Lecture Notes in Electrical Engineering

Volume 137

Dehuai Zeng (Ed.)

Advances in Control and Communication

Springer

Editor
Dehuai Zeng
Shenzhen University
Guangdong
China, People's Republic

ISSN 1876-1100 e-ISSN 1876-1119
ISBN 978-3-642-26006-3 e-ISBN 978-3-642-26007-0
DOI 10.1007/978-3-642-26007-0
Springer Heidelberg New York Dordrecht London

Library of Congress Control Number: 2011943876

Printed on acid-free paper

Springer is part of Springer Science+Business Media (www.springer.com)

ICEEE 2011 Preface

Electrical and Electronic Engineering is an exciting and dynamic field. Electrical engineering is responsible for the generation, transfer and conversion of electrical power, while electronic engineering is concerned with the transfer of information using radio waves, the design of electronic circuits, the design of computer systems and the development of control systems.

With success of ICEEE 2010 in Wuhan, China, and December 4 to 5, 2010, the second International Conference of Electrical and Electronics Engineering (ICEEE 2011) will be held in Macau, China, and December 1 to 2, 2011. ICEEE is an annual conference to call together researchers, engineers, academicians as well as industrial professionals from all over the world to present their research results and development activities in Electrical and Electronics Engineering along with Computer Science and Technology, Communication Technology, Artificial Intelligence, Information Technology, etc.

This year ICEEE is sponsored by International Industrial Electronics Center, Hong Kong. And based on the deserved reputation, more than 750 papers have been submitted to ICEEE 2011, from which about 282 high quality original papers have been selected for the conference presentation and inclusion in the proceedings based on the referees' comments from peer-refereed.

All the accepted paper are in the topics of Biotechnology, Power Engineering, Telecommunication, Control engineering, Signal processing, Integrated circuit, Electronic amplifier , Nano-technologies, Circuits and networks, Microelectronics, Analog circuits, Digital circuits, Nonlinear circuits, Mixed-mode circuits, Circuits design, Silicon devices, Thin film technologies, VLSI, Sensors, CAD tools, DNA computing, Molecular computing, Superconductivity circuits, Antennas technology, System architectures, etc. We expect that the conference and its publications will be a trigger for further related research and technology improvements in this importance subject.

We would like to express my deeply appreciations and thanks to Prof. Jun Wang for his high quality keynote speech, and to all contributors and delegates for their support and high quality contributions. Special thanks go to Springer Publisher.

We hope that ICEEE 2011 will be successful and enjoyable to all participants. We look forward to seeing all of you next year at the ICEEE 2012.

Dehuai Zeng

ICEEE 2011 Committee

Honorary Conference Chair

Jun Wang	The Chinese University of Hong Kong, Hong Kong

General Chairs

Jian Li	Nanchang University, China
Lei Yang	International Industrial Electronics Center, Hong Kong

Program Chair

Jin Chen	Wuhan University of Technology, China

Publication Chair

Dehuai Zeng	Shenzhen University, China

Program Committees

Yiyi Zhouzhou	Azerbaijan State Oil Academy, Azerbaijan
Garry Zhu	Thompson Rivers University, Canada
Ying Zhang	Wuhan Uniersity, China
Dehuai Zeng	Shenzhen University, China
Srinivas Aluru	ACM NUS Singapore Chapter, Singapore
Tatsuya Akutsu	ACM NUS Singapore Chapter, Singapore
Aijun An	National University of Singapore, Singapore
Qinyuan Zhou	Jiangsu Teachers University of Technology, China
Mark Zhou	Hong Kong Education Society, Hong Kong
Tianbiao Zhang	Huazhong Normal University, China

Contents

Future Information Technology and Computer Engineering

Control Fractional-Order Chaotic System to Approach any Desired Stability State via Linear Feedback Control

Zhang Dong and Zhou Ya-Fei

College of Electron and Electrical Engineering
Chongqing University of Arts and Science
Chongqing, 402160, China
{z3889,zhouyf}@163.com

Abstract. According to stability theory of fractional-order linear systems, we can obtain any desired stability state from fractional-order chaotic system via linear feedback control. The condition of any desired stability fixed point is established theoretically rigorous. The control parameters are irrelative with stability fixed point.

Keywords: stability theory of fractional-order linear systems, fractional-order systems, any desired stability state, linear feedback control.

1 Introduction

Many systems [1-8] are known to display fractional order dynamics, such as viscoelastic systems, dielectric polarization, electrode-electrolyte polarization and electromagnetic waves. But, the dynamics of fractional-order systems has not yet been fully studied. Recently, the chaotic dynamics of fractional-order dynamical systems have been investigated. It was shown that the fractional-order system can be produced a chaotic attractor. Chaos and hyperchaos in the fractional-order Rŏssler equations were studied, in which they showed that chaos can exist in the fractional-order Rŏssler equation with order as low as 2.4, and hyperchaos can exist in the fractional-order Rŏssler hyperchaos equation with order as low as 3.8.

On the other hand, chaotic synchronization and control have held the interest of many authors in the past few decades. Recently, the synchronization and control of fractional-order chaotic systems has attracted much attention [1-4,6-8], and many results have obtained. In this paper, a control method is presented for a class of fractional-order chaotic systems via linear feedback control, and we can obtain any desired targets from fractional-order chaotic systems. The control technique, based on stability theory of fractional-order systems, is simple and theoretically rigorous.

2 Control Theorem

Consider the following fractional-order chaotic systems,

$$\frac{d^q \mathbf{x}}{dt} = \mathbf{f}(\mathbf{x}) \tag{1}$$

D. Zeng (Ed.): Advances in Control and Communication, LNEE 137, pp. 1–7.
springerlink.com

Where $0 < q \le 1$ are the fractional orders, and $\mathbf{x} = (x_1 \quad x_2 \quad x_3)^{\mathrm{T}}$ are state variables and real numbers, and $\mathbf{f}(\mathbf{x}) = (f_1 \quad f_2 \quad f_3)^{\mathrm{T}}$ are nonlinear functions. There are many practical chaotic systems like system (1), for example, fractional-order Chua's chaotic circuit [1,4] and fractional-order Rŏssler chaotic system [1,4] etc.

Let $\mathbf{x}_0 = (x_{10} \quad x_{20} \quad x_{30})^{\mathrm{T}}$ is one state in the chaotic attractors of system (1), and is the desired state which it be need. According to the ergodic of chaotic system, system (1) can arrive to the point $\mathbf{x}_0$ at some times. But, system (1) is not stable at this point.

In order to make the point $\mathbf{x}_0$ is locally asymptotically stable state which it be need. We must add one control law for system (1). We can use a linear feedback control for system (1). We can obtain the following controlled system (2),

$$\frac{d^q\mathbf{x}}{dt} = \mathbf{f}(\mathbf{x}) - (\mathbf{k} + \mathbf{Df}(\mathbf{x})\big|_{\mathbf{x}_0})(\mathbf{x} - \mathbf{x}_0) - \mathbf{f}(\mathbf{x}_0) \tag{2}$$

Where $\mathbf{Df}(\mathbf{x})\big|_{\mathbf{x}_0}$ is the Jacobian matrix of system (1) at the point $\mathbf{x}_0$, and $\mathbf{k}(k_{ij} = k_i \ne 0, j = j; k_{ij} = 0, i \ne j)$ are control matrix and real numbers.

Certainly, the point $\mathbf{x}_0$ is a fixed point of controlled system (2). The goal in this paper is to make the point $\mathbf{x}_0$ stability by linear feedback control for system (2). Namely, the point $\mathbf{x}_0$ must be a stable fixed point for system (2). Therefore, we must choose suitable control matrix $\mathbf{k}$.

***Theorem*:** if $k_i > 0, i = 1,2,3$ for control matrix $\mathbf{k}$, then fixed point $\mathbf{x}_0 = (x_{10} \quad x_{20} \quad x_{30})^{\mathrm{T}}$ is locally asymptotically stable fixed point of system (2).

Proof

Because the point $\mathbf{x}_0$ is a fixed point of controlled system (2), so the Jacobian matrix of system (2) at the fixed point $\mathbf{x}_0$ is following,

$$D = \begin{pmatrix} -k_1 & 0 & 0 \\ 0 & -k_2 & 0 \\ 0 & 0 & -k_3 \end{pmatrix}.$$

So, the eigenvalues of the Jacobian matrix of system (2) at the fixed point $\mathbf{x}_0$ is $\lambda_i = -k_i (i = 1,2,3)$. Because $k_i > 0$, so $\lambda_i = -k_i < 0$. Therefore, the argument of the eigenvalue λ_i is $|\arg(\lambda_i(D))| = \pi > q\pi/2$. According to stability theory of fractional-order systems [4,6,8], we can obtain that the fixed point $\mathbf{x}_0$ is locally asymptotically stable. ■

This theorem indicates that we can obtain any desired stability state from fractional-order chaotic system via linear feedback control, and the control parameters $k_i\,(i=1,2,3)$ are irrelative with stability fixed point.

3 Applications

All the numerical simulation of fractional-order system in this paper is based on Ref [4,6,8]. It is a direct time domain approximation numerical simulation. Ref [7] has shown that using frequency domain approximation in the numerical simulations of fractional systems may result in wrong consequences. This mistake has occurred in the recent literature that found the lowest-order chaotic systems among fractional-order systems [7]. Now, we introduce the numerical solution of fractional differential equations in Ref [4,6,8]. Consider the following fractional-order system,

$$\begin{cases} d^{q_1}x/dt^{q_1} = f(x,y) \\ d^{q_2}y/dt^{q_2} = g(x,y) \end{cases},\ 0<q_1,q_2<1$$

with initial condition (x_0, y_0). Now, set $h=T/N$ $t_n=nh\,(n=0,1,2\cdots,N)$. The above system can be discretized as follows,

$$\begin{cases} x_{n+1} = x_0 + \dfrac{h^{q_1}}{\Gamma(q_1+2)}[f(x_{n+1}^p, y_{n+1}^p) + \sum_{j=0}^{n} \alpha_{1,j,n+1} f(x_j, y_j)] \\ y_{n+1} = y_0 + \dfrac{h^{q_2}}{\Gamma(q_2+2)}[g(x_{n+1}^p, y_{n+1}^p) + \sum_{j=0}^{n} \alpha_{2,j,n+1} g(x_j, y_j)] \end{cases}$$

where

$$\begin{cases} x_{n+1}^p = x_0 + \dfrac{1}{\Gamma(q_1)} \sum_{j=0}^{n} \beta_{1,j,n+1} f(x_j, y_j) \\ y_{n+1}^p = y_0 + \dfrac{1}{\Gamma(q_2)} \sum_{j=0}^{n} \beta_{2,j,n+1} g(x_j, y_j) \end{cases}$$

$$\alpha_{i,j,n+1} = \begin{cases} n^{q_i+1} - (n-q_i)(n+1)^{q_i}, & j=0 \\ (n-j+2)^{q_i+1} + (n-j)^{q_i+1} - 2(n-j+1)^{q_i+1}, & 1 \le j \le n \\ 1, & j=n+1 \end{cases}$$

$$\beta_{i,j,n+1} = \frac{h^{q_i}}{q_i}[(n-j+1)^{q_i} - (n-j)^{q_i}], \qquad 0 \le j \le n,\ i=1,2\,.$$

The error of this approximation is described as follows,

$$|x(t_n) - x_n| = o(h^{p_1}) \quad p_1 = \min(2, 1+q_1)$$

$$|y(t_n) - y_n| = o(h^{p_2}) \quad p_2 = \min(2, 1+q_2)$$

In the following, we take fractional-order Chua's chaotic circuit and fractional-order Rŏssler chaotic system for example.

3.1 Control of Fractional-Order Rŏssler Chaotic System to Any Desired Stability Fixed Point via Linear Feedback Control

Fractional-order Rŏssler chaotic system [1,4] is system (3),

$$\begin{aligned}\frac{d^q x_1}{dt^q} &= -(x_2 + x_3) \\ \frac{d^q x_2}{dt^q} &= x_1 + a x_2 \\ \frac{d^q x_3}{dt^q} &= 0.2 + x_3(x_1 - 10)\end{aligned} \tag{3}$$

Where q=0.9,a=0.4, the chaotic attractors [1,4] is shown as Fig.1.

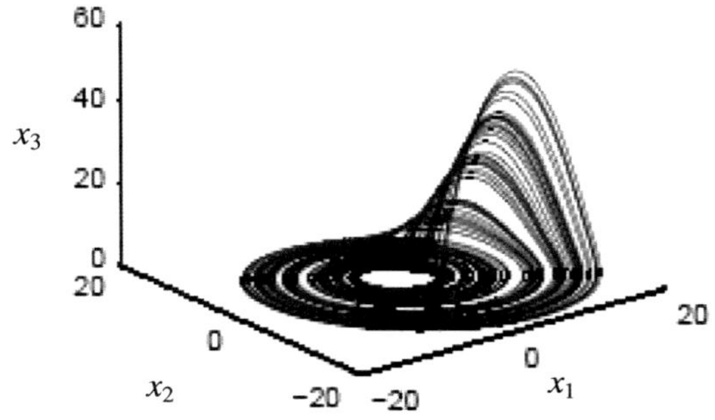

Fig. 1. Chaotic attractor of fractional-order Rŏssler system

Let $\mathbf{x}_0 = (x_{10} \quad x_{20} \quad x_{30})^{\mathbf{T}}$ is one state in the chaotic attractors of system (3), and is the desired stability fixed point which it be need. In order to make the point $\mathbf{x}_0$ is a locally asymptotically stable state, according to the above Theorem. We can obtain the following controlled system (4),

$$\begin{aligned}\frac{d^q x_1}{dt^q} &= -(x_2 + x_3) + u_1 \\ \frac{d^q x_2}{dt^q} &= x_1 + a x_2 + u_2 \\ \frac{d^q x_3}{dt^q} &= 0.2 + x_3(x_1 - 10) + u_3\end{aligned} \tag{4}$$

Where $u_i(i=1,2,3)$ is the linear feedback controller. According to the above Theorem, we can obtain the following,

$$\begin{pmatrix} u_1 \\ u_2 \\ u_3 \end{pmatrix} = \begin{pmatrix} -k_1 & 1 & 1 \\ -1 & -k_2 - a & 0 \\ -x_{30} & 0 & -k_3 - x_{10} + 10 \end{pmatrix} \begin{pmatrix} x_1 - x_{10} \\ x_2 - x_{20} \\ x_3 - x_{30} \end{pmatrix} + \begin{pmatrix} x_{20} + x_{30} \\ -x_{10} - ax_{20} \\ -0.2 - x_{30}(x_{10} - 10) \end{pmatrix}.$$

If $k_i > 0, i = 1,2,3$, then the point $\mathbf{x}_0 = (x_{10} \quad x_{20} \quad x_{30})^{\mathrm{T}}$ is a locally asymptotically stable state of system (4). Now, we choose $\mathbf{x}_0 = (10 \quad -10 \quad 5)^{\mathrm{T}}$, $k_1 = 1$, $k_1 = 2$ and $k_3 = 3$, Fig.2 shows the simulation result. Where $\varepsilon = [\sum_{i=1}^{3}(x_i - x_{i0})^2]^{1/2}$.

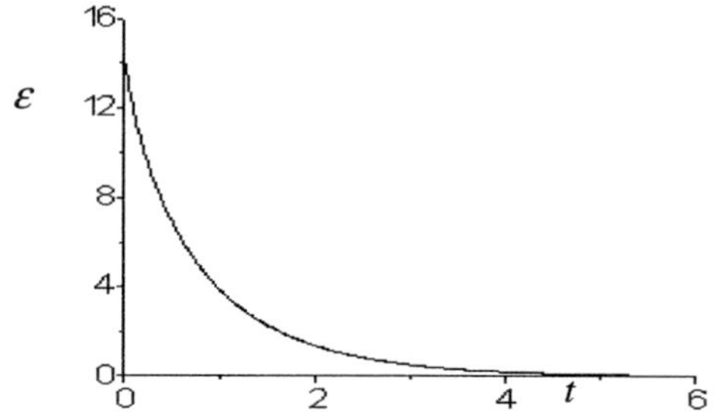

Fig. 2. The simulation result of fractional-order Rőssler chaotic system for $\mathbf{x}_0 = (10 \quad -10 \quad 5)^{\mathrm{T}}$

3.2 Control of Fractional-Order Chua's Chaotic Circuit System to Any Desired Stability Fixed Point via Linear Feedback Control

Fractional-order Chua's chaotic circuit [1,4] is system (4),

$$\begin{aligned} \frac{d^q x_1}{dt^q} &= a(x_2 + \frac{x_1 - 2x_1^3}{7}) \\ \frac{d^q x_2}{dt^q} &= x_1 - x_2 + x_3 \\ \frac{d^q x_3}{dt^q} &= -\frac{100}{7} x_2 \end{aligned} \qquad (5).$$

Where q=0.9,a=12.75, the chaotic attractors [1,4] is shown as Fig.3.

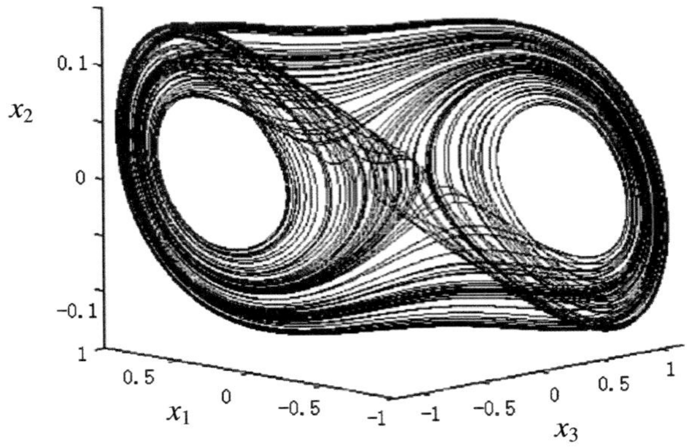

Fig. 3. Chaotic attractor of fractional-order Chua's circuit

Let $\mathbf{x}_0 = (x_{10} \quad x_{20} \quad x_{30})^{\mathrm{T}}$ is one state in the chaotic attractors of system (5), and is the desired stability fixed point which it be need. In order to make the point $\mathbf{x}_0$ is a locally asymptotically stable state, according to the above Theorem. We can obtain the following controlled system (6),

$$\begin{aligned} \frac{d^q x_1}{dt^q} &= a(x_2 + \frac{x_1 - 2x_1^3}{7}) + u_1 \\ \frac{d^q x_2}{dt^q} &= x_1 - x_2 + x_3 + u_2 \\ \frac{d^q x_3}{dt^q} &= -\frac{100}{7} x_2 + u_3 \end{aligned} \tag{6}$$

Where $u_i (i = 1,2,3)$ is the linear feedback controller. According to the above Theorem, we can obtain the following,

$$\begin{pmatrix} u_1 \\ u_2 \\ u_3 \end{pmatrix} = \begin{pmatrix} -k_1 - a(1-6x_{10}^2)/7 & -a & 0 \\ -1 & -k_2+1 & -1 \\ 0 & 100/7 & -k_3 \end{pmatrix} \begin{pmatrix} x_1 - x_{10} \\ x_2 - x_{20} \\ x_3 - x_{30} \end{pmatrix} + \begin{pmatrix} -a(x_{20} + \dfrac{x_{10} - 2x_{10}^3}{7}) \\ -x_{10} + x_{20} - x_{30} \\ 100x_{20}/7 \end{pmatrix}.$$

If $k_i > 0 (i = 1,2,3)$, then the point $\mathbf{x}_0 = (x_{10} \quad x_{20} \quad x_{30})^{\mathrm{T}}$ is a locally asymptotically stable state of system (6). Now, we choose $\mathbf{x}_0 = (-0.8 \quad 0.08 \quad 1)^{\mathrm{T}}$, $k_1 = 2$, $k_1 = 2$ and $k_3 = 6$, Fig.4 shows the simulation result.

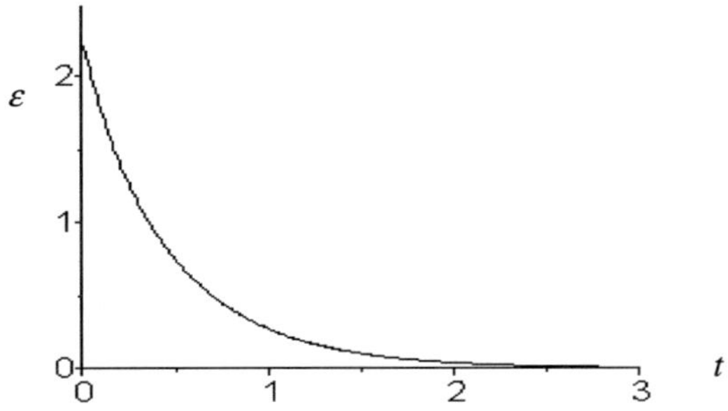

Fig. 4. The simulation result of fractional-order Chua's chaotic circuit system for $\mathbf{x}_0 = (-0.8 \quad 0.08 \quad 1)^{\mathrm{T}}$

4 Conclusions

According to stability theory of fractional-order linear systems, we establish one control method which it can obtain any desired stability state from fractional-order chaotic system via linear feedback control, and this method is theoretically rigorous. This control method is applied to fractional-order Chua's chaotic circuit and fractional-order Rŏssler chaotic system. Simulation results are the same as theoretical analyze results. It indicates that the control method in this paper is effective.

References

1. Li, C.G., Liao, X.F., Yu, J.B.: Synchronization of fractional order chaotic systems. Phys. Rev. E 68, 067203–067205 (2003)
2. Gao, X., Yu, J.B.: Synchronization of two coupled fractional-order chaotic oscillators. Chaos, Soliton and Fractals 26, 141–145 (2005)
3. Lu, J.G.: Synchronization of a class of fractional-order chaotic systems via a scalar transmitted signal. Chaos, Solitons and Fractals 27, 519–525 (2006)
4. Zhou, P.: Chaotic synchronization for a class of fractional-order chaotic systems. Chinese Physics 16, 1263–1266 (2007)
5. Mohammad, S.T., Mohammad, H.: A necessary condition for double scroll attractor existence in fractional-order systems. Physics Letters A 367, 102–113 (2007)
6. Zhou, P., Wei, L.J., Cheng, X.F.: A novel fractional-order hyperchaotic system and its synchronization. Chinese Physics B 18, 2674–2679 (2009)
7. Zhou, P., Wei, L.J., Cheng, X.F.: One new fractional-order chaos system and its circuit simulation by electronic workbench. Chinese Physics 17, 3252–3257 (2008)
8. Zhou, P., Cheng, X.F., Zhang, N.Y.: A new fractional-order hyperchaotic system and it's chaotic synchronization. Acta Physica Sinica 57, 5407–5412 (2008)

A Novel Fuzzy Entropy Definition and Its Application in Image Enhancement

Haitao Yu

School of Science
Xi'an University of Science and Technology
Xi'an, China
College of Marine
Northwestern Polytechnical University
Xi'an, China

Abstract. In order to improve the image enhancement quality and to reduce the processing time, a novel fuzzy entropy definition for image self-adaptive enhancement is proposed based on the exponential behavior of information-gain and a fuzzy domain partition method *M*. The proposed fuzzy entropy definition can avoid the defect of logarithmic one and makes the definition much reasonable and makes the physical meaning of the definition much evident due to exponential definition. And the partition method *M* enables the optimal enhancement for different images. The self-adaptive fuzzy parameters are gotten by enumeration method and classic genetic algorithm (GA) based on maximum entropy principle respectively. The experiment results show that processing time based on the new entropy definition is cut down a little on condition that the image enhancement quality is better or unchanged compared to that based on the existed entropy definition. And parameters optimization of GA costs less time than that of enumeration method for the simple optimization problem which the fuzzy domain partition method *M* is given for different images. The automatic acquisition of the partition method *M* is the next research.

Keywords: fuzzy entropy, definition, self-adaptive, image enhancement, exponential behavior.

1 Introduction

Image enhancement techniques play an important role for improving image quality in image processing, and are the basis of follow-up image analysis and pattern recognition. Owing to the complexity and relevance of the image itself, uncertainty and imprecision, that is the ambiguity, appear in image processing [1, 2]. As the fuzzy system can express various, imprecise, uncertain and inaccurate knowledge or information, many scholars introduced fuzzy theory into image processing and pattern recognition and achieved good results.

Currently, it becomes one of the obstacles for wide application of fuzzy image enhancement that some key parameters in fuzzy image enhancement are determined by

D. Zeng (Ed.): Advances in Control and Communication, LNEE 137, pp. 9–16.
springerlink.com

experiments based on different images. In order to achieve self-adaptive parameter selection, some scholars introduced the fuzzy entropy to image enhancement. But different fuzzy entropy definitions will lead to different processing results. The basic idea of self-adaptive fuzzy enhancement is briefly reiterated in part II. And after a brief analysis of the existed fuzzy entropy definitions in part III, a novel fuzzy entropy definition is proposed in part IV. The experiment results and conclusion will appear in Part V and Part VI, respectively.

2 The Basic Idea of Self-adaptive Fuzzy Enhancement

Fuzzy image enhancement mainly includes the following three steps: image fuzzification, membership modification and image defuzzification. Our fuzzy image enhancement is based on Pal and King's algorithm, whose description in detail can be found in [3]. Parameter F_d is usually obtained by experiments and will influence the final enhancement result.

In order to achieve self-adaptive parameter selection, some scholars introduced the fuzzy entropy to image enhancement. Its basic idea is as follows: firstly, the image is transformed from the gray domain to fuzzy domain by the membership function; secondly, the fuzzy entropy of the image is calculated within the limitation of parameter F_d and the optimal parameter F_d is obtained according to maximum entropy principle by enumeration method or genetic algorithm (GA); lastly, using the optimal parameter F_d obtained before, the fuzzy domain data is transformed to image gray domain by the inverse transformation of membership function.

3 A Brief Analysis of the Existed Fuzzy Entropy Definitions

Fuzzy entropy quantitatively reflects the fuzzy level of an image. And it signifies the average difficulty level of which a pixel can be considered to be a fuzzy subset element. At present, the fuzzy entropy definition has no unique one like the information entropy defined by Shannon. Different scholars have proposed different fuzzy entropy definitions, which will lead to different enhancement results.

Zadeh suggested a definition about the entropy of a fuzzy set which takes both distribution and membership into consideration in 1965. And his definition is a logarithmic one. De Luca and Termini proposed a quite different definition about the entropy of a fuzzy set in 1972. And this entropy has a nonprobabilistic feature. Kaufmann also defined entropy considering both the distribution and membership in 1980. But he takes the distribution in the fuzzy domain instead of that in the ordinary domain. And his definition is a natural logarithmic one. According to [4], each of the above fuzzy entropy definitions has some problems when using these fuzzy entropies to determine the fuzzy region. And our simulations also confirm the analysis. Due to the space limitations, simulations of the three definitions for self-adaptive image enhancement will not be given in the paper.

N.R. Pal and S.K. Pal [5] proposed a new nonprobabilistic definition based on the exponential behavior of information-gain in 1989. And the definition overcomes the inherent shortcoming of $\lim_{x\to 0}(\log x) = -\infty$. But it still has the shortcoming of De Luca and Termini's definition which is that the fuzzy entropy achieves maximum when fuzzy domain value equals to 0.5.

H.D. Cheng et al. [4] defined a novel fuzzy entropy based on the fuzzy domain partition method M which is task-dependent and may be equal partition or non-equal partition in 1999. And good image enhancement results were obtained using the definition. But their definition is still a logarithmic one. Y.M. Wang et al. [6] proposed a definition, still logarithmic one, based on a compensating parameter and the fuzzy domain partition method M of H.D. Cheng et al. in 2003.

4 A Novel Fuzzy Entropy Definition

As we can see that definitions in [4, 6] are based on logarithm and that definition in [5] is based on exponent. But the logarithm has the inherent shortcoming of $\lim_{x\to 0}(\log x) = -\infty$ for information-gain. According to [5], the analysis and the fact that the information gain approaches a finite limit when more and more image pixels are analysed strengthen the assertion that the gain in information is exponential in nature. And the fuzzy domain partition method M in [4] enables the optimal enhancement for different images. Thus, a novel fuzzy entropy definition is proposed as

$$H(A,N,M,\mu_A) = \frac{1}{N}\sum_{i=1}^{N}\{P_P(A_i)e^{[1-P_P(A_i)]} + [1-P_P(A_i)]e^{P_P(A_i)}\} \tag{1}$$

where

$$P_P(A_i) = \sum_{\mu_A(x)\in A_i} P(x) .$$

A is the fuzzy event, and N is the number of partitions of A in the fuzzy domain denoted as sub-events $A_1, A_2, \ldots, A_N$. The fuzzy domain partition method M is task-dependent which may be equal partition or non-equal partition. $\mu_A(x)$ is the membership function and $P(x)$ is the probability of x in the space domain. $P_P(A_i)$ means the probability summed in the space domain for the x (space domain) mapping into A_i (fuzzy domain) by the membership function $\mu_A()$. And it can be viewed as the probability of fuzzy events A_i, i=1,2,…,N, based on the membership function $\mu_A()$. Different membership functions will cause different values of $P_P(A_i)$. And it can be proved that the fuzzy entropy achieves maximum if and only if $P_P(A_i)$=0.5.

The ultimate difference between the definition of (1) and [4] is the information-gain behavior. According to [5], exponential behavior of information-gain can avoid the defect of logarithmic definition and makes the definition much reasonable and physical meaning of definition much evident. Another reason is that exponent arithmetic is faster than logaritmim for computer. Thus, exponent is adopted in (1). And it is expected that the image enhancement algorithm based on the novel fuzzy entropy definition costs less time than that based on [4].

The existing of partition method M is the common ground of (1) and [4]. And it enables the optimal enhancement according to maximum entropy principle for different characteristic images. But it also becomes its shortcoming that selecting of partition method M requires the priori knowledge of different images. The partition methods M in this paper are obtained by repeated experiments. And the automatic acquisition of the partition method M is the next research. The performance comparison experiment of the two definitions for image enhancement will appear in next part.

5 Experiment Results and Analysis

The original images are *lena*, *couple* and *template* widely used in image processing. Based on the basic idea described in part II, the optimal parameter F_d can be obtained according to maximum entropy principle by enumeration method or GA. If using enumeration method, the range of parameter F_d is from minimum of image gray level to maximum of that. If using GA, 8-bit binary string in Gray coding for image gray value of [0, 255] is adopted. And the fitness function is the novel fuzzy entropy definition as (1). Population size of 10, crossover probability of 0.95 of a single point crossover, mutation probability of 0.08, the generation gap of 0.9 and the termination generation of 10 are applied for reducing the processing time for the simple optimizing problem.

(a)

(b) (c)

Fig. 1. Comparison of processing result for *lena*: (a) original image, (b) enhancement result of five non-equal partition of (0.45, 0.35, 0.1, 0.05, 0.05) based on (1), (c) enhancement result of five non-equal partition of (0.45, 0.35, 0.1, 0.05, 0.05) based on [4]

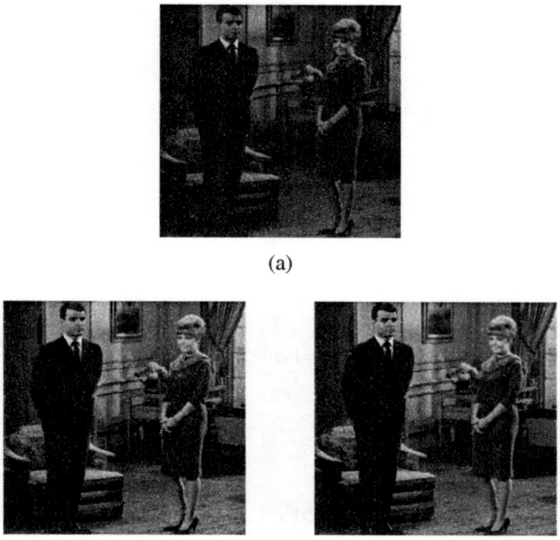

(a)

(b) (c)

Fig. 2. Comparison of processing result for *couple*: (a) original image, (b) enhancement result of five non-equal partition of (0.5, 0.3, 0.1, 0.05, 0.05) based on (1), (c) enhancement result of five non-equal partition of (0.5, 0.3, 0.1, 0.05, 0.05) based on [4]

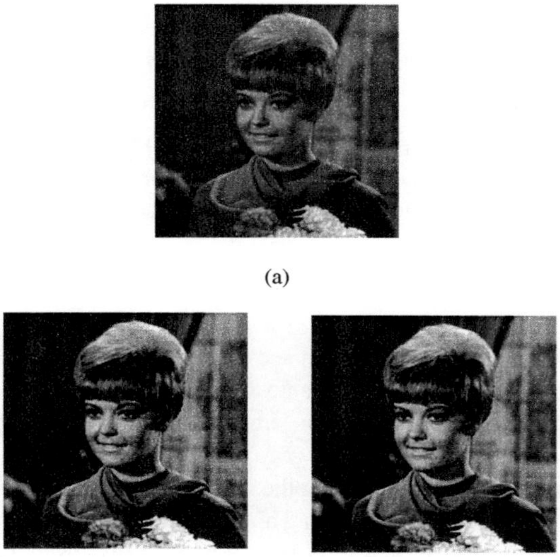

(a)

(b) (c)

Fig. 3. Comparison of processing result for *template*: (a) original image, (b) enhancement result of five non-equal partition of (0.48, 0.32, 0.1, 0.05, 0.05) based on (1), (c) enhancement result of five non-equal partition of (0.48, 0.32, 0.1, 0.05, 0.05) based on [4]

Table 1. Comparison of Parameter F_D Based on Entropy Definition of (1) and [4]

Entropy Definition	Parameter F_d		
	lena	*couple*	*template*
(1)	169	238	180
[4]	172	238	185

Table 2. Comparison of Processing Time Based on Entropy Definition of (1) and [4]

Entropy Definition	Processing Time (s)		
	lena	*couple*	*template*
(1)	3.2600	3.4200	3.4300
[4]	3.2750	3.4500	3.4650

Table 3. Comparison of Parameter F_D Based on Entropy Definition of (1) For Enumeration Method and GA

Parameter Optimization	Parameter F_d		
	lena	*couple*	*template*
Enumeration Method	169	238	180
GA	169	238	180

Table 4. Comparison of Processing Time Based on Entropy Definition of (1) for Enumeration Method and GA

Parameter Optimization	Processing Time (s)		
	lena	*couple*	*template*
Enumeration Method	3.2600	3.4200	3.4300
GA	1.3050	1.3400	1.4100

From Fig. 1, Fig. 2 and Fig. 3, we can see that almost similar enhancement in visual effects are obtained based on the fuzzy entropy definition of (1) and [4]. And the key parameters F_d based on them in Table 1 is subequal for *lena* and *template*, but equal for softimage *couple* which maximum of gray level is 238. Hence, almost the same enhancement results can be acquired using the fuzzy entropy definition of (1) and [4].

From the view of processing time in Table 2, the entropy definition of (1) is fast than that of [4]. The reason is that exponent arithmetic is faster than logaritmim for computer as pointed out in part IV. And it is expected that the image enhancement algorithm based on the novel fuzzy entropy definition costs less time than that based on [4]. At the same time, the exponential behavior of information-gain can avoid the defect of logarithmic definition and makes the definition much reasonable and physical meaning of definition much evident.

One may think that the processing time is cut down only a little. The time reduction for a gray image of 256×256 may be negligible. But it is noted that the difference will be evident when dealing with large amount of data.

If the non-equal partition method is given, the searching of optimal parameter F_d is a simple optimizing problem. And the stable solution can be obtained using classical GA in less evolutionary generations. From Table 3, we can see that the key parameter F_d is equal for the three images based on the definitions of (1) for enumeration method and GA. But optimization time using classical GA is cut down almost 60% compared to that using enumeration method according to Table 4.

It is noted that the processing time in this paper can be further reduced if using C or C++ programs due to the defect of MATLAB. And the automatic acquisition of the partition method M is the next research.

6 Conclusion

A novel fuzzy entropy definition for image self-adaptive enhancement is proposed based on the exponential behavior of information-gain and a fuzzy domain partition method M. The proposed fuzzy entropy definition can avoid the defect of logarithmic one and makes the definition much reasonable and makes the physical meaning of the definition much evident due to exponential definition. And the partition method M enables the optimal enhancement for different images. The self-adaptive fuzzy parameters are gotten by enumeration method and classic genetic algorithm (GA) based on maximum entropy principle respectively. The experiment results show that processing time based on the new entropy definition is cut down a little on condition that the image enhancement quality is better or unchanged compared to that based on the existed entropy definition. And parameters optimization of GA costs less time than that of enumeration method for the simple optimization problem which the fuzzy domain partition method M is given for different images. The automatic acquisition of the partition method M is the next research.

Acknowledgment. We thank for the sponsor of the Scientific Research Program Funded by Shaanxi Provincial Education Commission (Program NO.2010JK675) and the Education Reform Program Funded by Xi'an University of Science and Technology (Program NO.JG10066).

References

1. Li, H., Yang, H.S.: Fast and reliable image enhancement using fuzzy relaxation technique. IEEE Transactions on Systems, Man and Cybernetics 19(5), 1276–1281 (1989), doi:10.1109/21.44048
2. Yu, H.: Research on Image Enhancement Algorithms Based on Fuzzy Set Theory. Master Thesis of Xi'an Universtiy of Science and Technology (2005)

3. Pal, S.K., King, R.A.: Image Enhancement Using Smoothing with Fuzzy Sets. IEEE Transactions on Systems, Man and Cybernetics 11(7), 494–501 (1981), doi:10.1109/TSMC.1981.4308726
4. Cheng, H.D., Chen, Y.-H., Sun, Y.: A Novel Fuzzy Entropy Approach to Image Enhancement and Thresholding. Signal Processing 75(3), 277–301 (1999)
5. Pal, N.R., Pal, S.K.: Object-background segmentation using new definitions of entropy. IEEE Proceedings Computers and Digital Techniques 136(4), 284–295 (1989)
6. Wang, Y., Wu, G., Zhao, Y., Hu, J.: Image Enhancement Based on Fuzzy Entropy and Genetic Algorithm and Its Application to Agriculture 34(3), 96–98 (2003)

Automatic Security Analysis for Group Key Exchange Protocol: A Case Study in Burmester-Desmedt Protocol

Ren Chunyang, Wang Hongyuan, Zhang Zijian, and Liao Lejian

Beijing Key Lab of Intelligent Information, School of Computer Science and Technology
Beijing Institute of Technology
Beijing, China

Abstract. Canetti and Herzog have already proposed universally composable symbolic analysis (UCSA) for mutual authentication and key exchange protocols automatically without sacrificing the soundness of the cryptography. We want to extend their work to analyze group key exchange protocols. This paper takes the case of BD protocol with arbitrary participants against passive adversary (BD-Passive), and proves that BD-Passive is a secure group key exchange protocol. More specially, we (1) define the ideal functionality of BD-Passive; (2) prove the security property of BD-Passive in UC security framework by UCSA. Obviously, our work plays a new approach to prove group key exchange protocols automatically without sacrificing the soundness of the cryptography.

Keywords: Universally Composable Symbolic Analysis, Universally Composable, Ideal Functionality, Burmester-Desmedt protocol.

1 Introduction

Canetti [1] has proposed UC security framework a few years ago. He has analyzed practical protocols by protocol emulation. The main steps are as follow. All of the environment, protocol and adversary are viewed as PPT Interactive Turing Machine Instances (ITIs) in the beginning. Second, the ideal functionality should be established, which describes the execution of the ideal protocol which is believed to be secure and the abilities of the ideal adversary. Third, we represent the capabilities of the adversary by oracles. That is, an adversary can inquire its oracles to obtain the information which the parities output when the protocol is executed. Obviously, the type and number of oracles depend on the abilities of the adversary. Forth, the environment carries out an experiment. As long as the environment wants, it can ask the adversary to inquire any of the oracles, so as to get the respond what it is interested. Finally, the environment stops and makes the decision that whether it is interacting with the practical protocol and realistic adversary or with the ideal functionality and ideal adversary at the end of the experiment.

Researchers have focus on studying the approach which not only satisfies the soundness of cryptography, but also analyzes complex protocols automatically and always get the right answers in tolerable time. Canetti and Herzog [2] have proposed universally composable symbolic analysis (UCSA) to analyze mutual authentication and key exchange protocols. Furthermore, they has proved that the security properties

D. Zeng (Ed.): Advances in Control and Communication, LNEE 137, pp. 17–20.
springerlink.com

of the complex protocols can be satisfied, if they are decomposed to some simple single-session protocols which satisfy security properties respectively in UC security framework.

We want to extend their works to analyze group key exchange protocols. Therefore this paper takes the case of BD protocol with arbitrary participants against passive adversary (BD-Passive), and proves that BD-Passive is a secure group key exchange protocol in UC security framework by UCSA.

2 Ideal Functionality for Group Key Exchange Protocol

Here we first describe the ideal functionality for group key exchange protocol with arbitrary participants, before we discuss the security property of BD-Passive. More clarity, we describe the group key exchange functionality with arbitrary participants against passive adversary.

Definition 1. (Group Key Exchange Functionality with arbitrary participants $\mathrm{F}_{\text{n-GKE}}$)

$\mathrm{F}_{\text{n-GKE}}$ proceeds as follows, when parameterized with security parameter k , an ideal adversary S . In addition, the total number of the participants is arbitrary, therefore we define n as an arbitrary nature number.

(1) Initialization

Upon receiving an input $(\text{Establish-Key}, sid, gpid_i), i \in \{0,1,...,\text{n-1}\}$ from the participant p_i for the first time, record $(sid, p_i, gpid_i)$ and send this message to S . More specially, $gpid_i$ is a set including all of the three participant identifiers, while the participant p_i is the initiator.

Upon receiving an message $(\text{Establish-Key}, sid, gpid_i), i \in \{0,1,...,\text{n-1}\}$ from any other participants $p_j, j \in \{0,1,...,\text{n-1}\} \setminus i$, where $p_j \in gpid_i$, record $(sid, p_i, gpid_i)$ and send this message to S . Here the participant p_j is the responder. In addition, if there are already two recorded tuples $(sid, p_i, gpid_i)$, then choose $k \leftarrow \{0,1\}^*$ and recorded $[sid, gpid_i, k]$.

(2) Key Delivery

If S sends a message $(\text{deliver}, sid, p_i, gpid_i), i \in \{0,1,...,\text{n-1}\}$ where there is a recorded tuple $[sid, gpid_i, k]$, then send the message $(\text{Key}, sid, gpid_i, k)$ to the participant p_i immediately.

3 Analyze for Bd-Passive in Uc Security Framework

In this section, we prove BD-Passive is a secure group key exchange protocol in UC security framework.

Theorem 2 (Security Property of BD-Passive).
BD-Passive is a secure group key exchange protocol if 3-BD-Passive is a secure group key exchange protocol with three participants against passive adversary.

Proof: We prove this theorem by mathematical induction.

First of all, 3-BD-Passive has been proved secure in UC security framework automatically. Therefore, assume that BD protocol with n-1 participant against passive adversary ((n-1)-BD-Passive) is secure, so we have to prove that BD protocol with n participant against passive adversary (n-BD-Passive) is secure.

We prove it by contradiction. That is, if n-BD-Passive is not a secure group key exchange protocol, then (n-1)-BD-Passive is not a secure group key exchange protocol. Here we show the main idea.

If there exists an environment which can distinguish n-BD-Passive and $F_{n\text{-GKE}}$. Then either the secret key $g^{x_1x_2+x_2x_3+...+x_{n-1}x_n+x_nx_1}$ and random number can be distinguished or the secret key for all of the participants are not same.

If the former cannot be satisfy, we have the adversary can distinguish $g^{x_1x_2+x_2x_3+...+x_{n-1}x_n+x_nx_1}$ and random number. If we cannot distinguish (n-1)-BD-Passive and $F_{(n\text{-}1)\text{-GKE}}$, then we must distinguish $g^{x_1x_2+x_2x_3+...+x_{n-1}x_n+x_nx_1}$ and $g^{x_1x_2+x_2x_3+...+x_{n-1}x_1}$. It is equivalent to distinguish $g^{x_{n-1}x_n+x_nx_1}$ and $g^{x_{n-1}x_1}$. Obviously, it is indistinguishable if x_1, x_{n-1}, x_n are all of random members in the group. Otherwise, we can distinguish two random numbers $x_{n-1}x_n + x_nx_1$ and $x_{n-1}x_1$. So it is a contradiction.

If the latter cannot be satisfy, we have some secret keys are not same. Considered that the adversary is passive, so we have the group cannot satisfy the commutative law. Obviously, this group satisfies the commutative law. So it is a contradiction.

We complete the proof.

4 Conclusion

Based on that 3-BD-Passive has been proved secure in UC framework automatically, we have proved that BD-Passive is a secure group key exchange protocol in UC security framework. No doubt that our works provide a new thread to prove group key exchange protocol with arbitrary participants against passive adversary automatically without sacrificing the soundness of the cryptography.

References

1. Canetti, R.: Universally composable security: A new paradigm for cryptographic protocols. In: 42nd Annual Syposium on Foundations of Computer Science, pp. 136–145. IEEE Computer Society (2001)

2. Canetti, R., Herzog, J.: Universally Composable Symbolic Analysis of Mutual Authentication and Key-Exchange Protocols. In: Halevi, S., Rabin, T. (eds.) TCC 2006. LNCS, vol. 3876, pp. 380–403. Springer, Heidelberg (2006)
3. Burmester, M., Desmedt, Y.: A Secure and Efficient Conference Key Distribution System. In: De Santis, A. (ed.) EUROCRYPT 1994. LNCS, vol. 950, pp. 275–286. Springer, Heidelberg (1995)
4. Burmester, M., Desmedt, Y.: Efficient and secure conference key distribution. In: Lomas, M. (ed.) Security Protocols 1996. LNCS, vol. 1189, pp. 119–130. Springer, Heidelberg (1997)
5. Burmester, M., Desmedt, Y.: A secure and scalable group key exchange system. Information Processing Letters 94(3), 137–143 (2005)

A Virtual Organization Model Based on Semantic Web Services and Its Application in Supply Chain for Agricultural Product

Ruoyu Chen, Lejian Liao, and Zhi Fang

Beijing Laboratory of Intelligent Information Technology
School of Computer Science, Beijing Institute of Technology
Beijing, China
{twinsen,liaolj,fangz}@bit.edu.cn

Abstract. Cross-Organizational interoperability and coordination are core issues to Virtual Organization (VO) researches. We propose a model of VOs in which VO related concepts are modeled in an extension of the OWL-S service ontology and SWOCRL (Semantic Web Object Constraint Rule Language) rules. Within this framework, VO abilities can be seen as services provided; coordination between VOs can be seen as service invocation. SWOCRL rules are used as mediators between service descriptions and requests. A prototype system of our framework is proposed combining Constraint Handling Rules (CHR) and Prolog.

Keywords: Virtual Organization, OWL-S, SWOCRL, Supply Chain for Agricultural Product.

1 Introduction

Virtual Organization (VO)[1] is formed by a group of autonomous individuals and organizations based on their needs for each other's resources and problem solving abilities. The supply chain for agricultural product is dynamic and variant, so it is suitable to be modeled as VOs. With the increasing number of available web services in the open environment of the World Wide Web, we can model VOs under the semantic web services architecture and utilize the abundance of existing services. The coordination between VOs can be reduced to service discovery or composition in semantic web services.

We propose an VO modeling of supply chain for agricultural product based on extension of OWL-S[2] service ontology and SWOCRL (Semantic Web Object Constraint Rule Language) rules[3-4]. In our framework, VO concepts such as Organization, Activity, etc. are positioned under corresponding concepts in OWL-S ontology like Participant and Process. SWOCRL rules describe constraint relations between concepts in service ontology and are used as a bridge between service description and service requests. The reasoning of OWL-S ontology and SWOCRL rules are translated into CHR[5] rules. The intension of VO members is represented in service request that describes the desired resources or abilities provided by other members in VO.

The content of this paper is organized as follows. Comparison of related works is given in Section 2. To make the paper self-contained, Section 3 gives a brief introduc-

D. Zeng (Ed.): Advances in Control and Communication, LNEE 137, pp. 21–28.
springerlink.com

tion to the SWOCRL rule language, and then introduces the VO modeling for supply chain management based on our extended OWL-S ontology and SWOCRL rules. Section 4 discusses the translation from SWOCRL rules to CHR. The architecture of a prototype system is described in Section 5. We conclude our paper in Section 6.

2 Related Works

In[6], a VO modeling framework was proposed which was based on constraint logic. The modeling framework makes the full advantage of the representational and problem-solving powers of constraint logic programming, but with little concern for the interoperability and coordination between different organizations. The business-related concepts such as business activities and entities should be modeled in ontology as part of consensus between the organizational agents. The most important and widespread concepts in general VO include time, events, activities, interactions, services, resources and organizations. Process Specification Language (PSL)[7] is one of few work that construct enterprise ontology, including time, events, activities, and resources, in a systematic fashion. But PSL is written in KIF rules rather than description logics. The rules are not interoperable with current semantic Web formulation at conceptual level since the conceptual structure of the ontology is not explicit.

In our previous work[3-4], we have proposed an OWL based ontology for modeling VO concepts such as organization, activity, resources, etc. A constraint rule language SWOCRL is also proposed which is based on OWL and SWRL with constraint extension. This work extends the above ontology by modeling VO concepts under the OWL-S service ontology.

3 Vo Modeling of Supply Chain for Agricultural Products in Owl-S and Swocrl

3.1 SWOCRL: Backgrounds

SWOCRL (Semantic Web Object Constraint Rule Language) is a constraint language which is based on OWL and SWRL. It relies on conceptual structure defined by OWL and uses rules to infer object structures and to impose constraints on object attributes. SWOCRL modifies SWRL with some extensions and specializations. The abstract syntax of SWRL can be found in[8], the extension of SWOCRL syntax to SWRL is as follows, terminals are quoted, none terminals are italic and not quoted:

i-object ::= *i-variable* | *individualID* | *i-path-exp*
d-object ::= *d-variable* | *dataLiteral* | *d-path-exp*
constraint ::= *predicate-name* '(' { *d-object* } ')'
d-path-exp ::= *d-attribute* | *d-attribute* '.' *d-path-exp*
i-path-exp ::= *i-attribute* | *i-attribute* '.' *i-path-exp*

SWOCRL extends SWRL by allowing:

1) Some attributes and attribute paths as constraint variables. An RDF statement *rdfs:subPropertyOf(swocrl:constraintAttribute,*
owl:FunctionalProperty)

defines a special property *swocrl:constraintAttribute* for such constraint attributes. 2) Multi-attribute constraints as atoms in both antecedents and consequents of rules. Variables introduced in antecedents are taken to be universally quantified; variables introduced in consequents are taken to be existential ones.

SWOCRL specialize SWRL such that a SWOCRL rule only specifies assertions of just *one* class, featuring it as an object constraint language. Such specialization is desirable to circumscribe the complexity of rule reasoning. The specialization includes the following restrictions to the above syntax:

1) The first atom in the antecedent must be fixed as *class-name*(*i-variable*) which indicates the class that the rule asserts about. 2) For the following atoms, a unary class description atom must have an instantiated argument, i.e., either a constant individual, or an *i-variable* that occurs in a preceding atom; a binary property atom must have the first argument instantiated.

The operational semantics of SWOCRL is a forward inference process with constraint store accumulation in sense of concurrent constraint programming[9]. A SWOCRL rule is typically fired when an object is constructed or new information is added to it. The antecedent is checked against the object and the consequent is enforced if the check succeeds. The inference engine incrementally fills the values of object roles, constructs and accumulates constraints on object attributes, refines their ranges and derives their values. Note that a constraint in an antecedent performs an *asking* test while a constraint in the consequent performs a *telling* imposition in sense of CCP terms.

3.2 VO Modeling Based on OWL-S and SWOCRL

The VO model for supply chain of agricultural product includes concepts concern organizations, activities, product, transportation, and etc. Here we engraft these concepts into corresponding OWL-S concepts. Concept descriptions and associated SWOCRL constraint rules are given in DL-Form below. Simplifications are made for clarity and space limitation reasons.

In the supply chain of agricultural products, the major organizations include Producers, Resellers, Logistic Companies and Retailers such as Supermarket and Groceries. The role played by these organizations in the supply chain can be illustrated in Figure 1.

We view an organization as a service actor which corresponds to the class *Participant* in OWL-S.

$$Organization \subseteq Participant \cap \forall participate\text{-}in.\ Process$$

$$Producer \subseteq Organization \cap \forall participate\text{-}in.\ ProducerService$$
$$\cap \forall location.\ GeographicLocation$$

$$LogisticCo \subseteq Organization \cap \forall participate\text{-}in.\ LogisticService$$

$$Reseller \subseteq Organization \cap \forall participate\text{-}in.\ ResellerService$$
$$\cap\ \forall location.\ GeographicLocation$$

$$Retailer \subseteq Organization \cap \forall participate\text{-}in.\ RetailerService$$
$$\cap\ \forall location.\ GeographicLocation$$

$$Goods \subseteq =1 type.AgriculturalProduct$$
$$\cap = 1 productTime.Time \cap =1 bestAvaliableTime.Time$$
$$\cap =1 amount.Integer \cap =1 hasOwner.Organization$$
$$\cap =1 currentLocation.GeographicLocation$$

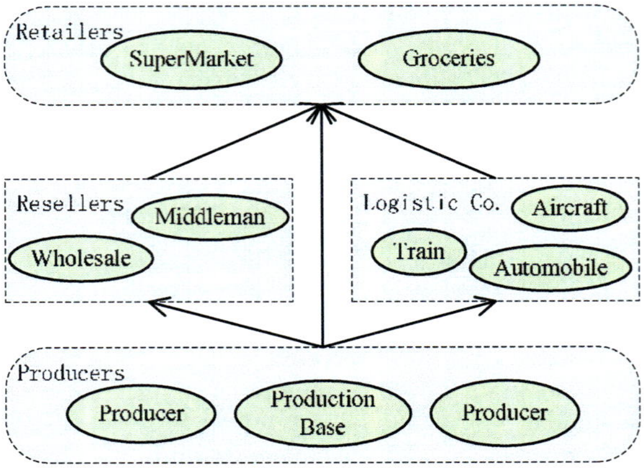

Fig. 1. Organizations in Agricultural Product Supply Chain

The composition structure and functioning for services provided by organizations can be represented as *Process*es that organizations involved in. Here takes *LogisticService* as an example.

$$
\begin{aligned}
&\textit{LogisticService} \subseteq \textit{Process}\\
&\quad \cap =1 \textit{performedBy. LogisticCo}\\
&\quad \cap =1 \textit{hasClient.Organization}\\
&\quad \cap =1 \textit{byTransportation. Transportation}\\
&\quad \cap =1 \textit{hasInputStartLocation.GeographicLocation}\\
&\quad \cap =1 \textit{hasInputDestination.GeographicLocation}\\
&\quad \cap =1 \textit{hasInputProductAmount.ProductAmount}\\
&\quad \cap =1 \textit{hasInputGoods.Goods}\\
&\quad \cap =1 \textit{hasInputStartTime.TimePoint}\\
&\quad \cap =1 \textit{hasPrecondition.}\\
&[\textit{Goods.currentLocation=LogisticService.hasInputStartLocation}]\\
&\quad \cap =1 \textit{hasOutputPrice.Price}\\
&\quad \cap =1 \textit{hasOutputETA.TimePoint}\\
&\quad \cap =1 \textit{hasEffect.}\\
&[\textit{Goods.currentLocation=LogisticService.hasInputDestination}]
\end{aligned}
$$

The properties *hasInputStartTime, hasOutputETA* are declared as constraint attributes:

hasInputStartTime : constraintAttribute
hasOutputETA : constraintAttribute

The above service description states that a logistic service provided by some logistic company can be invoked by offering start and destination locations as well as product type and amounts. The effect of invoking this service is that the goods are transported to the destination; this is expressed by the expression in property *hasEffect* of this service. The relation of the start time, estimated time arrival (ETA), and duration of a time interval can be represented as SWORCL rule:

Implies (
Antecedent(*LogisticService* (*X-ls*),
byTransportation(X-ls,X-tr),
hasInputStartTime(*X-ls, X-st*),
hasOutputETA(X-ls, X-eta),
hasInputStartLocation(*X-ls, X-sl*),
hasInputDestination(X-ls, X-dt))
Consequent(*time-sum*(*X-st, duration(X-tr, X-sl, X-dt), X-eta*)))

We assume the existence of function *duration* with start time, start location and destination as its parameters that compute the time interval.

4 Service Process and Swocrl Rules in Chr

Frühwirth[10] has developed a CHR constraint solver for OWL-DL, that is, OWL-DL reasoning rules in CHR. T-Box and A-Box assertions in OWL-DL can be translated into CHR constraints. In this section we discuss the representation of OWL-S service execution in CHR and conversion from SWOCRL rules to CHR.

4.1 Brief Introduction to CHR

CHR[5] is essentially a declarative concurrent committed-choice language consisting of multi-headed guarded rules that rewrite constraints into simpler ones until they are solved. There are three types of CHR as given in Figure 2.

$$\text{Simplification: } H \Leftrightarrow G|B \quad \forall\bar{x}(G \rightarrow (H \leftrightarrow \exists\bar{y}B))$$
$$\text{Propagation: } H \Rightarrow G|B \quad \forall\bar{x}(G \rightarrow (H \rightarrow \exists\bar{y}B))$$
$$\text{Simpagation: } H_1 \backslash H_2 \Leftrightarrow G|B$$
$$\forall\bar{x}(G \rightarrow (H_1 \wedge H_2 \leftrightarrow \exists\bar{y}\, H_1 \wedge B))$$

Fig. 2. Syntax and Semantics of CHR Rules

Operationally, a simplification rule replaces instances of the CHR constraints H by B provided that the guard G holds. A propagation rule instead just adds body B to H without removing anything. The hybrid simpagation rule removes matched constraints in H_2 but keeps constraints in H_1. More details about the syntax and semantics of CHR can be found in[5].

4.2 Service Process into CHR

The functionality provided by VO services are expressed by their IOPEs. A service can have any number of inputs representing the information that is required for the performance of the service. It can have any number of outputs, the information that the service provides to the requester. There can be any number of preconditions, which must all hold in order for the service to be successfully invoked. Finally, the service can have any number of effects, which specifies what changes to the state of the world are produced during the execution of the service. The evolution of a service execution can be captured by a propagation CHR which takes *Inputs* as its head, *Preconditions* as its guard, *Outputs* and *Effects* as its body. However, it should be noted that, when *Preconditions* and *Effects* have constraints in common, the CHR for this service should be a simpagation rule with the common constraints as its heads-to-be-removed part (H_2 part in Figure 2.). The execution of *LogisticService* in Section 3.2 can be defined as the following CHR:

(X_ls,X_sl)::hasInputStartLocation,
(X_ls,X_dt)::hasInputDestination,
(X_ls,X_pa)::hasInputProductAmount,
(X_ls,X_gs)::hasInputGoods,
(X_ls,X_st)::hasInputStartTime \ (X_cl=X_sl)
<=>
X_ls::'LogisticService', (X_gs,X_cl)::currentLocation |
(X_ls,X_price)::hasOutputPrice,(X_ls,X_eta)::hasOutputETA,
(X_cl=X_dt), serviceInvoked(PC,'LogisticService').

Note a constraint *serviceInvoked(Step,ServiceName)* was added into the constraint base after the *LogisticService* was invoked. This constraint records service invocation events as well as their order (by the self-incremental integer PC).

4.3 SWOCRL Translated into CHR

SWOCRL is based on OWL and SWRL, and relies on the conceptual structure defined by OWL. SWOCRL extends the rigid description-logic formulas of OWL with more flexible object constraints that are common in VO modeling. The basic syntax of SWOCRL has been described in Section 3.1. Every SWOCRL rule contains two parts: *Antecedent* and *Consequent*. Note that a constraint in *Antecedent* performs an ask operation while a constraint in the *Consequent* performs a tell operation to the constraint store in sense of CCP. The translation from SWOCRL to CHR is straightforward: atoms in antecedent form the Head part of translated CHR while consequent is the body of CHR. For example, the rule in Section 3.2 can be translated into the following CHR rule:

X_ls:: 'LogisticService',
(X_ls,X_tr):: byTransportation,
(X_ls, X_st):: hasInputStartTime,
(X_ls,X_eta):: hasOutputETA,
(X_ls, X_sl)::hasInputStartLocation,
(X_ls, X_dt)::hasInputDestination
==>X_duration = duration(X_tr, X_sl, X_dt)|
time-sum(X_st, X_duration, X_eta).

4.4 VO Coordination in Supply Chain for Agricultural Product

Coordination of organizations involved in Supply Chain of Agricultural Products can be seen as service invocations in our framework. The desire of organizations can be seen as requests for services described in above VO model.

Using the CHR rules for OWL and SWOCRL reasoning as well as service execution, a derived knowledge base can be built upon the CHR constraints translated from extended OWL-S ontology.

Service request should contain two parts of information: service inputs and desired goal state. In addition, the organization should have a profile describes personal information like credentials and credit that will be used for required preconditions for service invocation.

A special CHR rule can be composed and added into CHR rule base with the desired goal state as its head and guard. After obtaining the derived knowledge base, inputs in the service request are added into the knowledge base. Services that have their inputs and preconditions satisfied will execute and have the knowledge updated. It is obvious that when this rule is fired, the desired goal state is reached. Then in the rule body some cleanups and output can be done. This ends the coordination. If the information provided by the requester is not sufficient, CHR execution will stop and additional information can be added into knowledge base.

5 Prototype System Architecture

The framework proposed in this paper can be implemented combining SW toolkit such as Jena or Protégé and prolog systems such as SWI-Prolog or ECLiPSe. The architecture is shown in the following figure:

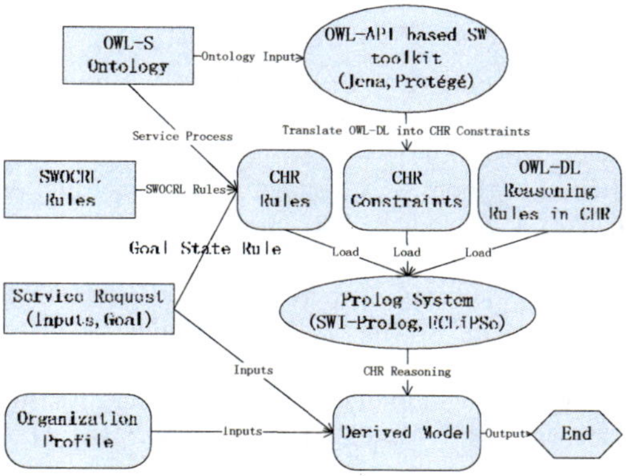

Fig. 3. Prototype System Architecture

First, the extended OWL-S ontology description is read into ontology base by SW toolkit. This ontology base can be translated into CHR constraints as described in [10].

Then the SWOCRL rules and service processes descriptions in OWL-S ontology are translated into CHR based on the CHR constraints generated in the first step.

All these CHR rules are loaded into some Prolog environment supports CHR; CHR constraints are then added into CHR constraint store. The execution of CHR rules will get a derived knowledge base.

A special CHR rule will be composed based on the goal part of service request and loaded into knowledge base. Then inputs are added into knowledge base until the goal state is reached and execution ends.

6 Conclusion

This paper presented a VO modeling framework combing an extended OWL-S service ontology and SWOCRL rules. The major concepts in VO such as organizations and activities are modeled as classes in OWL-S service ontology such as *Participant* and *Process*. The language SWOCRL allows the representation of object centered constraint rules. SWOCRL is especially suitable for describing Virtual Organization concepts under multi-agent environment, which is common in service-oriented applications. This combination utilizes the formal semantics offered by OWL-S service ontology and the expressive power offered by SWOCRL rules thus will have broad application prospects.

Based on the OWL-DL constraint solver in CHR[10], ontology description can be converted into CHR constraints. In addition, OWL-S offers formal semantics for service execution by their IOPEs. This service execution model and SWOCRL rules can be captured by propagation CHR rules. All these CHR rules and constraints are loaded into Prolog systems. After CHR reasoning, derived knowledge base is available for further manipulation.

Acknowledgment. This research is funded by the Natural Science Foundation of China (NSFC, Grant No.60873237) and partially supported by Beijing Key Discipline Program.

References

1. Foster, I.: The Anatomy of the Grid: Enabling Scalable Virtual Organizations. In: Sakellariou, R., Keane, J.A., Gurd, J.R., Freeman, L. (eds.) Euro-Par 2001. LNCS, vol. 2150, pp. 1–4. Springer, Heidelberg (2001)
2. Martin, D., Burstein, M., Hobbs, J., Lassila, O., McDermott, D., McIlraith, S., Narayanan, S., Paolucci, M., Parsia, B., Payne, T.R., Sirin, E., Srinivasan, N., Sycara, K.: OWL-S: Semantic Markup for Web Services
3. Lejian, L., Liehuang, Z.: Semantic Web Modeling for Virtual Organization: A Case Study in Logistics. In: Mizoguchi, R., Shi, Z.-Z., Giunchiglia, F. (eds.) ASWC 2006. LNCS, vol. 4185, pp. 602–608. Springer, Heidelberg (2006)
4. Lejian, L., Liehuang, Z., Jing, Q.: Ontological Modeling of Virtual Organization Agents. In: Shi, Z.-Z., Sadananda, R. (eds.) PRIMA 2006. LNCS (LNAI), vol. 4088, pp. 220–232. Springer, Heidelberg (2006)
5. Frühwirth, T.: Theory and practice of constraint handling rules. The Journal of Logic Programming 37(1-3), 95–138 (1998)
6. Chalmers, S., Gray, P.M.D., Preece, A.D.: Supporting Virtual Organisations Using BDI Agents and Constraints. In: Klusch, M., Ossowski, S., Shehory, O. (eds.) CIA 2002. LNCS (LNAI), vol. 2446, pp. 184–226. Springer, Heidelberg (2002)
7. Grüninger, M., Menzel, C.: The process specification language (PSL) theory and applications. AI Mag. 24(3), 63–74 (2003)
8. SWRL: A semantic web rule language combining OWL and RuleML: `http://www.w3.org/Submission/SWRL/`
9. Saraswat, V.A.: Concurrent constraint programming. The MIT Press (1993)
10. Frühwirth, T.: Description Logic and rules the CHR way. In: Proc. Fourth Workshop on Constraint Handling Rules, Porto, Portugal (2007)

Question Answering via Semantic Web Service Composition

Liu Wang, Lejian Liao, and Xiaohua Wang

School of Computer Science, Beijing Institute of Technology
Beijing Key Lab of Intelligent Information
Beijing, 100081, PRC

Abstract. Web question answering (WQA) and Semantic Web services are currently two separate fields in intelligent web research, but in practical application, they are usually highly correlative. This article combines the techniques in Natural Language Processing (NLP) and Hierarchical Task Network (HTN), and puts forward a WQA framework that is based on Semantic Web service composition. This framework uses NLP to analyze questions, describes the questions with predicate logic, maps them to relative Semantic Web services, and transforms question answering to a planning problem by constructing question domain and planning domain. It dynamically composites relative services making use of HTN system SHOP2, gets the planning series with Semantic Web services, and finally obtains the result that satisfied customers' need.

Keywords: WQA, Semantic Web services, dynamic service composition, OWL-S, NLP, HTN.

1 Introduction

Nowadays, web question answering system uses natural language processing and data mining techniques to find the best answer for user's questions, which has a considerable successfully match rate for WH questions. It extracts the answers from web page texts. But usually text is only a minor part of web resources, and massive data are saved in their database, which are hard to obtain by keyword searching, which usually exposes their interface to Web services to facilitate revoking by external programs [1]. There are some practical applications based on web question answering system to answer non-WH questions like "How to drive to Moscow from Beijing?" "How can I buy a second hand car with a price between 30 and 50 thousand dollars?" "I want to buy a science fiction movie disk that Schwarzenegger is protagonist, what shall I do?" When coming cross with non-WH questions, users will not be satisfied with the standard solutions that are just extracted from massive web pages or knowledge database, in other words, these solutions will not meet user's questions [2]. Users need extra information on the method to solve problems and advisements on solving steps. Web service is a kind of modular and self-expressive program described, published, located and revoked in web. It could exactly solve the problem raised above, but sole Web service is oriented constant problems and system integrations. It is

D. Zeng (Ed.): Advances in Control and Communication, LNEE 137, pp. 29–37.
springerlink.com

short of human-computer interaction and intelligent processing, and could not accept user's questions. When questions are complicated, users usually expect the system to provide a comprehensive solution rather than some simply extracted texts. Adopting WQA alone or just Semantic Web service composition is hard to satisfy users' various needs [3]. On this occasion we need to combine them together.

This paper puts forward a WQA framework that is based on Semantic Web services and Web services Compositions. The system analyzes the user's questions using NLP, gets their formalization descriptions and maps them to related Semantic Web services. After that the system gets domain problem satisfied HTN planning requirement and applies SHOP2 planner to conduct task planning [4] [5]. Finally the system obtains the sequence of Web services that meets user's needs.

The rest of the paper is organized as follows. Section 2 provides the conceptual model. Section 3 contains the framework. Section 4 covers the engine which will resolve nature language, discover Semantic Web services and express the logic formally. Section 5 provides conclusions and future work.

2 Conceptual Model

Let's illustrate with a case. A customer wants to buy all DVD directed by James Cameron, then there are four processing steps: First, query all movies directed by James Cameron, the Web service of querying movies by director is denoted as SearchMoviebyDirectorService. Second, according to the movies list, query stock of the DVD shop, the Web service of querying stock is denoted as SearchDvdbyNameService. Third, placing order by commodity ID. Web service for goods ordering is denoted as OrderService. Fourth, making payment of the order. Web service for payment with credit card is denoted as PayByCreditCardService.

The four steps are finished by WQA system sequentially, precondition is that users' questions have been processed by natural language, translated into expressions that are understandable by computer, after that, users demand is divided into subtasks and target task, then maps subtasks and target task to Web services respectively, then composite Web services by HTN, at last revoke Web services to achieve users' aim.

The difficulties in technique lie in three respects, transformation of questions to formalization description, mapping of users' tasks to Web services and the use of HTN. We use two-tuples to describe users' question, $Q(\Phi, \Psi)$.

Φ signify restrictive sets.
$\Phi:=\{\varphi_1, \varphi_2, \varphi_3 \ldots \varphi_n\} \mid n \in (1,2,3\ldots)$
Ψ signify target sets.
$\Psi:=\{\psi_1, \psi_2, \psi_3 \ldots \psi_n\} \mid n \in (1,2,3\ldots)$

For example, We can use $Q(\Phi,\Psi)$ to describe this question, "If the highway from Beijing to Shanghai is closed, how can I drive to Shanghai from Beijing?".

Φ:={HighwayBlock(Beijing, Shanghai), City(Beijing), City(Shanghai)},
Ψ:={Drive(Beijing, Shanghai)}.

So user's question can be formalized by natural language processing.

Natural Language Formalization Processing: analyze the grammar of users' input, divide sentences into words using strings marching method, analyze questions and grammar by manual regulation method, we only focus on some phrases, these includes noun phrase (np), verb phrase (vp), adjective phrases(ap), adverb phrases (advp), prepositional phrase (pp)…, among which np, vp and pp are the most important in understanding questions, according to the results of syntactic analysis and the presetting regulations, describes phrases with predicate logic, according to the generated sentence structure march predicate logic to users' question $Q(\Phi,\Psi)$.

The Application of Ontology in Semantic Web Service Mapping: according to specific ontology information, analyze the formalization descriptions introduced above with ontology, query Semantic Web service in relative domain, transform user's target task to Semantic Web service.

The Construction of HTN on Planning Questions: transform query and the mapping of Web service into formalization descriptions, format the functions set and methods set in HTN planning domain, finally format target task to questions domain.

When user raise a question, firstly extract user's questions and demands using natural language processing, get logic description, clarify its precondition and target. Secondly parse relative Semantic Web service into logic expression of function and method, add to planning domain, turn to planning questions, then run task planning program and get planning solution that meets user's need. Questions modeling are shown in Fig. 1.

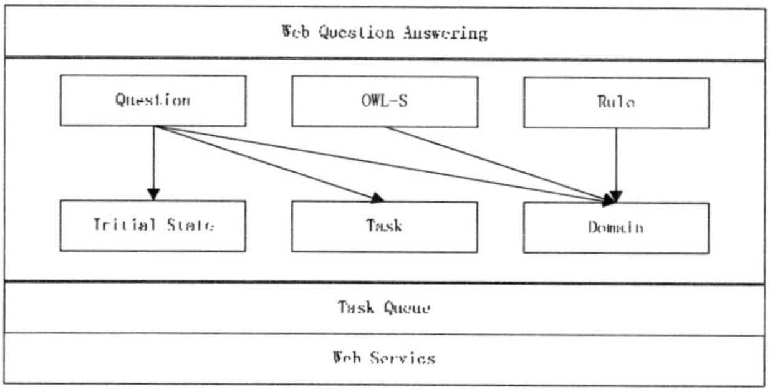

Fig. 1. Logical structure

Generally, we may use triple $P=(\sum, s_0, s_g)$ to describe a planning question. Among which $\sum=(S, A, \gamma)$ is a planning domain, S is a states set, A is an actions set, $\gamma$is a certain type state transition function, s_0 is the initial state of system, s_g is the target state of planning. In Fig. 1, user input questions to WQA, after natural language processing, user's questions come into being initial state, denote as s_0, target task description s_g and planning domain description, denote as $\sum$ which includes operation functions and methods. Semantic Web services in relative domain are parsed into planning domain description $\sum$, system defined rules or user defined rules are parsed into planning domain description $\sum$. Run planning on initial state, target state and planning domain, then output the tasks list, which is the revoking series of Web services, at last get the solution for user's question by revoking Web services.

3 Framework

3.1 Structure

In summary, we propose a framework based on the WQA system of Semantic Web service composition. This system answers the user's question by using the Semantic Web service composition and gives the resolution. The framework structure is shown in Fig. 2.

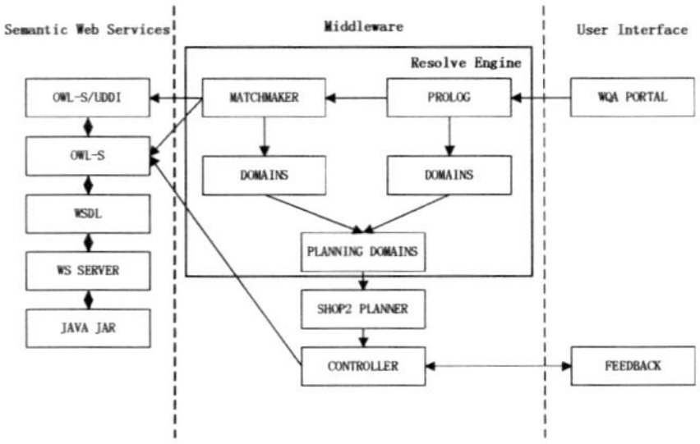

Fig. 2. The framework structure

In this framework, the resolve engine is the important module that combines WQA with Semantic Web services. The system receives the user's question and formalizes it by NLP. These formalization descriptions can be used to construct the ontology and map the Semantic Web services. These Web services are formatted to the domain descriptions that are suitable for HTN. The HTN planning system SHOP2 can use these descriptions to plan. So the user's question is transformed into the planning problem that is suitable for SHOP2. The system can use SHOP2 to process the plan problem and revoke the mapped Web services. Finally, the user's goal task can be accomplished.

There are three modules in this framework.

1) User Interface

This part provide user control portal. The user can ask the question through the WQA portal. Such that, "I want to buy a one-way ticket from Beijing to San Francisco on 9, July", "I want to buy a book of Harry Potter", "I would like to check my record of traffic violations", etc. When the system give options and solutions, the user can give his choice to make the system go for complete the services flow, or give some related advices for some more personal solutions. Such as, "price no higher than $10", "from 1, January to 1, March".

2) Middleware

This module plays a key role in the framework. It receives the question from the user interface and deals with it. The resolve engine parses the natural language that is input by the user and the related Semantic Web services into the domain problems that are suitable for HTN. As the resolve engine's output, a planning domain will be sent to SHOP2 planner. As the resolution, a sequence of semantics Web services will be given to the user. Then according to the user's feedback, the system will go on executing or go back for re-planning.

3) Semantic Web Services

This module is a foundation part that provides Semantic Web service description, performing and publishing. Other parts can revoke these services through the Internet. So the physical location of the services can be different. Any organization or person can publish and provide these services. Also the programming language that implements these services is different only if they follow the unified Semantic Web service protocol.

3.2 System Application

We will describe the whole process of the system to explain how to satisfy the customer's demand by using Semantic Web service composition. The process starts from the input of the user's question and do not end until the user gets the solution by system revoking Web services.

4) Step 1: Question input

The user inputs the question by using the WQA system interface. "I would like to have a self-driving travel by car from Beijing to Shanghai in 7 days, how to plan it and how to rent the car?"

5) Step 2: PROLOG process

This natural language question will be processed by PROLOG. According to the rules and syntactic, PROLOG parses this question into the logic descriptions that is suitable to the problem domain of SHOP2. Such as rent(car), location(Beijing), location(Shanghai), to(Shanghai), from(Beijing), duration(7).

6) Step 3: Semantic Web services analyze

The Semantic Web service matchmaker discovers and matches the related Semantic Web services. Different with [6], here the OWL-S services descriptions that represent the Semantic Web services can be transformed into the operators and methods that are suitable to SHOP2.

7) Step 4: Produce the planning problem

The outputs in step 2 are added to the problem domain of SHOP2. And the outputs in step 3 are added to the planning domain of SHOP2.All of these will be used in the next step.

8) Step 5: Plan

In this step, the SHOP2 module will process the planning problem and give a sequence of task. Such as, rentCar, bookCar, searchRoute, searchWeather, searchServicZone, searchSightSpot, searchHotel, bookHotel. These tasks correspond to a sequence of Web services. Performing the sequence will achieve the user's goal. That is "Rent a car and have a self-driving travel from Beijing to Shanghai in 7 days, how to plan it?"

9) Step 6: Services perform

The controller will handle the performing process of services. If there is a list that service returns for selecting, the controller will select the default selection and continue to perform the sequence of services. Finally, the resolution will be given to the user in visual format; also the alternative options are given. For example the renting car service will return a list of car rental agency. The controller will confirm the default option according to the score of the company and accomplish the remainder services. When the final resolution is given, also other options are given for user's reference.

4 Resolve Engine Model

The resolve engine in our framework combines WQA with Web services and describes the question as the planning problem which is suitable for SHOP2. The resolve engine should to do two things; one is NLP, which resolving the user's question meaning, finding out some related Semantic Web services, another thing is translating the above results into logical expressions. These logical expressions are the HTN domain description that will be processed by the HTN planer subsequently.

4.1 Natural Language Processing

Natural Language Processing (NLP) will be used in our model for question understanding of WQA. Using PROLOG in NLP has some advantages.

First, the natural language grammars can be written almost directly as PROLOG programs, which allow rapid prototyping of NLP systems. Second, semantic descriptions that use logic formalisms are readily produced in PROLOG because of PROLOG's logical foundations. Third, PROLOG is similar to the First Order Logic (FOL), so inviting of PROLOG should make it easily to reason with the knowledge base [7].

Using PROLOG, Grammar of natural language is described as the following rules.

```
sentence(S) :-
 nounphrase(S-S1),
 verbphrase(S1-[]).
 noun([car|X]-X).
 noun([route|X]-X).
 noun([hotel|X]-X).
 noun([sightspot|X]-X).
 verb([buy|X]-X).
 verb([search|X]-X).
 verb([book|X]-X).
 adjective([big|X]-X).
 adjective([brown|X]-X).
 adjective([lazy|X]-X).
 determiner([the|X]-X).
 determiner([a|X]-X).
 determiner([one|X]-X).
 nounphrase(NP-X):-
 determiner(NP-S1),
 nounexpression(S1-X).
 nounphrase(NP-X):-
 nounexpression(NP-X).
 nounexpression(NE-X):-
 noun(NE-X).
 nounexpression(NE-X):-
 adjective(NE-S1),
 nounexpression(S1-X).
```

```
verbphrase(VP-X):-
verb(VP-S1),
nounphrase(S1-X).
```

From these PROLOG scripts, the syntactic structure is settled. We can extend these structures to meet our real requirements. With these codes above, we can use DCG to parse the character list into word list. And then the word list will be transformed into predicate. For example, we can transform the command [rent, car] into rent(car). So user's sentences can be represented as functional commands and predicate commands. Such as "I want to rent car" can be represented as rent(car). "From Beijing to Shanghai" can be represented as location(Beijing), location(Shanghai), to(Shanghai), from(Beijing). "Seven days" can be represented as duration(7).

After processing the user's question with NLP, the Semantic Web matching module will search and discover the related Semantic Web services from UDDI, such as searchCarServer, bookCarServer, searchRouteServer, searchWeatherServer, searchHotelServer and bookRoomServer, etc. These services will be transformed into HTN planning domain subsequently [8].

4.2 HTN Planning Problem Constructing

In SHOP2, Planning problem is represented by a triple P(S, T, D), S is initial state, T is a task list and D is a domain description [9]. With these inputs, SHOP2 will return a plan P which is a sequence of operators that will be executed to make the state transit from S to D and achieve T.

We will use a HTN planning system SHOP2 to solve our problem. The objective of HTN is to produce a serial of actions which perform a specific function or task. For this sequence of actions, we will get a HTN planning domain for Web service dynamic composition. This planning domain includes operators (like atomic functions) and a set of methods which denotes how to decompose a task into subtasks. The function commands, predicate commands which we get in above section are correspond to a problem domain, the Semantic Web services are described as HTN planning domain. Then the problem of WQA is transformed to the problem of planning.

The functional commands and predicate commands using NLP in above section are corresponding to problem domain, which are given as following.

```
(defproblem problem task
  ( location(Beijing)
    location(Shanghai)
    from(Beijing)
    to(Shanghai)
    duration(7)
    vehicle(car) )
   ( (rent (car))
    (travel (Beijing, Shanghai)))
)
```

The problem domain denotes Beijing and Shanghai are the two location names. Beijing is start point, Shanghai is the end point. The time of duration is seven days. The task is car renting and travelling. These descriptions express the user's question that is "I would like to have a self-driving travel by car from Beijing to Shanghai in 7 days, how to plan it and how to rent the car?"

The planning domain will be extracted from the Semantic Web services. OWL-S describes the semantic of the Web services. These OWL-S marks will be resolved to the planning domain of SHOP2. As follows, we will introduce how to transform the input and output of OWL-S to the planning domain of SHOP2.

1) Semantic Web service: searchCarServer

Postcondition:
Service name: searchCar: searchCarServer
Input: location: Location, date: Date
Output: carinfolist: List<CarInfo>
Should be written into

```
(:operator
  (!searchCar ?location ?date)
  ((location ?loation)  (have ?date) )
  ()
  (have ?carinfolist)
)
```

2) Semantic Web service: bookCarServer

Precondition: searchCar: searchCarServer
Service name: bookCar: bookCarServer
Input: carinfo: CarInfo
Output: car: Car
Should be written into

```
(:operator
  (!bookcar ?car)
  (have ? carinfo))
  ()
  (have ?car)
)
```

3) OWL-S composition

if: have car information
then: involve bookCarServer
if: do not have car information
then: involve searchCarServer

```
(:method
   (buy ?car)
   (have ?carinfo)
   ((!bookcar ?car))
   (nothave ?carinfo)
   ((!searchCar ?location ?date)
   (!bookcar ?car) )
)
```

Here, we give an example to explain how to transform the OWL-S semantic description of querying and renting cars to the operators and the method set of SHOP2. In our works, all of the related Semantic Web services will be described. The resolving process and the result will be more complex. So, the question in WQA system and the Semantic Web services revoking are corresponding to the problem domain and the planning domain of HTN. We can accomplish the Semantic Web service dynamic composition by using SHOP2. Then we will get the solution of user's question which is put forward in WQA system.

5 Conclusion and Future Work

We have described a framework, which is combined by the WQA system and the Semantic Web service composition, supported by HTN and NLP. The resolve engine is provided as a bridge to connect the two systems. We have explained how to change the question into a problem domain and transform the basic OWL-S description into the planning domain; also, we have given an overview of project that change the WQA into planning problem and employ SHOP2 planner in combination with one or more Semantic Web services.

In summary, by the resolve engine, the framework allows the user get the dynamic services through Internet on demand, meet his need. Our component model allows developers to extend their individual function, to satisfy clients' requirements.

For future research, an efficient human feedback algorithm will be studied to improve the adaptability. Service protocol and contract shall be carried out in the system to restrict the services flow involving, Semantic Web service matching mechanism should be optimized by semantic to discover more accurate services in the Internet.

Acknowledgment. This work was supported by the Beijing Natural Science Foundation (Grant No. 4092037).

References

1. Kaisser, M.: Web question answering by exploiting wide-coverage lexical resources. In: Proceedings of the 11th ESSLLI Student Session, pp. 203–213 (2006)
2. Dumais, S., Banko, M., Brill, E., Lin, J., Ng, A.: Web question answering: Is more always better? In: The 25th Annual International ACM SIGIR Conference on Research and Development in Information Retrieval, SIGIR 2002 (2002)
3. Martin, D., Paolucci, M., McIlraith, S., Burstein, M., McDermott, D., McGuinness, D., Parsia, B., Payne, T., Sabou, M., Solanki, M., Srinivasan, N., Sycara, K.: Bringing Semantics to Web Services: The OWL-S Approach. In: Cardoso, J., Sheth, A.P. (eds.) SWSWPC 2004. LNCS, vol. 3387, pp. 26–42. Springer, Heidelberg (2005)
4. Nau, D., Au, T., Ilghami, O., Kuter, U., Murdock, J., Wu, D., Yaman, F.: SHOP2: an HTN planning system. J. Artif. Intell. Res. 20, 379–404 (2003)
5. Erol, K., Hendler, J., Nau, D.: Semantics for Hierarchical Task Network Planning. CS Technical Report TR-3239, University of Maryland (1994)
6. Wu, D., Sirin, E., Hendler, J., Nau, D., Parsia, B.: Automatic Web Services Composition Using SHOP2. Department of Computer Science University of Maryland (2004)
7. Park, S.-J., Kim, J.-H., Park, H.-G.: Reasoning relation among RDF/RDFS resources using PROLOG rules and facts. In: Cybernetics and Intelligent Systems (2004)
8. Sirin, E.: Combining description logic reasoning with ai planning for composition of web services. Ph.D. Dissertation, University of Maryland, College Park (2006)
9. Sirin, E., Parsia, B., Wu, D., Hendler, J., Nau, D.: HTN planning for web service composition using SHOP2. Journal of Web Semantics 1(4), 377–396 (2004)

Quality Evaluation Model Study of B2C E-Commerce Website

Zhiping Hou

School of Management, Guilin University of Technology, Guilin Guangxi, China
birdinmorning@163.com

Abstract. In order to solve the problem of B2C E-commerce website quality evaluation, the paper establishes an evaluation method based on Analytic Hierarchy Process and fuzzy comprehensive evaluation. Firstly, the paper describes the importance of website quality and builds a quality evaluation indexs system. Secondly, it determines indexs weight by using Analytic Hierarchy Process. Thirdly, the fuzzy comprehensive evaluation method is used to establish the B2C website quality evaluation model. Finally, according to questionnaires, the paper use the model to evaluate two B2C websites in Guangxi province, the case studies not only prove the rationality and scientificity of the evaluation model, but also put forward some proposals of website development.

Keywords: B2C Website, Quality Evaluation, AHP, Fuzzy Comprehensive Evaluation.

1 Introduction

In the internet times, B2C E-commerce website is not only a business interface directly to the customers, but also an important platform which enterprises can provide agile and high-quality services to the customers. Internationally, the United States, Japan and other developed countries, E-commerce and related research reached a certain scale. At present, China has emerged a multi-level and service diversity E-commerce websites, such as Taobao, Paipai, Dangdang, Amazon and other websites. In addition, Sina Mall, Sohu Mall and other portals open up shopping sites, Haier-online, Gome, Suning, and other traditional enterprises also carried out online sales. Therefore, the quality evaluation of B2C E-commerce website has important practical significance. In this paper, based on the perspective of quality evaluation, established B2C E-commerce website quality evaluation index system, and made the quality evaluation model of B2C e-commerce website.

2 Quality Evaluation Index System of B2c E-Commerce Website

Because the application of Internet in business less than 20 years, the quality research of business websites do not have mature theoretical foundation and framework, related disciplines and research areas to be further developed. DeLone & McLean put forward the information system success model when researched the evaluation factor of

D. Zeng (Ed.): Advances in Control and Communication, LNEE 137, pp. 39–47.
springerlink.com

information system effectiveness. The model include two key factors: information quality and system quality[1]. According to the D&M information system success model,

Mckinney et al proposed two corresponding dimension: website information quality(web-IQ) and website system quality(web-SQ), based on the factor of website user satisfaction[2]. Later, DeLone & McLean based on the original model, add "service quality" evaluation factor in the new information system success model[3]. Li & Amett visited the 1000 enterprise's 122 network managers, proposed system use, system quality, service quality, information quality were Website critical success factors[4]. After empirical analyzing, Negasha, Ryanb & Ighariab considered that there were positively correlated between information quality, system quality, service quality and the effectiveness of web-based customer support system[5].

In the 21st century, enterprises are facing an increasingly competitive market environment, how to provided knight sarvice become strategic issues, especially in E-commerce enterprise. Therefore, this paper select information quality, system quality, service quality three factors to evaluate B2C E-commerce website. The specific indexs are as follows:

a) Information Quality

i. Accuracy of the Information: *Information symbols expressed information is accurate[6].*
ii. Information Format Integrity: *The classification of B2C website information is scientific and rational, the information content is easy to understand.*
iii. Practicality: *Information practicality reflect that the information is to meet the actual needs of customers.*
iv. Comprehensive Information: *Comprehensiveness of the information include two meanings: first, coverage of information should be able to meet the needs of different users, second, the website should include extensive and comprehensive information.*
v. Information Update Rate: *Because of obvious time limit of information, beyond the time limit will lose the value of information[6]. So, the information update rate must be promptly.*
vi. Information Credibility: *The information credibility refers to the information is to be trustworthy and authority.*

b) System Quality

i. Transaction Security: *To ensure the inviolability of user's personal information and ensure transaction security.*
ii. Easy to Use: *The using convenience of website include three aspects: easy to manipulate,easy to retrieval, easy to release information.*
iii. Response Time: *The website can quickly respond to user request.*
iv. System Stability: *The error rate of website system is low and the performance is stability[7].*

v. Website Structure Reasonable: *Website navigation structure is clear and usable. The links provided by website is effective.*

vi. System Function: *The common function of B2C E-commerce website is complete, and it can meet the all transaction of E-commerce.*

c) Service Quality

i. Human Interface: *The original meaning come from the performance of physical equipment and service personnel. In the website service quality, it refers to the human design and use.*

ii. Personalized Recommendation: *According to the user's purchases, speculate the products that the user may be interest in, and recommendate the products to the user. Personalized recommendation has become an important part of E-commerce personalized service[8].*

iii. Reliable Service: *The website and the staff can provide the service accurately and effectively.*

iv. Interactivity: *Whether the user feedback comments can be received quickly. Whether the website has interactive functions, such as chat room, BBS.*

v. Customer Satisfication: *Customer satisfication in the entire transaction process.*

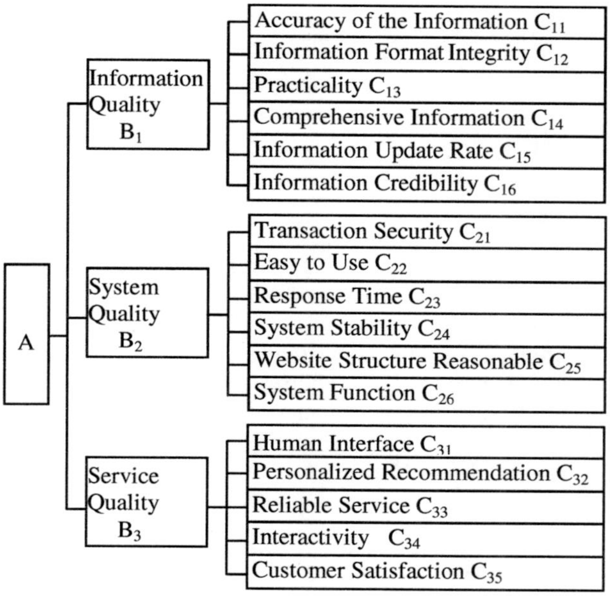

Fig. 1. Quality Evaluation Index of B2C E-Commerce Website

3 Determine Indexs Weight

This paper uses Analytical Hierarchy Process (AHP) to determine indexs weight. AHP is a kind of method which can translate semi-qualitative and semi-quantitative problems to quantitative problems.

a) The Basic Principle of AHP

i. Establish comparsion matrix A: *The first step is to establish the hierarchical structure, then compare relative importance of each indexs which belong to the same level index, finally, establish comparsion matrix A.*

ii. Calculate each row product of comparsion matrix M:

$$M_i = \prod_{j=1}^{n} a_{ij} \quad , \quad i = 1,2,\ldots\ldots,n \tag{3.1}$$

iii. Calculate Nth root of M_i:

$$\overline{w_i} = \sqrt[n]{M_i} \tag{3.2}$$

iv. Vector normalization:

$$w_i = \frac{\overline{w_i}}{\sum_{i=1}^{n} \overline{w_i}} \tag{3.3}$$

v. Consistency verify:

$$C.I. = \frac{\lambda_{\max} - 1}{n - 1}. \tag{3.5}$$

$$C.R. = C.I./R.I. \tag{3.6}$$

In the above formula, *C.I.* is coincidence index, *C.R.*is random coincidence ratio, *n* is order of comparsion matrix, *R.I.* is random coincidence index. λ_{max} is the largest eigenvalue of comparsion matrix, its calculation is as follow:

$$\lambda_{\max} = \frac{1}{n} \sum_{i=1}^{n} \frac{(AW)_i}{W_i} \quad , \quad i = 1,2,\ldots\ldots,n \tag{3.7}$$

When the *C.R.* is less than or equal to 0.10, that means the matrix has coincidence, otherwise should re-adjust the elements of comparison matrix until it reaches the coincidence.

b) Determine evaluation index weight

This paper invited 10 experts to score the indexs independently. By using the method of Delphi, the adjusted comparison matrix are as follow tables.

Table 1. Comparison Matrix: A-B

A-B	B_1	B_2	B_3	W
B_1	1	2	1	0.4126
B_2	1/2	1	1	0.2599
B_3	1	1	1	0.3276

n=3, $R.I.$= 0.58, λ_{max}=3.0598, $C.I.$= 0.0299, $C.R$=0.0516<0.1.

Table 2. Comparison Matrix: B_1-C_1

B_1-C_1	C_{11}	C_{12}	C_{13}	C_{14}	C_{15}	C_{16}	W_1
C_{11}	1	6	3	5	4	2	0.3936
C_{12}	1/6	1	1/3	1/2	1/2	1/4	0.0512
C_{13}	1/3	3	1	3	2	1/2	0.1579
C_{14}	1/5	2	1/3	1	1/2	1/3	0.0697
C_{15}	1/4	2	1/2	2	1	1/2	0.1043
C_{16}	1/2	4	2	3	2	1	0.2233

n=6, $R.I.$= 1.24, λ_{max}=6.1085, $C.I.$= 0.0217, $C.R$=0.0175<0.1.

Table 3. Comparison Matrix: B_2-C_2

B_2-C_2	C_{21}	C_{22}	C_{23}	C_{24}	C_{25}	C_{26}	W_2
C_{21}	1	2	3	2	4	3	0.3212
C_{22}	1/2	1	2	1/2	3	2	0.1685
C_{23}	1/3	1/2	1	1/3	3	2	0.1168
C_{24}	1/2	2	3	1	4	3	0.2550
C_{25}	1/4	1/3	1/3	1/4	1	1/2	0.0613
C_{26}	1/3	1/2	1/2	1/3	2	1	0.0772

n=6, $R.I.$= 1.24, λ_{max}=6.1742, $C.I.$= 0.0348, $C.R$=0.0281<0.1.

Table 4. Comparison Matrix: B_3-C_3

B_3-C_3	C_{31}	C_{32}	C_{33}	C_{34}	C_{35}	W_3
C_{31}	1	1/2	1/3	2	1/4	0.0994
C_{32}	2	1	1/2	3	1/2	0.1771
C_{33}	3	2	1	4	1/2	0.2685
C_{34}	1/2	1/3	1/4	1	1/5	0.0627
C_{35}	4	2	2	5	1	0.3924

n=5, $R.I.$= 1.12, λ_{max}=5.0649, $C.I.$= 0.0162, $C.R$=0.0145<0.1.

In the above table, W is the weight matrix of first level, W_1-W_3 are the weight matrix of second level.

4 Quality Evaluation Model of B2c E-Commerce Website

In fact, as it is difficult to define index's business, the specific score for each index is a very difficult task. Therefore, we can use fuzzy comprehensive evaluation method to evaluate B2C E-commerce website quality.

Using fuzzy comprehensive evaluation model for B2C E-commerce website quality evaluation, the specific steps are as follows:

i. Determine the evaluation degree domain: *Supposing the number of possible reviews are m, then the reviews set* $V=\{y_1,y_2,\ldots,y_m\}$.

ii. Evaluation of single index: *To evaluate each evaluation index* U_i, *then you can obtain single evaluation matrix* $R_i=(r_{ij,k})_{ni*m}$. *In this formula,* $i=1,2,\ldots,s$; $j=1,2,\ldots,n_i$; $k=1,2,\ldots,m$; $r_{ij\cdot k}$ *is membership degree of index* x_{ij} *to review* y_k.

iii. Give the weight of each index:

$$\omega_i=\{a_{i1},\ a_{i2},\ldots,\ a_{ini}\ \}.$$

This paper takes AHP to determine the index weight which have been completed in the chapter three.

ω =(0.4126, 0.2599, 0.3276)

ω_1 =(0.3936, 0.0512, 0.1579, 0.0697, 0.1043, 0.2233)

ω_2 =(0.3212, 0.1685, 0.1168, 0.2550, 0.0613, 0.0772)

ω_3 =(0.0994, 0.1771, 0.2685, 0.0627, 0.3924)

iv. Obtain the final reviews of U_i:

$$B_i=\omega_i\cdot R_i=(b_{i1},b_{i2},\ldots,b_{im}),\qquad i=1,2,\ldots,s.$$

v. Obtain fuzzy evaluation matrix: *Treat* U_i *as a single element,* B_i *is single index evaluation vector of* U_i, *obtain the fuzzy evaluation matrix* R.

$$R=\begin{pmatrix}B_1\\B_2\\\vdots\\B_j\end{pmatrix}=\begin{pmatrix}b_{11}&b_{12}&\cdots&b_{1m}\\b_{21}&b_{22}&\cdots&b_{2m}\\\cdots&\cdots&\cdots&\cdots\\b_{j1}&b_{j2}&\cdots&b_{jm}\end{pmatrix}$$

According to the index weight ω, you can obtain the final reviews vector of *U*.

$$B=\omega\cdot R=(b_1,b_2,\ldots,b_m)$$

vi. Calculate the final evaluation score: $G=B\times V^T$.

5 Case Study

Lo-go and Yoda are B2C E-commerce website located in Guangxi province. In 2005, Guangxi Boxin networks company created Lo-go website to provide a full range integrated information service for Guangxi local customers. Lo-go website aimed to create a safe, stable, honest E-commerce environment, the business of Lo-go website include B2B, B2C, C2C transaction. Yoda website, set up in 2008, intented to building the most outstanding online shopping platform on campus. The major business of Yoda website include information media, online mall, product experience center, etc.

In this paper, using the above method, according to questionnaires, invited 100 person in five aspects of undergraduate(50) major in E-commerce, postgraduate(20) major in business administrator, computer teacher(10), management teacher(10) and enterprise executives(10) to evaluate the two website's quality. The specific steps are as follows:

(i) Determine the evaluation degree domain: V=*{Very High, High, Medium, Low, Very Low }.*

(ii) Determine the evaluation membership matrix R: *The evaluation data of Lo-go and Yoda website are in table 5.*

Table 5. Expert Evaluation Data of Lo-go AND Yoda

INDEX		Very High	High	Medium	Low	Very Low
B_1	C_{11}	0.28, 0.22	0.31, 0.33	0.24, 0.21	0.13, 0.17	0.04, 0.07
	C_{12}	0.25, 0.28	0.43, 0.49	0.19, 0.14	0.11, 0.09	0.02,0
	C_{13}	0.23, 0.24	0.35, 0.32	0.22, 0.24	0.15, 0.17	0.05, 0.03
	C_{14}	0.25, 0.21	0.33, 0.29	0.26, 0.23	0.16, 0.24	0, 0.03
	C_{15}	0.19, 0.17	0.35, 0.31	0.29, 0.30	0.15, 0.17	0.02, 0.05
	C_{16}	0.24, 0.22	0.38, 0.37	0.27, 0.29	0.08, 0.10	0.03, 0.02
B_2	C_{21}	0.25, 0.27	0.32, 0.35	0.33, 0.36	0.07, 0.02	0.03,0
	C_{22}	0.27, 0.24	0.35, 0.30	0.25, 0.26	0.11, 0.14	0.02, 0.06
	C_{23}	0.28, 0.25	0.39, 0.35	0.27, 0.24	0.05, 0.12	0.01, 0.04
	C_{24}	0.26, 0.27	0.37, 0.32	0.28, 0.30	0.09, 0.07	0, 0.04
	C_{25}	0.28, 0.32	0.32, 0.33	0.25, 0.27	0.13, 0.08	0.02,0
	C_{26}	0.27, 0.23	0.38, 0.29	0.26, 0.32	0.05, 0.10	0.04, 0.06
B_3	C_{31}	0.22, 0.17	0.30, 0.32	0.32, 0.29	0.09, 0.15	0.05, 0.07
	C_{32}	0.24, 0.22	0.32, 0.30	0.31, 0.29	0.13, 0.15	0, 0.04
	C_{33}	0.25, 0.24	0.33, 0.28	0.29, 0.31	0.07, 0.09	0.06, 0.08
	C_{34}	0.22, 0.21	0.27, 0.30	0.35, 0.29	0.10, 0.13	0.06, 0.07
	C_{35}	0.27, 0.23	0.29, 0.24	0.32, 0.29	0.12, 0.19	0, 0.05

(iii) Fuzzy comprehensive evaluation of the level 1:

To the Lo-go Website:

ω_1 =(0.3936,0.0512,0.1579,0.0697,0.1043,0.2233)

$$R_1 = \begin{pmatrix} 0.28 & 0.31 & 0.24 & 0.13 & 0.04 \\ 0.25 & 0.43 & 0.19 & 0.11 & 0.02 \\ 0.23 & 0.35 & 0.22 & 0.15 & 0.05 \\ 0.25 & 0.33 & 0.26 & 0.16 & 0 \\ 0.19 & 0.35 & 0.29 & 0.15 & 0.02 \\ 0.24 & 0.38 & 0.27 & 0.08 & 0.03 \end{pmatrix}$$

$B_1 = \omega_1 R_1 = (0.250, 0.344, 0.248, 0.125, 0.035)$

In a similar way:

$B_2 = \omega_2 R_2 = (0.263, 0.351, 0.286, 0.082, 0.019)$

$B_3 = \omega_3 R_3 = (0.251, 0.306, 0.312, 0.104, 0.025)$

Lo-go Website fuzzy comprehensive evaluation matrix elements of level 1 are as follow:

$$R_{Lg} = \begin{pmatrix} 0.250 & 0.344 & 0.248 & 0.125 & 0.035 \\ 0.263 & 0.351 & 0.286 & 0.082 & 0.019 \\ 0.251 & 0.306 & 0.312 & 0.104 & 0.025 \end{pmatrix}$$

In a similar way, Yoda website fuzzy comprehensive evaluation matrix elements of level 1 are as follow:

$$R_{Yd} = \begin{pmatrix} 0.220 & 0.341 & 0.239 & 0.155 & 0.044 \\ 0.262 & 0.328 & 0.305 & 0.075 & 0.029 \\ 0.224 & 0.267 & 0.295 & 0.148 & 0.059 \end{pmatrix}$$

(iv) Fuzzy comprehensive evaluation of the level 2:

To the Lo-go Website:

$B_{Lg} = \omega \; R_{Lg}$

$$= (0.4126, 0.2599, 0.3276) \begin{pmatrix} 0.250 & 0.344 & 0.248 & 0.125 & 0.035 \\ 0.263 & 0.351 & 0.286 & 0.082 & 0.019 \\ 0.251 & 0.306 & 0.312 & 0.104 & 0.025 \end{pmatrix}$$

$= (0.254, 0.333, 0.279, 0.107, 0.028)$

To the Yoda website:

$B_{Yd} = \omega \; R_{Yd}$

$$= (0.4126, 0.2599, 0.3276) \begin{pmatrix} 0.220 & 0.341 & 0.239 & 0.155 & 0.044 \\ 0.262 & 0.328 & 0.305 & 0.075 & 0.029 \\ 0.224 & 0.267 & 0.295 & 0.148 & 0.059 \end{pmatrix}$$

$= (0.231, 0.313, 0.274, 0.132, 0.045)$

(v) Evaluation results analysis:

a) According to the information system evaluation results, the membership degree of Very High, High, Medium,Low, Very Low of the Lo-go website are 0.254, 0.333, 0.279, 0.107, 0.028 and the Yoda website's are 0.231, 0.313, 0.274, 0.132, 0.045. Thus, according to the principle of maximum membership degree, the fuzzy evaluation results of Lo-go and Yoda website are High.

b) Further processing of the above evaluation results, provision the review elements of the degree domain: $G \leq 0.1$ is Very Low, $G = 0.1 \sim 0.4$ is Low, $G = 0.4 \sim 0.7$ is Medium, $G = 0.7 \sim 0.9$ is High, $G \geq 0.9$ is Very High, and provision Column vector:

$$V = (1, 0.8, 0.6, 0.3, 0.1)$$

Final scores of Lo-go and Yoda website are as follows:

$$G_{Lg} = B_{Lg} \times V^T = (0.254, 0.333, 0.279, 0.107, 0.028) \times (1, 0.8, 0.6, 0.3, 0.1)^T = 0.722$$

$$G_{Yd}=B_{Yd}\times V^T=(0.231,0.313,0.274,0.132,0.045)\times(1, 0.8, 0.6, 0.3, 0.1)^T=0.689$$

According to the final evaluation scores, the website quality of Lo-go and Yoda obtain the reviews of High and Medium respectively. The results are a better description of actual performance of the two websites. Lo-go website has been established a long time, the business is more mature, but Yoda website is a new B2C website, the business to be further improved. Meanwhile, it is noteworthy that the two website do not obtain high scoring, the website quality are required to further improved.

6 Conclusion

In this paper, by establishing the quality evaluation indexs system of B2C E-commerce website, using AHP and fuzzy comprehensive evaluation method to build the B2C website quality evaluation model. According to the evaluation results of Lo-go and Yoda website, the evaluation method has many advantages: (1)The model considers information quality, system quality and service quality three factors and obtains an objective and credible evaluation result.(2)Comparison with the general evaluation, the method combine the quantitative and qualitative data and make the evaluation information more abundant.(3) This method is simple, easy to operate and promote.

Acknowledgment. This paper sponsored by scientific research project of Education Department of Guangxi Zhuang Autonomous Region, project number:201012MS106.

References

1. Christy, M.K., Matthew, C., Lee, K.O.: Consumer Satisfaction with Internet Shopping: A Research Framework and Propositions for Future Research. In: ICEC 2005, Xi'an, China, August 15-17, pp. 327–334. ACM (2005)
2. Mckinney, Yoon, Zahedi: The Measurement of Web-Customer Satisfaction: An Expectation and Disconfirmation Approach. Information Systems Research (3), 296–315 (2002)
3. DeLone, W.H., McLean, E.R.: The DeLone and McLean Model of Information Systems Success: A ten-year update. Journal of Management Information Systems (4), 9–30 (2003)
4. Li, C., Amett, K.P.: Exploring the Factors Associated with Web Site Success in the Context of Electronic Commerce. Information and Management 38(10), 23–34 (2000)
5. Negasha, S., Ryanb, T., Igbariab, M.: Quality and effectiveness in Web-based Customer support systems. Information and Management 40, 757–768 (2003)
6. Cao, R., Wu, J.: Information Quality and Evaluation Index System. Information Research (4), 6–9 (2002)
7. Kim, J., Lee, J., Han, K., Lee, M.: Business as Buildings: Metrics for the Architectural Quality of Internet Businesses. Information Systems Research 13(3), 239–254 (2002)
8. Li, J., Xu, Y., Wang, Y., Chu, C.: Strongest association rules mining for personalized recommendation. Systems Engineering Theory & Practice 29(8), 144–152 (2009)

A New Digital Watermarking Algorithm Based on NSCT and SVD

Xiong Shunqing, Zhou Weihong, and Zhao Yong

School of Mathematics & Computer Science Yunnan Nationalities University, YNU Kunming, China
{xsq324,ynzwh}@163.com, 23874260@qq.com

Abstract. Nonsubsampled Contourlet Transform(NSCT) take the advantages of multi-scale, multi-direction and translational invariance. In this paper, a new digital watermarking algorithm is proposed based on NSCT and Singular Value Decomposition(SVD). First, the low frequency component of image is got by apply Nonsubsampled Contourlet Transform to the original image. Second, the Singular Value Decomposition of low frequency component of image is taken. Last, the watermark information is embedded into the eigenvalue matrix of low frequency component of image. The result of experiment proves that the new algorithm can assure good imperceptibility for watermarked image, and the watermarks which is embedded using the algorithm have good robustness for familiar attacks, such as JPEG compression median filtering and added noise attack, especially good at anti-rotate attack, the NC of new algorithm can be increased by about 0.5 compared with watermarking algorithms based on DWT and SVD.

Keywords: Digital watermark, NSCT, SVD, Robustness.

1 Introduction

The purpose of digital watermarking is protecting the copyright information of the original digital medium. Digital watermarking algorithm contains two classes: spatial domain algorithm and transform domain algorithm. The transform domain algorithm has better robustness than spatial domain algorithm. The DFT、DCT and DWT are the commonly used transform in digital watermarking field. DCT is the kernel transform algorithm of JPEG compression standard. DWT is the kernel algorithm of JPEG2000 compression standard. But these algorithms base on above transform are hardly to resist the geometric attack.

The Contourlet transform proposed by M inh N. Do and Martin Vetterli [1,2] has advantage of multi-direction and anisotropy, catches the geometric features of the image very well, so the watermarking algorithm based on Contourlet transform has better robustness for resistance the common geometric attacking. But the Laplacian pyramid and the directional filter banks do down samplings, causing the Contourlet transform doesn't have translational invariance. The nonsubsampled Contourlet transform proposed by Do and Cunha[3,4] overcome the shortage of Contourlet transform.

D. Zeng (Ed.): Advances in Control and Communication, LNEE 137, pp. 49–57.
springerlink.com

This paper proposes a new watermarking algorithm based on the advantage of nonsubsampled Contourlet transform (NSCT), the new algorithm use both of the NSCT and SVD which decomposes the singular value of low-frequency sub band, take the advantage of the stabilization of singular value to enhance the robustness of algorithm.

2 Contourlet Transform

Contourlet transform [1], is also called Pyramid Directional Filter Banks(PDFB), is the expand of wavelet transform. It is a multi-scale, localization and multi-direction way to express the image, it is a very good two dimension expression of the image's geometry information [2].

2.1 Character of Contourlet Transform

Wavelet transform has a widespread use in image processing, and has best approximation to one dimension bounded variation function. But it can only capture limited information for the multi-dimension information. Contourlet has more primary function than wavelet, it can represent the curve using less variation coefficient. Contourlet transform has the real performance of two-dimensional representation of the image, and is more suitable for capturing the multi-dimension singular point. It has more useful performances as show in following:

- Catch the features of image in multi-scale;
- The coefficients of bands are nonlinear correlation after apply the Contourlet transform to image.
- The Contourlet transform has a redundancy rate of 4/3[2] because of the redundancy of Laplacian pyramid.

2.2 Discrete Contourlet Transform

Two dimension Discrete Contourlet transform is made up of Laplacian pyramid and directional filter banks:

(1) At first, multi-scale analysis of the image is taken by using Laplacian pyramid transformation, resolved to a set of band pass filter banks.

(2) Then remove the low-frequency portion, the singular value in same direction is converged to Contourlet coefficients after apply directional filter banks to high-frequency portion of image. As show in Figure1 is the Contourlet filter banks.

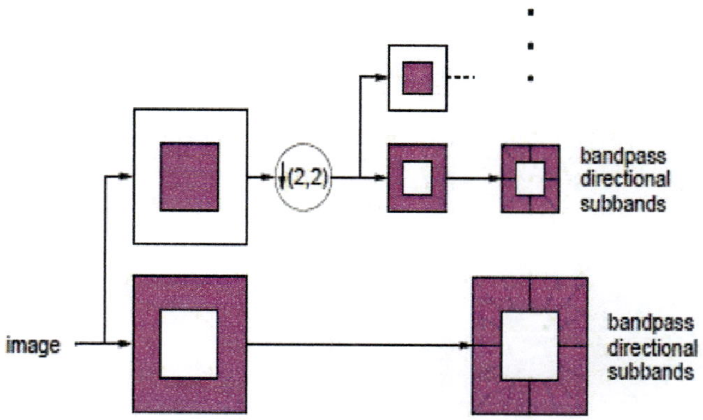

Fig. 1. Dispersed Contourlet transform filter banks

The figure 2 shows the three layer Contourlet transform coefficients of image Lena which after Laplacian pyramid transform, and every layer resolve to 4, 8 and 16 directions details in corresponding.

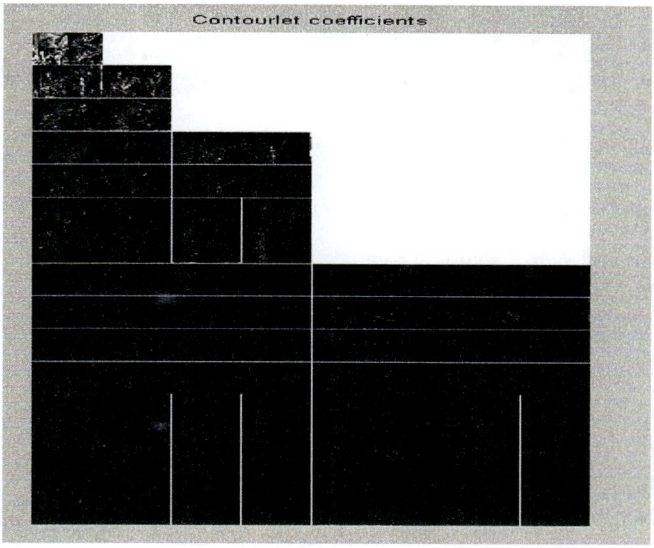

Fig. 2. Three layer Contourlet transform coefficients

2.3 Nonsubsampled Contourlet Transform (NSCT)

Nonsubsampled Contourlet transform (NSCT) has translation invariance, which the Contourlet transform doesn't have. Similar to Contourlet transform, first, NSCT uses nonsampled pyramid filter banks(NSPFT) to analysis image in multi-scale, and then, analysis the high-frequency sub-band in directional by nonsubsampled directional filter banks(NSDFB), the singular value in same direction is converge to one coefficient. If decompose the image in i-scale directional, we can catch 2^i coefficients matrix which have the same size of the original image. The Contourlet transform coefficients of 2 layer nonsmapled Contourlet transform of image Lena is show in Figure 3.

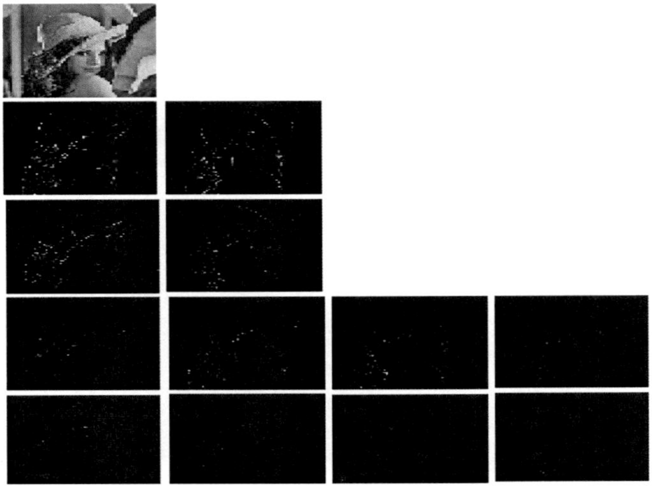

Fig. 3. Coefficients of 2-level nonsampled Contourlet transform in Lena

2.4 Singular Value Decomposition(SVD)

(1) Definition of matrix's singular value

A_{m*n} is an m*n matrix, U_{m*n} and V_{n*m} are unitary matrix, have following forms:

$$U^T AV = \begin{bmatrix} D & 0 \\ 0 & 0 \end{bmatrix} \tag{1}$$

$D = diag(d_1, d_2, \cdots, d_r)$,and $d_1 \geq d_2 \geq \cdots \geq d_r \geq 0$, $d_i (i = 1, 2, \cdots, r)$ is the singular value of matrix A, r is the rank of D.

(2) Stability of the matrix's singular value

Theorem: suppose, $A_{m \times n}, B_{m \times n} \in R^{m \times n}$, their singular value are: $\alpha_1 \geq \alpha_2 \geq \cdots \geq \alpha_n, \beta_1 \geq \beta_2 \geq \cdots \geq \beta_n$, So for every unitary unchanged nozzle $\|\bullet\|$, have following forms:

$$\|diag(\alpha_1 - \beta_1, \alpha_2 - \beta_2, \cdots, \alpha_n - \beta_n)\| \leq \|A - B\| \tag{2}$$

This theorem shows singular value is good at stability.

3 Watermark Embeding and Extracting

3.1 Ideology of This Paper's Plan

First, We get the low-frequency sub-band by apply NSCT to original image, As most of the power focuses on low-frequency sub-ban, and the low-frequency sub-band can able to meet the visual quality. So, embed the watermarking into the low-frequency sub-band can satisfy the invisibility of watermarking, and also has less influence in low-frequency sub-band that process the image by signal processing.

This paper combines nonsubsampled Contourlet transform with SVD, decompose the singular value of nonsampled Contourlet transform's low-frequency sub-bands, embed the watermark information in singular value. As the singular value has certain stability, that means if we do some tiny processing in singular value, which has embedded the watermarking, it doesn't have much change[3]. For the watermarking, it assures the robustness of watermarking algorithm. Because the nonsampled Contourlet transform not only has wavelet's multi-resolution and time-frequency modulation, but also has multidirectional, anisotropy and translation invariance, this algorithm can increase the ability of resistance of geometry attacking.

3.2 Watermarking's Embedding

(1) Apply the Arnold transform to watermark image for k times, the k is the secret key.

(2) Apply the nonsubsampled Contourlet transform to the original image with 3 layer, get the low-frequency and a group of high-frequency sub-bands.

(3) Decomposed the singular value of low-frequency sub-band matrix A with: $A=USV^T$

(4) Plus the information W which has been scrambled into S, get S': $S'=S + aW$;

(5) Decomposed the singular value matrix, which have been embedded the watermark :$S'=U_1S_1V_1^{T}$, S' replace the S, which is in the band, correspond with unitary matrix to re-configurate the low-frequency sub-band.

(6) Get the watermarked image by apply the inverse NSCT transform with the low-frequency sub-band and every band-pass band,

3.3 Watermarking's Extracting

(1) Apply nonsubsampled Contourlet transform to the watermarked image 3 times as above, extract the low-frequency sub-band.

(2) Decomposed the low-frequency sub-band matrix A' with SVD:$A'=U'S_1'V'^T$;

(3) Compute the matrix : $D'= U_1S_1'V^T$;

(4) Extract the singular values and subtract with the singular values of original image which doesn't embed the watermarking, get the watermark signal : $W'=(D'-S)/a$;

(5) Apply the Arnold transform to the watermark signal with T-k times, extract the watermark image.

4 Data Simulation

In our experiment, the gray level image Lena is used as the original image whose size is 512*512, and a 64*64 binary image is used as the watermarking image, the simulation experiments tool is Matlab 7.0, the correlation coefficient NC is used as the similarity standard between the original watermarking image and restore watermarking image, and the comparative and analysis experiment was done between our new algorithm and the algorithm based on the DWT and the SVD.

The watermarking image is scrambled with Arnold transform which make the watermarking information to be more safe and invisibility, and then the watermarking image is embed into the original image.

Fig. 4. Original image and watermarked image

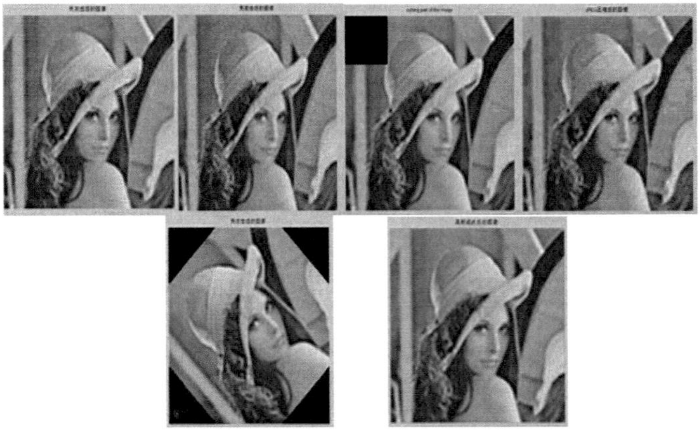

Fig. 5. The watermarked image is attacked by pepper-and-salt noise, Gaussian noise, cutting, JPEG compress, rotation and Gaussina filter.

The distortion of watermarked image is very small compare with the original image as show in figure 4, In order to prove the robustness and the ability of resistance of geometry attacking of the algorithm, several attacking experiments was taken in our paper. The result of watermarked image was attacked by pepper-and-salt noise, Gaussian noise, cutting, JPEG compress, rotation and Gaussian filter are shown in figure 5. And the NC value of corresponding restored watermarking image are shown in Table 1. The results of experiment show that the robustness resist the common geometry attacking of our algorithm is very good.

Table 1. The robustness of the new algorithm

Attack-Type	Pepper-and-Salt noise	Gaussian noise 0.005	Cutting		JPEG Compress 8%
			1/8	1/4	
NC	0.9573	0.9262	0.9613	0.9388	0.9996
Result of Test	OK	OK	OK	OK	OK

JPEG Compressing 10%	Rotating			Median filter 3*3	Gaussian filter
	10 degree	45 degree	60 degree		
0.9984	0.9167	0.8496	0.8359	1	0.9996
OK	OK	OK	OK	OK	OK

Table 2. The result of comparison the new algorithm with DWT and SVD algorithm

Attack-Type	Pepper-and-Salt noise 0.01	JPEG Compression			Cutting 1/8
		1%	8%	10%	
DWT	0.7569	0.5674	0.5567	0.5159	0.7475
SVD	0.9507	0.9972	0.9525	0.9525	0.9507
Our algorithm	0.9573	0.9996	0.9984	0.9926	0.9613

Rotation			Gaussian noise			Gaussian filter
10 degree	45 degree	60 degree	0.002	0.005	0.01	
0.3713	0.3346	0.3186	0.8376	0.8852	0.6195	0.8056
0.8056	0.7930	0.7138	0.9551	0.9060	0.8462	0.9839
0.9167	0.8496	0.8359	0.9741	0.9262	0.8816	0.9996

Furthermore, the contrast experiment was taken between our algorithm and the algorithm based on DWT and algorithm based on SVD. The DWT based algorithm and SVD based algorithm which we used are described as following: DWT based algorithm, decompose the original image by wavelet, embed the watermarking in low-frequency sub-graph; SVD based algorithm, decompose the original image's singular value, embed the watermark, the watermark image is also scrambled with Arnold transform as preprocessing.

The content of Table 2 show that when the DWT based algorithm is attacked by JPEG compression in 10%,revolved 10 degrees, pepper-and-salt noise in 0.01 and Gaussian noise in 0.002,the quality of extract watermarking is descend obviously. In contrast, when our algorithm is attacked by JPEG compression in 10%, revolved 60 degrees, pepper-and-salt noise in 0.01 and Gaussian noise in 0.002, the extracted watermarking is very well yet. The result of experiment show that for the common geometric attacking, the extracted watermarking with our algorithm has higher similar degree compare with other two algorithm. The algorithm proposed in this paper is not only has good robustness of resist in compression and noise attacking but also has the perfect ability of anti-rotation attacking.

5 Conclusion

In this paper, we propose an new algorithm of digital watermarking based on combining the nonsubsampled Contourlet transform and SVD, first apply the NSCT to the image and extract the low-frequency sub-band of image, and then decompose the low-frequency sub-band of image by SVD, finally embed the watermarking in the decomposed singular value. The experiment results show that the new algorithm has good ability in standing up to geometric attacking, especially rotation attacking.

Acknowledgment. This paper is partially supported by Yunnan Provincial Department of Education, Graduate Research Foundation (09J0061, 2010J071).

References

1. Do, M.N., Vetterli: Contourlets: a directional multiresolution image representation. In: Proceedings of 2002 International Conference on Image Processing, vol. 1, pp. I-357–I-360 (2002)
2. Do, M.N., Vetterli: The contourlet transform: an efficient directional multi-resolution image representation. IEEE Trans. on Image Processing 14(12), 2091–2106 (2005)
3. Zhou, J., Cunha, A.L., Do, M.N.: Nonsubsampled contourlet transform: construction and application in enhancement. In: Proceedings Of IEEE International Conference on Image Processing, Genoa, Italy, pp. 469–472 (2005)
4. Cunha, A.L., Zhou, J., Do, M.N.: The nonsubsampled contourlet transform: theory, design, and applications. IEEE Trans. Image Processing 15(10), 3089–3101 (2006)
5. Xu, C., Prince, J.L.: Snake, shapes, and gradient vector flow. IEEE Transaction on Image Processing 7(3), 359–369 (1998)
6. Liu, J., Liu, G., Li, A., Jin, H.: Watermarking Algorithm Besed On nonsampled Contourlet's Feature. Computer Engineering 35(17) (2009)

7. Wang, J., Lv, S.: Watermarking Algorithm Used Contourlet's Resist Geometry Attacking Ability. Journal of Sensing Technology 22(6), 883–886 (2009)
8. Chen, K., Wang, J.: An Watermarking Algorithm Based On Unite HVS And Contourlet. Journal of Computer-aided Design And Images 19(6) (2007)
9. Zhu, X., Xiao, L., Wu, H.: The Comparison of Robustness for Contourlet And Wavelet's Domain. Computer's Engineering And Application 45(9) (2009)
10. Feng, W.: Study in digital Watermarking new way. Master Thesis of Xi'an University of Science and Technology (2009)

Optimization of Harris Corner Detection Algorithm

Xiong Shunqing, Zhou Weihong, and Xia Wei

School of Mathematics & Computer Science Yunnan Nationalities University,
YNU Kunming, China
{xsq324,ynzwh}@163.com, xiawei025@sohu.com

Abstract. Corner detection is often influenced by noise when Harris corner detection algorithm is applied to digital images. In this paper, the response function of Harris corner detection algorithm is optimized to avoid the influence of k in response function, thus the precision of feature response function of target pixels is improved. The results of applying this method to detecting the feature point of face show that the anti-noise ability of this algorithm is greatly enhanced compared with Harris corner detection algorithm.

Keywords: Feature detection, Face recognition, Noise, Response function.

1 Introduction

Image feature detection is an important field of computer visual. Feature detection has been widely used in many aspects, such as image matching, objective description, object recognition etc. Human being identify objects and different people through the visual system, however, this simple process for human being is very difficult for computers, first computers need to analyze image data, then extract accurate and stable feature information to identify. The efficiency of feature points extracting directly affect precision rate and speed of image recognition. At present, the extraction feature points algorithm can be divided into two categories: one method is presented by Kitchen [1] based on contour line of image, the other one is the method based on image gray values, such as Susan method, Harris method etc. Harris corner detection algorithm doesn't affect by the posture of vidicon and light, it is a better algorithm compare with other algorithm, but it is very sensitive to noises.

2 Harris Corner Detection Algorithm

Harris corner detection operator is the improved operator based on Moravec operator[4]. Moravia operator is proposed by Moravec at 1977 which used gray variance of the pixels to extract feature points, first, a two value window is selected which uses the target pixel as central point, then calculate the sum value of the square of gray difference between the pixels at 0 degree, 45 degrees, 90 degrees, 135 degrees directions, the direction with minimum value is taken as the interest value of the pixel. Harris improved the algorithm by using Gaussian function instead of binary window function, Taylor expansion is used to get approximate value of all direction, and the interest value of

D. Zeng (Ed.): Advances in Control and Communication, LNEE 137, pp. 59–64.
springerlink.com

every target pixel is got. Assuming the coordinates of target pixel is (x, y), the distance in the direction x and y respectively is u and v, then, the formula of calculating the gray scale transformation of target pixel point is show as follows:

$$E_{x,y}=\sum_{x,y} w(x,y)[I(x+u,y+v)-I(x,y)]^2=\sum_{x,y} w(x,y)[I_x u+I_y v+o(u^2+v^2)]^2 \quad (1)$$

Written as matrix form:

$$M=\sum_{xy} w(x,y)\begin{bmatrix} I_x^2 & I_x I_y \\ I_x I_y & I_y^2 \end{bmatrix}=w(x,y)\otimes\begin{bmatrix} I_x^2 & I_x I_y \\ I_x I_y & I_y^2 \end{bmatrix} \quad (2)$$

$$E(u,v)=[u,v]M\begin{bmatrix} u \\ v \end{bmatrix} \quad (3)$$

Where $E_{x,y}$ is gray scale measurement in the window; $w_{x,y}$ is window function, generally defined as $w_{x,y}=e^{(x^2+y^2)/\sigma^2}$; I is image gray function, I_x is the difference in x direction, I_y is the difference in y direction, $o(u^2+v^2)$ is infinitesimal of the distance. Matrix M is the autocorrelation matrix of the target pixel, λ_1 and λ_2 is the two eigenvalue of the matrix, these two eigenvalue is positive proportion with the principal curvatures of the M matrix. When both of the two eigenvalue are smaller, indicating that the area near the target pixel is "flat areas"; one is big and the another one is small, indicating that the feature pixel is on the "edge"; when both of two eigenvalue are larger, indicating that moving in any direction will cause dramatic changes of gray. Harris' corner response function R is show as follows:

$$R=\det(M)-k(trace(M))^2 \quad (4)$$

Where $\det(M)=\lambda_1\lambda_2$ is the determinant of the matrix M , $trace(M)=\lambda_1+\lambda_2$ is the trace of the matrix, k ranges from 0.04 to 0.06. When R value of the target pixel is bigger than the given threshold, the pixel is taken as the corner point.

Although the Harris algorithm is improved with Moravec operator, but it still have following shortages:

(1) The quantity and quality of corner extraction largely depends on thresholds setting, if threshold is big, it will lose part of corners; if threshold is small, it will reduce the quantity of the corners, at the same time, the noise is more sensitive.

(2) It is difficult to control the window size of the Gaussian smoothing function in process, if the window is too small, is influence by noise largely; if the window is too big, it will make the offset to the corner, and increase the computational amount.

(3)If using infinite smooth Gaussian function as filter in order to reduce the influence of noise, a lot of corner information is lost because of the smooth function.

3 The Algorithm Design

The shortages of Harris corner detection algorithm can be optimized by improved formula (4). K in Formula (4) is in a range ;different k has different effect. if k pick the appropriate value, the algorithm can also achieve good results. The selection of the appropriate k need to consume a lot of time to debug and verify. To avoid this problem, we need to find a response function which doesn't affect by k and can replace the formula (4).

The response function R (4) is a function of the autocorrelation matrix M , and it's relevant to the two eigenvalue vectors λ_1 and λ_2 of matrix M. Assume the eigenvalue of autocorrelation matrix M of the target pixel is λ_1 and λ_2 $(\lambda_1, \lambda_2 \geq 0)$, if it meets the condition $\lambda_1 \approx \lambda_2$, then $R_{**} = \frac{(\lambda_1 - \lambda_2)^2}{(\lambda_1 + \lambda_2)^2}$ value tends to 0, and we have $0 \leq R_{**} \leq 1$. Her R_{**} is relevant to characteristic polynomial $\det(\lambda - M)$, and then we have $(\lambda_1 - \lambda_2)^2 = (\lambda_1 + \lambda_2)^2 - 4\lambda_1\lambda_2 = (trace(M))^2 - 4\det(M)$.

Suppose $R_* = 1 - R_{**}$, here we still have $0 \leq R_* \leq 1$. Thus we can get the expression about R_*:

$$R_* = \frac{4\det(M)}{(trace(M))^2} \tag{5}$$

Here using the above formula as the response function, when $R_* \to 1$, the point can be judged as corner point; when $R_* \to 0$, the point is the edge point. But there are still some shortcomings in this expression, R_* is different when the two eigenvalues are equal and not equal,. To avoid this problem, we usually plus a parameter in the denominator to get a more robust expression, as follows:

$$R_* = \frac{4\det(M)}{\varepsilon + (trace(M))^2} = \frac{4\lambda_1\lambda_2}{\varepsilon + (\lambda_1 + \lambda_2)^2} \tag{6}$$

Where ε is arbitrarily small positive number, there are no k in formula (6) , avoid the problem of random selection of k with response function. If one eigenvalue tends to 0, then R also tends to 0; if two eigenvalues are close and relatively small, R still tends to 0; if both of the two eigenvalue are big, R will get a little bit greater but still in interval [0,1]. The above optimized algorithm as follows:

Step 1: Calculate the autocorrelation matrix M of every pixel;

Step 2: Calculate corner response value R_* of every pixel;

Step 3: Search maximum value in the Gaussian window, if R_* is larger than the threshold then considered the pixel as corner.

In this paper, the original Harris corner response function formula (4) was replaces by new formula (6), not only doesn't change the autocorrelation between R and each pixel, retains the advantages of the original algorithm, but also has higher accuracy response value of the target pixel compared with the original algorithm. The result of simulation experiment show that the new algorithm has the ability of get more stable corners which miss-detected with old algorithm previously on the noise interference condition. The new algorithm can weaken the noise impact on the target pixel response values effectively.

4 Experimental Analysis

Our simulation experiment was carried out in the Matlab 7.0 platform, as show in Figure 1 is the feature points detected from original image. The left one is the result detected by Harris algorithm in reference[3], the right one is the result by our new algorithm. As we can see, our new algorithm obviously gets more stable corners by improving the accuracy of the response function compared to the original Harris algorithm.

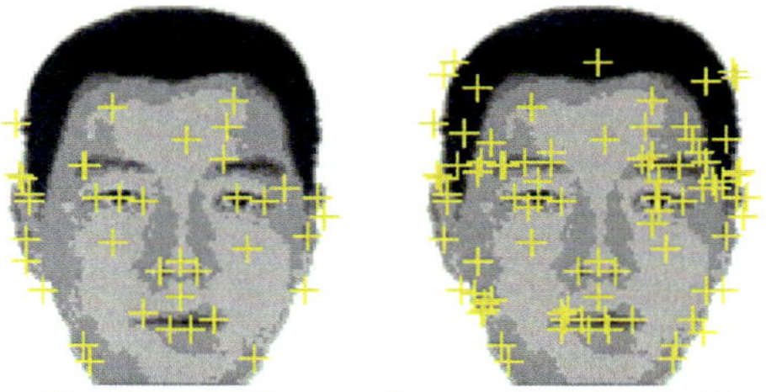

Harris algorithm Our improved algorithm

Fig. 1. The corner detected from the original image by two algorithm

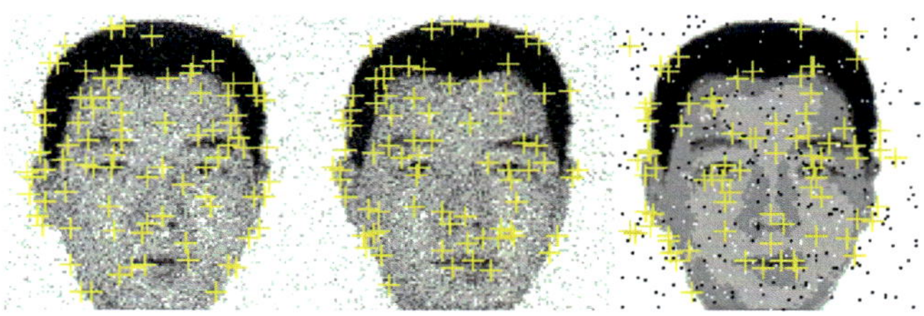

noise 1 noise 2 noise 3

Fig. 2. Corners detected by Harris algorithm from noised image

To compare the ability of anti-noise of the two algorithms, we add Gaussian noises with mean is 0, variance is 0.1 and mean is 0, variance is 0.02 to the original image respectively, and Salt and Pepper noise with density is 0.04. Then apply the two algorithms to the noised image respectively, as show in Figure 2 is the corners detected by original Harris algorithm, and Figure 3 is the corners detected by our new algorithm.

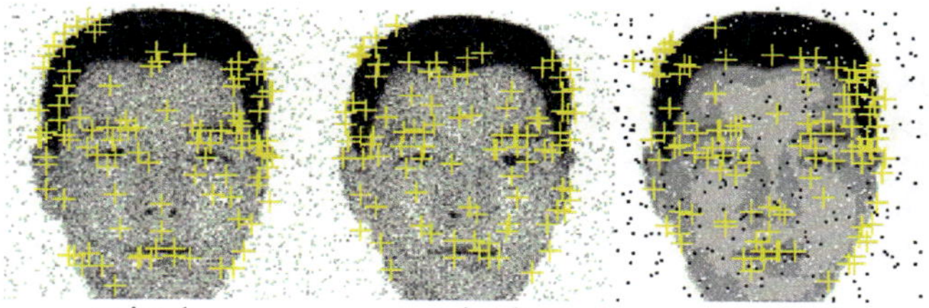

Fig. 3. Corners detected by our new algorithm from noised image

As we can see from Fig.2 and Fig.3, the number of corners detected from noised image increase rapidly compare with image without noise which applied Harris algorithm, but the effect of noise is much smaller with our new algorithm. The result can also be show in the following tables. As show in Table 1 is the number of corners detect by apply two algorithms to the face image with noise and without noise respectively. Table 2 is the impact rate of noise on the two algorithms. The rate is the extra number of corners bring from noise divide the number of original corners.

Table 1. The number of corners detected by two algorithms

Number of corners	Original image	Noise 1	Noise 2	Noise 3
Harris algorithm	48	90	98	97
Our algorithm	103	115	129	122

Table 2. The impact rate of noise with two algorithm

Impact rate	Noise 1	Noise 2	Noise 3
Harris algorithm	0.8750	1.0417	1.0208
The algorithm	0.1165	0.2524	0.1845

As we can see from Table 1 and Table 2, in the case of no noise, the number of corners detected by original Harris corner detection algorithm is small, and the number of corners detected are sharply increased after adding noise, the impact rate is big than 1, in comparison, the noise impact rate of our algorithm is no more than 0.3. The impact rate has been reduced a considerable amount compared to the original Harris algorithm.

To conclude, the impact of noise is reduced by using our new response function R, the corners detected are more stable in the same threshold, the ability of anti-noise of our algorithm is improved.

5 Conclusions

Feature point detecting is the premise of face recognition and play a important role on computer vision. Harris corner detection algorithm can not meet the requirement in many practical applications because it is easily affect by the noise, in this paper, an improved image feature detect methods is proposed by optimizing the response function of traditional Harris corner detection algorithm, the result of experiment show that the new detect algorithm have higher detection accuracy of target point and reduce the impact of noise on feature detection. Our future work is to apply this new algorithm to face recognition.

Acknowledgment. This paper is partially supported by Yunnan Provincial Department of Education, Graduate Research Foundation (09J0061, 2010J071).

References

1. Kitchen, L., Rosenfeld, A.: Gray Level Corner Detection. Pattern Recognition Letter 3(1), 95–102 (1982)
2. Smith, S.M., Brady, J.M.: SUSAN-A New Approach to Low Level Image Processing. International Journal of Computer Vision 23(1), 45–78 (1997)
3. Harris, C., Stephens, M.J.: A combined corner and edge detector. In: Proceedings of the 4th Alvey Vision Conference, pp. 147–151 (1988)
4. Mao, Y., Lan, M.: An improved method for Harris corner detection. Computer Technology and Development 5(5), 127–129 (2009)
5. Noble, A.: Descriptions of Image Surface, p. 45. Department of Engineering Science, Oxford University Press (1989)
6. Li, X., Zhao, Y.: Face classification algorithm based on feature. Microcomputer Applications 26(4), 60–64 (2010)
7. Qiu, J., Zhang, J., Li, K.: Image matching method based on harris and shift algorithm. Test and Measurement Technology 23(3), 271–274 (2009)
8. He, Y., Cai, G.: A new method of corner detection. Yunnan University of Nationalities Journal(Natural Science) 16(1), 45–47 (2007)
9. Xie, H., Wang, L.: The integration of classification of face images based on DT-CWT and SVM. Yunnan University of Nationalities Journal 19(5), 313–316 (2010)

Face Recognition Technology Based on Eigenface

Yan Xinzhong and Cui Jinjin

Detection Technique and Automation Device, College of Electronic Information and Automation Tianjin University of Science & Technology
Tianjin, China
yan_xz@tust.edu.cn, irics-c@163.com

Abstract. Ace recognition hasn't reached a satisfactory recognition ratio all the time, because it is influenced by light, expression and gestures etc. This article puts forward the face recognition technology based on statistic characteristics, meanwhile taking into consideration of characteristic information and structure information of face image. Test proved the feasibility of this method.

Keywords: ace recognition, statistic information, preprocessing, PCA(Principal Component Analysis).

1 Introduction

Face recognition, refers to especially the identification computer technology by analyzing and comparing face visual feature information. The process of face recognition is a comprehensive utilization of psychology, physiology, auto-detection technology and computer image processing technology. It is presently widely applied to border control, criminal identification and anti-terrorist. It also plays a significant role in civil attendance, identity authentication, bank and traffic etc [1].

Compared with other identification methods with human biologic characteristic, face recognition method is more friendly, convenient and covert. Owning to its great application prospect and incomparable superiority, face recognition increasingly becomes a highlight in current recognition and artificial intelligence.

The face recognition method can be divided into there categories according to the face representation form: one based on geometric characteristics, another based on statistic characteristics and one based on connectionism.

2 Face Recognition Principle

Generalized face recognition actually includes a series of technologies related to construct face recognition system which includes face image acquisition, face location, face recognition preprocessing, identity confirmation and identity search etc [2]. Simply put it, it's to judge if there's a face and the face can be found from an image. If it is judged to be a face, the face is classified, verified if it is a certain person, or find out it's characteristic to compare and distinguish it with the database face. People's daily behavior of comparing a face with the one a long time ago or a blurry one is seemingly simple; in fact it is a complex logic judgment function. The face recognition

D. Zeng (Ed.): Advances in Control and Communication, LNEE 137, pp. 65–70.
springerlink.com

system can be looked as a simulation of this behavior. And how to simulate and realize this process by using a quantitative logic calculation machine like computer is also very difficult. The basic framework of face recognition is shown in figure 1:

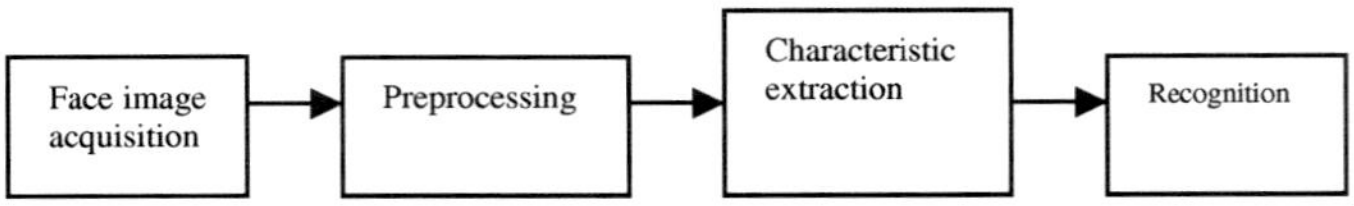

Fig. 1. The general process of face recognition

In figure 1, preprocessing module mainly improves the face image quality; including image contrast enhancemane, noise elimination etc. For face image with noise, it needs to eliminate the noise, filter interference, and raise the signal-noise ratio. For face image with weak information needs enhancements processing such as binary, gray variation etc. As for degenerative and blurry face image needs restoration processing. And distorted face image needs variation like geometric correction. Image processing aims at improving the quality of face image, changing face image to simplify people's observation and suitable for machine to identify it[3]. Characteristic extraction module extracts face characteristics. How to extract stable and reliable characteristics is the key to recognition system. Characteristic extraction is to extract shape information of certain meaning from solid model. The database has prestored face image known or related eigenvalue. The purpose of recognition is to match the image or characteristic to be recognized with the data in database. Recognition task divides into two categories: face recognition, confirming the image input is which one in the database, face verification, confirming the identity of somebody is true or not.

The successful face recognition methods includes at least the following elements [4]: image processing must reserve the useful information, restrain and separate the interference from outside; tiny face deformation should be taken into consideration when they are matched, and effect of face recognition can not be influenced under the premise of deformation; the proper representation method must be found out and it find out the differences between faces quickly and accurately.

3 Face Recognition Process

Face image preprocessing is a very important procedure among face recognition; it's indispensible especially for constructing face detection recognition system, because the face gesture, expression and actual light in the authentic environment greatly differ from images in database. Preprocessing eliminates the interference of hair and background, obtaining the face geometric invariance to certain degree.

Face characteristic acquisition algorithm adopted before: face recognition based on geometric characteristic is to extract face eigenvector from face organ structure to recognize face. This method is intuitionistic and simple with definite physical meaning. Characteristic extraction is not influenced by the number of classification. Face image is affected greatly by light, expression and facial details etc, leading to it's not good for exact characteristic point extraction. Besides, overlooking the field information of these characteristic points also causes the correct recognition ratio to much low [5]. Wavelet transform method extracts characteristic point from face image to solve the recognition problem by matching the characteristic point. This method accepts slight face shake and tiny expression changes. But the characteristic point extraction is not stable and has huge workload [6]. PCA method can get the principal component direction of face image distribution, using principal component of face image to represent face. The Eigen subface neglects the face set topological relations and face local characteristics. The recognition ratio becomes obviously low when light and gesture change [7]. To solve the aboving problems, this article sets out the face recognition technology based on statistic characteristics. Test proved its feasibility. Therefore we integrate principal components with the nonzero minimum components to recognize face; giving consideration to both the whole face information and local tiny difference improve the far. Thus the face recognition reliability is improved.

Face recognition method based on statistic characteristic not only aims at certain face characteristic. From the whole face standpoint, it uses statistic principles to extract some common face patterns from many face images to recognize faces. Because of face image complexity, face characteristics' display and description encounter great difficulties. Therefore methods based on statistic are attached more and more importance.

3.1 Face Eigenvector Analysis

PCA algorithm is a face recognition methods based on algebraic feature, global feature and K-L decomposition. This face recognition method based on PCA regards the face as a whole for the first time. Characteristic extraction is an important transformation of face recognition method from handmade definition to auto acquisition by statistic study. Simply put, its principle is to project a high dimension vector to a low dimension vector space through a special eigenvector matrix, represented as a low dimension vector without losing any information. That is, a high dimension vector accordance with the original one can be completely reconstructed through a low dimension vector and eigenvector matrix. Eigen face method is just to regard the image region with face included as a random vector. Accordingly, the orthogonal K-L base can be gotten by K-L transformation. The base accordance with the bigger eigenvalue has the shape similar to face, so it is called eigenface. This base linear combination can be used to describe, express and approach to face image, so face can be recognized and fused in this way. The recognition process is to project the face image to subspace composing of eigenface and compare it with the location of a known face in eigenspace to realize face recognition.

Let's suppose the face sample images from face training set are all obtained under the same light condition. It has a types of face, each type has b types of naked face sample x_i , $i = 1,2,3,...,$, $M = a \times b$ which has different gestures and expressions. The average value of naked face sample $\overline{x}$:

$$\overline{x} = \frac{1}{M}\sum_{i=1}^{M} x_i \tag{1}$$

in order to overcome the light effect to a certain degree, let me list the face sample set $\overline{x}_i$:

$$\overline{x}_i = x_i - \overline{x} \tag{2}$$

and calculate zero mean, covariance matrix of image serials $\overline{x}_i$ which calculate zero mean as follows:

$$C = \sum_{i=1}^{N_b} \overline{x}_i \overline{x}_i^T \tag{3}$$

When calculating the covariance matrix C's eigenvalue and eigenvector μ_i, find out the best projection space. When PCA transforms the high dimension vector to low dimension vector, variance of each component of low dimension vector is made to maximum. What's more, each component is not related to each other, so the best feature extraction can be reached.

The covariance matrix C's eigenvalue λ_i accords with eigenvector μ_i, the two has the following relation:

$$C\mu_i = \lambda_i \mu_i \quad u_i i = (0,1,\ldots,m-1,\ldots,i,\ldots,i+n-1) \tag{4}$$

The priority of principal component selection of PCA is usually determined by the corresponding eigenvalue. Bigger the eigenvalue, higher the priority. The PCA calculating process totally no need to set parameters deliberately or intervene calculation based on any empirical model. The last result only relates to data, independent to subscriber.

The eigenvector accordance with the former m eigenvalues only reflects the whole information distribution space of face sample from face training set. The Eigen vector accordance with the latter n minimum nonzero eigenvalues reflects the face sample's gesture and expression changes, which are the face local information. Therefore each face image can be projected to subspace composing of μ_0, μ_1,…, μ_{m-1}. Each face image accords with one point in subspace, equally each point in subspace accords with a face image. Owing to the similarity of these images with face, it is called eigenface. Any face image can be represented as a linear combination of this group of eigenface.

Different face recognition fuses the face holistic information eigenvector with minimum eigenvector, representing not only the principal eigenvector of the face holistic information discrepancy but also the nonzero minimum eigenvector of face local information discrepancy [8].

A. *Face recognition*

Project the difference

$$x' : x' = x_i - \overline{x} \tag{5}$$

between each naked sample x_i from face training set and face training set's average value $\overline{x}$, PCA space basis $\mu_A = (\mu_0, \mu_1, \ldots, u_{m-1})$ and minimum eigenvector space basis $u_B = (u_i, u_{i+1}, \ldots, u_{i+n-1})$

x'_i to PCA space basis and minimum eigenvector space basis, respectively representing project component of the face training set's holistic information in low dimension space, and that of face training local information in low dimension space.

Project the difference between face to be recognized x and face training set's average value $\overline{x}$, that is x' :

$$x' = x - \overline{x} \tag{6}$$

To μ_A and μ_B , respectively get the following formula:

$$\lambda_A = (\lambda_0, \lambda_1, \ldots, \lambda_{m-1}) = x' \times (\mu_0, \mu_1, \ldots, \mu_{m-1}) \tag{7}$$

$$\lambda_B = (\lambda_0', \lambda_1', \ldots, \lambda_{m-1}') = x' \times (\mu_0, \mu_1, \ldots, \mu_{m-1}) \tag{8}$$

The similarity of face holistic information's projection vector in low dimension space with training set D_A :

$$d_A^i = \lambda_A \times \lambda_A^i = \frac{\sum_{j=0}^{m-1} \lambda_j \lambda_{i,j}}{\sqrt{\sum_{j=0}^{m-1} \lambda_j^2} \sqrt{\sum_{j=0}^{m-1} \lambda_{i,j}^2}} \tag{9}$$

$$D_A = 1 - \exp(-d_A^i) \tag{10}$$

The similarity of coppresponding local information's projection vector in low dimension space with face sample from training set D_B :

$$d_B^i = \lambda_B \times \lambda_B^i = \frac{\sum_{j=0}^{m-1} \lambda_j' \lambda_{i,j}'}{\sqrt{\sum_{j=0}^{m-1} \lambda_j'^2} \sqrt{\sum_{j=0}^{m-1} \lambda_{i,j}'^2}} \tag{11}$$

$$D_B = 1 - \exp(-d_B^i) \tag{12}$$

The distance between face to be recognized and face sample numbered i from face training set is:

$$D^i = D_A^i + D_B^i$$

d_m is the minimum value of face to be recognized and face training set, that is the face serial number. Finally accomplish face recognition.

First, confirm that you have the correct template for your paper size. This template has been tailored for output on the US-letter paper size. If you are using A4-sized paper, please close this template and download the file for A4 paper format called "CPS_A4_format".

4 Conclusion

This article put out and realized a face recognition method based on statistic characteristics. This method combines principal Eigen component reflecting the holistic information with nonzero minimum eigenvector reflecting the tiny changes of image. Through this method, face statistic eigenvector can be obtained, so do the face statistic recognition. Using this method to research face recognition has not only the theory value but the application value. Through research, human vision mechanism, thinking mechanism as well as other behavior mechanism is expected to be revealed.

References

1. Zhang, H.: Digital Image Model Recognition Technology and Engineer-ing Practice, pp. 322–324. Post & Telecom Press (2003)
2. Liu, Z., Liu, L.: Face Recognition Research Based on Statistic Character-istic and Wavelet Decomposition Method. Computer Engineering and Application 6, 88–91 (2003)
3. Bartlett, M.S.: Face Recognition by Independent Component Analysis. IEEE Transactions on Neural Networks 13(6), 1450–1465 (2002)
4. Yang, M.H.: Kernel eigenfaces vs Kernel fisherfaces: Face recognition using Kernel methods. In: IEEE Conference Automatic Face and Gesture Recognition (AFGR), pp. 215–220 (2002)
5. Wang, J., Zhang, X.: Computer Image Recognition, pp. 161–165. China Railway Publishing House (1998)
6. Xiong, Z.: Face Recognition System Research. [doctorate thesis] Institute of Computing Technology Academy. Chinese Acedemy of Science, Beijing (1999)
7. Yao, F.: Face Characteristic Extraction Based on Wavelet Anylasis. Computer Engineering and Application 24(6), 32–35 (2000)

Generalized Diagonal Slices and Their Applications in Feature Extraction of Underwater Targets

Haitao Yu[1,2]

School of Science
[1] Xi'an University of Science and Technology
Xi'an, China
[2] College of Marine
Northwestern Polytechnical University
Xi'an, China
haitaoyu@xust.edu.cn

Abstract. In order to improve the classification accuracy of the radiated noise from underwater targets, generalized diagonal slices and their spectra of third cumulant are proposed by introducing an amendment quantity q into the original diagonal slice. The amendment quantity q represents the distance between the original diagonal slice and other slices parallel to it in bifrequency plane. And the generalized definition degrades into the original one when the amendment quantity q is set to zero. The physical meaning of the generalized diagonal slices spectra depicts lines parallel to the original diagonal slice. And two feature vectors are formed by summation or maximization of every generalized slice. It is validated by simulation that the target features in the bifrequency domain is drastically extracted by summation scheme and that the classification accuracy of 100% for feature vector obtained by summation scheme is steadily resulted by One-against-One (OAO) method of multi classification of Support Vector Machine (SVM).

Keywords: generalized diagonal slices, spectrum, bispectrum, feature extraction, underwater targets.

1 Introduction

Traditional signal processing methods are all based on the hypothesis that signals and noises are linear, stationary and Gaussian stochastic processes. P.L. Brockett et al. [1] believe that ambient records appeared to be both linear and Gaussian processes when examined over a period on the order of a minute, but were found to be nonlinear and non-Gaussian when examined over shorter time periods on the order of a second. And in each case the time series dominated by local shipping noise tested to be nonlinear and non-Gaussian over both short and long time periods. Whereas, Higher-Order Statistics (HOS) is applied extensively for it can not only detect and represent the nonlinear characteristic in signals and systems, but also restrain the impact of Gaussian colored noises from non-Gaussian signals [2]. But the large computation load and

D. Zeng (Ed.): Advances in Control and Communication, LNEE 137, pp. 71–79.
springerlink.com

output dimensions are the main factors of hindering its realistic applications. Many scholars have done quite a number of researches in dimension reduction based on HOS. Nagata proposed the concept of 1-D slices and their spectra in 1970. Many types of 1-D slices are possible, including radial, vertical, horizontal, diagonal, and offset-diagonal [3]. In the field of feature extraction of underwater targets, [4] and [5] decreased the computation load by the diagonal slice of bispectrum, but lost the other useful information in bifrequency plane. Reference [6] obtained the 100 sections in low frequency domain of bispectrum by uniformly-spaced sections paralleling to X-Z plane along Y axis and chose the maximum magnitudes of bispectrum in every section as the feature vector. It is feasible to choose the low frequency region of bispectrum matrix as the feature extraction region from the point of view of the characteristics of underwater acoustic signals. Whereas, it is found in simulations that the middle and high frequency regions still contains valid target features which can improve the classification accuracy.

Hence, generalized diagonal slices and their spectra of third cumulant are proposed by introducing an amendment quantity q into the original diagonal slice. The amendment quantity q represents the distance between the original diagonal slice and other slices parallel to it in bifrequency plane. And the generalized definition degrades into the original one when the amendment quantity q is set to zero. The physical meaning of the generalized diagonal slices spectra depicts lines parallel to the original diagonal slice. And two feature vectors are formed by summation or maximization of every generalized slice. It is validated by simulation that the target features in the bifrequency domain is drastically extracted by summation scheme and that the classification accuracy of 100% for feature vector obtained by summation scheme is steadily resulted by One-against-One (OAO) method of multi classification of Support Vector Machine (SVM).

2 Generalized Diagonal Slices and Its Spectrum

2.1 HOS and Bispectrum

Simply speaking, HOS [2] are the moments or cumulants higher than 2nd order and their spectra. For a k-th order stationary stochastic process $\{x(n)\}$, k-th order moment is defined as

$$m_{kx}(\tau_1, \tau_2, \ldots, \tau_{k-1},) = mom\{x(n+\tau_1), x(n+\tau_2), \ldots, x(n+\tau_{k-1})\}, \tag{1}$$

and k-th order cumulant is defined as

$$c_{kx}(\tau_1, \tau_2, \ldots, \tau_{k-1},) = cum\{x(n+\tau_1), x(n+\tau_2), \ldots, x(n+\tau_{k-1})\}, \tag{2}$$

where τ_i, i=1,2,…, k-1, is the time differences.

Third-order moment and third-order cumulant are respectively defined as

$$m_{3x}(\tau_1, \tau_2) = E\{\ x(n)\ x(n+\tau_1)\ x(n+\tau_2)\ \} \tag{3}$$

and

$$c_{3x}(\tau_1, \tau_2) = cum\{ x(n)\, x(n+\tau_1)\, x(n+\tau_2) \}$$
$$=E\{[x(n) - m_{1x}][x(n+\tau_1) - m_{1x}][x(n+\tau_2) - m_{1x}]\}. \quad (4)$$

For $\{x(n)\}$, its Power Spectral Density (PSD) is defined as the Fourier Transformation (FT) of the autocorrelation function. Hence, the spectra of k-th order moment and k-th order cumulant are defined as the k-1 dimensional FT of k-th order moment and k-th order cumulant. Higher-order cumulant and its spectrum have following advantages, but not the higher-order moment and its spectrum: (1) the higher-order cumulant of a Gaussian process is identically equal to zero even for a colored Gaussian process and its spectrum or itself provide a measurement of Gaussian bias and (2) higher-order cumulant and its spectrum can fully restrain the impact of Gaussian colored noises from non-Gaussian signals in theory. Hence, higher-order cumulant and its spectrum are applied widely.

Third-order spectrum, namely bispectrum, is the most common use in higher-order spectrum. It's the double FT of third order cumulant and its definition is as follows

$$B_x(\omega_1, \omega_2) = \sum_{\tau_1=-\infty}^{\infty} \sum_{\tau_2=-\infty}^{\infty} c_{3x}(\tau_1, \tau_2)\, e^{-j(\omega_1\tau_1+\omega_2\tau_2)} . \quad (5)$$

Let $X(\omega)$ be the Fourier Transformation of $x(n)$. Then bispectrum can be represented as

$$B_x(\omega_1, \omega_2)= X(\omega_1)\, X(\omega_2)\, X^*(\omega_1+\omega_2). \quad (6)$$

Actually, it can be the implement method in program. And the computation load is evidently large. Hence, the diagonal slice was proposed to save the computational time.

2.2 Diagonal Slice Definition and Its Spectrum

A 1-D slice of the k-th order cumulant is obtained by freezing (k-2) of its k-1 indexes. A diagonal slice is obtained by setting $\tau_i = \tau$, $i = 1, 2,\ldots, k$-1 [3].

The diagonal slice of third order cumulant is defined as

$$c_{3x}(\tau, \tau) = cum\{ x(n)\, x^2(n+\tau) \}. \quad (7)$$

And its spectrum is

$$B_x(\omega, \omega) = \sum_{\tau=-\infty}^{\infty} c_{3x}(\tau, \tau) e^{-j2\omega\tau} , \quad (8)$$

or

$$B_x(\omega, \omega)= X^2(\omega)\, X^*(2\omega) . \quad (9)$$

As we can see from the equations, the computation load is cut down compared to bispectrum. But diagonal slice and its spectrum lost the other useful information in bifrequency plane. Thus, the generalized diagonal slices and their spectra are proposed according to diagonal slice.

2.3 Generalized Diagonal Slices and Their Spectra

According to the habit of human eyes to classify the bispectrum, as shown in Fig. 1, one may differentiate signals from the vice-diagonal direction of the bispectrum easily. But only one line of bispectrum, for example, diagonal slice, provides limited information for signals classification for both human eyes and classifiers. Hence, the slices parallel to the diagonal slice should be included. And the generalized diagonal slices of third order cumulant are proposed as

$$c3x(\tau, \tau \pm q) = \text{cum}\{\ x(n)\ x(n+\tau)\ x(n+\tau \pm q)\ \} \tag{10}$$

where q is the amendment quantity of time differences τ and represents the distance between the original diagonal slice and other slices parallel to it in bifrequency plane. And its spectrum is

$$B_x(\omega, \omega \pm q) = \sum_{\tau=-\infty}^{\infty} c_{3x}(\tau, \tau \pm q) e^{-j(2\omega\tau \mp \omega q)}, \tag{11}$$

or

$$B_x(\omega,\ \omega \pm q) = X(\omega)\ X(\omega \pm q)\ X^*\ (2\omega \pm q)\ . \tag{12}$$

As we can see from (7) and (10), the generalized diagonal slices degrades into the diagonal slice when amendment quantity q=0. Equation (11) and (12) provide two different implement methods of spectrum for generalized diagonal slices in programs. And the physical meaning of the generalized diagonal slices spectra depicts lines parallel to the original diagonal slice. It provides more information for signal classification than original diagonal slice. As the generalized diagonal slices and their spectra are parts of third-order cumulant and bispectrum, the former inherit the properties of the latter. Because of the space limitation, the proof will be given in another paper.

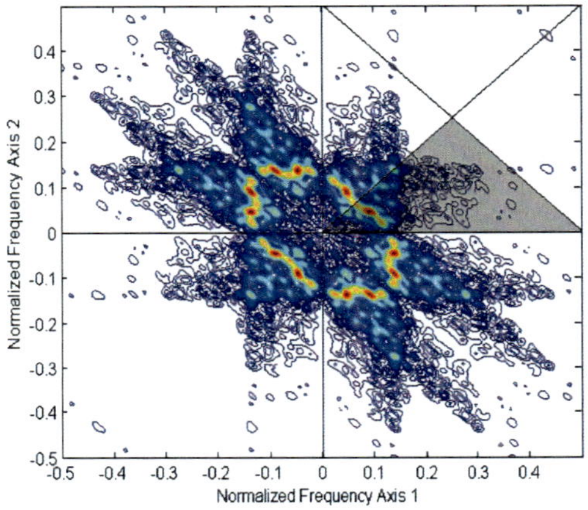

Fig. 1. Shadow region based on the symmetric characteristics of bispectrum

3 Feature Extraction Schemes Based on Generalized Diagonal Slices Spectra

Many scholars have proposed a large number of feature extraction schemes based on the original diagonal slice and its spectrum.

According to the symmetric characteristics of bispectrum, the features should be extracted in the following triangle, $\omega_2 \geq 0$, $\omega_1 \geq \omega_2$ and $\omega_1 + \omega_2 \leq \pi$, to cut down the computation load further, as shown in the shadow region of Fig. 1. And here two simple feature extraction schemes are proposed based on the generalized diagonal slices spectra. The summation or maximization of every slice is taken as the feature vector elements. Hence, two feature vectors E_1 and E_2 are formed by summation and maximization respectively. Maximization scheme is equivalent to selecting only the largest element. The summation scheme is equivalent to the energy accumulation. And it is expected that the summation scheme can effectively reduce the signal disturbance on the classification accuracy. Their separabilities are validated in the coming simulation.

4 Support Vector Machine (Svm)

SVM [7] was originally developed by Vapnik and his co-workers. And it is a new method for machine learning based on rigorous theory, Statistical Learning Theory (SLT) which is based on small samples. And it provides a new way for solving the nonlinear modeling problems. Some scholars believe that SVM is becoming the new research focus after ANN and that it will promote the further development of theories and technologies in machine learning.

Its basic thoughts is as follows. An optimal linear hyperplane is established based on structural risk minimization theory in feature space. And the optimizing linear hyperplane is solved by a convex programming problem. Thus, it can overcome the shortcomings of the hardness of network structure determinations, the slow convergence, the local minimum and the needs of amounts of data samples for training in ANN learning method. And the model based on SVM methods has the good generalization ability and can give its certain boundary, which is not available for any other learning methods.

Radial Basis Function (RBF) is efficient in most cases as the kernel function of SVM [8]. And the optimizing of the two parameters of RBF, C & γ, can be obtained by cross-validation and grid-search [8]. Hence, we choose RBF as the kernel function. And the optimizing for C & γ is conducted by 3-fold cross-validation and every 0.3 intervals grid-search in $[2^{-5}, 2^{2}]$. There are many multi-SVM methods. And classification accuracy of OAO method is better for most data sets [9-10]. Thus, OAO will be used for the classification of the radiated noise from underwater targets.

5 Experiment Results and Analysis

Data of 98,304 dots length for the radiated noise of underwater targets in three types, A, B and C, are obtained in various sea states respectively. And every class data are divided into 192 segments, the length of which is 512 dots. The former 96 segments of every class data serve as the training set. And the later 96 segments serve as the test set. And FFT of 256 dots for the short sample data of 512 dots is adopted to estimate the spectra of the generalized diagonal slices in this paper. Fig. 2 gives the normalized bispectra of the radiated noise of underwater targets in three types. And a certain similarities among them can be seen, which give rise to the inaccurate classification.

In order to validate the separabilities of the proposed feature vectors E_1 and E_2 based on the spectra of the generalized diagonal slices, feature extraction scheme in [6], represented as feature vector E_p, is used as comparison.

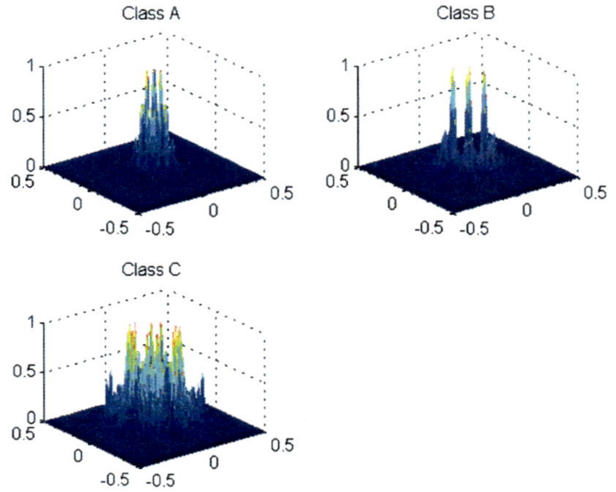

Fig. 2. Normalized bispectra of the radiated noise of underwater targets

Table 1. Classification Results Based on Feature Vectors E_1, E_2 and E_p by SVM for Test Sets

Feature Vectors	Class	A	B	C	Classification Accuracy (%)	Mean Classification Accuracy (%)
E_1	A	96	0	0	100	99.6528
	B	0	95	1	98.9583	
	C	0	0	96	100	
E_2	A	95	1	0	98.9583	99.3056
	B	0	95	1	98.9583	
	C	0	0	96	100	
E_p	A	95	1	0	98.9583	92.0139
	B	0	85	11	88.5417	
	C	0	10	85	88.5417	

Table 1. *(continued)*

Feature Vectors	Class	A	B	C	Classification Accuracy (%)	Mean Classification Accuracy (%)
E_1	A	96	0	0	100	100
	B	0	96	0	100	
	C	0	0	96	100	
E_2	A	96	0	0	100	98.9583
	B	0	94	2	97.9167	
	C	0	1	95	98.9583	
E_p	A	94	2	0	97.9167	89.9306
	B	0	88	8	91.6667	
	C	0	19	77	80.2083	

The classification results of training sets, as shown inClassification Results Based on Feature Vectors E1, E2 and Ep by SVM for Training Sets. Table 1, can reflect whether the parameters of classifiers are appropriate or not in a certain sense and can reflect the quality of the feature extraction schemes. It is apparent that the separabilities of feature vectors E1 and E2, which are formed by summation scheme and maximization scheme based on spectra of generalized diagonal slices respectively, are better than feature vector Ep formed by maximization scheme based on spectra of horizontal slices.

As we can see in Table 1, the classification accuracy of E1 is higher than that of E2 for training sets. Then classification accuracy of E1 higher than that of E2 for test sets is expected, as shown in Table 2. The possible reason is that maximization scheme is equivalent to selecting only the largest element but the summation scheme is equivalent to the energy accumulation. And it is expected that the separabilities of feature vector formed by summation scheme is higher than that by maximization scheme, even based on spectra of horizontal slices. And the simulation can verify the point. But because of the space limitation, the classification results for both its training sets and test set are not included in this paper.

The evident differences can be seen between class A and class B and C in Fig. 2. Thus, the feature vector which can fully distinguish class A from class B and C is expected. But the classification accuracy of feature vector Ep is only about 98% for class A for both training sets and test sets. It is noted that the classification accuracy of feature vectors E1 and E2 based on the spectra of the generalized diagonal slices for class A is 100%. It indicates that feature extraction scheme based on the generalized diagonal slices according to the habit of human eyes to classify the bispectrum is feasible.

The evident similarities can be found in class B and class C in Fig. 2. And the classification accuracy of Ep for class B and C is much lower than that of E1 and E2 for both training sets and test sets. It is noted that the classification accuracy of E1 for all class is 100% for test sets. It shows that feature vector extracted by the summation scheme based on the generalized diagonal slices according to the habit of human eyes to classify the bispectrum has the utmost classification accuracy.

From the view of the physical meaning, feature vector E1 is equivalent to the energy accumulation of the generalized diagonal slices spectra which depicts lines parallel to the original diagonal slice. The feature extraction scheme based on generalized diagonal slices makes full use of the information of bispectrum. At the same time, there is no risk of information loss and the calculation is simple. And it has high classification accuracy. But its feature vector dimension depends on the FFT points. If a large point of FFT is used, the feature vector dimension will also increase. And classification will cost more time. Hence, the next research is the dimension reduction of feature vector on condition that the classification accuracy is ensured.

6 Conclusion

In order to improve the classification accuracy of the radiated noise from underwater targets, generalized diagonal slices and their spectra of third cumulant are proposed by introducing an amendment quantity q into the original diagonal slice. The amendment quantity q represents the distance between the original diagonal slice and other slices parallel to it in bifrequency plane. And the generalized definition degrades into the original one when the amendment quantity q is set to zero. The physical meaning of the generalized diagonal slices spectra depicts lines parallel to the original diagonal slice. And two feature vectors are formed by summation or maximization of every generalized slice. It is validated by simulation that the target features in the bifrequency domain is drastically extracted by summation scheme and that the classification accuracy of 100% for feature vector obtained by summation scheme is steadily resulted by OAO method of multi classification of SVM. But the new feature vector dimension depends on the FFT points. If a large point of FFT is used, the feature vector dimension will also increase. Hence, the next research is the dimension reduction of feature vector on condition that the classification accuracy is ensured.

Acknowledgment. We thank for the sponsor of the Education Reform Program Funded by Xi'an University of Science and Technology (Program NO.JG10066).

References

1. Brockett, P.L., Hinich, M., Wilson, G.R.: Nonlinear and non-Gaussian ocean noise. J. Acoust. Soc. Am. 82(4), 1386–1394 (1987)
2. Zhang, X.: Modern Signal Processing, 2nd edn. Tsinghua University Press, Beijing (2002)
3. Nikias, C.L., Mendel, J.M.: Signal processing with higher-order spectra. IEEE Signal Processing Magazine 10(3), 10–37 (1993), doi:10.1109/79.221324
4. Li, Y., Feng, X., Fan, Y., Huang, J.: Extraction of low frequency line spectrum of ship radiated noise based on the 3/2 D spectrum. Acta Armamentarii 25(2), 238–241 (2004)
5. Shi, G., Hu, J.: Extraction and fusion of frequency-domain features from ship radiated-noise based on wavelet packet and 3/2 dimensional spectrum. Technical Acoustics 23(1), 4–7 (2004)

6. Peng, Y., Shen, L., Li, X., Wang, K.: Bispectrum based feature extraction and classification of radiation noise from underwater targets. Journal of Harbin Engineering University 24(4), 390–394 (2003)
7. Nello, C., John, S.-T.: An Introduction to Support Vector Machines and Other Kernel-based Learning Methods. Cambridge University Press (2000)
8. Hsu, C.-W., Chang, C.-C., Lin, C.-J.: A practical guide to support vector classification (April 3, 2010), `http://www.csie.ntu.edu.tw/~cjlin/libsvm/`
9. Yu, H., Zhao, H.: New research on multi-classification based on Support Vector Machines. Computer Engineering and Applications 44(7), 185–189 (2008)
10. Gou, B., Huang, X.: SVM multi-class classification. Journal of Data Acquisition & Processing 21(3), 334–339 (2006)

The BLAST Algorithm Based on Multi-threading in the DNA Multiple Sequence Alignment

Xiaojun Kang [1,2], LiYuan He[1], and LiJun Dong[2]

[1] College of Resources and Environment, Huazhong Agricultural University, Wuhan, P.R. China
[2] School of Computer, China University of Geosciences, Wuhan, P.R. China
xj_kang@126.com, heliyuan@mail.hzau.edu.cn

Abstract. DNA sequence alignment is the basis of DNA sequence analysis. The BLAST algorithm is one of the most classic and is widely used. In this paper, we propose to improve this algorithm based on multi-threading, in large amounts of data on the calculation of DNA multiple sequence alignment, we significantly improve the efficiency of the algorithm.

Keywords: DNA, multi-threading, BLAST.

1 Introduction

The human genome project is an international effort to identify the complete structure and function of the human genome. Life science data generated a flood of information. HGP project is a direct result of vast amounts of biological data obtained. With the advance of human genome project, many species, gene and protein data are has never been such high growth in the history of science. Currently, twenty kinds of bacteria have been measured. Now, the base amount doubles every 8 months. So far has completed more than 40 organism's whole genome sequencing, human genome sequencing 30 million bases has been completed. Up to now, registered in the GenBank DNA sequence database has more than 28 billion total base pairs. All of this constitutes a marine biological data.

Of massive data processing, making a single computer's memory capacity and computing speed show serious deficiencies. On the other hand, Large-scale data analysis will strongly depend on the super-computer. Therefore to develop a good parallel algorithm, using high-performance computers, is the only way to resolve this contradiction.

Sequence analysis is the basis of analysis of biological information, occupy the very important position in bioinformatics. Through sequence analysis, we can extract biological information which embedded in a simple sequence. Sequence alignment is the key of sequence analysis. By sequence alignment, we can determine the homology between species, to reveal the origin of life, the development of life has great significance.

The most common alignment compare between protein sequences and nucleic acid sequence.

D. Zeng (Ed.): Advances in Control and Communication, LNEE 137, pp. 81–84.
springerlink.com

By comparing the two sequences conserved between the sites of regional and conservative, Find the molecular evolutionary relationship between the two possible. Further alignment is a multiple protein sequence or nucleic acid sequence comparison; find their common conserved region, sites, etc.

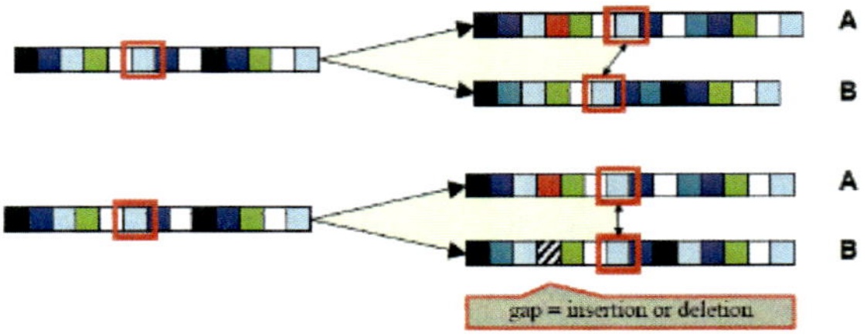

Fig. 1. Sequence alignment

2 The Blast Algorithm

The BLAST algorithm is a Basic Local Alignment Search Tool, it has been published by Altschul in 1990. BLAST algorithm can compare the homology between two nucleic acid or protein sequences. It can quickly find the homologous sequence between the two sequences, furthermore the algorithm scoring to determine the level of homology. When retrieved in BLAST, at first, we should list the name of database, gene sequence date, how to retrieve to invite the BLAST algorithm. After access to these parameters, BLAST read the sequence data in a structure, all records in the database mapping in the memory. At the last the algorithm began to compare the date from the first record to the last record in the database.

The algorithm can be described as:

(1) Set a minimum fragment length T, build the fragment index table for the query sequences.
(2) Find a space segment which can match the table from the database.
(3) Set the minimum score S of the match fragment.

```
>ref|NM_021257.3| Homo sapiens neuroglobin (NGB), mRNA
Length=1885

 Score = 3400 bits (3770),  Expect = 0.0
 Identities = 1885/1885 (100%), Gaps = 0/1885 (0%)
 Strand=Plus/Plus

Query  1    TTCCCAGGCCACCATAGCGGCTGGCGGAGGGAGCGCGCCTTGCTGGCCTGGAGGGGGC  60
            ||||||||||||||||||||||||||||||||||||||||||||||||||||||||||||
Sbjct  1    TTCCCAGGCCACCATAGCGGCTGGCGGAGGGAGCGCGCCTTGCTGGCCTGGAGGGGGC  60

Query  61   GGGGGCCGTGGCGGCTTTAAAGCGCCCAGCCCAggcgtcgcggggtggggcggctctggc  120
            ||||||||||||||||||||||||||||||||||||||||||||||||||||||||||||
Sbjct  61   GGGGGCCGTGGCGGCTTTAAAGCGCCCAGCCCAGGCGTCGCGGGGTGGGGCGGCTCTGGC  120

Query  121  ggctgcggggcgcagggcgcagcggccaagcggggTCCCCGGAAGCACAGCTGGGGTGTC  180
            ||||||||||||||||||||||||||||||||||||||||||||||||||||||||||||
Sbjct  121  GGCTGCGGGGCGCAGGGCGCAGCGGCCAAGCGGGGTCCCCGGAAGCACAGCTGGGGTGTC  180
```

Fig. 2. BLAST algorithm result

3 The Blast Algorithm Based on Multi-threading

Using multi-threading applications can make better use of system resources. Its main advantage is the full use of idle CPU time slice. Can use as little time respond to user requirements, making overall operating efficiency of process greatly improved. At the same time it also enhances the flexibility of the application. More importantly, since all the threads are the same process share the same memory Therefore, special mechanism for data transfer is not needed. We also not need the shared memory area or shared file. So that the coordination between different tasks and run, date exchange, resources allocation problems more easy to resolve

Step 1 and Step 2 in formula 2 are generally the same as formula 1,as we can see in 3.2.1, formula 1 just processes one set of data, while formula 2 processes both sets of data.

1) Divide the gens of Set 1 into 6 and write each share into a map, each used in one single thread.
2) Put the gens of Set 2 into a Map container, and establish a database according to the map container.
3) Execute step 4) to step 6) in each thread.
4) Write the number K (initialized with 1) gen to the file test.fa, which is used as the query, then try to find homologous gens in the database found in step 2).
5) Delete the homologous gens from the map, and write the gen in test.fa to a vector container.
6) Decide wether K=size (size=the number of gens in map),if so ,turn to step 7),else add 1 to K and turn to step 4).
7) Write the gens in 6 maps into file Carp11.fa one by one, and write the gens in the vector container into file Carp12.fa, and write the gens in the Map container into file Carp22.fa.
8) Now, the gens unique to Set 1 are stored in Carp11.fa, and the gens unique to Set 2 are stored in Carp22.fa, and the gens that Set 1 and Set 2 are in common are stored in Carp12.fa.

Different from the ordinary, we did not use public sequenced database, in other word, we have no standard references to compare with our source data. So, we use the source data to build the database for Blast, take the source sequence as its own reference. That is the creative point in our project .It can be said that Blast give us a judgment about weather this two sequence are from homologous gene, just as the symbol of "=".We use Blast in the follow scenarios:

1) Step 2 in formula 1, call Blast to reduce the amount of gene species.
2) Step 3 in formula 1, build database with the sorted by specific enzyme genes data, given a query sequence then search homologous genes in the database.
3) Step 3 in formula 2, call Blast to find common gens in two set of data

In our experiments, we use 3 group DNA data files to test, finally we get the following table:

Table 1. Efficiency comparison

File size	DNA group number	Clustering spend (Single-threaded)	Clustering spend (multi-threading)
143 M	95662	8h,25m	4h,35m
184 M	125520	8h,45m	5h,15m
19.2 M	15247	39m	26m

Experimental environment : Intel E5800, 320GB SATA, 4G RAM.

4 Conclusion

The BLAST algorithm has very broad applicability. We can not only use it to single-sequence alignment, but also we can do some multi-sequence clustering. In this paper from the experiments we know, after use the multi-thread the algorithm efficiency has significantly improved.

References

1. Altschul, S.F.: Gapped BLUST and PSI- BLAST:a new generation of protein database search program. Nucleic Acids Research 25(17), 3 389–3 402 (1997)
2. Wang, Z., Cao, X., Mao, L.: Research and Comparison of Great Efficiency Algorithm on DNA Sequence Alignment. Journal of Wuhan University of Technology 8 (2005)
3. Protein strueture Predietion and its applieation. Joumal of Fujian Agrieulture and Forestry University (Natural Science Edition) 35(3), 49-52 (2006)
4. Ogata, M.: Mammalian Acatalasemia: The Perspectives of Bioinfomatics and Genetic Toxicology. Acta Medica Okayama 63(2), 121–122 (2009)
5. Šerý, O.: Enhanced Property Specification and Verification in BLAST. In: Chechik, M., Wirsing, M. (eds.) FASE 2009. LNCS, vol. 5503, pp. 456–469. Springer, Heidelberg (2009)
6. Nie, J.S., Xu, J., Hofmayer, C.H., et al.: An Approach for Assessing Structural Upliftiong Using Blast Motions. Proceedings of The Asme Pressure Vessels And Piping Conference (8), 215–224 (2009)
7. Yang, Y., Zhang, J., Hoh, J., et al.: Efficiency of single nucleotide polymorphism haplotype estimation from pooled DNA. Proc. Nat. Acad. Set. 100(12), 7225–7230 (2003)
8. International Human Genome Sequencing Consortium.Initial sequencing and analysis of the human genome. Nature 409(6822), 860–921 (2001)

Research on TAGSNPS Selection Based on BLAST Algorithm

Xiaojun Kang[1,2], LiYuan He[1], and LiJun Dong[2]

[1] College of Resources and Environment, Huazhong Agricultural University, Wuhan, P.R. China
[2] School of Computer, China University of Geosciences, Wuhan, P.R. China
xj_kang@126.com, heliyuan@mail.hzau.edu.cn

Abstract. In this paper we showed a method on how to find a SNP using the BLAST algorithm. Which is more efficiency and reliable.

Keywords: TagSNPs selection, BLAST, DNA.

1 Introduction

Single Nucleotide Polymorphism(SNP)is a DNA set of polymorphism based on the single nucleotide variations at the genomic level. A small subset of informative SNPs, known as TagSNPs or htSNPs, is able to capture most variations in these haplotypes. However, it is expensive and hard to deal with massive data. So it is very necessary to develop mathematic and computational methods to select TagSNPs.

Our work is based on a new kind of biological experiments using second-generation DNA sequencing technology, through processing experimental data, to find out weather SNP exists in a great lot of base sequences. This project implements two solutions to search SNP in the source experimental data. In the first formula, we will statistic and compare the percentage of two types of homologous genes, which certainly are in the same trait. The other formula, we deal with two sets of genes from two phenotypes, filter out the genes unique to each carp and the common ones, because it is useful to make further analysis. In addition, this paper provides with a straightforward visual interface, users can easily set the parameters and choose one of the formulas to search SNP. During this data processing, comprehensive statistical data showed in UI will allow users to learn more about the source data.

The BLAST algorithm is a Basic Local Alignment Search Tool, it has been published by Altschul in 1990. BLAST algorithm can compare the homology between two nucleic acid or protein sequences. It can quickly find the homologous sequence between the two sequences, furthermore the algorithm scoring to determine the level of homology. When retrieved in BLAST, at first, we should list the name of database, gene sequence date, how to retrieve to invite the BLAST algorithm. After access to these parameters, BLAST read the sequence data in a structure, all records in the database mapping in the memory. At the last the algorithm began to compare the date from the first record to the last record in the database.

D. Zeng (Ed.): Advances in Control and Communication, LNEE 137, pp. 85–89.
springerlink.com

2 Overall Design

With the growing use of the second-generation DNA sequencing technology, previous approach for the enormously experimental data can not meet demand, so new kinds of approach are needed to try to data mining biological information hidden behind.

We implement tow schemes. In the first scheme, we calculate the gap of each base gene in two sets of Homology genes, then if the score is higher than the threshold, we suppose there is a SNP Occurs. In the second scheme, maybe not so intuitive, that is to say, it aims at filter out the genes unique to each carp and the common ones .More details will be talked about later, and we mainly talk about scheme 1.

The whole construct of this system as below:

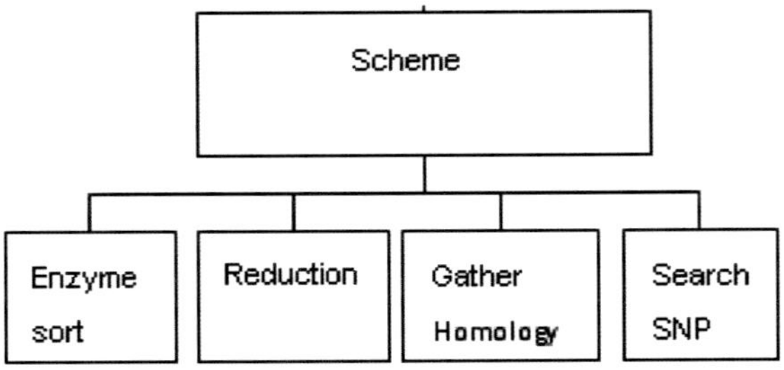

Fig. 1. The whole construct of the system

3 Algorithm and Detailed Steps

3.1 Enzyme Sort

In this biological experiment, genes are cut by the enzymes, that to say, all of the sequences in source data should be separated by which enzyme cut it down. So, this step just does this job. The Concrete steps are as below:

1) To achieve efficiency, use windows Memory mapping mechanism to map file from hard disk to memory.

2) Create output files, the number of files is the same as kinds of enzyme.

3) Traversal all of the source data in memory, match specific genes of each enzyme, then print the sequence into relative file.

4) Close all files, output file size and proportion to source data.

3.2 Reduction

Theoretically speaking, the two DNA sequences, if there is a continuous 18bp is the same, it can be Asserted that these two sequences are homologous. According to this hypothesis, we grouped according to restriction sites, and then there are **m** before the 18-bit (**m** <= 18, 18-**m** = Parallaxes is the potential difference) the same sequence is divided into a group and that group. All sequences are homologous sequences. Then grouped for homologous sequences using BLAST sequence alignment, to further reduce the number of groups of sequences (gene type number).

Specific steps are as follows:

1) Scan all of the sequence, if the first 18bp is the same, that divide it into a group, and record the number of sequence in each group;
2) Compare with the Paralaxe user given, if Parallaxes is not zero, and then go to 3);
3) Pick up a sequence in each group, and one of them as a reference, and then make contrast to the first 18bp between the reference and the rest sequences. If the same number of digits greater than or equal m, it will be clustered with the reference group;
4) Sort the groups by the number of sequence in each group, pick up a sequence in the first **N** group, which **N<M**, and the sequences in N groups should take iaP% of the source data (iaP means the input argument, can be input by user), to build up the database of Blast using the program makeblasédb.exe.
5) Get one of the sequences in the database just built as query, and call blast.exe to find its homologous sequences.
6) By repeatedly carry out 5), we can get the kinds of genes in the grass carp.

3.3 Gather Homology

Until now, we can search and divide all genes into groups using the kinds of genes in grass carp. As below:

1) Build database with the Enzyme sorted data, take one kind of genes as a query, and call blastn.exe to search its homologous sequences.
2) Output these homologous sequences to make up the SNP files.

3.4 Search SNP

From the above step, two set of homologous genes are generated, we should calculate each base gene in each kind of sequence, and set the threshold to identify SNP.

1) Traversal all file in each set of SNP file. get one file in each set.
2) Statistic the amount of base gene “A” ,“T” , “C” , “G” , “-” , “#” , and mark the number of sequences in the file , then, calculate the percentage of each base genes. Here, base gene “-” is generate by BLAST, it represent a blank here;”#” means the gene can not match by BLAST, we use “#” to the gene bit have not output form Blast.
3) Use absolute distance algorithm to calculate the score of each BP, if the score is higher than the threshold, that assert SNP bp found.

4 Program Display

In our program, we use MFC technology to produce visual interface that allows users to easily enter the relevant parameters from the interface, and you can choose between two different scenarios to find SNP. Meanwhile, the software implementation process step by step, each step can be executed on the interface by clicking "View Results” button to view the results, simple and convenient.

Some screenshots of the program as follows:

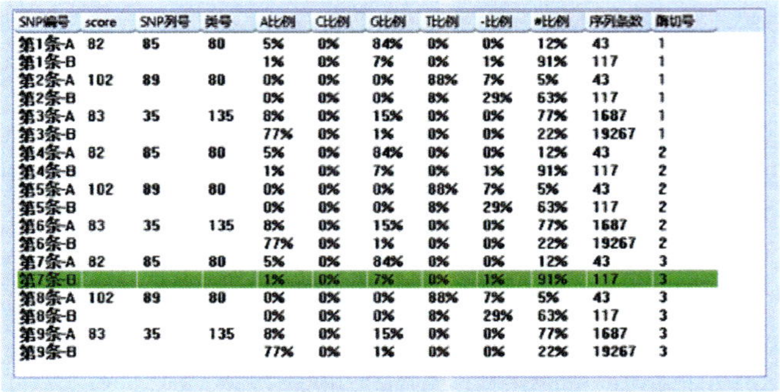

SNP编号	score	SNP列号	类号	A比例	C比例	G比例	T比例	-比例	#比例	序列条数	酶切号
第1条-A	82	85	80	5%	0%	84%	0%	0%	12%	43	1
第1条-B				1%	0%	7%	0%	1%	91%	117	1
第2条-A	102	89	80	0%	0%	0%	88%	7%	5%	43	1
第2条-B				0%	0%	0%	8%	29%	63%	117	1
第3条-A	83	35	135	8%	0%	15%	0%	0%	77%	1687	1
第3条-B				77%	0%	1%	0%	0%	22%	19267	1
第4条-A	82	85	80	5%	0%	84%	0%	0%	12%	43	2
第4条-B				1%	0%	7%	0%	1%	91%	117	2
第5条-A	102	89	80	0%	0%	0%	88%	7%	5%	43	2
第5条-B				0%	0%	0%	8%	29%	63%	117	2
第6条-A	83	35	135	8%	0%	15%	0%	0%	77%	1687	2
第6条-B				77%	0%	1%	0%	0%	22%	19267	2
第7条-A	82	85	80	5%	0%	84%	0%	0%	12%	43	3
第7条-B				1%	0%	7%	0%	1%	91%	117	3
第8条-A	102	89	80	0%	0%	0%	88%	7%	5%	43	3
第8条-B				0%	0%	0%	8%	29%	63%	117	3
第9条-A	83	35	135	8%	0%	15%	0%	0%	77%	1687	3
第9条-B				77%	0%	1%	0%	0%	22%	19267	3

Fig. 2. The result of our program

References

1. Šerý, O.: Enhanced Property Specification and Verification in BLAST. In: Chechik, M., Wirsing, M. (eds.) FASE 2009. LNCS, vol. 5503, pp. 456–469. Springer, Heidelberg (2009)
2. Zhang, K., Calabrese, P., Nordborg, M., Sun, F.: Haplotype block structure and its applications in association studies. The American Journal of Human Genetics
3. Huang, Y.T., Zhang, K., Chen, T., Chao, K.M.: Selecting additional tag SNPs for tolerating missing data in genotyping. BMC Bioinformatics 6, 263 (2005)
4. Crawfod, D.C., Nickerson, D.A.: Definition and clinical importance of haplotypes. Annu. Rev. Med. 56, 303–320 (2005)
5. Bafna, V., et al.: Haplotypes and informative SNP selection algorithms:don't block out information. In: Proceedings of 8th Pacific Symposium on Biocomputing (PSB 2003), pp. 466–477 (2003)

6. Zhang, K., Qin, Z., Liu, J., Chen, T., Waterman, M., Sun, F.: Haplotype block partitioning and tag SNP selection using genotype data and their applications to association studies. Genome Res. 14, 908–916 (2004)
7. Hinds, D.A., et al.: Whole-genome patterns of common DNA variation in three human populations. Science 307, 1072–1079 (2005)
8. International Human Genome Sequencing Consortium. Initial sequencing and analysis of the human genome. Nature 409(6822), 860–921 (2001)
9. Yang, Y., Zhang, J., Hoh, J., et al.: Efficiency of single nucleotide polymorphism haplotype estimation from pooled DNA. Proc. Nat. Acad. Set. 100(12), 7225–7230 (2003)

Research on Image Retrieval Based on Scalable Color Descriptor of MPEG-7

Wen Yong-ge and Peng Sheng-ze

School of Mathematics and Computer Science
Mianyang Normal University
Mianyang, China

Abstract. On the basis of MPEG-7 international standards, this paper explores the components semantics and syntax rules of the scalable color descriptor in detail, and further discusses the bitplane expression of image and the correlation theories of the color quantization in optimizing HSV space, it select a large number of images to carry on the empirical grouping research for the scalable and variable characteristics of coefficient in image SCD description, and carries on image content's similarity and time-consuming calculations in different coefficients, the experiment shows that SCD can obtain good performance both in the retrieval accuracy and the fast matching of image.

Keywords: MPEG-7, Scalable Color Descriptor, HSV optimization model, Calculation of similarity, Image matching.

1 Introduction

During the 1990s, Content-Based Image Retrieval (CBIR) is gradually becoming a research hotspot in the application of multimedia technology [1]. So far, the application range of the image retrieval technology also is involved in the fields of communication, medicine, military, transportation, public security, remote sensing, surveying and drawing, biology, industrial automation and office automation and so on.

In the visual features of image, the color is the most expressive visual feature in the low-level feature of image, and it is born with rotation invariance and scale invariance and shows the quite strong robustness [2], and the matching calculations based on the color characteristic are relatively simple, therefore, it becomes one of the most widely used features in the image retrieval system. The descriptors of color features in MPEG-7 standard mainly include four types, Dominant Color Descriptor, Scalable Color Descriptor, Color Layout Descriptor, and Color Structure Descriptor. And the Scalable Color Descriptor (SCD) is very suitable for the image-to- image matching and the retrieval based on color features, the retrieval accuracy increases with the increase of the number of bit in binary notation and the retrieval speed increases with the decrease of the value of coefficients; it comes from Haar transform of the histogram to HSV color space and can well achieve the description of the image color features, simultaneously, the size of descriptor is small, and easy to be adjusted according to the different demands [3], and it can flexibly deal with the accuracy and efficiency problems of image retrieval system.

D. Zeng (Ed.): Advances in Control and Communication, LNEE 137, pp. 91–98.
springerlink.com

2 Research Background

2.1 MPEG-7 Standards

MPEG-7 is "Multimedia Content Description Interface", it is the international standard which is recommended by ISO international organization and serves for the multimedia content description after MPEG-1, MPEG-2, and MPEG-4 standards, it aims to provide a whole set of technical standard for multimedia content description, and meets the wide range of multimedia applications and demands through describing various features of multimedia content [4]. The basic elements of media content description in MPEG-7 standards are:

(1) D (Descriptors), describe features, attributes, or groups of attributes of multimedia content.

(2) DS (Description Schemes), describe entities or relationships pertaining to multimedia content, it is composed of one or more D and DS and stipulates the structure and semantic relations between each element in DS.

(3) DDL (Description Definition Language), define Description Schemes, Descriptors, and Datatypes by specifying their syntax, and allows their extension, the Description Definition Language allows to modify and extend the existed description schemes.

(4) System tools, support delivery of descriptions about efficient storage, transmission, multiplexing of descriptions, description and content synchronization, file format, and manage and protect the intellectual property and so on.

The relationship between the elements of DDL, DS and D can be expressed by Figure 1.

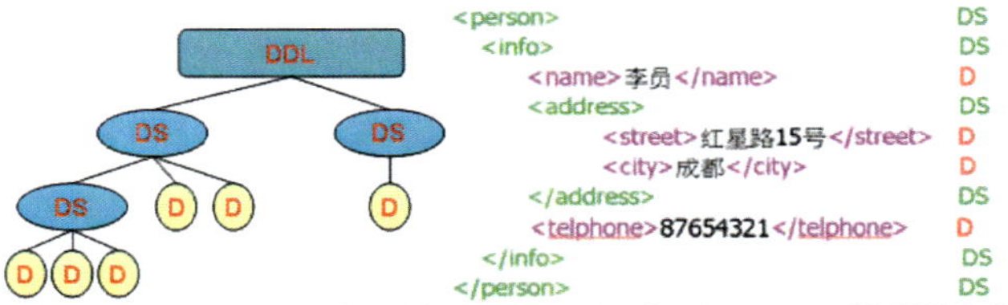

(a) The structure relationship between basic elements of MPEG-7
(b) The DDL representation instance corresponding to (a)

Fig. 1. The basic elements and constitution of MPEG-7

As in the figure above, the description scheme (DS) is composed of one or more D and DS, DS provides the structure and grammar of their relationships. DDL provides the description tools of MPEG, which includes descriptors and description schemes; it also allows to definite the extended DS in special applications at the same time.

2.2 HSV Color Model Optimization

The MPEG-7 standard supports the color space of RGB, YCbCr, HSV and HMMD and so on, and appoints HSV as the color space for the Scalable Color Descriptor (SCD) to carry on the color quantization [4]. When carries on the color histogram calculation, it is necessary to divide HSV color space into many small color intervals, each small interval corresponds to a bin of histogram, then it can get the color histogram by calculating the number of pixels for the color falling into each small interval. For example, the color histogram of a picture is constructed by 256 bins, and each value of the bin is expressed by 4 bits, then it needs 1024 bits to describe it, for retrieving on mass image information, both its storage capacity and match computation are seriously exhausted, the system speed also will be decreased greatly. The experiments shows that in the histogram representation of the natural image signal, the value of the adjacent histogram bin shows a large number of redundancies, in order to reduce the space occupied by the image description, first, take advantage of the function characteristics of evenly distributed and rapid convergence of Haar transformation, and carry on Haar conversion for the histogram in HSV color space, filter the bin which high-pass coefficient differential near 0, consequently, reduce the transformed coefficient; second, optimize HSV space model, carry on normalized processing to the areas which are close to black and white of the HSV color space, and default them as the same color, reduce the amount of calculation of spatial pixels, and further improve the processing speed of image. Optimizing HSV color model [5] is shown as the following Figure 2:

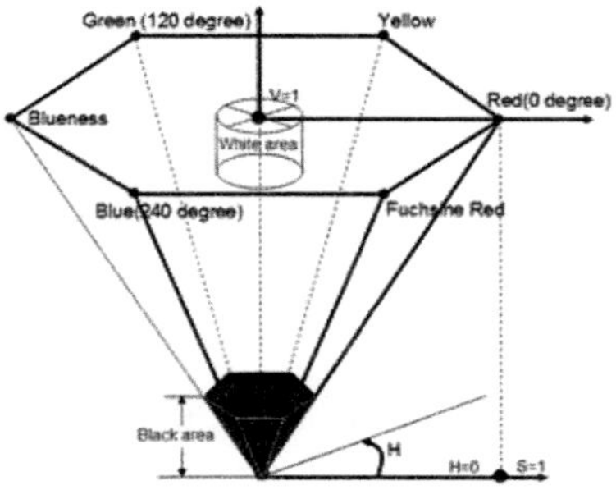

Fig. 2. Limits of the HSV color space

(1)Black area: all the V<15% colors are classified as black, set h=0, s=0, v=0;

(2)White area: all the s<10% and v>80% colors are classified as white, set h=0, s=0, v=1;

(3)Colored area: the other colors except the colors in (1) (2) areas, the values of H, S and V are unchanged.

3 Scd Descriptions to Images

The Scalable Color Descriptor specifies a color histogram in the HSV color space, which is encoded by a Haar transform, it is useful for image-to-image matching and retrieval based on color feature, and retrieval accuracy increases with the number of bits used in the representation [6]. In order to research the changes of SCD description accuracy of image, the system uses the JPEG image which is commonly used, in its DDL description, Haar transformation coefficient is given to be 72, 256, the number of bitplanes Discarded is given to be the parameter of 3, 6 and so on, its DDL grammar expression is shown as follows:

```
<MPEG7>
  <Descriptor type= "ScalableColorType">
  ......
    <attribute name="numOfCoeff" use="required">
           <enumeration value="72"/>
       <enumeration value="256"/>
    </attribute >
    <attribute name="numOfBitplanesDiscarded" use="required">
           <enumeration value="3"/>
       <enumeration value="6"/>
    </attribute >
  ......
  </Descriptor>
</MPEG7>
```

3.1 Bitplane Presentation of Image

The bitplane is a coding method for image. As for an image which color depth is expressed by 8 bits, regarding a plane of each bit as a layer in accordance with the hierarchical thinking, thus this picture is divided into 8 layers,that is to say, it is composed of 8 bitplanes; it is shown as the Figure 3. In general, using the bitplane 0 represents the lowest plane and the bitplane 7 represents the highest plane. Because it also includes amount of noise and interference signals except for the meaningful pixels in the image, therefore, the several highest bitplanes include the meaningful information that can be visiable, they are the key reference objects when constructing the index of image's color feature, and most of the low bitplanes are the image details and noise signal and have small contribution to the extraction and the match of the image color feature, but they will still increase system's vast computation when carries on quantization, so in order to reduce the amount of computation, they can be abandoned.

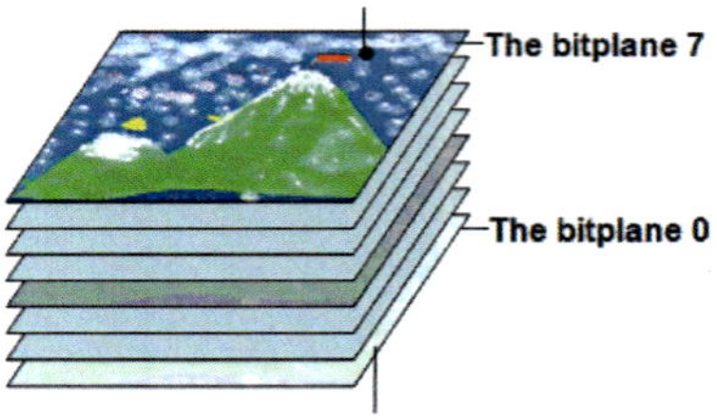

Fig. 3. The Bitplanes presentation of image

3.2 Color Quantification of HSV Space

The color histogram describes the proportion of different colors that account for the whole image, completely ignores the consideration of the rotation, scaling, deformation and other factors, and does not care the spatial location of each color, namely, it can not describe the targets or objects on the specific location of image, so it is suitable to describe those images which are difficult to carry on automated segmentation. Carrying on the process of quantification and reducing dimensions for image, which can simply and effectively calculate the color histogram of colored images. It uses unequal interval quantization strategy to carry on statistical quantification to image color in HSV color space, the three components with the hue H, the saturation S and the illumination V are respectively divided into several parts, each stand for 1 bin, it counts the number of each color's pixel according to the divided different color regions as the vector of image color feature, and establishes statistical histogram of each pixel and achieves the goal of quantifying image color.

When the system considers the balance between retrieval accuracy and efficiency, it mainly uses different coefficients such as 256, 72 and 32 respectively to carry on image retrieval based on SCD. When the coefficient is 256, the value range of color component H is 0 to 360, quantifyed 16 bins, the value range of S and V component are 0 to 1, quantifyed 4 bins respectively, it determines the specific corresponding relationship between coefficient and space division according to ISO/IEC FDIS 15938-3:2001(E) standards on MPEG-7, it is shown as Table 1.

Table 1. Bin numbers of reconstructed HSV color histograms

Number of coefficients	*Number of bins H*	*Number of bins S*	*Number of bins V*
32	8	2	2
72	8	3	3
256	16	4	4

4 Results and Conclusions

In the research process, according to the MPEG-7 standard, the programming realizes the SCD-based image retrieval system prototype, and it carries on the experimental measurements and calculations to the SCD scalable quantification coefficients via statistical methods. Based on the experimental results, this paper finally analyze and summarize to the similarity calculation of different coefficients, similarity sorting and the consumed time for image matching.

4.1 Calculation Model of Image Similarity

Currently, the usual method to calculate the image similarity is based on the vector space model, it take the visual features as points in vector space, by calculating the distance between two points to measure the similarity of image feature points, so that calculate the similarity between two images. According to the MPEG-7 international standard, scalable color descriptors uses the specified HSV color space, and quantifies the color features in HSV space, and calculates the image similarity according to color histogram after Haar transformation. Now, assuming that the color features in various practical applications are orthogonal and independent, and that the importance of each dimension is identical, at the same time, since the L1 distance algorithm is simple and efficient, which could be applied in the field of Haar transform, and bring a satisfactory retrieval accuracy [7], therefore, the L1 distance algorithm is applied in the programming experiment and shown as the following equation:

$$D_1 = \sum_{i=1}^{N} |A_i - B_i|$$

In the above equation, A_i and B_i represent the same color feature vector of two different images in the respective HSV space, N represents the number of bin in color histogram.

The corresponding main code segment to achieve L1 is as following:

```
for (int i = 0; i < secHaarHist.length; i++) {
      diff = Math.abs(secHaarHist[i] - haarTransHist [i]);
      diffsum += diff;
}
```

4.2 Image Retrieval and Conclusions

Compared with other color descriptors, the SCD descriptors have obvious advantages of flexibility in custom retrieval precision. When realizing the content-based image retrieval, this paper uses the step refinement method and implements the following three steps from the rough to the exact to get a predictive validation result:

(1) Firstly, set the coefficient of histograms as 32, and carry out the preliminary matching to the color features. Using 32-dimensional feature vector and the distance calculation method retrieve images from the image database. When the matching value between image in the database and the sample image is less than the threshold value, it reckons that the image does not meet the requirements and then test the next image. When the match value is greater than the threshold, the second step would be taken, that is, to take more accurate color coefficients from the color space.

(2) Secondly, using 72-dimensional feature vector and distance calculation method for image database retrieval, which is SCD-based universal quantization accuracy applied in the HSV color space, with the use of 8H * 3S * 3V's space division, it not only takes the computational complexity of matching into consideration, but also meets the system accuracy needs when it comes to image retrieval.

(3) Thirdly, using 256 dimensional feature vector and the Euclidean distance methods to make more accurate retrieval to these images, and then returns to the images which meet the conditions.

In the system, 100 images with different resolution are matched to computation. By the use of scalable quantization features of SCD, according to the different Haar coefficients to calculate the similarity between images. The results are shown in Table 2, and the consumed average time for the matching with the different coefficients is shown in Table 3.

Table 2. Similarity results with different numbers of Haar coefficients between pictures (part of data)

similarity matching	*Coef=256*	*Coef=72*	*Coef=32*
P0 and P1	1048	475	311
P0 and P2	974	391	281
P0 and P3	855	279	183
P0 and P4	1083	488	32

Table 3. The consumed average time results with different numbers of Haar coefficients between pictures (part of data)

Average consumed time(ms)	*Coef=256*	*Coef=72*	*Coef=32*
P0 and P1	204.725	201.95	195.62
P0 and P2	135.175	128.55	128.95
P0 and P3	151.95	143.375	137.9
P0 and P4	156.625	155.94	155.5
Total average time	162.119	157.454	154.493

From the above Table 2, Table 3, it can be seen that the greater the Haar coefficient value is, the greater the similarity would be, and the longer time it would take to match with another picture in normal circumstances; the smaller the coefficient value is, the smaller the similarity value would be, the less time it would take to make the similarity calculation. However, by studying the data in Table 2, it also can be seen that regardless of the size of coefficient values, the similarity sorting between images, namely, the similarity rank between images is basically stable. Therefore, when meet different application requirements, we can come to a conclusion, when it pays strict attention to the retrieval speed, firstly, take the smaller Haar coefficient as image feature vector to make the image retrieval. With the increasing requirements on the retrieval accuracy, it may be appropriate to improve the transformation coefficient, and make more accurate color information comparison without sacrificing the efficiency of the system at the same time.

The following Figure 4 is the sequence diagram of the system retrieval results. Take the picture 1 as an sample image among the following pictures, do the SCD description to 100 pictures with different degree similarities, extract the feature vector, and calculate the similarity degree and sort them. then the query results show that we can easily implement accurate and efficient image retrieval system by set the system coefficients and discarded bitplanes to meet different demands in terms of SCD description way.

Fig. 4. Retrieval images based SCD

5 Conclusion

MPEG-7 is a new multimedia content description standard promoted by ISO based on the previous three criteria of MPEG, which implements standardized descriptions for a variety of different types of multimedia information, and makes this descriptions link with the contents to achieve fast and efficient retrieval. SCD descriptor is a scalable description tool of color in MPEG-7, with the use the universality of the image color features and extraction conveniences, SCD description method can totally meet the actual image retrieval needs. Especially, in the case that implementing the different colors matching in details, it can customize a friendly and considerate retrieval service. Certainly, just like other descriptors, SCD can not solve all the issues of content-based multimedia retrieval but describe and extract the features from the viewpoint of the image color[8]. With the continuous deepening application study of SCD, combined with the main color descriptor, color distribution descriptor and color structure descriptor, as well as the integrated applied research on the descriptors of shape, texture, movement and location to realize powerful and highly practical image retrieval application system, which are entirely achievable.

Acknowledgement. Supported by Science Foundation of Sichuan Educational Committee(10ZC029), Foundation of Mianyang Normal University(MA2009013, MA2009012).

References

1. Rui, Y., Huang, T.S., Chang, S.-F.: Image Retrieval: Curren Techniques, Promising Directions and Open Issues. Journal of Visual Communication and Image Representation (10), 39–62 (1999)
2. Wang, H., Qin, T.: Image Retrieval Based on Combined Color Features in MPEG-7. Application Research of Computers 03 (2005)
3. Li, Z., Li, J., Yan, B.: Research of MPEG-7 Scalable Color Descriptor. Journal of the Graduate School of the Chinese Academy of Sciences 23(2), 192–197 (2006)
4. ISO/IEC FDIS 15938-3: Information technology – Multimedia content description interface – Part 3: Visual 1-8, 29–42 (2001)
5. Wang, T., Hu, S., Sun, J.: Image Retrieval Based on Color-Spatial Feature 13(10), 2031–2035 (2002)
6. ISO/IEC standard developed by Moving Picture Experts Group, MPEG-7 Overview, 52-58
7. Manjunath, B.S., Ohm, J.-R., Vasudevan, V.V., Yamada, A.: Color and Texture Descriptors. IEEE Transactions On Circuits And Systems For Video Technology 11(6) (June 2001)
8. Yang, Y., Zhang, J., Hoh, J., et al.: Efficiency of single nucleotide polymorphism haplotype estimation from pooled DNA. Proc. Nat. Acad. Set. 100(12), 7225–7230 (2003)

A Novel Helper Caching Algorithm for H-P2P Video-on-Demand System

Li Xia

State Key Laboratory of Networking and Switching Technology
Beijing University of Posts & Telecommunications
Beijing, China
lixiabupt@yeah.net

Abstract. In the field of Internet, the technology of P2P (peer-to-peer) has greatly enhanced the development of all kinds of video services. Particularly, the technology of P2P based on the mechanism of helpers has further exploited the potentialities of Internet. However, few researches related to utilizing the storage resources of helpers have been conducted. On the basis of P2P, this thesis aims at VoD (video-on-demand) service and proposes a novel algorithm making use of helpers to cache the content of VoD. According to the analysis and experiments, this kind of algorithm effectively reduces the server load and the average view delay of VoD system through utilizing the idle storage resources of helpers.

Keywords: P2P, VoD, H-P2P, streaming media, helper, caching.

1 Introduction

Video services on the Internet have developed rapidly in recent years. However, the video systems are confronted with many problems, for example, videos with not high quality and too much pressure on the server, because media transmission takes up too much bandwidth, the service time is quite long and the dynamic behaviors of peers are unpredictable. Through sharing bandwidth, the calculation and the storage resources on the Internet, the technology of P2P (peer-to-peer) completely changes the means of video service, significantly reduces the load of server, and improves the utility rate of the internet resources.

At present, the technology of P2P has been widely applied to video services. For example, among the live video projects, Coolstreaming [2] and PPLive [3] are rather successful. In the field of VoD (video-on-demand), the web like YouTube [4] has proved the efficiency and practicality of the technology of P2P. But in the P2P network, because of the limited upload bandwidth for the large number of internal peers and the inherent instability and unpredictability of peers, problems like system bottleneck still remain in P2P network.

Aiming at the shortcomings of P2P, in [5] Wang puts forward a helper transmission mechanism based on P2P so as to further exploit the resource potentialities of the Internet and improve the application capability of P2P, in which the concept of helper is first introduced to P2P. This mechanism names those peers uninterested in the delivered content as helpers. A helper overlay is added to the P2P network while

D. Zeng (Ed.): Advances in Control and Communication, LNEE 137, pp. 99–106.
springerlink.com

without changing those traditional P2P peer transmission means. Thus the mechanism is able to improve the capability of the P2P through using the idle resources of the helpers. Some later researches apply the helper mechanism to various services including file transmission, live video, and VoD, etc, and have achieved some research findings. In this thesis, the P2P system utilizing helpers is named H-P2P.

This paper proposes a content caching algorithm in helpers for P2P VoD system. According to the effect of video segments on view delay, this algorithm decides how to cache the segments in the helpers. In the third part of this thesis, by mathematical deduction, we conduct a qualitative analysis on how the video segments affect view delay. In the fourth part, based on the analysis made in the third part, we propose an algorithm making use of helpers to cache the content of VoD for the purpose of reducing the system's average view delay. And in the fifth part, with several simulation experiments, we have proved that this algorithm is able to efficiently reduce the view delay and the load of the media server.

3 Related Work

Wang first introduces the concept of helper to P2P system, making use of the idle peers uninterested in the delivered content of the system so as to further exploit the resource potentialities of the Internet and improve the application capability of P2P. However, there's a waste of resource in the system. Reference [6] brings socialized network into H-P2P and designs a system named tribler, which decides the choice of helpers on the basis of socialized relationship and makes the collector, i.e., the peers willing to become seeds, turn into seeds with the assistance of helpers, thus improving the performance of the P2P system. In [7], the authors develop the research by illustrating each mechanism of tribler in details under a practical system, namely, 2Fast, and introduce Promise mechanism and the way of measuring the contribution of peers according to the bandwidth, thus effectively holding back the "free ride" problem and building up fairness in the system. As for download of files, [8] proposes a content delivery mode based on H-P2P, and each helper only can download a very small segment. As shown in the model analysis and simulation experiment, one helper can nearly be equal to the capability of a seed. In [9], the authors look on some nodes which are equivalent to small servers as helpers, and these nodes are of high stability, quality and connection. In addition, the related parameters of these nodes are able to be changed by hand. Therefore, with these helpers giving priority to assisting the download of popular files, the system performance is able to be enhanced.

In terms of video service, in [10], Wang applies H-P2P to live video streaming system, in which, if introduced, helpers can allow peers to stream video at a rate above the average peer upload bandwidth, and through downloading a very small segment of the video in the form of erasure-coded parity packets, the server's burden is not to be increased. In [11], Wang and Zhang further expand the mechanism of H-P2P, and look into a novel P2P VoD architecture with using helpers to supply a scalable solution scheme. Besides, efficiency of the system has also been verified through packet level simulations. Under the content transmission mechanism of VoD, [12] brings in the method based on probability. Namely, the data piece close to the playback position has a comparatively higher download probability. As a result, the overall playback performance of H-P2P system is able to be improved to some degree. In [13], Zhang proposes a decentralized P2P VoD system, in which users and helper "servelets" cooperate in a P2P manner to transmit the video stream. Helpers are preloaded with only a small fraction of video data packets, and form into swarms, each of which serves partial video content. In [14], aiming at the system of VoD based on H-P2P, the authors put forward a data relay algorithm in which the helpers will give priority to relaying the segments with small bandwidth on average. The simulation experiments show that the streaming capacity of H-P2P VoD system can be greatly improved by adopting this algorithm. Reference [15] reckons that the streaming capacity is limited by the over-demanded video segments. Therefore, the researchers optimize helper assignment and rate allocation by a distributed algorithm, thus making the streaming capacity of VoD system increased effectively. In [16], aiming at the P2P VoD system, the researchers designed a novel grouping algorithm of helpers. This algorithm divides helpers into different groups according to the number of each video's users so as to reduce the average viewing delay of whole system, increase the utilization rate of helpers, and improve watch experience for users of the system.

Up till now, the researches on H-P2P have gained some achievements. But in these researches the usage of storage resources of helpers still remains in the stage of trying to take up the storage capacity as little as possible. However, as for the transmission capacity, most peers in the Internet are of comparatively sufficient storage capacity, so proper use of the helpers' storage capacity is acceptable for the helpers. If the storage capacity of many idle peers in the Internet can be used as helpers, it's bound to greatly improve and enhance the performance of P2P VoD system.

3 Model Analysis

Similar to [16], this paper mainly deals with how to effectively reduce the view delay of the video in demand by deducting and analyzing. In this section, we are not to consider some dynamic characteristics of the system like randomness of the peers, but just to carry on static analysis.

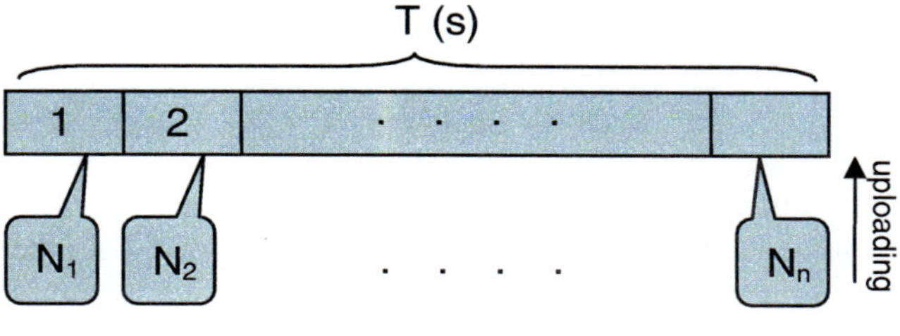

Fig. 1. Uploading segments of the video

As shown in Fig.1, we suppose a video with the length of T seconds is divided into n segments on average, and the bite size of each segment is m. Only when a peer has downloaded a segment completely can it claims of this segment to other peers in the network. When a user is watching the video, the array made up by the sum of the bandwidth of neighboring peers corresponding to each segment of this video, is $N_i(i = 1 \ldots n)$.

As for the No. i segment of the video, the time to download it is

$$TR_i = \frac{m}{N_i}. \tag{1}$$

We assume each user finishes watching the video from the beginning to the end. When the segment i is finished watching, the view delay caused by segment i itself is d_i, and the delay resulted from watching this video from the beginning to segment i is D_i. Then,

$$d_1 = D_1 = \begin{cases} 0 & \text{else} \\ TR_1 - \frac{T}{n} & TR_1 - \frac{T}{n} > 0 \end{cases} \tag{2}$$

$$d_i = \begin{cases} 0 & \text{else} \\ TR_i - \frac{T}{n} i - D_{i-1} & TR_i - \frac{T}{n} i - D_{i-1} > 0 \end{cases} \tag{3}$$

and

$$D_i = D_{i-1} + d_i \tag{4}$$

In which, i = 2…n.

As for the array d_i, we assume the number of elements which are not equal to 0 is h, and then the subscripts corresponding to the elements make up a new array t_j (j = 1…h). According to (1)(2)(3)(4), we can get

$$d_{t[j]} = \frac{m}{N_{t[j]}} - t[j]\frac{T}{n} - D_{t[j-1]} \tag{5}$$

and

$$D_{t[j]} = D_{t[j-1]} + d_{t[j]} \tag{6}$$

In which t[j] is t_j. Add (5) and (6), then get

$$D_{t[j]} = \frac{m}{N_{t[j]}} - t[j]\frac{T}{n} \tag{7}$$

We propose d_{fin} is the last element unequal to 0 of the array d_i. According to (7) we can infer that the total delay of watching this video depends on the segment fin. In this case, when a new helper joins in, if the caching segment of this helper is not the segment fin, the value of d_{fin} is not able to be decreased, thus not being able to reduce the total view delay of the video. Therefore, the helper is supposed to cache the No. fin segment of the video in priority.

4 Design of Algorithm

4.1 System Description

The viewer network of a VoD system based on H-P2P is a typical P2P streaming media network. In addition, when a viewer joins in this VoD system, the viewer is able to obtain a neighbor peer list and a neighbor helper list got from helper tracker. When the viewer needs the helper assist in transmitting data, some appropriate helpers are to be picked out from this helper list.

When a helper joins in or exits from the H-P2P system, it needs to register or logout on the helper tracker. Each helper is able to assist multiple viewers for content relay, and every viewer is also able to obtain the assistance from multiple helpers simultaneously. In the H-P2P system, any peer acts as either a viewer or a helper, but can't both be a viewer and a helper at the same time. No data between helpers is transmitted. Each helper is to contribute some storage space to cache the content of streaming media.

4.2 Practical Implementation

Since helpers are still of high dynamics and unpredictability, it's not economic to require helpers pre-cache any content, which may well lead to lots of additional consumption of bandwidth and storage resources. Therefore, in the algorithm of this thesis, all helpers can only cache the segments which have been relayed by themselves.

In the H-P2P system, any viewing peer's corresponding network surroundings, including neighbor peers and helpers, are always changing, and the value of its fin is accordingly changing. As for different viewing peers, the value of fin may well be different from each other. Thus, the integral performance of the system considered, only helpers give priority to caching the segment fin which repeats for the most times and update the caching content regularly can the average view delay of the system be reduced as much as possible. However, because it's impossible to get the real-time information about all of the peers, we have to adopt the means of local optimization. Based on the reasons above, we design the following helper caching algorithm:

When a peer starts viewing a video, it can figure out the segment fin at that moment according to the real-time available download bandwidth corresponding to each segment, and reports the value of fin to all helpers in its helper lists. Then the peer will regularly update fin and report the latest value of fin to helpers till the user finishes watching this video.

As for the helper, it uses an array R_i to store the times in turns that all segments were reported while these segments are taken as segment fin. At first, the helper will cache all segments relayed till the cache space is full. After the cache space is run out, when the helper keeps on relaying new segments, the helper will compare the value of R_i corresponding to new segments with that related to all segments which have been cached. If the value of R_i associated with a new segment is minimum, then the new segment will not be cached; otherwise, the new segment will be cached instead of the segment with the minimum value of R_i. In the course of comparison, if the values of R_i are equal, the segments will be replaced or cached according to the principle of FIFO.

5 Simulation Experiments

In this section, we will corroborate the effectiveness of the caching algorithm through some simulation experiments.

5.1 Simulation Environment

The simulation experiments base on PeerSim [17] which is a P2P network simulator. The video length is 600 seconds, and media streaming playing rate is 512kbps. The length of each segment is 10 seconds, and the number of users watching this video amounts to 500. The users will not drop out of the VoD system until they finish all of the video. All users and helpers have unlimited download bandwidth. The cache size of helpers is 5MB. Propagation delays are randomly assigned from the delay matrix (2500*2500) in the Internet measurement [18] and these numbers will change once every two seconds.

5.2 Simulation Result

In the simulation experiments, we compare the differences in average view delay rate of the system adopting caching algorithm of helpers and that without the algorithm in the first experiment, and compare server load rate adopting the algorithm and not in the second experiment. Here, we define average view delay rate of a video as mean of average of view delay per unit time by all of the users watching the video. And server load rate is the ratio of the bandwidth consumed by the server and the total bandwidth.

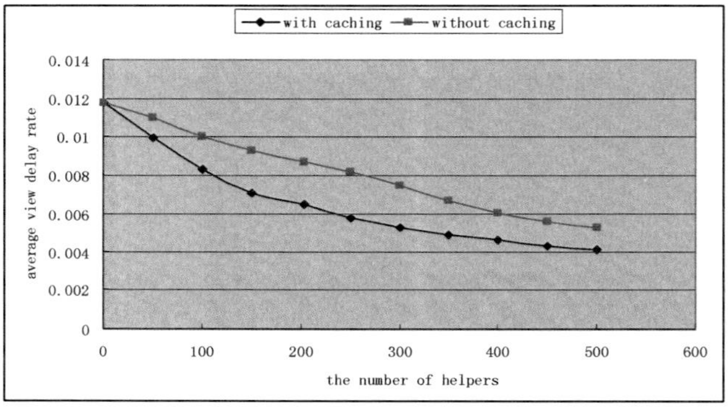

Fig. 2. Comparison of average view delay rate

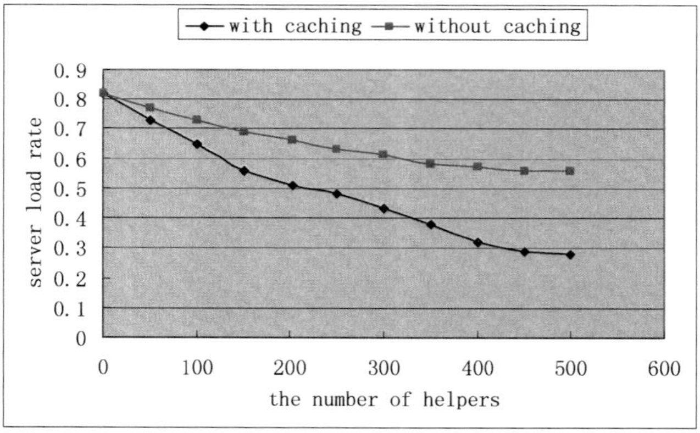

Fig. 3. Comparison of server load rate

It can be seen from Fig.2 and Fig.3 that after the caching algorithm adopted, the overall performance of the system is improved rather apparently, for which the effectiveness of this algorithm is proved. In addition, the two figures present that in the H-P2P VoD system, as the number of helpers increases, both the average view delay of the system and the load of the media server decrease to some degree, but when the helpers reach to a certain number, the system becomes saturated, and the effect of helpers on further enhancement of the system's performance is not great any longer.

6 Conclusions

Aiming at the H-P2P VoD system, this thesis puts forward an algorithm to cache video segments through using the storage capacity of helpers. This kind of algorithm locally assures that caching the segment of fin which repeats for the most time is to be given priority, thus effectively reducing the overall average view delay of the system and the bandwidth load of media servers. On the other hand, because the segment fin commonly lies in the back of a video, the algorithm can relieve the problem of comparatively long view delay in the back of the video. In the future research, we'll further analyze the boundary conditions of the system models so as to make them closer to the practical situation of the network, and take the dynamic characteristic of peers into consideration in order to improve the performance of H-P2P VoD system.

Acknowledgment. The work presented in this study is supported by the National Basic Research and Development Program (973program) of China (No. 2009CB320406); the National High Technology Research and Development Program (863program) of China (No. 2008AA01A317); the Foundation for Innovative Research Groups of the National Natural Science Foundation of China (No. 60821001); "HeGaoJi"-Important National Science & Technology Specific Projects (2009ZX01039-001-002-01).

References

1. Yiu, W., Jin, X., Chan, S.: Distributed storage to support user interactivity in peer-to-peer video streaming. In: Proc. of IEEE International Conference on Communication (ICC 2006), pp. 55–60. IEEE Press (June 2006), doi:10.1109/ICC.2006.254704
2. Zhang, X., Liu, J., Li, B., Yum, T.S.P.: Cool streaming/donet: A data-driven overlay network for efficient live media streaming. In: 24th Annual Joint Conference of the IEEE Computer and Communications Societies (INFOCOM 2005). IEEE Press (March 2005), doi:10.1109/INFCOM.2005.1498486
3. PPLive, http://www.pplive.com/
4. PPStream, http://www.ppstream.com/
5. Wang, J.: Enhancing collaborative content delivery with helpers, Master's Thesis. The University of British Columbia, unpublished (2004)
6. Pouwelse, J., Garbacki, P., Wang, J., et al.: TRIBLER: A Social-Based Peer-to-Peer System. Concurrency and Computation 20(2), 127 (2008)

7. Garbacki, P., Iosup, A., Epema, D., van Steen, M.: 2Fast Collaborative downloads in p2p networks. In: Sixth IEEE International Conference on Peer-to-Peer Computing (P2P 2006), pp. 23–30. IEEE Press (September 2006), doi:10.1109/P2P.2006.1.
8. Wang, J., Yeo, C., Prabhakaran, V., Ramchandran, K.: On the Role of Helpers in Peer-to-Peer File Download Systems: Design, Analysis and Simulation. In: Proc. of IPTPS (2007)
9. Xu, K., Yang, Y.H., Chen, T.: Improving BitTorrent Network's Performance via Deploying Helpers. In: EUC 2008 IEEE/IFIP International Conference, vol. 2, pp. 507–512 (December 2008), doi:10.1109/EUC.2008.96.
10. Wang, J., Ramchandran, K.: Enhancing Peer-to-Peer Live Multicast Quality Using Helpers. In: IEEE International Conference on Image Processing, pp. 2300–2303 (October 2008), doi:10.1109/ICIP.2008.4712251
11. Zhang, H., Wang, J., Chen, M., Ramchandran, K.: Scaling peer-to-peer video-on-demand systems using helpers. In: IEEE International Conference on Image Processing, pp. 3053–3056 (November 2009), doi:10.1109/ICIP.2009.5414399
12. Liang, C., Fu, Z., Liu, Y., Wu, C.W.: Incentivized peer-assisted streaming for on-demand services. IEEE Transactions on Parallel and Distributed Systems, 1354–1367 (September 2010), doi:10.1109/TPDS.2009.167
13. Zhang, H., Ramchandran, K.: A reliable decentralized peer-to-peer video-on-demand system using helpers. In: Picture Coding Symposium, pp. 1–4 (May 2009), doi:10.1109/PCS.2009.5167390
14. He, Y., Guan, L.: Improving the streaming capacity in P2P VoD systems with helpers. In: Proceedings of 2009 IEEE International Conference on Multimedia and Expo., pp. 790–793 (July 2009), doi:10.1109/ICME.2009.5202613
15. He, Y., Guan, L.: Solving streaming capacity problems in P2P VoD systems. IEEE Transactions on Circuits and Systems for Video Technology, 1638–1642 (November 2010), doi:10.1109/TCSVT.2010.2077553
16. Li, X., Zou, H., Zhao, X., Yang, F.: A grouping algorithm of helpers in peer-to-peer video-on-demand system. In: International Conference on Advanced Communication Technology, pp. 497–501 (2010)
17. Peersim, `http://peersim.sourceforge.net`
18. Saroiu, S., Gummadi, P., Gribble, S.: A measurement study of peer-to-peer file sharing systems. In: Multimedia Computing and Networking, pp. 156–170 (January 2002) doi:10.1117/12.449977

The Design and Implementation of a SMS-Based Mobile Learning System

Long Zhang[1,2], Linlin Shan[3], and Jianhua Wang[1]

[1] College of Computer Science and Information Engineering, Harbin Normal University, Harbin, Heilongjiang, China
[2] School of Computer Science and Technology, Harbin Institute of Technology, Harbin, Heilongjiang, China
[3] School of Art, Heilongjiang University, Harbin, Heilongjiang, China
zlwalkman@sina.com, shanlinlin1979@163.com

Abstract. Based on the detailed research on Mobile Learning and Short Message Service, the paper provides a Mobile Learning System, and perfects its design of structure and function. It also discusses about the critical stages and technology in this system, and completes it by means of Java. Finally the paper touches the problems of the system, and gives the solutions, which are very beneficial for subsequent research.

Keywords: Mobile Learning, Short message service, Java.

1 Introduction

Statistics released by Ministry of Industry and Information Technology show that up to February 2009, China's mobile phone users reached 660 million, and more than 100 million mobile Internet users. This tells us that, with the development of mobile technology, mobile equipment price reduction and performance improvement, Mobile Learning based on the wireless communication technology already has a mature soil in China, and it has inestimable application potential and huge market in the field of education and training. Mobile Learning will be triggered at any moment, and gradually entered the mainstream view of public.

In a broad sense, Mobile Learning refers to the learning activity with the help of mobile devices. The mobile devices comprise a mobile phone, electronic dictionary, Mp3 player, pocket dictionary and so on. From the perspective of educational technology, Mobile Learning relies on the mature wireless mobile network, Internet, and multimedia technology, and helps students and teachers to achieve the interactive teaching activities and the exchange of information, in the fields of education, science and technology. Mobile Learning allows anyone, at any time, any place to carry on the independent study, and thus contributes to the realization of life-long education and a learning society [1].

According to the development trends of Mobile Learning, the application of Mobile Learning is mainly divided into two categories: Online Mobile Learning Mode (OMLM) and Storage Based Mobile Learning Mode (SMLM). OMLM mainly relies on the mobile network, and freely access the Internet Education resources. The resource access is affected by mobile equipments and mobile communication network,

D. Zeng (Ed.): Advances in Control and Communication, LNEE 137, pp. 107–113.
springerlink.com

and also restricted by mobile communication network and Internet communication protocol. The present mobile communication protocol mainly has two forms: one is for the short message; the other is the connected online (real-time communication). So OMLM can be divided into two patterns: short message based OMLM (SMOMLM) and connection based OMLM (COMLM). SMOMLM has low operation cost, generally supported equipment and many other advantages, and mainly used in the learning activities and learning services that can communications with less data and simple text description. COMLM can direct accesses to teaching servers using mobile learning terminal, and then browses, queries and interacts with the learning resources. It mainly used in the learning activities with rich pictures, sounds, animations and other multimedia information materials. At present, domestic and foreign has launched the WAP, GPRS, 3G, wireless LAN and other connection based data services. It is worth mentioning that, 3G technology will make Mobile Learning unprecedented change in convenience and service quality, and ensure high quality teaching activities. SMLM stores e-book, multimedia courseware and other digital content in a portable or mobile device, and then learners can perform a learning activity whenever and wherever possible. SMLM is similar to the traditional sense of "mobile" in the process of learning. As can be seen, OMLM emphasizes real-time and interactive learning using wireless communication technology, and SMLM is a ubiquitous and practical learning ways.

From the above analysis we can see that, different forms of Mobile Learning have their characteristics. From the development perspective, OMLM will become the main direction of the development of distance education in the future [2].

However, there are still many new problems concerning Mobile Learning in our country. Though the popularity of the new technology provides a wider scope for the development of Mobile Learning, yet there isn't a practical and wide-used Mobile Learning System in existence because of many factors, including the variety of the mobile equipment and its functions, the limited capability, the small screen size, the low storage capacity, and the high cost of communication in China.

Short message service is the most popular means of mobile communication in china. It has many advantages, such as high capability of adaptation, easy operation, promptness and convenience, security, low cost, and separate sending. It is very fitful for the fragment learning in the mobile situations. So SMOMLM will be the best solution to Mobile Learning in China, which can be widely used.

Based on the detailed research on Mobile Learning and short message communication service, the paper provides Short Message Service Based Mobile Learning System (SMS-based MLS), and perfects its design of structure and function. It also discusses about the critical stages and technology in this system, and completes it by means of Java.

2 System Design and Implementation

2.1 System Structure

SMS-based MLS consists of six main functional modules: Student spatial module, teacher spatial module, administrator spatial module, short message monitor module, short message intelligent processing module and short message sending module. The

Student spatial module is an interactive interface between students and the system, and provides the students' role support. Teacher spatial module is an interactive interface between teachers and the system, and provides the teachers' role support. Administrator spatial module mainly completes the educational administration management and system maintenance work. Short message monitor module is used for monitoring whether a short information arrival. If any, it immediately notices the short message intelligent processing module for processing. Short message intelligent processing function module will distribute and process all kinds of short messages. Short message sending function module is used to send single or mass short message.

The users of the system need only to interact with the first three modules, and the rest three modules provide the corresponding services for the first three modules and invoked by the first three modules. The six modules all deal with the backstage database to store, update and search data.

2.2 Student Spatial Module

Here, student spatial module will be introduced in brief. In this module, students can go through two ways of interaction with the system. One is through the mobile terminal in Mobile Learning; the other is through the Internet computer or notebook for online learning. Students can only choose one course to study in a period of time.

(1) Register and Login. Students register for the system and login, then they can view and modify their registration information, or logout.

(2) Help. A specification for the student spatial module.

(3) Choosing Course. Students choose courses to study, and then set or modify ways of learning (including the learning mode, frequency and intensity).

(4) Message. Students can receive various types of educational administration information.

(5) Learning. According each student's learning situation, the system will choose to send for the students to study new knowledge, or review old knowledge, or test.

(6) Discussion. To discuss with the other students that study this course.

(7) Troubleshooter. To ask any question to the other students or teachers.

(8) Test. To test on the front of the knowledge learned, or one chapter knowledge l Troubleshooter earned.

Student spatial module is shown in figure 1.

2.3 The Format of Short Message Command

In order to support students by using short message service interaction with the system, we designed a series of short message command, and students can realize different functions through sending different commands. We put the first two characters of a short message as a command word, which descript the operation request between students and the system. The remainder as the command's parameters, is available to accept the require data. Short message command format is shown in figure 2:

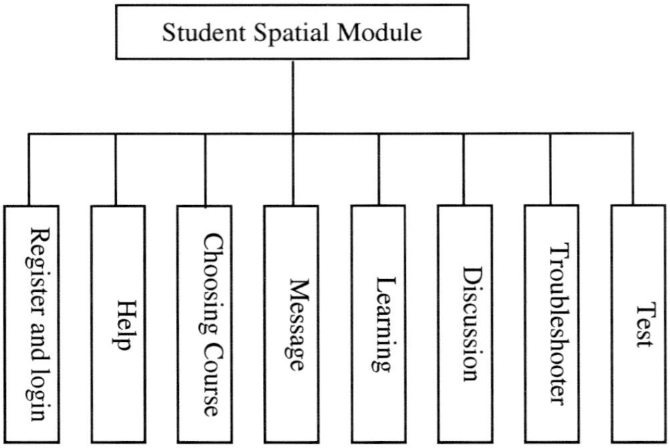

Fig. 1. Student spatial module

0	1	2	3	4	5	6	7	8	9	.	.	.	159

Fig. 2. Short message command format

The character at 0 and 1 position is a command word, the character at 2 and later position is the command parameter list.

At present, this system includes the following instructions. We can further expand the instruction set to enhance the functionality of the system.

(1) Register
Command format: zc nike_name
(2) Choosing course
Command format: dz course_name|course_id [Frequency] [Number]
(3) Discuss
Command format :: tl content
(4) Learning
Command format :: xx chapter_id [knowledge_point]
(5) Test
Command format :: ks [Number]
The "a|b" means that "a" and "b" choose either, and the "[a]" says that "a" is optional.

2.4 Short Message Monitor Module

Students can send a short message which is a specific command format to the system, and request the corresponding service. Short message monitoring module will regularly (every 60 seconds) scan wireless terminal devices which connect this sever. If a short message received, they are read into the buffer, and immediately processed. The algorithm is as follows:

(1) Firstly read the first new short message from the buffer zone.
(2) To decompose it and get the corresponding command word and parameter list.
(3) To determine whether it is lawful command word, if not, the sending module sends the help to the students.
(4) If it is a lawful command word, then turns to the corresponding command word processing functions.
(5) To read a new short message from the buffer zone, then jump 2, and continue processing until all of the new short messages has been processed.
Short message monitor module is showed in figure 3.

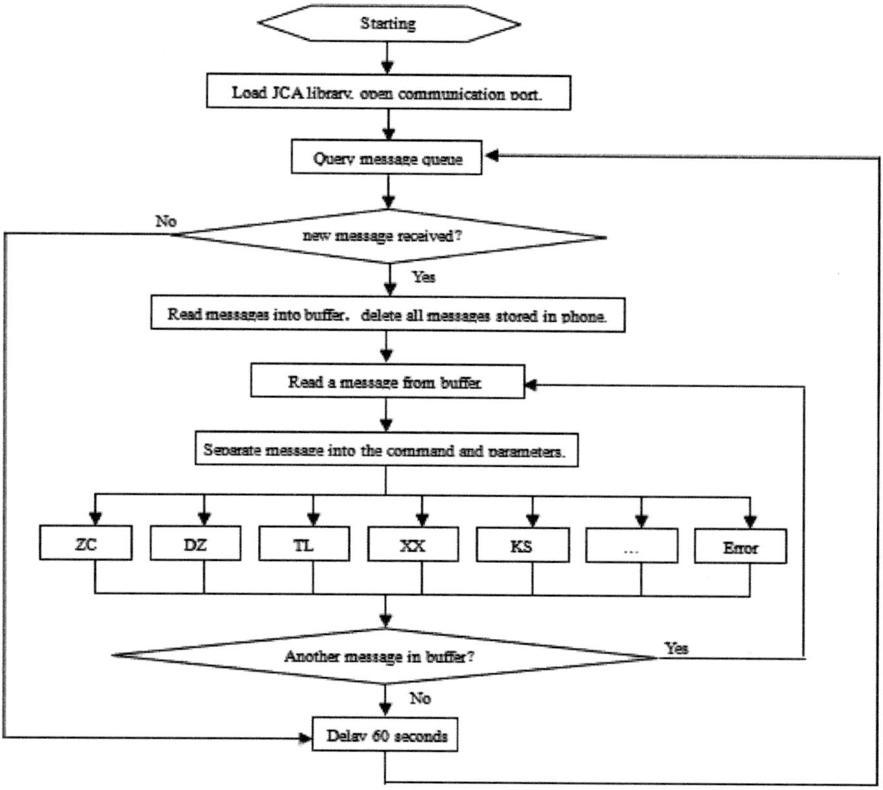

Fig. 3. Short message monitor module

2.5 System Implementation

Server Configuration: Windows 2000 Server, Apache Tomcat 4.0, JDK1.6.

Windows 2000 Server implements a good integration among the operating system, applications, network, communication and infrastructure services. Apache Tomcat 4 is the Web server, and it can support the JSP program. JDK1.6 is the development tools kit for Java. Serial connection to the server is a Siemens 3508i mobile phone, and infrared connection to the server is a Nokia 8850 mobile phone.

3 Summary

Mobile Learning is a union product which includes the mobile communication technology, network technology and modern education, and is also the trend of education technology development. The promotion and development of Mobile Learning will cause great changes in education technology. Mobile Learning System, will provide learners a learning environment whenever and wherever possible, and will contribute to the realization of life-long education and a learning society.

The work in this paper includes: (1) Design and implementation of serial communication based on short message. Short message module function is powerful, in support of the English short message, the PHS Short message, and automatic segmentation for super-long short message. (2) Analysis and Design of Mobile Learning System. To design the student short message command format, the student instruction in real-time monitoring module, the curriculum management module. (3) Finally, to realize a perfect short message service based Mobile Learning System, and practice in an actual teaching activities.

Design and development of Mobile Learning System is a very complicated work, there are many problems need to be further in-depth study and discussion. To build a safe and efficient, practical Mobile Learning System has a lot of work to do. The next step for the research mainly includes: (1) Development of mobile learning resource. Mobile Learning has many advantages, such as convenient, personalized teaching, rich interactive, situational relevance. According to these characteristics how to develop perfect mobile learning resource is a problem to be solved [3]. (2) To join the automatic answering function. Using word segmentation technology and intelligent semantic understanding, based on the received student questions, analysis short message content, query the database, finally give the correct answer or reasonable guidance. (3) Adding voice function. To increase the speech form of curriculum knowledge database, so that learners can not only through the text, but also through speech in learning and communicating. Students can freely send and receive voice messages. (4) To improve system processing intelligent level. It can automatically evaluate the students' learning situation, find the lack of knowledge points, and then generate adaptive knowledge content, to provide intelligent learning, review and test plan. (5) Joined to EMS and MMS, especially MMS. Text, sound, images, video are integrated together, will provide better support for Mobile Learning. (6) Utilizing short message channel, to further improve the system function. Because of research conditions, this system uses serial communication protocol to send and receive short message, and this led to the system processing speed is not high. So it is suitable for small range (such as campus environment). If you use the channel technology, not the existence of these problems, we can conduct more extensive social investigation, analysis user demand, and further improve the system function. We hope to develop more practical and perfect Mobile Learning System.

Acknowledgement. This study is supported by the science and technology project of the Education Department of Heilongjiang Province (11541093), the advanced research project of Harbin Normal University (10XYG-07) and the Heilongjiang Provincial Key Laboratory of Intelligence Education and Information Engineering.

References

1. Keegan, D.: The future of learning: From eLearning to mLearning (2004), http://learning.ericsson.net/mlearning2/project_one/index.html
2. Walcott, J.: An investigation into the use of mobile computing devices as Tools for Supporting Learning and Workplace Activities. In: Proc. of the 5th Human Centered Technology Postgraduate Workshop, pp. 265–268 (2001)
3. Zhao, G., Yang, Z.K.: Learning Resource Adaptation and Delivery Framework for Mobile Learning. In: Proc. of 35th Frontiers in Education Conference, pp. 19–22 (2005)

Summary of Digital Signature

Long Zhang[1,2], Linlin Shan[3], and Jianhua Wang[1]

[1] College of Computer Science and Information Engineering, Harbin Normal University, Harbin, Heilongjiang, China
[2] School of Computer Science and Technology, Harbin Institute of Technology, Harbin, Heilongjiang, China
[3] School of Art, Heilongjiang University, Harbin, Heilongjiang, China
zlwalkman@sina.com, shanlinlin1979@163.com

Abstract. With the increasing development of information application, the security of information has become a prominent problem. It also makes the digital signature technology has been rapid developed and applied. This paper introduces the concept, characteristics, related technologies of digital signatures, and the current research state of several types of digital signature. At last the use of digital signatures with various features is discussed.

Keywords: Digital Signature, Information security, Network security.

1 Introduction

Generally, in order to authenticate and verify the validity of transactions, a handwritten signature with or without an official or personal seal is used for signing a document or a convention between two countries in political, military or diplomatic activities, signing a contract or an agreement between two common companies or persons in commercial activities, or depositing money in or withdrawing money from a bank or making bank transferring.

As a result of the technical development and the demand of the information-based society, people want to sign documents, contracts or agreements remotely and quickly via the internet, thus digital signature was born in cryptology, especially on the basis of the fast development of public-key cryptology. As an important safety technique, digital signature plays a significant role in guaranteeing the integrality, privacy, and non-repudiation of data. With the technology of digital signature, people can realize contract or agreement signing, commodities purchase, bank transferring and information release and so on remotely within their doors, furthermore, the authorized persons can verify the authenticity and validity conveniently, thus to save time and resources effectively. Moreover, with the help of advanced technologies, the copying function which is impossible to be achieved in computation is realized.

The research on digital signature technology is attached importance to worldwide because of its major influence on the government, enterprises, common organizations and individuals at present and in the future. William Jefferson Clinton, the US president, officially signed the "Electronic signature Act" which is considered as a great legislation of the internet age in June 30, 2000, clearly admitting the legal force of

D. Zeng (Ed.): Advances in Control and Communication, LNEE 137, pp. 115–120.
springerlink.com

digital signatures, contracts and records. From 2000 to 2001, Ireland, Germany, Japan, Poland, etc. countries promulgated their own electronic signature act successively. In August 28, 2004, the Electronic signature Law of the People's Republic of China was adopted at the 11th Session of the Standing Committee of the 10th National People's Congress of the People's Republic of China and came in force in April 1, 2005.

2 Characteristics of Digital Signatures

The main differences of a digital signature and a handwritten signature are as follows:

(1) The connection with the document signed: a handwritten signature is a physical part of the document signed, and a digital signature is bound to the document signed as a mean to validate the signers.

(2) Validation methods: a handwritten signature is validated by comparing it with the original handwritten signature, and a digital signature is validated with a public proof technique, which enables anyone to do so.

(3) Copy protection ability: the copy document with a handwritten signature can be easily distinguish from the original document, but the all copies with a digital signature look just the same with the original, so an effective measure must be adopted to prevent the repeating use of digital signature information.

A common digital signature generally owns characteristics as follows:

Creditability: the receiver of the document believes the authenticity of the digital signature on the document and that the signer agrees its contents.

Unforgeability: no one can forge the signer's digital signature.

Non-reusability: a digital signature is an indivisible part of the signed document and cannot be moved to other documents.

Inalterability: the document cannot be altered since the moment it is signed, and once it is altered, its signature will change too and then the original one will not pass the validation and will be invalid.

Non-repudiation: the signer cannot deny his/her signature on the document after it was signed by him or her.

These characteristics enable a digital signature to have more advantages that a handwritten signature lacks, like convenience in use, and saving time and expenses, apart from the same functions with a handwritten signature.

3 Message Digest

The technology of digital signature is developed on the basis of cryptology. A digital signature can be obtained based on both public-key and private-key cryptology systems, and every type of signature scheme is closely connected to one or many cryptology systems. The research on digital signature technology mainly focuses on that of public-key cryptology system.

Since public-key cryptology system is essentially an arithmetical operation of larger integers based on trapdoor one way function, the encrypting speed of public-key cryptology system is very slow in practice, accordingly it's impracticable to make an digital signature for an entire piece of information with public-key cryptology arithmetic directly. To solve the problem, pre-processing can be made to large messages to be signed from which an eigenvalue in fixed size representing the message, and the eigenvalue is called a message digest.

A massage digest is a unidirectional arithmetic that can process message input strings in any length and produce pseudo-random codes to be outputted. A good message digest has the following features: (1) any slight change of a message will lead to a great change of the digest, namely avalanche effect; (2) it's impossible to restore a digest to a message; (3) it's impracticable in computation to find two messages with the same digest value.

4 Development and Types of Digital Signatures

The concept of digital signature was brought forward by Whitfield Diffie and Martin Hellman in 1976 with the purpose to make the signers impossible to deny their signatures after making digital signatures on electronic documents, and digital signatures have the same function with handwritten signatures.

With the development of cryptology, people put forward various cryptosystems meeting different needs, based on which many digital signature schemes have been proposed one after another. Rivest Shamir and Adleman put forward a digital signature scheme based on RSA public-key arithmetic in 1978, Shamir put forward a digital signature scheme based on identification in 1985, ELGamal put forward a public-key encryption arithmetic and a digital signature scheme based on discrete logarithms in 1985, Schnorr put forward an effective digital signature proper for smart cards in 1990, Agnew put forward an improved digital signature scheme based on discrete logarithms in 1990, NIST put forward the digital signature arithmetic (DSA) in 1991, and Scott Van stone first put forward the elliptic curve digital signature arithmetic (ECDSA) in 1992.

While people are studying the arithmetic of common digital signatures, the research scope is extended to include the research on special digital signatures, to meet the demand of special signatures used for specific situations in practice.

(1) Blind signatures [1, 3] are used when a user wants the signer to make a digital signature on a plaintext document, but does not want the signer to know its detailed contents. As a special digital signature, a blind signature should have the following three features compared with common digital signatures:

1) The signer cannot see the plaintext information;

2) The verifier can see the plaintext information, and can confirm the validity of the document only through the signature;

3) Both the signer and the verifier should not consider the signature to the blind message as a one to one correspondence.

In order to protect the fair benefit of the signer, especially to enable juridical authorities to track illegal and criminal acts like double spending and money laundering, Stadler et al [2] proposed fair blind signature scheme.

Blind signatures are mainly used in anonymous financial transactions on the internet, such as e-cash system [4-6], and anonymous electronic auction system.

(2) Threshold signatures own the same characteristics with secret sharing (t, n) threshold scheme. In a group with n members, only a group of not less than t members can represent the group to make a valid digital signature over a document. It is achieved through secret sharing, which means to divide the secret key into n shares, and a secret key can be reconstructed only when over t sub-keys combine together. Threshold digital signatures are well applied in key escrow technology. For instance, a person's private key is entrusted to n governmental departments, then a secret key can be reconstructed only when t of the n departments decide to implement monitoring.

(3) Proxy signatures, whose concept was proposed by Mambo, Usuda and Okamoto [7] in 1996, allow the key holder to give the authority to a third party to represent him/her to make a digital signature. Having a proxy digital signature like that a person resigns his/her seal to someone he/she trust and lets the person to exercise his/her rights. It was paid great attention to and widely researched once the concept was brought forward, because proxy digital signatures play an important role in practice.

On the basis of the classification scheme proposed by Mambo et al [7], proxy digital signatures are divided into types as follows: full proxy signatures, partial proxy signatures and proxy signatures with warrant. On this basis, S Kim et al [8] proposed partial proxy signatures with warrant.

Mambo et al [7] pointed out that proxy signature system should have basic prosperities as follows:

Unforgeability: besides the original signer, only the appointed proxy signer can represent the original signer to make a valid proxy signature.

Verifiability: from the proxy signature, the verifier believes that the original signer has agreed the signed document.

Undeniability: once a proxy signer makes a valid proxy signature for the original signer, he/she cannot deny the signature in front of the original signer.

Distinguishability: anyone is able to distinguish the proxy signature from the original signer's signature.

Inconformity of the proxy signer: the proxy signer should establish a valid proxy signature that can be checked.

Identifiablity: the original signer can confirm the identity of the proxy signer from the proxy signature.

B Lee et al [9] also pointed out that proxy signatures should have the feature of anti-misuse.

Since proxy signatures own an extensive application prospect, they have drawn much attention from the public and various proxy signature schemes suitable for different practical environments have emerged, such as threshold signature scheme [8, 10], proxy multi-signature scheme, anonymous proxy signature scheme, proxy blind signature scheme, etc., and they are widely discussed in various respects like safety transferring proxy keys via a public channel [12, 13, 14].

(4) Forward security l signature scheme [15]. Common digital signatures have a limit: if the key of a signer is leaked, all his/her signatures (both in the past and the future) are possible to be leaked. The forward security signature scheme has a great advantage that is the leaking of the current key will not affect the safety of the signatures made before.

(5) Group signatures [16] allow a member of the group to make a digital signature on behalf the entire group and the verifier is able to confirm the identity of the signer.

A good group signature scheme should meet the following security requirements:

Anonymity: when a group signature is provided, it's impossible in computation for any other person besides the group administrator to confirm the identity of the signer.

No connection: it's hard in computation to confirm whether two different signers are members of the same group when the signatures are not opened.

Anti- forgery: only group members can make a valid group signature.

Trackability: the group administrator has the authority to open a signature to confirm the signer's identity when necessary and the signer cannot hinder the opening of a legal signature.

Anti-circumvention and anti-attack: anyone including the administrator cannot make a valid group signature in the name of other group members.

Anti coalition attack: even many group members unite together, they cannot make a valid group signature that cannot be tracked.

After D.Chaum [16] brought forward the definition of group digital signatures and provided four schemes to implement it, people began to make more extensive researches on group signatures [17-20] resulting from their practicability. Rank multi-group signatures, blind group signatures, multi-group signatures, sub-group signatures, the group signatures meeting threshold natures and forward security group signatures have been proposed.

5 Further Research Directions

Although researchers have proposed many signature schemes, the current scheme needs further improvement more or less, now proxy signatures are taken as a case for explanation.

(1) It's possible to deny the fact of signing for most of current proxy signatures, but in practice, undeniability is very important. (2) Nearly all current proxy signature schemes are an interactive protocol of the original signer, the proxy signer and the receiver. The original signer participates in every proxy signing, which violates the original intention of proxy signature. (3) The research on using special signature schemes like blind signature, orientated signature, zero-knowledge signature and forward security signature to achieve proxy signature still needs to be furthered.

Acknowledgement. This study is supported by the science and technology project of the Education Department of Heilongjiang Province (11541093), the advanced research project of Harbin Normal University (10XYG-07) and the Heilongjiang Provincial Key Laboratory of Intelligence Education and Information Engineering.

References

1. Chaum, D.: Blind signatures systems. In: Proc. of Crypto 1983, Plenum, pp. 153–155 (1983)
2. Stadler, M.A., Piveteau, J.-M., Camenisch, J.L.: Fair Blind Signatures. In: Guillou, L.C., Quisquater, J.-J. (eds.) EUROCRYPT 1995. LNCS, vol. 921, pp. 209–219. Springer, Heidelberg (1995)

3. Camenisch, J.L., Piveteau, J.-M., Stadler, M.A.: Blind Signatures Based on the Discrete Logarithm Problem. In: De Santis, A. (ed.) EUROCRYPT 1994. LNCS, vol. 950, pp. 428–432. Springer, Heidelberg (1995)
4. Chaum, D., Fiat, A., Naor, M.: Untraceable Electronic Cash. In: Goldwasser, S. (ed.) CRYPTO 1988. LNCS, vol. 403, pp. 319–327. Springer, Heidelberg (1990)
5. Brands, S.: Untraceable Off-Line Cash in Wallets with Observers. In: Stinson, D.R. (ed.) CRYPTO 1993. LNCS, vol. 773, pp. 302–318. Springer, Heidelberg (1994)
6. Camenisch, J., Piveteau, J.M., Stadler, M.: An efficient fair payment systems. In: Advances in Cryptology, Proc. of Crypto 1995, pp. 88–94. Springer, Heidelberg (1996)
7. Mambo, M., Usuda, K., Okamoto, E.: Proxy signatures: delegation of the power to sign messages. IEICE Trans. Fundam., 1338–1354 (1996)
8. Kim, S., Park, S., Won, D.: Proxy Signatures. In: Han, Y., Quing, S. (eds.) ICICS 1997. LNCS, vol. 1334, pp. 223–232. Springer, Heidelberg (1997)
9. Lee, B., Kim, H., Kim, K.: Strong proxy signature and its applications. In: SCI 2001, pp. 603–608 (2001)
10. Zhang, K.: Threshold proxy signature schemes. In: Information Security Workshop, Japan, pp. 191–197 (1997)
11. Shum, K., Wei, V.K.: A strong proxy signature scheme with proxy signer privacy protection. In: Proc. of the 11th IEEE Int'l Workshops on Enabling Technologies: Infrastructure for Collaborative Enterprises (WETICE 2002), IEEE Press, New York (2002)
12. Lee, J.Y., Cheon, J.H., Kim, S.: An analysis of proxy signatures: Is a secure channel necessary? In: CTRST 2003, pp. 68–79 (2003)
13. Li, J.G., Cao, Z.F., Zhang, Y.C.: Improvement of M-U-O and K-P-W proxy signature schemes. Journal of Harbin Institute of Technology 9(2), 145–148 (2002)
14. Li, J.G., Cao, Z.F., Zhang, Y.C.: Nonrepudiable proxy multi-signature schemes. Computer Science and Technology 18(3), 399–402 (2003)
15. Bellare, M., Miner, S.K.: A Forward-Secure Digital Signature Scheme. In: Wiener, M. (ed.) CRYPTO 1999. LNCS, vol. 1666, pp. 431–448. Springer, Heidelberg (1999)
16. Chaum, D., van Heyst, E.: Group Signatures. In: Davies, D.W. (ed.) EUROCRYPT 1991. LNCS, vol. 547, pp. 257–265. Springer, Heidelberg (1991)
17. Petersen, H.: How to Convert any Digital Signature Scheme into a Group Signature Scheme. In: Christianson, B., Lomas, M. (eds.) Security Protocols 1997. LNCS, vol. 1361, pp. 177–190. Springer, Heidelberg (1998)
18. Camenisch, J.L., Stadler, M.A.: Efficient Group Signature Schemes for Large Groups. In: Kaliski Jr., B.S. (ed.) CRYPTO 1997. LNCS, vol. 1294, pp. 410–424. Springer, Heidelberg (1997)
19. Song, X.D.: Practical Forward Secure Group Signature Schemes. In: CCS 2001, pp. 225–235 (2001)
20. Popescu, C.: A Secure and Efficient Group Blind Signature Scheme. Studies in Informatics and Control 12(4), 78–82 (2003)

Optimal Selection of Working Fluid for the Organic Rankine Cycle Driven by Low-Temperature Geothermal Heat

Wang Hui-tao,Wang Hua, and Ge Zhong

Engineering Research Center of Metallurgical Energy Conservation and Emission Reduction, Ministry of Education
Kunming University of Science and Technology
Kunming, China
energywht@sina.com, wanghuaheat@hotmail.com, Zhongge@163.com

Abstract. To select the optimal organic working fluid for organic Rankine cycles driven by low-temperature geothermal heat, the exergetic analysis method and PR equation of state are utilized to calculate the exergy efficiency and other main thermal performances of low-temperature geothermal heat powered organic Rankine cycles using 10 kinds of dry fluids as working fluids. The results show that: As the critical temperature of working fluid increases, the evaporation pressure, condensing pressure, output power and exergy efficiency show a decreasing trend, but the cycle thermal efficiency and the final discharge (or rejection) temperature of the geothermal fluid show an increasing trend as a whole. Moreover, the organic Rankine cycle charged with R227ea (heptafluoropropane) gets the highest power output and exergy efficiency, and the evaporation and condensation pressures are in the appropriate range, therefore, R227ea appears to be the optimal working fluid for organic Rankine cycles used in low-temperature geothermal power plant.

Keywords: geothermal power generation, organic Rankine cycle, working fluid, exergetic analysis, equation of state.

1 Introduction

Geothermal is a renewable green energy of great reserves, the heat consisting in geothermal resources is 1.7 billion times as much as that consisting in coal reserves, the geothermal heat stored in the outermost layer just within 10km of the crust amounts to 12.6 x 10^{26}J, equivalent to 4.6 x 10^{16}tce. Since 1904, the Italians took the lead in Laderuiluo Solicitude to construct the world's first 550W rated geothermal power plant, during the sixties of twentieth century, United States, New Zealand, Japan and other 17 countries have successively built large-scale geothermal power stations. Statistics shows that there are several geothermal power plants rated more than 5x10^6 kW running all over the world in the late of 80s of twentieth century, and the geothermal power output reached 6.8 million kW in 1995, which accounts 16% annual growth. Because environmental pollution and energy shortage problems have plagued

D. Zeng (Ed.): Advances in Control and Communication, LNEE 137, pp. 121–129.
springerlink.com

the world, many countries have regarded geothermal development and utilization as an important strategic solution to develop new energy and to improve the environment [1]. China is located in two of the world's great geothermal zone, rich in geothermal resources, the proven geothermal reserves of china equal to $4.626\text{x}10^{11}$tce, but, only one hundred thousandth of which has been used, the potential for development is very great [2]. With the growing demands of electricity required by national economy development and our increasing awareness on environmental protection, the need for power generated by geothermal heat will become stronger and stronger. For Organic Rankine cycle (ORC) can achieve a good efficiency [3], ORC technology is appropriate to be applied to convert the low-temperature geothermal heat into electricity. The world's existing low-temperature geothermal power plants using organic Rankine cycle consumed lots of organic fluids having great ability of ODP and a strong effect of GWP as working fluids [4,5], such as R113, R11, etc. A large number of organic fluids can be produced up to now, therefore, the optimal selection of environment-friendly alternative working fluids for low-temperature geothermal heat powered organic Rankine cycle has become a top priority and a research focus. In this paper, exergy analysis and PR (Peng-Robinson) equation of state are applied to establish the thermodynamic method of working fluid selection for low-temperature geothermal heat powered organic Rankine cycle.

2 Exergy Analysis of the Low-Temperature Geothermal Powered Orc

In order to simplify the analysis, assuming that the process being pressurized by pump and expanding in turbine are isentropic, ignoring the external heat loss from pipes and equipments, neglecting the fluid flow resistance in heat exchangers and pipes. Fig.1 is the flow chart of the geothermal heat powered ORC; Fig.2 is the T-S diagram of the geothermal heat powered ORC using dry fluid being saturated vapor at the inlet of expander.

The specific power output is:

$$w_{net} = i_4 - i_5 - (i_2 - i_{1'}) \tag{1}$$

Thermal efficiency is described as follows:

$$\eta_{th} = \frac{w_{net}}{i_4 - i_2} \times 100\% \tag{2}$$

According to the first law of thermodynamics:

$$m_g \bar{c}_p \left(T_{in} - T_m\right) = m_{wf} \left(i_4 - i_3\right) \tag{3}$$

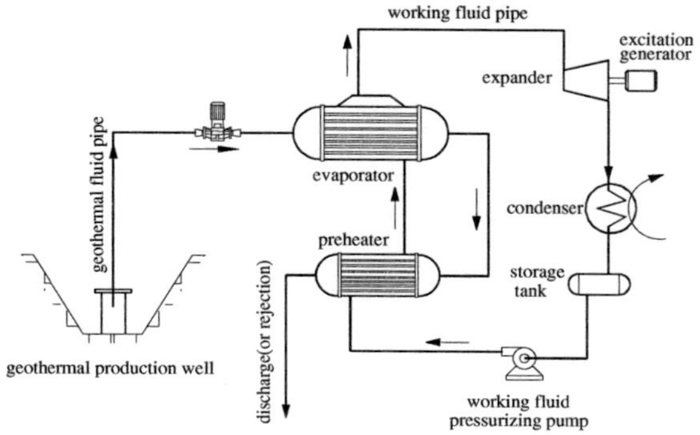

Fig. 1. Flow chart of the geothermal heat powered ORC

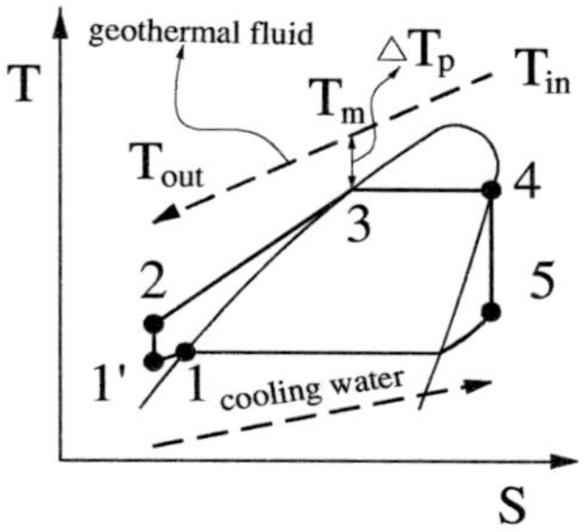

Fig. 2. T-S diagram for the geothermal heat powered ORC

$$m_g \bar{c}_p (T_m - T_{out}) = m_{wf} (i_3 - i_2) \tag{4}$$

To meet the heat transfer requirement of temperature difference, assuming that the heat transfer pinch point temperature difference in the evaporator is ΔT_p, then:

$$T_m = T_3 + \Delta T_p \tag{5}$$

Combining (3) - (5) to yield:

$$m_{wf} = \frac{m_g \bar{c}_p (T_{in} - T_3 - \Delta T_p)}{i_4 - i_3} \tag{6}$$

$$T_{out} = T_3 + \Delta T_p - \frac{i_3 - i_2}{i_4 - i_3} (T_{in} - T_3 - \Delta T_p) \tag{7}$$

The net power output is:

$$P_{out} = m_g \bar{c}_p (T_{in} - T_3 - \Delta T_p) \frac{(i_4 - i_5) - (i_2 - i_{1'})}{i_4 - i_3} \tag{8}$$

According to the second law of thermodynamics, the exergy input carried by geothermal fluid is:

$$E_{in} = \int_{T_0}^{T_{in}} \left(1 - \frac{T_0}{T}\right) \bar{c}_p m_g dT$$

$$= m_g \bar{c}_p \left[(T_{in} - T_0) - T_0 \ln \frac{T_{in}}{T_0} \right] \tag{9}$$

Exergy efficiency is deduced as follows:

$$\eta_{Ex} = \frac{T_{in} - T_3 - \Delta T_p}{(T_{in} - T_0) - T_0 \ln \frac{T_{in}}{T_0}} \frac{(i_4 - i_5) - (i_2 - i_{1'})}{i_4 - i_3} \tag{10}$$

Where, i is the specific enthalpy of each state point, J/kg; m_g, m_{wf} represents respectively mass flow rate of geothermal fluid and working fluid, kg/s; T_0, T_3 stands respectively for the ambient temperature and evaporation temperature of working fluid, K; $\bar{c}_p$ is the average constant pressure heat capacity of geothermal fluid during the cooling down process from inlet temperature T_{in} to ambient temperature, J/ (kg·K).

3 Preselection of Working Fluids

Generally speaking, the selection of working fluid for organic Rankine cycle should meet the following requirements as far as possible [6, 7]:

1) Environmental impacts. Many organic fluids have different values of ODP and GWP, the fluids without ODP and with low value of GWP should be used, such as HFC, HC, FC and other halogenated hydrocarbons.

2) Chemical stability. It is vital to select the appropriate chemically stable working fluid according to the temperature and other conditions of heat source.

3) Security. Including toxicity, flammability and corrosion property. In order to avoid corrosion of pipes and equipments and to avert poisoning accidents caused by improper operation, the working fluids should be of non-toxicity and low toxicity.

4) Appropriate critical parameters. Normal boiling point and freezing temperature of working fluids.

5) Flow and heat transfer characteristics. The fluids of higher heat transfer coefficient and lower viscosity should be used as far as possible.

6) Price and availability. Working fluids should be inexpensive and easy to buy. Basically, the working fluids can be classified into three categories. Those are dry, isentropic, and wet depending on the slope of the saturated vapor *T–s* curve (d*T*/d*s*) to be positive, infinite, and negative, respectively [8-10]. The working fluids of dry or isentropic type are more appropriate for ORC systems. This is because dry or isentropic fluids are superheated after isentropic expansion, thereby eliminating the concerns of impingement of liquid droplets on the turbine blades. Moreover, the superheated heat transfer surface is not needed. For this reason, this paper comparatively calculated the cyclic performances of ORC using the candidate ten dry or isentropic fluids. Table I lists the physicochemical properties of these 10 fluids.

Table 1. Main Physicochemical properties of some dry fluids [11]

Fluids	critical temperature /K	critical pressure /KPa	ODP	GWP_{100}	security
R227ea	374.89	2929	0	2900	A2
R236fa	398.07	3200	0	6300	
R600a	407.85	3640	0	20	A3
R236ea	412.44	3502	0	710	
R600	425.16	3796	0	20	A3
R245fa	427.20	3640	0	820	A2
Neo-C_5H_{12}	433.80	3202	0		
R601a	460.40	3384	0		A3
R601	469.60	3374	0	11	A3
n-hexane	507.40	2969	0		A3

4 Calculation of Low-Temperature Geothermal Heat Powered Orc

4.1 Calculation of Working Fluid'S Thermodynamic Parameters

The preselected candidates are all non-polar organic fluid, PR equation of state can be applied to calculate their thermodynamic parameters, according to reference [12], and PR equation of state can ensure a sufficient accuracy.

PR equation of state:

$$p = \frac{RT}{v-b} - \frac{\alpha(T)}{v(v+b)+b(v-b)} \tag{11}$$

$$\alpha(T) = \alpha(T_c)\alpha(T_r, \omega) \tag{12}$$

$$\alpha(T_C) = {0.45724 R^2 T_c^{\ 2}} / {P_c} \tag{13}$$

$$\alpha(T_r, \omega) = \left[1 + k(1 - T_r^{0.5})\right]^2 \tag{14}$$

$$k = 0.37464 + 1.54226\omega - 0.26992\omega^2 \tag{15}$$

$$b(T) = b(T_c) = {0.07780RT_c}/{p_c} \tag{16}$$

Where, p is pressure, Pa. v is specific volume, m^3/kg. T is temperature, K. R is gas constant, J/ (kg·K). ω is working fluid's eccentric factor. Z is compression factor. p_c is critical pressure, Pa. T_c is critical temperature, K. $T_r = T/T_c$ is reduced temperature, dimensionless.

Fugacity coefficient ϕ can be derived as follows:

$$\ln\phi = (Z-1) - \ln(Z-B) - \frac{A}{2\sqrt{2}B}\ln\frac{Z+(1+\sqrt{2})B}{Z+(1-\sqrt{2})B} \tag{17}$$

Wherein, $A = \frac{\alpha p}{R^2T^2}$ $B = \frac{bp}{RT}$

The residual function method is used to calculate working fluid's specific enthalpy and entropy. The residual functional equation can be deduced as follows:

$$a_r = a^* - a = RT\left[\ln(Z-B) - \frac{A}{2\sqrt{2}B}\ln\frac{Z+(1-\sqrt{2})B}{Z+(1+\sqrt{2})B}\right] \tag{18}$$

$$s_r = s^* - s = -\frac{\partial a_r}{\partial T} = -R\ln(Z-B) + \frac{P\beta}{2\sqrt{2}RTB}\ln\frac{Z+(1-\sqrt{2})B}{Z+(1+\sqrt{2})B} \tag{19}$$

$$h_r = a_r + T \bullet S_r + RT(1-Z) \tag{20}$$

$$\beta = \frac{\partial\alpha}{\partial T} = -\frac{0.45724R^2T_Ck \bullet \left[1 + k(1 - T_r^{0.5})\right]}{p_C T_r^{0.5}} \tag{21}$$

According to the definitions:

$$s = s^* - s_r \tag{22}$$

$$h = h^* - h_r \tag{23}$$

Where, the superscript * indicates the corresponding thermodynamic properties of ideal gas at the same temperature and pressure.

$$s^*_{(p,T)} = s_{(p_0,T_0)} + \int_{T_0}^{T} c_p^{\ 0} \frac{dT}{T} - R\ln\frac{p}{p_0} \tag{24}$$

$$h^*_{(p,T)} = h_{(p_0,T_0)} + \int_{T_0}^{T} c_p^{\ 0} dT \tag{25}$$

Where, s (p_0, T_0), h (p_0, T_0) were calculated at the base state (p_0, T_0). In accordance with ASHRAE convention, their values should meet the requirements that the specific entropy and enthalpy of the saturated liquid is respectively 1.0000 kJ/ (kg·K) and 200 kJ/kg at T_0=273.15K; c_{p0} is constant pressure ideal gas heat capacity of working fluid, solely varying with temperature, increasing with the increase of temperature, its polynomial can be fitted as formulation (26) using experimental data:

$$c_p^{\ 0} = d_0 + d_1 T + d_2 T^2 + d_3 T^3 \tag{26}$$

4.2 Results

Calculation conditions are defined as follows:

Inlet temperature of geothermal fluid t_{in}=95°C, working fluid's temperature at the inlet of expander t_3=70 °C, specific heat capacity $\bar{c}_p$ is assumed to be a fixed value 4.2kJ/(kg·K), geothermal mass flow rate m_g =100 kg/s. Considering the geothermal power plant is mainly running in winter or during the transition seasons, ambient temperature t_0=15°C, working fluid's condensation temperature t_{cond} is designed to be 22°C, sub-cooling of working fluid at the exit of condenser t_{subc}=1°C, heat transfer pinch point in evaporator ΔT_p=8°C, no regenerator is considered in this paper.

Table 2 lists the calculated values of evaporation pressure p_2, condensation pressure p_{cond}, geothermal fluid's discharging (or rejection) temperature T_{out}, net power output P_{out}, thermal efficiency η_{th} and exergy efficiency η_E.

As can be seen from Table 2, as the working fluid's critical temperature increases, evaporating pressure, condensation pressure, output power and exergy efficiency decrease as a whole, however, thermal efficiency and geothermal fluid's discharging (or rejection) temperature show an increasing trend. ORC with n-hexane gets the minimum of power output and exergy efficiency, and, condenser is at a very low sub-atmospheric pressure state, it's necessary to mount vacuum maintaining system. However, ORC with R227ea (heptafluoropropane) gets the maximum of power output and exergy efficiency, and, evaporation pressure is in the low-mid pressure range, no strict requirements for metal's pressure bearing, the pressure in condenser is also higher than atmospheric pressure, which can effectively prevent air leakage in, therefore, R227ea is the optimal selection.

Table 2. Some calculated results

Fluids	p_2 /MPa	p_{cond} /MPa	T_{out} /°C	P_{out} /kW	η_{th} / %	η_E / %
R227ea	1.491	0.412	64.2	1422.3	11.04	36.14
R236fa	0.988	0.245	68.7	1271.9	11.50	32.32
R600a	1.088	0.319	70.0	1242.2	11.87	31.57
R236ea	0.794	0.185	70.3	1230.6	11.90	31.27
R600	0.810	0.221	71.1	1206.3	12.05	30.65
R245fa	0.606	0.134	70.9	1206.7	11.97	30.66
Neo-C_5H_{12}	0.584	0.155	70.4	1215.7	11.79	30.89
R601a	0.356	0.083	71.7	1176.5	12.05	29.90
R601	0.283	0.061	72.0	1167.0	12.12	29.67
n-hexane	0.105	0.018	72.3	1160.2	12.19	29.48

5 Conclusions

From the above, the following conclusions can be drawn:

1) As the working fluid's critical temperature increases, evaporating pressure, condensation pressure, output power and exergy efficiency decrease as a whole, however, thermal efficiency and geothermal fluid's discharging (or rejection) temperature show an increasing trend.
2) R227ea is the optimal working fluid for ORC powered by low-temperature geothermal heat.

Acknowledgment. The study presented in this paper is financially supported by National Natural Science Foundation of China and Yunnan Natural Science Foundation (Grant Nos. 51066002, U0937604, 2008KA002).

References

1. Lund, J.W., Freeston, D.H., Boyd, T.L.: Direct application of geothermal energy: 2005 Worldwide review. Geothermics 34(6), 691–727 (2005)
2. China's New Energy and Renewable Energy White Paper. China Planning press, Beijing (1999) (in Chinese)
3. Yari, M.: Exergetic analysis of various types of geothermal power plants. Renewable Energy 35(1), 112–121 (2010)
4. Kenneth, H., Williamson Richard, P.: Geothermal Power Technology. Proceedings of the IEEE 89(12), 1783–1792 (2001)
5. Bertani, R.: World geothermal power generation in the period 2001–2005. Geothermics 34(6), 651–690 (2005) (in Chinese)
6. Wang, H., Wang, H.-T.: Organic Rankine cycle Technologies to convert low-temperature waste heat into electricity, Beijing
7. Wang, H., Wang, H.: Selection of working fluids for low-temperature solar powered organic Rankine cycles. Power Engineering 29(3), 87–91 (2009) (in Chinese)

8. Hung, T.C.: Waste Heat Recovery of Organic Rankine Cycle Using Dry Fluids. Energy Conversion & Management 42(5), 539–553 (2001)
9. Liu, B.-T., Chien, K.-H., Wang, C.-C.: Effect of Working Fluids on Organic Rankine Cycle for Waste Heat Recovery. Energy 29(8), 1207–1217 (2004)
10. Mago, P.J., Chamra, L.M., Srinivasan, K., et al.: An Examination of Regenerative Organic Rankine Cycles Using Dry fluids. Applied Thermal Engineering 28(8-9), 998–1107 (2008)
11. Aalto, M.M.: Correlation of Liquid Densities of Some Halogenated Organic Compounds. Fluid Phase Equilibria 141(1-2), 1–14 (1997)
12. Steven Brown, J.: Predicting performance of refrigerants using the Peng-Robinson Equation of State. International Journal of Refrigeration 30(8), 1319–1328 (2007)

Development Tendency of the Embedded System Software

Chen Jiawen

Information Engineering Department,
Jilin Liaoyuan Vocational and Technical College,
Jilin Liaoyuan, China
jllycjw@yahoo.cn

Abstract. Currently, the embedded system is rapidly developing. The embedded technology has been applied in every field of the social life. The embedded software is facing new opportunities. Its development is of more diversity, intelligent and humanization. The embedded system application software development need strong development tools and operating system support.

Keywords: Embedded software, Network, Intelligent.

1 Introduction

The third tide of the world information industry, the internet of things, has come quietly after the computer and the World Wide Web. The embedded system technology takes a vital role in the internet of things. The embedded system is described as a special computer system which is concentrated in application with tailored software and hardware, and meet the serious comprehensive requirements of the application system for function, reliability, cost, structure and power consumption, which is composed of embedded hardware and software, the former is the support, and the latter is soul. Almost all embedded product need software to provide various and tailored function. As the core of the embedded product, the embedded software takes more and more important role in the industry development.

2 Embedded Software Connotation

The embedded software is one kind of the computer software, composed of program and text. It can be divided into system software, support software and application software. The embedded system was original defined as the computer which was used to control equipment and installed inside the equipment, or the special software and hardware which was embedded in other system. The embedded system is usually required real time response capability, while not required complex user interface, even keyboard support, displayer, serial port and hard disk, secondary development of user. This system is extensively used in device, industrial control equipment, electric lift, SPC exchange, microwave equipment, traffic light and household appliances. In

D. Zeng (Ed.): Advances in Control and Communication, LNEE 137, pp. 131–135.
springerlink.com

recent years, the embedded system has new meaning and various access devices, such as handheld PC, wireless mobile net phone, set top box, family gateway, net TV, net vehicle bone box and intelligent house hold appliances. At the same time, the embedded system is gradually divided into system software, support software and application software.

3 Embedded Software Development Situation

Since the 20th century, the embedded operating system began to grow vigorously. Facing the Internet and specific application of the embedded operating system is the important development trend. Embedded operating system along with the development of embedded system is now more mature. In long-term application and business competition of industrial control and other related areas, some relatively successful Embedded operating system have gradually formed, including the mainstream Windows CE, Palm OS, Embedded VxWorks pSOS, Linux, and OS - 9, etc. Analysis of these systems show that they provided powerful function, similar to the desktop operating system. But the embedded system are lack of effective application and network connection function. And the application development support ability of this system is relatively weak. Thus, the application of a new generation special embedded operating system is needed.

Microsoft's Windows CE is a representative embedded real-time operating system, derived from the desktop operating system. It is a concise Windows 95,which is not a good RTOS from the viewpoint of technology. First, RTOS pays much attention to individuation, and Windows CE is not open OS, which make it difficult for the third party to tailor the product. Secondly, RTOS aimed high efficiency, energy-saving, while Windows CE is clumsy and occupy too much RAM, which application function is huge. Thirdly, Windows CE does not consider the height of the system in the kernel structure design.

Embedded database technology is widely used. Along with the mobile communication technology progress, mobile data processing is achieved more requirements. The embedded database technology has already got more attention from the academic and industrial and military and civilian areas. In foreign countries, Sybase dominated in the mobile database fields, and has been applied in medical insurance, financial services, and retail, transportation and government, etc. Oracle has put much resources in mobile enterprise applications market, provide complex mobile e-commerce solutions, including technology, application, online services and consulting service.

In China, the embedded database system, independently developed by National People's Congress JinCang company, has been applied to military mobile information system, e-medicine, personal marketing assistant. Neusoft group has also developed the embedded database system, OpenBASE Mini.

In embedded systems, the application software is the most active element. The application software often determines the performance, location and characteristics of the product. In recent years, due to the rapid development of embedded products, some kinds of the application software have been born in the market, such as browser, E-mail, word processing software, communication software, multimedia software, personal information processing software, intelligent human-machine interaction software, and other industry application software.

4 Embedded Software Applications

Embedded software can be applied in the household market, industrial market, commercial market, communications market and defense markets. The product, applied the embedded operating system, is various. And the potential market is tremendous. Therefore, the embedded software is facing the rare market opportunities in the next 10 years.

4.1 Information Household Application

The consumer and family health electronic equipment, such as STB, WebTV, network refrigerator, air conditioning network, will be rapidly developed in the next few years. The personalized, regional and seasonal change trend of the household electronic equipment provided the development space for embedded operating system. It is estimated that the next 10 years is the golden age of information household appliances. For example, there will be 100-150M set of STB produced in 10 years. The total market demand for STB will reach from 90 billion to 150 billion RMB. Information intelligent household is the direction of future development, which will be developed rapidly in a few years in China.

4.2 Medical Instruments Field

The application of medical instruments, such as embedded pacemakers, embedded radiation monitoring equipment and analysis ASOS equipment, need support from ASOS. Some kinds of test equipment, such as muscle, discrete dynamic current cardiography luminosity chemical analysis, spectrophotometer etc, need to use high-performance, specialty of embedded system to improve the accuracy and speed. Existing monitor function and performance will be greatly improved.

4.3 Intelligent Vehicle Field

Communication, information, navigation, entertainment and all kinds of auto safety electronic system in an on-board box will become the next generation and future automobiledevelopment direction.

As wireless communication and GPS technology mature and are extensively applied, the on-board box will become the next generation and future automobile development direction, which integrates communication, information, navigation, entertainment and all kinds of auto safety electronic system. Due to the market demand, the car carries enough box will be the focus of recent development in this field, and the market scale in future years will increase quickly.

4.4 The Intelligent Transportation Field

Because of the demand of environment, the intelligent transportation system (ITS) is becoming the rapid developing backbone industry in the new century. Specific application of the embedded operating system will be key technology of the intelligent comprehensive intersection control machine, interactive system, the new road car parking system, highway information monitoring and integrated management system. Its application will ensure the intelligent transportation system with low cost and high performance, and greatly improve the reliability of the system and the level of intelligence.

Embedded software has received attentions from many related industry. Some companies are studying on it, such as ASOS system Co. LTD.. To pursue high performance and low cost, the embedded technology will become key technology of many products and has wide prospect.

5 Embedded Software Development Trend

In recent years, the development of embedded system will have the following development trends

1 the embedded products will rapidly develop, with the Internet application mutually promote. The embedded products will be one of the main terminals of the internet. A lot of software in the service of embedded products will appear in the internet, which have special contents servicing the embedded products.

2 with the rapid development of the microelectronics technology, Chip function become more powerful and SOC (System) will become trend on Chip, which will not only can reduce cost, product volume, but also increase the reliability of the product. At the same time, the software and hardware will combine closely. Their boundaries become more ambiguous. The embedded often appears in hardware style. This form can improve the real-time properties and enhance the maintainability.

3 The wireless communication products will become an important applications of embedded software. On one hand, wireless products will depend on the chip technology and embedded software to improve performance. On the other hand, the current embedded products will add the wireless communication function.

4 The embedded operating system will develop in coordination with the embedded application software. The important role of embedded system software includes embedded application software. The embedded application fields are various. To meet all kinds of requirements, it is necessary to pay full attention to the development of application software.

5 The embedded operating systems have been developed on various hardware platforms. Along with the wide application of the embedded system, the information exchange and sharing of relative resources increased. Thus, the matter of standard will also becomes more and more serious. It drew more attention from industry how to establish the relevant standards.

The development of embedded system needs to choose embedded processors and embedded operating system. The embedded application software will be developed through cross-compilation and links. After commissioning, the embedded system will be conducted system test by the memory analysis tools, performance analysis tools and analytical tools. The development of the embedded system application software needs strong development tools and operating system support. It should support small electronic equipment to fulfill small size, low cost and power consumption and to provide exquisite multimedia human-machine interface. The combination of the embedded system and the Internet is a future development trend.

References

1. United by eclipse, embedded system design (January 2007)
2. Eclipse focus: Motorola joins eclisp, propses Tml project, embedded computing design (October 2006)
3. Wang, S.: Safe choice, 3–4 (2007)
4. Fink, M.: Linux and open source in the application of commercial economy. Tsinghua university press
5. He, X.Q.: Mobile phone challenge embedded Linux. Electronic Products to the World (December 2006)
6. He, X.Q.: The latest development of embedded software. In: Shenzhen Embedded System Meeting (October 2006)
7. http://www.mvista.com
8. Do-it-yourself linux? Jim Ready Enterprise Open Source Journal (December 11, 2006)
9. http://www.linuxdevices.com
10. http://www.bmrtech.com

Numerical Simulation on the Coal Feed Way to 1000MW Ultra-Supercritical Boiler Temperature Field

Liu Jian-quan[1], Bai Tao[1], Sun Bao-min[1], and Meng Shun[2]

[1] Key Laboratory of Condition Monitoring and Control for Power Plant Equipment,
North China Electric Power University,
Bing Jing, China
[2] Combustion Science and Research Institute, School of Energy Science and Engineering,
Harbin Institute of Technology,
Harbin, Heilongjiang Province, China
ljq8981@sina.com, sbm@ncepu.edu.cn,
{hdbaitao,mengshunhit}@126.com

Abstract. The factors influenced the distribution of tempera -ture and velocity field was numerically studied with realizable k-ε model in a 1000 MW ultra supercritical boiler. The results indicate that the ring way of pulverized coal feeding feed powder can form the short and wide flame shape, combustion stability is better; the center way can form the long and narrow flame shape, combustion stability is worse. The concentration NO emissions decrease slightly. The simulation results can coincide with the practical combustion process. All of these can provide a theoretical basis for the HT-NR burner design and operation.

Keywords: HT-NR3 burner, numerically studing, Realiz-able k-ε model, Combustion.

1 Introduction

Ultra-supercritical units, for the steam pressure and steam temperature rises, the thermal efficiency of domestic ultra-supercritical units increase nearly 10% than the average level. In this paper, the introduction of 1000MW Ultra-supercritical boiler HT-NR3 burner temperature field was the research object (China's first design coal for Shenhua 1000MW supercritical swirling coal combustion boiler). The boiler burner adopted Babcock and Hitachi advanced NR (NOx reduce) technology. The flame NOx reduction technology down part burns of the furnace was instead with central fuel burners to achieve oil-free ignition, this burner is the main burner for the boiler. Currently, the characteristic of this type of boiler burner combustion is relatively small. On this to study and fully understand the characteristics of the combustion technology, guidance unit operation is of great significance.

Actual operation experiment can not measure the change of temperature, and the workload is huge[1-3]; Numerical simulation can reflect the combustion process in detail that has been widely used to study the combustion process[4-6]. In this paper, numerical simulation of combustion process was selected to calculate, Experimental

D. Zeng (Ed.): Advances in Control and Communication, LNEE 137, pp. 137–144.
springerlink.com

results were compared with that calculating, the calculating showed relatively accurate by comparing.

2 Simulation and Calculation Object

2.1 Experiment Equipment Cases

The width depth and height respectively of boiler furnaces wide are 33.9734 m × 15.5584m × 64.600m. The boiler is reheat, supercritical pressure and swing combustion operation Bunsen boiler, single chamber, balanced ventilation, solid-state discharge slag, π-type layout, The capacity to design evaporation is 3033t / h. Shenhua coal is the design bituminous coal, burning coal analysis in Table 1, the boiler is equipped with 6 mills, Corresponding to total of 48 HT-NR3 low NOx pulverized coal burners, the burner structure show in Figure 1, before and behind the hedge wall total arrange three layer. HT-NR3 staged single port combustion burner means outside the pale of the pulverized coal concen-tration distribution in favor of the formation of NO reduction zone; the nitride can be quickly converted into gas, reducing substances generated within the flame acceleration peak for NO.

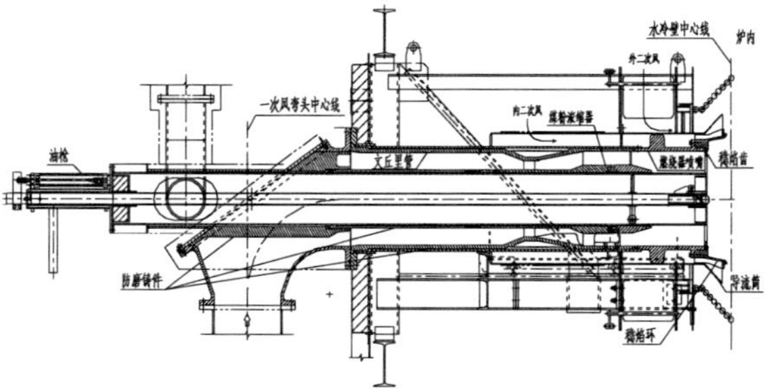

Fig. 1. Diagram of the HT-NR3 burner

Table 1. Approximate and elementary analysis of the coal

elementary analysis					Approximate analysis			$Q_{net.ar}$/ (MJ/kg)
C_{ar}/%	H_{ar}/%	O_{ar}/%	N_{ar}/%	S_{ar}/%	A_{ar} /%	M_{ad}/%	V_{daf}/%	
61.88	3.40	10.728	0.80	0.44	9.10	3.68	34.08	23.47

2.2 Mathematical Model and Calculation Method

1) The computational domain and mesh.

Computational domain is used unstructured tetrahedral grids. For full furnace simulation, the refined grid with the greatest density of cells is constructed near the burner region consider the flow field changes more violent, and ensure the center air, primary air and the inner or outer secondary air at different grid, to avoid pseudo-diffusion. Accounting to the limitations of computer hardware, the burners can be simplified. Considering the selection of inlet parameters are consistent with single burner simulation results, the total number of meshing about 1.116 million, the mesh of whole furnace burner computational domain show in Figure 2.

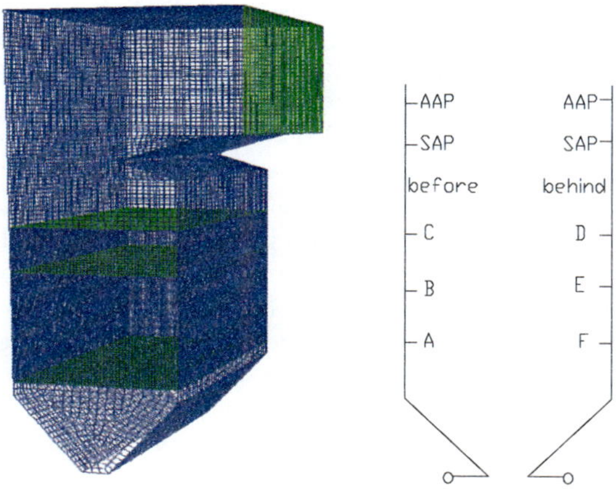

Fig. 2. Arrangement and Calculation domain or grid numbers of the total burners

2) Mathematics and geometry model.

Full-size geometry model is used in the Numerical simulation of the whole furnace, the same sizes of selection geometry with real furnace. The numerical simulation calculation is used three-dimensional steady-state. The simulation of gas phase turbulent flow can be achieved to use k-ε model, (realizable k-ε model). The unified form of (1):

$$div(\rho \upsilon \phi) = div(\Gamma_{\Phi} \nabla \phi) + S_{\phi} \tag{1}$$

In equation: ϕ is general dependent variable; r_{Φ} is transport coefficients; S_{Φ} is source term; is air density; ρ is velocity vector; υ is velocity vector Log-law wall function is used to treatment the near wall. Radiation is described to use the P-1 model. The devolatilization process is modeled by single step model that is used to the combustion simulation. The turbulent combustion of gas phase is modeled by Mixture fraction -the probability density function (Mixture-Reaction/PDF), char combustion is used a diffusion/kinetics model. The coal particles follow random tracks (Stochastic Tracking) model, size distribution follows the Rosin-Rammler distribution. The Inlet and export condition are used pipe value and full development conditions [7-12].

3) Boundary conditions.

This calculation of the rated load design conditions, based on rated load design conditions into 5-layer burner, we calculated the rated load conditions into A, B, C, E, F burner, excess air coefficient is 1.14, the boiler coal capacity of 104kg / s, one of which wind is 194.8kg / s, the total amount of secondary air 733 kg / s, AAP and SAP total air volume of 276.7 kg / s, the main combustion zone of secondary air is 456.5

kg / s . Primary air temperature is 350K, the secondary air temperature is 619K. Other conditions under wind load and excess air ratio based on the amount of wind in the calculation.

3 Numerical Simulation Results and Analysis

3.1 Analysis of the Whole Burner Flow Field Distribution in the Furnace

Numerical simulation results of HT-NR3 burner center and original structure to the powder outlet flow field show in Figure 3 (a) and Figure 3 (b). Change the primary air way to the burner, the flow field change greatly. original structure, burner swirl flow field show rotational flow characteristic step by step, the flow field center to a larger recirculation region, the flow of the exit flow field lines are also upturned tail, the jet expansion angle is relatively large; Center to the burner case, burner smaller recirculation flow, burner flow field show line characteristics step by step, jet expansion angle is small, show similar characteristics of the Line jet.

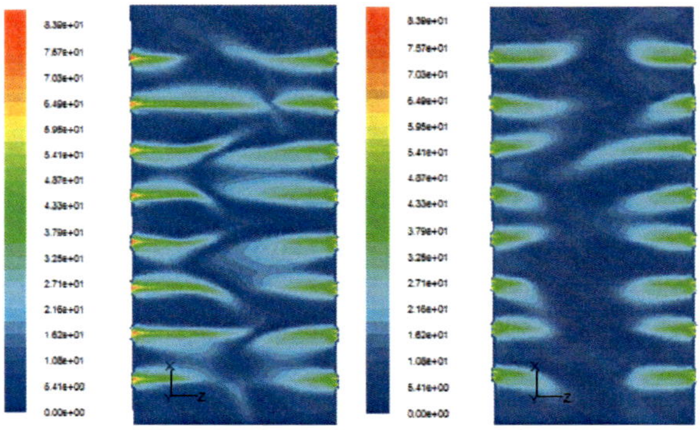

(a) 19768. 7mm (b) 25588. 5mm

Fig. 3. Burner velocity contrast on the furnace

3.2 Characteristics of the Furnace Temperature Distribution

1) Horizontal section comparative analysis of the whole furnace temperature field.

Put into A, B, C, D, E, F total 6 layer burners, boiler coal capacity is 104kg/s, primary air capacity is 194.8kg/ s, the secondary air flow is 733 kg/s, excess air coefficient is 1.14, basic working conditions in accordance with the air. The combustion condition show in Figure 4 at 19768.7 mm height (the lower burner region) department, 25588.5 mm height (the middle burner region) at the level of cross-section temperature field. Can be seen that in the original structure, at the time air vents from 400mm, flue gas temperature up to 1390°C, the area around the burner nozzle at the temperature of 350mm from the Wall is 1155°C, the secondary air expansion angle is 105°, concentration of pulverized coal in which the ambient temperature far more than its ignition temperature, flame temperature and the highest decay faster. Center to the powder case, the primary air nozzle at the distance of 400mm, the flue gas temperature rises to 1350°C, the temperature of the area around the burner nozzle from Wall 350mm is 1100°C, the secondary air expansion angle is 100 °, the flame distance from the nozzle increased, the vents near the water wall, the flue gas temperature decreased, the phenomenon of the furnace flame is not significant deviation, the highest flame temperature slow decay. Reason for the center to the powder case, the actual operation of the burner during the rotation of its impact outside the secondary air force weakened.

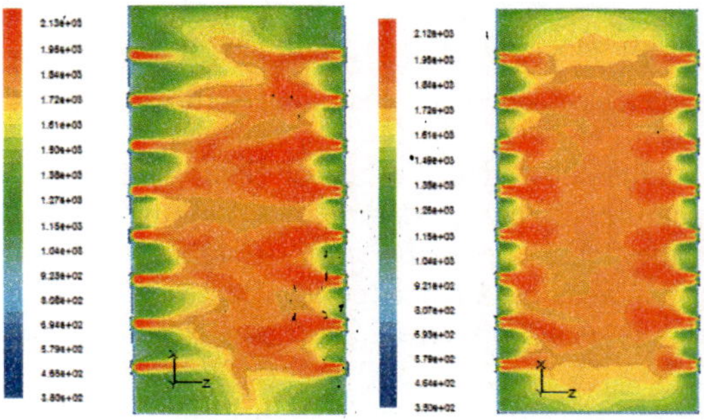

(a) y=19768. 7mm (b) y=25588. 5mm

Fig. 4. Temperature contrast on furnace burners

2) Vertical section comparative analysis of the whole furnace the temperature field.

Put into A, B, C, D, E, F total 6 layer burners, boiler coal capacity is 104kg / s, primary air capacity is 194.8kg /s, the secondary air flow is 733 kg/s, excess air coefficient is 1.14, basic working conditions in according with the air. The condition show in Figure 5, the furnace with the vertical cross section (here to take x = 1841.5mm, X-axis center set at the center of the furnace width) velocity field and temperature field. Flow field in rotating ,the role of the hedge, in 19768.7mm height, forming two rows of tapered cylindrical high temperature zone, and the central part of the furnace bottom temperature is showed, the temperature field and flow field show good synergy; With the high increase in the 25588.5mm furnace at the middle-row tapered column is increasingly, becoming the junction of the high tempera-ture furnace, the temperature field and flow field synergy decline; In 31408.3mm high position, the flame maximum temperature near the center of the furnace, the role of flow rise, the high temperature at the top furnace show the features extending. Take the bottom burner as the research object, compared to trends before and after the wall that can be found, burner temperature is lower than the upper part of the same layer temperature region, The results of comparative temperature field and temperature field analysis in the horizontal section is consistent.

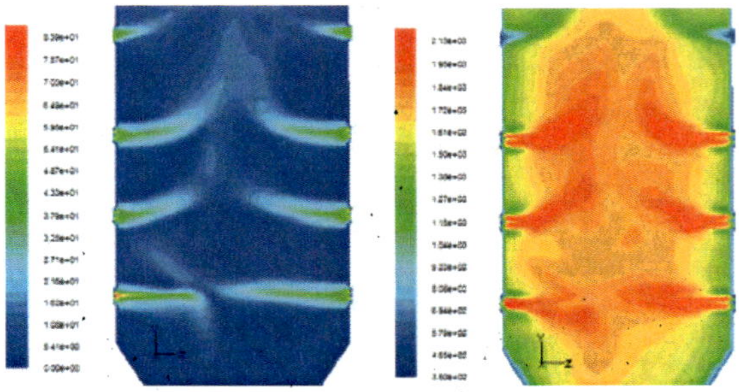

(a) velocity field (b) temperature field

Fig. 5. Velocity and temperature fields contrast on vertical cross-section (X=1841.5)

The simulation results show that the burner flame show great effect to the coal feed, the original structure, the burner flame rapid combustion, there is a slight deflection phenomenon in the flame, the temperature rise rapid. Near wall region, the flame stability is effectively increasing. Center fuel can increase the flame rigid; the phenomenon of flame deflexion is not obviously, the flame stability deteriorates [9-15].

3.3 Characteristics of the Furnace No Distribution

Put into A, B, C, D, E, F total 6 layers burner, boiler coal capacity is 104kg/s, primary air capacity is 194.8kg/s, the secondary air flow is 733kg/s, excess air coefficient is 1.14 at 19768.7mm height (the bottom burner region) department, show in Figure 6(a). The bottom foreside burner is Centre Powder and rear burner is Original structure. Taking the bottom burner as the research object, compared foreside burner outlet region No distributing trends with the rear wall burner that can be found, burner outlet region No was lower than the rear part of the same layer, But not very obvious. Take the vertical cross section (here to take x=1841.5mm, X-axis center set at the center of the furnace width) as the research object, show in Figure 6(b). Comparing bottom foreside burner outlet region No distributing trends with the rear wall burner that can be found the same case. The comparison result show that all conditions concentration of the NO emission decrease is not obvious.

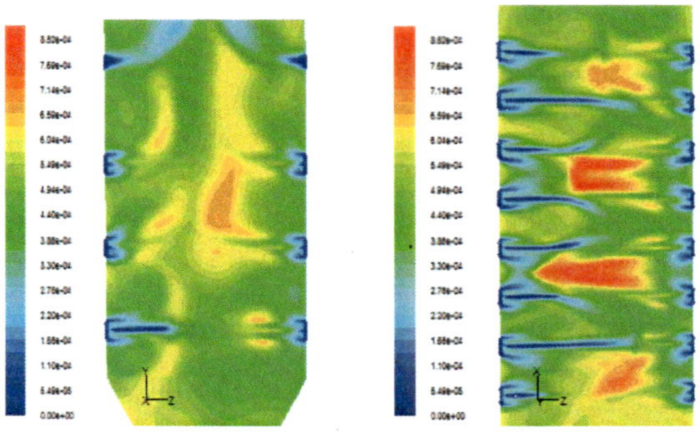

(a) No distribution (x=1841.5) (b) No distribution (y=19768.7)

Fig. 6. NO distribution contrast on furnace burners

4 Actual Industry Experiment Study

4.1 Combustion Experiment Study

Put into the burner operation, select the observation hole beside A8 burner to take on temperature field (thermal calculation results shown in Figure 4 (a)) measurements, using armored nickel-chromium - nickel silicon thermocouple temperature on the burner field measurements, and data with the corresponding numerical simulation of temperature conditions were compared. The maximum numerical simulation results and measuring deviation was 20%, the trend consistent with the numerical results. Select the hole near C8 burner to see the fire, through the high temperature endoscope Cyesco portable to watch burner nozzle and the water wall near the observation area, coal quality is bad, combustion stability is good when the original structure; coal quality is better ,when the center coal flour can prevent slagging phenomenon[11-15].

4.2 NO Emission Characteristics of Burners Combustion

Experiment power load is 1000MW. Put into A, B, C, E, F total 5 layer burners, boiler coal capacity is 104kg/s supremacy capacity is 194.8kg / s, the secondary air flow is 733 kg / s, excess air coefficient is 1.14, Accordance with the basic working conditions ; other conditions burner air and excess air ratio is based on the amount of air in the calculation. 60% of the power load ,put into total 3 layer burners; 80% of the power load ,put into 4 layer burners; 100% of the load put into 5 layers burner, measured concentration of NO emission, All this show in Table 2. Comparison shows that all conditions concentration of the NO emission decrease is not obvious.

Table 2. Furnace outlet NO effects under different power load and different burners

Power /MW	*Burner operation mode*	*NO discharge Φ_{NOx}/ (mg.m^{-3})*	*Burner operation mode*	*NO discharge Φ_{NOx}/ (mg.m^{-3})*
1000	ABCDE	291	BCDEF	293
800	ABCE	264	BCEF	267
600	ABE	185	BEF	190

5 Conclusion

- Simulation and experimental results are good qualitatively inosculated with the actual operating results, Indicating that the use of CFD on Numerical calculations are feasible in the furnace burner.
- In the original structure, the burner flame rapid combustion, there is a slight flame deflection phenomenon; temperature rises up near wall region that can effectively increase the flame stability.
- Center to feed powder can increase the flame rigid, no obvious deflection phenomenon of the flame, the flame stability deteriorated.

- On HT-NR3 burner, center to feed powder compare with the original structure, NO emission concentration is not obvious decreasing.

References

1. E. Mineral impurities in coal combustion behavior, problem and remedial measures. Hemisphere Publishing Corporation, New York (1985)
2. Schnell, U., Kaess, M., Brodbek, H.: Experimental and numericalinvestigation of NOx formation and its basic inter-dependencies onpulverized coal flame characteristics. Combust. Sci. and Tech. 93(1-6), 91–109 (1993)
3. Ji, C.C., Cohen, R.D.: An investigation of the combustion of pulverizedcoal-air mixture in different combustor geometries. Combustion and Flame 90(3-4), 307–343 (1992)
4. Anagnostopoulos, J.S., Sargianos, N.P., Bergeles, G.: The prediction ofpulverized Greek lignite combustion in axisymmetric furnaces. Combustion and Flame 92(3), 209–221 (1993)
5. Faltsi-Saravelou, O., Wild, P., Sazhin, S.S., et al.: Detailed modelling ofa swirling coal flame. Combust. Sci. and Tech. 123(1-6), 1–22 (1997)
6. Gorres, J., Schnell, U., Hein, K.R.G.: Trajectories of burning coalparticles in highly swirling reactive flows. Int J. Heat and Fluid Flow 16(5), 440–450 (1995)
7. Yu, M.J., Baek, S.W., Kang, S.J.: Modeling of pulverized coal combustion with non-gray gas radiation effects. Combust. Sci. and Tech. 166(1), 151–174 (2001)
8. Zhou, L.X.: Theory and numerical modeling of turbulent gas-particle flows and combustion. Science Press, CRC Press, Beijing, Boca Raton (1994)
9. Liu, Z., Yan, W., Gao, Z., et al.: The effect of the micro-pulverized coal fineness on nitric oxide reduction by reburning. Proceedings of the CSEE 23(10), 204–208 (2003) (in Chinese)
10. Zhang, J., Sun, R., Wu, S.-H., et al.: An Experimental and numerical study on swirling Combution Process in a 200MW pulverized coal fired boiler. Proceedings of the CSEE 23(8), 215–220 (2003)
11. Ubhayakar, S.K., Stickler, D.B., Von Rosenberg, C.W., et al.: Rapiddevolatilization of pulverized coal in hot combustion gas. In: Proc. of 16th Symp(Int) on Combustion, Pittsburgh, pp. 427–436 (1976)
12. Hassan, M.A., Hirji, K.A., Lookwood, F.C., et al.: Measurements in apulverized coal-fired cylindrical furnace. Experiments in Fluids 3(3), 153–159 (1985)
13. Cheng, J., Zeng, H., Xiong, W., et al.: Research and test for reducing NOx emission of A 300MW lean coal-fired boiler. Proceedings of the CSEE 22(5), 157–160 (2002)
14. Li, Z., Feng, Z., Wangyang, et al.: Study on dual vertical dense/lean combustion of pulverized coal in order to decrease NOx emission and stabilize combustion. Proceedings of the CSEE 23(11), 184–188 (2003)
15. Dong, H., Cao, X.-Y., Niuzhi-gang, et al.: The characteristic of NO release for the chars and volatiles of bituminous during combustion. Journal of China Coal Society 30(1), 95–99 (2005)

Research on the Coking Features of 1900t/h Supercritical Boiler with LNASB Burner Vents

Liu Jian-quan[1], Bai Tao[1], Sun Bao-min[1], and Wang Hong-tao[2]

[1] Key Laboratory of Condition Monitoring and Control for Power Plant Equipment, North China Electric Power University, Bing Jing, China
[2] Boiler Science and Research Institute, North China Electric Power Science and Research Institute generating co. LTD, Bing Jing, China
ljq8981@sina.com, hdbaitao@126.com, sbm@ncepu.edu.cn, wanghthbdky@sohu.com

Abstract. In this paper 1900t/h supercritical boiler with swirl burners as research object, dropping coke, pressure fluctuating and large coke blocks plug slag discharge port frequently in this boiler. Combining with the practical situation of slagging, ribbon and actual method of measurement was used to measure and analysis the secondary air flow and swirl intensity. Experimental results can find the slagging factors of boiler burner nozzles. All this can provide some theoretical basis for adjusting and improving burner to solve the slagging problems.

Keywords: boiler coking, combustion adjust, data analysis, flow characteristics.

1 Introduction

Currently, the slagging character of shenhua coal is studied seldom. Less experience about the coal suitability of large load boiler in the subject area can be referred. In this paper 1900t/h supercritical boiler with swirl burners is the research object, after the boiler put into operation, large slag block break Slag conveyor into, plug the boiler slag mouth, leading to the pressure fluctuations and fire off in the furnace frequently .In order to solve the vicious slagging phenomenon, first, according to the instructions to adjust the thermal condition, but the slagging can not be eliminated. Combining with the characteristic of burner mouth slagging, according to the characteristics of near the burner outlet region, cold flow field and thermal dynamic properties is similar [1-3], based on the actual burner as a experiment model, finding out the vicious burner nozzle slagging reason by using of ribbon method combination with instruments to measure, providing some theoretical basis for the domestic large-capacity supercritical coal-burning boiler LNASB burner malignant Slagging problems.

D. Zeng (Ed.): Advances in Control and Communication, LNEE 137, pp. 145–153.
springerlink.com

2 Experiment Section

2.1 Experiment System Introduction

The experiment equipment is 1900t/h Supercritical swirl boiler LNASB burner, cold model experiment system shown in Figure 1, primary and secondary air come from fan through pipes throttle adjustment, flow meters and other devices entered into LNASB burner.

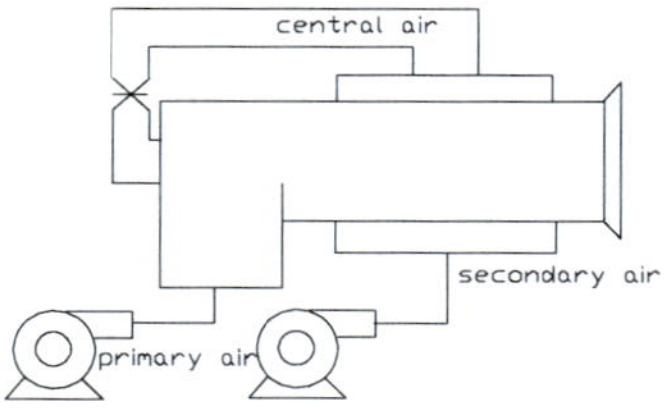

Fig. 1. LNASB burner cold modeling experiment system

2.2 Experiment Equipment Introduction

The experiment of equipment technologies come from Mitsui Babcock Energy (Mitsui Babcock Energy Limited), and with this to design and manufacture. The boiler is single chamber, balance ventilation, solid slag, π-type layout. Shenhua coal is the design coal. Boiler with 6 mills, corresponding to 6 layer and total 30 LNASB (Low NO Axial Swirl Burner) pulverized coal burners. The air come from the burner including primary air, inside and outside secondary air, Primary air through bend pipe then into the burner volute, last through the coal collector and BANJO into the furnace duct. Inner secondary air swirl strength can be axial adjusted; outer secondary air swirl strength can not. During the operation, Boiler drop slag and slag mouth being blocked frequently .During examination find a lot of slag around the burner nozzle, A lot of slag heap outlet the second air nozzle, all this show in Figure 2.

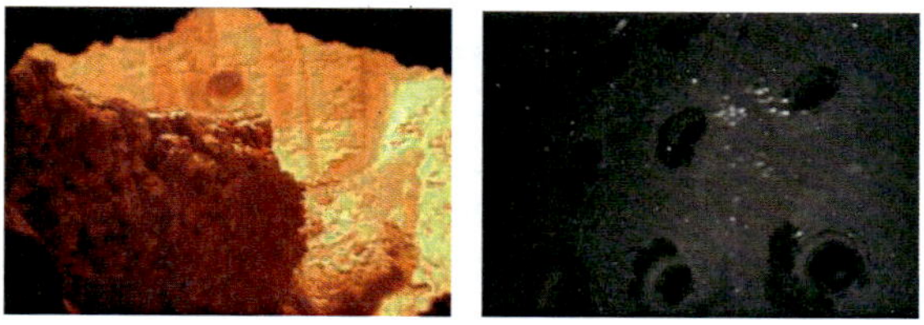

Fig. 2. The picture of large coke block plug tap mouth and burner mouth coking

2.3 Experiment Methods Introduction

Adopt original burner as the actual model in the experiment, satisfy the principle of geometric similarity. For the furnace reaching or exceeding 1.8×105 is satisfied with requirements, the state air movement is similar with it. Experiment is satisfied with the burner jet similarity, Ensure the same burner air distribution as operation condition air distribution manner first, secondly followed by cold jets the same Re number as the operation condition, Maintain momentum ratio of cold jets the same as operation condition, In line with the similar requirements of boundary conditions [4-6].

Experiment method was used combination of ribbon and instruments measure, Select C3 and A3 burner with a combination of ribbon observation and measuring instruments first, Maintain the burner flow field in normal conditions, Find the most suitable flow conditions to avoid burner muzzle slag, Then actual measuring powder deviation and combustion system, find the deterioration slagging reasons.

3 Experiment Results and Analysis

3.1 Analysis of the Burner Primary Air Changes

In order to study the burner jet recirculation zone and secondary air expansion angles, when the primary air ratio changed, find the burner primary air vent slag characteristics. According to actual operating flow conditions, The experiment is divided into 6 working conditions to measure, with inner secondary air 19m/s and outside the secondary air 38m/s. The experiment data show in Table 2, The burner nozzle flow characteristics show in Figure 3.

Table 1. Boiler burner primary air experiment Data Sheet

Condition	*CM11*	*CM12*	*CM13*	*CM14*	*CM15*
Primary air velocity/m.s	0	15	18	30	38
Primary air ratio	0	9.7/%	11.8/%	19.6/%	24.5/%
Secondary air ratio	100/%	90.3/%	88.2/%	80.4/%	75.5/%

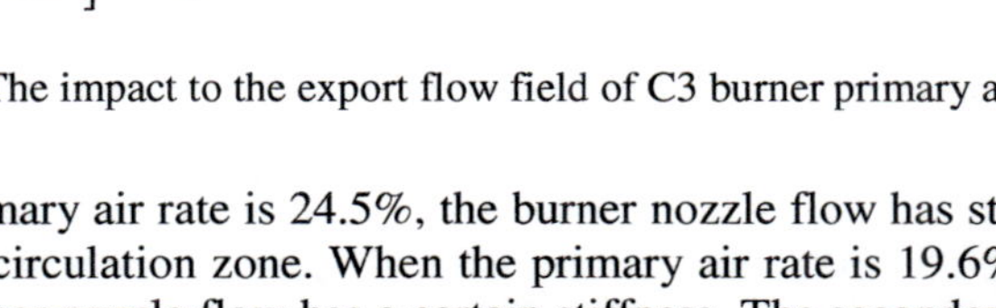

Fig. 3. The impact to the export flow field of C3 burner primary air change

When the primary air rate is 24.5%, the burner nozzle flow has strong stiffness and no significant recirculation zone. When the primary air rate is 19.6% (design primary air rate), the burner nozzle flow has a certain stiffness, The secondary air recirculation zone and the extension angle is better, when the primary air rate dropped from 19.6% to 11.8%, the recirculation region and second air expansion angle increases; primary air rate to 9.7% (50% of design primary air rate), The recirculation region migration occurred, The secondary air expansion angle is larger; when the primary air rate from

9.7% to 0 (primary air is fully closed), The recirculation zone extend and show a larger angle. Experiment results show that LNASB burner recirculation zone and extension angle of primary air with a larger decrease, Small rate of primary air can improve the stable character of the burner flame, But the primary air ratio too small will cause recirculation zone deflection. Larger primary air rate can reduce the expansion angle and the secondary air recirculation zone width, Avoid outside the recirculation zone and the flue gas sweep wall phenomenon [7-12]. Increasing the distance from the nozzle, nozzle slag can be avoided.

3.2 Analysis of the Burner Inner Secondary Air Changes

The inner and outside secondary air blade angle was 55 ° and 15 °of LNASB boiler burner, The inner and outer secondary air blade angle are not adjustable. the inner secondary air duct can be axis adjusted, moved forward, the secondary air swirl strength increases, whereas decreases; outside the secondary air swirl strength is not adjustable. By adjusting the outside secondary air damper to control the inner and outer secondary air volume, By adjusting the inner secondary air damper to control the inner and outer secondary air ratio, the outside secondary air can not be adjusted separately. According to the different ignition. That was divided into the oil ignition and ion ignition burner, The later without central air.

1) Analysis of the Centre air burner inner secondary air change.

C3 oil burner is selected as the research object; primary air was maintained with velocity of 29.5m/s. The experiment is divided into 6 working conditions to measure data, experiment conditions show in Table 3, the burner nozzle jets show in figure4 and figure 5.

Table 2. C3 burner inner secondary air experiment Data Sheet

Condition	*CM21*	*CM22*	*CM23*	*CM24*	*CM25*	*CM26*
Inner/ outside secondary air velocity /m/s	25/32	25/32	25/32	25/32	18/45	0/48
Inner/ secondary air swirl strength	100%	50%	30%	20%	20%	20%

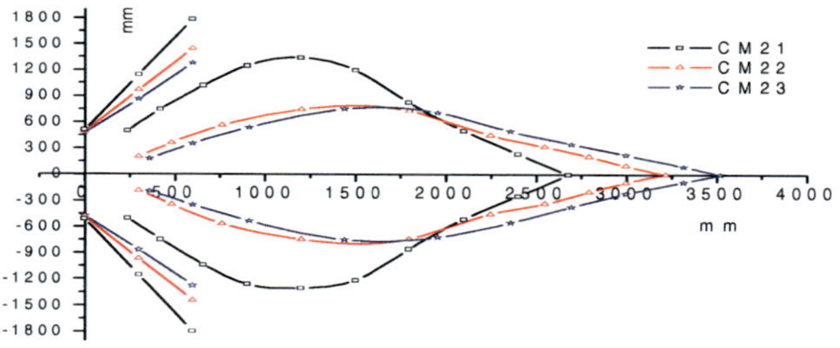

Fig. 4. The impact to the export flow field of C3 burner inner secondary air change (1)

By analyzing 6 experiment conditions show that ,when the secondary air swirl strength was adjusted to 20%, the inner secondary air damper to maintain 60%, outside the secondary air baffle to maintain 100% (condition CM25) condition, the secondary air dispersion between the 90~ 105°. Center recirculation region is the right size, Without gas brushing wall phenomenon; The length of the central recirculation zone is right size, the maximum axial velocity gradual decayed to 0 about 2.2D (the design burner primary air mouth), The recirculation zone roots from the burner nozzle about 0.25D, The burner secondary air rotary relatively strong to benefit staged combustion, That is the best theoretical conditions to prevent slag[9-11].

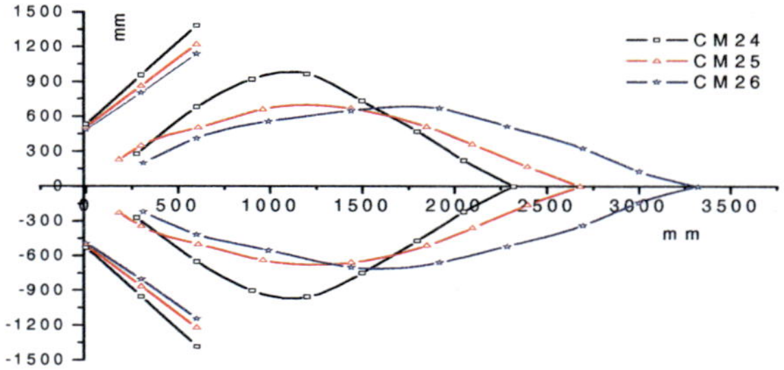

Fig. 5. The impact to the export flow field of C3 burner inner secondary air change (2)

Reduced inner secondary air appropriately under the same total amount of the secondary air, while appropriate increased inner secondary air swirl strength, Can reduced extension angle, can avoid the outside recirculation zone and the phenomenon of flue gas to sweep wall, Increase recirculation zone distance from the nozzle, Increasing the central air can also prevent slagging.

2) Analysis of without Centre air burner inner secondary air Changing.

C3 burner is selected as the research object, primary air is maintained as a velocity of 29.5m/s, The experiment is divided into 6 working conditions to measure data, experiment conditions show in Table 4, the burner nozzle jets show in figure 3 and figure5.

Table 3. A3 burner inner secondary air experiment Data Sheet

condition	*CM31*	*CM32*	*CM33*	*CM34*	*CM35*	*CM36*
Primary air velocity	29	29	29	29	29	43
inner secondary air swirl strength/m/s	19/38	19/38	19/38	25/32	19/36	19/38
Inner / outside second air velocity	50%	100%	0	50%	0	50%

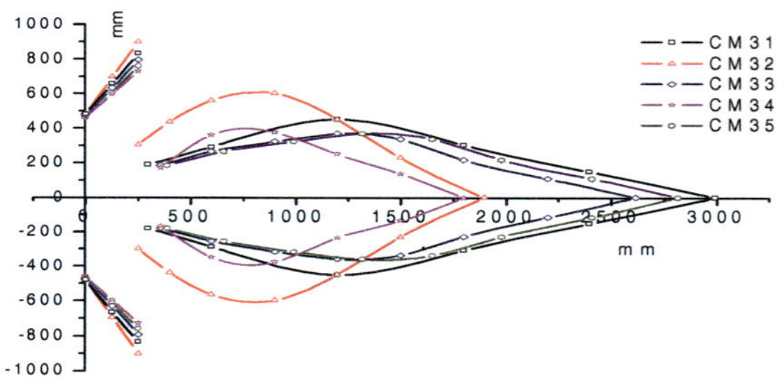

Fig. 6. The impact to the export flow field of A3 burner inner secondary air changing

By analyzing 6 experiment dates of flow conditions show that, Burner Air jets without central air is different with the conventional burner. The secondary air swirl strength to the minimum, the secondary air to keep 70% open cases (condition CM33), secondary air diffusion angle is 95～100 °; centers recirculation region is the right size, without the phenomenon of air brush wall; Maximum axial velocity gradual decay to 0 about 2.2D distance, the roots of the recirculation zone from the burner nozzle has a distance about 0.25D; The burner secondary air rotary relatively strong. Axial velocity and tangential velocity benefit staged combustion, This case is the best theoretical condition to prevent the burner nozzle slagging[9-11]; increase the external central air in condition CM35, Increasing central air can increase the recirculation root distance from the burner nozzle , increasing the maximum axial velocity attenuation distance. The results showed that, Reference the method to central air burner adjustment, When the secondary air swirl strength be adjusted small, appropriate to reduce the inner secondary air Increased outside secondary air, This can avoid the outside recirculation zone and the phenomenon of flue gas to sweep wall, Increased recirculation zone distance from the nozzle, Increasing the central air can also preventive role slagging.

Results show that: LNASB burner inner secondary air rotational intensity and amount of air impact strength to the jet, with the volume and the swirl strength increases, the recirculation zone is essentially the same shape, maximum width increases, the length decrease; when the inner secondary air swirl strength increases to a certain extent, Its impact on the size of recirculation zone was significantly enhanced. The amount of inner secondary air and rotational intensity is subjected to external restrictions on the outside secondary air swirl strength, When the secondary air flow is reduced to 0 and the secondary air swirl strength minimum, The recirculation zone also is obvious.

3.3 Analysis of Combustion System Deviation Primary Air and Coal Powder

3) Numerical measurements results of the primary air.

The primary air pipe air and coal flow fineness powder was determined. Coal fineness (R_{90}) measurements show in Figure 6, primary air and powder flow measurement results shown in Figure 7. Curve can be seen in the figure, there is a big deviation of primary air velocity and pulverized coal fineness. According to the experiment results in certain conditions (primary air as 25m / s), by adjusting the coal pipe and mill shrinkage folding doors to the tailgate, Measure air and coal flow fineness powder by the back pipe with installing in the coal by pipeline, and adjust coal pipe air velocity and pulverized coal fineness in the same condition.

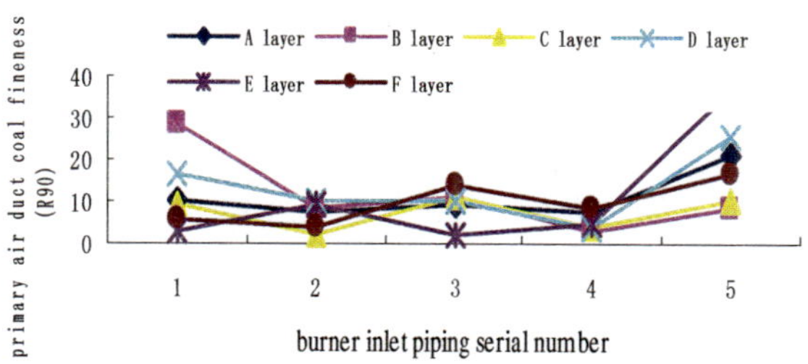

Fig. 7. Burner primary air pipe coal fineness deviation

4) Bias results of the burners primary air and coal powder.

The boiler combustion system was examined in cold air cases. There is a large wearing capacity deviation of different burner primary hairdryer, there is a local serious wear and tear area, Its location lie in the burner primary mouth along the spiral rotation to 270°~300°. In the pulverized coal duct 300mm at the front of the collector, the primary air duct was divided into 8 points and averaged wear volume equidistant along the circumference, the results of wearing bias show in Figure 8.

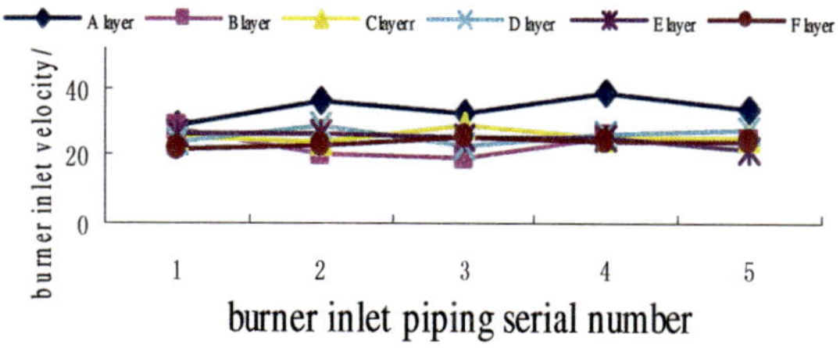

Fig. 8. Burner primary air pipe air and coal velocity deviation

There is a big air and coal uniformity coefficient in the same burner primary duct. It shows that there is a big air and powder deviation about the single burner. At the same time there is a big air and coal uniformity coefficient in the different burners primary duct. Although this data can not show the air and powder proportional uneven, But still show that there is a big air and powder deviation in the difference burners.

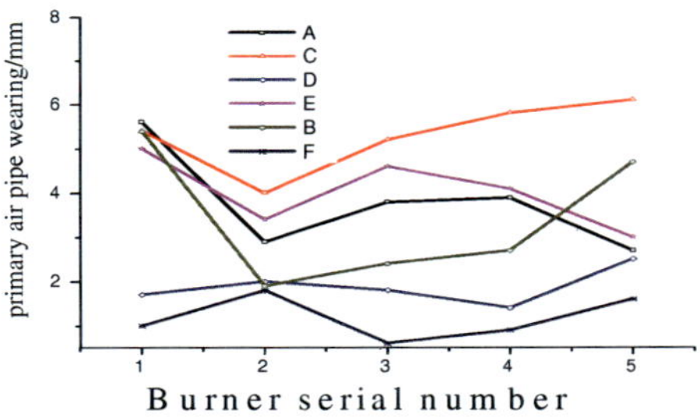

Fig. 9. Primary air pipe wear comparison of different burners

4 Experiment Results and Analysis

The burner air and powder measuring position must be actual response the real working conditions, and to be adjusted leveling, otherwise easily lead to high heat load of the burner, cause nozzle slag. The same layer burner uneven distribution of pulverized coal .Easily lead to single burner overload; Single burner pulverized coal uneven distribution along the circumferential direction, the single burner will be caused local high load, both easy cause burner slag.

LNASB burner inner secondary air flow and swirl impacted strength on the jet, with the inner secondary air volume and the swirl strength increases, the recirculation zone is essentially the same shape, maximum width increases, the length has decreased slow. When the secondary air swirl strength and the air increase to a certain extent, its impact on the size of recirculation zone was significantly enhanced. LNASB inner secondary air swirl strength and air flow rate on the export to be the limited with outside secondary air, when the inner secondary air flow is reduced to 0 and swirl strength is minimum. The recirculation zone is also very obvious. LNASB central air burner reasonable adjustment can increase the flame distance from the nozzle to reduce the vent slagging tendency should not be ignored.

References

1. Gu, M., Zhang, M., Fan, W., et al.: The effect ofthe mixing characters of primary and secondary air on NOx formationin a swirling pulverized coal flame. Fuel 84(8), 2093–2101 (2005)
2. Zhang, J., Sun, R., Wu, S., et al.: An experimental and numerical study on swirling combustion process in a 200MWpulverized coal fird boiler. Proceedings of the CSEE 23(8), 215–220 (2003)
3. Vander Lans, R.P., Glarborg, P., Dam-Johansen, K.: Influence of process parameters on nitrogen oxide formation in pulverized coal burners. Prog. Energy Combust. 23(3), 349–377 (1997)
4. Changfu, Zhou, Y.: Effect of operation parameters on theslagging near swirl coal burner throat. Energy Fuels 20(11), 1855–1861 (2006)

5. Chen, Y., Yan, W., Shi, H.: Research on furnaceslagging of 330MW boiler of dalate power station. Proceedings of the CSEE 25(11), 79–84 (2005)
6. Anacleto, P.M., Fernandes, E.C., Heitor, M.V., et al.: Swirl flow structureand flame characteristics in a model lean premixed combustor. Combustion Science and Technology 175(8), 1369–1388 (2003)
7. Pu, Y., Zhang, J., Zhou, L.X.: Measurement of velocity fields for turbulent combustion in a swirl combustor. Chinese J. Theoretical and Applied Mechanics 35(3), 341–347 (2003)

Study on Fractal-Like Dissociation Kinetic of Methane Hydrate and Environment Effect

Xu Feng[1,2,*], Wu Qiang[2], and Zhu Lihua[2]

[1] Key Lab. for the Exploitation of Southwestern Resources & the Environmental Disaster Control Engineering, Ministry of Education, Chongqing University, Chongqing, China
[2] School of Safety Engineering & Technology, Heilongjiang Institute of Science & Technology, Harbin, China
xufeng79_79@163.com

Abstract. By using observation method, the dissociation process of methane hydrate at normal temperature and pressure is studied, and it is discovered which the dissociation of methane hydrate isn't occurred synchronously at whole surface, but heterogeneous characteristic of dissociation is emerged. The dissociation rate of hydrate is slow at normal temperature and pressure, and average dissociation rate is $2.15\times10^{-4}m^3\cdot h^{-1}$. The dissociation kinetic equation of methane hydrate is proposed. It is revealed that the dissociation of methane hydrate at normal pressure is a pseudo first-order reaction. The dissociation rate coefficient of methane hydrate in different time is calculated, and the result shows that the rate coefficient is relative to reaction time, namely the dissociation kinetics of methane hydrate does not tally with the typical kinetics law, and the dissociation reaction kinetics of methane hydrate is fractal-like. Lastly, the environment effent of hydrate dissociation is also analyzed.

Keywords: methane hydrate, dissociation, fractal-like, environment.

1 Introduction

Gas hydrates are crystalline, non-stoichiometric, clathrate compounds. They are formed by certain gases when contacted with water under favourable temperature and pressure conditions. Gas components like CH_4, C_2H_6, C_3H_8, i-C_4H_{10} and n-C_4H_{10} besides other gases like N_2, CO_2 and H_2S form hydrates[1,2]. The gas hydrate with a methane molecule percent more than 99% is generally called methane hydrate. By some estimates, the energy locked up in methane hydrate deposits is more than twice the global reserves of all conventional gas, oil, and coal deposits combined. But no one has yet figured out how to pull out the gas inexpensively, and no one knows how much is actually recoverable. Because methane is also a greenhouse gas, release of

* Visiting Scholar of of Key Lab. for the Exploitation of Southwestern Resources & the Environmental Disaster Control Engineering in Chongqing University, Ministry of Education, China.

D. Zeng (Ed.): Advances in Control and Communication, LNEE 137, pp. 155–161.
springerlink.com

even a small percentage of total deposits could have a serious effect on Earth's atmosphere. There have been a series of work reported on the dissociation kinetics of methane hydrate[3-5]. However, little research has been done on the fractal-like dissociation of methane hydrate. Gas hydrate's surface is heterogeneous, and dissociation process of hydrate belongs to heterogeneous phase reaction. So, this heterogeneous characteristic of gas hydrate can be described by using fractal geometry theory. Since 1975, when Mandelbrot introduced the term "fractal", the theory and application of fractal geometry have been developed very quickly. The application of fractal geometry has great influence on natural science[6]. Kinoshita[7] studied the adsorption of Sr^{2+} in the activated charcoal particles volume, and obtained following equation:

$$c_{out} \propto t^{-\alpha} \tag{1}$$

where, c_{out} is concentration of effluent solution, t is adsorption time, and α is a constant. The α is correlated with the constant $1/n$ in Freundlich formula:

$$\alpha = \frac{1}{1-\frac{1}{n}} \tag{2}$$

The relationship between $1/n$ and fractal dimension D is shown in Eq.(3) [8]:

$$\frac{1}{n} = D - 2 \tag{3}$$

and thus,

$$c_{out} \propto t^{-\frac{1}{3-D}} \tag{4}$$

Xu et al[9] studied the adsorption characteristics of rock in Pb-Zn mine of Yunnan to Pb^{2+}, and found that Eq(1) is adapted to describe the adsorption kinetics of rock to Pb^{2+}, that is, the adsorption kinetics is fractal-like. Wang et al[10] researched the adsorption process of dye compounds onto granular activated charcoal, and the result showed that the adsorption process has fractal-like kinetic characteristic, and there exists exponential relation between the rate coefficient k and reaction time t, and the effective reaction order relating to fractal dimension. The research of Liu et al showed that the dissolution process of rock salt is also fractal-like[11]. Thus it can be seen that the heterogeneous phase reaction kinetics does not tally with the classical kinetics law, and rate coefficient k is relative to reactive time. The rate constant or order of reaction is related to spectral dimension. Kinetics of reaction is fractal-like.

Based on the above analysis, we discuss the fractal-like kinetics characteristics of methane hydrate dissociation at normal pressure and the environment effent of hydrate dissociation. This paper will provide a new method for research on hydrate dissociation.

2 Observing the Dissociation Behavior of Methane Hydrate

In order to study the kinetic behavior of methane hydrate dissociation, the surface modality of methane hydrate in the dissociation process at normal temperature and pressure is investigated by method of observation. It is observed by naked eyes that the gas bubbles are produced on the surface of the hydrate when the dissociation reaction just start; subsequently, increase and assemble on the surface of the hydrate solid. The dissociation of hydrate is not occurred synchronously at whole surface of the hydrate, and heterogeneous characteristic of dissociation is emerged. That is, the dissociation rate is fast for some areas, and is slow or none for other areas. The nonuniform kinetic behavior can be caused by the different gas storage density on each parts of solid hydrate. The results of molecular dynamics simulation illustrat that high cavity occupancy is propitious to the stability of the hydrate crystal structure[12,13].

The methane from dissociation releases into the atmosphere through the holes of hydrate, and the methane release rate is nonuniform in whole dissociation process. The dissociation rate of methane hydrate is slow at normal temperature and pressure, for example, the methane hydrate with 150.7cm^3 synthesized is decomposed at 19℃ and 1.05×10^{5}Pa after divided into many parts, and the average dissociation rate, which is calculated based on the time of complete dissociation and the volume of hydrate, is 2.15×10^{-4}m^3·h^{-1}.

3 Fractal-Like Kinetic of Methane Hydrate Dissociation

3.1 Kinetic Equation for Methane Hydrate Dissociation

Kim et al.[14] studied the dissociation kinetic of methane hydrate by semi-batch stirred tank reactor, and they indicated that the dissociation rate equation of methane hydrate can be described with Eq.(5).

$$\frac{dn_H}{dt} = k_d A_s (f_e - f) \tag{5}$$

where, n_H is the amount of residual hydrate(mol), k_d is the dissociation rate constant(mol·m^{-2}·MPa^{-1}·min^{-1}), A_s is the surface area of a particle(m^2), f_e is the three phase equilibrium fugacity of the gas(Pa), f is the fugacity of the gas(Pa), $f_e - f$ is the dissociation driving force is defined as the fugacity difference.

Tian et al.[15]obtained the kinetic model of methane hydrate dissociation at normal pressure based on Eq.(5):

$$-dn_H / dt = kn_H \tag{6}$$

where, n_H is the amount of residual hydrate(mol), k is the apparent dissociation rate constant(min^{-1}). We know that the dissociation of hydrate is not occurred synchronously at whole surface of the hydrate by the above analysis. So, it is too ideal that the kinetic model of methane hydrate dissociation at normal pressure is given with Eq.(6). In addition, the n_H in Eq.(6) isn't a quantity which is measured by experiment. We think that it is more equal to describe the dissociation kinetic equation with the amount of releasing methane which can be measured by experiment. Thus the relationship between volume of dissociation gas and dissociation time of hydrate can be expressed by,

$$V = V_0[1 - \exp(-kt)]^n \tag{7}$$

where, n is a experiential constant, k is rate constant ($\min^{-1}$), V is the volume of dissociation gas, V_0 is the limit volume.

The data, which were listed in literature[15], are fitted with Eq(7), and the regression results are listed in Table1.

Table 1. Regression parameters of Eq.(7) under different temperature

Temperature [℃]	0	2	4	6
V_0[L]	61.281	61.188	58.470	61.464
K [$\min^{-1}$]	0.0237	0.0385	0.0561	0.0584
n	0.809	0.862	0.901	0.804
r	0.9975	0.9997	0.9997	0.9994

It is thus obvious that the reliability of relative coefficient in various temperature exceeds 99%. It indicates that Eq.(7) is adapted to describe the quantitative relationship between the volume of dissociation gas and time. The dissociation rate is expressed by the amount of releasing methane in unit time. Thus, Eq.(7) can be transformed into following form:

$$\frac{dV}{dt} = knV\left[\left(\frac{V_0}{V}\right)^{\frac{1}{n}} - 1\right] \tag{8}$$

As seen in Table1, values of n approach to 1. If n=1, Eq.(8)can be rewritten into:

$$\frac{dV}{dt} = k(V_0 - V) \tag{9}$$

Both Eq.(6) and Eq.(9) are kinetic equation for methane hydrate dissociation under ideal state, and their form is also similar. Thus, it can be proved indirectly that Eq.(6) and Eq.(9) are logical as kinetic models of methane hydrate dissociation at normal pressure.

The relationship between apparent dissociation rate constant k in Eq.(8) and temperature can be expressed by Arrhenius equation,

$$k = A\exp\left(-\frac{\Delta E}{RT}\right) \tag{10}$$

or

$$\ln k = \ln A - \frac{\Delta E}{RT} \tag{11}$$

where, A is the preexponential factor ($min^{-1}\cdot MPa^{-1}$). Based on the apparent dissociation rate constants at different temperature, ln$A$ and ΔE are calculated by Eq.(11), and they are lnA = 39.38 and ΔE = 97.66 $kJ\cdot mol^{-1}$.

3.2 Fractal-Like Kinetic Characteristics

For classical kinetics, the rate constant k is independent of time. However, the rate constant depends on the time in most heterogeneous phase reaction, and it can be described by Eq.(12)[16-18]

$$k = k_1 t^{-h} \tag{12}$$

where, k_1 is the constant, in agreement with the classical kinetics result, h is a constant that measures the degree of local heterogeneity.

The dissociation reaction of methane hydrate belongs to heterogeneous phase chemical reaction, and the kinetics of most heterogeneous phase reaction does not tally with the typical kinetics law. So, the dissociation kinetic of methane hydrate is fractal-like. Eq.(8) and Eq.(9) are experiential formulas that describe the quantificational relation between the amount of methane from hydrate dissociation and dissociation time. Eq.(9) is regard as a pseudo first-order reaction. Thus, the rate constants k (or rate coefficients) of the methane hydrate dissociation reaction, which was described in literature[15], are calculated according to the equation of first-order kinetic. The results show that these k values might increase or decrease as temperature increases.

In order to explore the quantitative relation between rate coefficient and time, the data calculated are regressed by using Eq.(12), and the results are listed in Table2.

Table 2. Regression results

t [℃]	0	2	4	6
h	2.0381	2.3841	2.1808	2.3949
k_1 [min^{-1}]	10.118	16.568	7.0964	8.2035
r	0.96081	0.9476	0.9811	0.95675

As seen in Table2, the relative coefficients in various temperature are very high. It indicates that Eq.(12) is adapted to describe the relation between the dissociation rate coefficient of methane hydrate and time. It means that dissociation kinetics of methane hydrate at normal pressure is fractal-like. Since h is larger than that in literature[18], the spectral dimension can't be obtained. But it is affirmative that the process of methane hydrate dissociation at normal pressure is fractal-like.

4 Environment Effect of Methane Hydrate Dissociation at Normal Pressure

Gas hydrates are huge potential green energy resource. The total amount of gas hydratereaches $0.84\times10^{18}m^3$ in global. The dissociation reaction of hydrate at normal pressure generally occurs in the process of utilizing gas hydrate. Once gas hydrates decompose largely, the trapped methane is released, greenhouse effect is occurred, which damage the people surrounding climate. Our experiment shows that the dissociation rate of hydrate is slow, which clearly verified the environment security of gas hydrate utilization.

5 Conclusions

(1) The dissociation of methane hydrate isn't occurred synchronously at whole surface, but heterogeneous characteristic of dissociation is emerged.

(2) The dissociation reaction of methane hydrate belongs to heterogeneous phase chemical reaction, and the dissociation kinetic of methane hydrate is fractal-like.

(3) The dissociation rate of hydrate is slow, so the utilization of gas hydrate has environment security.

Acknowledgment. This work was supported by Visiting Scholar Foundation of Key Lab. for the Exploitation of Southwestern Resources & the Environmental Disaster Control Engineering in Chongqing University, Ministry of Education, China.

References

1. Zhong Min, H., Hai Hua, W., Hai Yan, W.: Oil-Gas Field Surface Engineering 20, 21–25 (2004) (in Chinese)
2. Dendy, S.E.: Clathrate hydrate of natural gases. Marcel Dekker Inc., New York (1998)
3. Chang Yu, S., Guang Jin, C., Tian Min, G., et al.: Journal of Chemical Industry and Engineering(China) 53, 899–903 (in Chinese)
4. Ying Mei, W., Qing Bai, W., Peng, Z., et al.: Natural Gas Geoscience 2, 245–248 (2009) (in Chinese)
5. Xiao Xia, H., Jin Song, Y., Ying Hai, M., et al.: Natural Gas Geoscience 16, 818–821 (2005) (in Chinese)
6. Long Jun, X., Dai Jun, Z., Xue Fu, X.: Coal Conversion 18, 31 (1995) (in Chinese)
7. Kinoshita, M., Harada, M., Sato, Y., et al.: AICHE J. 43, 2187–2193 (1997)
8. Long Jun, X., Le Guan, G., Xue Fu, X.: Coal Conversion 23, 91 (2000) (in Chinese)
9. Long Jun, X., Xue Fu, X., Chuan Long, M.: Journal of Chongqing University(Natural Science Edition) 23, 60 (2000) (in Chinese)
10. Yi Li, W., Fu Ling, Y., Dong Sheng, W.: Acta Scientiae Circumstantiae 25, 643–649 (2005) (in Chinese)
11. Cheng Lun, L., Long Jun, X., Xue Fu, X.: Colloids and Surfaces A:Physicochemical and Engineering Aspects 201, 231–235 (2002)
12. Chun Yu, G., Li Ying, D., Qing Zhen, H., et al.: Acta Phys.-Chim. Sin. 24, 595–600 (2008) (in Chinese)

13. Li Ying, D., Chun Yu, G., Yue Hong, Z., et al.: Computers and Applied Chemisny 24, 569–574 (2007) (in Chinese)
14. Kim, H.C., Bishnoi, P.R., Heidemann, R.A.: Chem. Eng. Sci. 42, 1645–1653 (1987)
15. Long, T., Shuan Shi, F., Wen Feng, H.: Journal of Wuhan University of Technology 28, 23–26 (2006) (in Chinese)
16. Klymko, P.W., Koplamn, R.: J. Phys. Chem. 87, 4565 (1983)
17. Koplamn, R.: J. Stat. Phys. 42, 185 (1986)
18. Koplamn, R.: Science 241, 1620 (1988)

Repair Geometric Model by the Healing Method

Di Chi[1], Wang Weibo[2], Wen Lishu[3], and Liu Zhaozheng[4]

[1] Colledge of Mechanical Engineering, Northeast Dianli University, Jilin, China
[2] School of Accounting, Zhejiang Gongshang University, Hangzhou, China
[3] General Education Department Neusoft Institute of Information, Dalin, China
[4] Jilin Petrochemical Company, Jilin, China
di.yichi@qq.com, {wangweibo,liuzhaozheng}@sohu.com, wenlishu@neusoft.edu.cn

Abstract. Repairing the panel CAD geometric model is an important pre-processing step of the finite element analysis, and widely used in the computer-aided geometric design field. This paper focused on the surface healing method of panel CAD model, which was proposed and analyzed by NURBS technology. Through the pipeline method determines the matching-curves border fast firstly, make the curves discrete points sets and on the basis of the characteristics of the matching-curves method, the surfaces were stitched. On the basis of the algorithm, a surface stitching program module was developed. The validity of algorithm was verified by healing automobile front fender parts.

Keywords: Auto panels, Geometric model, Surface healing, pipeline method.

1 Introduction

The geometry of auto body parts CAD model is very complex [1], thus, in its design and format conversion often have some data errors, the more common, such as the lack of the complete topology information of a CAD model which led to adjacent entities in the public entities may have different functions expressed, so that solids discontinuity showed in the graphical display; Another example is the current existence of CAD model data storage has many formats such as IGES, STEP, SAT, etc., during the file format conversion the model reconstruction would occur errors or missing information. The errors of Model data will result in subsequent data processing in difficulties, such as the automatic finite element mesh generation, process design and other difficult to mold design, therefore, repair the panel's CAD model is very important.

There are two ways to repair CAD models by now which are mesh-based data recovery method [2, 3] and geometric model-based data recovery method[4-8]. The former method generate the finite element mesh of panel CAD model firstly, followed by the grid identification and node merging approach to eliminate the errors in the grid data, at last the legitimate mesh model generated. The main purpose of this

D. Zeng (Ed.): Advances in Control and Communication, LNEE 137, pp. 163–169.
springerlink.com

restoration is to generate finite element model consistent with the requirements of FE, less to CAD model repairing. The latter method aimed to CAD models directly, repairing the geometric error or making up topology missing, such as: Chang Su presents a method through the adhesive initial trimmed parametric surface to generate an "unleaked" grid model[7]; Barequet used local curve matching technology to search the defects of CAD model and made up the gaps between surfaces by optimized triangle patches[5]. Currently the main problem of this class method is cumbersome and complex judgment of the match entity before repair the CAD model, which is the main problem this article settled.

This paper studied a healing method to make up the gaps between surface patches based on CAD models and the general organization of the paper is as follows: Section 2 proposes the match curve search method; Section 3 presents the hilling method based curve-merged technology.

2 Surface Healing Based on the Geometric Model

The main purpose of surface healing is eliminate error information of geometric model to generate consistent finite element model as well as up build the complete topology information of model. According to the characteristics of CAD model, the geometric-based healing method is generally as that: firstly, we find the adjacent surface information of each surface by the features of it; then to discrete each surface boundary curve, by estimate the spacing of discrete points we can determine the match curve which would be merged later, and then confirm the topological relations of the match curves according to the relative discrete points; finally, we merge the discrete points of the match curves to regenerate the public curve to eliminate the gaps between surfaces. The process shown in Figure 1.

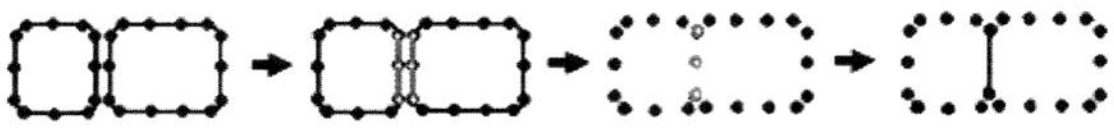

Fig. 1. Traditional healing method

Generally, the deficiencies of the stitching method is that there would be a great of calculation to process the distance of the number of discrete points to judge whether the given curves are match and the topological relations based on the given tolerance. This method is easy to deal with simple models, but for the auto panels which constituted by the thousands of surfaces, curves and points, the efficiency of method is not ideal. In addition, because of the complexity of panel models, the problems that whether the discrete points of boundary curves of adjacent surface can corresponding to each other and the given tolerance can satisfy the margin requirements of all match curves judgment would be difficult to solve. All the questions are inevitably affected the overall efficiency of the classic algorithm, increasing the possibility of error generated.

By comprehensive consideration, if we can confirm the match curve directly by the geometric algorithm, and then combine them by discrete points which shown in Fig.2, the overall algorithm is easier to handle, and efficiency can be raised.

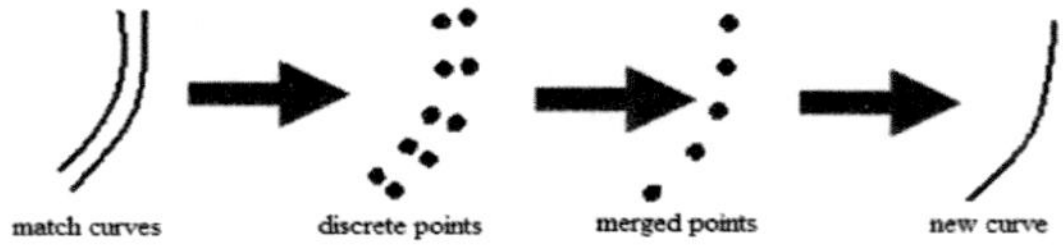

Fig. 2. Healing method based on the geometric method

3 Searching Match Curve

Searching the match curves is a key step in the process of surface stitching because that if we confirm the match curves, that is the location of the gap between the surface is determined. Judging from geometric characteristics, the necessary and sufficient conditions of two curves matching are: 1) the two lines belonging to two adjacent surfaces; 2) the two lines may be in separating, overlapping or intersecting state; 3) the coedge of two curves is in the opposite direction, and at least uniform of the part and minimum distance between them. According to these three conditions, we can determine whether the two curves match.

According to condition 1, we first determine the adjacent surfaces information of the given surface so that the match curve of each boundary curve of surface can be found in the adjacent surfaces, this treatment method can effectively reduce the searching range and improve the searching efficiency. By the classical bounding box algorithm the adjacent surfaces is easy to confirm.

According to condition 2 and 3, the problem for searching match curve can be expressed as solving the problem: l_1 is a given curve, which belongs to surface S_0, adjacent surfaces collection of S_0 is $\{S_i \mid 1 \le i \le n\}$, boundary curves collection of S_i is $\{l_{i,j} \mid 1 \le j \le m\}$, to solve a and b satisfied that $l_{a,b}$ is the match curve of l_1.

In this paper, "pipeline method" is used to solve the problem above, as shown in Figure 3, the algorithm works as follows: First, it's known the B-spline expression from reading the geometric model data as that:

$$p_1(u) = \sum_{i=0}^{n} d_i N_{i,k}(u) \tag{1}$$

where d_i is control vertices, $N_{i,k}(u)$ is B-spline basis functions and $u \in [0,1]$. Then make the start point of l_1 as the sweep circle center, radius as ε and l_1 as the sweep line to generate pipeline surface s_1 which can express as:

$d(x, p_1(u)) = \varepsilon$ ($x \in S^i_{si}$, $u \in [0,1]$) ;

$$\text{It's } \begin{cases} (x - p_1(u)) \cdot p_1'(u) = 0 \\ |x - p_1(u)| = \varepsilon \end{cases} ;(u \in [0,1]) \tag{2}$$

Simplify 2, we get s_1 :

$$s_1(u,v) = p_1(u) + \varepsilon[\cos v \cdot N_1(u) + \sin v \cdot B_1(u)] \quad (u \in [0,1],\ v \in [0,2\pi]) \tag{3}$$

$N_1(u) = \dfrac{\dot{p}_1 \otimes \ddot{p}_1}{|\dot{p}_1 \otimes \ddot{p}_1|}$, $B_1(u) = \dfrac{\dot{p}_1 \otimes \ddot{p}_1}{|\dot{p}_1 \otimes \ddot{p}_1|} \otimes \dfrac{\dot{p}_1}{|\dot{p}_1|}$ are the principal normal vector and binormal vector of $p_1(u)$ respectively.

Do curve $l_{i,j}$ the same treatment, we can get the equation of $s_{i,j}(u,v)$:

$$s_{i,j}(u,v) = p_{i,j}(u) + \varepsilon[\cos v \cdot N_{i,j}(u) + \sin v \cdot B_{i,j}(u)] \tag{4}$$

Simultaneous equation 3 and 4, that is by substitution of parametric surface $s_1(x(u,v), y(u,v), z(u,v))$ into 4, all that is needed to judge whether there has solution. If there has solution, l_1 is the match curve of $l_{i,j}$, let $a = i, b = j$ to solve the problem; if there has not solution, let $j = j+1$, if $j = m$, then let $i = i + 1$, cycle process above until problem solved.

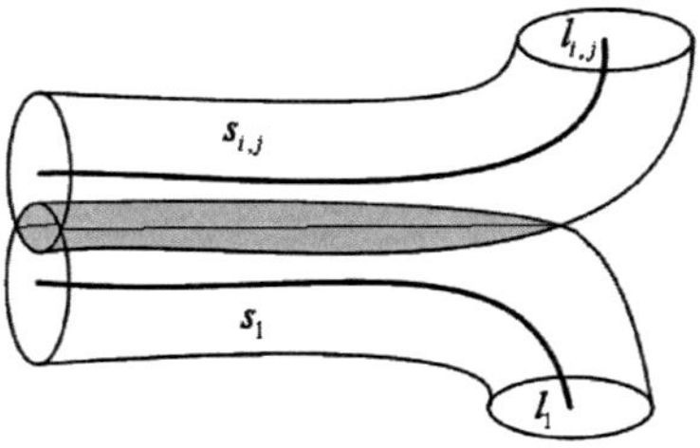

Fig. 3. The pipeline method to judge the matching curves

In summary, the match curve searching algorithm is as follows:

Step 1. Read the surface information of geometric model.

Step 2. If the surface information is correct, the loop direction is counter-clockwise in accordance with the right-hand rule, turn out every boundary curve in this direction and give it the globally unique ID value.

Step3. Use the bounding box algorithm to determine the adjacent surfaces of each surface and save it in the vector.

Step4. Make two cylindrical surfaces of the given boundary curve and its maybe match curve, and then determine whether they intersecting. If the equation has solution, the given curve and its match curve are sure, then breaks the current loop, mark and saves them, do the next curve; If there is no solution, then do the same deal to the rest curves of the adjacent surface, repeat Step 4, until all the curves of the current surface are determined the matching curves; if loop termination, but there has no match curve be found which indicates that the curve is the boundary curve of geometric model.

Step5. Repeat steps 4, until each curve of CAD model has been confirmed its matching.

In general, most of the curves and its matching of the geometric model are full-match, shown in Figure 4, that is the two roughly equal length and curve shape is similar to doing so before the column surface intersection, be the first to judge the first two curves present end point is less than the tolerance distance a, in order to effectively reduce the search complexity of the algorithm. Another match the curve of each curve may not be unique, but at this point still only use circle string algorithm to judge once.

4 Merge with the Match Curve

If two curves are match, the relationship of which is shown in Figure 4, where the top curve is in the current processing, the following is its match curve. In practice, most of the two curves would be stitched are full-match as shown in Figure 4 (a), the others are part-match, which are very complicated, but it is noticed that the part-matched curves always can come down to “T” type topology, which is shown in Figure 5, then we can map the endpoint of the processing curve(denote by P) to the other curve(denote by M), the mapping point also called match point, interrupt the curve M on the point and merge the full-match part and P as one curve, while the remaining part of M does not match the curve P can re-merger with other match, thus we complete the surface stitch.

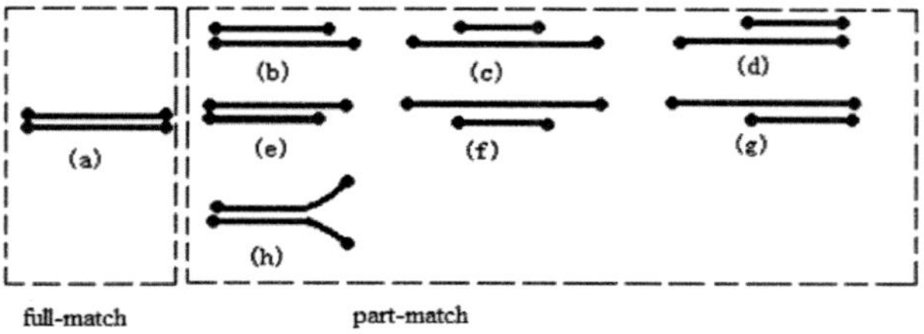

Fig. 4. The topological relationship of matchings

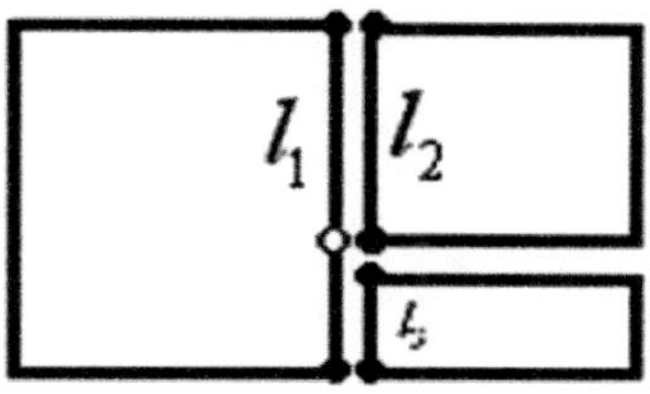

Fig. 5. “T” type relationship

5 Some Examples

To verify the effectiveness of the algorithm, we draw box-model shown in Figure 6 (a), in which the longest side is 75.599 mm, the smallest edge is 32.786 mm, the maximum surface gap is 3.518 mm, therefore, take tolerance parameter as 4.00 mm, after surface stitched, geometric model shown in Figure 6 (a) (model shown using KMAS system platform). In order to fully demonstrate the algorithm effect, we also take a typical body panels-fender part as shown in Figure 7 to stitch, for better analysis of the results, the calculated data listed in Table 1.

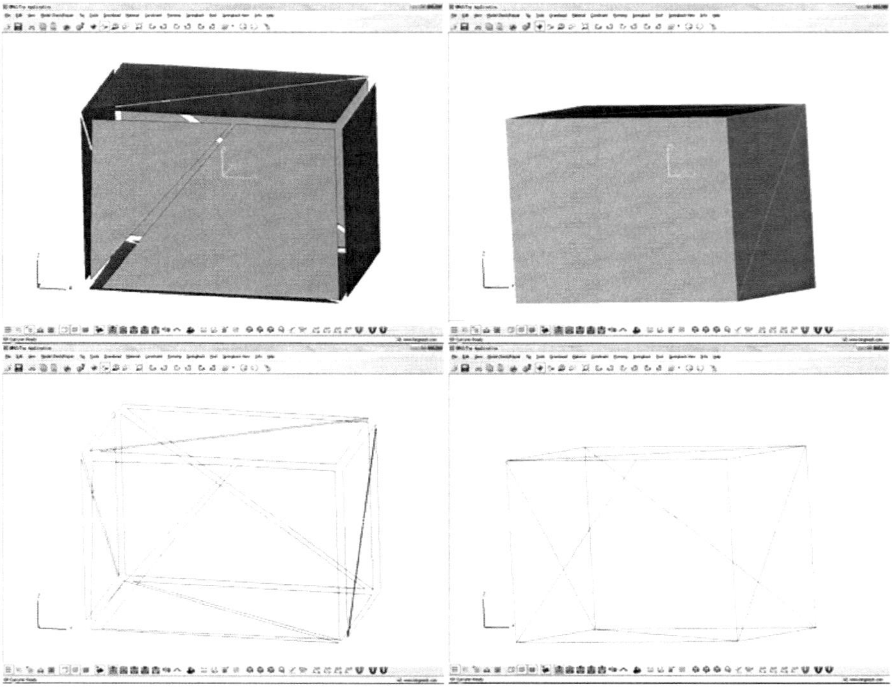

a. Pre-healing b. After healing

Fig. 6. The surface healing example of cube model

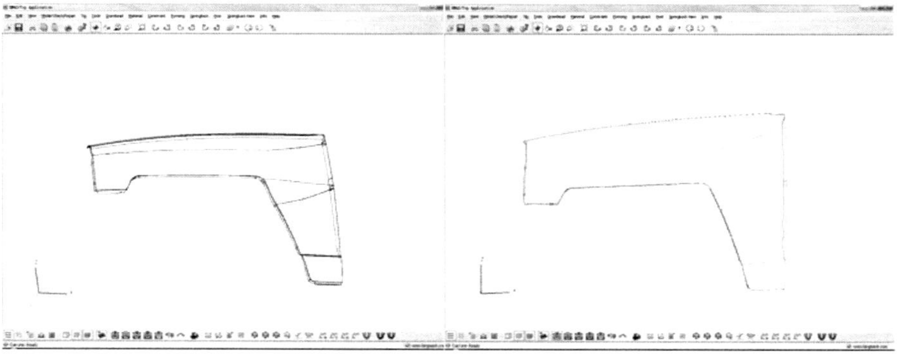

a. Pre-healing b. After healing

Fig. 7. The surface healing example of fender model

Table 1. Data of Healing Process

Model	**Process Data**					
	Data Format	***Process Time***	***Status***	***Surface Num***	***Curve Num***	***Vertex Num***
Box	IGES	5.7 s	Pre	12	37	37
			After	12	18	9
Fender	IGES	21.4s	Pre	97	356	287
			After	97	317	225

6 Conclusion

The most common data error of auto panel geometrical model is that there often have gaps between surface patches for various reasons. This has hampered the finite element model generation and die face design. This paper discusses a surface stitch method based on the CAD model, proposed a "pipeline method" to judge the match curves quickly and complete the surface stitch by feature-based method to merge the match curves. It's developed the surface stitching program modules based on the algorithm, through the auto front fender part processing, the algorithm is effective and robust.

Acknowledgment. The research is funded by the Doctor Scientific Research Foundation of Northeast Dianli University(BSJXM-200906), and we use software KMAS^{TM} to provide technical supports.

References

1. Gong, K., Hu, P.: Method for generating additional surface in die face design of automotive panel. Journal of Jilin University(Engineering and Technology Edition) 31(2), 63–66 (2006)
2. Mihailo, R., Djordje, B., Surya, H.: CAD-based triangulation of unordered data using trimmed NURBS models. Journal of Materials Processing Technology 107(1), 60–70 (2000)
3. Volpinf, O., Sheffert, A., Bercovier, M., et al.: Mesh simplification with smooth surface reconstruction. Computer-Aided Design 30(11), 875–882 (1998)
4. Ribo, R., Bugeda, G., Onate, E.: Some algorithms to correct a geometry in order to create a finite element mesh. Computers and Structures 80(16), 1399–1408 (2002)
5. Barequet, G., Kumar, S.: Repairing CAD Models. In: Proc. IEEE Visualization 1997, pp. 363–370 (1997)
6. Gill, B., Micha, S.: Filling gaps in the boundary of a polyhedron. Computer Aided Gemetric Design 12(2), 207–229 (1995)
7. Zhang, S., Shi, F.: Repair and Stitch in Multiple Trimmed Free Surfaces. Journal of Computer Aided Design & Computer Graphics 17(4), 699–703 (2005)
8. Barequet, G.: Using geometric hashing to repair CAD objects. IEEE Journal of Computational Science and Engineering 4(4), 22–27 (1997)

Fracture Characteristics of Archean-Strata in Jiyang Depression and the Meaning for Oil-Gas Reservoir*

Li Shoujun[1,2], He Miao[1,2,**], Yuan Liyuan[1,2], Yin Tiantao[1,2], Jia Qiang[3], Zhao Xiuli[2], and Jin Aiwen[2]

[1] Shandong Provincial Key Laboratory of Depositional Mineralization & Sedimentary Minerals,
Shandong University of Science and Technology,
Qingdao, China
[2] College of Geological Science & Engineering,
Shandong University of Science and Technology,
Qingdao, China
[3] College of Geo-Resources & Information,
China University of Petroleum (Huadong),
Qingdao, China
hemiaogeology@qq.com

Abstract. We study the fracture characteristics of archean-strata based on the outcrops, drilling, well test and log data in Jiyang Depression and Western Shandong. Through field work, core sample survey, literature review, laboratory study and experimental observation, We finally discuss the main factors that control the fracture development in the study area——lithology, tectonic stress, and overlying strata. We study the tectonic stress through numerical modeling in detail in this paper, which is a comparatively new research field.

Keywords: Fracture Characteristics, Archean-Strata, Jiyang Depression, lithology, tectonic stress, overlying strata.

1 Introduction

The archean-strata in Jiyang Depression is Taishan Group which is correlated with the same strata in Western Shandong [1-3]. The characters are complex, and the seismic data is disordered, so the degree of natural gas exploration is lower. With deepening development of hydrocarbon exploration, deep layers have become key target and popular domain for hydrocarbon exploration[4]. So systematical research on fracture characteristics of Archean-Strata−−one of the basic geological problems−− is seriously needed.

* This study was supported by the Science & Technology Leading Project of SINOPEC (S0-PT-1091D010Z) and the Postgraduate Science &Technology Innovation Foundation of SDUST(YCA100203).
** Corresponding author.

D. Zeng (Ed.): Advances in Control and Communication, LNEE 137, pp. 171–177.
springerlink.com

2 Results and Disscussion

2.1 The Development Characteristic of the Outcrops-Fractures

1) Fracture groups: We select eight pionts among the measuring data of Taishan Group in the outcrop area. Then 499 fracture direction were statistically analyzed(Fig.1, Fig.2). Fractures are well developped on the strikes of SSE, at the dip angle of 220°~230°and 50°~60°. It is agree with that the orientation of the fault——SSE. Then followed by that Perpendicular to fault strike. The remainder is diagonal to fault trend. Many groups intersect With each other, then formed the fractures grid system. That make the fracture reservoir a connection.

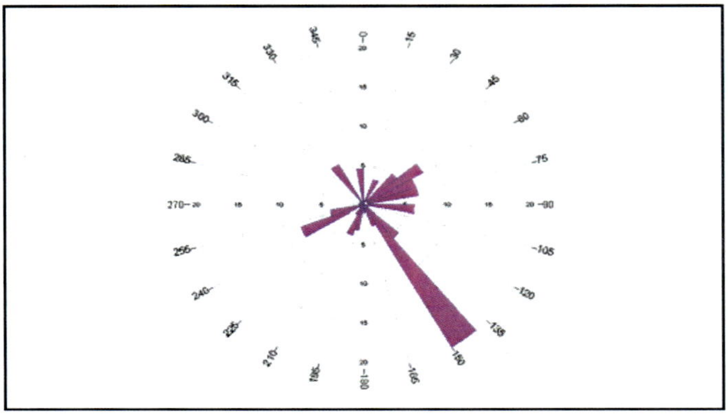

Fig. 1. Rose diagrams of Taishan-Group-fracture direction in outcrop area

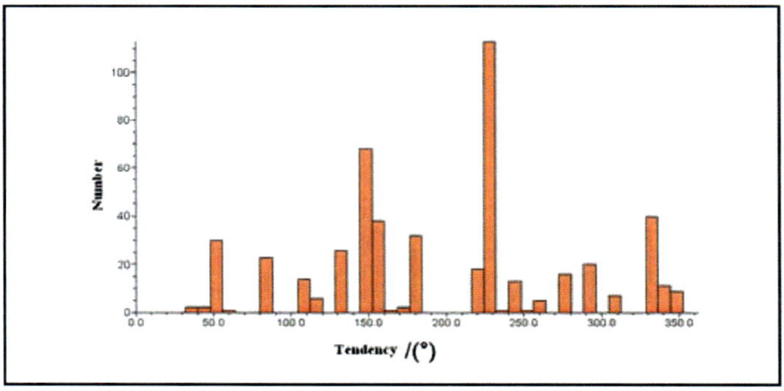

Fig. 2. Distribution of Taishan-Group-fracture tendency in outcrop area

2) Fracture dip: The development of fracture dip is mainly concentrated on the 40°~90°, accounting for 90.38% of the total dips of statistics. At 58°~82° is particularly concentrated (Fig.3), accounting for 76.15% of the total dips of statistics.

reservoir drilling bases on vertical wells in High-angle fractures, drilling data revealed the extent of fracture development may be much lower than the actual development of underground fracture system.

3) Fracture width: Micro-fractures develop well has the higher density. There are more than 600 micro-fractures in straight-line distance of 1m(Fig.4).

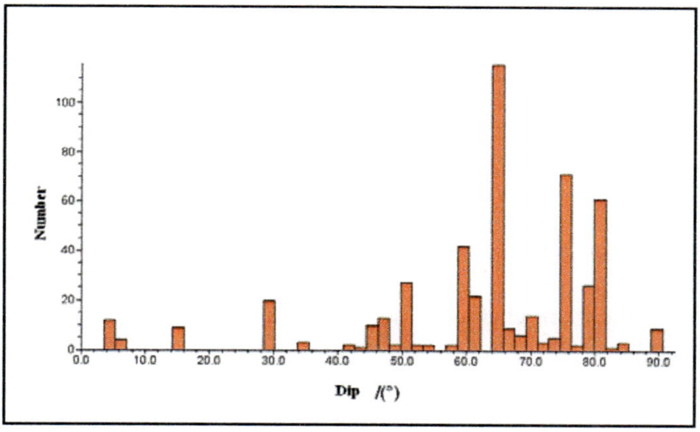

Fig. 3. Distribution of Taishan-Group-fracture dip in outcrop area

Fig. 4. Distribution of fracture dip of Taishan Group Field points

2.2 Influencing Factors on Fracture Development

According to the study on 15 Archean core fracture in Jiyang Depression, In the foundation of the predecessor's study[5], we study the main factors that control the fracture development in the study area are following:

4) Lithology: According to the archean fracture development statistical data in the Chenggu-Zhuangxi region, we can obtain the relationships between fracture development and lithology——Correlation with the development-degree of fractures be from high to low, the lithology is in turn: cataclastic rock, moyite, adamellite, gneissic granite, gneissic diorite, mylonite, lamprophyre rock, diorite, biotite granulite, and plagioclase amphibolite.

5) Tectonic stress

The degree of the fragmentation depends on characteristic of the stress, depth and length of action time. If stress action is too weak, it would not have sufficient fractures. If too hard, it would break rock into pieces, the mineral into powder, and even the recrystallization which blocks the fracture; only when the stress intensity is moderate, the rock coincided fragmentation, and very few pieces of grain. And then fractures would become the favored access for oil migration and the favorable accumulation environment for oil-gas. Dongying depression is the important part of Bohai Bay congruent basin.it experiences multistage evolution include the Paleozoic, Mesozoic and Caenozoic, tectonic Movements of the tectonic cycles which provide the power for metamorphism of Archaeozoic Qianshan Internal Fragmentation, and form rich fractures.

According to the main characteristics of the study area, we selected the ANSYS software to simulate and analyse the two-dimensional stress field of metamorphic rocks'geostatic stress in Chengdao – Zhuanghai region.

This simulation of two-dimensional stress field mainly takes Chengdao - Zhuanghai regional strata as the research object. The EW direction is X axis, and NS direction is Y axis. This model includes 24 faults in work area, we make them into the north-east and south-east direction according to their directions, and the secondary and tertiary faults according to fracture sizes.

According to the nature of the study area, this Modeling selects the free-form-meshing, bases on triangulation point unit, and focuses on dividing the fault zone in the bottom of target layer), the key point and so well by human intervention. Rock mechanics parameters used in the modeling process are showed in Tab.1.

Table 1. Rock mechanics parameters used in the modeling process

Number	Media type	Elastic modulus	Poisson ratio
1	Second-order fault	20	0.18
2	Third-order fault	30	0.2
3	Granitic gneiss	40	0.25

The main force in the study area is horizontal compression and principal stress is nearly east -west direction and south-north direction. Respectively, principal stress of east -west direction is about 120Mpa, principal stress of south-north direction is about 70Mpa. The southern boundary of work area was constrainted, no displacement and rotation.

Established tectonic stress model in Chengdao - Zhuanghai region of Jiyang Depression (Fig.5), according to the figure, it shows that the range of the stress distribution mainly concentrated near the well point, and two north-south stress banded high values area; the maximum stress in work area is located in the Northwest region of Chengbei-303 wells, and some edge and peripheral location of the partial fault. The distribution range of stress concentration has its limitations and therefore it is the fracture development center.

Fracture developed degree is controlled by the tectonic stress field for structural fractures in the rock layers which have the same lithological characters, structural fractures are relatively developed in the tectonic stress field with the high value area. Statistics of unit-thickness liquid-producing capacity of some wells in Chengdao region and tectonic stress analysis show that the larger is simulate principal stress in the region, the relatively higher is unit-thickness liquid-producing capacity. Liquid-producing capacity of Archean wells depends mainly on the development of layer fractures. Hence the higher is the tectonic stress fields, relatively the better developed the fractures, so these regions become the favorable accumulation environment for oil-gas.

Fault is often the rupture zone, the Archean fracture development zone of the North China has been studied by the predecessors who think that the two sides are closely related. There grows many faults in the study area, It is no doubt that there are the fracture zones near these fault zones. Control "Mountain" faults play a leading role on the fracture, and all levels of the Archean faults work for fracture development.

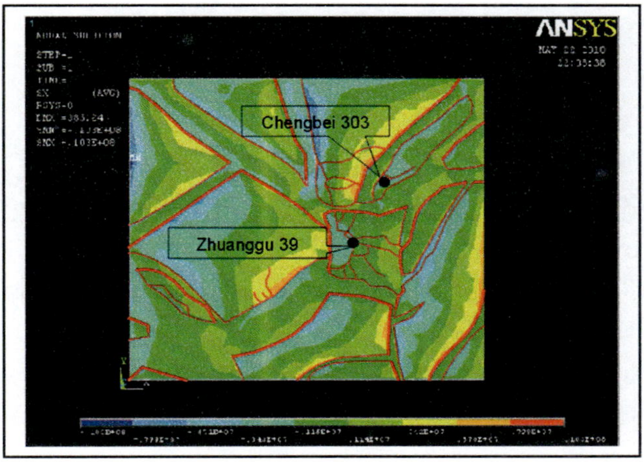

Fig. 5. Tectonic stress analysis on Archaean strata in Chengdao-Zhuanghai region

The study shows that fractures mainly developed near the faults,. Fig.6-A is serpentinized hornblende schist in the 1 core of the Chenggu-19 well (1760m-1762m), the mirror of the faults indicate the faults are presence, Fig.6-B shows that the core is seriously damaged.

Granite and gneiss are brittle highly, so the areas where the tectonic activities are intense easily develops fractures(Fig.6-C). Such as serpentinized hornblende schist and so on , are also easy to produce fractures (Fig.6-B).

6) Overlying strata

When the overlying strata were developped in Paleozoic, it is carbonaterock. After recrystallization of fractures infilling, the fractures were made to be bigger and well opened(Fig.6-D). When the overlying strata are clastic rock developed in Mesozoic and Cenozoic. The top surface of Archeozoic strate weathered seriously, fractures and microfractures developped well. They are filled with argillaceous infill and dripping(Fig.6-E). The Archaean reservoir rocks are silicates formed in high temperature and mixing melting condition, so the rocks lack the pores. The main reservoir spaces are pore and fracture reformed from the early fracture —structural fracture(Fig.6-F). The reformed procession is mainly caused by alteration of unstable melanocratic mineral and dissolution of mineral such as quartz.

A B C

D E F

A: The faults mirror of the Serpentinized amphibole serpentinized hornblende schist.

B: Cataclasite.

C: Hole and calcite.

D: Fractures in hornblende granitic gneiss and with gas.

E: Filling fractures.

F: Unfilling fractures.

Fig. 6. Dveloppment of fractures

3 Conclusions

Lithology is the basic factors that affected fracture development. The fractures develop better in cataclastic rock than in Plagioclase amphibolite.

Tectonic stress plays a dominate role. Where the structural stress is high with other same condition, there always develop fractures densely. The closer to the fault, the better Fractures develop. While the farther away from the fault, the fractures become fewer and fewer. So the fields with higher tectonic stress, always are the favorable accumulation environment for hydrocarbon.

When the overlying strata are carbonaterock developped in Paleozoic and clastic rock developped in Mesozoic and Cenozoic, the overlying strata are in unconformable contact with the Taishan Group. Fractures are filling with carbonaterock and clastic rock (mudstone). The Pacific fault zones develop fractures in this area, and have the certain phenomenons of corrosion or air slaking, with good reservoir properties.

Acknowledgment. This study was supported by the Science & Technology Leading Project of SINOPEC (S0-PT-1091D010Z) and the Postgraduate Science & Technology Innovation Foundation of SDUST(YCA100203).

We thank the Geophysical Research Institute in Shengli Oil-field of Sinopec for the provision of seismic data.

References

1. Zhang, Z., Liu, M., et al.: Rock Strata in Shandong Province. China University of Geosciences Publishing House, Wuhan (1996)
2. Bureau of Geology and Mineral Resources of Shandong Province. Regional Geology of Shandong Province. Geological Publishing House, Beijing (1991)
3. Wang, S.: The Stratigraphic Division of the Taishan Group in Western Shandong and the Characteristics of its Protoliths. Geological Bulletin of China 9(2), 140–146 (1992)
4. Hu, H.: Deep Hydrocarbon Reservoir Formation Mechanism Survey. Petroleum Geology and Oilfield Development in Daqing 6, 24–26 (2006)
5. Wang, R., Jin, Q., Dai, J., Zhang, J.: Distribution Rule and Methods of Evaluation of Buried-hill Oil-gas Reservoir Space. Geological Publishing House, Beijing (2003)
6. Zhang, J., Jin, Q.: Characteristics of fractures and their hydrocarbon reservoir meanings for the Archean outcrop in Laiwu area, Shandong province. Petroleum Geology and Experiment 25(4), 371–374 (2003)

Influencing of Liquid Film Coverage on Marangoni Condensation

Jun Zhao[1], Bin Dong[1,2], and Shixue Wang[1]

[1] School of Mechanical Engineering, Tianjin University, Tianjin, China
[2] School of Vehicle & Motive Power Engineering, Henan University of Science and Technology, Luoyang, China
Zhaojun@tju.edu.cn, Db8003@yahoo.com.cn

Abstract. In this paper, the liquid film coverage influencing the Marangoni Condensation heat transfer characteristics for steam-ethanol mixture vapor on a vertical surface were investigated. The condensation heat transfer coefficients were calculated according to the two methods respectively. The diffusion resistance of the concentration boundary layer was considered in the first method and was neglected in the second one. Then the calculation values were compared with the experimental results. It shows that the calculation values are higher, close to and lower than the experimental results respectively in the very low, the lower and the higher concentration area when considering the diffusion resistance. It suggests that the diffusion thermal resistance can be neglected in the very low concentration area but must be taken into consideration in the higher concentration area.

Keywords: Marangoni condensation, liquid film coverage, heat transfer coefficient (HTC), surface subcooling, mixture vapor.

1 Introduction

Energy system with medium-low level heat sources, such as geothermal, ocean heat, factory waste heat, have been widely used in energy industries including steam power, petrochemical industry and refrigeration. How to improve the efficiency of the mixture vapor condensation heat transfer process is a common problem in heat transfer through phase change of condensers in these energy systems [1]. There has been considerable research on the Marangoni condensation of mixture vapor. The most recent motivation for this has been the development of energy utilizing efficiency in the new thermodynamic and refrigeration cycles.

During surface condensation of vapor mixtures of a positive system [2], such as steam-ethanol system, where the surface tension of the high-boiling point component is larger than that of the low-boiling point component, the liquid-phase surface tension difference is formed between the gas-liquid interfaces. This resulted in the uneven distribution of thickness of the condensation liquid film, and finally led to Marangoni condensation which is similar to dropwise condensation [3]. The

D. Zeng (Ed.): Advances in Control and Communication, LNEE 137, pp. 179–186.
springerlink.com

characters of Marangoni condensation heat transfer had been observed by Markovich[4], Fujii[5], Utaka[6] and so on. Utaka and his co-workers [6] integrated the relationships of the condensation HTC changing along with the degree of subcooling (the temperature differences between the mixture vapor and the condensation surface) at the condensation surface into characteristic curves. As illustrated in Fig.1, the degree of subcooling at the condensation surface has great effect to Marangoni condensation heat transfer coefficients. The characteristic points B, C and D on the curve stand for the beginning soaring point of HTC, ending soaring point of HTC and maximum condensation HTC point respectively.

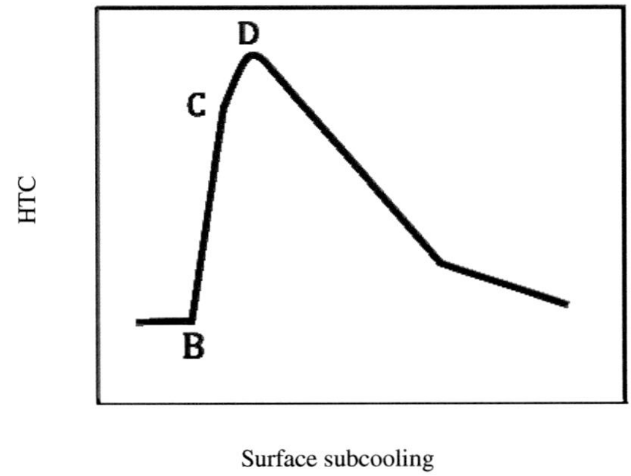

Fig. 1. Heat transfer characteristic curve of Marangoni condensation

In this paper, the condensation heat transfer coefficients were obtained by using the liquid film coverage combined with the liquid film conductive heat resistance and concentration boundary layer diffusion resistance. Then comparisons of the calculated values and the experimental results were made. Further, analyses on the characters of Marangoni condensation heat transfer process were made.

2 Experimental and Analytical Methods

The experimental results in this paper were obtained within a rectangular heat transfer surface with the size of 10mm×20mm, under the working condition which the ethanol mass concentration (*C*) is 0.02, 0.07, 0.20, 0.43 respectively. The vapor pressure is 0.1 Mpa and the vapor velocity is 0.4m/s.

The physical model of the mixed vapor filmwise condensation is showed in Fig.2. With assumptions that the process is one-dimension steady state heat conduction, the

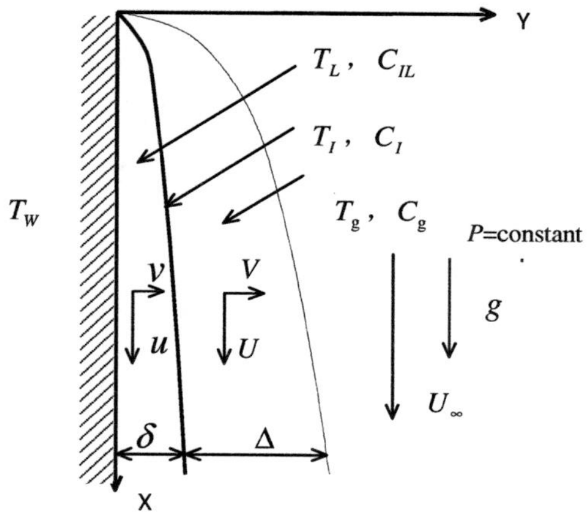

Fig. 2. Model of mixed vapor filmwise condensation

conductive resistance of liquid film and diffusion resistance of concentration boundary layer are the main resistance, the overall resistance R is as follows:

$$R = 1 / k = R_1 + R_2 \tag{1}$$

Where k is the HTC, R_1 is the conductive resistance of liquid film and R_2 is the diffusion resistance of the concentration boundary layer.

A. Calculation of the conductive resistance

$$R_1 = (T_\mathrm{I} - T_\mathrm{W}) / q \tag{2}$$

Where q is the heat flux, T_I is the vapor-liquid interface temperature and T_W is wall temperature.

If definite the liquid film coverage as:

x = (the area of heat transfer – the area of the droplet) /the area of heat transfer surface (3)

When ignoring the heat resistance of concentration boundary layer, then

$$R_1 = 1/k = x\delta \ / \lambda \tag{4}$$

Where λ is the liquid film heat conductive coefficient, δ is the film thickness.

B. Calculation of diffusion resistance

Definite the diffusion resistance as:

$$R_2 = (T_\infty - T_\mathrm{I}) / q \tag{5}$$

The partial Nusselt number of vapor side is

$$Nu_{cx} = \frac{\alpha_{cx} x}{\lambda_V} = -\Theta'_{FVI}\, \mathrm{Re}_{Vx}^{1/2} \tag{6}$$

Given U_∞、 T_∞、 C_∞、 T_W, combining with equation (2), (4), (5) and (6) will get R_2. More Details can be found in paper [7].

3 Results and Discussions

3.1 The Relationship between Liquid Film Coverage and Condensation HTC

In the condition of various concentrations, the curves of condensation HTC and liquid film coverage changing along with characteristic points are given respectively in Fig.3 and Fig.4. As showed in the two figures, along with the change of direction B-C-D, the HTC goes up gradually, while the liquid film coverage goes down. It indicates that Marangoni condensation destroys the condensation film and leads to the notable increase of HTC and the decrease of liquid film coverage meanwhile. When the liquid film coverage reaches the minimum value point, the HTC reaches the peak point. In addition, when in the extremely low concentration, the liquid film coverage is very low especially at the point D. The reason is that the concentration boundary layer is difficult to form in the extremely low concentrations. Thus the Marangoni condensation emerged upon the liquid film and the HTC increases greatly. This shows that the condensation heat transfer should be studied as the main aspect when the concentration is extremely low. The results at the point D also indicate that the liquid

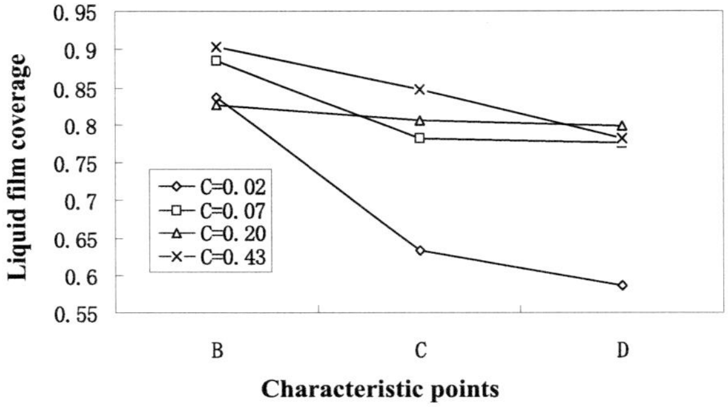

Fig. 3. The condensation heat transfer coefficients variation with the characteristic points in different concentrations

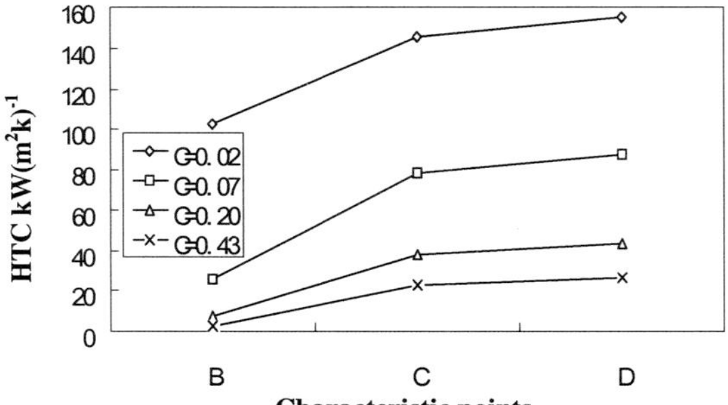

Fig. 4. Liquid film coverage values variation with characteristic points in different concentrations

film coverage of different concentrations are approaching equal under different concentration in high concentration area, while the HTC differs evidently. The main reason is that in different concentrations the ratio of the diffusion resistance and the overall resistance varies greatly. Along with the concentration increasing the diffusion resistance proportion rises rapidly. Though the liquid film coverage is roughly same, the condensation HTC decreases evidently.

3.2 Comparison

Fig.5 gives the comparison of the calculated values and the experimental results of condensation HTC in different concentrations. The figure gives the method to figure out condensation HTC using the liquid film coverage and film thickness when ignoring the concentration boundary layer diffusion resistance and mainly considering the liquid film conductive resistance. The result shows that the calculation values are greater than the experimental results, but in the extremely low concentration, they are well-agreed.

The main reason is that the influence of diffusion resistance differs according to the variation of concentrations. When considering the cooperation effect of the liquid film conductive resistance and the concentration boundary layer diffusion resistance, the calculated values are well-agreed with the experimental results. However, they differ a lot by only taking the liquid film conductive resistant into consideration, especially in the condition C=0.07 Fig. 5 (b). But in the condition C=0.02 as show in Fig. 5 (a),

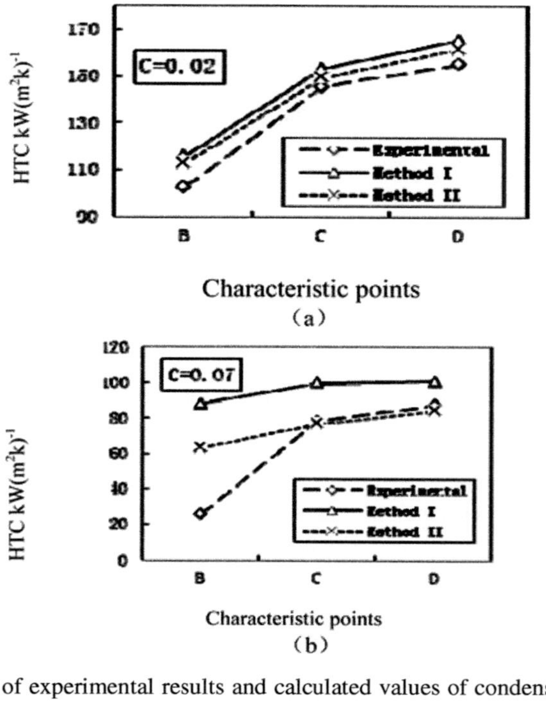

Fig. 5. Comparison of experimental results and calculated values of condensation heat transfer coefficients

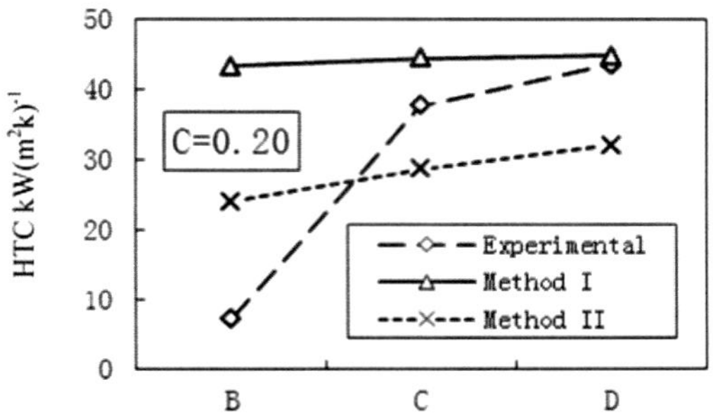

(C)

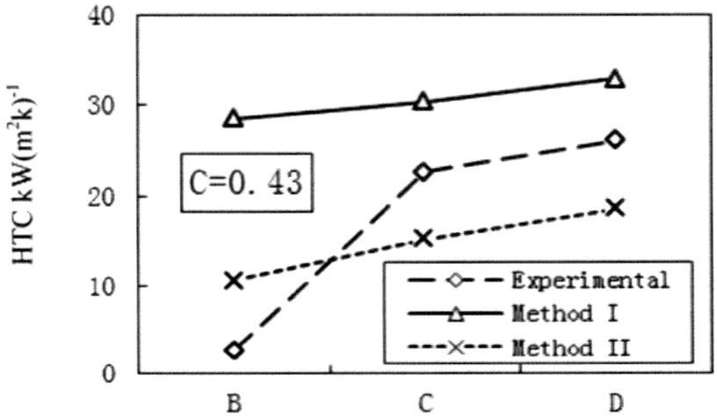

(d)

Fig. 5. *(continued)*

the differences become indistinctive. The reason is the effect of the diffusion resistance is weak when the concentration is low.

Fig.6 illustrates the ratio of diffusion resistance and the conductive resistance changing with the characteristic points. In the condition C=0.02, the effect of concentration boundary layer diffusion resistance is weaker than that of liquid film conductive resistance, so the diffusion resistance could be ignored. But in the conditions C=0.2 and C=0.43, the calculated values obtained by the second method are smaller than the experimental values. Fig.6 also shows that the diffusion resistance is approaching and finally greater than the liquid film conductive resistance in concentration C=0.20 and C=0.43. So the effect of diffusion resistance should be taken into consideration. But the calculated values are lower than the experimental results, the reason is that the disturbance in the concentration boundary layer was ignored. The sweeping effect of liquid drop will lead to the decrease of diffusion resistance. This phenomenon will become even important when the diffusion resistance takes a large proportion. Under the high concentration condition, the sweeping effect of liquid drop on the diffusion resistance will weaken and then the calculated values can be more similar to the experimental results.

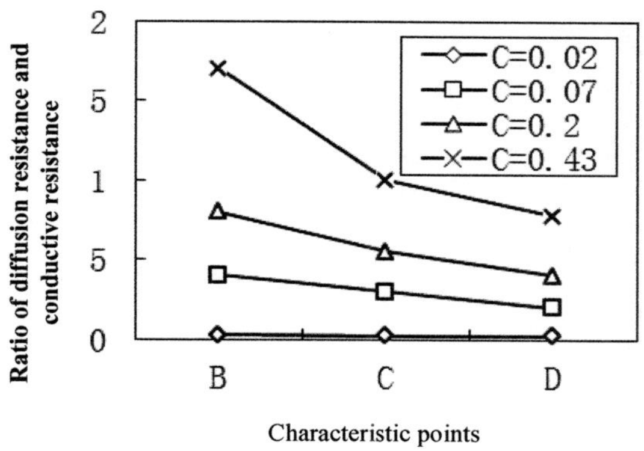

Fig. 6. Ratio of diffusion resistance and the conductive resistance varies with the characteristic points

4 Conclusions

A. At the characteristic points of the condensation heat transfer curve, the condensation HTC is generally contrary to that of liquid film coverage. When the liquid film coverage comes to the minimum, the HTC reaches its peak point.

B. When the concentration is extremely low, the diffusion resistance can be ignored because of its weaken effect to the heat transfer. Meanwhile, the Marangoni condensation HTC is very high, thus, the process should be mainly studied under the condition where the concentrations are extremely low.

C. The effect of diffusion resistance should be taken into consideration when the concentrations are relatively high. When considering the cooperation effects of the liquid film conductive resistance and the concentration boundary layer diffusion resistance, the calculated values are close to the experimental results in the low concentrations especially at c=0.07. But when the concentrations are relatively high, the inaccuracy is great. It's caused by ignoring the weakening effect on the diffusion resistance by the liquid drop sweeping. Therefore, the effect of diffusion resistance and the effect of liquid drop sweeping should be considered at the same time when the concentrations are relatively high.

Acknowledgment. The authors would like to acknowledge the Science and Technology Project of Tianjin, China (Grant No. 08JCYBJC26000). We would like to thank Professor Yoshio Utaka from Yokohama National University for providing the experimental data.

References

1. Dennyve, Jusjonisv, J.: Effects of forced flow and variable properties on binary film condensation. International Journal of Heat and Mass Transfer 15, 2143–2152 (1972)
2. Ford, J., Missen, R.: On the conditions for stability of falling films subject to surface tension disturbance: the condensation of binary vapors. Canadian Journal of Chemistry Engineering 48, 309–312 (1968)

3. Scriven, L., Sternling, C.: The Marangoni effects. Nature 187, 186–188 (1960)
4. Mirkovich, V., Missen, R.: Non-filmwise condensation of binary vapors of miscible liquid. Canadian Journal of Chemistry Engineering 39, 86–87 (1961)
5. Fujii, T., Koyama, S., Shimizu, Y.: Gravity controlled condensation of an ethanol and water mixture on a horizontal tube. Transactions of JSME, Series B 55, 210–215 (1989)
6. Utaka, Y.: Measurement of condensation characteristic curves for binary mixture of steam and ethanol vapor. Heat Transfer-Japanese Research 24, 57–67 (1995)
7. Fujii, T.: Theory of Laminar Film Condensation, pp. 78–93. Springer, Heidelberg (1991)

Study on the Droplet Size Distribution of Marangoni Condensation

Jun Zhao[1], Bin Dong[1,2], and Shixue Wang[1]

[1] School of Mechanical Engineering, Tianjin University, Tianjin, China
[2] School of Vehicle & Motive Power Engineering, Henan University of Science and Technology, Luoyang, China
zhaojun@tju.edu.cn, db8003@yahoo.com.cn

Abstract. In this paper, the droplet size distribution of steam-ethanol mixture Marangoni condensation on a vertical surface was investigated. Based on the experimental images, the droplet radius under nine experimental conditions was measured and then the droplet sieve accumulative weight fractions were acquired. The regression curve was fitted out from the scatterplot chart of the droplet sieve accumulative weight distribution under different working conditions. It is proved that the droplet size is mostly in the range from 0 to 0.5m and the droplet size distribution shows a good agreement with the Rosin-Rammler model. In a fixed mass fraction, along with the increase of surface subcooling, the droplet sieve accumulative weight fraction of one certain droplet radius increased greatly.

Keywords: Marangoni condensation, steam-ethanol, mixture vapor, dropletsize distribution.

1 Introduction

There has been considerable research on the condensation heat transfer of mixture vapor. The most recent motivation for this is to improve the energy utilizing efficiency of heat exchangers in the energy systems. In most cases the focus has been on the diffusion process in the vapor phase which results in the so-called mass-transfer resistance and diminution of the heat transfer. The vapor phase convection-with-diffusion process in forced and free convection of binary mixtures is now well understood [1]. For certain binary mixtures, a mode of condensation whose appearance resembles that of dropwise condensation of a pure vapor on a hydrophobic surface, has been observed [2-3]. Most notably it has been found that, notwithstanding the vapor phase diffusion resistance and with quite small velocity (0.4 m/s), vapor side heat transfer coefficient enhancement up to around 8 times can be obtained by adding very small amounts (0.5% or less) of ethanol to the boiler feed water [3].

So-called Marangoni condensation may occur when the more volatile constituent has the smaller surface tension e.g. a steam–ethanol mixture. When Marangoni

D. Zeng (Ed.): Advances in Control and Communication, LNEE 137, pp. 187–194.
springerlink.com

condensation occurs the effect of composition on surface tension presumably outweighs the effect of temperature and the pressure gradient resulting from change of interface curvature. Irregular modes of condensate of uneven thickness may appear which result in the "pseudo-dropwise".

The description of droplet size distribution (DSD) is the key to simulating Marangoni condensation heat transfer. Gose[4] and Tanasawa[5] have attempted to model the growth and coalescence for the entire range of droplet size on a fixed area. Glicksman and Hunt [6] divide the dropwise condensation cycle in to a number stages, starting with values of nucleation site density up to $10^8/cm^2$. Although many research works on droplet size distribution had been conducted early and a general droplet size distribution function is obtained, the DSD on Marangoni condensation is less well understood till now. DSD of Marangoni condensation has become the main obstacle to the direct numerical simulation of Marangoni condensation heat transfer. It is the aim of the present work to put forward the describing of the DSD.

In this paper, the DSD of steam-ethanol mixture Marangoni condensation on a vertical surface was studied. Based on the experiment images, using statistical analysis method and mathematics distribution function, the DSD character was discussed.

2 Experimental Apparatus

A copper heat transfer block devised specifically for investigating phenomena with large heat flux and high-heat transfer coefficients was used. The condensing surface had an area of 10mm×20 mm.

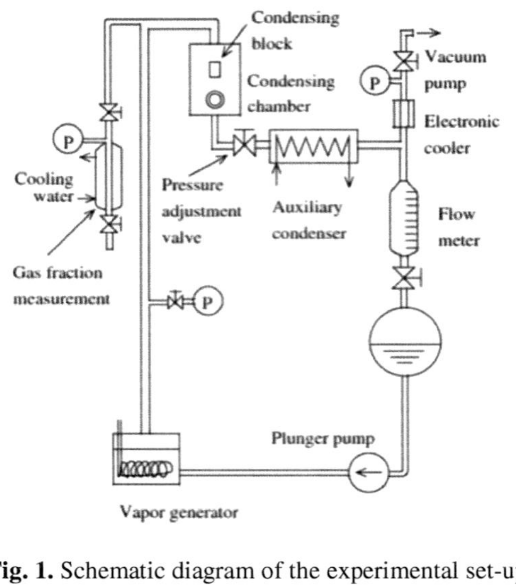

Fig. 1. Schematic diagram of the experimental set-up

Fig.1 shows the schematic diagram of the experimental apparatus. After passing through the condensing chamber in which the heat transfer block is placed, the vapor generated in the steam generator is condensed almost entirely in the auxiliary condenser. The condensate is returned to the vapor generator by the plunger pump via the flow measurement equipment. The vapor flow is in the same direction in which gravity acts. Noncondensing gas is continuously extracted by the vacuum pump near the outlet of the auxiliary condenser. The inlet of the vacuum pump is cooled by an electronic cooler to maintain a constant concentration in the vapor mixture, by maintaining low-vapor pressure. The loop was divided into a high-pressure part and a low-pressure part bounded by the pressure adjusting valve and the return pump. The vapor pressure of the high-pressure-side is maintained at approximately 1 kPa above atmospheric pressure. The concentration of non-condensing gas in the vapor mixture is measured before and after the experiment. Another heat transfer block for the vapor concentration measurement is attached in the condensing chamber located downstream of the main heat transfer block. After the vapor condition reaches the steady state, the condensation characteristic curves were measured continuously using a quasi-steady measurement in which the temperature of the cooling water was changed very slowly for a fixed concentration and fixed velocity of vapor. The aspect of condensate was observed and recorded through the glass window of the condensing chamber using a CCD camera, and the transition points of the condensate aspect were determined using these photographs.

3 Experimental Results and Discussion

3.1 Condensate Images

During the experimental process, a large number of condensate images acquired by the visualization system to analyze the condensate shape and factors influencing the DSD. The statistics of droplet radius and occupied area in all the working conditions were obtained by direct measurement. The non-circular condensate parts were expressed by equivalent radius. All the Statistics were collected in the working condition of ethanol mass fraction (C = 0.02, 0.07, 0.20), pressure (P = 0.1Mpa) and mixture vapor velocity (U = 0.4m/s). Fig.2 depicts some condensate images obtained from the experiment.

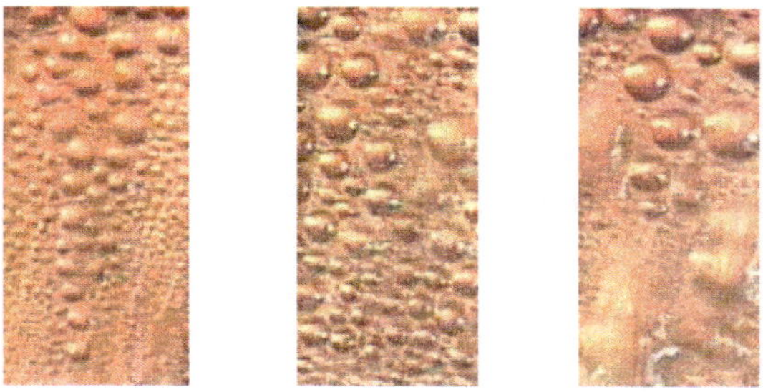

C = 0.02 P = 0.1Mpa U = 0.4m/s

Fig. 2. Schematic of droplet size distribution on different surface subcooling

3.2 Rosin-Rammler Model

Rosin-Rammler model [7]: When droplet size data can be represented closely by a mathematical expression, the maximum of useful information can be revealed. Also, it allows ready graphical representation and offers opportunities for interpolation, extrapolation, and comparison among systems. Various two-parameter mathematical models and expressions have been developed, ranging from the well-established normal and log-normal distributions to the Rosin-Rammler and the Gates-Gaudin-Schumann models. Numerous 3 and 4 parameter models have also been proposed for greater accuracy in describing DSD. But these wide spread applications have been limited due to their greater mathematical complexity. However, the Rosin-Rammler model described by Djamarani and Clark proves relatively suited and efficient. The Rosin-Rammler distribution function (RRDF) is expressed as

$$F(r) = \exp\left[-\left(r/a\right)^b\right] \tag{1}$$

$F(r)$ is the droplet sieve accumulative weight fraction (%), r is the droplet radius (mm), a is the median droplet size (mm), and b is a measure of the spread of droplet radius (mm). The applicability of RRDF can be determined by a curve fitting the actual sieve size data of droplets of a sample. A least square regression analysis can be carried out to fit the data points and the correlation coefficient can be used to estimate the goodness of the fit. From (1), it is clear that when b and a are specified, the only size distribution of the droplets is created. Equation (1) can be varied as a logarithmic form as follows:

$$\ln\left(-\ln F(r)\right) = b \times \ln r - b \times \ln a \tag{2}$$

It can be seen from (2) that the regression will be a line in the $\ln r \sim \ln\{-\ln[F(r)]\}$ coordinates provided that droplet size satisfies RRDF. The exponent b and the median radius a can be specified as the slope and the intercept of the inclined line, respectively. The density function of DSD can be written as the differential coefficients of (1):

$$f(r) = -\frac{b}{a^b} \times r^{b-1} \exp\left[-\left(r/a\right)^b\right] \tag{3}$$

From (3), it is easy to obtain the accumulative weight fraction of the droplets with a radius between two arbitrary representative values r_1, r_2:

$$F(r_2 - r_1) = \int_{r_1}^{r_2} f(r)dr \tag{4}$$

3.3 Droplet Size Distribution

The droplet radius under nine experimental conditions was measured and then the droplet sieve accumulative weight fractions were acquired. We can find that the droplet size is a discontinuous variable and most droplet radius is below 1mm. Appropriate droplet size interval should be large enough to contain a certain number of drop and at the same time the interval shouldn't be too large to ignore the droplet size details. After comprehensive consideration, we select the droplet size interval as 0.05mm. For unity, the droplet size interval above 1mm also takes 0.05mm. Table 1 shows the statistical result under the condition of $C = 0.02$, $\Delta T = 2.17$K.

Table 1. Seive droplet accumulative weight fraction ($C = 0.02$ $\Delta T = 2.17$K)

r mm	F(r)	*r* mm	F(r)
0.05	0.999	0.45	0.688
0.1	0.978	0.5	0.637
0.15	0.933	0.55	0.613
0.2	0.904	0.6	0.584
0.25	0.890	0.65	0.447
0.3	0.855	0.7	0.406
0.35	0.803	0.75	0.350
0.4	0.751	1.4	0

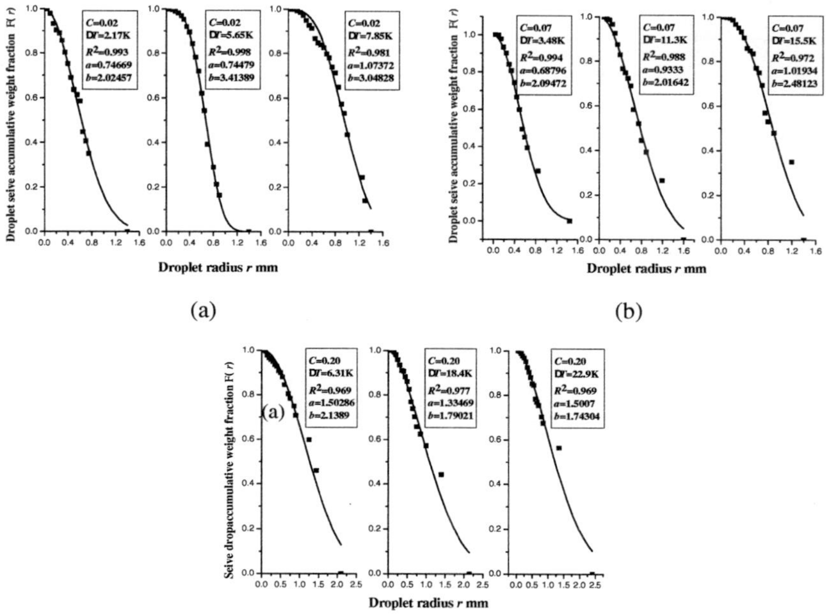

(c)

Fig. 3. The DSD regression curve

According to the Rosin-Rammler distribution function's requirements, from (2) and droplet sieve accumulative weight fraction statistic values, the droplet sieve accumulative weight distribution under different working conditions can be obtained. And then fit out the regression curve as show in Fig. 3. In Fig. 3, the parameter *a*, *b* in (1) and the related coefficient R^2 can also be acquired. As can be seen from Fig. 3, all the working conditions except the condition of C=0.20, ΔT= 6.31℃ and the condition C=0.20, ΔT= 22.9℃(as show in Fig. 3(c)) the data regression are good. It indicates that the DSD shows a good agreement with RRDF. From Fig. 3 it still can be seen that in a fixed mass fraction, along with the increase of subcooling, the droplet sieve accumulative weight fraction of one certain droplet radius increased. It can also be observed in Fig.3 that along with the increase of ethanol mass fraction, the droplet sieve accumulative weight fraction increased.

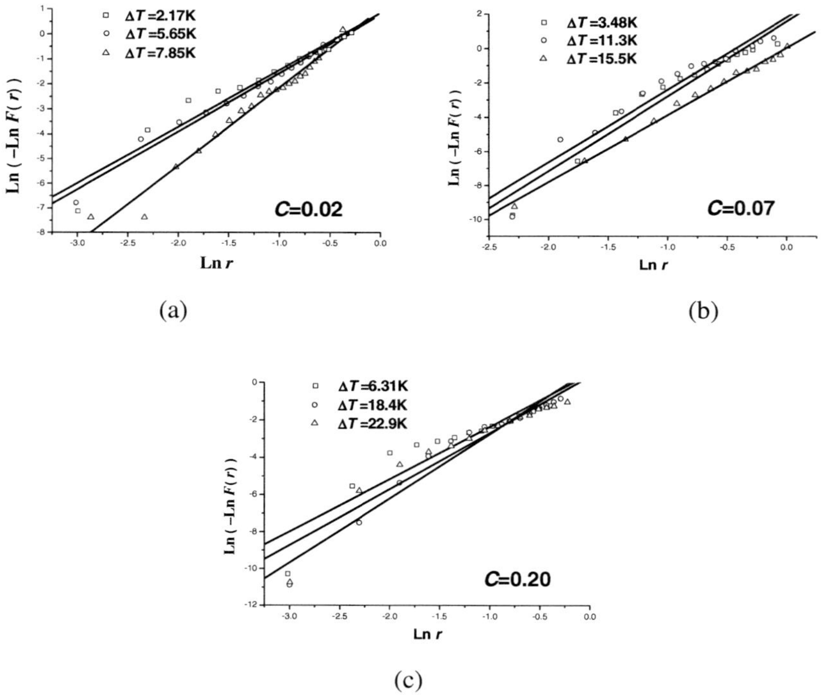

Fig. 4. Linear regression under bi-logarithm coordinate

Fig. 4 is the regression curve of the DSD drew in the bi-logarithm coordinates according to (2). We can see that the regression curve of the DSD is a straight line which proved the DSD matched with the RRDF well. The slope of the regression curve indicates the uniformity of the DSD. It is observed that the maximum slope in Fig. 4 was under the condition of $C = 0.02$, $\Delta T = 7.85$K which means the uniformity of the droplet distribution is the best among the whole working conditions.

Table 2. Related coefficient of regression

Ethanol mass fraction *C*	**Subcooling Δ*T* K**	**Related coefficient *R***
0.02	2.17	0.970
0.02	5.65	0.992
0.02	7.85	0.988
0.07	3.48	0.970
0.07	11.3	0.966
0.07	15.5	0.997
0.20	6.31	0.926
0.20	18.4	0.974
0.20	22.9	0.953

Table 2 shows the relative coefficients of the regression curve under all the experimental working conditions. All the relative coefficients except the condition C=0.20, ΔT= 6.31°C are very close to 1 which again proved the accordance of the DSD of this experiment with the RRDF.

Fig. 5 is the droplet number statistics for each droplet size interval and then the percentage of the total droplet number was acquired. It is found that the droplet size is mostly in the droplet size from 0 to 0.5mm which occupied 80% of the total droplet number. The droplet size outside the range from 0 to 0.5mm is relatively few. It can also be observed that the maximum value of the droplet number percentage is between 0.1 mm and 0.3mm. With the subcooling reduction, the droplet number between 0.1mm and 0.3 mm increased greatly. That is with the increase of subcooling, the relative number of the droplet whose average droplet radius is 0.2 mm increased.

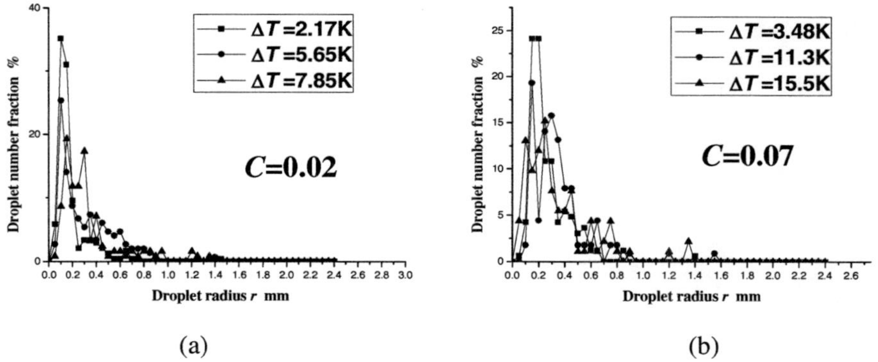

Fig. 5. The droplet radius distribution

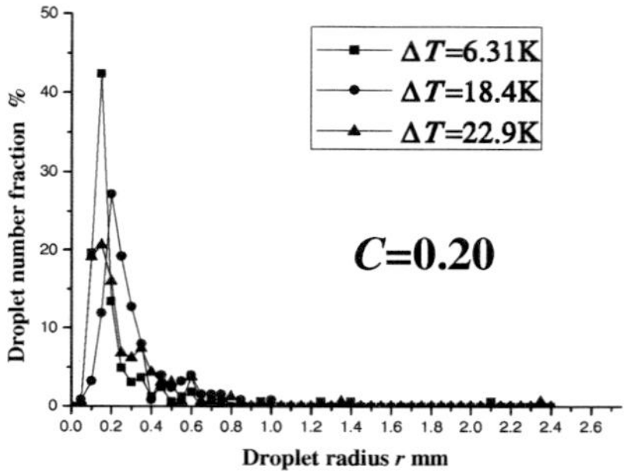

(c)

Fig. 5. (*continued*)

4 Conclusions

The DSD of steam-ethanol mixture Marangoni condensation on a vertical surface was measured through the experimental condensate images. For all the experimental working conditions, the DSD shows a good agreement with the RRDF. In a fixed mass fraction, along with the increase of subcooling, the droplet sieve accumulative weight fraction of one certain droplet radius increased greatly. Through the droplet number analysis, the droplet size is mostly in the range from 0 to 0.5mm.

Acknowledgment. The authors would like to acknowledge the National Major Project of Scientific and Technical Supporting Programs of China during the 11th Five-year Plan Period (Grant No. 2006BAJ03A06). We would like to thank Professor Yoshio Utaka from Yokohama National University for providing the experimental data.

References

1. Fujii, T.: Theory of Laminar Film Condensation, pp. 78–93. Springer, Heidelberg (1991)
2. Mirkovich, V., Missen, R.: Study of condensation of binary vapors of miscible liquids. Canadian Journal of Chemistry Engineering 39, 86–87 (1961)
3. Utaka, Y., Wang, S.: Characteristic curves and promotion effect of ethanol addition on steam condensation heat transfer. International Journal of Heat and Mass Transfer 47, 4507–4516 (2004)
4. Gose, E.E., Mucciardi, A.N., Baer, E.: Model for dropwise condensation on randomly distributed sites. International Journal of Heat and Mass Transfer 10, 15–22 (1967)
5. Tanasawa, I., Tachibana, F.: A synthesis of the total process of dropwise condensation using the method of computer simulation. In: Proceedings of the 4th International Heat Transfer Conference, vol. 6, Paper Cs. 1. 3 (1970)
6. Glicksman, L.R., Hunt, A.W.: Numerical simulation of dropwise condensation. International Journal of Heat and Mass Transfer 15, 2251–2269 (1972)
7. Peng, Z.B., Liang, K.F.: Numerical simulation and experimental study of liquid-liquid jet-flow atomization. Journal of Engineering for Thermal Energy and Power 22, 205–212 (2007)

Research on Market Influence of Wind Power External Economy and Its Compensation Mechanism

Yu Shunkun, Zhou Lisha, and Li Chen

School of Economy and Management, North China Electric Power University, Beijing, China
{ysk21,cressa26,lichenbj}@126.com

Abstract. The market influence of the external economy of wind power is analysed, firstly. The results show that the "market failure" is caused by the external economy of wind power. To compensate the losses caused by the external economy of wind power, the Government established the corresponding compensation mechanism, mainly including three types of promotion policies. Then the compensation mechanism of the external economy of wind power has been researched and compared. It showed that three kinds of promotion policies have their own advantages and disadvantages. According to the policy adjusted objectives, a variety of policies shoud be used together, in order to achieve the good results of implementation.

Keywords: the external economy of wind power, market failure, compensation mechanism, wind power promotion policies.

1 Introduction

The external economy means the external influnence which economic activities of econcomic subjects have on others. The existence of the external economy leads the difference between individual and social marginal benefits. It also leads the difference between individual and social marginal costs. As the economic subjects making decisions, they usually only consider the benefits and costs which have a direct impact on themselves, and neglect the costs and benefits that are not directly related to themselves. The dicision biases of economic subjects are caused by these differences. Thus, when ecnonomic subjects making decisions, they should not only consider the internal economy of things, but also consider the external economy to avoid the biases. Only in this way, the scientific and accurate decision-making results can be obtained.

2 Impact Analysis of the External Economy of Wind Power

2.1 Structural Characteristics of the Electricity Market in China

There have being the supply-side market and the demand-side market in the electricity market, as shown in Figure 1.

D. Zeng (Ed.): Advances in Control and Communication, LNEE 137, pp. 195–203.
springerlink.com

Figure 1 (a) shows, traditional power plants can sell all the power which the basic level can provide in the current market supply and demand. This point has reached the equilibrium price ***PS**** and the equilibrium quantity ***QS****. There is also a small part of supply relying on wind power to meet more electricity demand Q_0. Because of the high cost of wind power generation, Q_0 has not yet reached the optimum level of wind power supply. Then the Figure 4 (b) shows the power companies purchase power from the traditional power plants ***S1*** and wind farms ***S2*** and sell the power to the end-users of demand side. During this process, We obtained the market equilibrium price ***PD**** and the equilibrium quantity ***QD*=QS*+ Q0***.

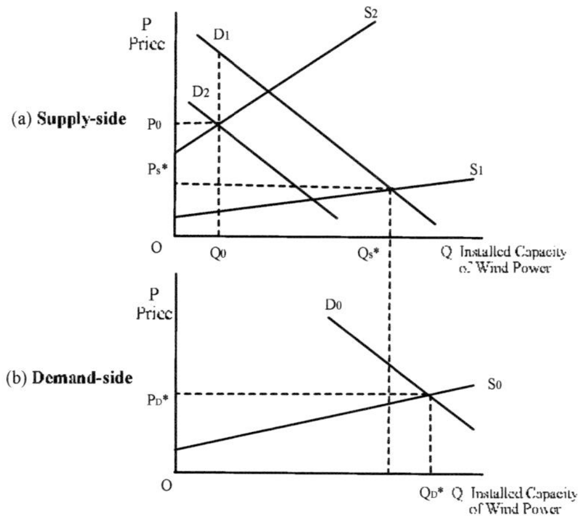

Fig. 1. The market strcture of China's power market supply-side and demand-side

*Where: **S1**- the supply curve of traditional energy plants; **S2** – the supply curve of wind farms; **D1** – the demand curve faced by traditional power suppliers; **D2** - the demand curve faced by wind power suppliers; **S0** – the supply curve of Power Grid Corporation in the demand-side market; **D0** – the demand curve of Power Grid Corporation in the demand-side market.*

2.2 Influnence on the Market Allocation of the External Economy of Wind Power

Within the market mechanism, whether producing or using wind power is based on the maximum personal benefit principle, as the Figure 2 shown. Because of the strong external economy, the social marginal benefit of wind power ***MSR*** is more than the personal marginal benefit ***MR***. From the Figure 2, we can see that the balanced installed capacity is decisided by E_1 which is the intersection of ***MR*** and ***MC***. The corresponding point Q_1 is the balanced capacity of wind power. However, in the view of the whole society, the social optimum balanced installed capacity of wind power is decisided by E_2, which is the intersection of ***MSR*** and ***MC***. The corresponding point Q_2 is the social optimum ballanced capacity. Therefore, the balanced installed capacity Q_1 is less than the social optimal balanced installed capacity Q_2 within the market mechanism, the difference is $Q_2 - Q_1$. Meanwhile, the poersonal marginal benefit of wind power is less

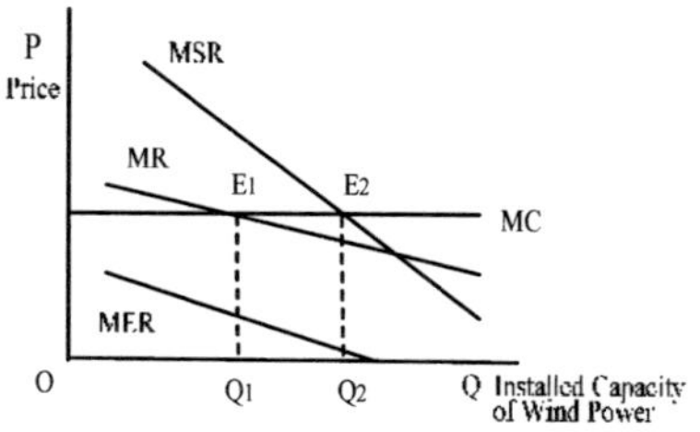

Fig. 2. External economy of wind power and the balanced analysis chart

Where: MR - the personal marginal benefit curve of wind power; MC - the personal marginal cost curve of wind power; MSR - the social marginal benefit curve of wind power; MER - the external marginal benefit curve of wind power.

than its social marginal benefit, so the level of development of wind power can not achieve the optimal level required by society, that is to say, it doesn't achieve Pareto optimal state. There is "market failure" situation in the wind power industry.

Furthermore, the adverse effects of the external economy of wind power to the market allocation are shown in Figure 3. Assume that there are only wind power and traditional power in the market. The total market installed capacity is a fixe value ***AB***. If there is no external economy, the equilibrium point is decided by ***F*** which is the intersection of MR_1 and MR_2. At this time, $MR_1 = MR_2$. It means achieving the optimal allocation of social resources.

When there is the external economy, the equilibrium point of wind power and traditional power should be decided by ***G*** which is the intersection of MSR_1 and MSR_2. At this time, $MSR_1 = MSR_2$. At this time, the installed capacity ***QFB*** allocated by market spontaneously is less than the social optimal allocation $Q_G B$. Their difference is ***QGQF***. As achieving the social optimal allocation, the installed capacity of traditional power should be controlled at the Q_G point and the quantity is $Q_G A$. But it will lead to the price difference $\triangle P$ between the personal marginal benefit P_2 of traditional power producers or consumers and the personal marginal benefit P_1 of wind power producers or consumers. Since $P_2 > P_1$, $\triangle P = P_2 - P_1$. As wind power can not internalized, it will make Q_G move to Q_G, reducing the production or consumption of wind power and

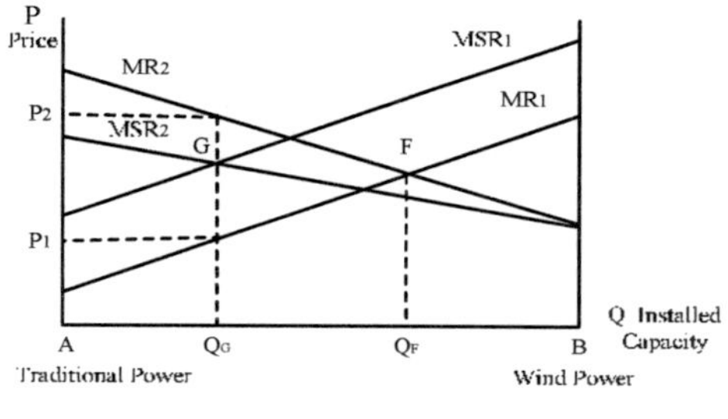

Fig. 3. The market distribution model of wind power and traditonal power

Where: ***MR1****-the personal marginal curve of wind power;* ***MSR1****-the social marginal curve of wind power;* ***MR2****-the personal benefit curve of traditional power;* ***MSR2****-the social marginal benefit curve of traditional power.*

increasing the production or consumtption of traditional power. Thus it deviates from the optimal equilibrium point ***G*** of social resources allocation.

In order to protect the environment and promote the sustained and healthy development of wind power industry, the Government needs to take appropriate measures to compensate producers or consumers' losses caused by the external economy of wind power.

3 Compensation Mechanism of the External Economy of Wind Power

Usually, the basic idea to solve "market failure" caused by external economy is to transform social costs or benefits to personnal ones with related systems or policies. Our Government adopted three key policies to compensate the losses caused by the external economy of wind power. They are economic promotion policies, market forced policies and internalization policies of external economiy. The compensation mechanism of the external economy of wind power is constituted with these three policies.

3.1 Economic Promotion Policies

The economic promotion policies adopted by government are always including subsidies, taxes and other forms. As Figure 4 shows, in the short term, the increasing demand of wind power means reducing demand of traditional power. The change range is the same, we set $\triangle Q$. Before using the economic promotion policies, due to the high costs, wind power is only used to meet regional demand. Thus, Q_0 is the market equilibrium quantity and P_0 is the market equilibrium price. The total market trading volume is $QD^*=QS^*+Q_0$.

To promote the development of wind power, the Government has taken the price-based subsidies. Assuming the subsidy rate is t. The result of subsidy is reducing the costs of wind power and increasing sales $\triangle Q$. Obviously, if there is no government subsidies, the price which wind power producers can afford is P_1 and the price which the power grid companies can afford is P_2. Since $P_1>P_2$, the power grid companies can not consume wind power $\triangle Q$. With government subsidies, if the company should increase consumption $\triangle Q$, it must satisfy four conditions: (1) Purchase amount and price should be at D_1; (2) Sales volume and selling prices must be at S_2; (3) Purchase and supply are equal, the difference between purchase price and supply price is t. That is, to make wind power market increase sales $\triangle Q$, the Government needs to provide the amount t of subsidies.

Shown in Figure 4, the implementation of economic promotion policies, the wind power market increased sales $\triangle Q$, D_2 moved to the D_1 position. In the traditional power supply market, the trading volume decreased $\triangle Q$, D_1 moved to the D_2 location, the transaction price decreased The producer surplus of wind power increase the area of the triangle e. In the traditional power plants, the producer surplus of those who are still sold power are f (the area of regional $ABCPS^*P_3$ between PS^* and P_3). The power

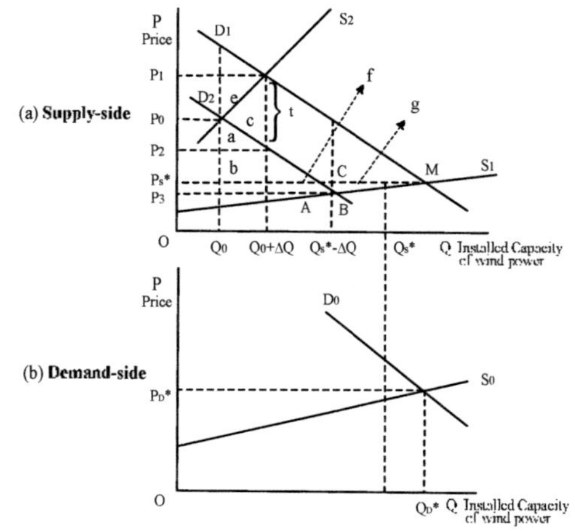

Fig. 4. The effects of government subsidies to the power market supply-side and demand-side

plants which have lost power market lost ***g*** (the area of the triangle ***BCM***). Therefore, the losses of the producer surplus of traditional power plants are $(f+g)$.When power grid companies purchased traditonal power, the consumer surplus increased the area of ***f*** beacause of its lower price. But the consumer surplus loss is ***b*** (the area between ***P****** and P_2) when purchasing the wind power $\triangle Q$. Thus, the total variable quantity of the consumer surplus of the powr grid companies is $(f-b)$. At the same time, the government provided subsidies, so it undertaked the difference between purchase price and sale price of wind power. The government lost the area of ***e***, ***a*** and ***c***, which can be expressed as $-(e+a+c)$.

After the implementation of economic promotion policies, the variable quantity of totoal social welfare L_1 is the sum of profit and loss of plants (including traditional power plants and wind farms), power grid companies and government. That is $L_1=e-(f+g)+(f-b)-(e+a+c)=-(a+b+c+g)$. This shows that the total social welfare have reduced $(a+b+c+g)$ as the implementation of government subsidies for wind power. If the government subsidies aren't effective, power companies can not purchase wind power at price P_2 which is lower than the traditional power price ***P****.The companies' losses ***b*** for purchasing power can not be zero. According to the maximum principle, wind power market can not automatically increase sales. This requires the Government to introduce a number of mandatory market policies, which are in conjunction with economic promotion policies.

3.2 Market Enforced Policies

These policies mean to transfer a mandatory part of the traditional power to wind power with laws and regulations, such as market quotas. Quota system usually requires the market share of wind power have a certain proportion, as shown in Figure 5.

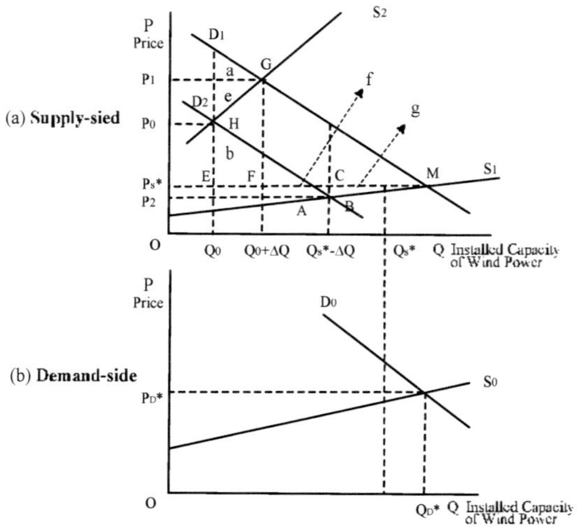

Fig. 5. The impact of quata system on the power market supply-side and demand-side

The supply of wind power increased $\triangle Q$, while the supply of traditional power plants decreased ($QS^*-\triangle Q$). The price of traditional power is reduced to P_2 and the price of wind power is P_1. Within the new level of market transactions, the producer surplus and consumer surplus are also changed. The producer surplus of wind power increased the area of ***e***. The producer surplus of traditional power decreased the area of $(f+g)$. Because the price elasticity of S_1 is small, the area of ***f*** and ***g*** is not large. Due to the lower price, the consumer surplus of power grid companies increased the area of ***g*** when they purchased the traditional power. But their consumer surplus decreased the area of ***a***, ***e*** and ***b*** (trapezoid ***EFGH***) when purchasing wind power$\triangle Q$. Therefore, the totoal variable quantity of the consumer surplus of grid companies is $(g-e-a-b)$. As the electricity demand is price rigidity, the change from PS^* to P_2 is very small, ***g*** is less than $(e+a+b)$, so the consumer surplus of power grid companies decreased.

The totoal social welfare is $L_2=e-(f+g)+(f-e-a-b)=-(a+b+g)$. This shows that the implementation of market forced quota policies reduced the total social welfare $(a+b+g)$. If the market quota is less than $\triangle Q$, the price of wind power will be lower than P_1, but the end result will not change when this price is higher than the fixed cost of wind power ($P_1>P_2$). If the market quota is more than $\triangle Q$, it needs other measures use together to achieved market transactions.

3.3 Internalization Policies of External Economy

In order to achieve the internalization of external economy, it needs to achieve the internalization of the external costs of traditional power. There are many ways to achieve the internalization, such as imposing on sulfur and carbon taxes and other environmental taxes.

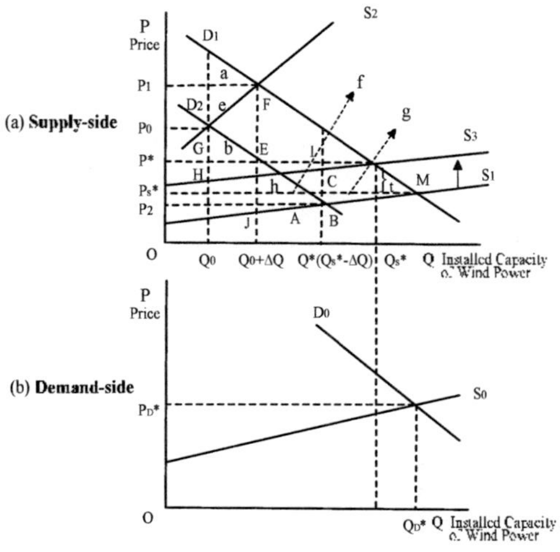

Fig. 6. The effects of internalization policies of external economy to power market supply-side and demand-side

As shown in Figure 6, set t is environmental tax imposed by per unit of electricity quantity.The costs of traditional power increased because of imposing on the taxes, which causing S_1 move to the S_3 position. Shown in Figure 8, the moving height of S_1 is t. At this time, the new market equilibrium price is P^* and the new market equilibrium quantity is Q^*. P^* is more than PS^* which traditonal power plants can undertake at before imposing on taxes. The difference (P^*-PS^*) between the two is t which is handed in the government. Since the price of traditional power increased, power grid companies have to reduce the consumption $\triangle Q$ of it. Then the companies increase the consumption of wind power, so that the wind power sales increased $\triangle Q$ and it is sold at P_1. At this time, the producer surplus of wind power increased the area of e. The producer surplus of traditonal power decrease the area of $(f+g)$. Due to the higher price of traditonal power, the consumer surplus of power grid companies decreased the area of h (the area of the rectangle ***EIBJ***) when they purchase the traditional power. The consumer surplus of power grid companies lost the area of a, e and b (the area of the rectangle ***EIBJ***) when they buy wind power $\triangle Q$. Therefore, the totoal comsumer surplus of grid companies decreased $(h+a+e+b)$.In addition, as imposing tax, the totoal increased benefits of the government are the area of h and f.

The totoal social welfare is $L_3=e-(f+g)-(h+a+e+b)+(h+f)=-(a+b+g)$. This shows that the implementation of internalization policies reduced the total social welfare $(a+b+g)$. Among them, power grid companies and traditional power plants afford the environmental costs together and the revenue of government increased $(h+f)$. This means that, the losses of power grid companies increased further by using the internalization policies. However, the government can use the revenue to solve the environmental problems. Meanwhile, it will reflect the cost advantage of wind power on the environmental pollution-free and help to improve the public's awareness of protecting environment.

4 Comparison of Three Policies within the Compensation Mechanism

In summary, the inplementation costs and effects of the three policies are different, as shown in Table 1.

Table 1. The market impact of the three main promotion policies in short-term

Policies	Economic Promotion Policies	Market Forced Policies	Internalization Policies of External Economy
the Increased Amount of Wind Power Market Share	$\triangle Q$	$\triangle Q$	$\triangle Q$
Producer Surplus of Wind Power	E	e	e
Producer Surplus of Traditional Power	$-(f+g)$	$-(f+g)$	$-(f+g)$
Consumer Surplus of Power Grid Companies	Minimum	Medium	Maximum
Changes of Government Revenue	Reduction	None	Increase
Changes of the Totoal Social Welfare	$L1$	$L1+c$	$L1+c$

We can see that no matter what kind of policies is used, the totoal social welfare will decrease. Among them, the loss of totoal social welfare of market forced policies and internalization policies are the same, which are $(a+b+g)$. As the policies intervention of government decrease market efficiency, the loss of economic promotion policies is the most, which is $(a+b+c+g)$. The environmental and social benefits of wind power are more than the economic benefit, so the loss of economic benefit is the result of adjusting the "market failure". Among them, the policy-oriented of economic promotion policies is strong, but it will reduce government revenue. The government guidance of market forced policies is strong and these policies encourage investment. Internalization policies transfer environmental signals and increase the government revenue.

5 Conclusion

First, within the frame of three promotion policies, the increased amount of wind power producer surplus equals to the decreased amount of traditional power producer surplus. It means with the development of wind power, the interests of conventional power plants will be subject to be damaged; meanwhile, with the certain supply-demand relationship, the increased or decreased amount of power producers has nothing to do with policy choices, but the increased market share of wind power producers.

Second, in order to support the development of wind power, power grid companies will give up a portion of consumer surplus. But within different promotion policies, these portions of consumer surplus are different. Within the economic incentive policy, the loss of power grid companies will be transferred to the government; within the market enforcement policy, power grid companies bear a large amount loss; within the internalization policy of the external economic, the environmental tax promote the price of traditional power, power grid companies entail a part of environmental costs, and then bear a greater loss.

Third, all these three polices have positive effect on the development of the society. In these three policies, the internalization policy of external economic is the optimal policy. The implementation of environmental tax policies has both the social significance and the environmental significance, it can reflect the relationship between environment and energy consumption, and then encourage traditional energy companies to speed up the use of environmental technology. However, this policy increases market loss, so it increases the difficulty of policy implementation.

In short, the three promotion polices within the compensation mechanism their respective advantages and disadvantages. They must be used in combination according to the internal and external circumstances, and considering the short-and long-term effect, this can achieve a good result.

References

1. Gao, H.: Western Economics, vol. 3, pp. 112–207. China Renmin University Press, Beijing (2007)
2. Nicola, A.: Economic policy principles: Value and Technology, pp. 113–252. China Renmin University Press, Beijing (2001); Wang, G., Liu, Q. (translated)
3. Gu, E.: China's wind power industry to develop new strategies and non-grid wind power theory, pp. 52–186. Chemical Industry Press, Beijing (2006)
4. Jiang, L.: 2008 Summary of domestic and international wind power development. Power Technology Economy 21(2), 12–15 (2009)
5. Xiao, Z.: Nuclear power price of the external economy. China Power Enterprise Management 19, 22–23 (2008)
6. Su, H.H.: Network externality effect of the micro-economic analysis. Industrial Technology & Economy 26(6), 129–132 (2007)

Research on the Evaluation of External Economy of Wind Power Project Based on ANP-Fuzzy

Yu Shunkun, Zhou Lisha, and Li Chen

Economics and Management School
North China Electric Power University
Beijing, China
{ysk21,cressa26,lichenbj}@126.com

Abstract. The evaluation system has the interaction among the indexes, as well as the evaluation of the target is difficult to define precisely. Therefore, the methods of ANP and Fuzzy are adopted to evaluate comprehensively the external economy of wind power project. Firstly, determing the index weight with ANP can overcom AHP's flaws which can not reflect the interaction among the indexes. In this way, it reflects the interaction among the indexes more objective. Then build the evaluation model with Fuzzy method based on ANP to assess the external economy with fuzziness. And by means of Super Decisions software, the effectiveness and scientific of the evaluation model is verified through an example.

Keywords: The external economy of wind power project, analytic network process, fuzzy comprehensive evaluation, Super Decisions software.

1 Introduction

When evaluating the economy of wind power and conventional power, some pepole are more likely to only consider the internal economic factors. In fact, if taking the impact on the environment and other external economic factors into account, the overall economy of wind power is superior to other traditional power supply. In the view of environmental impact, wind energy is clean renewable energy, while the traditional fossil fuel emissions large amount polution on the environment. Therefore, when ecnonomic subjects making decisions, they should not only consider the internal economy of things, but also consider the external economy to avoid the biases. Only in this way, the scientific and accurate decision-making results can be obtained. The scientific and effective external economic evaluation of wind power projects will help to improve the efficiency of wind power and promote the sustained development of wind power industry.

2 Anp Method Introduced

2.1 The Concept of ANP

The Analytic Network Process (ANP) developed by Thomas Saaty in 1996 is a decision technique to deal with the network structures with dependence and feedback

D. Zeng (Ed.): Advances in Control and Communication, LNEE 137, pp. 205–217.
springerlink.com

relationship in it. ANP is a new practical decision method which is developed on the basis of AHP. ANP is particularly useful for complex decision system which has internal dependence and feedback relationship among the elements [1]. The ANP describes the relationship among elements in the system with network structure, not with a simple hierarchical structure, so the interaction of elements can be demonstrated more accurately [2]. Therefore, ANP is a more effective decision method compared with the AHP. Figure 1 shows a typical ANP structure.

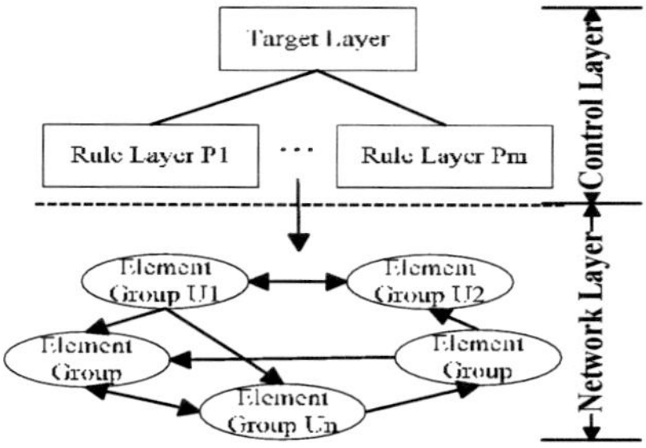

Fig. 1. A typical ANP structure

2.2 Steps of ANP

• Establish elements and element sets through systematic analyzing for decision problems. We should estimate that if these elements are independent or interdependent. Here, conference, expert inquiry or brainstorming can be used.

• Establish ANP matrix. Determine control layer and analyze the mutual influence between each two elements according to control criteria. Confirm the relationship among criteria, elements and element sets.

• Establish ANP super matrix. Elements in ANP control layer are set to be $P_1, P_2, \ldots, P_m$. Element sets in network layer are set to be $U_1, U_2, \ldots, U_N$. And, there are $u_{i1}, u_{i2}, \ldots, u_{in_i}$ $(i=1,2,\ldots,N)$ in each U_i. Regard the elements of the control layer P_s $(s=1,2,\ldots,m)$ as criteria, and regard the elements u_{jk} $(k=1,2,\ldots,n_j)$ of U_j as subordinate criteria. Make a comparative analysis for the influence between each two u_{jk} in the element set U_i to establish matrix P_s. The data reflecting the relative importance of each two elements in the marix can be obtained by consulting corresponding experts. In order to quantify the judgment matrix, ANP model uses rule of nines proposed by Saaty to scale. Then use latent root method to obtain the priority vectors $(w_{i1}^{(jk)}, w_{i2}^{(jk)}, \cdots w_{in_i}^{(jk)})^T$. According to consistency test, if the above eigenvectors meet compatible condition, they are the weights of elements. A matrix can be established by using all priority vectors. It is denoted as:

$$W_{ij} = \begin{bmatrix} w_{i1}^{(j1)} & w_{i1}^{(j2)} & \cdots & w_{i1}^{(jn_j)} \\ w_{i2}^{(j1)} & w_{i2}^{(j2)} & \cdots & w_{i2}^{(jn_j)} \\ & & \vdots & \\ w_{in_i}^{(j1)} & w_{in_i}^{(j2)} & \cdots & w_{in_i}^{(jn_j)} \end{bmatrix}$$

The column vectors W_{ij} of the matrix represent the comparative importance of the priority vectors $u_{i1}, u_{i2},\ldots, u_{in_i}$ in U_i with the $u_{j1}, u_{j2},\ldots, u_{jn_j}$ in U_j. If the elements of U_i are not affected by the element of U_j, W_{ij} is equal to 1. Put all the interactive priority vectors together in the network layer, we can obtain a super matrix under the control layer W:

$$W = \begin{array}{cc} & \begin{array}{cccc} U_1 & U_2 & \cdots & U_N \\ u_{11}\cdots u_{1n_1} & u_{21}\cdots u_{2n_2} & \cdots & u_{N1}\cdots u_{Nn_N} \end{array} \\ \begin{array}{cc} U_1 & \begin{array}{c} u_{11} \\ \vdots \\ u_{1n_1} \end{array} \\ U_2 & \begin{array}{c} u_{21} \\ \vdots \\ u_{2n_2} \end{array} \\ & \\ U_N & \begin{array}{c} u_{N1} \\ \vdots \\ u_{Nn_N} \end{array} \end{array} & \begin{bmatrix} W_{11} & W_{12} & \cdots & W_{1N} \\ W_{21} & W_{22} & \cdots & W_{2N} \\ \vdots & \vdots & \ddots & \vdots \\ W_{N1} & W_{N2} & \cdots & W_{NN} \end{bmatrix} \end{array}$$

Each element in this matrix is another matrix. And the column sum in this matrix is equal to 1. The super matrixes should be normalized so as to calculate conveniently. Weighting the elements of the super matrix can get a weighted super matrix $\overline{W} = (\overline{W})_{N\times N}$.

In the formula $\overline{W}_{ij} = a_{ij} \times W_{ij}$, a_{ij} is a weighted factor, $i=1,2,\ldots, N; j=1,2,\ldots, N$.

- In order to reflect the interdependent relationship among elements, the super matrix W needs a stability treatment. That is to calculate each super matrix's relative limit priority vector:

$$W^{\infty} = \lim_{k\to\infty}(1/N)\sum_{k=1}^{N}\overline{W}^{k}$$

If the limit is converging and unique, the values of corresponding rows of original matrix are the stable weights of the evaluation indices. The core of determining weights with ANP is solving super matrix. The calculation process is very complex. This paper adopts Super Decisions software in the example to solve super matrix. This software is developed on the basis of ANP theory. It is a powerful tool.

3 Fuzzy Comprehensive Evaluation Method Introduced

3.1 The Concept of Fuzzy Comprehensive Evaluation Method

In real life, there are some fuzziness which have no clear boundaries and clear classification. If the evaluation process involved in this type of fuzzy factor, we have to use the fuzzy comprehensive evaluation method (Fuzzy). The basic idea of Fuzzy is to use fuzzy linear transformation theory and maximum degree principle. Multi-level fuzzy comprehensive evaluation is followed by repeated operation of this synthesis, from the lowest level to the highest level, until the final results [3].

3.2 Steps of Fuzzy

• Determine Object Sets $O=\{ o_1, o_2,\ldots, o_l \}$,Factor Sets $U=\{ U_1, U_2,\ldots, U_m \}$,Reviews Sets $V=\{ v_1, v_2,\ldots, v_n \}$. Among them, $\bigcup_{i=1}^{m} U_i = U, U_i \cap U_j = \Phi$, $i \neq j$.

• Establish the weight distribution vectors of evaluation factors. Factor sets are set to be U. Weight vectors are set to be W, $W=\{ w_1, w_2,\ldots, w_m \}$, $0 \le w_i \le 1$, $\sum_{i=1}^{m} w_i = 1$. In this paper, the weight of all factor sets are calculated and determined by using Analytic Network Process (ANP).

• The fuzzy comprehensive evaluation matrix R is obtained by evaluating each factor.

$$R = (r_{ij})_{m\times n} \begin{bmatrix} r_{11} & r_{12} & \cdots & r_{1n} \\ r_{21} & r_{22} & \cdots & r_{2n} \\ \vdots & \vdots & & \vdots \\ r_{m1} & r_{m2} & \cdots & r_{mn} \end{bmatrix}$$

The fuzzy evaluation matrix R is a sinlge-factor evaluation matrix. r_{ij} in R reflect the various grades of membership of objects. Among them, r_{ij} can be obtained by fuzzy statistical method. That is:

$$r_{ij} = f_{ij} \Big/ \sum_{j=1}^{n} f_{ij}$$

f_{ij} is the total number of the i factor evaluated to v_j in the j grades.

• The results of the first level comprehensive evaluation are obtained by fuzzy transforming the various factors.

$$B_1 = W_1 \circ R_1 = (b_{11}, b_{12}, \cdots, b_{1n})$$
$$B_2 = W_2 \circ R_2 = (b_{21}, b_{22}, \cdots, b_{2n})$$
$$\vdots$$
$$B_m = W_m \circ R_m = (b_{m1}, b_{m2}, \cdots, b_{mn})$$

As above, " $\circ$ " indicates a broad fuzzy composition. B_s (s=1,2,…,m)are the results of Fuzzy about evaluation object s factor. Using the maximum membership degree principle, the optimal evaluation result is the remark grade v_j corresponding to the maximal b_{sj}.

According to the needs of practical problems, we can select a specific method to evalue b_{sj} (j=1,2,…,n). Fuzzy operators commonly used are $M(\wedge,\vee)$ model, $M=(\bullet,\vee)$ model and $M(\bullet,+)$ model. In order to take into account the impact on various factors, the weighted average type operators are used in this paper. It is $b_{sj} = \sum_{i=1}^{m} w_i \cdot r_{ij}$, among which, j=1,2,…,n.

• The factor layers are evaluated by the second level comprehensive evaluation. By $B_1, B_2,\ldots, B_m$, the single factor evaluation matrix can be obtained in U, U={ $U_1, U_2,\ldots, U_m$ }.

$$R_U = \begin{bmatrix} B_1 \\ B_2 \\ \vdots \\ B_m \end{bmatrix} = \begin{bmatrix} b_{11} & b_{12} & \cdots & b_{1n} \\ b_{21} & b_{22} & \cdots & b_{2n} \\ \vdots & \vdots & & \vdots \\ b_{k1} & b_{k2} & \cdots & b_{mn} \end{bmatrix}$$

If the corresponding weight vector of $U_1, U_2,\ldots, U_m$ is W_U={ $w_1, w_2,\ldots, w_m$ }, the comprehensive evaluation of U which is on behalf of relative things is $B_U = W_U \circ R_U = (b_1, b_2, \cdots, b_n)$.

• Analyse the result of Fuzzy. According to the certain standards, Fuzzy results can be transferred into scores. By analyzing, the results of sorting can be more clarity. The final determination of the evaluation object can be obtained.

4 Case Study

4.1 Establish Evaluation Index System

There are many factors which affect the external economy of wind power project. The Delphi method is adopted in this paper. After three rounds of consultation, the opinions of 30 experts which are about the important indexes are obtained. According to these opinions, the important first, second and third level indexes are selected through the ranking method. Table 1 shows the evaluation index system.

4.2 Determine the Weights of Evaluation Indexes Based on ANP

As the evaluation indexes of the external economy of wind power project are not independent and the elements affect each other, a network model of the elements is constructed with ANP method. First, we analyze and list out the associated relationship of the indices. Then, an ANP structure model is built to evaluate the indices. Figure 2 shows this.

Second, the third level elements are paired compared on the basis of the three first level criterion indexes (u_1, u_2, u_3) and the second level sub-criterion indices in the control layer. Third, judgment matrix is built based on expert opinions to judge whether to accept the judgment matrix according to the result of consistency test. At last, the index weights are determined by the weighted super matrix and the limit matrix. We adopt the rule of nines to score the relationship among the indexes. Below, figure 3 shows an example of the evaluation indices judgment matrix score interface in SD software.

Table 1. The evaluation index system of the external economy of wind power project

Object Sets	**Factor Sets** First Level Index	**Sub Factor Sets** Second Level Index	Third Level Index
External Economy of Wind Power Project ***u***	Econom-y Impact **u1**	Impact on Country's Economic Development Goals u11	The Electricity Capacity of Generators, the Status of the Grid u111
			The Effect to Improve the Economic Structure, the Distribution of Productive Forces and the National Economy u112
			The Environmental Benefits and Social Benefits u113
			The Impact on the National Economy and the Sustainable Development of Energy and Electricity u114
		Impact on Regional Economic Development u12	The Effect to improve the Economic GDP of the Project Area u121
			The Impact on the change of the Production Structure of the Project Area u122
			The Impact on the Land Use Adjustment and the Land Value-added of the Project Land u123
		The Impact on Department's Economic Development u13	The Contribution to the Agriculture Economic Development u131
			The Contribution to the Industry Economic Development u132
			The Contribution to the Economic Development of the Tertiary Industry u133
		The Impact on Scientific and Technological Progress u14	The Technological Advancement Degree and Application Used by the Project u141
			The Effect to Promot Scientific and Technological Progress of the Nation, Department and Local Place u142
	Social Impact **u2**	The Impact on the Living Standard and Life Quality of the Project Area u21	The Impact on Per Capita Income Increase u211
			The Effect to Enhance the Living Conditions u212
			The Effect to Improve the Living Standard u213
		The Mutual Adaptation between the project and the Local Community u22	Local Government Support for the Project u221
			The Attitude of Local Residents for the Project u222
			Compensation for Damaged Groups u223
			Range of the Beneficiaries and Their Reflect u224
			The Impact on Local Community Development u225
			The Impact on Women, Ethnic and Cultural Practices u226
		Employment Impact u23	The Effects on Short-term Employment u231
			The Impact on the Long-term Direct Employment u232
			The Impact on the Long-term Indirect Employment u233
			The Contribution to the Employment Rate and the Displaced Rate u234

Table 1. *(continued)*

	Environmental Impact **u3**	The Project Status and the Environment Report Compared u31	The Impact on Water Quality u311
			Tbe Impact on Soil Erosion u312
			Safety Control of Engineering Geology u313
			Impact on Construction Waste u314
			Noise Impact u315
			Electromagnetic Impact u316
			Impact on Rare Animals and Plants u317
		Environmental Management u32	Environmental Monitoring and Management u321
			Execution of Environmental Systems, Policies and Provisions u322
			Management of Environmental Protection Funds, Equipment and Instruments u323
			Environmental Technology Management and Staff Training u324

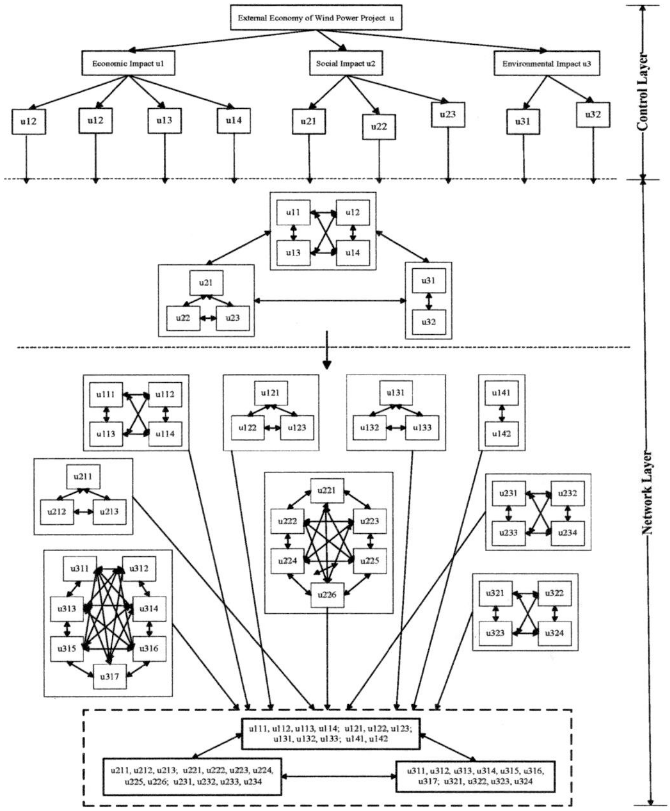

Fig. 2. The ANP_structure model of the evaluation index system of wind power project's external economy

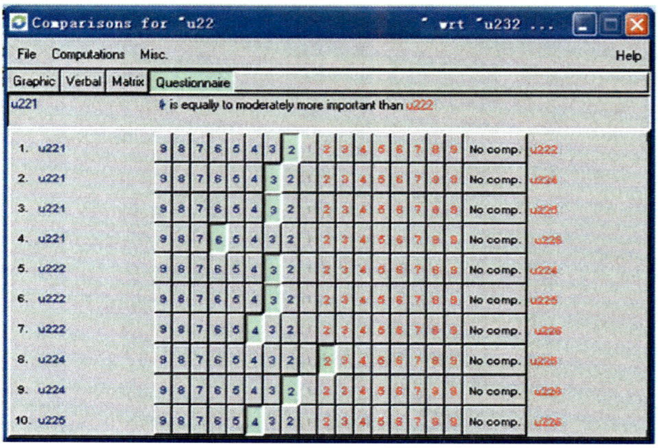

Fig. 3. An example of the evaluation indexes judgemet matrix mark interface in SD software

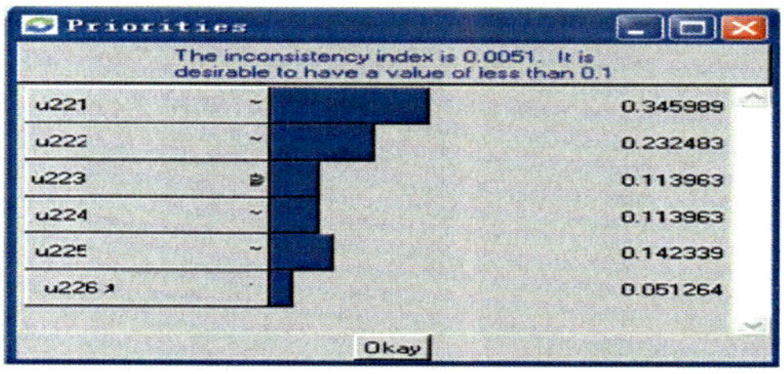

Fig. 4. An example of the index consistency check and the weigh of third grade indexes

We input the relationship judgment matrix of each evaluation index into Super Decisions software respectively. After consistency test, the super matrix *W*, the weighted super matrix $\overline{W}$ and the limit super matrix W^{∞} can be obtained. Figure 4 shows the result of consistency test and the weights of the first grade indexes. The result of consistency test for the judgment matrix is 0.0051, less than 0.1000. Therefore, this judgment matrix should be accepted.The weigh of each third grade evaluation index and its corresponding limit is shown in table 2.

Then, the second and the first level elements are paired compared on the basis of the three first level criterion indices (u_1, u_2, u_3) in the control layer. Then, the judgment matrix is built based on expert opinions to judge whether to accept the judgment matrix according to the result of consistency test. At last, the index weights of the second and the first levle are determined by the weighted super matrix and the limit matrix. Obtaining the weight of each first level index is the same way as the second level indexes. We can obtain the weigh of each first grade evaluation index is: W= (0.3108, 0.1958, 0.4934).The weigh of each second grade evaluation index is:

W_1= (0.4182, 0.1205, 0.2707, 0.1906), W_2= (0.3333, 0.3333, 0.3333), W_2= (0.6667, 0.3333). The weigh of each third grade evaluation index is: W_{11}= (0.2699, 0.2598,

0.2109, 0.2594), W_{12}= (0.3078, 0.3751, 0.3171), W_{13}= (0.3074, 0.5347, 0.1579), W_{14}= (0.6544, 0.3456), W_{21}= (0.3695, 0.3404, 0.2901), W_{22}= (0.3952, 0.2510, 0.0683, 0.0749, 0.1728, 0.0378), W_{23}= (0.2814, 0.3998, 0.2105, 0.1083), W_{31}= (0.0695, 0.1121, 0.1283, 0.0597, 0.1685, 0.1966, 0.2653), W_{32}= (0.3053, 0.3318, 0.1235, 0.2394).

Table 2. The weigh of each third grade evaluation indexes and its corresponding limit

Index	Weight	Limit
u111	0.2699	0.0314
u112	0.2598	0.0302
u113	0.2109	0.0245
u114	0.2594	0.0302
u121	0.3078	0.0358
u122	0.3751	0.0436
u123	0.3171	0.0369
u131	0.3074	0.0357
u132	0.5347	0.0622
u133	0.1579	0.0184
u141	0.6544	0.0716
u142	0.3456	0.0378
u211	0.3695	0.0430
u212	0.3404	0.0396
u213	0.2901	0.0337
u221	0.3952	0.0459
u222	0.2510	0.0292
u223	0.0683	0.0079
u224	0.0749	0.0087
u225	0.1728	0.0201
u226	0.0378	0.0044
u231	0.2814	0.0293
u232	0.3998	0.0416
u233	0.2105	0.0219
u234	0.1083	0.0113
u311	0.0695	0.0066
u312	0.1121	0.0106
u313	0.1283	0.0122
u314	0.0597	0.0057
u315	0.1685	0.0159
u316	0.1966	0.0187
u317	0.2653	0.0252
u321	0.3053	0.0336
u322	0.3318	0.0365
u323	0.1235	0.0136
u324	0.2394	0.0264

4.3 Fuzzy Comprehensive Evaluation of Indexes

We select 50 inquiry objects in the form of questionnaire, such as experts, business managers, power users and the local residents. They judge the economic rate of the external economy of wind power project.

Table 3. The index weight and the evaluation result of each factor set (the third grade evaluation indexes)

Factor Sets		Sub Factor Sets (Third Level Index)	Evaluation Results				
First Level Index	Second Level Index		Very Good	Good	General	Poor	Very Poor
u1	u11	u111	0	0	14	33	3
		u112	10	26	10	4	0
		u113	12	22	12	4	0
		u114	30	17	3	0	0
	u12	u121	12	24	12	2	0
		u122	8	26	13	3	0
		u123	15	17	15	3	0
	u13	u131	0	0	8	23	19
		u132	9	12	19	10	0
		u133	0	0	4	20	26
	u14	u141	17	21	9	3	0
		u142	14	27	7	2	0
u2	u21	u211	1	2	10	31	6
		u212	4	21	17	7	1
		u213	6	17	18	7	2
	u22	u221	15	20	14	1	0
		u222	16	17	10	7	0
		u223	22	11	11	6	0
		u224	7	26	15	2	0
		u225	24	19	5	2	0
		u226	0	3	5	19	23
	u23	u231	9	20	12	9	0
		u232	12	21	13	4	0
		u233	9	16	16	7	2
		u234	2	22	18	8	0
u3	u31	u311	27	20	3	0	0
		u312	19	22	9	0	0
		u313	8	17	21	4	0
		u314	23	19	8	0	0
		u315	17	23	9	1	0
		u316	0	12	18	13	7
		u317	0	12	24	14	0
	u32	u321	14	19	14	3	0
		u322	25	21	4	0	0
		u323	19	19	12	0	0
		u324	18	22	10	0	0

The questionnaires are fully recovered. We determine the first level evaluation set of customer satisfaction is $V=\{v_1, v_2, v_3, v_4, v_5\}=$ (very good, good, general, poor, very poor) and the measurement scale vector is $\boldsymbol{H}=\{5, 4, 3, 2, 1\}$. A questionnaire with 5 values indicates that the index is very good, the same as other values. The evaluation result of each sub-factor set is obtained on the basis of sorting the results of questionnaire inquiry, as shown in Table 3. A single factor fuzzy evaluation matrix $\boldsymbol{R_i}$ is obtained by single-factor-evaluating each element in sub-factor set, i=1,2,...,5. With fuzzy statistical method, the fuzzy evaluation matrix $\boldsymbol{R_{11}}$ of u_{11} is:

$$R_{11}=\begin{bmatrix}0.0000 & 0.0000 & 0.2800 & 0.6600 & 0.0600\\ 0.2000 & 0.5200 & 0.2000 & 0.0800 & 0.0000\\ 0.2400 & 0.4400 & 0.2400 & 0.0800 & 0.0000\\ 0.6000 & 0.3400 & 0.0600 & 0.0000 & 0.0000\end{bmatrix}$$

We use the weighted average type operators $M(\bullet,+)$ on fuzzy calculation for $\boldsymbol{R_{11}}$ and obtain the comprehensive evaluation vector $\boldsymbol{B_{11}}$of $\boldsymbol{u_{11}}$:

$$\boldsymbol{B}_{11}=\boldsymbol{W}_{11}\circ\boldsymbol{R}_{11}=(0.2582,0.3161,0.1937,0.2158,0.0162)$$

Similarly, the comprehensive evaluation vector of u_{11}、u_{12}、u_{13}、u_{14}、u_{21}、u_{22}、u_{23}、u_{31}、u_{32} is obtained respectively.

$$\boldsymbol{B}_{12}=\boldsymbol{W}_{12}\circ\boldsymbol{R}_{12}=(0.2290,0.4506,0.2665,0.0538,0.0000)$$
$$\boldsymbol{B}_{13}=\boldsymbol{W}_{13}\circ\boldsymbol{R}_{13}=(0.0962,0.1283,0.2650,0.3115,0.1989)$$
$$\boldsymbol{B}_{14}=\boldsymbol{W}_{14}\circ\boldsymbol{R}_{14}=(0.3193,0.4615,0.1662,0.0531,0.0000)$$
$$\boldsymbol{B}_{21}=\boldsymbol{W}_{21}\circ\boldsymbol{R}_{21}=(0.0694,0.2564,0.2941,0.3174,0.0628)$$
$$\boldsymbol{B}_{22}=\boldsymbol{W}_{22}\circ\boldsymbol{R}_{22}=(0.3224,0.3653,0.2194,0.0755,0.0174)$$
$$\boldsymbol{B}_{23}=\boldsymbol{W}_{23}\circ\boldsymbol{R}_{23}=(0.1888,0.3955,0.2778,0.1294,0.0084)$$
$$\boldsymbol{B}_{31}=\boldsymbol{W}_{31}\circ\boldsymbol{R}_{31}=(0.1854,0.3318,0.3162,0.1390,0.0275)$$
$$\boldsymbol{B}_{32}=\boldsymbol{W}_{32}\circ\boldsymbol{R}_{32}=(0.3845,0.4076,0.1895,0.0183,0.0000)$$

According to $\boldsymbol{B_{11}}$, $\boldsymbol{B_{12}}$, $\boldsymbol{B_{13}}$, $\boldsymbol{B_{14}}$, the fuzzy comprehensive evaluation matrix $\boldsymbol{R_1}$ of $\boldsymbol{u_1}$ in the second level factor sets is obtained.

$$\boldsymbol{R}_1=\begin{bmatrix}\boldsymbol{B}_{11}\\ \boldsymbol{B}_{12}\\ \boldsymbol{B}_{13}\\ \boldsymbol{B}_{14}\end{bmatrix}=\begin{bmatrix}0.2582 & 0.3161 & 0.1937 & 0.2158 & 0.0162\\ 0.2290 & 0.4506 & 0.2665 & 0.0538 & 0.0000\\ 0.0962 & 0.1283 & 0.2650 & 0.3115 & 0.1989\\ 0.3193 & 0.4615 & 0.1662 & 0.0531 & 0.0000\end{bmatrix}$$

Similarly, the comprehensive evaluation vector $\boldsymbol{B_1}$ of $\boldsymbol{u_1}$ is:
$$\boldsymbol{B}_1=\boldsymbol{W}_1\circ\boldsymbol{R}_1=(0.2225,0.3092,0.2165,0.1912,0.0606)$$

Then, the comprehensive evaluation vector of u_2、u_3 is obtained respectively.

$$\boldsymbol{B}_2=\boldsymbol{W}_2\circ\boldsymbol{R}_2=(0.1935,0.3390,0.2637,0.1741,0.0295)$$
$$\boldsymbol{B}_3=\boldsymbol{W}_3\circ\boldsymbol{R}_3=(0.2518,0.3571,0.2740,0.0988,0.0183)$$

The fuzzy comprehensive evaluation matrix $\boldsymbol{R}$ of $\boldsymbol{u}$ in the first level factor sets is obtained according to $\boldsymbol{B_1}$, $\boldsymbol{B_2}$, $\boldsymbol{B_3}$.

$$\boldsymbol{R}=\begin{bmatrix}\boldsymbol{B}_1\\ \boldsymbol{B}_2\\ \boldsymbol{B}_3\end{bmatrix}=\begin{bmatrix}0.2225 & 0.3092 & 0.2165 & 0.1912 & 0.0606\\ 0.1935 & 0.3390 & 0.2637 & 0.1741 & 0.0295\\ 0.2518 & 0.3571 & 0.2740 & 0.0988 & 0.0183\end{bmatrix}$$

Similarly, the comprehensive evaluation vector $\boldsymbol{B}$ of $\boldsymbol{R}$ is obtained: $\boldsymbol{B}=\boldsymbol{W}\circ\boldsymbol{R}=(0.2313,0.3387,0.2541,0.1423,0.0336)$.

5 Results Analysis

The evaluation results of the first, second and third level, which are generated by fuzzy operation of various indexes, are shown in Table4, Table 5 and Table 6 below.

Table 4. The first grade evaluation result of fuzzy comprehensive evaluation

The Second Level Indexes	The First Level Evaluation Results				
	Very Good	Good	General	Poor	Very Poor
u11	0.2582	0.3161	0.1937	0.2158	0.0162
u12	0.2290	0.4506	0.2665	0.0538	0.0000
u13	0.0962	0.1283	0.265	0.3115	0.1989
u14	0.3193	0.4615	0.1662	0.0531	0.0000
u21	0.0694	0.2564	0.2941	0.3174	0.0628
u22	0.3224	0.3653	0.2194	0.0755	0.0174
u23	0.1888	0.3955	0.2778	0.1294	0.0084
u31	0.1854	0.3318	0.3162	0.1390	0.0275
u32	0.3845	0.4076	0.1895	0.0183	0.0000

We evaluate the external economy of wind power project with the principle of maximum membership degree. The maximum membership degree of u_{11} index is 0.3161. The corresponding remark is "good". It indicates that the wind power project has well economic in promoting national economic development goals. Similarly, this project' remarks corresponding to the maximum membership degree are all "good" in u_{12}, u_{14}, u_{22}, u_{23}, u_{31}, u_{32}. It means that the wind power project has well economic on these aspects. However, the remarks corresponding to the maximum membership degree are "poor" in u_{13} and u_{21}. It shows that wind power is still in the early development stage. Its influence on the industrial sector is not significant. Meanwhile, it shows that the construction and operation of wind power project has little effect on the living standard and life quality of project area residents.

The second evaluation results and their corresponding scores, which are generated by fuzzy operation of various second evaluation indexes, are shown in Table 5. Among them, the remark set V= (very good, good, general, poor, very poor) = (100, 80, 60, 40, 20). The score and rank result of each index is as follows:

Table 5. The second grade evaluation result, the scores and the ranking of fuzzy comprehensive evaluation

The First Level Index	The Second Evaluation Results					Score	Rank
	Very Good	Good	General	Poor	Very Poor		
u1	0.2225	0.3092	0.2165	0.1912	0.0606	68.836	3
u2	0.1935	0.3390	0.2637	0.1741	0.0295	69.846	2
u3	0.2518	0.3571	0.2740	0.0988	0.0183	74.506	1

The maximum membership degree of u_1, u_2 and u_3 is respectively 0.3092, 0.3390 and 0.3571. The corresponding remarks are all "good". It shows that the wind power project has well economic effect on economy, society and enviroment. But we can see that the external economy of enviromental impact of this wind power project is the best because of its compared highest score. Its score is 74.506, close to the good level which is 80 points. The economy of social and enviromental impact is respctively ranked second and third.

The final evaluation result and its corresponding score, which is generated by fuzzy operation of various first evaluation indexes, are shown in Table 6.

Table 6. The final evaluation result and the total scores of fuzzy comprehensive evaluation

Index	The Final Evaluation Result					Total Scores
	Very Good	Good	General	Poor	Very Poor	
u	0.2313	0.3387	0.2541	0.1423	0.0336	**71.836**

We can see the maximum membership degree of the external economy of this wind power project is 0.3387. Its corresponding remark is "good". Its score is 71.836l, colse to the good level which is 80 points. It indicates that this wind power project has a strong external economy.

6 Conclusions

It has great importance to improve the efficiency of wind power through the external economic evaluation. As the interdependent and feedback relationship existing among the indexes of the external economy of wind power project, ANP method is introduced into this paper instead of AHP to overcome the defect that traditional AHP technique can not reflect the relationship among the indexes. The Fuzzy method is introduced into this paper to overcome the problem that the indexes has no clear boundaries and clear classification In a word, this evaluation model based on ANP-Fuzzy is more realistic and comprehensive for the external economy evaluation of wind power project. Meanwhile, it is easy to use this model to calculate and analyze. It is more scientific for enterprises to make decisions for wind power.

References

1. Saaty, T.L.: The Analytical Network Process:Decision Making with Dependence and Feedback. RWS Publications, Pittsburgh (2001)
2. Saaty, T.L.: Decision Making with Dependence and Feedback. RWS Publications, Pittsburgh (1996)
3. Zhou, L.-S., Li, C., Yu, X.-H.: A post Evaluation technique for engineering project investment based on ANP-ENTROPY-TOPSIS. In: International Conference on Engineering Management and Service Sciences (September 2009)

Liuhang Formation and Its Characteristics of Fracture Development in Western Shandong and Jiyang Depression*

He Miao[1,2,**], Li Shoujun[1,2], Tan Mingyou[3], Han Hongwei[3], Guo Dong[3], Wang Huiyong[3], Jia Qiang[4], Yin Tiantao[2], and Yuan Liyuan[2]

[1] Shandong Provincial Key Laboratory of Depositional Mineralization & Sedimentary Minerals
Shandong University of Science and Technology,
Qingdao, China
[2] College of Geological Science & Engineering,
Shandong University of Science and Technology,
Qingdao, China
[3] Geophysics Institute of Shengli Oilfield Company Limited,
SINOPEC,
Dongying, China
[4] College of Geo-Resources & Information,
China University of Petroleum (Huadong),
Qingdao, China
hemiaogeology@qq.com

Abstract. With deepening development of oil-gas exploration, deep layers have become key target and popular domain for oil-gas exploration[4]. So systematical research on fracture characteristics of Liuhang Formation in Taishan Group in Archean is seriously needed. We do research on Liuhang Formation and its characteristics of fracture development based on the outcrops, drilling, well test and log data in Jiyang Depression and Western Shandong. Through field work, core sample survey, literature review, laboratory study and experimental observation, We finally discuss the development of Liuhang Formation and its fracture characteristics in Jiyang Depression and Western Shandong. We study the fracture characteristics in Liuhang formation with higher-accuracy than in Taishan Group, which have not been found in the study area.

Keywords: Fracture Characteristics, Liuhang-Formation, Jiyang Depression, lithology, Fracture groups, Fracture dip, Fracture density, Facial porosity.

1 Introduction

Liuhang Formation is the top strata of Taishan Group developed in the Archean. Liuhang formation in Jiyang Depression can be correlated with the one in Western

* This study was supported by the Science & Technology Leading Project of SINOPEC (S0-PT-1091D010Z) and the Postgraduate Science &Technology Innovation Foundation of SDUST(YCA100203).

** Corresponding author.

D. Zeng (Ed.): Advances in Control and Communication, LNEE 137, pp. 219–225.
springerlink.com

Shandong. So we study it and its fracture development not only by drilling, well test and log data in Jiayang Depression, but also the outcrops—the direct, abundant, accurate observation—in Western Shandong. In the outcrops, the strata develop in Liuhang in Xintai, Dongniujia Village, Panchegou, Xiannanyu in Taian City, Xiangshan in Laiwu City, Huoguan of Zhangqiu City. The Holostratotype section with the thick of 622.4m, locates 300 meters south to Xinanyu Village of Jiaoyu town in Taian City.

2 Results and Disscussion

2.1 Strata

1) Strata in outcrops: Liuhang Formation is in conformable contact with the schistose biotite amphibole-oligoclase-granulite containing trondhjemite gravel belongs to Shancaoyu Formation (Fig.1, A&B). And is in unconformable contact with Mantou Formation of Lower Cambrian Series (Fig.1, G&H). It is composed of micron-anorthosite-amphibolite, chlorite schist, biotite granulite, amphibole-biotite-granulite, Sericite-quartz-schist, neutral and acidic metamorphic volcanic breccia, and sedimentary conglomerate, with iron content of tremolite quartzite(Fig.1, C-F). Palimpsest structure is seldom seen in anorthosite-amphibolite. Blasto crystal fragment texture and blastopsammitic texture are clear in granulite. Magmatic rock with acidic component is their main virgin rock, Magmatic rock with neutral component comes second. The composition of gravels are complex, there are two broad categories: Volcano dust and sedimented clast — such as plagioclase granite, altered quartz diorite, pegmatite, vein quartz and felsite.

2) Strata in Jiyang Depression: Among the wells in Jiyang Depression Liuhang Formation developed in Zheng-401, Chenggu-19, with the main rock is hornblende schist, amphibolite, granitic gneiss, metamorphogenic crystal tuff(Fig.1, I), with lamprophyre intrusion, granodiorite and biotite adamellite intrusion (Fig.1, J). In Chenggu-19 the Liuhang Formation develops 1760 meters below the ground and is at least 1402 meters long. In Zheng-401 it develops 1675.37 meters below the ground and is at least 15.97 meters long.

2.2 The Development Characteristic of the Outcrops-Fractures

We choose the data of fracture development in one typical station in the study area, through laboratory study and analysis we obtain a result about fracture groups, dip and density. We also do research on the facial porosity based on the observation of Chenggu19 and Zheng401. The characters are as follows:

1) Fracture groups: There are faults to the right of the study area, the downthrown side developed middle and upper cambrianseries Ordovicianseries. While the upthrown block developed Liuhang Formation. 54 fracture direction were statistically analyzed(Fig.1). Fractures are best developped on the strikes of NEE, with 52 fractures at the angle of 60°~65°, perpendicular to fault strike. While on the strikes of SSE, 22 fractures are developed at the angele of 175°, It is agreed with that the orientation of the fault——SSE. These groups intersect With each other, then formed the fractures grid system. That make the fracture reservoir a connection.

A&B: Schistose biotite amphibole-oligoclase-granulite containing trondhjemite gravel belongs to Shancaoyu Formation.

C: Quartziferous granulite. D: Micron-anorthosite-amphibolite.

E: Weathering front of Serpentinite-talc-schist

F: schistosityplane of Chlorite- tremolite-schist

G&H: Unconformable contact with Mantou Formation

I: Metamorphogenic crystal tuff (plane polarized light, length 1.2mm).

J: Biotite adamellite (plane polarized light, length 2.9mm).

Fig. 1. Characters of Liuhang Formation

2) Fracture dip: The distribution of fracture dip has a dissymmetrical bimodal feature(Fig.2). The development of fracture dip is mainly concentrated on the 60°~90°, accounting for 72.60% of the total dips of statistics. At 60°~76° is particularly concentrated (Fig.2), accounting for 58.90% of the total dips of statistics. While there are few fractures from 60°~90°. We predict that it is caused by shear stress. Vertical and high-angle oblique fractures developed well in Liuhang Formation(Tab.1). They can incise the rock mass into clintheriform and columanr(Fig.3). The Reservoir drilling bases on vertical wells in High-angle fractures, drilling data revealed the extent of fracture development may be much lower than the actual development of underground fracture system. which belongs to vertical fractures.

3) Fracture density: Fracture density is the fracture number that there developed in unit area. Micro-fractures develop well and has the higher density. In the schist of schist-granulite interbeds there are more than 600 micro-fractures every 1m. Fracture density is controlled by boundary faults control. the near by the fault, the density is high. But near the intrusion, the fracture density can result in great changes. Invades body can make the fracture development degree of get high improvement.

4) Fractures in Jiyang Depression and Facial porosity: Facial porosity is the accumulative total surface area in unit area. The facial porosity can reach maximum 24%, mainly concentrated in the 10% ~ 20%. Liuhang Formation in Jiyang Depression mainly developed unfilled fractures with oil and oil patch. Fractures are mostly less than 1 mm wide, and micro-fractures develop well, there always have catactastic structure(Fig. 5), detrital material content can accounts for 26-34%. The well is very near to fault, so the well development of fractures are mainly caused by the tectonic stress.

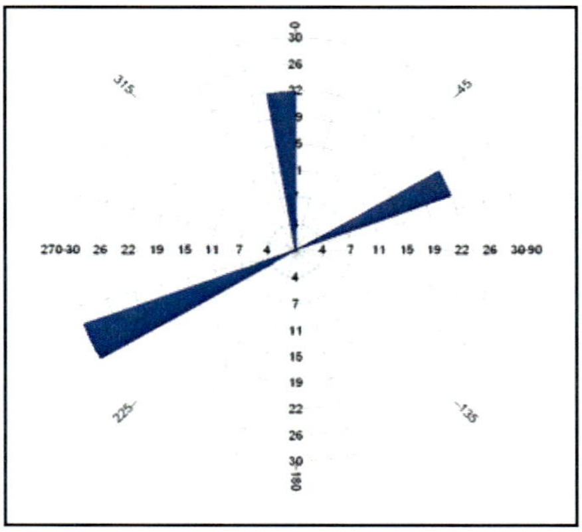

Fig. 2. Rose diagrams of Liuhang-Formation-fracture direction in outcrop area

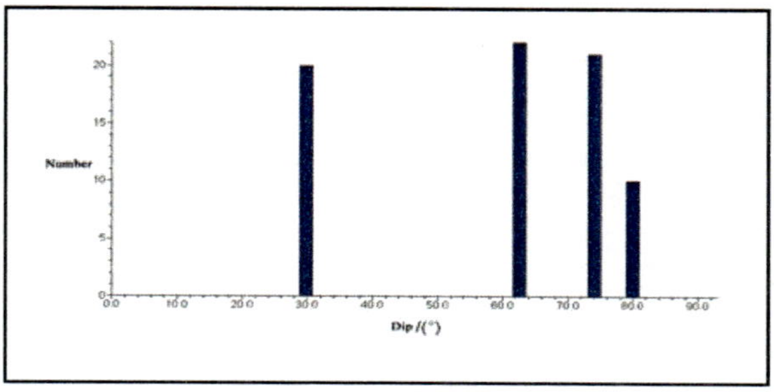

Fig. 3. Distribution of Liuhang-Formation-fracture dip in outcrop area

Table 1. Distribution of fracture developed angle

Fractures	**Level**	**Low-angle oblique**	**High-angle oblique**	**Vertical**
Angle	<5°	5°~30°	30°~70°	>70°
Accounting(%)	0	27.40	30.14	42.46

Fig. 4. Development of fracture of Liuhang Formation in outcrops

2.3 Governing Factors on Fracture Development

According to the study on the characters of fracture in Western Shandong and Jiyang Depression, combine with the predecessor's study we obtain the governing factors on fracture development——lithology, tectonic stress and overlying strata.

Lithology is the basic factors that affected fracture development. In cataclastic rock the fractures develop well while in Plagioclase amphibolite the fractures develop not so well.

The development of fractures also depends on the characteristic of the stress, depth of function and length of action time. Tectonic stress plays a dominate role in the fracture development. Where there is high tectonic stress, where there are always fractures densely developed. Always The closer to the fault, the better Fractures develop. While the farther away from the fault, the few fractures developed.

When the overlying strata were formed in Paleozoic, it is easy to develop well opened vuggy porosity. Always the top surface of Archeozoic strate weathered seriously. Fractures and microfractures developped well in Taishan Group with clastic rock of Mesozoic and Cenozoic overlying.

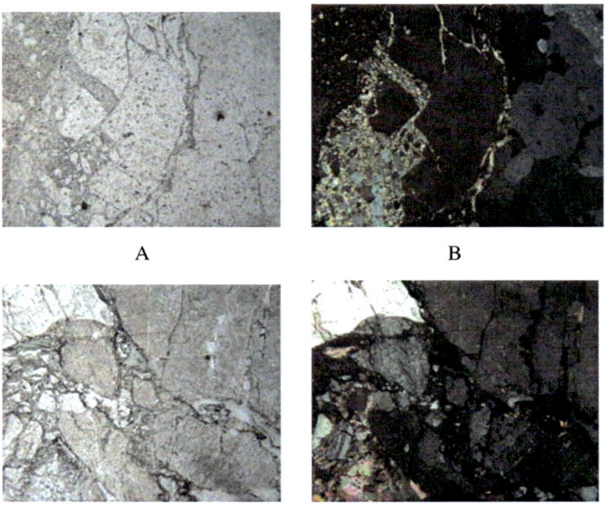

A B

C D

A&C: Plane polarized light.

B&D: crossed polarizers.

Fig. 5. Catactastic structure and micro-fractures in Liuhang Formation

3 Conclusions

Liuhang Formation is composed of micron-anorthosite-amphibolite, chlorite schist, biotite granulite, amphibole-biotite-granulite, Sericite-quartz-schist, neutral and acidic metamorphic volcanic breccia, and sedimentary conglomerate, with iron content of tremolite quartzite. The formation always have catactastic structure.

In Liuhang Formation, the fractures develops well and can be the reservoir for oil-gas. In Shengli oil field the development of fracture is mainly controlled by regional tectonic activities. Also be affected by the lithology. The closer to big fault, the better develop the fractures and hydrocarbon pore volume. It always is the main part for oil-gas reservoir.

Outcrop observation shows that the reservoirs are mainly caused by high-angle and Vertical fractures, so in this reservoirs deviated wells and horizontal wells. can meet more fractured reservoir, which is beneficial to the efficient development.

Acknowledgment. This study was supported by the Science & Technology Leading Project of SINOPEC(S0-PT-1091D010Z) and the Postgraduate Science &Technology Innovation Foundation of Shandong University of Science and Technology (YCA100203).

We thank the Geophysical Research Institute in Shengli Oil-field of Sinopec for the provision of seismic data.

References

1. Bureau of Geology and Mineral Resources of Shandong Province. Regional Geology of Shandong Province. Geological Publishing House, Beijing (1991)
2. Hou, G., Li, J., Jin, A., et al.: New Comment on the Early Precambrian Tectono-Magmatic Subdivision and Evolution in the Western Shandong Block. Geological Journal of China Universities 10(2), 239–249 (2004)
3. Hu, H.: Deep Hydrocarbon Reservoir Formation Mechanism Survey. Petroleum Geology and Oilfield Development in Daqing 6, 24–26 (2006)
4. Wang, D., Jin, Q., Dai, J., Zhang, J.: Distribution Rule and Methods of Evaluation of Buried-hill Oil-gas Reservoir Space. Geological Publishing House, Beijing (2003)
5. Wang, S.: The Stratigraphic Division of the Taishan Group in Western Shandong and the Characteristics of its Protoliths. Geological Bulletin of China 9(2), 140–146 (1992)
6. Zhang, J., Jin, Q.: Characteristics of fractures and their hydrocarbon reservoir meanings for the Archean outcrop in Laiwu area, Shandong province. Petroleum Geology and Experiment 25(4), 371–374 (2003)
7. Zhang, Z., Liu, M., et al.: Rock Strata in Shandong Province. China University of Geosciences Publishing House, Wuhan (1996)
8. Lemon, A.M., Jones, N.L.: Building solid models from boreholes and user-defined cross-sections. Computers & Geosciences 29, 547–555 (2003)
9. Deutsch, C.V., Journel, A.G.: GSLIB:geostatistical software library and user' s guide. Oxford University Press, New York (1992)
10. Ehlen, J., Harmon, R.S.: Geo Comp99: CeoComputation and the geosciences. Computers & Geosciences 27(8), 1–2 (2001)
11. Jessel, M.: Three dimensional geological modeling of potential-field data. Computers & Geosciences 27(4), 455–465 (2001)
12. Lemon, A.M., Jones, N.L.: Building solid models from bore-holes and user-defined cross-section. Computers and Geosciences 29(3), 547–555 (2003)
13. Simon, W.H.: 3D Geoscience modeling: Computer Techniques for Geological Characterization. Springer, New York (1994)

The Long-Range Monitoring System of Water Level Based on GPRS Network

Zhang Yi-Bing

The Measure and Control Technology Institute,
Taiyuan University of Technology,
Taiyuan, Shanxi, 030024, China
600600106@163.com

Abstract. Through a combination of detection technology and GPRS communication technology, rainfall stations and watershed hydrological stations in each observation point will be used as the system terminal. Those data collection terminals are composed of sensor unit and microcontroller system, mainly in charge of on-site water level, flow, and rainfall and other signal acquisition, software filtering, processing, storage and display. Real-time data collected will be stored, or transmitted through the RS232 and GPRS network to the data terminals, after post-processing by the central computer, provided to upper management. This system has been achieved a long-range, wide range of data transmission, and provided a scientific way for flood control decision.

Keywords: GPRS water information Monitoring system.

1 Introduction

Traditional water information system has characters of short transmission distance because monitor stations are usually located in bad environment which makes the construction and layout is difficult to be set up. The wireless station can cover at most dozens of mile. As social and economic development and population increase, the losses by the same order of magnitude flood will be more and more. This enhances flood control projects, non-engineering facilities, improving disaster prevention and resilience, with a very importance and urgency. The development of mobile communication and Internet technology has brought deep change to society. GPRS wireless data transmitting technology ripens step by step to make it be applied in many trades, and it offers a kind of new means for the data transmission of testing system. According to the characteristic of distributed data collection system, through the combination collection equipment of field data and GPRS wireless communication , we can link mutually the intelligent instruments that spreads in every collection point and can complete certain task independently. When certain condition comes, the system can realize the response immediately, long-range online detection. In this way it can save plenty of manpower and material resources. It is the outcome of Internet technology and communication technology and detection technology. Because valley water-information observation and prediction has the characteristic of

D. Zeng (Ed.): Advances in Control and Communication, LNEE 137, pp. 227–234.
springerlink.com

broad regional distribution, we adopt the GPRS wireless communication, and the data of the hydrometric stations and rainfall stations will be delivered to the data handling center that locates in distant place, so the remote online water-information Monitoring system is formed. The system utilizes the mobile communication network which covers most areas of the whole country to realize wireless data transmission, monitoring long-range online. In real time, quickly and accurately capture the representative points within the catchment rainfall, hydrological information, and passed to the relevant flood control department, to provided a scientific basis for decision making for the flood forecasting warning, the reservoir for flood control and flood mitigation etc.

2 GPRS Summary

GPRS is the abbreviation of General Packet Radio Service. It is the technical standard established by European telecommunications standard association (ETSI).It is a kind of packet radio exchange technology based on GSM system, and it offers the connection of end to end and wireless IP of broad region. Comparatively the way of dialing of relative GSM is the way of circuit exchange data delivering. GPRS adopts the packet exchange technology, so it have the excellence of " real time online" , "charge according to quantity" , "shortcut registration","high speed transmission". Because GPRS is based on the extensive applied protocol of TCP/IP, it offers the powerful and functional agile solving scheme of data transmitting. We can use the system everywhere only if the place is covered by the GSM network. GPRS network is the system which we increase SGSN (supporting node of service GPRS), GGSN (supporting node of gateway GPRS) to the GSM network foundation. The PCU located in BSC side is used to separate the data service and voice service, and control the distribution of wireless resource. The function of SGSN node is more similar with MSC/VLR, the functions of it ,such as, encryption , lesson right, IMEI inspection , moving management , counting data management , divide into groups road from with transmit , logic link management and connection the functions such as GSM net Yuan, GGSN node has network , receive to enter , grouping route , transmitting ,logical link management and connecting GSM network or so. And the GGSM node has the functions suck as connecting network, dominating and transmitting the grouping route, connecting the other IP network .It is similar with the traditional gateway. That using GPRS communication technology to transmit data information is made use of existing resource, that the network of mobile communication that can cover the most of nation-wide areas. So it establishes for user the data transmission of celerity, convenient, low cost passageway of data transmission, it realizes wireless data transmission, long-range online detection. Its data transmission speed: 57.6 kb/s, the transmission speed of next a tape, GPRS business can reach 384 kb/s. GPRS uses standard SIM card to stock disposition information ,such as machine card , baud rate of serial port , the checking way , server IP address and joint port number and etc. From above-mentioned GPRS characteristic, we can find out, GPRS network is suit for the real-time transmission which is frequently, small data.

3 The Structure of On-Line Detection Monitoring System

The system is made up of three parts: the long-range on-line Monitoring, transmission and workstation of data processing centre. The system software is divided into two parts, the first is the equipment application of remote on-line Monitoring, which is programmed by assembly language of SCM, and the second is the program of running on server, which is programmed by VC++, and combines with the provided saddlebag by the GPRS manufacture.

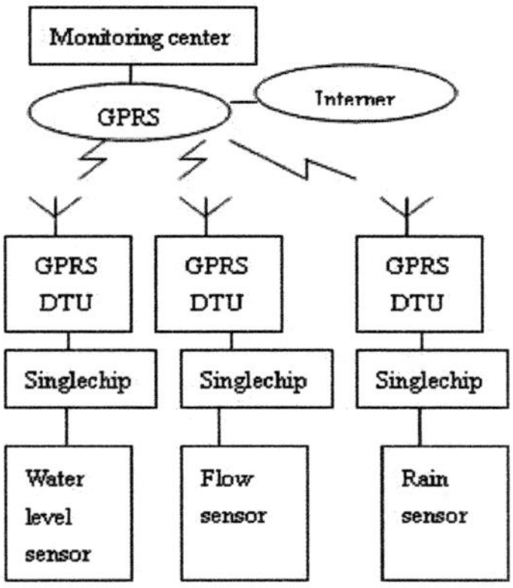

Fig. 1. System structure

l) Long-range on-line Monitoring: In the part, we have the hydrological station, rainfall station which work on the field as the terminal of the system, and they are fixed on the observation stations. The system is made up of sensor unit and SCM, and the main function is the Monitoring of field water-level, flux, rain full, and software filtration and management and storage and display. The SCM system has signal converting, 12bit A/D converting, serial EEPROM, digital display, keyboard, controlling circuit and interface circuit of RS232 asynchronism. The result of real-time Monitoring can be stored or transmitted to the data terminal of GPRS by RS232 port.

2) Transmission: We make use of GPRS wireless data terminal unit(DTU) , and it adopt the wireless DDN data terminal . System core modules from the main controller DTU GPRS wireless communication module Rabbit3000 and Siemens MC391i as core part, mainly by the following components:

Central control part adopts semiconductor company was the industrial control chip Rabbit3000 as microprocessors. This processor has rich resources, can match the interface software development platform Dynamic C can realize online programming! Debug! Simulation and guarantee the GPRS terminal powerful and good maintainability and scalability.

GPRS module USES MC391i. MC391i support standard set of instructions and standard 3VSIM card, its good temperature properties and stable job performance. Its serial 3.3 V interface levels) and Rabbit3000 serial port C with standard 9 line way (3.3 V interface levels) connected. RS232 / RS485 interface part adopts MAX3232CSE and MAX3485 on level conversion.Each block needs SIM card when using ,and it has the unique ID which is similar with the mobile telephone . The terminal supports GSM/GPRS, and it accords with the standard of ETSI GSM Phase 2+. It adopts the real-time clock, supports encrypting algorithm of A5/1& A5/5, supporting the special dummy data traffic network of high-speed, on-line forever, data transmitting lucidity. Rabbit3000 microprocessor internal except for storage download on the program of the ROM (with) beyond campaigns for, no other built-in data storage area and program storage area, so the user need according to oneself project requires add-in storage area and data storage area. Structure the simplest was by a Flash memory chips (with/CS0 interface) and a RAM memory chips (with/CS0 interface) composition. In practical application, the smallest Flash capacity is 128K, RAM capacity is 32K. Was also can support smaller chips, but these small sram is obsolete, so there is no support.

Although was supported by the code biggest can reach 1-m, but in most applications 250K, equivalent to the best less than 10000-20000 bar C statement. This was not only reflects the streamline code, also accord with most embedded application size.

In usually memory model, data space must with root code, stack and XPC window sharing a64k space. Normally, this kind of practice is put apart 40K or less of a potential data space. XPC Windows need to 8K, stack need 4K, and most system should have at least 12K root code. For the vast majority of embedded application speaking, these data space enough.

Was restricted to directly access C variables for about 44K memory, in a storage in Flash RAM and data between between isolated. To most of the embedded application speaking, these memory space already enough.

Maybe some applications require additional data memory to store huge an array or a list, so the dynamic C provides for extension data memory type of support, it expanded additional data memory quantity to outweigh the 1M.

Whether you need a stack memory depends on the type of application, especially whether to use the beat multitasking, then each task needs to have his stack. Because the stack in a 16-bit address space has its own section, therefore it is convenient to use memory to achieve a lot of stack. When environment accident of change occurs, STACKSEG register will be made corresponding change, with the stack segment mapped to RAM for a part, this part of the RAM contains will run new tasks associated stack. Usually the stack size 4K, enough for a stack (usually 4) provide space. When need greater than 4K stack, also can expand stack segment. If only need a stack, can leave out the whole stack segment and will this single stack into data segment. This kind of method applies to 32KRAM only and do not need multiple stack system.GPRS DTU block has the user data interface of 20 pins, in order to supply power to the block and exchange the data .The connection of external data wire and the way of data exchange are similar with the RS232. At the same time, most types of the series support the power of +7.5DCV--+26DCV, so it is very convenience to fix in more places. When the connection of terminal is over, we can manage the GPRS terminal through the foundation, management, debug, so it is easy to collocate the interfix parameter before using, neatly change the parameter during debug, and upgrade software and simply test. The system adopts the way of center to multi-spot .There are many devices of automatic water-level detection, they bags the data by their own GPRS data terminal, connect the wireless GPRS network through GPRS interface, and then transfer to internet by mobile service, finally reach the data processing working station by gateway and route. It is necessary to point out that the data transmission which is the terminal of GPRS to data business center is discontinuity, and we can adjust the interval time of data transmission according to the needs, so it debases the fee of wireless transmission in a certain extent.

3) Data processing center: Data processing center is the dominating part of the whole system, which contains one pedestal main server and three other data processing servers. Windows Server 2003 OS and SQL Server 2000 database software are both installed in the main sever, and it also has a fixed IP address in the internet. All the data that comes from water monitor station and rain monitor station transports to this main server through the network on the first step. In view of system stability, only database software is allowed to run on main server; in order to reduce data processing load of main server, three other data processing servers are established. This set, including one main server and other three data processing server, is similar to the data server center in intranet. Although the data sever center contains many servers, there is only one IP address to internet by using the network address translation technology (NAT). The NAT ports reflect function connects the main server to every particular host in intranet, so the interview to the main server through a destined port is provided by a destined host in intranet. For instance, we can hide a host's inner IP address when it provides the www service to outward. In our system, data processing servers handle the information that comes from different port respectively, and then send the data back to the SQL database in main server; the on-line server is in charge of receiving the data from the line. In the end, the www host reserves data from the SQL database in main server, and displays the water information on the web page timely, so the superior department can get all information through the internet.

4 The Realization of System Function

There are two different way to build GPRS network, outside net method and inside net method. The outside net method, such as figure 2 shown, is stable and the GPRS DTU (Data Transmission Unit) is simple to design. But besides the connect problem with dynamic IP address, the system safety is not ensured. The inside net method is shown in figure 3. By this way, all important parts of the system are inside the GPRS network, without any contact to the internet. It is safe, convenient and cheap to set up, therefore, we choose the inside net method to build our system. But it is still not perfect; some problem still needs to improve.

l) Communication protocols: Communication protocols reflect the function of the whole system, so the design of communication protocols is very important. Generally speaking, supervise-control system software mostly adopts three layer C/ S structures. This kind of structure separates the presentation layer, the session layer and the data link layer. We can design a particular module handling the data exchange between supervise-control center and terminal on the session layer. This module means efficiency, safety and change to extend. In this article, the communication protocols between data processing center and remote data terminal unit (DTU) is the communication protocols between message assistant (MA) and remote data terminal unit (DTU). In addition, this communication protocol is an application layer protocol, it shields the difference among data deliver in wireless network, can be in good use in GPRS network etc. Three is no relationship between this protocol and transport layer 、 network layer or data link layer.

2) Hosts break and information exchange: Because of the characteristics of the GPRS network, break happens constantly when the main server of the data server center receives information from equipments. So some precondition that CMCC provides a fixed IP address in GPRS network for main server is required. For in MH2MH network, if the main server did not have a fixed IP address, there would be no effective way to rebuild the system after a host break. In consideration of the characteristics of the system, it is an efficient way to adopt the UDP protocol for the communication between message assistant and remote terminal unit (RTU). To ensure the stability, every time of the communication includes one request and one corresponding response. The response failure maybe due to the packet loses in transfer process or some other problem in GPRS network. If there was no corresponding response, the system would send request message again instantly and constantly. If the resend request way did not work, we should check the network and reconnect, or use the standby communication system to solve.

5 System Software Design

Software design using Dynamic C, it is the Z - World a company developed the special IDE, it contains a integrated C compiler, editor, the link editor, loader and debugger, has replaced expensive circuit simulators, simplifies the use of user. Because the software design requirements of GPRS

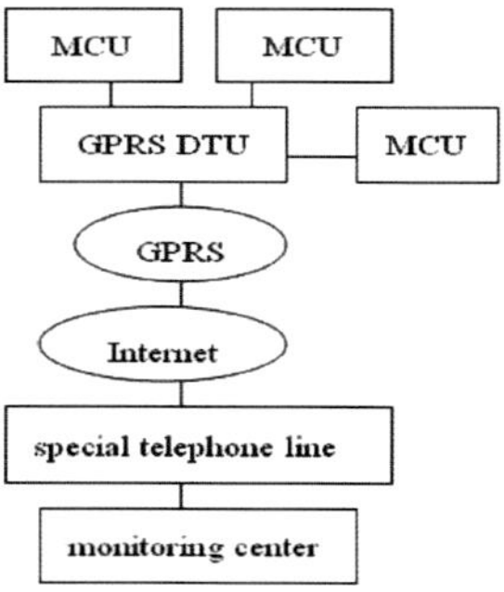

Fig. 2. The outside net method

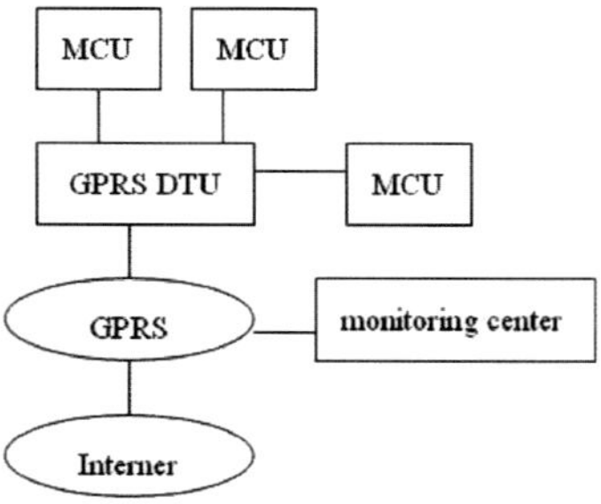

Fig. 3. The inside net method

terminal Rabbit3000 at work to executing simultaneously two or more tasks, and each task has certain relevance between each other, so need to adopt more mature real-time operating system. Real-time operating system flexibly allocation of resources (CPU, memory, etc.) for each task, according to the different situations of each task execution of different operating. Usually real-time operating system kernel divided into two types of Non - before type (preemptive, also called a collaborative multitasking, Cooperative Multi Tasking) kernel and wonderful - type (preemptive) kernel. The wonderful type kernel is through the central processor will timer time into time slice, then according to the different situations of each task assigned time slice. Each task priority of the same, according to their own needs to apply for or abandonment of the CPU use, mutual cooperation sharing a CPU. Wonderful type kernel is CPU according to various task priority different determine which task access control of the central processor, making the task response time to optimization. Dynamic C itself built-in two real-time operating system kernel, wonderful type kernel USES is transplantation, the II kernel muon COS kernel is wonderful type by Dynamic C provides Multi Tasking functions. - Because GPRS Modem software function modules for task response time request is not too high, and most of the time by time slice round-robin method parallel operation, does not exist priority level problem, thus GPRS DTU reactive software design multi-back Multi Tasking technology as design basis . the terminal software system mainly divided into six working modules. System initialization module, network login module, connect module, network state detection module, the data transmission module and watch-dog module.

First of all, according to the need for initialization parameter Settings, and monitoring whether need to modify Settings, if need be, then pass the super GPRS TU display Settings menu in Chinese, according to clew setting, and will be kept in Flash parameters, otherwise, in accordance with the initialization parameter configuration system, the initialization each I/O ports, serial ports, and the corresponding global variables. Initialize Siemens MC391i module and dial-up through the PPP agreement login GPRS networks, through monitoring DCD feet potential and module return values determine login results. If there is a failure, automatic weighing dial to ensure system reliable login GPRS networks and real-time online. GPRS networks login successfully automatically after run this module connects to the server, sends GPRS directory terminal address, the user password, waiting for login information such as the result. If successful, criterion according to login set keep alive time automatically send keep alive after login information, otherwise overtime give up and automatic reconnecting with, so as to ensure the reliable connection "directory server connection, according to the directory server instruction execution corresponding operation.

6 Conclusion

This system is applicable to the ambulated, separate, no man guarding and dynamic monitor-control system. It can be expanded in many industries such as electric power, Water conservancy, petroleum, communication, environmental protection etc.

References

1. Feng, L.-J., Zhang, M., Fan, J.-F.: The Communications Realization of Remote Monitoring System based on GPRS. Microprocessors (3), 36–38 (2010)
2. Yin, L., Zhao, X.-Q., Sun, X.-H.: A Design of Remote Monitoring Software for Fiber Optic Sensing Net. A Design of Remote Monitoring Software for Fiber Optic Sensing Net (9), 69–71 (2010)
3. Luo, W., Wang, L.N., Xiao, K.: Remote monitoring and software upgrade of embedded system based on GPRS. Application of Electronic Technique 31(53), 159–162 (2010)
4. Chen, X., Liu, Z., Han, Z.: Design and Implementation of Remote Monitoring System Based on GPRS. Network Security Technology & Application (10), 44–45, 67 (2010)
5. Zhang, W., Wang, H., Cheng, P.: Remote Intelligent Monitoring System of Street Lights Based on GPRS. Application Research of Computers 27(9), 2104–2106 (2010)
6. Zhao, X.-Q.: The Designed of Remote Sewage Monitor System Based on GSM. Microcomputer Information 8(5), 91–92, 85 (2010)
7. Tong, W.-Y., Liu, C.-M., Zhao, G.-C.: Water Quality Parameters Monitoring System of Data Long-distance Transmission Based on GPRS. Automation & Instrumentation 6(7), 52–55 (2010)
8. Jiang, Y.-L.: Data Transmission and Management of Long-distance Monitoring System. Instrumentation Technology 35(8), 43–45, 48 (2009)

Chat Analysis to Understand Students Using Text Mining

Yao Leiyue and Xiong Jianying

Department of Software Research,
JiangXi Bluesky University,
Nanchang China 330098
ylyyly2001@163.com, special8212@sohu.com

Abstract. Network communication has been the main method in people communication. The goal is to help educators grasping the students' social and psychological status and personal characteristics by mining conversation contents of a student group. In this paper, the text mining is used to find hot topics in chat groups, personal language behavior, personal motivation, and members of the emotions involved in the overall performance trends and the personal emotional expression trend. By using word frequency analysis, co-occurrence word frequency analysis, and lexical emotional similarity analysis, experimental results show that the method can quickly and effectively grasp the characteristics of subjects for students to develop strategies provide the basis for education.

Keywords: Chat analysis, Text mining, word orientation, Student behavior.

1 Introduction

People interactive each other all the life, and with the development of information technology, to communicate through the network has become very popular, and is the mainstream of communication. Network communication includes receiving and sending e-mails, distance learning, online searching, cooperating in collaborative environments by chatting, or browsing the tremendous amount of online information. Currently, there are several chat tools available on the internet, and the emergence of such tools has realized the communication among the internet users from all over the world. Some of them are very popular such as ICQ, hotmail. In China, most Internet users prefer to use Tencent's QQ, it has also become an important tool for office and entertainment [1]. QQ chat text is conductively to a certain degree of monitoring, such as to aid in crime detection or even crime prevention. The children and young teenagers chats their thoughts and desires in a relatively short period of time, and expressed a certain literary form. In particular, university students is a popular, for a critical stage of psychological development, the analysis of the conversation text of them is conducive to help educators understanding the social psychology, personality status, hot issues among them, and make best education management for students [2].

The motivation is that the current chat content analysis techniques are basically manual [3], which is difficult, costly, and time consuming. A text mining techniques for the problem of automatic monitoring of QQ chat group is presents in this paper. By use text mining, these research including group membership, individual characteristics, hot issues, and emotions recognition among the group could been done.

D. Zeng (Ed.): Advances in Control and Communication, LNEE 137, pp. 235–243.
springerlink.com

The paper is organized as follows; section 2 describes the text mining and an overview of the chat room or group chat content monitoring problem. Section 3 gives an overview of our study and our experimental results. The experimental evaluation is discussed in section 4. Finally, section 5 concludes this paper and discusses possible extensions.

2 Related Work

Text mining is a rich semantic analysis process of text to understand the content and meaning it contains. It has become an important field of data mining. Existing text mining techniques rely on more structured, formal corpuses containing research papers, abstracts, reports, etc. Online chatting differs in a number of ways from everyday face-to-face conversation, both qualitatively and quantitatively. Text chat is a dynamic text, the article inputted by user each time is not as complete as would not necessarily follow strict grammatical structure, is a context-sensitive dynamic text. The user may express a specific theme by few words. The topics may be totally irrelevant, and chat text is formed by a short text. Approaches toward understanding the dynamics of chat conversation are limited, and as usage grows the need for automated analysis increase. From this point of view, it becomes necessary to analyse these conversations and to understand the characteristic of the speakers.

1) Summarizing conversation topic: In text mining applications, determination of conversation topic is one of the important study areas. Most of the studies are based on classification of news texts, and others' researches related to the determination of text writer's characteristics.

2) Understanding the user behavior: Radford examine the communication and information seeking preferences of the Internet users. They also compare traditional libraries and the Internet as the means for an information repository and emphasize the fact that the internet is starting to become an alternative for text based communication.

3) Investigation of chat user attributes: Gender variations was examined in Web logs using logistic regression techniques. However, the authors can not find any conclusive results binding the users genders and web writings. In their work, herring and danet examine several aspects of the language use in the Internet. They assert that gender is reflected in online discourse in every language they studied.

4) Understanding social and semantic interactions: The social and semantic relationships extracted from chat conversations can lead to a better understanding of human relations and interaction. Social clusters and conversational content are understood automatically.

5) Others research: Monitoring chat room conversations, extracting interesting information, authorship attribution and so on.

3 Methold of Chat Mining

3.1 Goal of Mining

We can obtain lots of information about the group and each people in the group from the conversation. Members of the group also have different responsiveness to conversation content. By using the text mining, group membership and individual characteristics can be analyzed by speech frequency and word frequency; the hot issues can be conducted by high frequency words and occurrence words; and understand emotions and personal feelings using the text orientation recognition.

3.2 Data Gathering

The conversation data were gathered from chat mediums by using QQ messenger log files. These data were subjected to pretreatment, and were prepared for data mining. In the pre-treatment of the data, basic steps of the data mining were taken into consideration. The formatted conversation which includes name, id, content, time is as follow:

```
罗青林(962566099)        非英语专业研究生演讲比赛开始报名，欢迎大家踊
考虑。预赛时间为2010年11月15日晚上6点半至9点半   2010-11-7 11:16:15
詹宏陆(403328235)        怎么几位大侠都未睡呀      2010-11-7 23:12:40
詹宏陆(403328235)        都是性情中人啊  2010-11-7 23:13:15
```

Fig. 1. Formatted conversation text

3.3 Mining Process

1) Exclude formatted data: the nickname, QQ number, speeches are formatted data, the conversation content was extracted, and the feature of text processing is required before this part of the data extracted. Text chat is a form of informal text, so there will be some sign language, network language. The definition of some commonly used languages in Table 1 and Table 2, and some other face figures as fig 3 defined in the QQ file can be read directly.

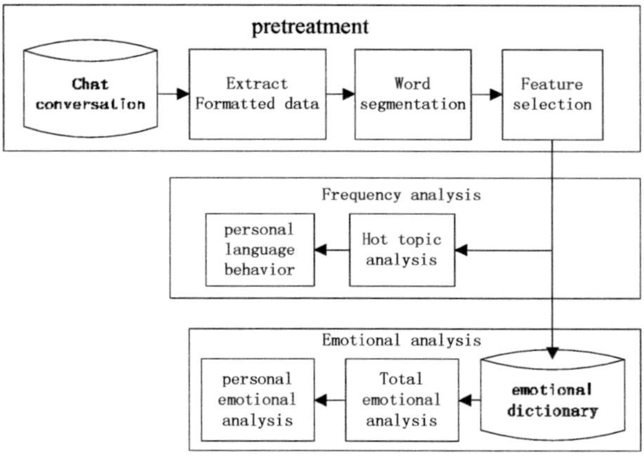

Fig. 2. Mining Process Flow

Fig. 3. QQ face

Table 1. Definition of network slang

slang	meaning
886, 88	goodbye
Lp	wife
Lg	husband
MMD, NND, TMD	curse
…	…

Table 2. Definition of signs

Signs	meaning
☺,:),:)),:D, :d	langhing
:-O	surprised
@>>--->---	rose
?_?	what
…	…

In order to speed up the efficiency of word segmentation, stop words can be filtered firstly. Segmentation algorithm is Reverse Maximum Method (RMM) which is a simple and effective method for Chinese word segmentation based on dictionary. Then filter characteristic words using defined nonsense words. Then conversation content is collected by speaker, and the storage data structure is as fig 3, and name is nickname, id is QQ account number, conFeature is collection of feature word list.

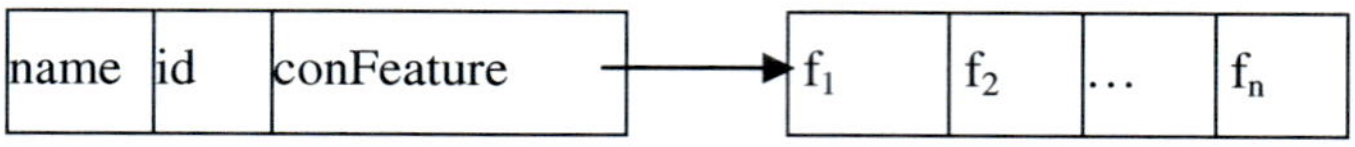

Fig. 3. Data structure

Algorithm described as follows:

```
WHILE(!endof(textFile))
    {
    read(textFile) /*read one conversation record*/
    extract(Name,Id,Time) /*extract the format data*/
    extract(Content) /*extract the unformatted data*/
    ConText=filter(Stoplist) /*filter by stoplist words*/
    segmentation(ConText)
    f=filter(InsignificanceList)
    addto(f,ConFeature)
    }
```

2) Analysis using word frequency: TF is word frequency, F is feature set, the hot words can be found by TF rank. When the TF beyond a threshold, it can be used as a hot word, is also means high frequency words. The hot words list can be defined as $T(t_1,t_2\ldots t_n)$. N(t,f) define as the occurrences of the hot words and other words, of which $t \in T$, $f \in F$. then select the high occurrences value to produce hot topic sentence.

3) Emotional recognition: Emotional orientation of words includes words polarity, intensity and context model analysis. The polarity and strength of words can be defined in dictionary.

If analyze emotional tendency to all chatters by all words, social and psychological tendencies of the group can be understood. And if analyze emotional tendency to each chatter by his own words, the individual's social and psychological tendencies can be got. By comparing all individual's incline to the overall degree, the differences DIFF can be calculated as follows.

$$DIFF=EMP-EMT \tag{1}$$

$$EM = \sum_{k=1}^{t} X_k \tag{2}$$

EMP means individual's emotional orientation, and EMT means the overall emotional orientation which is calculated as the sum of all the characteristics of emotional words.

If DIFF is positive, then the individual holds a positive attitude on the issue than the whole, greater value, and the greater the individual differences.

If DIFF is negative, then the individual on the issue of negative attitudes held higher than the overall, greater value, the greater the individual differences.

If DIFF is equal to or close to 0, then the issue held by individuals is consistent with the general.

4 Results and Discussion

Total size of the conversations is 2.37MB. The conversation which included the chat content in 2 months has 41 talkers. 1839 conversation records were gathered from these mediums.

4.1 Basic Analysis

Speech frequency: From the speech frequency of the 49 talkers, only several people performs actively, most of people have little interest, the result illustrate that the cohesion of the group is lower.

Characteristic words: Through segmentation, the word frequency is greater than 3 times a 501. After the definition of the word meaningless word filters are 408 features, combined with HOWNET Chinese dictionaries, 182 nouns separated by a synonym for the word combination are features of 146 seeds as a hot topic the search words.

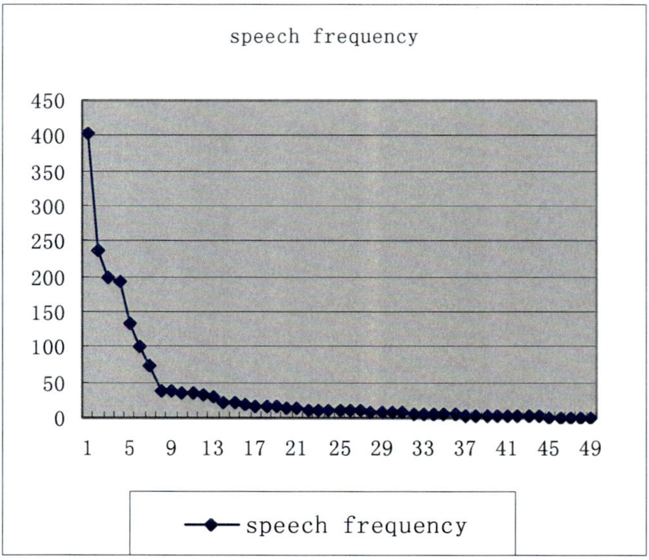

Fig. 5. Speech frequency distribution

4.2 Hot Topic

Select as term frequency greater than 10 hot words, a total of 41, through co-occurrence analysis, we found that the 41 nouns with a total of 106 key words are distributed scatter as follows:

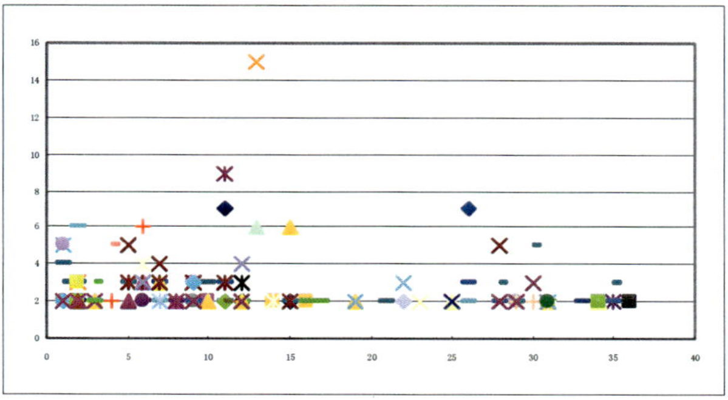

Fig. 6. Words Co-occurrence Distribution

If connecting the hot words with those words that have high Co-occurrence, the topic can be deduced. Such as teacher call the roll, paper issued, economics research, game theory research, Chinese politics status. For the group is made up by doctoral student, so the topics are in line with the doctoral student population characteristics.

4.3 Overall Emotion Analysis

The different emotions results are as Table 3. From the overall distribution data, the overall group community emotion lay particular stress on positive,(35.2% vs 13.21%). Neutral attitude very clear(51.59%).

Table 3. Tendency distribution

Emotion	records	distribution
positive	653	35.20%
neutral	957	51.59%
negative	245	13.21%

The different polarity level of positive emotions segment results are as Table 4.

Table 4. Positive emotion polarity distribution

polarity	records	distribution
Common(0~10)	500	26.95%
Middle(10~20)	116	6.25%
High(>20)	37	1.99%

The different polarity level of negative emotions segment results are as Table 5.

Table 5. Negative emotion polarity distribution

polarity	records	distribution
Common(-10~0)	188	10.13%
Middle(-20~-10)	48	2.59%
High(<-20)	3	0.16%

In further emotion classification, most of the word's polarity of positive and negative is common; and the high level words occupied very small. It hints that the sociality of this group is normal, during this period.

4.4 Difference between Individual and Overall Emotion

There is existed great difference between individual and overall in positive and neutral attitude from the individual emotion minus overall, as fig 7 shows the neutral and positive two lines fluctuate badly. It is means that the attitude towards some topics has little consistency. But to the negative attitude, as fig 7 shows the flat negative lines. It implicated that more people express negative towards topics has consistency.

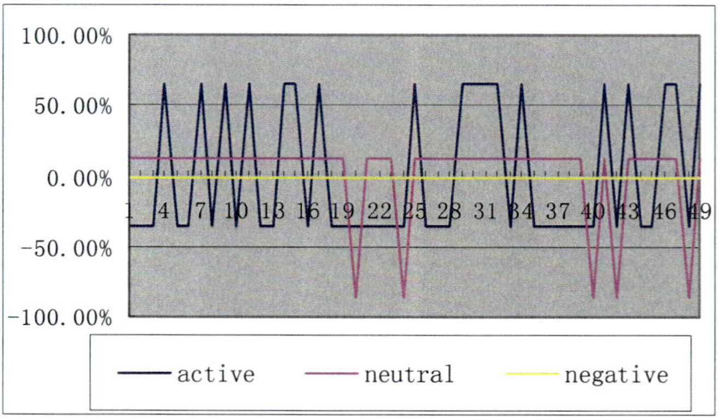

Fig. 7. Speech frequency distribution

5 Conclusion

Development of information technology makes network communication become one of the mainstream communications. In China, many point to point chat tools provide group chat, such as Tencent's QQ which is an important tool in office work and entertainment. Especially for students group, their thoughts and desires often express when chatting in a relatively short period of time, is a certain product of literary form of a campus cultural phenomenon. It is beneficial to formulate education management from the analysis of the text. In this paper, word frequency analysis, document analysis, and lexical emotional recognition are used in the mining of chat content, in order to understand personal language behavior, personal motivation, and members of the emotions involved in the overall performance trends and the personal emotional expression trend. Experimental results show that the method can quickly and effectively grasp the characteristics of subjects for students, and to develop strategies provide the basis for education.

References

1. Tian, F.: Hot QQ: a new social intercourse. China Youth Study (3), 16–19 (2003)
2. Li, X., Ma, L.: Psychological Perspective from QQ chat behavior of college students. Higher Science Education (6), 72–75 (2009)
3. Meehan, A., Manes, G., Davis, L., Hale, J.: Packet sniffing for automated chat room monitoring and evidence preservation. In: Proceeding of the 2001 IEEE, Workshop on Information Assurance and Security (June 2001)
4. Li, S.Y., He, W.: A study on the method of feature selection in chat text. Computer Science (5), 202–204 (2007)
5. Bengel, J., Gauch, S., Mittur, E., Vijayaraghavan, R.: ChatTrack: Chat Room Topic Detection Using Classification. In: Chen, H., Moore, R., Zeng, D.D., Leavitt, J. (eds.) ISI 2004. LNCS, vol. 3073, pp. 266–277. Springer, Heidelberg (2004)

6. Argamon, S., Sacic, M., Stein, S.: Style mining of electronic message for multiple authorship discrimination:First results. In: Proceddings of the Ninth ACM SIGKDD International Conference on Knowledge Discovery and Data Mining, pp. 475–480 (2003)
7. Dickey, M.H., Burnett, G., Chudoba, K.M., Kazmer, M.M.: Do you read me? Perspective making and perspective taking in chat communities. Journal of the Association for Information Systems 8(1), 47–70 (2007)
8. Walther, J.B., Bunz, U., Bazarova, N.N.: Rules of virtual groups. In: Proceedings of the 38th Annual Hawaii International Conference on System Sciences (2005)
9. Zheng, R., Li, J., Chen, H., Huang, Z.: A framework for authorship identification of online messages: Writing-style and classification techniques. Journal of the American Society for Information Science and Technology 57(3), 378–393 (2006)
10. Aas, K., Eikvil, L.: Text categorisation: A survey, tech. rep., Norwegian Computing Center (June 1999)
11. Kerrand, D.S., Murthy, U.S.: Divergent and convergent idea generationin teams: A comparison of computer-mediated and face-to-face communication. Group Decision and Negotiation 13, 381–399 (2004)
12. Liao, X., Cao, D., Fang, B.: Research on Blog Opinion Retrival Based on Probabilistic Inference Model. Journal of Computer Research and Development 46(9), 1530–1536 (2009)

Pricing Mechanism on Carbon Emission Rights under CDM

Jiangshan Bao and Jinjin Zuo

The School of Economic and Management,
Beijing University of Technology, Beijing, China
by1258@163.com, zuojinjin_2045@126.com

Abstract. Cost difference between developing countries and developed countries is basic condition of Clean Development Mechanism(CDM).This paper, based on cost difference between developing countries and developed countries, aims to construct bilateral transaction model in the condition of price limits and without price limits, through the bilateral transaction game between developing countries and developed countries, coming to the conclusion that forced quotation police takes important influence trading price and trading volume in national CDM market, and advances some suggestion when China faces the opportunity and challenge.

Keywords: CDM, carbon emission rights, pricing mechanism, bilateral transaction model, bayes game.

1 Introduction

"Kyoto Protocol" marks the emergence of the economic era of carbon emissions, which created a new product (greenhouse gas emissions) in the world and was quickly commercialized. According to statistics of World Bank, from 2005 to 2008, average annual growth in global carbon trading reached 126.6%.Although the global financial crisis in 2008 resulted in gliding of projects based on CDM, yet secondary market remained active all the time; quota-based transactions still maintain a rapid growth momentum, the annual trade volume reached 126.35 billion dollars and increased 100.5% ,compared to 63.01 billion dollars in 2007 .The World Bank estimated global carbon trading will amount to 1,500 billion dollars by 2012, which is expected to become the world's largest market and over the oil market. In the global carbon trading, the EU emissions trading system has been dominant. In 2008, the EU emissions trading system trading amounted to 91.91 billion dollars, trading of carbon dioxide equivalent volume was 3.09 billion tons, far more than the EU emissions trading system and the average level of global carbon trading. Under the current CDM projects, China is the main supply-side transactions in 2008, accounting for 84% share of the world, India and Brazil ranked second and third, respectively, of global proportion of 4%, 3%.

2 The Review of Domestic and Foreign Research

Bahn (2001) [1] imitated ERUs market of national emissions trading by using the MARKAL-MACRO model, assessed CO2 emissions trading of ERUs market in

D. Zeng (Ed.): Advances in Control and Communication, LNEE 137, pp. 245–252.
springerlink.com

European countries, and considered to reduce emissions marginal cost through cooperation in the different States. Bosello and Roson (2002) [2] studied the impact of social welfare between two trading countries because of emissions trading system, and took the idea that the research of climate change policy should focus on equity and efficiency, fairness involved in the distribution of emissions reduction target between countries, efficiency involved in the potential national income by using flexible mechanisms to trade .Rehdanz and Tol (2005) [3] introduced game theory into bilateral trade between developed countries, and analyzed the results of market trading rules and the impact of environmental benefits. Kemfert (2004) [4] and Bernard (2007) [5] use game theory and general equilibrium theory of the emissions trading market, the simulation analysis, that the emissions trading market will be the main supplier of China and Russia. Julien Chevallier (2008) [6] studied pricing model under the EU emissions system in all sectors of bilateral transaction in the first phase (from 2005 to 2007).

In China, research on pricing mechanism of carbon emissions is still in its infancy. Chen Wenying and Wu Zongxin (1998) [7] took the fairness and efficiency of carbon allocation mechanism into consideration, determined the price of carbon trading based on the global carbon trading conduct, and analyzed impact of carbon emissions trading on China and impact of different carbon credit allocation mechanism on the earnings carbon trading. Yu Tianfei (2007) [8] introduced option mechanism into the initial allocation of emission rights, and took advantage of B-S model to study on option pricing of carbon emissions. Hong Juan and Chen Jing (2010) [9] analyzed the future price and spot price relationship between main two commodities (EUA and CER) and relations between futures prices of commodities. Through the establishment of VAR model and Granger causality test and impulse response function method, the author made empirical analysis of the international carbon emissions trading market price. The paper, from the microeconomic point, discussed bilateral mechanism of carbon emissions trading between the developed and developing countries, and confirmed that bilateral limit policies of international carbon trading could have a significant impact on the equilibrium price.

3 Carbon Pricing Model of Bilateral Trade

Current carbon emissions trading roughly divided into two categories, one is based on the quota trading, buyers purchase emission reduction quota distributed by the administrator in the "cap and trade" system, such as the allocation units (AAU) under the Kyoto Protocol, or the EU Emissions Trading Scheme (EU ETS) under the EU quota (EUAs); the second is project-based transactions, the buyers buy emission credits from projects confirmed the reduction of greenhouse gas emissions. The most typical example is certified emission reductions (CERs) and Emission Reduction Units (ERUs) under the CDM and JI.As can be drawn from table 1, carbon trading volume in clean development mechanism reached 389 million tons in 2008, and creates profits of 6519 million dollars, the second turnover CDM up to 26277 million dollars. In the emissions market, the EU Emissions Trading Scheme (EU ETS) dominant market share in 2008, trading accounted for 98% of the total, and the transactions volume show the growing trend.EUA and CER have been two main types of commodities in international carbon emissions trading market. In Figure 1, prices of EUA and CER present change of V-shaped model because of turmoil in the financial crisis in the world economy, carbon trading shows stagnation. From May 2009, the global carbon trading prices keep steady-state trend. Therefore, the global carbon trading market has become interconnected unity, developing countries as a supplier of CERs market, whether CERs price change will affect the world trading or not? Through the following establishment of bilateral trade model and analyze the reasons.

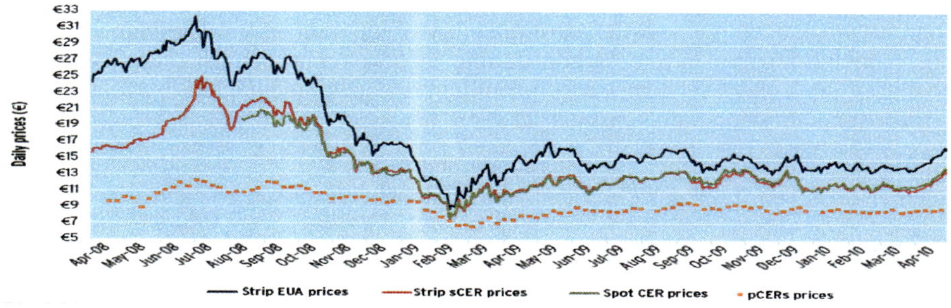

Fig. 1. Carbon emissions trading price, Source: World Bank Report 2010

Table 1. The scale of global carbon emissions trading

		2007		2008	
		Volume (million tons)	Turnover (million dollars	Volume (million tons)	Turnover (million dollars
Trading conditions in different types of projects	Primary CDM	552	7433	389	6519
	JI	41	499	20	294
	Voluntary market	43	263	54	397
	sum	636	8195	463	7210
Secondary CDM					
sum		540	5454	1072	26277
Emissions Quota Market	EU ETS	2060	49065	3093	91910
	New South Wales, Australia	25	224	31	183
	Chicago Climate Exchange	23	72	69	246
	RGGI	na	na	65	246
	AAUs	na	na	18	211
sum		2108	49361	3276	92859
total		2984	63007	4811	126345

Source: World Bank Report 2009.

3.1 Bilateral Trading Model under No Price Limit

Cost of reducing emissions of carbon dioxide depends on a variety of marginal substitution possibilities, such as various alternative fuels and alternative energy-intensive products capabilities. The greater capabilities of the potential alternative, the lower cost to meet targets of specific emission reduction, usually take "top down" and "bottom up" model to assess the substitutability of the options and mitigation costs. In this paper, from the starting carbon trading in developing countries and developed countries, the author establishes cost function in regional i, the model is as follows:

The total emission reduction cost function:

$$TCD_i = \frac{1}{3}a_i Q^3 + \frac{1}{2}b_i Q_i^2 \tag{1}$$

The marginal emission reduction cost function:

$$MC_i = a_i Q^2{}_i + b_i Q_i \tag{2}$$

Where, Q represents region's emission reductions, a_i and a_i represent the parameters of the marginal cost functions. Make the following assumptions: There are developed country (A)and developing country(B), A need to take emission reduction targets, but there is no clear emission reduction obligations for B, A and B state trade CERs with each other in accordance with emission reduction of CDM projects, both pursue for maximize profits and cost minimization, that is: A wants to minimize the cost of reducing emissions to achieve emission reduction targets, B wants to gain maximum profit in the transaction, and gains more than financial support and technology transfer, according to (1) and (2), establish equations is:

$$\min C_A = \frac{1}{3}a_A Q^3{}_A + \frac{1}{2}b_A Q^2{}_A + p_A q_A \tag{3}$$

$$s.t. Q_A + q_A \geq T_A \tag{4}$$

$$\max G_B = p_B q_B - (\frac{1}{3}a_B Q_B^2 + \frac{1}{2}b_B Q_B^2) \tag{5}$$

$$s.t. Q_B \geq Q_A \tag{6}$$

Where, C_A represents abatement costs of A country, Q_A and Q_B represents respectively that emissions of A and B countries, p_A represents the purchase price of country A , p_B represents he sale price of country B, q_A represents purchase of country A , q_B represents the sale amount of B country , T_A represents emission reduction targets of A country, G_B represents trading profits of the country B, a_B and b_B are the parameter values of marginal abatement costs. When reach balance, the developed and developing countries have reached the optimum profit, now, has the condition: $p_A = p_B = p^*$ and $q_A = q_B = q^*$, A's domestic emission reductions is $T_a - q^*$, B's emission reductions is q^* .

3.2 Bilateral Trading Model under Price Limit

Make the following assumptions: developing countries have the lowest price v, when only the transaction price p> v, there is the transaction, or there is no transaction; for developed countries, only if the marginal abatement costs (MCA) is higher than the price-fixing v, they will purchase emissions from developing countries t, or no trading. So the critical point of both transactions is:

$$MC_A = a_A Q^2{}_A + b_A Q_A = v \tag{7}$$

When $MCA > V$, developed countries will choose to complete the emissions reduction targets in the territory, not for trading.

When $MCA < V$, there are three cases:

When $MCA > V > MCB$, Limit price is higher than the marginal cost of emissions in developing countries, the transaction price p = v, trading volume is decided by the equation (2), (4) and (7) jointly.

When $MCA > MCB > V$, Limit policies have no effect on the developing and developed countries, the actual transaction price is equal to the transaction costs in developing countries, $a_B Q^2{}_B + b_B Q_B = p \quad q = Q_B$

When $MCA < MCB$, there is no trading, CDM mechanism is meaningless.

3.3 Bayes Game Analysis of Bilateral Trade

In carbon emissions trading market, market participants includes the supply side (developing countries) who provide emissions commodities, the demand side (developed countries) who buy emissions commodities, supply and demand sides through bargaining, and ultimately determine the transaction price, the process is actually a Bayesian game process. Assumptions: in linear equilibrium case, emissions for the units of emissions goods, the cost of goods is provided by developing countries is c, priced is Ps, and the price is a linear function of cost, namely: $P_S = P_S(c) = a_s + b_s c$; Assessment value of developed countries is v, bid price is Pd, and the bid price is a linear function of the value of v, that is: $P_d = P_d(v) = a_d + b_d v$, also assume: c and v obey uniform distribution [0, 1]. When $p_s \le p_d$, then the sides of the transaction price is $p = (p_s + p_d)/2$, the effectiveness of the developed countries as $U_s = (p_s + p_d)/2 - c$, the utility of the buyer is $U_d = v - (p_s + p_d)/2$; $p_s > p_d$, no transaction occurs, the effectiveness of the two sides are zero. Conditions of profit maximization for parties transaction is:

For developed countries,

$$\max\left\{\left[v - \frac{1}{2}(p_d + E(p_s(c) | p_d \ge p_s(c)))\right] * \Pr ob(p_d \ge p_s(c))\right\} \tag{8}$$

For developing countries,

$$\max\left\{\left[\frac{1}{2}(p_s + E(p_d(v)|p_d(v) \geq p_s) - c)\right] * \Pr ob(p_d(v) \geq p_s)\right\} \tag{9}$$

Satisfying: $p_s(c)$ is subject to $[a_s, a_s + b_s c]$ with uniform distribution, $p_d(v)$ is subject to $[a_d, a_d + b_d v]$ with the uniform distribution. Then,

$$\Pr ob(p_d \geq p_s(c)) = \Pr ob(p_d \geq a_s + b_s c)$$
$$= \Pr ob(c \leq (p_d - a_s)/b_s) = (p_d - a_s)/b_s$$

$E(p_s(c)|p_d \geq p_s(c))) = \dfrac{\frac{1}{b_s}\int_{a_d}^{a_d+b_d(v)} xd_x}{2\Pr ob(p_d \geq p_s(c))} = \dfrac{1}{2}(a_d + p_d)$ Take it into (8); it can obtain the first-order condition:

$$p_d = \frac{1}{3}a_s + v \qquad E(p_d(v)|p_d(v) \geq p_s) = \frac{1}{2}(p_s + a_d + b_d)$$

Take it into (9); it can obtain the first-order condition:

$$p_s = \frac{1}{3}(a_d + b_d) + \frac{2}{3}c$$

Simultaneous can be obtained:

$$p_s(c) = \frac{1}{4} + \frac{2}{3}c \qquad p_d(v) = \frac{1}{12} + \frac{2}{3}v$$

This can be linear equilibrium,

$$p = \frac{1}{2}(p_s + p_d) = \frac{1}{6} + \frac{1}{3}(c + v)$$

The effectiveness of the developed countries is $U_d = \frac{2}{3}v - \frac{1}{3}c - \frac{1}{6}$

The effectiveness of the developing countries is $U_s = \frac{1}{6} + \frac{1}{3}v - \frac{2}{3}c$

The total effectiveness of all countries is $U = v - c$

4 Inspiration of Carbon Price Mechanism on Carbon Trading In China

4.1 Establish Domestic Carbon Trading Platform and Actively Involve in International Carbon Trading Market

China, as the world's leading emitter of greenhouse gases and largest supplier of global carbon emission, are suppose to make full use of CDM mechanism to promote sustainable economic and social development. in one hand, expand CDM financing, on another hand, speed up the domestic energy industry and energy-intensive traditional industry transformation, thereby increase its ability of dealing with adverse influence of climate change, but also can take advantage of CDM carbon emissions to promote the production of CERs trade for the domestic sale of foreign exchange and increase foreign exchange reserves so that enhance China's international competitiveness. However, China failed to establish a national carbon trading platform in the international carbon trading market. CERs for sale in China only enter the international carbon trading market by the international carbon funds, or through the involvement of the World Bank and other institutions. Therefore, the establishment of a domestic carbon trading platform has become China's further integration into the international carbon trading market, the need for more ways to profit.

4.2 Establish and Improve Mechanisms for Initial Allocation of Emission Rights

Carbon emissions trading is the most important issue in the initial allocation of emission rights, different distribution methods will generate different environmental and economic effects, thus affect the competitiveness of enterprises, cause a fair question. Current international allocation of carbon emission rights have: equal per capita allocation method and the contraction and convergence approach, allocation based on historical emissions, or based on the latest data distribution and sale distribution and mixing distribution. The international allocation of carbon emissions on the distribution is both economic and also political game; the role of each country varies. For China and other developing countries, in order to take full advantage of the advantage, to ensure that the technical conditions of use should be taken based on the historical per capita emissions equivalent way.

4.3 Explore the Potential of Domestic Capital and Expand Financing Channels

CDM achieves sustainable development by providing funds and technology to developing countries to assist them in achieving emission reduction targets, co-funding for developing countries should also become the main source of income under the emission reduction activities of CDM. China has means of expanding finance: First, through government direct investment, which is very suitable for CDM investments that have large external effects. Second one is through financial channels, which is reflected the fluctuations by the price of CERs, the investor's decision-making can be more rational. When resolve the financing channels, China can improve negotiations status with foreign partners for CDM projects and get a higher price advantage.

Acknowledgment. This work is partially supported by Science & Technology Innovation Platform Grant # 0110005466009 and National Academic Team Project of BJUT Grant # 01100054R7001to Jingwen Li and Education Skills Project Grant # 01100054R6002 to Jun Guan.

References

1. Bahn, O., Barreto, L., Büeler, B., Kypreos, S.: A Multi-regional MARKALMACRO Model to Study an International Market of CO2 Emission Permits: A detailed analysis of a burden sharing strategy among the Netherlands, Sweden and Switzerland. PSI technical paper, Villigen (1998)
2. Bosello, F., Roson, R.: Carbon Emissions Trading and Equity in International Agreements. Environmental Modeling and Assessment 7(1), 29–37 (2002)
3. Rehdanz, K., Tol, R.S.J.: A No Cap but Trade Proposal for Greenhous Gas Emission Reduction Targets for Brazil, China and India. Working Papers FNU-68, Research unit Sustainability and Global Change, Hamburg University (revised July 2005)
4. Kemfert, C.: Climate coalitions and international trade: assessment of cooperation incentives by issue linkage. Energy Policy 32(4), 455–465 (2004)
5. Bernard, A.L., Vielle, M.: Measuring the welfare cost of climate change policies: A comparative assessment based on the computable general-equilibrium Model GEMINI-E3. Environmental Modeling & Assessment 8(3), 199–217 (2003)
6. Chevallier, J.: Strategic Manipulation on Emissions Trading Banking Program with Fixed Horizon. Economics Bulletin, Access Econ. 17(14), 1–9 (2008)
7. Chen, W., Wu, Z.: Allocation of carbon emissions and carbon emissions trading. Tsinghua University (Natural Science) (December 1998)
8. Yu, T.: Carbon emissions trading market research. Nanjing Forestry University (2007)
9. Hong, J., Chen, J.: Empirical Study on Price of international carbon emissions trading. China Price (January 2010)

Analytic Solutions of an Iterative Functional Differential Equation Near Resonance

Lingxia Liu

Department of Mathematics, Weifang University, Weifang,
Shandong 261061, P.R. China
llxmath@126.com

Abstract. This paper is concerned with a functional differential equation with a deviating argument depending on the state derivative. By reducing the equation with the *Schröder* transformation to another functional differential equation with proportional delay, an existence theorem is established for analytic solutions of the original equation. In this paper, we obtain results of analytic solutions in the case of α at resonance and the case of α near resonance under the Brjuno condition.

Keywords: Functional differential equation, Analytic solution, Brjuno condition, Resonance.

1 Introduction

Functional differential equations of the form $x'(t) = f(x(\sigma(t)))$ have been studied by many authors. However, when the delay function $\sigma(t)$ is state dependent, $\sigma(t) = x(t)$, relatively little is known. Cooke in [1] studied the equation $x'(t) = x(t - x(t))$. In [2], analytic solutions of the state dependent functional differential equations

$$x'(z) = x^{[m]}(z), m \geq 2 \tag{1}$$

are found, where $x^{[m]}(z) = x(x^{[m-1]}(z))$. As in [2], our strategy remains to reduce the equation (1) with the *Schröder* transformation

$$x(z) = y(\alpha y^{-1}(z)) \tag{2}$$

to the auxiliary equation

$$\alpha y'(\alpha z) = y'(z)y(\alpha^m z), \tag{3}$$

Where iteration of the unknown function is not involved but an indeterminate complex α needs to be discussed. We need to find invertible analytic solutions of the equation (3) for possible choices of α. For technical reason, in [2], the constant α given in the *Schröder* transformation, is required to fulfil that α is off the unit circle $(0 < |\alpha| < 1)$ or α is on the circle with the Diophantine condition: $|\mu| = 1$, μ *is not*

D. Zeng (Ed.): Advances in Control and Communication, LNEE 137, pp. 253–260.
springerlink.com

a root of unity, and $\log\frac{1}{|\mu^n-1|}\le T\log n,\quad n=2,3,\cdots$ *for some positive constant* T. The Diophantine condition requires α to be far from all roots of unity. In this paper, we break the restriction of the Diophantine condition and study the situation that the constant α is a root of unity in **C** considered in [3], and those α near resonance under the Brjuno condition. In order to find analytic solution of (3), we assume that (3) satisfies the initial value conditions

$$y(0)=\alpha, \tag{4}$$

and α satisfies one of the following conditions

(H1) $\mu=e^{2\pi i\theta}$, where $\theta\in$ **R\Q** is a Brjuno number ([4, 5]), i.e., $B(\theta)=\sum_{k=0}^{\infty}\frac{\log q_{k+1}}{q_k}<\infty$, where $\{p_k/q_k\}$ denotes the sequence of partial fraction of the continued fraction expansion of θ, and is said to satisfy the Brjuno condition.

(H2) $\alpha=e^{2\pi iq/p}$ for some integers $p\in\mathbf{N}$ with $p\ge 2$ and $q\in\mathbf{Z}\setminus\{0\}$, and $\alpha\ne e^{2\pi il/k}$ for all $1\le k\le p-1$ and $l\in\mathbf{Z}\setminus\{0\}$.

We observe that α is on the unit circle in (H1) and (H2). In the case of (H2) α is a p-th unit root (or called p-order resonance), while the case (H1) contains a part of α near resonance, i.e., neither being roots of the unity nor satisfying the Diophantine condition.

The Brjuno condition in (H2) provides such a chance for us. As stated in [6], for a real number θ we let θ denote its integer part and let $\{\theta\}=\theta-[\theta]$. Then every national number θ has a unique expression of the Gauss' continued fraction

$$\theta=a_0+\theta_0=a_0+\frac{1}{a_1+\theta_1}=\ldots,$$

denoted simply by $\theta=[a_0,a_1,\ldots,a_n,\ldots]$, where a_j's and θ_j's are calculated by the algorithm:

(a) $a_0=[\theta],\ \theta_0=\{\theta\}$ and

(b) $a_n=\left[\frac{1}{\theta_{n-1}}\right],\theta_n=\left\{\frac{1}{\theta_{n-1}}\right\}$ for all $n\ge 1$.

Define the sequences $(p_n)_{n\in\mathrm{N}}$ and $(q_n)_{n\in\mathrm{N}}$ as follows

$$q_{-2}=1,\ q_{-1}=0,\ q_n=a_nq_{n-1}+q_{n-2};\qquad p_{-2}=0,\ p_{-1}=1,\ p_n=a_np_{n-1}+p_{n-2}.$$

It is easy to show that

$$p_n/q_n=[a_0,a_1,\ldots,a_n].$$

Thus, for every $\theta\in$ **R\Q** we associate, using its convergence, an arithmetical function $B(\theta)=\sum_{n\ge0}\frac{\log q_{n+1}}{q_n}$. We say that θ is a Brjuno number or that it satisfies Brjuno condition if $B(\theta)<+\infty$. The Brjuno condition is weaker than the Diophantine condition. For example, if $a_{n+1}\le ce^{a_n}$ for all $n\ge 0$, where $c>0$ is a constant, then is a

Brjuno number but is not a Diophantine number. So, the case (H1) contains both Diophantine condition and a part of μ near resonance.

$$\theta = [a_0, a_1, \ldots, a_n, \ldots]$$

In this paper, considering the Brjuno condition instead of the Diophantine one, we discuss not only the case (H1) but also (H2) for functional differential equation (1).

2 Auxiliary Equation Case (H1)

In order to discuss the existence of analytic solutions of the auxiliary equation (3) under (H1), we need to introduce Davie's lemma. Let $\theta \in R\setminus\theta$, and $(q_n)_{n\in\mathbf{N}}$ be the sequence of partial denominators of the Gauss's continued fraction for θ as in the Introduction. As in [6], let

$$A_k = \{n \ge 0 \mid \|n\theta\| \le \frac{1}{8q_k}\},\ E_k = \max(q_k, \frac{q_{k+1}}{4}),\ \eta_k = \frac{q_k}{E_k}.$$

Let A_k^* be the set of integers $j \ge 0$ such that either $j \in A_k$ or for some j_1 and j_2 in A_k, with $j_2 - j_1 < E_k$, one has $j_1 < j < j_2$ and q_k divide $j - j_1$. For any integer $n \ge 0$, define

$$l_k(n) = \max\left((1+\eta_k)\frac{n}{q_k} - 2,\quad (m_n\eta_k + n)\frac{1}{q_k} - 1\right),$$

Where $m_n = \max\{j \mid 0 \le j \le n,\ j \in A_k^*\}$. We then define function $h_k : \mathbf{N} \to \mathbf{R}_+$ as follows:

$$\begin{cases} \dfrac{m_n + \eta_k n}{q_k} - 1, & \text{if} \quad m_n + q_k \in A_k^*, \\ l_k(n), & \text{if} \quad m_n + q_k \notin A_k^*. \end{cases}$$

Let $g_k(n) := \max\left(h_k(n), \left[\dfrac{n}{q_k}\right]\right)$, and define $k(n)$ by the condition $q_{k(n)} \le n \le q_{k(n)+1}$. Clearly, $k(n)$ is non-decreasing. Then we are able to state the following result:

Lemma 1. (Davie's Lemma [7]) Let

$$K(n) = n\log 2 + \sum_{k=0}^{k(n)} g_k(n)\log(2q_{k+1}).$$

Then

(a)There is a universal constant $\gamma > 0$ (independent of n and θ) such that $K(n) \le n\left(\sum_{k=0}^{k(n)} \dfrac{\log q_{k+1}}{q_k} + \gamma\right)$,

(b) $K(n_1) + K(n_2) \le K(n_1 + n_2)$ for all n_1 and n_2, and

(c) $-\log|\alpha^n - 1| \le K(n) - K(n-1)$.

Now we state and prove the following theorem under Brjuno condition. The idea of our proof is acquired from [7].

Theorem 1. Suppose (H1) holds, then for any complex number $\eta \neq 0$, the equation (3) has an analytic solution $y(z)$ in a neighborhood of the origin such that $y(0)=\alpha$ and $y'(0)=\eta$.

Proof. We assume Eq. (3) has a power series solution of the form

$$y(z)=\sum_{n=0}^{\infty} b_n z^n,\ b_0=\alpha,\ b_1=\eta\neq 0. \tag{5}$$

Substituting series (5) of y in (3) and comparing coefficients, we obtain

$$(\alpha^{n+1}-\alpha)(n+1)b_{n+1}=\sum_{k=0}^{n-1}(k+1)\alpha^{m(n-k)}b_{k+1}b_{n-k},\ n=1,2,\cdots \tag{6}$$

in a unique manner. Furthmore, we have

$$\left|\frac{(k+1)\alpha^{m(n-k)}}{(n+1)(\alpha^{n+1}-\alpha)}\right|\leq\frac{1}{|\alpha^n-1|},\ n\geq 1,\ 0\leq k\leq n-1.$$

It follows from (6) that

$$|b_{n+1}|\leq\frac{1}{|\alpha^n-1|}\sum_{k=0}^{n-1}|b_{k+1}||b_{n-k}|,\ n=1,\ 2,\ \cdots. \tag{7}$$

In order to construct a governing series of (7), we define a sequence $\{B_n\}_{n=1}^{\infty}$ by $B_1=|\eta|$ and

$$B_{n+1}=\sum_{k=0}^{n-1}B_{k+1}B_{n-k},\ n=1,2,\cdots. \tag{8}$$

If we define

$$G(z)=\sum_{n=1}^{\infty}B_n z^n, \tag{9}$$

then

$$\begin{aligned}G^2(z)&=\sum_{n=2}^{\infty}(B_1B_{n-1}+B_2B_{n-2}+\cdots+B_{n-1}B_1)z^n\\&=\sum_{n=1}^{\infty}(B_1B_n+B_2B_{n-1}+\cdots+B_nB_1)z^{n+1}\\&=\sum_{n=1}^{\infty}B_{n+1}z^{n+1}\\&=G(z)-|\eta|z.\end{aligned}$$

So we have

$$G(z)=\frac{1}{2}\pm\frac{\sqrt{1-4|\eta|z}}{2}.$$

Because $G(0)=0$, so

$$G(z)=\frac{1}{2}-\frac{\sqrt{1-4|\eta|z}}{2}.$$

We obtain $G(z)$ is convergence in $|z|<\frac{1}{4|\eta|}$. Hence there is a constant $T>0$ such that

$$B_n<T^n,\ n=1,2,\cdots. \tag{10}$$

Now we prove

$$|b_n|\le B_n e^{K(n-1)},\ n=1,2,\cdots,$$

where $K:N\to R$ is defined in lemma 1. In fact $b_1=|\eta|=B_1$, we assume that $|b_j|\le C_j e^{K(j-1)}$, $j\le n$. From Lemma1 and (7) we have

$$\begin{aligned}|b_{n+1}| &\le \frac{1}{|\alpha^n-1|}\sum_{k=0}^{n-1}B_{k+1}B_{n-k}e^{K(k)+K(n-k-1)}\\ &\le \frac{e^{K(n-1)}}{|\alpha^n-1|}\sum_{k=0}^{n-1}B_{k+1}B_{n-k}\\ &\le \frac{e^{K(n-1)}}{|\alpha^n-1|}B_{n+1}.\end{aligned}$$

Note that

$K(l_1-1)+K(l_2-1)+\cdots+K(l_t-1)\le K(n-2)\le K(n-1)\le K(n)+\log|\alpha^n-1|$, then $|b_{n+1}|\le B_{n+1}e^{K(n)}\le T^{n+1}e^{K(n)}$.

Note that $K(n)\le n(B(\theta)+\gamma)$ for some universal constant $\gamma>0$, then $|b_{n+1}|\le T^{n+1}e^{n(B(\theta)+\gamma)}$, that is,

$$\limsup_{n\to\infty}\left(|b_{n+1}|^{\frac{1}{n}}\right)=\limsup_{n\to\infty}\left(T^{\frac{n+1}{n}}e^{B(\theta)+\gamma}\right)=Te^{B(\theta)+\gamma}.$$

This implies that the convergence radius of the series (5) is at least $\min\{\frac{1}{4|\eta|},\left(Te^{B(\theta)+\gamma}\right)^{-1}\}$. This completes the proof.

3 Auxiliary Equation Case (H2)

Let $\{D_n\}_{n=1}^{\infty}$ be a sequence defined by $D_1 = |\eta|$ and

$$D_{n+1} = \Gamma \sum_{k=0}^{n-1} D_{k+1} D_{n-k},\ n = 1, 2, \cdots, \tag{11}$$

where $\Gamma = \max\left\{1, \left|\alpha^i - 1\right|^{-1}, i = 1, 2, \cdots, p-1\right\}$.

Theorem 2. Suppose (H2) holds, let $\{b_n\}_{n=0}^{\infty}$ be determined recursively by $b_0 = \alpha, b_1 = \eta$ and

$$(\alpha^{n+1} - \alpha)(n+1) b_{n+1} = V(n, \alpha),\ n = 1, 2, \cdots, \tag{12}$$

where

$$V(n, \alpha) = \sum_{k=0}^{n-1} (k+1) \alpha^{m(n-k)} b_{k+1} b_{n-k}.$$

If $V(sp, \alpha) = 0$ for all $s = 1, 2, \cdots$, then equation (3) has an analytic solution $y(z)$ in a neighbourhood of the origin such that $y(0) = \alpha$, $y'(0) = \eta$, and $y^{(sp+1)}(0) = (sp+1)!\eta_{sp+1}$, where all $\eta_{sp+1}'s$ are arbitrary constants satisfying the inequality $|\eta_{sp+1}| \le D_{sp+1}$ and the sequence $\{D_n\}_{n=1}^{\infty}$ is defined in (11). Otherwise, if $V(sp, \alpha) \ne 0$ for some $s = 1, 2, \cdots$, then the equation (3) has no analytic solution in any neighborhood of the origin.

Proof. As in the proof of Theorem 1, we seek a power series solution of (3) of the form (5), then the equality in (6) is indispensable. If $V(sp, \alpha) \ne 0$ for some number s, then the equality in (6) does not hold for $n = sp$. This is because $\alpha^{sp+1} - \alpha = 0$, then such a circumstance equation (3) has no formal solutions. When $V(sp, \alpha) = 0$ for all natural numbers, for each s the corresponding b_{sp+1} in (6) has infinitely many choices in **C**, that is, the formal series solution (5) define a family of solutions with infinitely many parameters. Choose $b_{sp+1} = \eta_{sp+1}$ arbitrary such that

$$|\eta_{sp+1}| \le D_{sp+1},\ s = 1, 2, \cdots, \tag{13}$$

where D_{sp+1} is defined by (11). Now we prove the power series solution (5) converges in a neighborhood of the origin. Note that $\left|\alpha^n - 1\right|^{-1} \le \Gamma$ for all $n \ne sp$, then

$$|b_{n+1}| \le \Gamma \sum_{k=0}^{n-1} |b_{k+1}| |b_{n-k}|,\ n \ne sp,\ s = 1, 2, \cdots.$$

So we have

$$|b_n| \le D_n,\ n = 1, 2, \cdots. \tag{14}$$

Now we prove that $\{D_n\}_{n=1}^{\infty}$ is convergence in a neighborhood of the origin.

Let

$$V(z)=\sum_{n=1}^{\infty}D_nz^n,\ D_1=|\eta|,$$

where $\{D_n\}_{n=1}^{\infty}$ is defined in (11), then

$$\begin{aligned}V^2(z)&=\sum_{n=2}^{\infty}(D_1D_{n-1}+D_2D_{n-2}+\cdots+D_{n-1}D_1)z^n\\&=\sum_{n=1}^{\infty}(D_1D_n+D_2D_{n-1}+\cdots+D_nD_1)z^{n+1}\\&=\frac{1}{\Gamma}\sum_{n=1}^{\infty}D_{n+1}z^{n+1}=\frac{1}{\Gamma}V(z)-\frac{1}{\Gamma}|\eta|\,z.\end{aligned}$$

Then

$$V(z)=\frac{1}{2\Gamma}(1\pm\sqrt{1-4M\,|\eta|\,z}).$$

But $G(0)=0$, so

$$V(z)=\frac{1}{2\Gamma}(1-\sqrt{1-4M\,|\eta|\,z}).$$

We can obtain that $V(z)$ is convergence in $|z|<\dfrac{1}{4\Gamma|\eta|}$. From (14) we see that the series (5) converges in a neighbourhood of the origin. This completes the proof.

4 Existence of Analytic Solution

Theorem 3. Suppose the conditions of Theorem 1 and 2 are satisfied, then equation (1) has an analytic solution

$$x(z)=y(\alpha y^{-1}(z)) \tag{15}$$

in a neighbourhood of α, such that $x(\alpha)=\alpha,\ x'(\alpha)=\alpha,$ where $y(z)$ is an analytic solution of (3).

Proof. In Theorem 1 and Theorem 2 we have found a solution $y(z)$ of (3) in the form (5), which is analytic near the origin. Since $y(0)=\alpha, y'(0)=\eta\neq 0,$ the function y^{-1} is also analytic near α. Thus $x(z)=y(\alpha y^{-1}(z))$ is analytic, moreover, $x^{[k]}(z)=y(\alpha^k y^{-1}(z))$ and

$$x'(z)=\frac{\alpha y'(\alpha y^{-1}(z))}{y'(y^{-1}(z))}.$$

As in [2],from (3) we have

$$x'(z) = \frac{\alpha y'(\alpha y^{-1}(z))}{y'(y^{-1}(z))} = y(\alpha^m y^{-1}(z)) = x^{[m]}(z).$$

Because $y(0) = \alpha, y'(0) = \eta \neq 0$, then

$$x(\alpha) = y(\alpha y^{-1}(\alpha)) = y(0) = \alpha,$$

$$x'(0) = \frac{\alpha y'(0)}{y'(0)} = \alpha.$$

This completes the proof.

References

1. Cooke, K.L.: Functional differential systems: some models and perturbation problems. In: Inter. Symp. Diff. Eqs. Dynamical Systems, Puerto Rico, CA (1965)
2. Si, J.G., Li, W.R., Cheng, S.S.: Analytic solutions of an iterative functional differential equation. Computer Math. Applic. 33(6), 47–51 (1997)
3. Si, J.G., Zhang, W.N.: Analytic solutions of a nonlinear iterative equation near neutral fixed points and poles. J. Math. Anal. Appl. 284(1), 373–388 (2003)
4. Bjuno, D.: Analytic form of differential equations. Trans. Moscow Math. Soc. 25, 131–288 (1971)
5. Marmi, S., Moussa, P., Yoccoz, J.-C.: The Brjuno functions and their regularity properties. Comm. Math. Phys. 186(2), 265–293 (1997)
6. Carletti, T., Marmi, S.: Linearization of Analytic and Non-Analytic Germs of Diffeomorphisms of (ℂ, 0). Bull. Soc. Math. 128, 69–85 (2000)
7. Davie, M.: The critical function for the semistandard map. Nonlinearity 7, 219–229 (1994)

Local Analytic Solutions of a Functional Differential Equation

Lingxia Liu

Department of Mathematics, Weifang University,
Weifang, Shandong 261061, P.R. China
llxmath@126.com

Abstract. Analytic solution are studied for an iterative functional differential equation. For technical reasons, in previous work the constant μ given in the *Schröder* transformation, is required to satisfy that μ is off the unit circle S^1 or lies on the circle with the Diophantine condition. In this paper, we study results of analytic solutions in the case of μ at a root of the unity and the case of μ near resonance under the Brjuno condition. Analytic solutions of neutral differential-difference equation are also discussed.

Keywords: Functional differential equation, Analytic solution, Power series, Brjuno condition, Resonance.

1 Introduction

Nonlinear science is one of the most important topics in today's science. The theory of iterative dynamical systems plays an important role in nonlinear science. Many mathematical models in physics, mechanics, biology and astronomy can be reducing to an iterative functional differential equation. The two-body problem in a classic electrodynamics, some population models, some models of blood cell productions are given in the form of iterative functional differential equations. In the last few years, there has been growing interest in studying the existence of solutions of functional differential equations with state dependent delay [1-10]. However, there are only few papers dealing with a second-order iterative functional differential equation. In [8], Petahov considered the existence of solutions of the equation $x''(t) = ax(x(t))$. In [9], the authors studied the existence of analytic solutions of the equation $x''(t) = x(az + bx(z))$. In [10] , the authors considered the existence of analytic solutions to the differential equation.

$$x''(z) = (x^m(z))^2 \tag{1}$$

in the complex field, where $x^m(z)$ denotes the mth iterative of the function $x(z)$.

Existence of local analytic solutions is studied in this paper. As in [10] , we make the *Schröder* transformation $x(z) = y(uy^{-1}(z))$ to reduce the equation (1) to the auxiliary equation

$$\mu^2 y''(\mu z)y'(z) = \mu y'(\mu z)y''(z) + (y'(z))^3 (y(\mu^m z))^2 \text{ , } z \in c \text{ ,} \tag{2}$$

D. Zeng (Ed.): Advances in Control and Communication, LNEE 137, pp. 261–269.
springerlink.com

in which iteration of the unknown function is not involved but an indeterminate complex μ needs to be discussed. In [10], the existence of analytic solutions for Eq. (1) is closely related to the position of an indeterminate constant μ. For technical reason, in [10], the constant μ given in the *Schröder* transformation, is required to fulfill that μ is off the unit circle S^1 or μ is on the circle with the Diophantine condition: $|\mu|=1$, *μ is not a root of unity, and* $\log\frac{1}{|\mu^n-1|}\le T\log n,\quad n=2,3,\cdots$ *for some positive constant* T. The complex μ has been discussed in the condition:

$(i)\ 0<|\mu|<1$

$(ii)\ |\mu|=1$, μ is not a root of unity, and $\log\frac{1}{|\mu^n-1|}\le k\log n,\quad n=2,3\cdots$.

In this paper, we break the restriction of Diophantine condition and discuss not only those u at resonance, i.e., at a root of the unity but also those u near resonance under the Brjuno condition considered in [7]. Now we discuss the analytic solutions of Eq. (2) in the following conditions:

(H1) $\mu=e^{2\pi i\theta}$, where $\theta\in\mathbf{R}\backslash\mathbf{Q}$ is a Brjuno number [11,12], i.e., $B(\theta)=\sum_{k=0}^{\infty}\frac{\log q_{k+1}}{q_k}<\infty$, where $\{p_k/q_k\}$ denotes the sequence of partial fraction of the continued fraction expansion of θ, and is said to satisfy the Brjuno condition.

(H$_2$) $\alpha=e^{2\pi iq/p}$ for some integers $p\in\mathbf{N}$ with $p\ge 2$ and $q\in\mathbf{Z}\backslash\{0\}$, and $\alpha\ne e^{2\pi il/k}$ for all $1\le k\le p-1$ and $l\in\mathbf{Z}\backslash\{0\}$.

Observe that μ is on S^1 in the case of (H1) and (H2). The difficulties are encountered for μ on S^1, as mentioned in the so-called "small-divisor problem" (seen in [13] p.146 and [14]). In [15] the case of (H2), where μ is a root of the unity, was also discussed for a general class of iterative equations. Since then, we have been striving to give a result of analytic solutions for those μ "near" a root of the unity, i.e., neither being roots of the unity nor satisfying the Diophantine condition.

The Brjuno condition in (H1) provides such a chance for us. As stated in [16], for a real number θ we let θ denote its integer part and let $\{\theta\}=\theta-[\theta]$. Then every national number θ has a unique expression of the Gauss' continued fraction $\theta=a_0+\theta_0=a_0+\frac{1}{a_1+\theta_1}=\ldots$, denoted simply by $\theta=[a_0,a_1,\ldots,a_n,\ldots]$, where a_j 's and θ_j 's are calculated by the algorithm: **(a)** $a_0=[\theta]$, $\theta_0=\{\theta\}$ and **(b)** $a_n=\left[\frac{1}{\theta_{n-1}}\right]$, $\theta_n=\left\{\frac{1}{\theta_{n-1}}\right\}$ for all $n\ge 1$. Define the sequences $(p_n)_{n\in\mathbf{N}}$ and $(q_n)_{n\in\mathbf{N}}$ as follows: $q_{-2}=1, q_{-1}=0, q_n=a_nq_{n-1}+q_{n-2}$;

$p_{-2}=0, p_{-1}=1, p_n=a_np_{n-1}+p_{n-2}$. It is easy to show that $p_n/q_n=[a_0,a_1,\ldots,a_n]$. Thus, for every $\theta\in\mathbf{R}\backslash\mathbf{Q}$ we associate, using its convergence, an arithmetical function $B(\theta)=\sum_{n\ge 0}\frac{\log q_{n+1}}{q_n}$. We say that θ is a Brjuno number or that it satisfies Brjuno condition if $B(\theta) < +\infty$. The Brjuno condition is weaker than the Diophantine condition. For example, if $a_{n+1} \le ce^{a_n}$ for all $n \ge 0$, where $c > 0$ is a constant, then $\theta = [a_0, a_1, \ldots, a_n, \ldots]$ is a Brjuno number but is not a Diophantine number. So, the case (H1) contains both Diophantine condition and a part of μ ``near" resonance. In this paper, considering the Brjuno condition instead of the Diophantine one, we discuss not only the case (H1) but also (H2) for Eq. (2).

A change of variable further transforms (2) into the differential-difference equation

$$g''(\omega+\beta)\cdot g'(\omega) - g'(\omega+\beta)\cdot g''(\omega) = [g'(\omega)]^3[g(\omega+m\beta)]^2, \tag{3}$$

Where α is a complex constant. The solutions of this equation have properties similar to those of (2). If Eq. (3) has an invertible solution $g(\omega)$, then we can show that $x(z) = g(g^{-1}(z)+\beta)$ is an analytic solution of (1).

2 Analytic Solutions of Auxiliary Equation

The existence of an analytic solution of (2) in the initial condition

$$g(0) = s,\ g'(0) = \eta \ne 0,\ 0 < |\eta| \le 1 \tag{4}$$

is discuss. First, we discuss the existence of analytic solutions of Eq. (2) under the Brjuno condition which is weaker than the Diophantine condition. To do this, we need to introduce Davie's Lemma. First, we recall some facts in [15] briefly. Let $\theta \in \mathbf{R}\backslash\mathbf{Q}$ and $(q_n)_{n\in\mathbf{N}}$ be the sequence of partial denominators of the Gauss's continued fraction for θ as in the Introduction. As in [14], let

$$A_k = \{n \ge 0 \mid \|n\theta\| \le \frac{1}{8q_k}\},\ E_k = \max(q_k, \frac{q_{k+1}}{4}),\ \eta_k = \frac{q_k}{E_k}.$$

Let A_k^* be the set of integers $j \ge 0$ such that either $j \in A_k$ or for some j_1 and j_2 in A_k, with $j_2 - j_1 < E_k$, one has $j_1 < j < j_2$ and q_k divide $j - j_1$. For any integer $n \ge 0$, define

$$l_k(n) = \max\left((1+\eta_k)\frac{n}{q_k} - 2,\ (m_n\eta_k + n)\frac{1}{q_k} - 1\right),\quad \text{Where}\quad m_n = \max\{j \mid 0 \le j \le n,\ j \in A_k^*\}.$$

We then define function $h_k : \mathbf{N} \to \mathbf{R}_+$ as follows:

$$\begin{cases} \dfrac{m_n + \eta_k n}{q_k} - 1, & \text{if}\quad m_n + q_k \in A_k^*, \\ l_k(n), & \text{if}\quad m_n + q_k \notin A_k^*. \end{cases}$$

Let $g_k(n) := \max\left(h_k(n), \left[\dfrac{n}{q_k}\right]\right)$, and define $k(n)$ by the condition $q_{k(n)} \le n \le q_{k(n)+1}$. Clearly, $k(n)$ is non-decreasing. Then we are able to state the following result:

Lemma 1. (Davie's Lemma [15]) Let

$$K(n)=n\log 2+\sum_{k=0}^{k(n)} g_k(n)\log(2q_{k+1}).$$

Then

(a) There is a universal constant $\gamma>0$ (independent of n and θ) such that

$$K(n)\le n\left(\sum_{k=0}^{k(n)}\frac{\log q_{k+1}}{q_k}+\gamma\right),$$

(b) $K(n_1)+K(n_2)\le K(n_1+n_2)$ for all n_1 and n_2, and

(c) $-\log|\alpha^n-1|\le K(n)-K(n-1)$.

Now we state and prove the following theorem under Brjuno condition. The idea of our proof is acquired from [14].

Theorem 1. Assume that (H1) holds, then for the initial condition (4), Eq. (2) has an analytic solution of the form

$$y(z)=\sum_{n=0}^{\infty} b_n z^n,\ b_0=s,\ b_1=\eta \tag{5}$$

in a neighborhood of the origin.

Proof. As in [10], we rewrite (2) in the from

$$\frac{\mu y''(\mu z)y'(z)-y'(\mu z)y''(z)}{[y'(z)]^2}=\frac{1}{\mu}y'(z)[y(\mu^m z)]^2,$$

or

$$\left(\frac{y'(\mu z)}{y'(z)}\right)'=\frac{1}{\mu}y'(z)[y(\mu^m z)]^2.$$

If $y'(0)=\eta\neq 0,$ then Eq. (2) is reduced equivalently to the integro-differential equation

$$y'(\mu z)=y'(z)[1+\frac{1}{\mu}\int_0^z y'(t)(y(\mu^m t))^2dt]. \tag{6}$$

Substituting to $y(z)$ their power series (5) respectively in (6). We see that the sequence $\{b_n\}_{n=2}^{\infty}$ is successively determined by the condition

$$(\mu^{n+1}-1)(n+2)b_{n+2}=\sum_{i=0}^{n}\sum_{j=0}^{n-i}\sum_{k=0}^{n-i-j}\frac{(i+1)(j+1)\mu^{m(n-i-j)}}{n-i+1}b_{i+1}b_{j+1}b_k b_{n-i-j-k},\ n=0,1,2,\cdots \tag{7}$$

in a unique manner.

We need to show that the subsequent power series (5) converges in a neighborhood of the origin. First of all, note that $|\mu|=1$, then

$$|b_{n+2}|\le\frac{1}{|\mu^{n+1}-1|}\sum_{i=0}^{n}\sum_{j=0}^{n-i}\sum_{k=0}^{n-i-j}|b_{i+1}||b_{j+1}||b_k||b_{n-i-j-k}|,\ n=0,1,2,\cdots, \tag{8}$$

thus if we define a sequence $\{B_n\}_{n=0}^{\infty}$ by $B_0=|s|$, $B_1=1$, and

$$B_{n+2}=\sum_{i=0}^{n}\sum_{j=0}^{n-i}\sum_{k=0}^{n-i-j}B_{i+1}B_{j+1}B_kB_{n-i-j-k},\ n=0,1,2,\cdots.$$

Now if we define

$$G(z)=\sum_{n}^{\infty}B_nz^n, \tag{9}$$

then $G(z)$ is continuous and satisfies the equality

$$G^4(z)-2|s|G^3(z)+|s|^2G^2(z)-G(z)+(z+|s|)=0. \tag{10}$$

As in [10], let $R(z,\varphi)=\varphi^4-2|s|\varphi^3+|s|^2\varphi^2-\varphi+z+|s|$ for (z,φ) from a neighborhood of $(0,|s|)$. Since $R(0,|s|)=0$, $R_{\varphi}'(0,|s|)=-1\neq 0$, there exists a unique function $\varphi(z)$, analytic on a neighborhood of zero, such that $\varphi(0)=|s|$, $\varphi'(0)=|\eta|$ and satisfying the equality $R(z,\varphi(z))=0$. On account of (9) and (10), we have $G(z)=\varphi(z)$. Then the power series (9) converges on a neighborhood of the origin, which implies that the power series (5) also converges in a neighborhood of the origin. So there is a constant $T>0$ such that $B_n<T^n$, $n=1,2,\cdots$.

Now we prove $|b_n|\le B_ne^{k(n-1)}$, $n=1,2,\cdots$, where $K:N\to R$ is defined in Lemma 1. In fact, $|b_0|=|s|=B_0$, $|b_1|=|\eta|\le 1=B_1$, we assume that $|b_j|\le B_je^{k(j-1)}$, $j\le n$. From Lemma 1 and (8)

$$\begin{aligned}|b_{n+2}|&\le\frac{1}{|\mu^{n+1}-1|}\sum_{i=0}^{n}\sum_{j=0}^{n-i}\sum_{k=0}^{n-i-j}B_{i+1}e^{K(i)}B_{j+1}e^{K(j)}B_ke^{K(k-1)}B_{n-i-j-k}e^{K(n-i-j-k-1)}\\&\le\frac{1}{|\mu^{n+1}-1|}\sum_{i=0}^{n}\sum_{j=0}^{n-i}\sum_{k=0}^{n-i-j}B_{i+1}B_{j+1}B_kB_{n-i-j-k}e^{K(n-2)}\\&=\frac{e^{K(n-2)}}{|\mu^{n+1}-1|}B_{n+2}.\end{aligned}$$

Because $K(n)$ is non-decreasing, so $e^{K(n-2)}\le e^{K(n)}$, then $|b_{n+2}|\le\frac{e^{K(n)}}{|\mu^{n+1}-1|}B_{n+2}$. Note that $K(n)\le K(n+1)+\log|\mu^{n+1}-1|$, then $|b_{n+2}|\le e^{K(n+1)}B_{n+2}$, so we have $|b_n|\le B_ne^{K(n-1)}$, $n=1,2,\cdots$. From Lemma 1 we have $K(n)\le n(B(\theta)+\gamma)$ for some universal constant $\gamma>0$, then $|b_n|\le\Gamma^ne^{(n-1)(B(\theta)+\gamma)}$, $n=1,2,\cdots$, i.e., $\limsup_{n\to\infty}(|b_n|)^{\frac{1}{n}}\le\limsup_{n\to\infty}(Te^{\frac{n-1}{n}(B(\theta)+\gamma)})$ $=Te^{B(\theta)+\gamma}$. This implies that the convergence radius of the series (5) is at least $(Te^{B(\theta)+\gamma})^{-1}$. The proof is complete.

The following theorem is devoted to the case where μ is a root of unity. The idea of our proof is acquired from [17].

Theorem 2. Suppose that (H_2) holds. Suppose that for $b_0 = s,\quad b_1 = \eta$, and the system

$$(\mu^{n+1}-1)(n+2)b_{n+2} = \sum_{i=0}^{n}\sum_{j=0}^{n-i}\sum_{k=0}^{n-i-j}\frac{(i+1)(j+1)\mu^{m(n-i-j)}}{n-i+1}b_{i+1}b_{j+1}b_k b_{n-i-j-k}\ ,\ n = 0, 1, 2, \cdots, \tag{11}$$

has a solution $\{b_n\}_{n=0}^{\infty}$ such that $b_{lp+1} = 0$ and

$$\sum_{i=0}^{lp-1}\sum_{j=0}^{lp-i-1}\sum_{k=0}^{lp-i-j-1}\frac{(i+1)(j+1)\mu^{m(n-i-j)}}{n-i+1}b_{i+1}b_{j+1}b_k b_{n-i-j-k} = 0,\quad l = 1, 2, \cdots.$$

Then the initial value problem (2) and (4) has an analytic solution of the form

$$y(z) = s + \eta z + \sum_{n \neq lp-1,\, l \in N} b_n z^n\ ,\ \ N = \{1, 2, 3, \cdots\}$$

in a neighborhood of the origin.

Proof. If $\{b_n\}_{n=0}^{\infty}$ is a solution of system (11) such that $b_{lp+1} = 0$, then $y(z) = \sum_{n=1}^{\infty} b_n z^n$ is the formal solution of the auxiliary equation (2) . Let

$$\Gamma = \max\left\{\frac{1}{|\alpha - 1|}, \frac{1}{|\alpha^2 - 1|}, \cdots, \frac{1}{|\alpha^{p-1} - 1|}\right\},$$

from (8) we have

$$|b_{n+2}| \le \frac{1}{|\mu^{n+1} - 1|}\sum_{i=0}^{n}\sum_{j=0}^{n-i}\sum_{k=0}^{n-i-j}|b_{i+1}||b_{j+1}||b_k||b_{n-i-j-k}|,$$

$$\le \Gamma\sum_{i=0}^{n}\sum_{j=0}^{n-i}\sum_{k=0}^{n-i-j}|b_{i+1}||b_{j+1}||b_k||b_{n-i-j-k}|,\quad n = 0, 1, 2, \cdots$$

for all $n \neq lp$, where $l = 1, 2, \cdots$. In order to construct a majorant series, we consider the equation

$$F(z, \psi) = \psi^4 - 2|s|\psi^3 + |s|^2\psi^2 - \frac{1}{\Gamma}\psi + \frac{1}{\Gamma}(z + |s|) = 0$$

For (z, ψ) from neighborhood of $(0, |s|)$. Since $F(0, |s|) = 0$ and $F_{\psi}'(0, |s|) = -\frac{1}{\Gamma} \neq 0$, there exists a unique function $\psi(z)$, analytic in a neighborhood of zero, such that $\psi(0) = |s|$, $\psi'(0) = 1$. So $\psi(z)$ can be expanded into a convergent power series. Write $\psi(z) = \sum_{n=0}^{\infty} c_n z^n$, then $c_0 = |s|,\ c_1 = 1$ and

$$c_{n+2} = \Gamma\sum_{i=0}^{n}\sum_{j=0}^{n-i}\sum_{k=0}^{n-i-j}c_{i+1}c_{j+1}c_k c_{n-i-j-k}\ ,\ \ n = 0, 1, 2, \cdots.$$

Furthermore, $|b_n| \le c_n$, $n = 0, 1, 2, \cdots$. In fact, $|b_0| = |s| = c_0$, $|b_1| = |\eta| \le 1 = c_1$, for inductive proof we assume that $|b_j| \le c_j$ for $j \le n+1$, when $n+1 \ne lp$, i.e., $n \ne lp-1$, we have

$$|b_{n+2}| \le \Gamma \sum_{i=0}^{n} \sum_{j=0}^{n-i} \sum_{k=0}^{n-i-j} |b_{i+1}||b_{j+1}||b_k||b_{n-i-j-k}|$$

$$\le \Gamma \sum_{i=0}^{n} \sum_{j=0}^{n-i} \sum_{k=0}^{n-i-j} c_{i+1} c_{j+1} c_k c_{n-i-j-k} = c_{n+2}$$

as required. By convergence of the series of $\psi(z)$, we see that the series $y(z) = \sum_{n=0}^{\infty} b_n z^n$ converges uniformly in a neighborhood of the origin. The proof is complete.

3 Analytic Solutions

Having knowledge about the auxiliary equation (2) and (3), we are ready to give analytic solutions of (1).

Theorem 3. Suppose one of the conditions in Theorem 1-2 is fulfilled. Then equation (1) has a solution of the form $x(z) = y(\mu y^{-1}(z))$ in a neighborhood of s such that $x(s) = s$, $x'(s) = \mu$, $x''(s) = s^2$, where $y(z)$ is an analytic solution (2).

Proof. By Theorem 1-2, we can find an invertible analytic solution $y(z)$ of the auxiliary equation (2) in the form of (5) such that $y(0) = s,\ y'(0) = \eta \ne 0$. Let $x(z) = y(\mu y^{-1}(z))$, which is also analytic in a neighborhood of S. From (2), it is easy to see $x^m(z) = y(\mu^m y^{-1}(z))$, $x'(z) = \dfrac{\mu y'(\mu y^{-1}(z))}{y'(y^{-1}(z))}$ and

$$x''(z) = \frac{\mu^2 y''(\mu y^{-1}(z)) \cdot y'(y^{-1}(z)) - \mu y'(\mu y^{-1}(z)) \cdot y''(y^{-1}(z))}{\left[y'(y^{-1}(z))\right]^3}$$

$$= \left[y(\mu^m y^{-1}(z))\right]^2 = (x^m(z))^2.$$

Thus, we have $x(s) = y(\mu \cdot 0) = y(0) = s$, $x'(s) = \dfrac{\mu y'(0)}{y'(0)} = \mu$ and $x''(s) = (y(0))^2 = s^2$. The proof is complete.

Theorem 4. Suppose that $\mu = e^{\beta}$ and one of the following conditions is fulfilled:

(A_1) $\Re\beta = 0,\ \Im\beta = 2\pi\theta$ and $\theta \in \mathbf{R}\backslash\mathbf{Q}$ is a Brjuno number.

(A_2) $\Re\beta = 0,\ \Im\beta = \dfrac{2\pi q}{p}$ and $q,\ p$ satisfy the assumptions of (H2).

Then equation (1) has a solution of the form $x(z) = g(g^{-1}(z) + \beta)$ in a neighborhood of S, where $x(\omega)$ is an analytic solution of functional (3) in the half-plane $\Omega_\kappa = \{\omega\colon \Re\omega < \ln\kappa, -\infty < \Im\omega < +\infty\}$ for certain constant $\kappa > 0$.

Proof. First of all, we note that if β satisfies (A_1) or (A_2), then $\mu := e^{\beta}$ satisfies the corresponding condition (H1) or (H2). From Theorem 1-2, we know that there is a positive κ such that Eq. (2) has an analytic solution $y(z)$ in a neighborhood of the origin $U_\kappa = \{z : |z| < \kappa\}$. Let the variable z be connected with ω by the equation $z = e^{\omega}$, and then we have that $z \in U_\kappa$ if $\omega \in \Omega_\kappa$. Define an analytic function $g(\omega)$ for $\omega \in \Omega_\kappa$ by the equation $g(\omega) = y(e^{\omega})$. We assert that $g(\omega)$ satisfies (3) when $e^{\beta} = \mu$. In fact, $g'(\omega) = e^{\omega} y'(e^{\omega})$, $g''(\omega) = y''(e^{\omega})e^{2\omega} + y'(e^{\omega})e^{\omega}$, then

$$\begin{aligned}
&g''(\omega+\beta)g'(\omega) - g'(\omega+\beta)g''(\omega) \\
&= \left[y''(e^{\omega+\beta})e^{2(\omega+\beta)} + y'(e^{\omega+\beta})e^{\omega+\beta}\right]\cdot y'(e^{\omega})e^{\omega} \\
&\quad - y'(e^{\omega+\beta})e^{\omega+\beta}\cdot\left[y''(e^{\omega})e^{2\omega} + y'(e^{\omega})e^{\omega}\right] \\
&= y''(e^{\omega+\beta})\cdot y'(e^{\omega})e^{3\omega+2\beta} + y'(e^{\omega+\beta})y'(e^{\omega})e^{2\omega+\beta} \\
&\quad - y'(e^{\omega+\beta})\cdot y''(e^{\omega})e^{3\omega+\beta} - y'(e^{\omega+\beta})y'(e^{\omega})e^{2\omega+\beta} \\
&= y''(e^{\omega+\beta})\cdot y'(e^{\omega})e^{3\omega+2\beta} - y'(e^{\omega+\beta})\cdot y''(e^{\omega})e^{3\omega+\beta} \\
&= \mu^2 y''(\mu z)\cdot y'(z)z^3 - \mu y'(\mu z)\cdot y''(z)z^3 \\
&= [y'(z)]^3\left[y(\mu^m z)\right]^2\cdot z^3 = [y'(z)\cdot z]^3\left[y(\mu^m z)\right]^2 \\
&= [g'(z)]^3\cdot[g(\omega+m\beta)]^2 .
\end{aligned}$$

Since $y'(0) = \eta \neq 0$, the function $y^{-1}(z)$ is analytic in a neighborhood of the point $y(0) = s$. Thus $g^{-1}(z) = \ln y^{-1}(z)$ is analytic in a neighborhood of s . If we now define $x(z)$ by $g(g^{-1}(z)+\beta)$, then from (3) we have the proof is complete.

$$x'(z) = \frac{g'(g^{-1}(z)+\beta)}{g'(g^{-1}(z))},$$

$$\begin{aligned}
x''(z) &= \frac{g''(g^{-1}(z)+\beta)\cdot g'(g^{-1}(z)) - g'(g^{-1}(z)+\beta)g''(g^{-1}(z))}{(g'(g^{-1}(z)))^3} \\
&= g^2(g^{-1}(z)+m\alpha) = (x^m(z))^2 .
\end{aligned}$$

References

1. Eder, E.: The functional differential equation x'(t)=x(x(t)). J. Diff. Eq. 54, 390–400 (1984)
2. Feckan, E.: On certain type of functional differential equations. Math. Slovaca 43, 39–43 (1993)
3. Stanek, S.: On global properties of solutions of functional differential equation x'(t)=x(x(t))+x(t). Dynamic Sys. Appl. 4, 263–278 (1995)
4. Jackiewicz, Z.: Existence and uniqueness of solutions of neutral delay- differential equations with state dependent delays. Funkcial. Ekvac. 30, 9–17 (1987)
5. Wang, K.: On the equation x'(t)=f(x(x(t))). Funkcialaj Ekvacioj 33, 405–425 (1990)
6. Si, J.G., Li, W.R., Cheng, S.S.: Analytic solutions of an iterative functional differential equation. Computer Math. Applic. 33, 47–51 (1997)

7. Liu, L.X.: Local analytic solutions of a functional equation $\alpha z+\beta x'(z)=x(az+bx''(z))$. Appl. Math. Compu. 215, 644–652 (2009)
8. Si, J.G., Wang, X.P.: Analytic solutions of a second-order functional differential equation with a state derivative dependent delay. Colloquium Math. 79, 273–281 (1999)
9. Si, J.G., Wang, X.P.: Analytic solutions of a second-order functional differential equation with a state dependent delay. Result. Math. 39, 345–352 (2001)
10. Si, J.G., Wang, X.P.: Analytic solutions of a second-order iterative functional differential equation. Compu. Math. Appli. 43, 81–90 (2002)
11. Bjuno, A.D.: Analytic form of differential equations. Trans. Moscow Math. Soc. 25, 131–288 (1971)
12. Marmi, S., Moussa, P., Yoccoz, J.-C.: The Brjuno functions and their regularity properties. Comm. Math. Phys. 186, 265–293 (1997)
13. Berger, M.S.: Nonlinearity and Functional Analysis. Academic Press, CA (1977)
14. Siegel, C.L.: Vorlesungenuber Himmelsmechanik. Springer, CA (1956)
15. Si, J.G., Zhang, W.N.: Analytic solutions of a nonlinear iterative equation near neutral fixed points and poles. J. Math. Anal. Appl. 283, 373–388 (2003)
16. Carletti, T., Marmi, S.: Linearization of Analytic and Non-Analytic Germs of Diffeomorphisms of ($\mathbb{C}$, o). Bull. Soc. Math. France 128, 69–85 (2000)
17. Bessis, D., Marmi, S., Turchetti, G.: On the singularities of divergent majorant series arising from normal form theory. Rend. Mat. Ser. 9, 645–659 (1989)

Time to Maximum Rate Calculation of Dicumyl Peroxide Based on Thermal Experimental Analysis

Zang Na

Department of Fire Protection Engineering Chinese People's Armed Police Force Academy
Langfang, China
zangnaduoduo@163.com

Abstract. The calculation about the kinetics and thermal decomposition of dicumyl peroxide (DCPO) is required for safety concerns, because of its wide applications and accident cases. The isoconversional kinetic study on the thermal behaviour of DCPO has been carried out by differential scanning calorimetry. The apparent activation energy Ea and pre-exponential factor A are not constant during the reaction. Application of accurate decomposition kinetics enabled the determination of the time to maximun rate under adiabatic conditions (TMRad) with a precision given by the confidence interval of the predicitons. This analysis can then be applied for the examination of the effects of the surrounding termperature for safe storage or transportation conditions

Keywords: Dicumyl peroxide (DCPO), Differential scanning calorimeter (DSC), Thermodynamic, Time to maximum rate (TMRad).

1 Introduction

In recent years, the evaluation of chemical hazards has become important in the chemical industry because explosion accidents with regards to organic peroxide have been reported [1]. These accidents are mostly attributable to thermal hazards, such as runaway reactions or thermal decomposition of reactive chemicals. Dicumyl peroxide (DCPO, C_6H_5C $(CH_3)_2OOC$ $(CH_3)_2C_6H_5$) is widely used as a cross-linking agent for polyethylene (PE), ethylene vinyl acetate (EVA) copolymer, ethylene-propylene terpolymer (EPT), and also as a curing agent for unsaturated polystyrene (PS) [2]. Dicumyl peroxide has high commercial value, whereas the inherent decomposition character can release large amounts of thermal energy and spark off high pressure during runaway excursion, resulting in fires or explosions. Dicumyl peroxide requires inherently higher design during manufacturing, transportation, storage, and even disposal. Thermal decomposition or runaway reaction accidents occasioned by dicumyl peroxide and cumene hydroperoxide (CHP) (the raw materials of DCPO) have also been an important issue, as displayed in Table 1.

Important parameters, such as self-heating decomposition temperature (SADT), temperature of no return (T_{NR}), and adiabatic time to maximum rate (TMR_{ad}) and so on, are applied to dictate vessel temperature for transportation. The thermodynamics of DCPO have been studied [2-4]. In all DSC experiments carried out in the reaction is fully exchanged with the surrounding environment, thus not influencing the

D. Zeng (Ed.): Advances in Control and Communication, LNEE 137, pp. 271–277.
springerlink.com

Table 1. Thermal Explosion Incidents Caused By Dcp & Chp

Year	Chemical	Site	Deaths/injuries	Position
1988	DCPO	Taiwan	0/0	reactor
1991	CHP	USA	1/0	reactor
1992	CHP	Taiwan	0/0	reactor
1999	DCPO	Taiwan	0/0	reactor
2002	CHP	USA	0/0	reactor
2003	DCPO	Taiwan	0/2	reactor
2005	DCPO	Taiwan	0/0	reactor
2008	DCPO	Taiwan	0/0	reactor
2010	CHP	Taiwan	0/0	reactor

reaction rate. However, the real conditions of transportation and usage of DCPO are real with the samples in kg or Mg-scale, or in an ideal adiabatic system, all heat lives in the sample. Therefore. During up-scaling two important factors have to be considered: (i) the isoconversional methods of determination of the kinetic parameters, (ii) the effect of heat accumulation in the energetic system, as the sample mass is increased by a few orders of magnitude compared to the thermoanalytical experiments.

This study aimed to calculate the kinetic parameters of DCPO based on the isoconversional methods. Then using the kinetic parameters obtained from experiments, the adiabatic time to maximum rate (TMR_{ad}) can be calculated based on the advanced kinetics.

2 Experimental

2.1 Samples

The DCPO of 99 mass%, white crystals was directly purchased from Beijing Chemical Co., and DCPO was stored at room temperature in moisture-proof box. The melting point of DCPO was about to be 40℃.

2.2 Differential Scanning Calorimeter

DSC was used to measure the difference between the heat flows to a sample and a reference pan, performed on a Mettler TA8000 system coupled with a DSC821e measuring cell that could withstand relatively high pressure. Temperature-programmed screening experiments were performed with DSC. The heating rates (β) selected for the temperature-programmed were 4, 6, 8 10°C min^{-1}. The range of temperature rise was chosen from 30 to 300℃. The mass of sample roughly 8 mg was used for acquiring the experimental data. The test cell was sealed manually by a special tool equipped with Mettler's DSC.

3 Results and Discussion

3.1 Thermal Analysis

In order to determine the thermal hazard of 99 mass% DCPO, we used DSC under various heating rate to estimate the thermal hazard of 99 mass% DCPO and determine kinetics, as shown in Fig. 1. Figure 1 shows thermal curves of decomposition of 99 mass% DCPO with various types of heating rate (β=4, 6, 8 and 10°C min^{-1}) by DSC. From Fig. 1, we determined that the initial reaction of DCPO was endothermic when temperature approached 40°C and DCPO decomposed at about 140 °C.

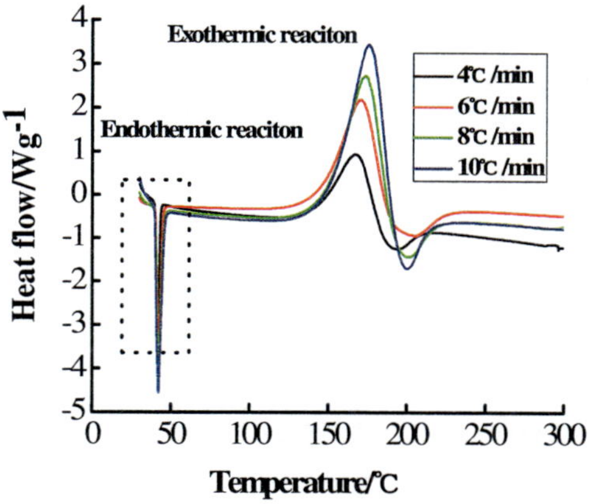

Fig. 1. DSC curves of DCP at different heating rate

3.2 Kinetic Analysis

3.3 Differential Method of Friedman

Friedman analysis [5] based on the Arrhenius equation, applies the logarithm of the conversion rate $d\alpha / dt$ as a function of the reciprocal temperature at different degrees of the conversion α :

$$\ln \frac{d\alpha}{dt} = \ln[A_i f(\alpha_{i,j})] - \frac{E_i}{R} \frac{1}{T_{i,j}} \tag{1}$$

where *i*: index of conversion, *j*: index of the curve and $f(\alpha_{i,j})$ the function dependent on the reaction model that is constant for a given reaction progress $\alpha_{i,j}$ for all curves j. As $f(\alpha)$ is constant at each conversion degree α_i , the method is so-called isoconversional. The ration $-E_\alpha / R$ and $\ln[A_\alpha f(\alpha)]$ are the slope and the intercept with the vertical axis of the plot of $\ln(d\alpha / dt_\alpha)$ *vs.* $1/T_\alpha$, respectively. It is possible to make kinetic predictions at any temperature profile $T(t)$, from the values of E_α and $[A_\alpha f(\alpha)]$ extracted directly form the Friedman method by suing the following expression:

$$d\alpha / dt_\alpha = [A_\alpha f(\alpha)] \exp\left(-\frac{E}{RT(t_\alpha)}\right) \tag{2}$$

where t_α, $T(t_\alpha)$, E_α and A_α are the time, temperature, apparent activation energy and pre-exponential factor, respectively, at conversion α.

The results of isoconversional differential Friedman method are depicted in Fig.2 and Fig.3, the dependence of *E* and *A* on the reaction progress. From Fig.2, it can be seen that the decomposition of DCP usually have a multi-step nature, the kinetic parameters are not constant during the reaction. The isoconversional analysis enables a more accurate determination of the kinetic characteristics than simplified kinetic approaches.

3.4 Kinetics at Tones Scale

Due to the ideal heat transfer in isothermal or non-isothermal experiments, all heat evolved during the reaction is exchanged with surrounding and decomposition takes place at the constant temperature or under constant heating rate. If the reaction with the sample is tones scale, the situation is radically different. We should consider the adiabatic conditions that are usually assumed when working with more than one cubic meter of the substance. For example, the scenario of the cooling failure of the batch reactor is assumed as a adiabatic conditions. All heat evolved during reactions is accumulated in the system what leads to, firstly very slow, and later, very fast increase of the sample temperature and self-heat rate which can result in runaway process. The adiabatic induction time is defined as the time which is needed for self-heating of the sample from the starting temperature to the time of maximum rate (TMR_{ad}). The precise determination of the time to maximum rate under adiabatic conditions is necessary for the safety analysis of many technological processes [6-7].

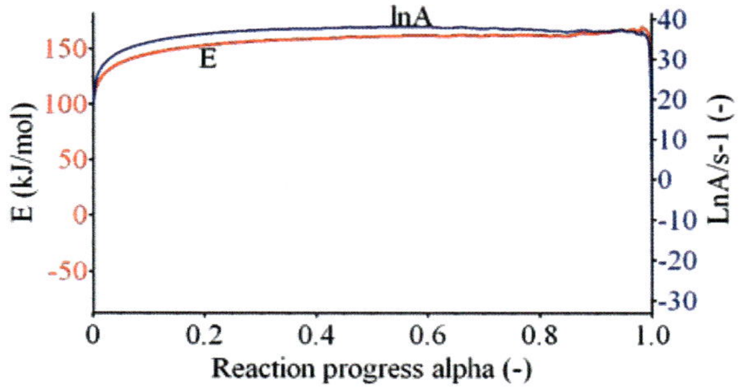

Fig. 2. Activation energy and pre-exponential factor as a function of the reaction progress

For an arbitrarily chosen starting temperature, one can predict a runaway reaction will occur assuming the ideal adiabatic conditions without any heat transfer to of from the experimental material. Heat balance over the sample inside the vessel may be expressed by the equation (3) [8]:

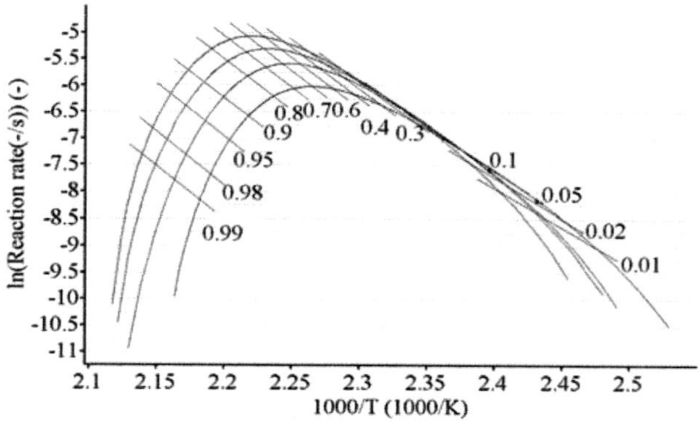

Fig. 3. Friedman analysis- the logarithm of the reaction rates as a function of the reciprocal temperature

$$\frac{dT}{dt} = \frac{1}{\phi} \Delta T_{\text{ad,real}} \frac{d\alpha}{dt} \tag{3}$$

with the adiabatic temperature rise:

$$\Delta T_{\text{ad,real}} = \frac{\Delta H}{C_{\text{p,s}}} \tag{4}$$

The ϕ factor:

$$\phi = \frac{M_c C_{\text{p,c}} + M_s C_{\text{p,s}} + M_x C_{\text{p,x}}}{M_s C_{\text{p,s}}} \tag{5}$$

with *M*: mass, C_p: specific heat, T : temperature,indices *c*, *s*, *x*: container, sample, solvent respectively , ϕ is thermal inetia.

Using the presented above equation (3) describing the heat balance under experimental conditions and kinetic description of the process equation (2) one can now predict the reaction progress $\alpha(t)$ and the rate $d\alpha / dt$ as well as the development of the temperatures $T(t)$ and dT / dt, the adiabatic induction times at any selected starting temperature. After solving both above presented differential equations we will obtain the thermal stability diagram, depicting the dependence of the adiabatic induction time TMR_{ad} on the starting temperature (Fig.4). A TMR_{ad} of 24h is reached for starting temperatures of 71.87 ℃. Due to the fact the value of the heat release applied in these calculations has been determined with the standard deviation of ca.10%, the confidence interval (determined for 95% probability) for 71.87 ℃ is obtained between 20 h and 27 h.

The simulation of the sample temperature (confidence interval for 95% probability) under adiabatic conditions for the starting temperature of 71.87 ℃ is presented in Fig.5. Depending on the decomposition kinetics and ΔT_{ad}, the choice of the starting temperature strongly influences the adiabatic induction time.

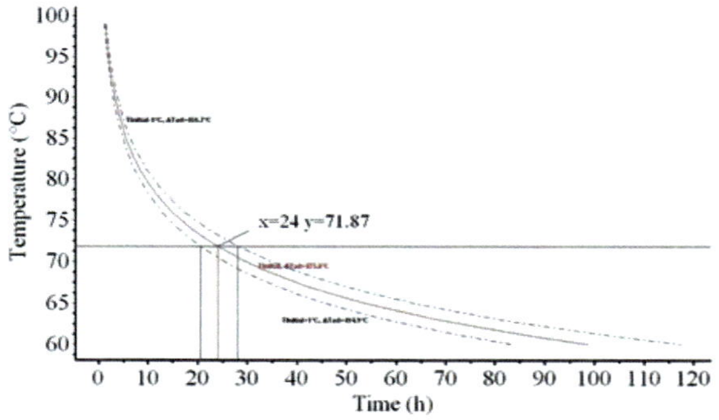

Fig. 4. Thermal safety diagram of the DCPO

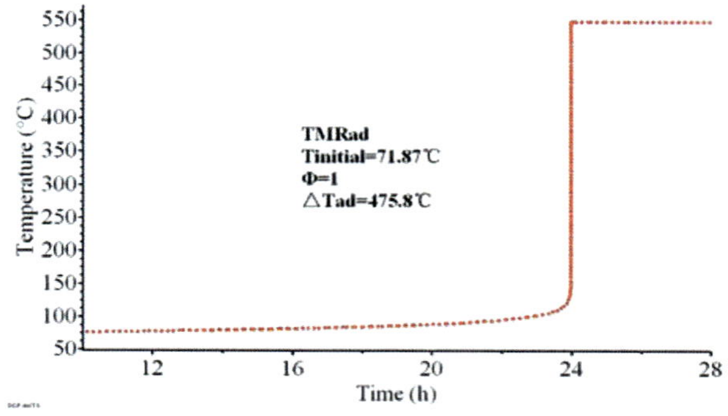

Fig. 5. Change of the sample temperature under adiabatic conditions

4 Conclusion

The isoconversional method-Friedman analysis has been used for the kinetics of DCPO. The apparent activation energy E_a and pre-exponential factor A depends on the reaciton progress which proves the thermal decomposion of DCPO is a multi-stage process. Combined the kinetics with heat balance, it allows the successful simulation of the thermal behaviour under adiabatic conditions such as the determination of *TMR*ad at arbitrary temperature. TMR_{ad} of 24h of DCPO is reached for starting temperatures of 71.87 ℃. It is demonstrated that DCPO is properly maintained at low temperature during transportation and storage; it will not occur runaway reaction. However, it is assumed to result in accidents under higher temperature, improper temperature control, human error, cooling failure and so on. The calculated data can give a reference of the intrinsic safety of DCPO during manufacturing, transportation, storage, and even disposal.

References

1. Sun, J.H.: Evaluation on Thermal Hazards of Chemical Substances, p. 3. Science Press, Beijing (2005)
2. Wu, K.W., Hou, H.Y., Shu, C.M.: Thermal Phenomena Studies for Dicumyl Peroxide at Various Concentration by DSC. Journal of Thermal Analysis and Calorimetry 83, 41–44 (2006)
3. Shen, K.S.J., Wu, S.H., Chi, J.H., Wang, Y.W., Shu, C.M.: Thermal Explosion Simulation and Incompatible Reaction of Dicumyl Peroxide by Calorimetric Technique. Journal of Thermal Analysis and Calorimetry 102, 569–577 (2010), doi:10.1007/s10973-010-0916-4
4. Hou, H.Y., Liao, T.S., Duh, Y.S., Shu, C.M.: Thermal Hazard Studies for Dicumyl Peroxide by DSC and TAM. Journal of Thermal Analysis and Calorimetry 83, 167–171 (2006)
5. Budrugeac, P.: Differential Non-linear Isoconversional Procedure for Evaluating the Activation Energy of Non-isothermal Reactions. Journal of Thermal Analysis and Calorimetry 68, 131–139 (2002)
6. Pastré, J., Wörsdörfer, U., Keller, A., Hungerbühler, K.: Comparison of different methods for estimating TMRad from dynamic DSC measurements with ADT 24 values obtained from adiabatic Dewar experiments. Journal of Loss Prevention in the Process Industries 13, 7–17 (2000)
7. Roduit, B., Borgeat, C., Berger, B., Folly, P., Andres, H., Schädeli, U., Vogelsanger, B.: Up-scaling of DSC Data of High Energetic Materials Simulation of Cook-off experiments. Journal of Thermal Analysis and Calorimetry 85, 195–202 (2006)
8. Roduit, B., Dermaut, W., Lunghi, A., Folly, P., Berger, B., Sarbach, A.: Advanced Kinetics-Based Simulation of Time to Maximum Rate under Adiabatic Conditions. Journal of Thermal Analysis and Calorimetry 93, 163–173 (2008)

Thermal Stability Analysis of Dicumyl Peroxide

Zang Na

Department of Fire Protection Engineering
Chinese People's Armed Police Force Academy Langfang, China
zangnaduoduo@163.com

Abstract. To prevent fire and explosion accident caused by dicumyl peroxide (DCPO), the adiabatic stability of DCPO was investigated by an accelerating rate calorimeter (ARC) in this paper. Comparison of test results of ARC with differential scanning calorimetry (DSC) was carried out. The test results show that the initial exothermic temperature for DCPO was 379.47 K, the maximum exothermic temperature was 439.82K, maximum self-heating rate and the time to maximum self-heating rate were 3.44 K·min-1 and 608.16 min respectively, and the maximum pressure produced by unit mass was 7.74 MPa·g-1. The high thermal hazard and powerful explosion of DCPO were demonstrated by the results of DSC and ARC. Finally, The kinetic parameters such as apparent activation energy and pre-exponential factor of the thermal decomposition of DCPO were calculated pressure kinetic model, and the temperature of no return (TNR) and self accelerating decomposition temperature (TSADT) of DCPO with some specific package were predicted. The studies have revealed that DCPO is sensitive to thermal effect, and the thermal decomposition is hazardous due to its serious pressure effect.

Keywords: Dicumyl peroxide (DCPO), Adiabatic decomposition, Pressure, Accelerating rate calorimeter, Thermal stablility.

1 Introduction

DCPO ($C_{18}H_{22}O_2$), one of the OPs, has caused several thermal fire and explosions because of its unstable peroxy bonds (-O-O-). DCPO is white crystals under room temperature and is widely used as vulcanizing agent of natural rubber and synthetic rubber, initiator of polymerization and cross-linking agent of polyethylene resin. Sodium carbonate and phenol as two typical catalysts are commonly used in the manufacturing process of DCP [1]. Small-scale tests are necessary and mandatory for loss prevention and damage control.

The thermokinetic parameters for dicumyl peroxide were studied by DSC, ATM and VSP2, respectively [2-4]. The thermal decomposition mechanism of dicumyl peroxide was also studied by using the Satava-Sestak method, Borchardt-Daniels method and Flynn-Wall-Ozawa method [4]. In this paper, the study of the decomposition of dicumyl peroxide is performed by accelerating rate calorimeter (ARC), and the kinetic model method is used to calculate kinetic parameters from the experimental data.

D. Zeng (Ed.): Advances in Control and Communication, LNEE 137, pp. 279–286.
springerlink.com

2 Experimental

2.1 Instrument and Samples

Accelerating Rate Calorimeter (ARC) used is a product of the Columbia Scientific Industries, Austin, TX, USA. Experimental materials were purchased from Sinopharm Chemical Reagent Beijing Co., Ltd. DCPO of 99 mass%, white crystals, was stored at room temperature, below 25℃. The samples and main test conditions were shown in Table 1. The experimental mode of Heat-Wait-Search and the slope sensitivity of 0.02 °C·min^{-1} were used for all samples in ARC tests.

Table 1. Test Conditions Of Dcpo

sample	sample mass m/g	bomb mass m/g	start temperature ℃
DCPO-1	1.0236	8.7894	70
DCPO-2	1.0218	8.7836	70

2.2 Results and Analysis

ARC curves of DCPO are given in Fig. 1-Fig. 4 and the tested results are shown in TABLE 2.

From Fig.1, it is seen that the decomposition of DCPO stats at 379.47K (temperature rise rate of 0.0027K·min^{-1} >experimental slope sensitivity of 0.002 K·min^{-1}). The initial exothermic temperature is low which is lower than the experimental test of 396 K based on DSC [2-3]. With the increase of the temperature, a large amount of gases generate and the pressure rises sharply. The pressure rise rate changes with the temperature rise rate. The temperature rise rate of sample and bomb sample is continuously improved by the heat evolved from the thermal decomposition. The maximum self-heat rates are 2.15 K·min^{-1} and 3.44 K·min^{-1},

Table 2. Measured Thermal Decomposition Characteristic Data Of Dcpo

characteristic parameters	**DCPO-1**	**DCPO-2**
thermal inertia factor	5.49	5.41
initial exothermic temperature of system/℃	106.40	115.59
initial self-heat rate of system/℃·min^{-1}	0.19	0.19
final temperature of system/℃	429.91	433.38
adiabatic temperature rise of system/℃	323.53	327.79
maximum self-heat rate of system/℃·min^{-1}	2.15	3.44
time to maximum self-heat rate/min	639.07	608.16
maximum pressure per sample mass/MPa·g^{-1}	7.74	3.37

respectively. The temperature rises accompany with the pressure. The maximum pressure per mass of DCP is 7.74 MPa·g^{-1}. From Fig.3, the first experimental test of logarithm of temperature rise rate vs. reciprocal of temperature is mainly approximately linear between 393 K and 443 K. (the second experiment is between 383 K and 433 K). It demonstrates that the thermal decomposition of DCPO corresponds to Arrhenius equation. It is shown in Fig.2 and Fig.4 that the system has little gas before the initial exothermic temperature. The pressure rises as exponent type.

In this case, dispersive points are observed, which may be caused by the low sensitivity of ARC. The sensitivity of ARC for heating-rate is 0.02 K·min^{-1}. In other words, ARC regards the case as the beginning of an exothermic reaction when the heating-rate of a system is larger than 0.02 K·min^{-1}. To avoid the influence of these occasional factors in calculating the kinetic parameters, several data at the beginning of the reaction should not be used.

3 Calculation of Kinetic Parameters

3.1 Pressure Kinetic Model

According to the pressure rise rate equation [5]:

$$m_{\mathrm{p}} = \frac{dp}{dt} = k(\frac{p_{\mathrm{f}} - p}{\Delta p})^n \Delta p c_0^{n-1} \tag{1}$$

k_{p}^* can be written as

$$k_{\mathrm{p}}^* = k c_0^{n-1} \frac{m_{\mathrm{p}}}{(\frac{p_{\mathrm{f}} - p}{\Delta p})^n \Delta p} \tag{2}$$

By using eqn (2), the kinetic parameter may be calculated easily. Replacing the c with α, the reaction rate can be expressed as

$$\frac{d\alpha}{dt} = kf(\alpha) \tag{3}$$

where symbolizes the kinetic model. α can be given as

$$\alpha = \frac{p - p_0}{\Delta p} \tag{4}$$

Differentiating eqn (4) with respect to t and comparing with eqn (3), it is obtained

$$m_{\mathrm{p}} = \frac{dp}{dt} = \Delta p \cdot Ae(-E_{\mathrm{a}} / RT) f(\frac{p - p_0}{\Delta p}) \tag{5}$$

Then the following equation can be deduced

$$\ln \frac{m_{\mathrm{p}}}{f(\frac{p - p_0}{\Delta p})} = \ln(A \cdot \Delta p) - \frac{E_{\mathrm{a}}}{R} \cdot \frac{1}{T} \tag{6}$$

The plot of $\ln \frac{m_{\mathrm{p}}}{f(\frac{p - p_0}{\Delta p})}$ vs T^{-1} should give a straight line with a slope providing the kinetic model is correctly chosen.

Because the order of the decomposition of DCPO is not defined ,three different values 0, 1 and 2 are selected to fit the experiment data according to eqn (2). The results of the linear regression are listed in Table 3. The data shown in Table 3 present that the linearity is the best when the reacion order is one. The corresponding value of the apparent activation energy is 158.4 $kJ{\cdot}mol^{-1}$.

In order to compare with other methods, the results obtained form different methods are listed in Table 4. From Talbe 4, it can be seen that the activation energies obtained from DCPO are in good agreement., but there is a samll diffetent in the values obtained from reference[11]. This may be due to the purity of experimental samples.

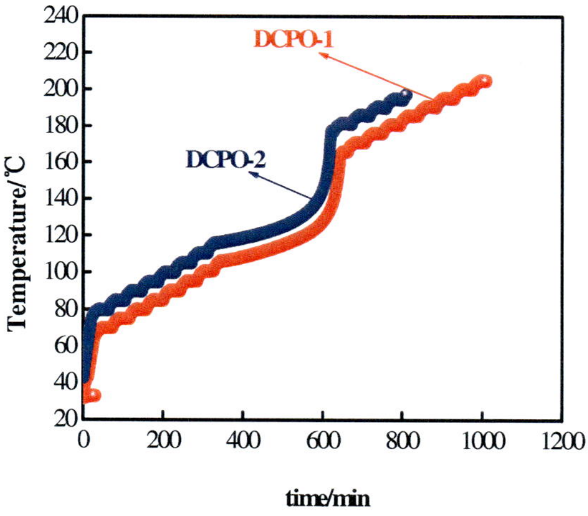

Fig. 1. Curves of temperature vs. time

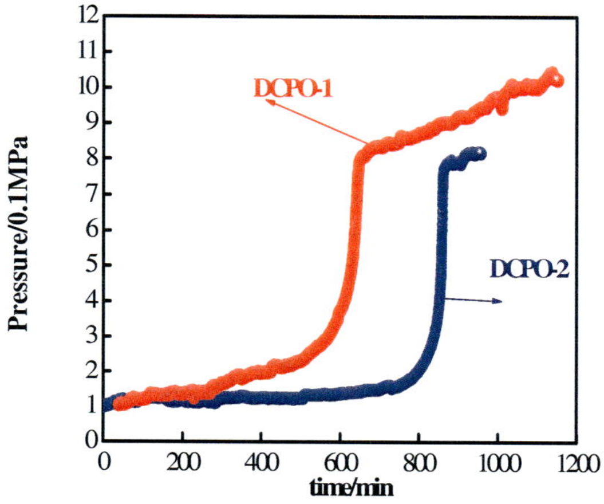

Fig. 2. Curves of pressure vs. time

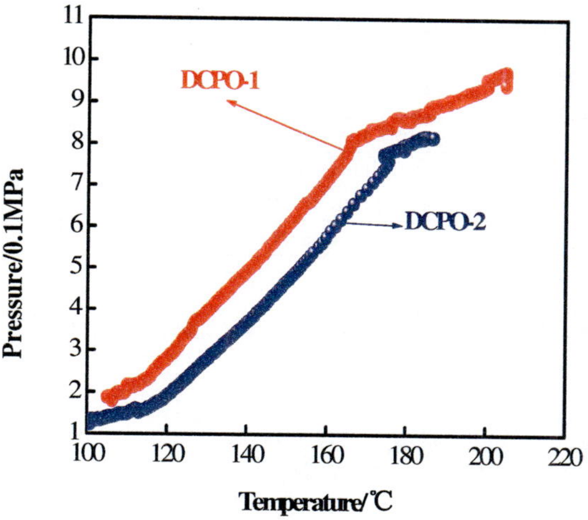

Fig. 3. Curves of pressure vs. temperature

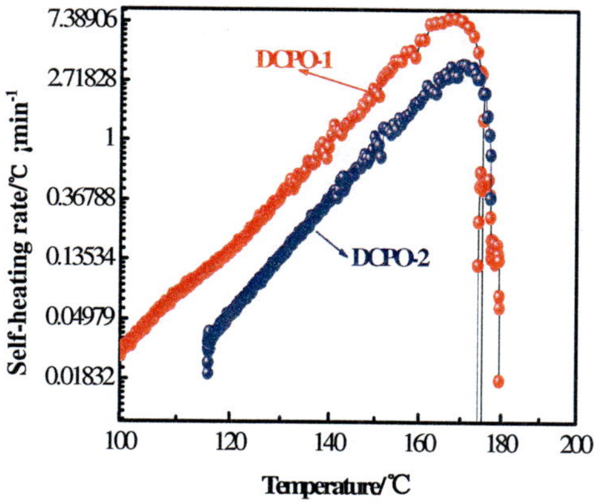

Fig. 4. Curves of temperature rise rate vs. temperature

Table 3. Regression Results Of Kinetic Parameters For Dcpo

reacion order	ln A	R^*	E_a **/ kJ·mol^{-1}**
$n=0$	35.42	0.9797	112.3
$n=1$	42.09	0.9953	158.4
$n=2$	37.15	0.9812	120.6

Table 4. Resultts Of Calculated Kinetic Parameters Of Dcpo Using Different Methods

	methods	E_a **/ kJ·mol^{-1}**
1	pressure kinetic model	158.4
2	Borchardt-Daniels[6]	140
3	Kissinger's method[7]	134.5
4	Built-in model of DSC[8]	117

4 Evaluation of Self Accelerating Decomposition Temperature (Tsadt)

In the thermal hazard analysis of materials, temperature of no return T_{NR} and self accelerating reaction temperature T_{SART} are two important parameters. Self accelerating decomposition temperature (SADT) is the lowest ambient temperature at which the temperature increase of a chemical substance is at least 6 K in a specified commercial package during a period of seven days of less. The SADT is a very important parameter for assessing the safety management of reactive substances in storage, transportation and usage. In this study, SADT of DCPO is calculated from the results of accelerating rate calorimeter (ARC).

The curve of time to maximum rate vs. temperature is given in Fig.5. According to the real heat release condition and the curve of time to maximum rate vs. temperature, eqn (7) gives the time constant of the exothermic reaction system. The no return temperature of DCPO is obtained in Fig. 5. Then, the SADT can be calculated from eqn (8).

$$\tau = \frac{M \bar{C}_V}{hS} \tag{7}$$

$$T_{SADT} = T_{NR} - \frac{RT_{NR}^2}{E_a} \tag{8}$$

The SADT of DCPO filled with 25 kg drum (the drum: 29cm in diameter, 33 cm in filled height) is calculated under the assumption of Semenov's model. The heat release coefficient is 10 kJ·m^{-2}·K^{-1}·h^{-1}. The surface is calculated as $S = \pi \times 29 \times 33 = 3004.98$. According to eqn (7) : $\tau = \frac{M \bar{C}_V}{hS} = \frac{25 \times 1.5}{10 \times 0.3} = 12.5$. $T_{NR} = 395.2$ K. (in Fig. 5) According to eqn (8), SADT can be calculated as follows.

$$T_{SADT} = T_{NR} - \frac{RT_{NR}^2}{E_a} = 386.9 \text{ K}$$

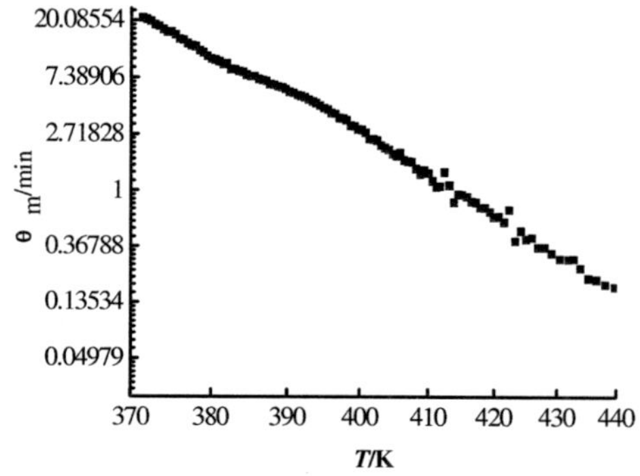

Fig. 5. Curves of time to maximum rate vs. temperature

5 Conclusion

The adiabatic experiment of DCPO has been made. There is much of gas during the thermal decomposition. The pressure rises with temperature as a type of exponential relationship. The maximum pressure per mass is 7.74 MPa·g^{-1}. It would wear down the pressure container especially may occur fire and explosion accidents. Some proper measures are taken to prevent the pressure system from hazardous accidents in the course of production, storage, transportation and usage.

According to the ARC experimental tests, the apparent activation energy E_a and pre-exponential factor A can be obtained based on the pressure kinetic model, it is 158.4 kJ·mol^{-1} and 1.89×10^{18}, respectively. Combined the kinetics with heat balance, it allows the successful simulation of the thermal behaviour under adiabatic conditions such as the self accelerating decomposition temperature SADT. T_{SADT} of DCPO is reached for 386.9 K. It is demonstrated that DCPO is properly maintained at low temperature during transportation and storage; it will not occur runaway reaction. However, it is assumed to result in accidents under higher temperature, improper temperature control, human error, cooling failure and so on. The calculated data can give a reference of the intrinsic safety of DCPO during manufacturing, transportation, storage, and even disposal.

References

1. Wu, K.W., Hou, H.Y., Shu, C.M.: Thermal Phenomena Studies for Dicumyl Peroxide at Various Concentration by DSC. Journal of Thermal Analysis and Calorimetry 83, 41–44 (2006)
2. Shen, K.S.J., Wu, S.H., Chi, J.H., Wang, Y.W., Shu, C.M.: Thermal Explosion Simulation and Incompatible Reaction of Dicumyl Peroxide by Calorimetric Technique. Journal of Thermal Analysis and Calorimetry 102, 569–577 (2010), doi:10.1007/s10973-010-0916-4
3. Hou, H.Y., Liao, T.S., Duh, Y.S., Shu, C.M.: Thermal Hazard Studies for Dicumyl Peroxide by DSC and TAM. Journal of Thermal Analysis and Calorimetry 83, 167–171 (2006)

4. Budrugeac, P.: Differential Non-linear Isoconversional Procedure for Evaluating the Activation Energy of Non-isothermal Reactions. Journal of Thermal Analysis and Calorimetry 68, 131–139 (2002)
5. Qian, X., Liu, L., Feng, C.: Calculating Apparent Activation Energy of Adiabatic Decomposition Process Using Pressure Data. Acta Physico-Chimica Sinica 21, 134–138 (2005)
6. Chen, F., Wu, J., Ding, A.: Thermokinetic Studies of Dicumyl Peroxide. Journal of China University of Mining & Technology 38, 846–850 (2009)
7. Wu, K.-W., Hou, H.-Y., Shu, C.-M.: Thermal phenomena studies for dicumyl peroxide at various concentrations by DSC. Journal of Thermal Analysis and Calorimetry 83, 41–44 (2006)
8. Shen, S.-J., Wu, S.-H., Chi, J.-H.: Thermal explosion simulation and incompatible reaction of dicumyl peroxide by calorimetric technique. Journal of Thermal Analysis and Calorimetry 102, 569–577 (2010)
9. Pastré, J., Wörsdörfer, U., Keller, A., Hungerbühler, K.: Comparison of different methods for estimating TMRad from dynamic DSC measurements with ADT 24 values obtained from adiabatic Dewar experiments. Journal of Loss Prevention in the Process Industries 13, 7–17 (2000)

Modeling and Simulation of High-Pressure Common Rail System Based on Matlab/Simulink

Haitao Zhi[1], Jianguo Fei[1], Jutang Wei[1], Shuai Sun[2], and Youtong Zhang[2]

[1] Department of Engineering Southwest Forestry University Kunming, Yunnan Province, China
[2] Clean Vehicles Laboratory Beijing Institute of Technology Beijing, China
zhi_haitao@163.com, sunshuai1@bit.edu.cn

Abstract. High-pressure common rail system can further improve the economy, power and emissions performance of diesel engines. In this paper we build a air path model which based on Matlab/simulink. Through the simulation we get the main curve of injection system and provide the basis for the optimization of high pressure injection system.

Keywords: Fuel, Simulation, Common rail, Matlab/simulink.

1 Introduction

The increasingly serious energy crisis has become the focus of internal combustion engine industry worldwide, but also let the engine have more and more users. Compared with the gasoline engine it has many advantages: it can reduce 20-25% of the CO_2 emissions, have more advantage in acceleration when speed is lower, and let the average fuel consumption lower than 25-30%. This can provide more fun to drive. Therefore, someone boldly predicted the development tendency of the engine of global vehicle production, and divide the proportion of diesel engines of the world's vehicle production by region. However, compared with gasoline engine, the emission control of diesel is a difficult point. In order to meet emissions standards, advanced fuel injection system of diesel --- High-pressure common-rail technology become the industry focus. The simulation of high-pressure common rail system features can guide the design of the system's structural parameters, and provide design ideas and experimental calibration for the whole software and hardware design of Electronic Fuel Injection System. It has been widely used in the design and performance studies of diesel fuel injection system.

2 Structure and Principle of High-Pressure Common Rail System

The common-rail system consists of several modules like the high pressure pump, fuel metering unit, pressure control valve and injector. The common rail fuel system provides fuel for injection is stored at high pressures in one chamber for all cylinders. This chamber is called the common rail. Common-rail systems enable fuel to be stored at high pressures independently of engine speed.

D. Zeng (Ed.): Advances in Control and Communication, LNEE 137, pp. 287–294.
springerlink.com

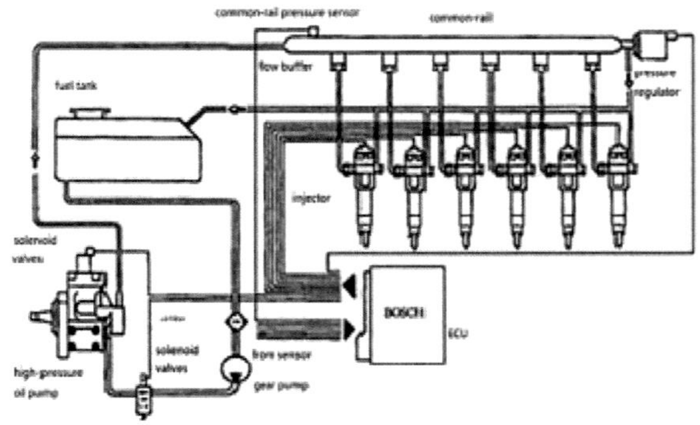

Fig. 1. The basic principle diagram of high-pressure common rail system

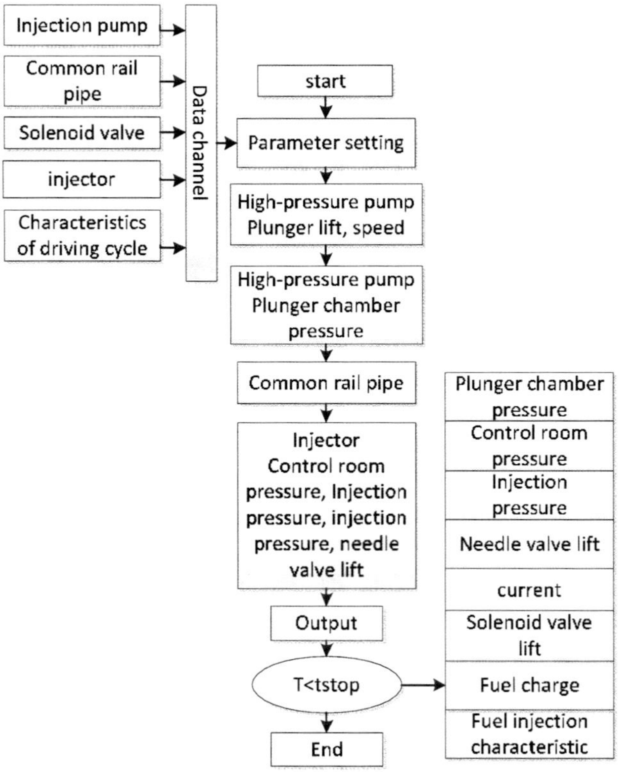

Fig. 2. The simulation of high-pressure common rail system

2.1 Rail

Fuel for injections is stored at high pressures in the common rail. The fuel is supplied by the high-pressure pump. Injectors take the injected fuel from the rail. Additionally rail leakage is taken into account.

The rail is modeled as a chamber with a constant volume according to the following equation:

$$p_{Rail} = \frac{E_{Fuel}}{V_{Rail}} \int q(t)dt = \frac{1}{C_H} \int \sum q_i(t)dt \tag{1}$$

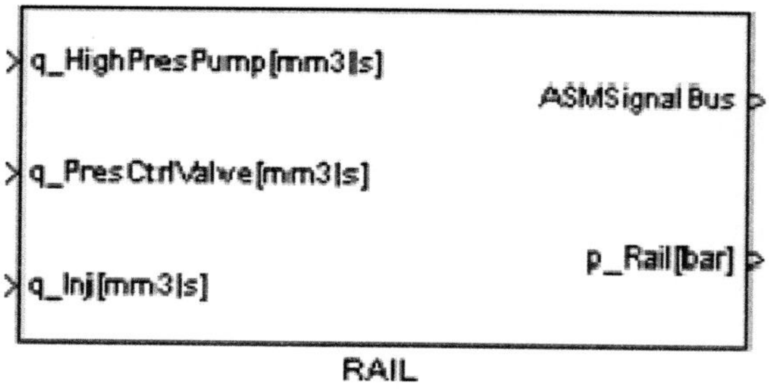

Fig. 3. The simulation figure of RALL

The RAIL block calculates the pressure in the rail as a function of the fuel mass flows out of and into the rail.

$$Q_{zg} = \frac{V_{gg}}{E}\frac{dP_{gg}}{dt} + Q_{kzsi} + Q_{cyqi} + Q_{gg\text{-}h} \tag{2}$$

The flow from common rail pipe to oil slot is:

$$Q_{cyqi} = \xi S_{cyqi} c_{cyqi} \sqrt{\frac{2}{\rho}\left|P_{gg} - P_{cyq}\right|} \tag{3}$$

Circulation area in the equation is:

$$\mathrm{S_{cyqi}} = \pi \mathrm{d}_9^2/4 \tag{4}$$

P_{cyq} is pressure in the oil slot.

$$\xi = \begin{cases} 1 & P_{gg} \geq P_{cyq} \\ -1 & P_{gg} < P_{cyq} \end{cases} \tag{5}$$

The flow from common rail pipe to control chamber is:

$$Q_{kzsi} = \zeta S_z c_z \sqrt{\frac{2}{\rho}\left|P_{gg} - P_{kzs}\right|} \tag{6}$$

P_{kzs} is pressure in the control room.

$$\zeta = \begin{cases} 1 & P_{gg} \geq P_{kzs} \\ -1 & P_{gg} < P_{kzs} \end{cases} \tag{7}$$

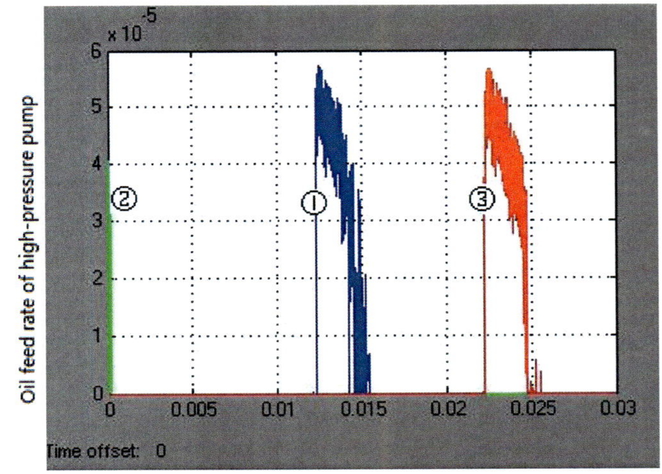

Fig. 4. The changing situation of fuel delivery of high-pressure., pemp

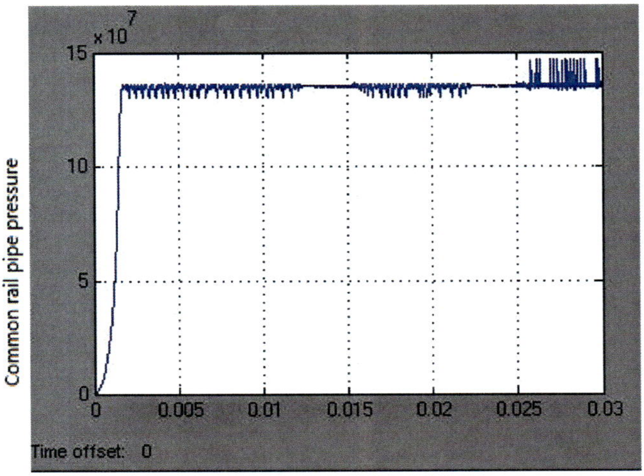

Fig. 5. The changing situation of pressure in common rail

2.2 High-Pressure Pump

The high-pressure pump provides fuel to the common rail. It is driven by the crankshaft at a specific transmission ratio. Newer pumps may contain a fuel metering unit (also called a volume control valve) which decreases the amount of fuel which is pumped to the rail. This is used at high engine speeds to decrease energy losses.

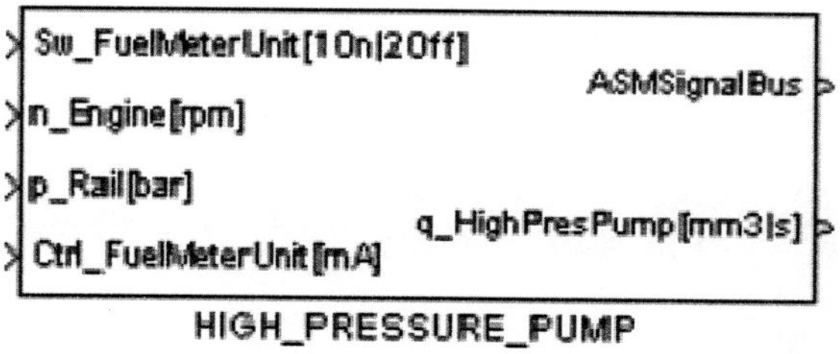

Fig. 6. The simulation figure of HIGH_PRESSURE_PUMP

The HIGH_PRESSURE_PUMP block calculates the maximum fuel flow to the rail as a function of pump speed and pump volume. This maximum flow is decreased by an efficiency map. It depends on pump speed and rail pressure.

The fuel metering unit is modeled as a map according to the control signal and pump speed. The fuel flow through the fuel metering unit is not transmitted to the rail. The fuel metering unit can be deactivated by an external switch.

Fuel continuity equation in plunger chamber is:

$$Q_z = Q_{zV} + Q_{zg} + Q_{zx} + Q_{zh} \tag{8}$$

Fuel flow through high pressure pump (into rail):

$$Q_{zg} = \xi \mu_c F_c \sqrt{\frac{2}{\rho} \times \left| P_z - P_{gg} \right|} \tag{9}$$

Instantaneous oil mass pressured into the fuel plunger chamber is:

$$Q_z = S_z \frac{dh_z}{dt} \tag{10}$$

Change rate of compression oil mass caused by pressure in the plunger chamber is:

$$Q_{zV} = \frac{V_z}{E} \frac{dP_z}{dt} \tag{11}$$

Flow from plunger chamber to low-pressure oil road is Q_{zh} (mm^3/s)

$$Q_{zh} = \lambda \mu_h F_h \sqrt{\frac{2}{\rho}\left|P_z - P_h\right|} \tag{12}$$

$$\lambda = \begin{cases} 1 & P_z \geq P_h \\ -1 & P_z < P_h \end{cases} \tag{13}$$

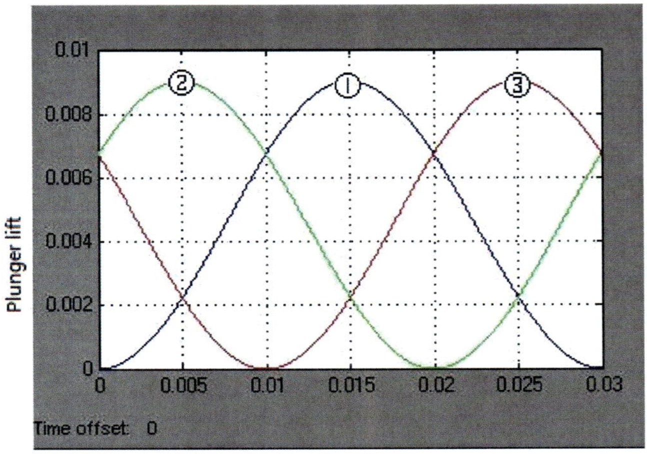

Fig. 7. The changing situation of lift of three plunger

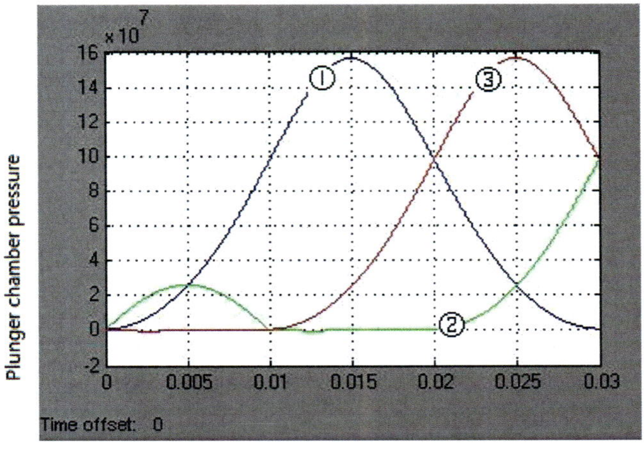

Fig. 8. The changing situation of pressure in three plunger chamber

2.3 Injector

Injector is the most critical and the most complex component of high-pressure common rail system. According to the control signal that ECU send out, by controlling the opening and closing of the solenoid valve, it inject fuel in the High-pressure fuel rail into the engine combustion chamber with the best injection timing, Fuel injection quantity and Fuel injection rate.

In the INJECTOR block, injected fuel mass is calculated according to injection time and rail pressure. The number of injections and number of cylinders have to be specified.

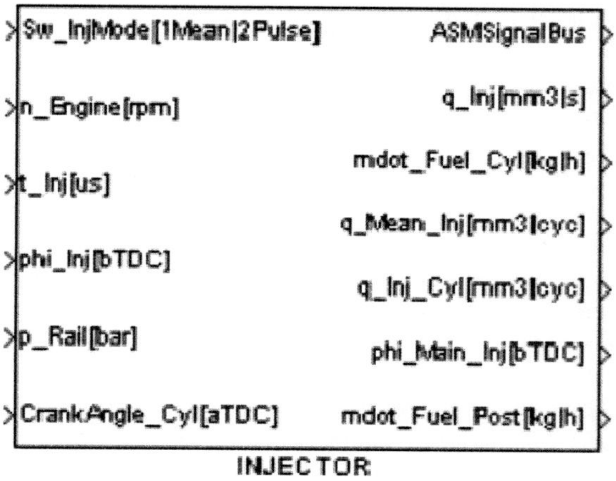

Fig. 9. The simulation figure of INJECTOR

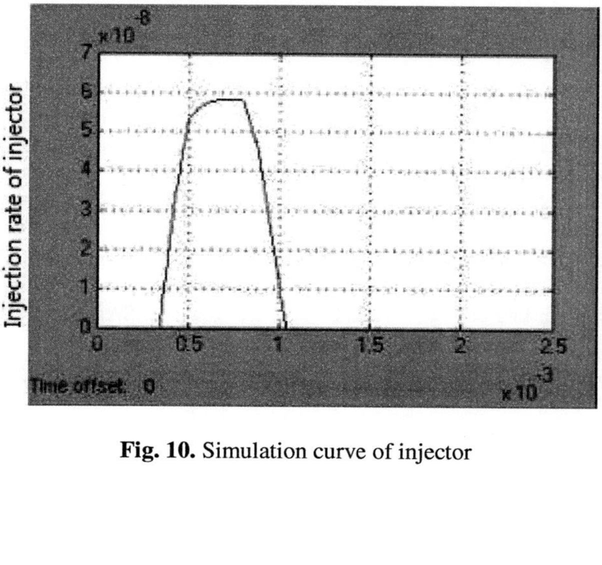

Fig. 10. Simulation curve of injector

Two modes for the injector model can be chosen by external switch: mean injector fuel flow and pulse-wise injection. In pulse-wise injection mode, the entire cylinder fuel mass (for all pulses for multiple injections) is taken out of the rail in one sample time. The injectors provide fuel for combustion to the cylinders. The fuel is taken from the rail. The injected fuel mass depends on the length of injection pulses and on rail pressure. For Diesel Particulate Filter (DPF) regeneration, the calculation of post injection mass flow mdot_Fuel_Post is included and provided by a new out port. Divide injector in two parts: solenoid valves and hydraulic, and then analysis, modeling, simulation partly.

3 Experiment and Simulation

Combining sub-models of high-pressure oil pump, high-pressure oil pump, injector built in above paragraphs we can obtain integrated model of high-pressure common rail system. After completing parameter settings of the model partly, it means that we have completed settings of the whole high-pressure common rail system. After setting parameters reasonably, click it to run and the whole system start simulation. Click the parameters homologous oscilloscope that has simulation results you want to obtain, and it will show changing process of the parameter along with time. The simulation results and experimental data are consistent.

In view of the structure characteristics and request for calculation of high-pressure common rail system, we done abstract and simplification, and thought over the main factors affecting veracity of model as entirety as possible, then built the simulation model based on the calculating assumption. It proves that Simulink's modularization program design method is suit to each process to building the model of high-pressure common rail system. In subsequent studies, the emphasis is to further improve the structure of the model, optimize the control strategy and optimize the parameters.

References

1. Gebert, Barkhimer (eds.): An Evaluation of Common Rail Hydraulically Intensified Diesel Fuel Injection System Concepts and Rate Shapes. SAE paper (1998)
2. Woermann, R.J., Theuerkauf, H.J., Heinrich, A.: A Real-Time Model of a Common Rail Diesel Engine. SAE paper (1999)
3. Wickman, Tanin, Senecal (eds.).: Methods and Results From the Development of a 2600 Bar Diesel Fuel Injection System. SAE paper (2000)
4. Morselli, R.: Energy Based Model of a Common Rail Injector. IEEE (2002)
5. Amoia, Ficarella, Laforgia (eds.).: A Theoretical Code to Simulate the Behavior of an Electro-Injector for Diesel Engines and Parametric Analysis. SAE paper (1997)

Information Technology, Education and Scocial Capital

Huaiwen Cheng

Laboratory of Watershed Hydrology and Ecological
Cold and Arid Regions Environment and Engineering Research Institute, C A S,
Lanzhou, China
chenghuaiwen@yahoo.com.cn

Abstract. This paper discusses the relations between social capital and education, and between social capital and information technology (IT). Social capital can foster educational achievement at family, school/community and region/nation level. Information technology may play a central role in the creation of social capital.

Keywords: social capital, education, information thchnology.

1 Introduction

Social capital as an influential factor for children's educational achievement was first introduced by Coleman. Coleman suggests that in addition to parental educational attainment and income, another equally important determinant of the well-being and educational development of children is the level of connectedness among the child and family, friends, community and school [1]. According to Coleman, this connectedness is a product of social relationships and social involvement, it generates social capital. In the context of education, social capital in the forms of parental expectations, obligations, and social networks that exist within the family, school, community and region are important for student success [2-3]. These variations in academic success can be attributed to parents' expectations and obligations for educating their children [4]; to the network and connections between families whom the school serves [5]; to the disciplinary and academic climate at school [6]; and to the cultural norms and values that promote student efforts [7]. The concept of social capital is a useful theoretical construct for explaining the disparities in students' educational performance among different nations [6]. But, not all social capital is good. There are negative impacts of social capital for educational achievement [8].

Information technology (IT), recently, develops rapidly with increasing the forms of social capital, and it brings up new era in the construction and development of social capital [9]. However, there are no all-sided discussions of social capital, education and IT, though there are some of relation between social capital and education. This paper summarizes how social capital fosters educational achievement and information technology shapes social capital and will provide an all-sided relationship among social capital, education and information technology.

D. Zeng (Ed.): Advances in Control and Communication, LNEE 137, pp. 295–302.
springerlink.com

2 Social Capital and Education Achievement

2.1 Family Social Capital

Family social capital is defined by the resources that individuals may access through social ties [10](Frank & Yasimoto,1998). It includes family background, family structure, parent involvement with their child's school [3]. Family social capital impacts children academic performance [11].

A child's family background affects his or her goals and attitudes toward learning opportunities and school. Families' help students focus on and finish the work assigned at school—a value placed upon the institution itself [12]. Familial background offers social, financial, and human capital as resources for youth. Social capital serves as the medium through which children access their parent's financial and human capital [3]. A person's actions are shaped by social context and not simply by the financial and human resources available to them. The social capital developed within the family has wide ranging effects on the student's opportunities, development, and behavior [3].

In terms of social capital, children are not seen separately from their parents. Social capital in the lives of children is seen as a bi-product of their parents' relationships with others and as a result, their own social capital networks are rendered invisible [13]. Accordingly, for children social capital is seen as an asset that they can utilize later in life, rather than during their childhood. Thus, the primary concern of children is not with having social capital themselves, but with the ability of adults to transfer the stock of social capital to them, and then later in life, the children can cash in on their generated social capital [3]. Marrow noted that research fails to explore how children actively generate their own social capital, and even make links for their own parents' social capital [14].

Family structure is one of important factors impacting children's educational achievement. By examined relationships between family structure and student outcomes, researchers found that in relation to two-parent families, children living in single parent homes have higher drop-out rates, are more vulnerable to pressure from their friends to engage in deviant behaviors, have lower academic achievement [15]. Thus, the benefits of having a two-parent home have been shown to be greater for elementary school students than for high school students. It is also noted that young adults from single family homes are also more likely to enter low-social status occupations. Mulkey *et. a*l. concluded, "Consistent with previous studies, this analysis shows that the effect of a single parent upbringing on the evaluation of students in school-whether by standardized tests or by grades, is small" [16]. Similarly, Dronkers found that children from two-parent families have more school success than children from one parent families [17].

It is generally accepted that the quality of parent–child interactions during childhood has an important association with school success and the eventual social-status attainment of young adults. The home environment is the most powerful factor in determining the school learning of students, their level of school achievement, their interest in school learning, and the number of years of schooling they will receive [3].

Parental involvement is important in the children's academic performance. Some researchers who reviewed various parental influences related to academic outcomes, have concluded that parental involvement is consistently found to be positively related to better grades in school [5]. There is a meaningful relationship between parental involvement and children's academic success. Parental expectations and beliefs about education are among the most influential family variables that can affect children's achievement at school [18]. On the whole, parental expectations and beliefs appear to be important for children's academic success. At the same times, parental involvement can influence student achievement through instructional interactions between parents and children. This type of involvement reflects collaborative learning and the parental sharing of information and a task-related process which enables the child to learn effectively and assume appropriate responsibility for his/ her learning [5].

Therefore, family social capital can impacts children' educational achievement.

2.2 School and Community Social Capital

Social capital in school and community involves in networks that linked parents, children, teachers and community embedded school.

The close contact between parents, teachers and children benefits children academic. Smaller schools tend to have better academic results and fewer disturbances than larger schools, though it appears possible to mitigate these size effects through other forms of social capital building [6]. There are better close contact between parents, teachers and children in smaller schools than in larger schools, because the networks are more easily constructed in smaller space or groups than in larger space or groups.

The researchers, in their studies on school effect for the students' educational attainments, found that much of positive influence might lie in community in which the school was embedded rather than in the school itself [19]. Parent-school connectivity and the negative impact of high residential mobility have been strongly implicated. The average socioeconomic status of a child's class or school has an effect on his or her outcomes, even after taking account of ability and socioeconomic status [20]. Neighborhood characteristics predict educational outcomes, and the strength of these effects rivals that associated with more commonly cited family and school factors. These neighborhood effects were explained by neighborhood social capital, with collective socialization having the strongest influence. If having only one biological parent in the household tends to lead to lower educational achievements in children, then a community with a high proportion of such single-parent and step-parent families will, on average, have a poorer educational record.

Community social capital is more important to educational achievement than characteristics. Both the structure and process of community social capital have an impact on high school students' educational achievements [6]. Community social capital was particularly important in helping some students to excel, for wider social networks help to take you extra mile.

2.3 Region and Nation Social Capital

There is a strong relationship between region social capital and educational attainment. Putnam found US states with higher levels of social capital at the aggregate level achieve consistently better academic results[Putnam,2000]. The US is not alone in showing this remarkable relationship between social capital and educational attainment at the national level. A similar very high correlation between educational attainment was found in UK [6].

OECD had tried to explain national differences in educational attainment by differences in social capital. It is slightly surprising that the authors of the OECD report on the role of social capital did not attempt to model this relationship, especially as the OECD gathers the relevant data on cross-national differences in educational performance [21]. However, for industrialized nations, these with lower educational achievement are the nations with the lower level of social capital. And a remarkable correlation between national literacy levels and levels of social trust (social capital), if controlling for social trust entirely eliminated the effect of GDP per capita, confirming that social trust was the dominant variable in explaining national educational attainment [6].

So, the problem of interpreting causal direction for national social capital and educational achievement arises. Do look educational achievements as a result of national high social capital, or do people trust each other more as a result of their high levels of education or equity more generally? "This is an issue that we cannot resolve now ", Halper say [6].

2.4 Negative Social Capital

For educational achievement, not all social capital is good capital. Moore *et. al.* found the negative association between social capital and educational attainment [22].The way individual accessed social capital is important for understanding how social capital impacts education. In network terms, a person's network of family and friend is likely more dense than their acquaintance network. As we known, family and friends, i.e., strong ties, become even more important and necessary for one's access to resources, than weak bridging ties in one's network. This poses a double burden on marginalized groups whose family and friends are also likely of low socioeconomic status with few personal resources. Moreover, access through strong ties may impose additional demands and pressures which potentially less likely develop when accessing resources through weak ties. These pressures could lead to the educational achievement of negative consequences of social capital through the range of mechanisms that Portes identifies: ''restricted opportunities,'' ''excessive demands,'' ''limited freedoms,'' and ''down-leveling pressures'' may all operate to reduce the sense of mastery of individuals with relatively higher social capital in a disadvantaged network [23]. Moreover, when the factions are formed among children duo to social capital, the social capital will have negative impact for their educational achievement.

At the same times, parent's social capital sometimes is not good for children's educational achievement, in particular in China. Parents are self-serving on their children education, and they can use all social relationship and network to invest in their children education, in order to search a good job for children. Moreover, the ties of friendship are recognized by people in China, and many things are accomplished by it. So, the children whose parents have high level of social capital will think that they have more chance in future society. Then, they have no power for learning, leading to bad children educational achievement.

3 Information Technology as Creator of Social Capital

Recently, Information Technology (IT) develops rapidly with the development of telecommunications that provides Internet and other various types of information system with global technology infrastructure by connecting computers and other communications systems in the world [24].

The role of IT in the creation of social capital was found by Lin. He thought that IT increased the forms of social capital and made social capital transcend national or local community boundaries [9]. The emerging of e-networks was promoted by the development of IT.E-networks are defined as the social networks in cyberspace, and specifically on the Internet. E-networks have become a major avenue of communication globally since the early 1990s; an overview of their extent and scope is informative here. E-networks create social capital in the sense that they carry resources that go beyond mere information purposes. E-commerce is a case in point. Many sites offer free information, but they carry advertisements presumably enticing the user to purchase certain merchandise or services. They also provide incentives to motivate users to take actions. The Internet has also provided avenues for exchanges and possible formation of collectivities [25]. These "virtual" connections allow users to connect with others with few time or space constraints. Access to information in conjunction with interactive facilities makes e-networks not only rich in social capital, but also an important investment for participants' purposive actions in both the production and consumption markets.

Now, researchers currently very interested in how IT can be used to build and maintain one's social capital. Especially IT is interesting in this aspect as they offer a new way of handling relationships in an offline and online environment. Researchers found that the IT supports the formation and maintenance of weak ties, increasing the bridging social capital of its users [26]. For example, IT users can use their large number of friends in internet or e-network to help them to get a job or information about issues that they and their immediate friends do not possess. Ellison *et al.* investigated the role of Facebook to build and maintain social capital [27]. Their findings show that students who use Facebook more intensely report higher bridging social capital as well as higher bonding social capital. This shows that Facebook was used to maintain both loose acquaintances and close friends. In addition, findings showed that Facebook usage was especially beneficial for students reporting a low satisfaction and low self-esteem, as those who reported low self-esteem also reported higher bridging social capital if they used Facebook more intensely [28]. So, we can see that IT can help to build and maintain social capital.

IT provides an equalizing opportunity in the access to social capital. Given the easy, low-cost access to IT that is being provided to more and more people around the world, the abundance and flow of information, the multiplicity of alternative channels as sources and partners, and the increasing need for and gratification of almost instantaneous exchanges, power differentials will be smaller. The development of IT and the emergence of social, economic, and political networks in IT signal a new era in the construction and development of social capital.

4 Conclusion

In conclusion, social capital at family, school and community, regional and national level has a significant impact on educational achievement, while information technology play a important role in the creation of social capital (Fig.1).

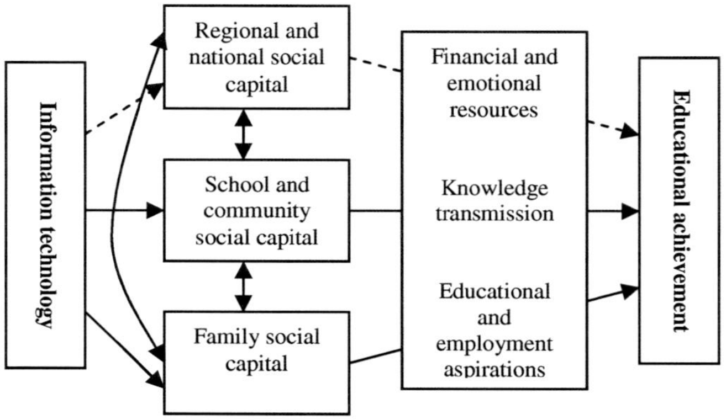

Fig. 1. The relationship among social capital, educational achievement and information technology [6]. Note: Strength of lines roughly indicaties strength of direct relationship.

At family level, social capital of child and parent can improve educational achievement. Higher levels of child-parent contact generally lead to higher educational aspirations and achievements. Parents' social capital is the support children receive from the rest of the family, their friendship networks and their relationship with the children's school, and can also positively affect the children's educational outcome. Differences in family social capital help to explain differences in educational achievements across family types, such as the lower attainment of children form single-parent and step-parent families; across social class; and across ethnic groups [6].

At the school and community level, the positive effects are found. Some school types, notably small school, appear to perform significantly better. Stronger parent-school relationships and parent-parent relationships appear to help explain the effect of school for children educational achievements. The school social capital - teacher-teacher relationships – may be one important factor in explaining differences of educational attainment between schools. But many so-called school effects are really community effects. The low average social capital of the community adds to the children's educational disadvantage.

At the region and nation level, a startlingly strong relationship between social capital and educational achievement. Across US states and UK nations, measures of social capital are highly correlated with educational attainment. But we don't interpret direct cause why educational achievements as a result of national high social capital.

IT helps to build and maintain social capital. It can increase the forms of social capital and made social capital transcend national or local community boundaries. The development of IT brings up a new era of social capital.

References

1. Coleman, J.S.: Social capital in the creation of human capital. American Journal of Sociology 94(supplement), 95–120 (1988)
2. Huang, L.H.: Social capital and student achievement in Norwegian secondary schools. Learning and Individual Differences 19, 320–325 (2009)
3. Schlee, B.M., Mullis, A.K., Shriner, M.: Parents social and resource capital: Predictors of academic achievement during early childhood. Children and Youth Services Review 31, 227–234 (2009)
4. Entwistle, D.R., Alexander, K.L.: Family type and children's growth in reading and math over the primary grades. Journal of Marriage and the Family 58, 341–355 (1996)
5. Hoover-Dempsey, K.V., Sandler, H.M.: Why do parents become involved in their children's education? Review of Educational Research 67, 3–42 (1997)
6. Halper, D.: Social Capital. Polity Press, UK (2005)
7. Thomson, R., Henderson, S., Holland, J.: Making the most of what you've got? Resources, values and inequalities in young women's transitions to adulthood. Educational Review 55, 33–46 (2003)
8. Moore, S., Daniel, M., Gauvin, L., Dube, L.: Not all social capital is good capital. Health & Place, publishing (2009)
9. Lin, N.: Social Capital- A Theory of Social Structure and Action. Cambridge University Press, Cambridge (2005)
10. Frank, K.A., Yasimoto, J.Y.: Linking action to social structure with a system: Social capital within and between subgroups. American Journal of Sociology 104, 642–686 (1998)
11. Mullis, R.L., Rathge, R., Mullis, A.K.: Predictors of academic performance during early adolescence: A contextual view. International Journal of Behavioral Development 27, 541–548 (2003)
12. Coleman, J.S., Hoffer, T.: Public and Private Schools: The Impact of Communities. Basic Books, New York (1987)
13. Leonard, M.: Children, childhood, and social capital: Exploring the links. Sociology 39, 605–622 (2005)
14. Marrow, V.: Conceptualizing social capital in relation to the well being of children and young people: A critical review. The Sociological Review 47, 744–765 (1999)
15. Majoribanks, K.: Family and School Capital: Towards a Context Theory of Students' School outcomes. Kluwer Academic Publishers, MA (2002)
16. Mulkey, L.M., Crain, R.L., Harrington, A.J.C.: One-parent households and achievement: Economic and behavioral expectations of a small effect. Sociology of Education 65, 48–65 (1992)
17. Dronkers, J.: The changing effects of lone parent families on the educational attainment of their children in a European welfare state. Sociology 28, 171–191 (1994)
18. Fan, X., Chen, M.: Parental involvement and students achievements: A meta-analysis. Educational Psychological Review 13, 1–22 (2001)
19. Ainsworth, J.W.: Why does it take a village? The Mediation of neighborhood effects on educational achievement. Social Forces 81, 117–152 (2002)
20. Willms, J.D.: Three hypotheses about community effects relevant to the contribution of human and social capital to sustaining economic growth and well-being. In: OECD International Symposium (March 2000)
21. OECD, The Well-Being of Nations: The Role of Human and Social Capital, OECD, Paris (2000)

22. Moore, S., Shiell, A., Hawe, P., Haines, V.A.: The privileging of communitarian ideas: Citation practices and the translation of social capital into public health research. American Journal of Public Health 95, 1330–1337 (2005)
23. Portes: Social capital: Its origins and applications in modern sociology. Annual Review of Sociology 24, 1–24 (1998)
24. Ahluwalia, P., Varshney, U.: Composite quality of service and decision making perspective in wireless networks. Decision Support Systems 46, 542–551 (2009)
25. Watson, N.: why we argue about virtual community: A case study of the Phish. Net Fan community. In: Virtual Culture, Sage, London (1997)
26. Onath, J., Boyd, D.: Public displays of connection. BT Technology Journal 22, 71–82 (2004)
27. Ellison, N., Steinfeld, C., Lampe, C.: Spatially bounded online social networks and social capital: the role of Facebook. In: Proceedings of the Annual Conference of the International Communication Association (2006)
28. Ulrike, P., Raj, A., Panayiotis, Z.: Age differences in online social networking – A study of user profiles and the social capital divide among teenagers and older users in My Space. Computers in Human Behavior 25, 643–654 (2009)

Analysis of Antibacterial Activities of Antibacterial Proteins/Peptides Isolated from Serum of *Clarias Gariepinus* Reared at High Stocking Density

Wang Xiaomei, Dai Wei, Chen Chengxun, Li Tianjun, and Zhu Lei

Tianjin Key Laboratory of Aqua-Ecology & Aquaculture,
Department of Fisheries Science, Tianjin Agricultural University,
Tianjin, 300384, China
xiaomeiwem@yahoo.com

Abstract. Antibacterial proteins are an important part of the innate immune system for all animals. The aim of the present work was to identify and characterise antibacterial proteins/peptides in serum of *Clarias gariepinus*. Antibacterial proteins/peptides were isolated from the serum of *Clarias gariepinus* using ammonium sulfate precipitation and further isolation on Sephadex G-50 column chromatography. Two fractions, namely, AP 1 and AP 2 associated with two absorption peaks at 280 nm were obtained. The antibacterial activity of AP 1 and AP 2 was tested against *Escherichia coli*, *Aeromonas hydrophila* and *Edwardsiella tarda*. AP 1 exhibited strong inhibitory activity against *E. coli*, *A. hydrophila* and *E. tarda*, and inhibited *E. tarda* best. AP 2 had also higher inhibitory activity against *E. tarda* than against *A. hydrophila*, but it had not inhibitory activity against *E. coli*. Protein profiles of AP 1 and AP 2 were determined by Laemmli SDS-PAGE system. The result showed that the AP 1 had a broad range of proteins/peptides with molecular weights ranging from about 9.5 kDa to over 66 kDa, and a 27 kDa protein/peptide was the most abundant than others, while the AP 2 had only 1 distinct protein band, which molecular weight was about 3.8 kDa.

Keywords: Clarias gariepinus, serum, antibacterial proteins/peptides, antibacterial activity, electrophoresis.

1 Introduction

A large variety of antimicrobial proteins/peptides are found in various organisms [1]. These proteins/peptides are broad-spectrum with potent antimicrobial activity against a wide range of microorganisms, including bacteria, viruses, fungi and protozoan parasites [2]. Such broad-spectrum proteins/peptides offer the distinct advantages of nonspecificity and rapid response, which favored their investigation and exploitation as potential new antibacterial agents.

Clarias gariepinus is native to the River Nile of Africa and introduced to China in 1981. It has become an economically important aquaculture species in China due to its fast growth, high disease resistance and hypoxia tolerance. Fish rely heavily upon innate or nonspecific immune mechanisms for initial protection against infectious agents [3]. The property of high disease resistance of this fish might be attributed to its

D. Zeng (Ed.): Advances in Control and Communication, LNEE 137, pp. 303–309.
springerlink.com

innate immunity system. It was well known that antibacterial proteins/peptides are a key component of innate immunity of fish and present in following organs or tissue, such as skin mucus, skin, gills, intestinal, serum [2-6]. In this study, the antibacterial activities of potential antibacterial proteins/peptides in serum of *C. gariepinus* reared at 200 kg/m^3 were detected, which will not only enhance our understanding of innate immune system of this fish, but also lay a foundation for feed additives development.

2 Materials and Methods

2.1 Fish

Healthy *Clarias gariepinus*, obtained from Deren aquaculture center, Tianjin, China, were reared in static tanks at a stocking density of 200 kg/m^3 at temperature of 25°C. Fish were fed with commercial diet until they reached a mean weight 554.7±31.5 g at the time of sampling.

2.2 Sample Collection

Blood was drawn from vessel in vascular arc of caudal peduncle. Serum was prepared according to the paper [7]. Serum was collected and pooled from 25 fish individuals for proteins/peptides isolation. Before sampling, experimental fish were anesthetized with a sub-lethal dose of tricaine methanesulphonate (MS222, sigma).

2.3 Sample Extraction

Serum was mixed with ice-cold PBS buffer (pH 6.0) in a ratio of 1:5. The mixture was treated in water bath at 70°C for 20 min, and stirred frequently at the same time. High molecular weight proteins were removed by centrifugation at 10, 000 rpm for 25 min at 4°C. The supernatant was treated with solid ammonium sulfate (AS) to reach a concentration of 70% saturation. The mixture was left overnight at 4°C and centrifuged at 10, 000 rpm for 30 min at 4°C. The precipitated proteins/peptides were collected. The precipitates were redissolved in sterile deionized water and dialyzed in 1000 MWCO (molecular weight cut-off) dialysis bag against sterile deionized water for 24 h at 4 °C with tenfold changes of dialysis water. Crude protein/peptide extracts obtained were lyophilized and kept at -20°C.

0.1 g lyophilized crude proteins/peptide extract was dissolved in 4 mL of 0.05 mol/L ammonium acetate buffer solution (pH5.5). The redissolved sample was subject to a column of Sephadex G-50 (1.0×60 cm) and eluted with 0.05 mol/L ammonium acetate buffer solution at room temperature. Fraction of 3 mL each was collected on fraction collector (Shanghai Luxi Fraction Collector, China) at a flow rate of 0.3 mL/min. The absorbance at 280 nm was measured to monitor the proteins/peptides during the chromatography separation. The proteins/peptides associated with the absorption peaks were pooled together and lyophilized, respectively. The lyophilized peak samples were kept at -20 °C until used.

2.4 Antibacterial Activity Test

Escherichia coli, *Aeromonas hydrophila* and *Edwardsiella tarda* obtained from micro lab of Tianjin Agricultural University, China, were used as the main test bacteria. They were grown in LB medium at optimal temperature to logarithmic phase, harvested by centrifugation at 5, 000 rpm for 15 min at 4°C, washed twice with sterilized physiological saline and resuspended, finally counted using a Neubauer hemocytometer.

Antibacterial activity of lyophilized protein/peptide extracts separated on Sephadex G-50 was tested using the agar diffusion method described by Hellio et al. with some modifications [8]. The lyophilized different fractions separated on Sephadex G-50 were resuspended in sterile water to give concentration of 25, 50, 75 and 100 mg/mL for one peak sample and only 10 mg/mL for less amount peak sample. Twenty milliliters of 1.5% (w/v) agar in LB medium were poured into sterile Petri dishes and formed a layer. The bacterial suspension was added aseptically to 5 mL agar medium at 45°C to a final concentration of approximately 10^6 cfu/mL, then mixed well and poured immediately to form the second layer. After hardening, sterile Oxford cups (6 mm inner diameter) were placed on this bilayer medium. 200 μL of resuspended samples were pipetted into Oxford cups and allowed to diffuse for 24 h at 4°C. After incubation for 24 h at appropriate temperatures for bacteria growth, such as *E. coli* at 37°C, *A. hydrophila* and *E. tarda* both at 31°C, the antibacterial activity was evaluated by measuring the diameter of the inhibition zone. 200 μL of sterile water was used for the control for all assays and showed no inhibition of bacterial growth. All inhibition assays and controls were carried out in triplicate.

2.5 Electrophoresis

The homogeneity of pooled fractions was determined using 15% SDS-substrate acrylamide gel, according to the method of Xia Qichang et al. with a slight modification [9]. Acrylamide bisacrylamide stock solution (30% T, 2.67% C mixture) was used. 0.001 g of lyophilized pooled fractions was dissolved in 50 μL sterile water. 7.5 μL of lyophilized proteins/peptides solution and 7.5 μL of protein loading buffer [2% SDS (wt/vol), 5% mercaptoethanol (vol/vol), 20% glycerol (vol/vol), 0.02% bromophenol blue, 100 mM Tris-HCl, pH 6.8] were mixed, boiled for 10 min, and then loaded into each well of the gel. Protein markers were used to estimate the molecular weights of proteins. Electrophoresis started with an initial constant current of 11 mA for 30min, and then ran at constant current of 24 mA until the dye front migrated to 1 cm from the bottom of the gel. After that, the gel was stained with 0.025% Coomassie brilliant blue R-250 in 40% methanol and 10% acetic acid for 4 h and destained in 10% methanol and 10% acetic acid until clear bands were observed.

3 Results and Discussion

3.1 Sample Extraction

Fish are totally dependent upon the innate immune system to help in the maintenance of homeostatic integrity against pathogenic or opportunistic microbial invaders [10]. Antibacterial proteins/peptides have been recognized to be important one of factors in innate immunity and to be the first line of defence against pathogenic microorganism. Some results in study on antibacterial proteins/peptides of fish were achieved [4-6, 11-15].

In this paper, extraction of proteins/peptides from serum of *C. gariepinus* on Sephadex G-50 column chromatography were obtained and showed that there were two optical density peaks (Fig.1). The proteins/peptides associated with these two optical density peaks were pooled, and termed as AP 1 and AP 2 in the order.

3.2 Antibacterial Activity of Crude Proteins/Peptides Extract, AP 1 and AP 2

In this paper, crude proteins/peptides extract from serum of *C. gariepinu* was examined for their antibacterial activity against *A. hydrophila*, *E. tard* and *E. coli*. The sample at concentration of 100 mg/mL exhibited antibacterial activity against all the three tested bacterial strains, but the diameters of inhibition zone were the same as that of Oxford cup (results didn't shown), which was consistent with the research on *Paralich thysolivaceus* [11], earthworm [16], deer leukocytes [17] and *Bullacta exarata* [18]. Proteins and peptides could be separated on Sephadex columns [19]. In order to further test the antibacterial activities of antibacterial proteins/peptides from serum of *C. gariepinu*, the crude proteins/peptides extract from serum in this study was subject to a column of Sephadex G-50 and the elution samples, namely, AP 1 and AP 2 were obtained.

The antibacterial activity against *A. hydrophila*, *E. tard* and *E. coli* of AP 1 and AP 2 were also detected. As shown in Table 1, AP 1 at concentration of 25, 50, 75 and 100 mg/mL exhibited antibacterial activity against all the three tested bacterial strains, but it was observed that the antibacterial activity of proteins/peptides differed among the three tested bacterial strains. At the same concentration,

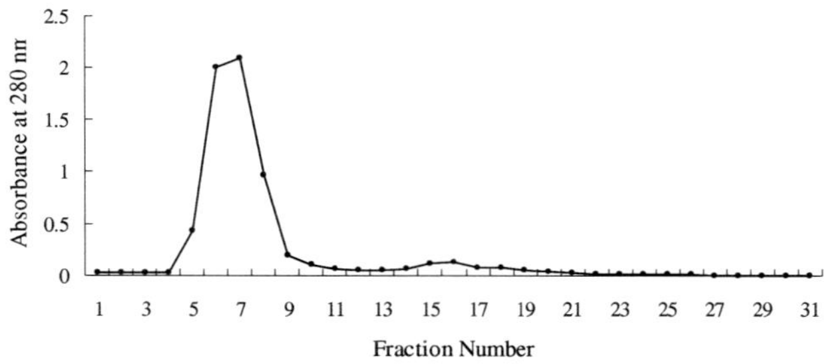

Fig. 1. Absorbance at 280 nm for each fraction eluted off from Sephadex G-50 chromatography column

Table 1. Diameter of Inhibition Zone of AP 1 and AP 2

Elution peak (mg/mL)	Diameter of antibacterial zone (mm)		
	E. coli	*A. hydrophila*	*E. tarda*
AP 1 (25)	8.04	8.38	13.97
AP 1 (50)	8.51	17.20	20.28
AP 1 (75)	9.38	18.62	22.03
AP 1 (100)	11.56	21.17	26.50
AP 2 (10)	0.00	8.19	8.73

AP 1 exhibited the biggest inhibition zone to *E. tard* and the smallest inhibitory zone to *E. coli.* AP 2 exhibited bigger inhibition zone against *A. hydrophila* than against *E. tard*, but it didn't exhibit inhibition zone to *E. coli*, which might be due to that the concentration of AP 2 was too low to inhibit the growth of *E. coli*. Despite this, these results indicated that the same proteins/peptides isolated on Sephadex G-50 showed difference of antibacterial activity against different bacterial strains (Tab. 1). The composition and amount of antibacterial components produced in the same tissue has been observed to change in response to ecological and physiological condition changes, such as handling stress and stages of growth [20, 21]. The stress coming from high density culture conditions might influence the antibacterial activity of *C. gariepinus* in present study. The antibacterial activity of AP 1 and AP 2 isolated from crude proteins/peptides in serum of *C. gariepinus* on Sephadex G-50 was more effective compared with that of crude proteins/peptides, which was consistent with previous research findings. Which might be due to that crude protein included proteins/peptides, salts, and other a small molecules (not protein), however, these components of nonprotein could be removed from proteins by Sephadex Chromatography.

3.3 Electrophoresis

The protein profiles of AP 1 showed a broad range of proteins/peptides, with molecular weights ranging from about 9.5 kDa to over 66 kDa. Abundant protein/peptide with molecular weight of 27 kDa was observed in AP1. Abundant 27 kDa protein/peptide was also observed in AP1 from all epithelial tissues and skin mucus of this fish, *C. gariepinus* in our previous study [22]. Lemaître et al. reported the isolation of two antibacterial proteins of 27 and

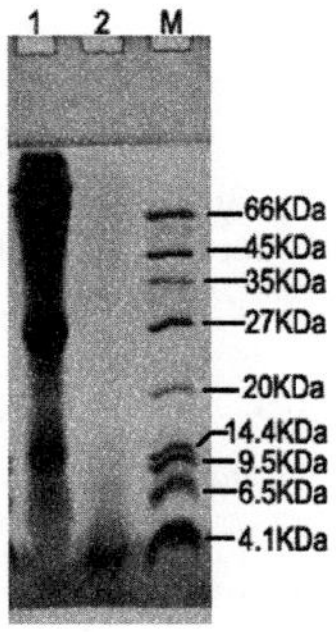

Lane 1: AP 1; Lane 2: AP 2; Lane M: protein marke

Fig. 2. SDS-PAGE Electrophoresis patterns of the AP 1 and AP 2

31 kDa from the skin mucus of carp, *Cyprinus carpio* [5]. Protein of 27 kDa in *C. gariepinus* and in *C. carp* might be the same antimicrobial proteins expressed in different fish species, which will be further confirmed. Only one band (3.8 kDa) was observed in the protein profile of AP 2, which suggested that antibacterial activity of AP2 could be attributed to the 3.8 kDa peptide. Further work would be under progress in this laboratory for further isolation of AP 1 on Sephadex 25 column chromatography and determination of amino acid composition of the both 27 kDa from AP 1 and 3.8 kDa from AP 2 proteins/peptides.

Acknowledgment. This work is supported by Tianjin natural science foundation of China (Grant No 08JCZDJC19000).

References

1. Brogden, K.A., Ackermann, M., McCray Jr., P.B., Tack, B.F.: Antimicrobial Peptides in Animals and Their Role in Host Defences. International Journal of Antimicrobial Agents 22, 465–478 (2003), doi:10.1016/S0924-8579(03)00180-8
2. Robinette, D., Wada, S., Arroll, T., Levy, M.G., Miller, W.L., Noga, E.J.: Antimicrobial Activity in the Skin of the Channel Catfish *Ictalurus punctatus*: Characterization of Broad-Spectrum Histone-Like Antimicrobial Proteins. Cellular and Molecular Life Sciences: CMLS 54, 467–475 (1998), doi:10.1007/s000180050175
3. MagnadÓttir, B.: Innate Immunity of Fish (overview). Fish & Shellfish Immunology 20, 137–151 (2006), doi:10.1016/j.fsi.2004.09.006
4. Subramanian, S., Ross, N.W., MacKinnon, S.L.: Comparison of Antimicrobial Activity in the Epidermal Mucus Extracts of Fish. Comparative Biochemistry and Physiology Part B: Biochemistry and Molecular Biology 150, 85–92 (2008), doi:10.1016/j.cbpb.2008.01.011
5. Lemaître, C., Orange, N., Saglio, P., Saint, N., Gagnon, J., Molle, G.: Characterization and Ion Channel Activities of Novel Antibacterial Proteins from the Skin Mucosa of Carp (*Cyprinus carpio*). European Journal of Biochemistry 240, 143–149 (1996), doi:10.1111/j.1432-1033.1996.0143h.x
6. Ruangsri, J., Fernandes, J.M., Brinchmann, M., Kiron, V.: Antimicrobial Activity in the Tissues of Atlantic Cod (Gadus morhua L.). Fish and Shellfish Immunology 28, 879–886 (2010), doi:10.1016/j.fsi.2010.02.006
7. Wang, Y., Wu, Z., Pang, S., Zhu, D., Feng, X., Chen, X.: Effect of fructooligosaccharides on non-specific immune function in *Carassius auratus*. Acta Hydrobiologica Sinica 32, 488–492 (2008) (in Chinese)
8. Hellio, C., Pons, A.M., Beaupoil, C., Bourgougnon, N., Gal, Y.L.: Antibacterial, Antifungal and Cytotoxic Activities of Extracts from Fish Epidermis and Epidermal Mucus. International Journal of Antimicrobial Agents 20, 214–219 (2002)
9. Xia, Q., Zeng, R.: Protein Chemistry and Proteomics, pp. 80–91. Science Press, Beijing (2004) (in Chinese)
10. Smith, V.J., Fernandes, J.M.O., Jones, S.J., Kemp, G.D., Tatner, M.F.: Antibacterial Proteins in Rainbow Trout, *Oncorhynchus mykiss*. Fish & Shellfish Immunology 10, 243–260 (2000), doi:10.1006/fsim.1999-0254
11. Wang, H., An, L., Yang, G.: Extraction and Some Characterization of Antibacterial Peptides from Different Tissues of Japanese Flounder (*Paralichthys olivaceus*). Fisheries Science 26, 87–90 (2007) (in Chinese)
12. Su, J., Lei, H., Huang, P., Xiao, T.: Antibacterial activities of antimicrobial peptides in the intestine of grass carp. Journal of Hunan Agricultural University (Natural Sciences) 35, 162–164 (2009) (in Chinese)
13. Oren, Z., Shai, Y.: A Class of Highly Potent Antibacterial Peptides Derived from Pardaxin, a Pore-Forming Peptide Isolated from Moses Sole Fish Pardachirus marmoratus. European Journal of Biochemistry 237, 303–310 (1996), doi:10.1111/j.1432-1033.1996.0303n.x
14. Conlon, J.M., Sower, S.A.: Isolation of A Peptide Structurally Related to Mammalian Corticostatins from The Lamprey *Petromyzon marinus*. Comparative Biochemistry and Physiology Part B: Biochemistry and Molecular Biology 114, 133–137 (1996), doi:10.1016/0305-0491(95)02132-9

15. Park, C.B., Lee, J.H., Park, I.Y., Kim, M.S., Kim, S.C.: A Novel Antimicrobial Peptide from The Loach, Misgurnus anguillicaudatus. FEBS Letters 411, 173–178 (1997), doi:10.1016/S0014-5793(97)00684-4
16. Cui, D., Zheng, Y., Wang, Y., Zhang, L.: Purification of antibacterial peptides from earthworm. Journal of Dalian Institute of Light Industry 23, 265–269 (2004) (in Chinese)
17. Chen, L., Wang, Y., Yu, P.-L.: Extraction of antimicrobial peptides from deer leukocytes and identification of their bactericidal activities. Journal of Henan Agricultural University 39, 406–409 (2005) (in Chinese)
18. Li, Y., Su, X., Li, T.: Study on antimicrobial peptides from Bullacta exarata. Journal of Oceanography in Taiwan Strait 24, 145–149 (2005) (in Chinese)
19. Casimpoolas, N., Kenney, J.: Rapid Analytical Gel Chromatography of Proteins and Peptides on Sephadex Microbore Columns. Journal of Chromatography A 64, 77–83 (1972), doi:10.1016/S0021-9673(00)92950-9
20. Agarwal, S.K., Banerjee, T.K., Mittal, A.K.: Physiological adaptation in relation to hyperosmotic stress in the epidermis of a fresh-water teleost *Barbus sophor* (Cypriniformes, Cyprinidae): a histochemical study. Zeitschrift für Mikroskopisch-anatomische Forschung 93, 51–64 (1979)
21. Zuchelkowski, E.M., Lantzm, R.C., Hinton, D.E.: Effects of Acid-Stress on Epidermal Mucous Cells of The Brown Bullhead *Ictalurus nebulosus* (LeSeur): A Morphometric Study. The Anatomical Record 200, 33–39 (1981), doi:10.1002/ar.1092000104
22. Wang, X., Dai, W., Xing, K., Li, T., Wang, X.: Antibacterial Activities of Antibacterial Proteins/Peptides Isolated from Organs and Mucus of *Clarias gariepinus* Reared at High Stocking Density. In: Proc. IEEE Symp., Intenational Conference on Cellular, Molecular Biology, Biophysics and Bioengineering 2010, IEEE Press (2010) (in press)

Prediction Model of Iron Release When the Desalinated Water into Water Distribution System

Wang Jia, Tian Yi-mei, and Liu Yang

School of Environmental Science and Engineering,
Tianjin University, Tianjin, China
ymtian_2000@yahoo.com.cn

Abstract. A trial network was constructed in the laboratory to simulate the water quality condition of real water distribution network. Three models were established based on experimental data and regression analysis method, which are total regression model, stepwise regression model and principal component regression model. The relationship between iron concentration and related water quality indexes was described quantitatively and the iron concentration was predicted by the reliable statistical models. The result shows as for prediction accuracies of these models, the principal component regression model is a little better than total regression model, and is much better than stepwise regression model. The accuracies of three models are 72.03%, 65.02% and 72.92% respectively.

Keywords: iron release, prediction model, water quality stability, desalinated water, regression analysis.

1 Introduction

With the sustainable development of cities, the application of desalinated water will be the development trend of multi-water resources utilization in many cities, especially in coastal cities. However, "red water" or "color water" [1]often occurs when desalinated water is transported and distributed as municipal water in water distribution systems due to the corrosion of metal pipes and iron release, and often leads to the user's strong dissatisfaction. In order to control the release of iron and ensure water quality, iron concentration must be predicted timely and accurately in pipe network. But the research on the prediction model of iron release is very few [2]. Models which have been established are mainly in the form of power exponent according to water quality indexes and regarding the corrosion rate or color as the dependent variable [3]. Unified prediction model of iron release are not established until now because of the differences between the structure of the pipe network, water quality and the treatment process in different places.

In order to establish strongly applicable and highly confident prediction model of iron release amount, three methods were selected to establish the mathematical model of iron concentration to predict total iron concentration in water.

D. Zeng (Ed.): Advances in Control and Communication, LNEE 137, pp. 311–318.
springerlink.com

2 Sample Date Source

2.1 Simulation Experiment

Simulation experiment was carried out in the laboratory to study the iron release mechanism and establish prediction model in pipe network after mixing desalinated water and tap water. A more integral replaced steel pipe of a water plant in Tianjin was selected to build the pipe network. The water plant effluent and high-quality desalinated water were studied to simulate water quality conditions in the actual water supply network which was produced by the method of multi-effect distillation in condition of the low temperature.Water quality characteristics: alkalinity and total hardness were relatively high; pH was between 7 and 8.5; the chemical stability of water was guaranteed basically; water plant chose aluminum as the coagulant to reduce iron concentration effectively. Desalinated water features: the water quality was better than tap water; alkalinity, hardness and various ions were very low; pH was about 7.8 in general while sometimes pH also occurred 6.5, leading to the strong corrosion on the network.

2.2 Method and Sample Data

The tube must be sealed and vertical. Effluent and influent was respectively from the bottom of pipe and from the 40 cm section below the liquid level by the method of siphon. Static immersion test was made in the tube with tap water and desalinated water mixture inside. Related water quality indicators were monitored continuously such as the iron concentration, pH, temperature, alkalinity, total hardness, sulfate and chloride concentration. According to hydraulic retention time in the actual pipe network, the experimental period was set to range from 8 to 24 hours. Every 1 or 2 hours we had to obtain a sample with the quantity of water ranging from 100 to 200 ml.

The influence factors of the iron release mainly include hydraulic operating conditions and water quality conditions. Seven influence factors such as pH, temperature, residence time, alkalinity, total hardness, Cl^- and SO_4^{2-} concentration can be determined according to this study.

After the experimental monitoring data was analyzed and screened, the singular points were removed. Then 92 groups of the monitoring data were selected from 107 groups of data as the sample data to establish the prediction model. The results of correlation analysis between the influence factors and iron release were shown in Table 1 based on the data.

Table 1. Correlation Analysis between the Influence Factors and Iron Release

Projects	ΔFe (mg/L)	Residence time (h)	Temperature (°C)	pH	Total Hardness (mg/L as CaCO3)	Alkalinity (mg/L as CaCO3)	Cl^- (mg/L)	SO_4^{2-} (mg/L)
ΔFe	1.000							
Residence time	0.806	1.000						
Temperature	0.254	0.100	1.000					
pH	-0.297	-0.154	-0.050	1.000				
Total hardness	-0.259	-0.087	-0.006	0.499	1.000			
Alkalinity	-0.171	-0.051	-0.685	-0.284	-0.110	1.000		
Cl^-	0.014	-0.033	0.212	-0.299	-0.206	0.026	1.000	
SO_4^{2-}	0.170	-0.020	-0.007	0.104	-0.189	0.068	-0.218	1.000

3 Prediction of Iron Release Amount

In this study, the empirical and statistical models of iron concentration were established in the pipe network based on correlation analysis between iron release mechanism and influence factors. The three methods of multiple linear regression, stepwise regression and principal component analysis were selected under the condition in which the characteristics of each prediction methods were combined and data acquisition conditions were taken into account. The models can predict the total iron release and describe quantitatively the relationship between the iron concentration in water and each related indicators.

3.1 Total Regression Model

First the method of the multiple linear regression was selected to establish the whole regression model in the form as follows:

$$Y = k \cdot X_1^{\alpha} \cdot X_2^{\beta} \cdot X_3^{\chi} \cdot X_4^{\delta} \cdot X_5^{\varepsilon} \cdot X_6^{\varphi} \cdot X_7^{\gamma} \quad (1)$$

Where: Y was a dependent variable, namely iron concentration in pipe network; X_1, X_2, X_3, X_4, X_5, X_6, X_7 were the selected model variables, respectively, namely pH, temperature, residence time, alkalinity, total hardness, chloride and sulfate concentration; α, β, χ, δ, ε, φ, γ were fitting coefficients of the independent variables in the model, respectively.

To facilitate the nonlinear model fitting, equation was taken logarithm on both sides and the model parameters were obtained by regression fitting. The prediction model of iron concentration was obtained as follows:

$$Fe = 3.25 \times 10^{10} \frac{T^{0.101} \cdot [Cl^-]^{0.477} \cdot [SO_4^{2-}]^{0.691} \cdot HRT^{0.371}}{pH^{2.952} \cdot Hardness^{5.458} \cdot Alk^{0.051}} \quad (2)$$

Where: Fe was iron release, mg / L; T was the temperature of the mixed water in the pipe network, °C; Alk and Hardness were bicarbonate alkalinity and total hardness, mg / L; $[Cl^-]$ and $[SO_4^{2-}]$ were chloride and sulfate concentration, mg / L; HRT was the retention time, h.

The multiple correlation coefficients of the model were two parts which R^2 was 0.76 and F was 40.93. $F_{0.99}(7, 84)$ was 2.86 in the table. Because F was larger than $F_{0.99}(7, 84)$, the model reached the level that α was 0.01 significantly, which had guiding significance to a certain [4]. Based on result of total regression model, the iron release amount was negatively correlated with pH, alkalinity and total hardness. However, the amount was also positive correlated with the residence time, temperature, chloride and sulfate concentration, in which the result was consistent with previous studies [5—8].

15 groups of experimental data which were not involved in establishing the model were selected to calculate the total iron concentration. The total regression model was tested by the way of comparing the differences between the real value and the predictive data. The results were shown in Figure 1.

3.2 Stepwise Regression Model

In total regression model, due to the introduction of more independent variables, not only unreliable model variables were easy to lead to error accumulation, but also the stability of the model might degrade [9]. To explore the model error when a small amount of independent variables with weak partial correlation coefficients was removed and got a simpler model equation to improve the accuracy of prediction model, the method of stepwise regression was continuously selected to establish prediction model.

Stepwise regression is a regression method that screens the independent variable gradually. The independent variables are introduced one by one during establishing the regression equation, and the partial correlation coefficient is tested statistically. If the independent variable has a significant role, it can be remained in the regression equation, otherwise the introduction was stopped. The original variables are tested at the same time to guarantee whether they are still significant after the introduction of new variables. If they are not significant, they should be removed. This continues until there is no new variables can be introduced, and no old variable can be removed. Then the stepwise regression equation is established in the end.

In this study, stepwise function in MATLAB toolbox was selected to carry out regression analysis to obtain model parameters, which can reject non-significant variables. The eventual prediction model of iron concentration was established as follow:

$$Fe = 1.04\times10^{13}\frac{[Cl^-]^{0.495}\cdot[SO_4^{\ 2-}]^{0.704}\cdot HRT^{0.378}}{pH^{3.244}\cdot Hardness^{5.7}} \tag{3}$$

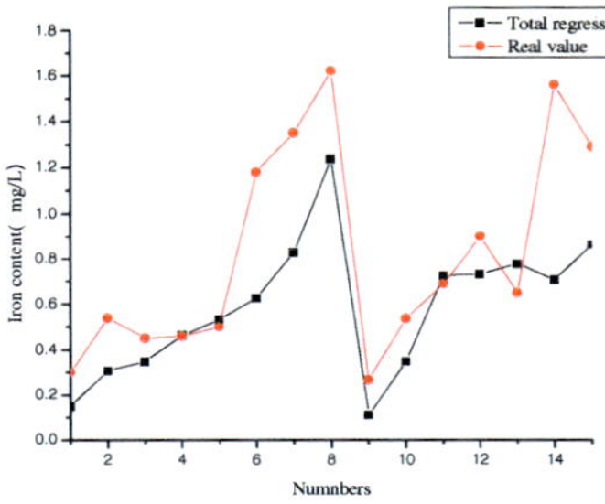

Fig. 1. Comparison of predicting values of total regression model and real data

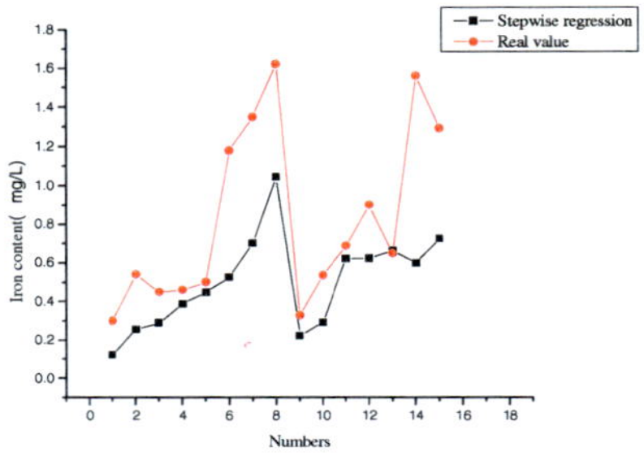

Fig. 2. Comparison of predicting values of stepwise regression model and real data

Table 2. Correlation analysis between the influence factors and iron release

Independent Variables	Residence Time(h)	pH	Total Hardness (mg/L)	Cl^- (mg/L)	SO_4^{2-} (mg/L)
t Score	13.070	-4.553	-5.305	5.155	2.264
p Score	0.000	0.000	0.000	0.000	0.026

The multiple correlation coefficients of the model included two parts which R^2 was 0.756 and F was 53.21. Because F was larger than $F_{0.99}(7, 84)$, the model reached the level that α was 0.01 significantly, which had guiding significance to a certain [4]. It can be seen that the stepwise regression model rejected two independent variables including alkalinity and water temperature. Alkalinity was removed mainly because of the overlap of alkalinity and pH on the influence of iron release to a certain extent, leading to the result that pH was filtered out automatically as a model of the independent variables due to its larger significant in the process of the stepwise regression analysis; The reason of water temperature rejection might be caused by its unobvious changes. Model retained significant influence factors on iron release, including residence time, pH, total hardness, Cl^- and SO_4^{2-}, to weaken the effect of independent variables with weak correlation and mutual overlapping influences. Coefficient and the comparison results were shown in Table 2 and Figure 2 respectively.

3.3 Principal Component Regression Model

The prediction accuracies of the total regression model and the stepwise regression model were not high as shown in Figure 1 and Figure 2. To search for a more accurate prediction model of iron concentration in the network, prediction model by the principal component analysis continued to be established.

Principal component analysis [10] is the method that researches the internal structure of the correlation matrix of original variables to find several important and independent integrated variables, which are linear combinations and retain the main information of the original variables. From the statistical point, the idea of reduction dimension in principal component analysis is to find a linear function $y = \sum_{i=1}^{n} \alpha P_n$ by the data matrix X given which can reflect the changes in these indexes best and is the linear combination of $P_1, P_2, \ldots, P_n$.

Princomp function of MATLAB toolbox was used to analyze principal component in this study. It is considered that they retain the basic information of the original water quality when the accumulation contribution rate of new variables reached 85% or more. The results of principal component analysis were shown in Table 3 and Table 4.

The cumulative contribution of the first three variables was 98.27% as shown in Table 3, which has focused the main information of original variables. Therefore, regression was used to establish the model that reflected relationship between iron concentration and the principal components according to the first three principal components.

$$\mathrm{LogFe} = 0.6074 + 0.648P_1 - 0.518P_2 - 0.33P_3 \tag{4}$$

$$\mathrm{Fe} = 4.05\frac{T^{0.0168}\cdot[Cl^-]^{0.057}\cdot[SO_4^{2-}]^{0.090}\cdot HRT^{0.6993}}{pH^{0.05}\cdot Alk^{0.461}\cdot Hardness^{0.286}} \tag{5}$$

Table 3. Eigenvalue and Proportion of the 7 Principal Components

Component	P_1	P_2	P_3
Eigenvalue	0.2829	0.1459	0.0773
Proportion (%)	54.93	28.33	15.01
Cumulative (%)	54.93	83.26	98.27

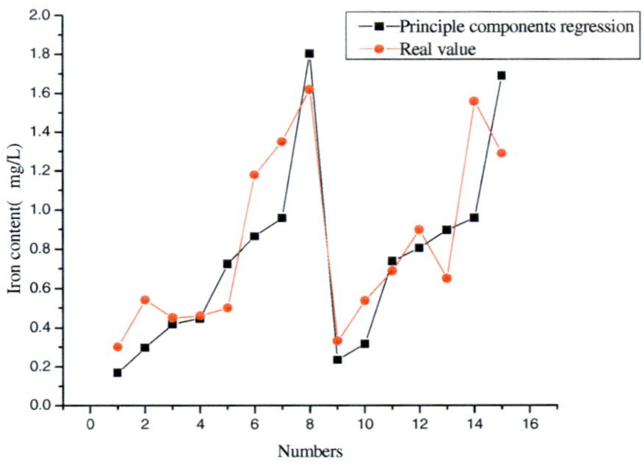

Fig. 3. Comparison between predicting values of principal component regression model and real data

The multiple correlation coefficients of the model included two parts which R^2 was 0.756 and F was 53.21. Because F was larger than $F_{0.99}(7, 84)$, the model reached the level that α was 0.01 significantly, which had guiding significance to a certain [4]. It can be seen from Figure 3 that the model by principal component analysis predicted more accurately.

Table 4. Error analysis of prediction models

Real Values (mg/L)	**Relative Error In Each Model**		
	Total Regression	***Step Regression***	***Principal Component Regression***
0.301	-50.77	-59.08	-44.09
0.540	-43.52	-52.57	-45.07
0.45	-23.14	-35.88	-7.10
0.46	0.50	-15.76	-2.84
0.5	5.94	-10.65	44.90
1.18	-47.08	-55.54	-26.71
1.35	-38.81	-48.1	-29.01
1.62	-23.73	-35.57	11.30
0.331	-19.47	-32.82	-28.95
0.537	-35.22	-45.64	-40.73
0.688	5.18	-9.900	7.45
0.899	-18.85	-30.89	-10.46
0.65	19.35	1.71	37.93
1.56	-54.80	-61.68	-38.50
1.29	-33.19	-43.89	31.09
Mean Percentage Absolute Error（%）	27.97	35.98	27.08

3.4 Model Validation and Comparison

As for prediction accuracy of these models, the principal component regression model was a little better to total regression model, and was much better than stepwise regression model; The accuracy of total regression model was about 72.03%; Stepwise regression model based on the total regression model removed insignificant variables, leading to the result that the model was simpler and less data was required to be monitored, but its efficiency declined to only 64.02%; The average prediction accuracy of the two models was both low generally; The prediction accuracy of principal component regression model had been improved, reaching 72.92%, so it had better application.

4 Prediction of Iron Release Amount

In this paper, models were established to predict the iron release in the form of power exponent by utilizing monitoring data in pipe network, including the total regression model, stepwise regression model and principal component regression model. The accuracies of three models were 72.03%, 65.02% and 72.92% respectively; The stepwise regression model which based on the total regression model removed insignificant variables, making the model simpler and more easily calculated while its efficiency decreased; The accuracy of principal component regression model had been improved a little to have better application.

Acknowledgment. The study was supported by the National Water Pollution Control and Management of Science and Technology Project of China (No. 2008ZX07317-005) and the Tianjin Science Association Key Project (No.033113111).

References

1. Sarin, P., Snoeyink, V.L., Bebee, J., Kriven, W.M., Clement, J.A.: Physico-chemical characteristics of corrosion scales in the old iron pipes. Water Research 35, 2961–2969 (2001)
2. Mutoti, G., Dietz, J.D., Imran, S.A., Taylor, J., Cooper, C.D.: Development of a novel iron release flux model for distribution systems. J. Am. Water Works Assoc. 99, 102–111 (2007)
3. Pisigan, R.A., Singley, J.E.: Effect of the Water Quality Parameters on the Corrosion of Galvanized Steel. J. Am. Water Works Assoc. 77, 76–82 (1985)
4. Ji, S.D., Zhang, Y.H.: Mathematical statistics, 1st edn., pp. 284–287. Tianjin University Press, Tianjin (2008)
5. Niu, Z.B., Wang, Y., Zhang, X.J.: Effect on Iron Release in Drinking Water Distribution Systems. Environmental Science 28, 2270–2274 (2007)
6. Savoye, S., Legrand, L., Sagon, G., Lecomte, S.: Experimental investigations on iron corrosion products formed in bicarbonate/carbonate-containing solutions at 90 ℃. Corrosion Science 43, 2049–2064 (2001)
7. Larson, T.E., Skold, R.V.: Corrosion and Tuberculation of Cast Iron. J. Am. Water Works Assoc. 49, 1294–1301 (1957)

8. Sarin, P., Snoeyink, V.L., Lytle, D.A., Kriven, W.M.: Iron Corrosion Scales: Model for Scale Growth, Iron Release, and Colored Water Formation. J. Envir. Engrg. 130, 364–373 (2004)
9. Zhou, W.F., Li, M.: Discussion on shortcomings of Stepwise Regression Analysis. Northwest Water Power 4, 49–50 (2004)
10. Ye, X.F., Wang, Z.L.: Principal Component Analysis in the Evaluation of Water Resources. Journal of Henan University 37, 276–279 (2007)

Study on Biodegradability of Acrylic Retanning Agent DT-R521

Xuechuan Wang, Yuqiao Fu, Taotao Qiang, and Longfang Ren

Key Laboratory of Chemistry and Technology for Light Chemical Industry,
Ministry of Education, Shaanxi University of Science and Technology, Xi'an 710021, China
wangxc@sust.edu.cn, fyuqiao@126.com

Abstract. The methods of respiratory curve and COD_{30} were adopted to evaluate the biodegradability of acrylic retanning agent DT-R521, which was used as the substrate of microorganism. The main results obtained by the method of respiratory curves were as follows: When the concentration of substrate was 0~3000mg/L, the respiratory curves of the substrate were under the curve of endogenous, which showed that the substrate has inhabiting effect on microorganism. When the concentration of the substrate was 1500mg/L, the utilization of the substrate by microorganism was the most. The main results obtained by the method of COD_{30} were as follows: when sludge concentration was 1000mg/L and pH was 7, the biodegradation rate of 500mg/L DT-R521 was 26.21%. Under above condition, the biodegradation rates of DT-R521 with other concentrations were much smaller. Moreover, as indicated in the respiratory curves, the effect of sludge concentration, salinity and co-metabolism on the biodegradability of the substrate was obvious. Under the experimental conditions, when the concentration of the substrate was 1500mg/L, the sludge concentration was 3000mg/L and salinity was 0.7%, the biodegradability of the substrate was the best. The glucose was as co-metabolism substrate and when its concentration was 300mg/L, the biodegradability of the substrate was obviously increased.

Keywords: acrylic retanning agent DT-R521, biodegradability, COD_{30}, respiratory curve.

1 Introduction

The use of green chemicals has been an important part of clean production in leather industry. As an important parameter for evaluating the environmental friendliness of organic chemicals, biodegradable property parameter has been widely acknowledged. However, the study on the biodegradability of leather chemicals was limited. Wang, et al. (2004) and Jiang, et al. (2005) have emphasized the importance of leather chemicals biodegradability study. Kylie Catterall et al (2003) have studied the biodegradability of Ferricyanide-Mediated using a mixed microbial consortium. With the acrylic retanning agent patent of Rohm & Hass in America was published by Netherlands in 1966, development and application of acrylic retanning agent started. Acrylic retanning agent is polymer, copolymerization mainly with (methyl) acrylic acid, assisted by other vinyl monomers. Compared with other synthetic and natural

D. Zeng (Ed.): Advances in Control and Communication, LNEE 137, pp. 319–325.
springerlink.com

tanning agent, acrylic retanning agent exhibit unique properties, such as increasing absorb amount of chromium, leather products fullness and softness, grain tightness as well as good elasticity, so as to was emphasized by leather manufacturers and researchers. H.Pan et al. (2004) have emphasized the characteristics of acrylic acid. Up till now, the development of acrylic retanning agent is still broad prosperity. Nevertheless, the biodegradation of acrylic retanning agent is still at an initial stage. Granted by Tingjiang Fine Chemical Co., Ltd., our studying team elected acrylic retanning agent DT-R521 (a kind of acrylic retanning agent with middle and small molecular weight is sticky liquid with light yellow) as the research object and statistically studied its biodegradability, which is of significance for the healthy development of the company.

2 Experimental

2.1 Chemicals and Mineral Salt Medium

$C_{12}H_{22}O_{11}$ (purity 99%), $(NH_4)_2FeSO_4$ (purity 99.5%) and $K_2Cr_2O_7$ (purity 98.5%) were purchased from Tianjin Chemical Reagent Co., Ltd, China. A mineral salt medium (MSM) used for the growth of microorganisms contained (mg/L) NH_4Cl, K_2HPO_4 $FeSO_4{\cdot}7H_2O$ 45, $CaCl_2$ 45, $MgSO_4$ 30, $ZnCl_2$ 30 at pH 7.0. This MSM didn't contain source of carbon and energy. All the chemicals were of analytical grades and were purchased from Tianjin Chemical Reagent Co., Ltd, China.

2.2 Microorganism Cultivation

The activated sludge sample was collected from the sludge compression workshop of Bei Shi Qiao Wastewater Treatment Plant in Xi'an, China. Appropriate amount of collected activated sludge was transferred to a culture flask filled with 2000mL of sterile MSM containing $C_{12}H_{22}O_{11}$ as the sole source of carbon and energy. The microorganism was incubated day by day until the microorganism characteristics were detected as follows. The collected activated sludge became yellow flocculent sludge, the SV_{30} was about 30%, and the MLSS was about 2500mg/L, which indicated that the activity of microorganism was good enough for following experiment of organic biodegradability determination.

2.3 Evaluation of the Biodegradability of Acrylic Retanning Agent DT-R521

2.3.1 The Biodegradability of Acrylic Retanning Agent DT-R521

As presented by Kong et al. (2004) and Zhang, et al. (1998), the methods of organic chemical biodegradability evaluation include respiratory curves, COD_{30}, et al. The studies of Yao et al. (2006) and Dong, et al. (2003) showed that co-metabolism is an effective method to the refractory compounds.

2.3.1.1 Respiratory Curves. Respiratory curve was mainly used to evaluate the initial adaptability of the microorganism to the substrate. BOD of the different concentration substrates (500mg/L, 1000mg/L, 1500mg/L, 2000mg/L and 3000mg/L) was measured using BOD analyzer and other measurement conditions were pH at 7.0, 20°C, prepared mineral salt medium (MSM), 1000mg/L fully aerated fresh activated sludge. Moreover, The BOD of blank sample, whose substrate concentration was 0 mg/L, was measured by BOD analyzer with the same measurement conditions. During each sample BOD measurement, the BOD value in every day was recorded, and the respiratory curves was obtained as Fig 1.

2.3.1.2 COD_{30}: COD30 was mainly used to evaluate the ultimate biodegradability of the substrate, which reflected the maximal degree of substrate utilization by the microorganism. 250ml substrate with different concentration (500mg/L, 1000mg/L, 1500mg/L, 2000mg/L and 3000mg/L) were transferred to five conical flasks. Other test conditions were controlled as follows. The pH was 7.0, mineral salt medium (MSM) through adding appropriate mineral salts in the conical flasks, 1000mg/L fully aerated fresh activated sludge. The five conical flasks were put in constant temperature shock incubator with the temperature at 20°C. The test was lasted for 30 days, The COD value of each sample was determined and recorded every 5 days. The COD30 curves were obtained as Fig 2.

2.3.2 Factors Affecting the Biodegradability of Acrylic Retanning Agent DT-R521

The method of respiratory curves was adopted to evaluate the effect of sludge concentration, pH, salinity and co-metabolism on the substrate biodegradation. Firstly, the substrate biodegradability was determined at sludge concentration 500mg/L, 750mg/L, 1000mg/L, 2000mg/L, 3000mg/L with other determination conditions as follows: 1500mg/L substrate, mentioned salt medium conditions, pH 7.0 and salinity 0%. Secondly, the substrate biodegradability was determined at pH 6, 6.5, 7, 7.5, 8, 8.5 with 1500mg/L substrate, 3000mg/L sludge concentration, 0% salinity and the same determination conditions. Thirdly, the substrate biodegradability was measured at salinity 0%, 0.1%, 0.5%, 0.7%, 0.9% with 1500mg/L substrate, 3000mg/L sludge concentration and at pH 8. Finally, the substrate biodegradability was measured at glucose concentration 0mg/L, 300mg/L, 600mg/L, 900mg/L, 1200mg/L, 1500mg/L with 1500mg/L substrate, 3000mg/L sludge concentration, at pH 8 and salinity 0.7%.

3 Results and Discussion

3.1 The Biodegradability of Acrylic Retanning Agent DT-R521

3.1.1 Respiratory Curves

Fig. 1 showed that the substrate concentration had great influence on the substrate biodegradability. Under the experimental conditions, the respiratory curves of substrate were all under the endogenous respiration curve, which indicated that the tested different concentration of substrate has inhabiting effect on microorganism and could not be degraded by the microorganism. When the concentration of the substrate was 1500mg/L, the inhabiting effect of the substrate by microorganism was the least. The next were 1000mg/L, 500mg/L. When the concentration of the substrate was 3000mg/L, the inhabiting effect was the most obvious.

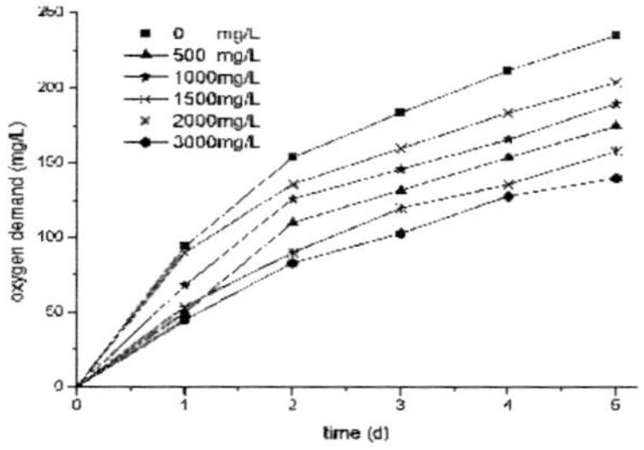

Fig. 1. The relationship between the substrate concentration and biodegradability

3.1.2 COD_{30}

As shown in Fig. 2, in the condition of pH at 7.0, 20°C, when the substrate concentration was in the range of 500~3000mg/L, the ultimate biodegradation rates were slow. The biodegradability of the substrate with the concentration of 500mg/L was the best and the ultimate biodegradation rate was 26.21%. The following was 1500mg/L, 1000mg/L, 2000mg/L, 3000mg/L. When the substrate concentration was 3000mg/L, the ultimate biodegradation rate only was 11.96%. When the substrate concentration was 3000mg/L and 1000mg/L, the inflexion points of 5d and 10d indicated that the biodegradation was false. That is to say, substrate firstly was adsorbed on the surface of microorganism and then released, this phenomenon was most significant when the substrate concentration was 3000mg/L.

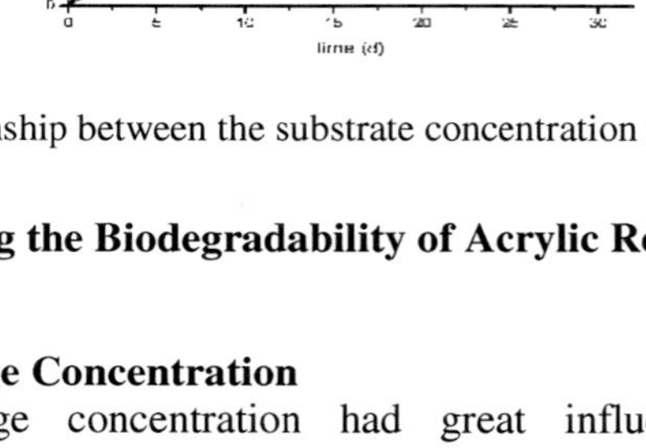

Fig. 2. The relationship between the substrate concentration and biodegradability

3.2 Factors Affecting the Biodegradability of Acrylic Retanning Agent DT-R521

3.2.1 Effect of Sludge Concentration

Fig. 3 showed sludge concentration had great influence on the substrate biodegradation. With the increase of inoculation amount, the substrate biodegradability became better as the sludge concentration was among 500～3000mg/L. The more the inoculation amount, the more the substrate degraded by microorganism. When the inoculation concentration increased to 3000mg/L, the substrate biodegradation rate increased.

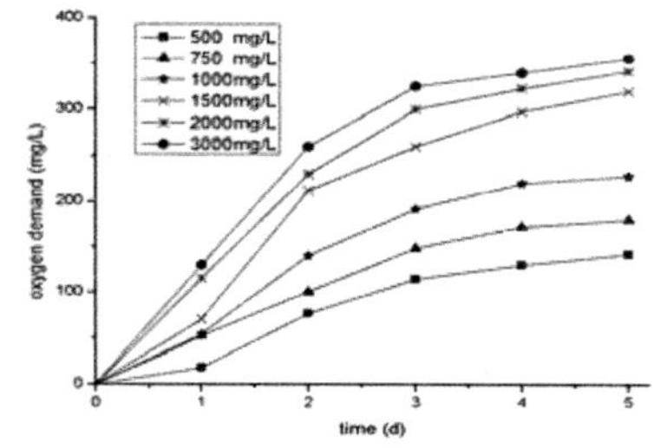

Fig. 3. The relationship between sludge concentration and the substrate biodegradability

3.2.2 Effect of pH

Fig. 4 showed that the hydrogen ion concentration of the culture medium greatly influenced the substrate biodegradability. The reason was pH limited the activity of the microorganism. Under the test conditions, the optimum pH for the substrate biodegraded was 8. When the pH was 8.5, the respiratory curve of substrate was lower than that of pH was 7.5, which indicated that strong alkaline medium had bad influence on the biodegradability of the substrate. When the pH was 6, the respiratory curve was the lowest, which indicated that weak alkaline medium had bad influence on the biodegradability of the substrate too.

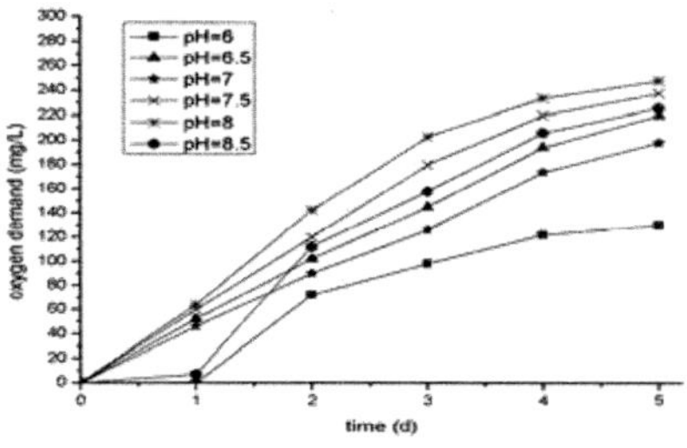

Fig. 4. The relationship between pH and the substrate biodegradability

3.2.3 Effect of Salinity

Fig. 5 showed the salinity of the culture medium has great influence on the substrate biodegradability. During the process of microorganism using the substrate as the sole source of carbon, appropriate mineral salts should be provided for the normal activity of the microorganism. Under too high or too low salinity, the microorganism cells would dehydrate or hydrate and the microorganism activity would be inhibited as a result of the inapplicable osmotic pressure environment. Under the test conditions, when the additional NaCl was in the range of 0%-0.7%, the substrate biodegradability became better with the increase of salinity. However, when the additional NaCl was above 0.7%, the substrate biodegradability became worse with the increase of salinity. The optimum salinity for the substrate biodegraded by the microorganism was 0.7% additional NaCl based on the prepared MSM.

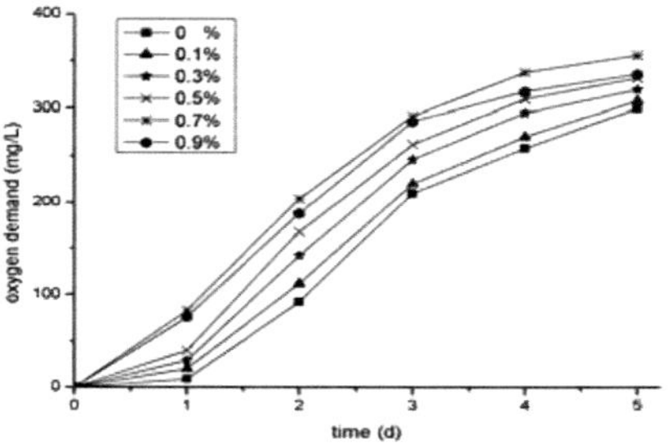

Fig. 5. The relationship between salinity and the substrate biodegradability

3.2.4 Effect of Co-metabolism

Fig. 6 showed glucose concentration had great influence on the substrate biodegradability when it was as co-metabolism substrate. Compared with blank sample, when the glucose concentration was in the range of 0~1500mg/L, the substrate respiratory curves became increase then decrease. As indicated in Fig. 6, the optimum glucose substrate was 300mg/L. the respiratory curves of substrate with the concentration of 1200mg/L and 1500mg/L were lower than that of the blank sample, which indicated that when the glucose concentration was too high, the respiratory of the microorganism was inhabiting effected because the COD of the microorganism was large.

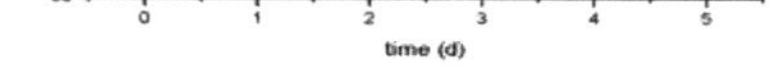

Fig. 6. The relationship between glucose concentration and the substrate biodegradability

4 Conclusions

Acrylic retanning agent DT-R521 was used as the substrate of microorganism, when sludge concentration was 1000mg/L, pH was 7 and with no additional NaCl, the biodegradation rate of 500mg/L DT-R521 was 26.21%. The methods of respiratory curve showed that the substrate had inhabiting effect on microorganism and could not were biodegraded by microorganism. The result of the method of respiratory curve was consistent with that of COD_{30}. The difference was the optimum concentration. The result of COD_{30} method was 500mg/L and that of the respiratory curves was 1500mg/L. In order to increase accuracy, other evaluation method could be used. Moreover, as indicated in the respiratory curves, the effect of sludge concentration, salinity and co-metabolism on the biodegradability of the substrate was obvious. Under the experimental conditions, when the concentration of the substrate was 1500mg/L, the sludge concentration was 3000mg/L and salinity was 0.7%, the biodegradability of the substrate was the best. The glucose was as co-metabolism substrate and when its concentration was 300mg/L, the biodegradability of the substrate obviously increased.

Acknowledgements. This research was supported by National Natural Science Foundation of China (20876090), the Program of Science and technology projects in Xianyang City (XK0909-3) and the Graduate Innovation Fund of Shaanxi University of Science and Technology.

References

1. Wang, X.C.: Journal of Shaanxi University of Science and Technology 22(3), 161–163 (2004)
2. Jiang, Z.P., Yang, H.W., Sun, L.X.: Environmental Science 6, 11–13 (2005)
3. Catterall, K., Zhao, H., Pasco, N.: Development of a Rapid Ferricyanide-Mediated Assay for Biochemical Oxygen Demand Using a Mixed Microbial Consortium. Anal. Chem. 75, 2584–2590 (2003)
4. Pan, H., Zhang, J.X., Dang, H.X.: Journal of Henan University (Natural Science Edition) 34(1), 38–42 (2004)
5. Saravanabhavan, S., Thanikaivelan, P., Raghavarao, J.: Reversing the Conventional Leather Processing Sequence for Cleaner Leather Production. Environ. Sci. Technol. 40, 1069–1075 (2006)
6. Sun, D.H., Wang, Z.Y., Shi, B.: Leather Science and Engineering 15(2), 16–19 (2005)
7. Kong, F.X., Yin, D.Q., Yan, G.A.: Environmental Biology, pp. 202–209. Higher Education Press (2000)
8. Zhang, X.J., Di, F.P., He, M.: Environmental Science 19(5), 25–28 (1998)
9. Yao, J., Zhao, Y., He, M.: Environmental Science and Technology 29(3), 11–13 (2006)
10. Dong, C.J., Lv, B.N., Chen, Z.Q.: Chemical Industry Environment Protection 23(2), 82–85 (2003)

Relationship between Audible Noise and UV Photon Numbers

Mo Li, Yuan Zhao, and Jiansheng Yuan

Department of Electrical Engineering,
Tsinghua University, Beijing 100084, China
limo10@mails.tsinghua.edu.cn

Abstract. When corona occurs at hardwire fittings or high voltage line, the surrounding gas will be ionized. As a result, the electromagnetic energy would release through light, audible noise, or other different ways. This paper analyzes the relationship between the photon numbers measured by ultraviolet imager (UVI) of different fittings and the audible noise. Analyses show that the A-weighted audible noise changes with the photon number exponentially. This exponential relationship is related to the kind of hardwire fittings with different exponential curve fittings coefficients. According to this exponential relationship, the number of photons of different hardwire fittings could be used ro calculate the A-weighted audible noise, or judge the contribution of one ionized hardwire fittings in the whole noise level of electric transmission line.

Keywords: UVI measured photon numbers, audible noise, electromagnetic environment, corona, hardwire fittings.

1 Introduction

Recently, with the enhancement of transmission line voltage in China, the electromagnetic environment issues such as radio disturbance and audible noise drawm more and more attention. [1]For the high-voltage lines, audible noise has become one of the main factors in determining wire structure and analyzing of construction costs. As a result, it is necessary to measure the audible noise accurately. [2]

For the hardwire fittings of transmission line, many experiments on corona noise of a series of hardwire fittings were made, and noise spectrum distribution and ultraviolet imager measured photon numbers of some different hardwire fittings on different voltage were received.

As for the measurement method of audible noise, sound level meter was mainly used. However, the requirements for a sound level meter to maintain its accuracy are very harsh. For example, the influence of wind speed could not be avoided and it is hard to get rid of the background noise when the noise level to be tested is low. [3] Using the ultraviolet image technology, the photon number produced by corona discharge of electrical apparatus could be detected to make a quantitative analysis of the discharge strength. The ultraviolet imager (UVI) could accurately position the discharge point, and has a good anti-interference ability, so it is used more and more widely. [4]Based on the experiment data, the relationship between UVI measured photon numbers and audible noise is studied, and estimation methods of audible noise level with photon numbers are given.

D. Zeng (Ed.): Advances in Control and Communication, LNEE 137, pp. 327–333.
springerlink.com

2 Characteristics of Audible Noise and Photon Numbers Changing with Voltage

Frequency spectrum of audible noise can be measured by gradually increasing voltage, and A-weighted noise level with different voltages can be figured out using A-weighted network. Plus, along with the ultraviolet (UV) photon numbers at the same voltage level measured by UVI, we could receive the characteristics of audible noise and photon number changing with voltage.

2.1 Introduction of Test Conditions and Test Objects

The laboratory in the ultra high voltage (UHV) ac test base is cylindrical tank structure with ellipsoidal cap and the ac test power is YDTCW-6000 kVA / 3 × 500 kV test transformer. The voltage regulator is column regulator with a capacity of 4800kVA, and the voltage output is continuously adjustable.

Under these test conditions, the test objects were arranged by either suspension string or strain string.

The test objects including XGF-5X suspension clamp, which is used to suspend wires to insulator strings and the main componets are pylons, U-shaped screw and hull; FJZ240 Two Split Wire Spacer, which is used to maintain a certain geometric arrangement of split sub-conductor wires; and FDN-30YXL Performed Wire Damper, which is used to reduce the vibration of conducting wires and ground wires which is shaped like dumbbell.

2.2 Characteristics and Changing Rules of Audible Noise

At the altitude of 19 m, frequency spectrum of audible noise was measured of different fittings at various voltages using a sound level meter. Then A-weighted noise level at different voltages can be figured out using A-weighted network. [5]

The audible noise level at different voltage of three hardwire fittings is shown in Fig. 1. The noise level with voltage is monotonically and roughly linearly. There are differences among A-weighted noise level of each hardwire fittings, but the trends are almost the same.

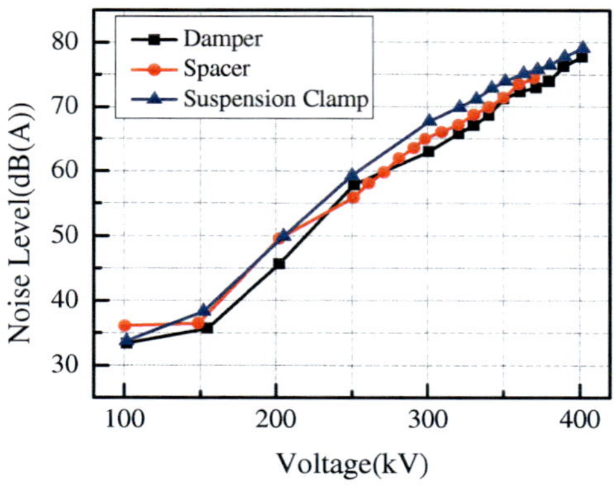

Fig. 1. Audible Noise at Different Voltage

2.3 Characteristics and Changing Rules of UV Photon Numbers

When voltage level is low, the photon number is almost zero because there is a Thompson dark process at the beginning of the discharge and the instrument could not detect the extremely weak light.[6] Afterwards, with the increasing level of corona, photons begin to appear and increase rapidly. The photon numbers of different fittings with increasing voltage is shown in Fig. 2.

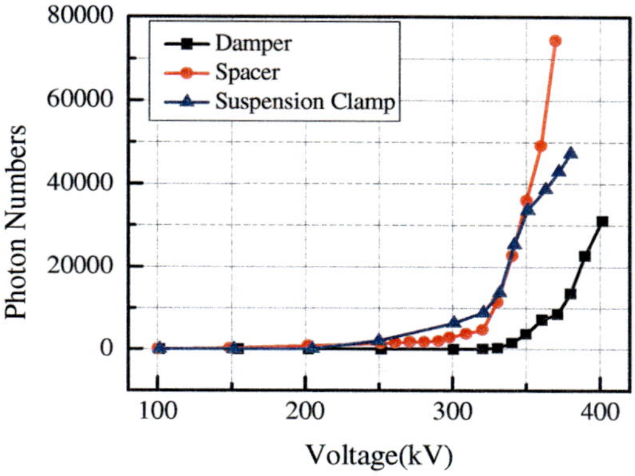

Fig. 2. Photon Numbers at different Voltage

UVI measured photon numbers increase in approximately exponential form with voltage rising while different fittings have different discharge inception voltage. All the photon numbers increase exponentially, but the coefficients are not the same because the shape and discharge point of these three fittings are different.

3 Relationship between Audible Noise And Uv Photon Numbers

According to the analysis above, a connection of A-weighted noise level and UV photon numbers can be established at the same voltage level and the characteristics can be analyzed.

3.1 Exponential Fitting Curve of Different Hardwire Fittings

The UV photon numbers changes with A-weighted noise level of three hardwire fittings are shown in Fig. 3, Fig. 4 and Fig. 5. The noise levels of the three fittings are all in the range from 64 dB to78 dB. UV photon numbers increase with A-weighted noise level approximately exponentially but the coefficients are not the same. In the fitted curves, the differences of fitted value and measured noise level are no more than 0.5 dB.

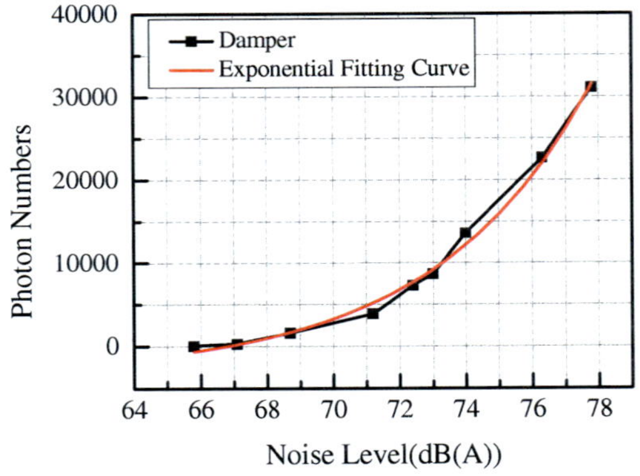

Fig. 3. Relationship between Photon Numbers and Audible Noise of FDN-30YXL Performed Wire Damper

With L and N representing photon numbers and noise level, the fitted equations of FDN-30YXL Performed Wire Damper is (the unit of N is dB)

$$N = -3316.2 + 0.00197e^{-L/4.6615} \tag{1}$$

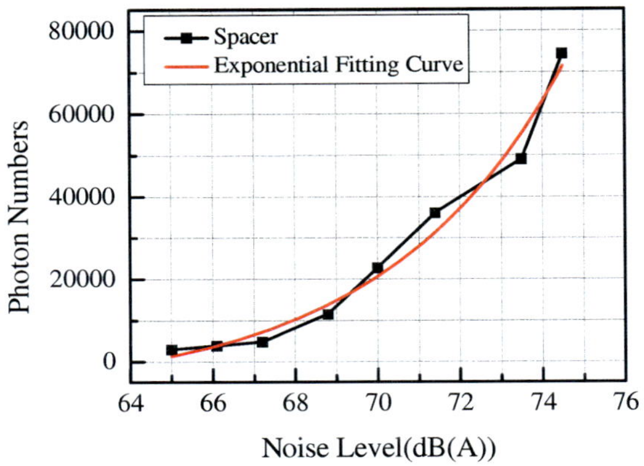

Fig. 4. Relationship between Photon Numbers and Audible Noise of FJZ240 Two Split Wire Spacer

With L and N representing photon numbers and noise level, the fitted equations of FJZ240 Two Split Wire Spacer is

$$N = -7634.2 + 0.00317e^{-L/4.3739} \tag{2}$$

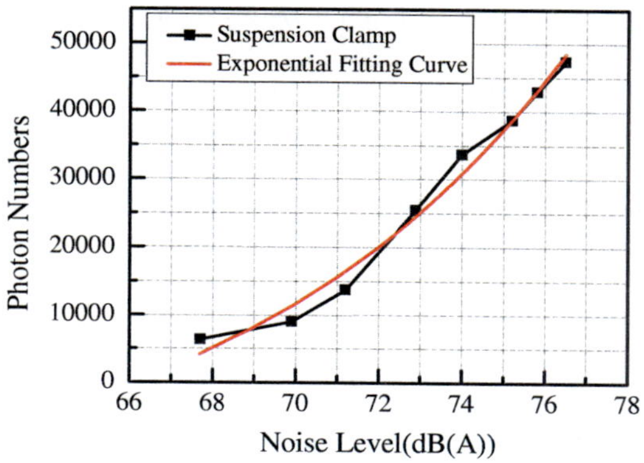

Fig. 5. Relationship between Photon Numbers and Audible Noise of XGF-5X suspension clamp

With L and N representing photon numbers and noise level, the fitted equations of XGF-5X suspension clamp is.

$$N = -18514.1 + 5.579e^{-L/8.1439} \tag{3}$$

3.2 Piecewise Linear Fitting of Different Hardwire Fittings

Using the properties of exponential function, piecewise linear fitting can be used so that we can do linear computation seperately in different intervals. Piecewise linear fittings were made between A-weighted noise level and UV photon numbers of three different hardwire fittings. The noise level was piecewise fitted from 65 to 71 and from 71 to 78, respectively. With L and N representing photon numbers and noise level, the piecewise linear fitted equations of FDN-30YXL Performed Wire Damper is

$$\begin{cases} N = 728.24L - 48238.25 & 64 \le L \le 71 \\ N = 4159.55L - 293791.75 & 71 < L \le 78 \end{cases} \tag{4}$$

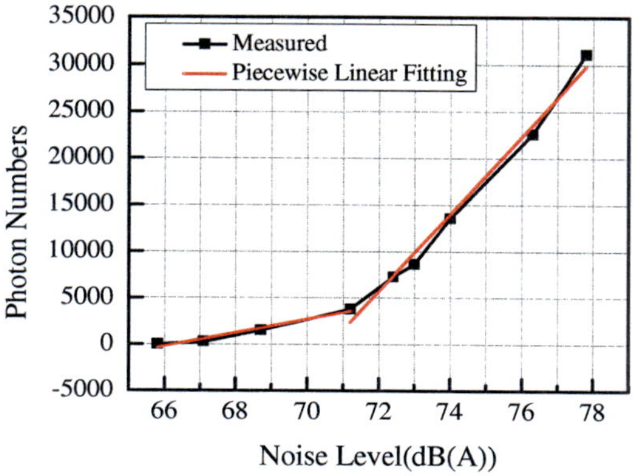

Fig. 6. Piecewise Lenear Fitting of between Photon Numbers and Audible Noise of FDN-30YXL Performed Wire Damper

Specific fitted curve is shown in Fig. 6. With piecewise linear fitting method, the differences of fitted value and measured noise level are no more than 0.1dB. Noise level of other fittings could also be divided at 71 and fitted respectively. The fitted curves of FJZ240 Two Split Wire Spacer and XGF-5X suspension clamp are shown in Fig. 7 and Fig. 8.

With L and N representing photon numbers and noise level, the piecewise linear fitted equations of FJZ240 Two Split Wire Spacer is.

$$\begin{cases} N = 2215.18L - 142136.3 & 64 \le L \le 69 \\ N = 10044.66L - 680853.7 & 69 < L \le 78 \end{cases} \tag{5}$$

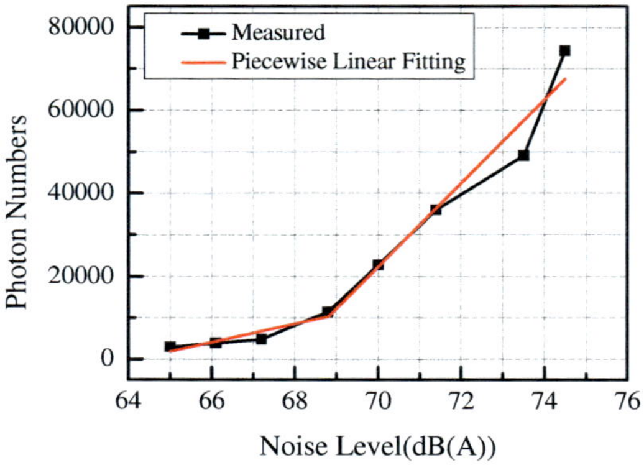

Fig. 7. Piecewise Lenear Fitting of between Photon Numbers and Audible Noise of FJZ240 Two Split Wire Spacer

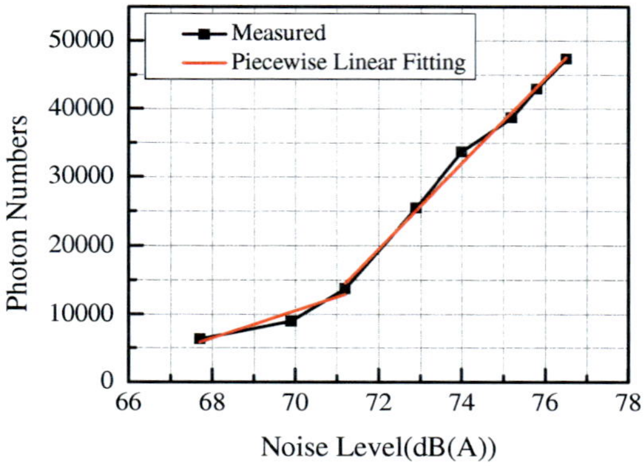

Fig. 8. Piecewise Lenear Fitting of between Photon Numbers and Audible Noise of XGF-5X suspension clamp

With L and N representing photon numbers and noise level, the piecewise linear fitted equations of XGF-5X suspension clamp is

$$\begin{cases} N = 2003.67L - 129745.72 & 64 \le L \le 71 \\ N = 6225.33L - 428674.69 & 71 < L \le 78 \end{cases} \tag{6}$$

4 Conclusions

In this paper, the characteristics of audible noise and photon numbers changing with voltage is given and the curve fittings of A-weighted noise level and UV photon numbers of three hardwire fittings are made using exponential function and piecewise linear function. We could draw conclusions as follows:

(1) The noise level increases with voltage monotonically and roughly linearly. UVI measured photon numbers increase in an exponential form with voltage rising, but the photon numbers of different hardwire fittings are not the same.
(2) Audible noise level of hardwire fittings could be estimated by photon numbers using the exponential or piecewise linear fitted equations.
(3) Furthermore, audible noise produced by different hardwire fittings can be distinguished by their photon numbers during onsite measurement.

Acknowledgment. This work is supported by State Grid Electric Power Research Institute.

References

1. Shu, Y., Huang, D., Ruan, J., Hu, Y.: Construction of UHV Demonstration and Test Projects in China. In: Power and Energy Engineering Conference (APPEEC 2009), pp. 1–7. IEEE Press (March 2009), doi:10.1109/APPEEC.2009.4918237
2. Wan, B.-Q., Xie, H.-C., Zhang, G.-Z., Zhang, X.-W.: Electromagnetic environment of 1000kV UHV AC substation. In: Electromagnetic Compatibility (APEMC 2010), pp. 1413–1416. IEEE Press (April 2010), doi:10.1109/APEMC.2010.5475850
3. Lundquist, J.: Results from AC transmission line audible noise studies at the Anneberg EHV test station. IEEE Transactions on Power Delivery 5, 317–323 (1990), doi:10.1109/61.107291: (Asea Brown Boveri HV Switchgear and Ludvika)
4. Liu, Y.-P., Wang, H.-B., Chen, W.-J., Yang, Y.-J., Tang, J.: Test Study on Corona Onset Voltage of UHV Transmission Lines Based on UV Detection. In: High Voltage Engineering and Application (ICHVE 2008), pp. 387–390. IEEE Press (November 2008), doi:10.1109/ICHVE.2008.4773954
5. EPRI, "Transmission line reference book 345kV and above", 2nd edn., pp.259–319. Eelcrtci Power Research Insitutte, California (1982)
6. High Voltage Engineering and Application, (ICHVE (2008)

Analysis and Design of the Coils System for Electromagnetic Propagation Resistivity Logging Tools by Numerical Simulations

Yuan Zhao, Mo Li, Yueqin Dun, and Jiansheng Yuan

Department of Electrical Engineering,
Tsinghua University,
Beijing 100084, China
yuan-zhao06@mails.tsinghua.edu.cn

Abstract. The coils systems of electromagnetic propagation resistivity logging (EPRL) tools use coils as transmitting and receiving antennas for a fixed frequency. The Numerical Mode Matching (NMM) method is applied in the simulation of electromagnetic fields with multi-layer structure domains to study the key design parameters of logging tools, which contain the source frequency, the transmitter-receiver spacing, and the receiver interval. A critical requirement in the design of the coils system is to produce different desired signals for different typical ground model. In this paper, the three key design parameters are analyzed, and the principles and steps of coils system design are given. The parameters must be reasonable values to achieve the desired effect. The simulation results also show that depth of investigating is related with the source frequency, the transmitter-receiver spacing, and the receiver interval.

Keywords: electromagnetic propagation resistivity logging tools, resistivity, coils system, numerical mode matching method, numerical simulation.

1 Introduction

The electromagnetic propagation resistivity logging is an important part of geophysical exploration. The well-logging tool usually contains one or two transmitting antennas and two receiving antennas. A critical requirement in the design of the coils system is to produce different desired signals for different typical ground model.

In this paper, the Numerical Mode Matching (NMM) method, which is proposed by Chew et al. [1-3], is used to solve this problem, and it has been shown to be more efficient. The efficiency of the NMM method is based on the idea that a higher dimensional problem can be reduced to a series of lower dimensional problems. Based on NMM method, codes for calculating the EPRL responses in layer structure soil has been developed.

The paper is constructed as follows. Part II explains the principles and steps of coils system design. Part III provides a design example and its analysis. Finally, the conclusion is given in the last section of the paper.

D. Zeng (Ed.): Advances in Control and Communication, LNEE 137, pp. 335–341.
springerlink.com

2 Approach of Coils System Design

The simulated coils system consists of one transmitter and two receivers [4]. It can be applied to detect physical property parameters of the soil or medium surrounding the borehole, mostly the resistivity. The key design parameters of logging tools are the source frequency, the receiver interval, and the transmitter-receiver spacing [5]. Fig. 1 shows, that the source frequency *f* is current's frequency in the transmitter, the receiver interval ΔL is the spacing between receivers, and the transmitter-receiver spacing *L* is the spacing between the transmitter and midpoint of two receivers.

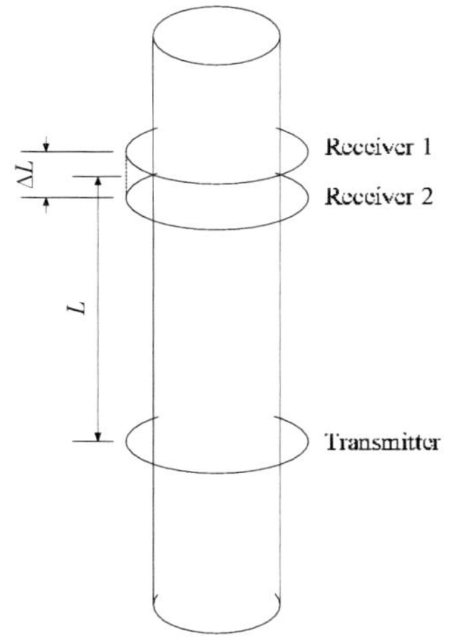

Fig. 1. The coils system model of the EPRL tool

In this section, these three key design parameters would be described by the earth structure and required precision. The earth structure includes earth resistivity, dielectric constant, and mud resistivity. The required precision includes depth of investigation, and thickness of the thinnest formation that can be detected.

2.1 The Source Frequency f

1) Reducing the effect of dielectric constant

The main purpose of electromagnetic propagation resistivity logging is to measure the resistivity of the earth. It is required to minimize the effect of the dielectric constant, so the source frequency *f* should be,

$$f \ll \frac{1}{2\pi\varepsilon\rho_{\max}} \tag{1}$$

where $\rho_{\max}$ is the maximum resistivity, and ε is dielectric constant of the earth.

The required precision of minimum phase difference

For the required precision of phase difference, the phase difference of the two receivers should be greater than the minimum phase difference of the instrument. So the source frequency *f* should also be,

$$f > \frac{\pi\rho_{\max}}{\mu}\cdot\left(\frac{\Delta\phi_{\min}}{180\cdot\Delta L}\right)^2 \tag{2}$$

where $\Delta\phi_{\min}$ is the minimum phase difference, and μ is permeability.

From (1) and (2), the source frequency *f* can be determined by,

$$\frac{\pi\rho_{\max}}{\mu}\cdot\left(\frac{\Delta\phi_{\min}}{180\cdot\Delta L}\right)^2 < f \ll \frac{1}{2\pi\varepsilon\rho_{\max}} \tag{3}$$

In addition, the source frequency also affects depth of investigation and resolution of horizontally layered formations. Numerical simulation shows that reducing the source frequency can increase the depth of investigation, but will reduce the resolution of horizontally layered formations.

2.2 The Receiver Interval ΔL

2.1.1 The Minimum Electromagnetic Wavelength

As we know, the phases of two receivers' voltages change both in [0, 360] (°), so the phase difference of two receivers' voltages also changes in [0, 360] (°). If the receiver interval is greater than minimum electromagnetic wavelength, the real phase difference may be greater than 360 (°), but the tool can only detect the range [0, 360] (°). So the phase difference detected at this time is not the true phase difference. In order to make sure that the phase difference is the single-valued function of resistivity, it is needed to ensure that the receiver interval is less than the minimum electromagnetic wavelength. That is,

$$\Delta L > \lambda_{\min} \tag{4}$$

where $\lambda_{\min}$ is the minimum electromagnetic wavelength.

2.1.2 The Required Precision of Minimum Phase Difference

In the case of certain electromagnetic wavelength, the smaller the receiver interval, the smaller the phase difference is. Considering the accuracy of phase difference measurement, the phase difference can not be too small. So,

$$\Delta L > \frac{\lambda_{\max}\cdot\Delta\phi_{\min}}{360} \tag{5}$$

where $\lambda_{\max}$ is the maximum electromagnetic wavelength.

2.1.3 The Minimum Thickness of the Horizontally Thin-Layer That Can Be Detected

Simulation results show that the minimum thickness of the horizontally thin-layer that can be detected is almost the same as the receiver interval. This is one criterion to determine the minimum the receiver interval. So the receiver interval should be as follows,

$$\Delta L > d \tag{6}$$

where d is the minimum thickness of the horizontally thin-layer that can be detected.

From (4), (5) and (6), the receiver interval ΔL can be determined by,

$$\max\left(\frac{\lambda_{\max} \cdot \Delta\phi_{\min}}{360}, d\right) < \Delta L < \lambda_{\min} \tag{7}$$

2.3 The Transmitter-Receiver Spacing L

2.3.1 The Required Minimum Voltage of Receivers

The transmitter-receiver spacing directly affects the voltage of receivers. In order to get large enough signals in receivers, the transmitter-receiver spacing can not be too large. The required minimum voltage signals can fix the limit of the transmitter-receiver spacing.

2.3.2 Depth of Investigation

The transmitter-receiver spacing affects depth of investigation. The numerical simulation shows that increasing the transmitter-receiver spacing can increase depth of investigation. Also we know that the smaller resistivity is, the shallower the depth of investigation is. When determining the transmitter-receiver spacing, we can calculate in condition of the smallest resistivity.

The steps of designing coils system of electromagnetic propagation resistivity logging tools are: firstly, the initial frequency is determined by (3), and then the receiver interval, and the transmitter-receiver spacing. An appropriate frequency is selected to consider both depth of investigation and resolution of horizontally layered formations. If the depth of investigation and resolution of horizontally layered formations does not meet the design requirements, we can adjust the frequency according to the above principles until the meet all design requirements.

3 Results and Analysis

In this part, a design example is given. There are several assumed conditions. The main ground resistivity change in the range: $[\rho_{\min}, \rho_{\max}] = [0.1, 100]\ (\Omega \cdot \mathrm{m})$. In the borehole, there is KCL mud, whose resistivity is $0.4\ (\Omega \cdot \mathrm{m})$. The minimum thickness of the horizontally thin-layer that can be detected is: d=0.30 (m). The required depth of investigation is 0.25 (m). Based on these condition, the source frequency, the receiver interval, and the transmitter-receiver spacing are obtained by above steps.

3.1 The Source Frequency *f*

The source frequency need to satisfy (1), that is,

$$f \ll \frac{1}{2\pi\varepsilon\rho_{\max}} = 1.8\times10^{8}\ (\text{Hz}) \tag{8}$$

Also the source frequency should satisfy (2), that is,

$$f > \frac{\pi\rho_{\max}}{\mu}\cdot\left(\frac{\Delta\phi_{\min}}{180\cdot d}\right)^2 = 8.6\times10^{4}\ (\text{Hz}) \tag{9}$$

So we choose the most common frequency of electromagnetic propagation resistivity logging, which is 2 (MHz).

3.2 The Receiver Interval ΔL

From (4) we can see, the receiver interval is less than the minimum electromagnetic wavelength, that is,

$$\Delta L < \lambda_{\min} = \frac{2\pi}{\sqrt{\frac{\omega\mu}{2\rho_{\min}}}} = 2\sqrt{\frac{\pi\rho_{\min}}{f\mu}} = 0.71\ (\text{m}) \tag{10}$$

Assuming that the required precision of minimum phase difference is 1 ($^{\circ}$). So the receiver interval can be designed as (5), that is,

$$\Delta L > \frac{\lambda_{\max}}{360} = \frac{1}{180}\sqrt{\frac{\pi\rho_{\max}}{f\mu}} = 0.062\ (\text{m}) \tag{11}$$

The receiver interval is almost as the minimum thickness of the horizontally thin-layer that can be detected. That is,

$$\Delta L > d = 0.30\ (\text{m}) \tag{12}$$

From (10), (11) and (12), the receiver interval should be in the range: $0.062\ (\text{m}) < \Delta L < 0.30\ (\text{m})$. In order to make sure the measurement accuracy and resolution, the receiver interval can be 0.2 (m).

3.3 The Transmitter-Receiver Spacing *L*

Firstly, the voltage signals can not be too small. Based on above parameters, the amplitude attenuation and phase difference curve are calculated at resistivity of $0.1\ (\Omega\cdot\text{m})$ and $100\ (\Omega\cdot\text{m})$, shown in Fig.2 and Fig.3. To ensure the measurement signals are large enough, the transmitter-receiver spacing can be taken as less than 0.85 m.

In addition, the smaller resistivity is, the shallower the depth of investigation is. So we calculate the depth of investigation at resistivity of $0.1\ (\Omega\cdot\text{m})$. If we take the transmitter-receiver spacing L=0.75 (m), the depth of investigation for amplitude attenuation is 0.51 (m), the depth of investigation for the phase difference is 0.27 (m). to meet the requirements.

In summary, if the source frequency takes 2 (MHz), the receiver interval takes 0.20 (m), and the transmitter-receiver spacing takes 0.75 (m), the coils system of electromagnetic propagation resistivity logging tools can meet all the design requirements.

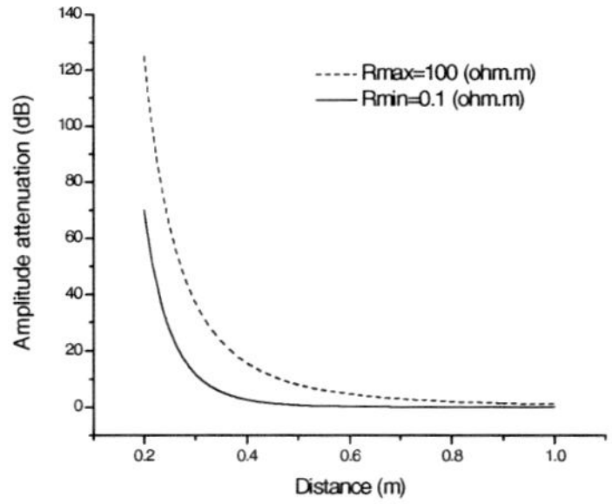

Fig. 2. The amplitude attenuation coefficient curve.

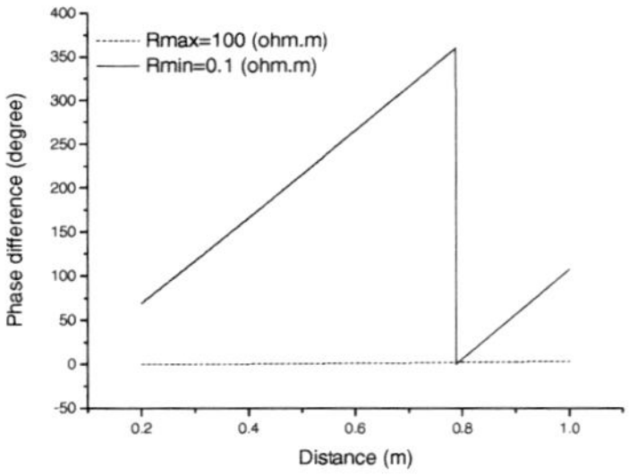

Fig. 3. The phase difference curve

4 Conclusion

In this paper, the design parameters of electromagnetic propagation resistivity logging tools including the source frequency, the receiver interval, and the transmitter-receiver spacing, are studied, and the principles and steps of coils system design are given. The source frequency should be: $\pi\rho_{max}/\mu\cdot(\Delta\phi_{min}/180\cdot\Delta L)^2 < f \ll 1/2\pi\varepsilon\rho_{max}$, the receiver interval should be: $\max(\lambda_{max}\cdot\Delta\phi_{min}/360, d) < \Delta L < \lambda_{min}$, and the transmitter-receiver spacing should meet requirements of depth of investigation at smallest resistivity.

The simulation results also show that the lower frequency can increase the depth of investigating, but will reduce the investigating characteristic of thin layers. Reducing the receiver interval can improve investigating characteristic of thin layers, but if it is too small that will affect the resolution of the phase difference. Increasing the transmitter-receiver spacing can increase the depth of investigating, but too larger spacing will make the signals are too small. The parameters must be reasonable values to achieve the desired effect.

References

1. Chew, W.C., Barone, S., Anderson, B., Hennessy, C.: Diffraction of axisymmetric waves in a borehole by bed boundary discontinuities. Geophysics 49, 1586–1595 (1984), doi:10.1190/1.1441567

2. Chew, W.C.: Modeling of the dielectric logging tool at high frequencies theory. IEEE Trans. on Geoscience and Remote Sensing 26, 382–387 (1988), doi:10.1109/36.3041
3. Chew, W.C., Nie, Z.P., Liu, Q.H., Anderson, B.: An efficient solution for the response of electrical well logging tool s in a complex environment. IEEE Trans. on Geoscience and Remote Sensing 29, 308–313 (1991), doi:10.1109/36.73673
4. Zhang, Y., Liu, C., Shen, L.C.: The performance evaluation of MWD logging tools using magnetic and electric dipoles by numerical simulations. IEEE Trans. on Geoscience and Remote Sensing 34, 1039–1044 (1996), doi:10.1109/36.508421
5. Xing, G., Wang, H., Ding, Z.: A new combined measurement method of the electromagnetic propagation resistivity logging. IEEE Trans. on Geoscience and Remote Sensing 5, 430–432 (2008), doi:10.1109/LGRS.2008.919817

Does Investment Efficiency Have an Influence on Executives Change?

Lijun Ma and Tianhui Xu

School of Business,
Renmin University of China,
Beijing, China
malj324@163.com, xutianhui30411@126.com

Abstract. The effectiveness of executives change is an important aspect of the effectiveness of corporate governance mechanism, research in this area has both theoretical and practical significance. This article studies the influence of executives change and corresponding economic consequences from the point of investment efficiency and enterprise property. The results show that, overall, little evidence prove that executives are replaced because of low investment efficiency; no remarkable improvement on investment efficiency can be seen after the change of executives. This paper's study enriches the literature of executives change research from a new angle, providing reference meaning for the practice of our country.

Keywords: executives change, investment efficiency, sensitive, state-owned companies, non-state-owned companies.

1 Introduction

The effectiveness of executives change play an important role in the effectiveness of corporate governance mechanism, effective alteration mechanism will help to improve the company's value. Domestic and foreign literature mainly concern about the affecting factors and economic consequences of executives change. For affecting factors, existing literature focus mainly on firm performance, equity structure, characteristics of the board of directors, control market, laws and regulations, the competition in product market and executives features. These studies have gained plenty of achievement.

This paper takes the data of China's listed company from 2001 to 2006 as research samples. The result indicates two aspects: firstly, there's little chance that executives are replaced because of low investment efficiency; secondly, the change of executives doesn't mean a remarkable improvement on investment efficiency. Considering the significant differences between state-owned enterprises and private enterprises on corporate governance, we analyzed these two kinds of companies separately. This paper has the following research contributions: firstly, the paper studies the question of executives change from the perspective of investment efficiency, and further more, studies how the executives alteration affect investment efficiency and the corporate governance; secondly, we provide reference evidence for the design of effective incentive and restraint mechanisms in our country to regulate executives' behavior.

D. Zeng (Ed.): Advances in Control and Communication, LNEE 137, pp. 343–347.
springerlink.com

2 Research Hypotheses

Managers generally have the tendency to build business empire (Jensen, 1993), they may put all available funds into investment, thereby causing excessive investment; and may on the other hand conduct insufficient investment because of the agency problem (Holm strong and Weiss, 1985). These two kinds of agency problems depict opposite enterprise behavior, but both of them will cause damage to stockholders' benefits. Therefore, it's necessary for shareholders to seek for effective governance mechanism to ease the agency problems. However, since most of the listed companies in China are state-controlled directly or indirectly, most executives' are appointed and dismissed by the government department at all levels, so more executives are responsible for the government, who masters their appointed qualification; this finally cause the state-owned to diversify their goals in order to meet with the government's requirements, such as increasing employment opportunities, stabilizing social environment, investing in emerging industry, increasing the local tax and the growth of GDP, etc. Moreover, the listed companies in China are mainly controlled by the first majority shareholders especially the state shares, lacking checks and balances from other big shareholders; such equity structure against managers to accept the supervision and restraint from multiple ownership in wider range, which is not conducive to perfect corporate governance, and more unfavorable to improve corporate value. The policy burdens and government intervention has distorted state-owned companies' internal management. The Reduction of State-owned Shares Group (2001) reported in their research paper that the function of external governance mechanism such as the manager talent market, capital market and the corporate control market was restricted, since the proportion of state-owned shares were too high and couldn't circulate. Kato and Long's (2005) study found that for China's state-owned listed company, the sensitivity between company performance and CEO alteration is lower than that of foreign investment holding company, and only when the ownership convert from state-owned into private, can we get a significant negative correlation between firm performance and the CEO alteration.

Enterprise's investment efficiency is a very important index that can reflect the agency problems between shareholders and the managers, and it directly determines the future development prospect of the enterprises. For a company, if the investment efficiency is low, the shareholder's benefits will be undoubtedly damaged, in this case, to replace the incompetent executives can yet be regarded as a kind of good governance mechanism. Therefore, if a enterprise' executive is replaced because of low investment efficiency, its company governance mechanism can be considered as more effective; Conversely, the governance mechanism will be think as inefficient or invalid. According to the above analysis, the corporate governance of state-owned enterprise is poorer, which, obviously limits the function of the governance mechanism that allow companies to change their executives when facing a low investment efficiency.

Compared with the state-owned enterprises' managers, private enterprises' managers are faced with greater market pressure, therefore, the later have greater motivation to conduct effective investment to improve enterprise performance, in order to obtain the faith of the market. Xiaodong Xu and Xiaoyue Chen's(2003) research found that the company whose biggest shareholder are non-state-owned has higher corporate value and stronger profitability, and are more flexible on company operation and effective on corporate governance, at the same time, its senior management also confront with more supervision and motivation from the internal enterprise and external market. Zhengyu Zhao, Zhishu Yang and ChongEn Bai (2007) found that seeing from the contrast of state-owned and non-state-owned, company performance's positive incentive effect is more remarkable in those non-state-owned than in state-owned companies, that's to say, corporate governance effectiveness performs better in non-state-owned corporations. Therefore, we can reasonably expect that when the enterprise's investment efficiency is low, private companies are more easily to replace executives than the state-owned. Based on the above analysis, the paper puts forward the following hypothesis:

Compared with the state-owned companies, the worse the investment efficiency of non-state-owned companies is, the more the possiblities of changing executives are.

3 Variable and Model

In this article, we use the following LOGISTIC regression model to exam the hypothesis.

$$\begin{aligned} Turnover = a_0 + a_1 ABSinvt + a_2 Ownership \\ + a_3 Grow + a_4 ROA + a_5 Size + a_6 Top1 + a_7 Age \\ + a_8 Cash + a_9 Number + a_{10} Holder + a_{11} Meeting + \varepsilon \end{aligned} \quad (1)$$

$$\begin{aligned} Turnover = \gamma_0 + \gamma_1 ABSinvt + \gamma_2 Cengov + \gamma_3 Private + \gamma_4 Cengov * ABSinvt \\ + \gamma_5 \text{Private} * \text{ABSinvt} + \gamma_6 Grow + \gamma_7 ROA + \gamma_8 Size + \gamma_9 Top1 \\ + \gamma_{10} Age + \gamma_{11} Cash + \gamma_{12} Number + \gamma_{13} Holder + \gamma_{14} Meeting + \varepsilon \end{aligned} \quad (2)$$

In model(1),the explained variable is Turnover which reflects the executives change of companies. Turnover takes "1" when there is a change to the chairman of the board or the general manager, otherwise, takes "0". ABSinvt, the explanatory variable, is the absolute value of investment efficiency. If the regression coefficient of ABSinvt is significantly positive, it indicates that the bigger the deviation of investment is in enterprises, the more likely the executives are going to get changed. Referring to researches related and current situations in china, this paper has also controlled some other factors in the model above.

4 Empirical Results

Table 1 reports the regression results of model 1 and model 2 .In model 1, ABSinvt's regression coefficient is positive but not significant, indicating that enterprises' investment efficiency does not have significant influences on the change of executives. Moreover, the results of model 2 tell that Private*ABSinvt's regression coefficient is significantly positive, showing that compaired with the state-owend companies', executives change is more sensitive to investment efficiency. In conclusion, results above confirm the hypothesis we put forward.

Table 1. Results of LOGISTIC Regression

		Turnover		Turnover	
		Coefficient	*Wald*	*Coefficient*	*Wald*
Intercept		0.2031	0.0967	0.1767	0.0707
ABSinvt	+	0.2858	0.1886	-0.4708	0.2970
Ownership	−	-0.0347	0.2873		
Cengov	−			0.1992	3.6617*
Private	+			0.0048	0.0029
Cengov*ABSinvt	−			0.9810	0.2589
Private*ABSinvt	+			2.4206	2.5825*
Grow	?	0.0179	0.2342	0.0184	0.2488
ROA	−	-1.2764	15.3503***	-1.2897	15.5242***
Size	−	0.0261	0.0705	0.0302	0.0942
Top1	−	-0.4071	5.8304**	-0.4237	6.2692**
Age	?	0.0184	3.4681*	0.0189	3.6323*
Cash	?	-0.0618	0.8603	-0.0681	1.0384
Number	−	0.0136	1.3649	0.0105	0.7985
Holder	−	-0.0395	13.1448***	-0.0372	11.6169***
Meeting	+	0.0143	2.7511*	0.0147	2.8965*
Effects of Year		Controlled		Controlled	
Effectsof Industry		Controlled		Controlled	
R-Square		0.0328		0.0349	
Chi-Square		216.6890***		230.8403***	
Size of sample		6495		6495	

Notes: ***stands for a 1% level of significance,while **for 5%,and * for 10%.

In addition, we carried the steady examination from the following aspects. Firstly, referring to researches related, we used the sensitive coefficient of "investment - cash flow" to measure the investment efficiency. Then we carried on the steady examination of the results above which ended up with nothing changed. Next, we rejected samples in which executives changes were caused by controlling transfer, expiration of the term of office, retirement.etc. According to results of this new regression analysis, the conclusion did not changed.

5 Conclusion

The effectiveness of executives change is an important aspect of the effectiveness of corporate governance mechanism. The change is not only an extreme restraint to managers, but also a correction to bad performance. This article not only inspects executives change on the aspect of investment efficiency, but also conducts a further study on influences on investment efficiency of executives change. Results indicate that, compaired to the state-owned companies, non-state-owned companies are more sensitive on "executives change-investment efficiency", but the efficiency does not have a remarkable improvement after the change of executives.

References

1. Beatty, R.P., Zajac, E.J.: CEO Change and Firm Performance In Large Corporations: Succession Effects and Manager Effects. Strategic Management Journal 8, 305–317 (1987)
2. Denis, D.J., Denis, D.K., Sarin, A.: Ownership Structure and Top Management Turnover. Journal of Financial Economics 45, 193–221 (1997)
3. Eric, C.C., Sonia, M.L.W.: Chief Executive Officer Turnovers and the Performance of China's Listed Enterprises. Working paper. Hong Kong Institute of Economics and Business Strategy and The University of Hong Kong (2004)
4. Holmstrom, B., Weiss, L.: Managerial Incentives, Investment and Aggregate Implications. Review of Economic Studies 52, 403–426 (1985)
5. Kang, J., Shivdasani, A.: Firm performance, corporate governance and top executive turnover in Japan. Journal of Financial Economics 38, 29–58 (1995)
6. Kato, T., Long, C.: CEO Turnover, Firm Performance, and Corporate Governance in Chinese Listed Firms. Working paper (2005)
7. Firth, M., Fung, P.M.Y., Rui, O.M.: Firm Performance, Governance Structure, and Top Management Turnover in a Transitional Economy. Journal of Management Studies 43, 1289–1330 (2006)

How to Attract More Service FDI in China?

Changhai Wang, Yali Wen*, and Kaili Kang

College of Economics and Management,
Beijing Forestry University, Beijing, 100083, China
wangchanghai1982@163.com

Abstract. In today's economical environment, service industry plays an important role in a country's development. This paper researches the role and impact of TNCs in global service industry. It analyses the investment environment of service industry in China as a case to reflect how a government can alter their policy and institutional framework to cope with recent changes in the industry as well as attract more TNCs to invest into its country, which would help China to realize the sustainable development.

Keywords: FDI in services, China, strategy.

1 Introduction

1.1 The Growth FDI in Services and Its Implications

In today's economical environment, service industry plays an important role in a country's development, while FDI in services is a key access to achieve advanced knowledge or gain capital for production. There are some features for the FDI's shifting: First, the patterns of FDI have been changing with a growth in services and mixing; Second, there is changing trend in distribution among the host and home countries in terms of shifting the share of outward and inward; Third, there is relative lower transnationalization in service industry compared to manufacture industry, and which might be different in different countries; Last, it is quite common and crucial that non-equity forms of investment are in services industry. There are a large number of TNCs invest in services of developing countries. Furthermore, TNCs tend to take M&A model as a speedy and practical model to enter into host countries. Also, there are some drives to expand TNCs activities in services which are in terms of ownership-specific advantages, location-specific advantages, and internationalization advantages. Finally, there are some impacts brought by TNCs in services to developing countries, which embody in financial resources, balance of payments, services provisions, competition and crowding out, technology, exports and employment[1].

1.2 The Offshoring of Corporate Service Functions

In recent years, many companies establish their affiliates abroad, such as Infineon Technologies set up three of their new centers in Dubin, Kista and Munich in July

* Corresponding author.

D. Zeng (Ed.): Advances in Control and Communication, LNEE 137, pp. 349–356.
springerlink.com

2003; In the same year, British Telecom set up two call centers in India, etc. There are several factors that explain the growth of offshoring, traces the role of FDI in the process, and explores the influences for host and home countries: A. many services, such as haircut, couldn't be traded between parties located in different countries because of technical constrains, customs and habits, etc. However, new information and communication technologies(ICTs) make the information-centered set of services possible to be traded in several ways. But there are still limits to offshoring: one is technical limitation because many services functions cannot be digitized; The other one are legal factors, which might limit the globalization of IT-enabled services. Furthermore, there are also some limits to the providing of educated employees. B. The offshoring in services has been growing quite fast since the year 2001, and still be expected to accelerate in the foreseeable future. However, the offshoring in IT and other electronic industries are just at the beginning stage. TNCs still play important roles in offshoring. C. When it comes to the determinants how offshoring is undertaken, we usually decide to use different models based on different features in a host country. There are four types of factors affect the decisions, which are controlling of that activity, level of internal interaction, the availability of capable local firms, and larger scale activities. TNCs have good advantages to provide their services globally. D. If we observe the competitiveness drives to corporate offshoring, we can find the FDI plays the most important roles in offshoring, which influences offshoring in the following two ways: one is through captive offshoring; the other way is that providers in special services establish foreign affiliates to serve foreign clients. At the same time, cost reduction and improved quality is also very crucial to offshoring. While, we also found that the US offshoring companies are much more than the European offshoring companies.E. On one hand, FDI-oriented offshoring brings a lot of economical benefits and an impact on labor force such as job creation, higher wages, skills upgrading to host countries; On the other hand, offshoring provides some benefits to home countries, which include cost reduction, quality improving, competitiveness increasing, more productive and higher value shifting, using exports revenues on imports advanced products, etc[1].

1.3 National Policies

International rule-making decides the national policies, and at the same time, the policies in both host and home countries also decide FDI development and its process. For the host countries, the policies on services are quite crucial for the benefits from FDI, which can embody in the following aspects: the countries' opening up to FDI in services, basic infrastructure to attract investors, promoted activities as well as the implementation of various of fiscal, financial, and other incentives to attract FDI, also export processing zones(EPZ) might be used in some export-oriented production of goods to attract FDI, adequate infrastructures and appropriate skills play key roles in determining whether to invest in a country, regulatory issues occurred between host countries and home countries, in particularly in data protection and intellectual property, are key points that should be considered, policies in improving local capabilities such as improve skills, establish linkages with local institutions, and upgraded technologies, etc. For the home countries, there are some policies or measures that used by home countries to attract or benefit from FDI, such as to provide information, encourage technology, transfer risk, offering incentives, etc, which might aim to limit the ability of other countries to attract FDI. The offshoring on services in the US sets off a strong argument that lots of advanced jobs lost from being taken place by foreign employees through offshoring. While there are also some responses from European countries, in UK, Trade Unions do not take a protectionist approach in concerning about the impact from offshoring, but regarding a three-pronged approach: early consultation to influence decisions; questioning of the case for offshoring; and avoidance for compulsory redundancies. In some other European countries, they seem not to be against offshoring either. It's possible for developed countries to welcome offshoring, but at the same time they also concern how to meet the challenges of adapting in this process.

1.4 Interaction of National and International Policies

When it comes to the international policies, there is a broad used definition of international investment agreements(IIAs) that address the investment issues, which is covered by three following approaches: the investment-based approach, the service-based approach and the mixed approach. However, there are still some policy challenges raised by the multifaceted reality, for instance in some cases, policies might be overlapped, inconsistent, and gapped, which might raise to conflicts. Further more, the complexity and ambiguity of the rules to services might compromise the clarity of the system and make it difficult to negotiate through the other rules. Also, there will be huge amount of potential benefits brought by IIAs, which can offer a stable, predicable framework for attracting FDI and benefitting from it. Meanwhile, there are challenges leaded by the potential benefits, which strikes a balance between using IIAs to attract FDI, benefiting from it, and preserving the flexibility used to the application of national development strategies in services. In conclusion, there is a complex interaction between the national regulations and international policies, on one hand, countries need to strengthen their capabilities to attract more FDI, which on the other hand can also help link developing countries to global value chain. Finally, the key point to reduce the negative effects is to pursue the right policies within a broader development strategy.

2 Case Study—How to Attract More Service FDI in China?

2.1 Introduction

Since the reforming and opening policies in 1980s, China's FDI has been boasting the largest amount of FDI inflow of all developing countries and China has been ranked the top three FDI recipients with a FDI inflow of $72 billion [2]. 90 percent of FDI in China are brought by Greenfield investment. FDI plays an important role in contributing China's economical development, in particularly in service industry [3].It has been expected that China's services industry will be the most attractive sector to foreign investment as well as the largest level of opening up among all the fields since China entered WTO. At the same time, there has been a rapid progress in China's services development since the past decades. GDP in services had percentages of 33.7 among the whole Chinese GDP in 2002, which increased to 12.3 percent compared to GDP in services at the beginning of implementation of reforming and opening policies in China. While FDI in services contributed a lot to the process of development in China's services. Therefore, it is quite significant to learn how is the current situation, problems existing of Chinese FDI in services' and what are the determinants of the attractiveness to Chinese FDI in services?[4]

2.2 Background of Chinese Service FDI

As we can see in Table 1, the FDI mainly focused on manufacture industry at the beginning of implementation of reforming and opening policies, and there is hardly ever FDI in service sector. Service FDI has been increasing since the late 1990s. The amount of China's FDI in services increased since 1997, then following a sharp decrease in 1999 to 2000, and there is a steady increase after the year of 2000. The development of China's FDI in services followed the shape of "W" since the past decade. In 1997, the amount of FDI in services was $12.1 billion, which increased to $13.5 billion in the year of 1998, and then fluctuated to $10.5 billion in 2000, and after that following with a steady increase because of China entering into WTO in 2001. However, the increasing rate of service FDI went down to 23.6% in the year 2008, which might be affected by the global financial crisis.[5]

Table 1. The total amounts of China's FDI in services Unit: $ billion

Year	1997	1998	1999	2000	2001	2002
SFDI	12.06	13.51	11.83	10.46	11.18	12.25
Increased Rate(%)	-17.9	12.0	-12.5	-11.5	6.85	9.56
Year	2003	2004	2005	2006	2007	2008
SFDI	13.33	14.05	14.92	21.14	30.70	37.95
Increased Rate(%)	8.77	5.44	6.19	41.69	45.50	23.60

Source: National Bureau of Statistics 2009.

Proportion of China's FDI in services had an increase trend from 26.65% in 1997 to 29.34% in 1999 among the total FDI. However, the proportion of China's service FDI was lower than the previous one since 2000, which can be seen in the **table2**. In 1980s, FDI in industry accounted for 60.3% among the whole FDI, which FDI in agriculture and services accounted for 2.9% and 36.8% respectively. When in 1990s, the ratio of FDI in services decreased to 31.4%, while the ratio of FDI in industry increased to 66.7%. The ratio of China's FDI in services continued to decrease and fluctuated with an average level of 23.2% in 2004, then went up substantially from 2005[5].

Table 2. Ratio of China's service FDI accounts for China's total FDI Unit: %

Year	1997	1998	1999	2000	2001	2002
Ratio (SFDI/FDI)	26.65	29.72	29.34	25.70	23.85	23.23
Year	2003	2004	2005	2006	2007	2008
Ratio (SFDI/FDI)	24.9	23.2	24.7	30.4	41.0	41.1

Source: National Bureau of Statistics 2009.

When it refers to the distribution of services FDI in China, we found that there is an unbalance in the area of FDI distribution. Foreign investors focused more on basic infrastructures, industrial supporting capabilities, etc, when they chose their investing areas, because of the special nature of services industry. While eastern area of China has the advantages mentioned above, where has been the most centralized area for investing in services by foreign investors. On the opposite, the investment in Middle-west of China is comparatively small because of the poor basic infrastructures. Moreover, foreign capitals in services preferred to big cities as well as middle-sized cities because the higher level of urbanization, the stronger capabilities of attracting foreign investment[6].

With the rapid development of economic integration, there are four economic areas becoming the major invested areas in China, which are Zhujiang triangle of economic area, Changjiang triangle of economic area, Huanbohai economic area and north-east economic area. These four areas have become the top areas of attracting foreign investment in services industry and also the most developed area in economic development in China because of the strong economic capabilities and good investment environment[7].

There are two main categories of investment in China from the view of investment sources: one of which are from Hongkong, Macao, Taiwan and other countries from south-eastern Asia; The other categories of investment are some companies from the US, Japan and European countries. Different investors have different preferences in the invested areas[7].

The Greenfield investment are the dominating ways of China's service FDI, also there is an increasing trend in merging and acquisitions. In China, foreign investors have more and more interest in companies' M&A, and which have been the leading way of FDI in services [1]. Since there are overproduction in domestic manufacture and less and less lands for industry, also the acquisition environment has been matured, TNCs accelerated to M&A to Chinese companies in services as well, especially in financial, insurance, tourist and retailing, etc. TNCs tended to invest in the way of M&A, such as The Hongkong and Shanghai Banking Corporation has some shares of Bank of Communications, Deloitte merged Chinese accounting companies, Morgan Stanley bought Yongle Electric Appliances Company, etc. From the whole point of view, the leading way of FDI is still Greenfield, however, Mergers and Acquisitions might be the dominating way of FDI in services in China[8].

2.3 Problems Existing and Strategies to Increase the Attractiveness to Chinese Service FDI

1) Problems existing in the current service FDI in China

FDI makes up for the lack of capital input in China's service industry and promotes to develop the service sectors to some extent. However, China's openness policies in service industry to foreign investment are still at the beginning stage. The level and the quality of using service FDI in China should still be improved compared with trends of international direct investment. Therefore, some aspects should be improved.

First, the small scale of using service FDI , the unsteady inflow foreign capital, and the unbalanced distribution in service FDI, enlarges the structural differences among the service industry in China. Until the end of 2003, the proportion of foreign investment items among the first, second and third industry were respective 2.87%, 75.26% and 21.87%, which suggested that the scale of using FDI was still small, there were obvious differences among the industrial distribution of flow foreign capital[4].

Second, the structure of service FDI is not reasonable, which couldn't improve the capabilities of using service FDI. In the 1980s, there was serious unbalanced distribution among the third industry in terms of lack shares of FDI in education, science studying and culture industry; While there were higher shares of FDI in real estates, basic infrastructures and other services. In 2000s, FDI in real estates and social services still kept a high position. Until 2003, the most FDI used in service was in real estates, which accounted for 61.7% among the total FDI, following by the renting and business services with 10.3%. FDI in finance, insurance, trades and information industries was much lower than the average level in the world, and also lower than that of in the other developing countries[4].

Last but not least, service FDI still shows their preference to the developed areas of China. Most of the service FDI is input in the eastern developed areas, while little in the middle-west area.

2) Strategies to attract service FDI in China

With the increasing share of services in economic activities, the liberalization of FDI policies as well as the intensive competitive pressures in service markets, there are more and more opportunities for Chinese service FDI increased. Within this background, the localization-specific advantage of China, which in terms of China's local advantages and good FDI policies, determines the extent and patterns of services FDI[1]. Therefore, how to cope with the problems existing in the service markets and make good use of the localization-specific advantages is a key issue to decide whether Chinese government can attract more service FDI.

a: Improving China's policies on FDI

Since China's entering WTO in 2000, Chinese government has made relative laws and regulations according to their promises and the WTO requirements, which includes the equal positions for foreign firms with domestic firms in terms of being equally treated in policies. However, there are big differences between demanding to FDI and the rapid development of Chinese services in the real investment environment. Since 2006, Chinese government fulfilled all of their promises in WTO to enlarge the openness not only in field of finance, insurance, trading, retailing, etc, but also include communications, conferences, tourist, and other services industry that used to be in low level of opening[9].

Therefore, Chinese government should increase some laws and regulations that were not existing to the FDI and improve some laws and regulations which have been existing but still not matured in the laws system according to the opening process of service FDI; At the same time, to some laws and regulations which are lack of maneuverability, Chinese government should make necessary individual laws and regulations to a further improving [10].

b. Improve China's investment environment in service

Investment environment includes not only the basic infrastructures, but also the marketing system, intellectual property, administrations, laws and regulations, and human resources, etc. At the current situation, the basic infrastructures in China should be improved on one hand, and the reformation of marketization, improvement of laws and regulation environment, public service environment and intellectual property protection environment, etc, should be enhanced on the other hand, in particularly to strengthen government service orientation, services to markets actors. At the same time, it should be encouraged to construct more basic infrastructures for service industry and noted to cover the least developed areas and middle-west areas[10].

c. Optimizing the internal service structures.

Chinese government should adjust the internal service structures to develop modern service industry from a point of the practical situations of Chinese economies and services. For one thing, FDI in labor intensive industry should be encouraged to give full play to the advantage of low cost of Chinese labor. For the other thing, catching opportunities that current international manufactures have been transferred to China, which leads to linkages to promote the development of the service industry, such as the outplacement services might be improved, international logistics might be developed, and international E-business might be enhanced, etc. Moreover, to attract more FDI in some intellectual intensive industry such as information technology and Biotechnologies, which will be the leading industry in the future, will make the Chinese internal structures tend to be reasonable [11].

d. Liberalization gradually

Since many of China's service industries are at the beginning of the market development, which has weak competition, therefore, gradual policies of protection are needed to boost the openness of service markets. In the mean time, gradual liberalization as well as the opening of marketization is proposed by GATTS. China should utilize the preference treatment provided to developing countries and take into account of China's actual situations to attract FDI and develop China's service industries [12].

3 Conclusion

China has made great progress in attracting service FDI since its opening up to foreign investment. And until now, China has been the largest recipient of FDI among all the developing countries [13]. However, there are still some problems existing in attracting service FDI in the process of development, such as the scale of service FDI is still not big enough, the structure of service FDI is not reasonable, as well as the area distribution is seriously unbalanced. Correspondingly, there are some improving measures that how to amend China's policies on FDI, and how to create a better investment environment[14], etc, are given.

The aim of this study is to learn the shift of Chinese service FDI in the past decade as well as the current situations, analyze some problems facing by Chinese government in service FDI, and also to put forward some facts that Chinese government should improve in the future in order to attract more service FDI.

Acknowledgment. This study was funded by Chinese National Natural Science Foundation (Project Number:70803005) and the mining management Project of WWF-China. In particular, the authors thank Mr. Shi Jian and Mr. Si Kaichuang and Mr. Hu Chongde for their help in data collection.

References

1. WIR. World Investment Report: The shift towards services. pp. 97–237
2. UNCTAD, Investment Brief. Rising FDI into China: The facts behind the numbers (2009), `http://www.unctad.org/Templates/Search.asp?intItemID=2441&lang=1&frmSearchStr=investment+brief+China&frmCategory=all&section=whole` (accessed February 09, 2010)
3. Guoqiang, L.: China's Policies on FDI: Review and Evaluation 315, 318 (2006)
4. Feng, Y.: China's service industry makes use of foreign capital: the current situation, problems and determinants analysis. Study of The World Economics (1), 4–6 (2006)
5. National Bureau of Statistics, Foreign Direct Investment by Sector (2009), `http://www.stats.gov.cn/tjsj/ndsj/2008/html/R1717e.htm` (accessed February 09, 2010)
6. OECD, Directorate for Financial, Fiscal and Enterprises Affairs working Main determinants and impacts of foreign direct investment on China's economy, pp. 8–9 (2000a)
7. Xianbin, H.: The impacts and determinants analysis of service FDI in China, 12–14 (2008)
8. OECD, Directorate for Financial, Fiscal and Enterprises Affairs working Main determinants and impacts of foreign direct investment on China's economy, p. 10 (2000b)
9. China Trade in Services, Laws by sectors (2010), `http://tradeinservices.mofcom.gov.cn/en/b/news_101220.shtml` (accessed February 09, 2010)
10. Yi, F.: Study on promote service FDI in China. China FDI 2009 189, 28 (2009)
11. Lei, Y.: Policies on how to attract service FDI. Marketing Modernization 579, 76 (2009)
12. Jianhua, W.: Trends and policies to service FDI after China entered into the WTO. Issues of International Business and Trade 7, 7–8 (2003)
13. Fung, K.C.: FDI in China: Policy, trend, and impact, 17 (2002)
14. China's Outward FDI: Past and Future, pp. 315–316

Pretreatment of Micro-polluted Raw Water by the Combined Technology of Photocatalysis-Biological Contact Oxidation

Yingqing Guo[1], Changji Yao[1], Erdeng Du[1], Chunsheng Lei[1], and Yingqing Guo[2]

[1] School of Environmental & Safety Engineering
Changzhou University
Changzhou, China
[2] College of Environmental Science and Engineering
Tongji University
Changzhou, China
water8820@163.com

Abstract. This paper gives a pilot-scale study on the pretreatment effect of micro-polluted raw water by taking advantage of the combined technology of photocatalysis—biological contact oxidation. The research shows that: turbidity removal ratio of the combined technology reaches more than 70% and the average removal ratio of COD_{Mn} reaches 22.8%; the average NH3-N removal ratio of photocatalytic oxidation is only 11.9%, while biological contact oxidation reaches 29.5%, and of which the average TP removal ratio is 28.5%. A preferable effect can be acquired in processing micro-polluted raw water by photocatalysis—biological contact oxidation.

Keywords: photocatalysis, biological contact oxidation, TiO_2, micro-polluted, series operation.

1 Introduction

Photocatalytic oxidation, a special photosensitive oxidation, with its sensitizer of the n-type semiconductor, is a new water-processing technology appeared in the late 30 years and possesses such outstanding advantages as low energy consumption, easily operated and no secondary pollution. Under the irradiation of ultraviolet and solar rays, the photocatalytic reaction of TiO_2 occurs and produces a strong oxidation capacity that is able to degrade many organic contaminations which are hard to be decomposed into inorganic substances like CO_2, H_2O, etc[1].

Biological contact oxidation is a procedure during which oxygenated water flows circularly on the surface of artificially synthetic media in a pond and contaminant substrates that are biochemically available can be degraded through flocculating absorption and oxidation of the biofilm of the media. There is a large variety of organisms in the biofilm, such as bacterium, fungus, filamentous bacteria, protozoan, metazoan and others, constituting a relatively stable eco-system [2].

D. Zeng (Ed.): Advances in Control and Communication, LNEE 137, pp. 357–363.
springerlink.com

During the procedure of this technology that combines nano-TiO_2 photocatalytic oxidation and biological contact oxidation together, macromolecular contaminants in raw water that are hard to biodegrade can be oxidized by pretreatment of photocatalysis to improve the biodegradability of raw water. Then, increase the total removal ratio and decrease the producing of by-products in subsequent chlorination by giving full play to the strong capacity of removing organic substances of subsequent biological treatments so as to improve the safety of drinking water.

2 Experiment Section

2.1 Experimental Materials

The TiO_2/PP compound media is prepared from the mixture of TiO_2 powder and polypropylene (PP) by means of twin-screws extruding. The polypropylene media is pure plastic without TiO_2, of which the specification is $\Phi 50 \times 50$mm, the specific surface area is 278m2/m3, the void ratio is 96.6%, the density is 0.907g/cm3, and the packing coefficient is 8000/m^3. The light source of photocatalysis is ultraviolet lamp with a power of 18W and a wavelength of 254nm.

2.2 Experimental Apparatus

The apparatus is composed of two parts: photocatalytic reaction zone and biological contact oxidation zone. The raw water flows into the photocatalytic zone which is made of organic glass. A gaseous diffusion system is equipped in the bottom of the reactor and two quartz socket tubes, where the ultraviolet light source is placed, are in the center. The external part of photocatalytic reaction zone is muffled by aluminum foil which not only prevents the entry of other lights, but also improves the utilization percent of ultraviolet light. The liquid, mixed by the compound media and the water sample to be processed, is between quartz socket tubes and the reaction zone. Let the air to the bottom of the mixed liquid by a pneumatic pump, which functions not only as aerating so that suspended media in the reaction zone can be fully fluidized, but also as preventing the simple compounding of electron and cavity by using O_2 in the air as the electron acceptor of oxidant. In the reacting process, the dissolved oxygen concentration maintains the invariable range of 3.0~3.5 mg/L. After the dynamic natural biofilm formation of polypropylene media through continuous water feeding, combine the photocatalytic oxidation reactor with the biological contact oxidation reactor in series. Refer to figure 1 for the experimental apparatus.

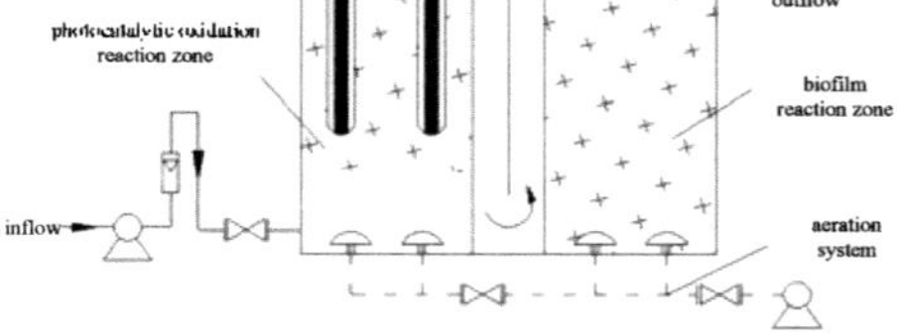

Fig. 1. Schematic diagram of experiment setup

2.3 Quality of the Raw Water

The raw water is taken from a lake of Changzhou in Jiangsu Province which has high concentrations of COD_{Mn} and NH_3-N. Analyze the raw water sample and conclude a water quality index in table1.

Table 1. The characteristics of raw water

Items	pH	Turbidity NTU	COD_{Mn} (mg/L)	NH_3-N (mg/L)	UV_{254} (cm^{-1})	TP (mg/L)	Chl-a (mg/L)
Range	6.52~8.17	25~30	10.57~11.03	0.62~0.82	0.373~0.382	0.252~0.314	2.11~4.63
Average value	7.35	27.5	10.80	0.72	0.378	0.283	3.37

2.4 Biofilm Formation of Polypropylene Media

Biofilm formation methods of polypropylene media majorly include artificial biofilm formation and natural biofilm formation through dynamic culturing. The former needs to introduce cultures and add nutrients that promote the growth of microorganisms in the water, so that they are able to aggregate and grow on the media to form a biofilm; the latter neither need the introduction of cultures nor the adding of nutrients. It requires only a continuous water inflow for the apparatus so that microorganisms will assemble and grow on the media to form a biofilm.

Giving consideration to the low pollutant concentration of slightly contaminated water, which is the object of this experiment, natural biofilm formation through dynamic culturing is adopted. Inlet the water continuously and add an aeration system to the bottom of the media so as to shorten the incubation time of the biofilm. The water temperature maintains 20～30℃ throughout the experiment. During the initial stage, the quantity of inflow is 120L/h, the air water ratio is 0.50:1, and the hydraulic retention time (HRT) is 60min. Starting from the second day of reaction, determine COD_{Mn} concentrations of both inflow and outflow of the reactor. If the COD_{Mn} concentration of outflow remains in a certain range and the removal ratio is 25～30% since the 15th day, biofilm is successfully formed.

3 Experiment Results and Analysis

3.1 The Turbidity Removal Effect of the Combined Technology

It can be seen from figure 2 that both photocatalytic oxidation and biological contact oxidation have a preferable effect of turbidity removal. When the apparatus functions stably, the turbidity removal ratio of the combined technology remains above 70%. With the increase of reacting time, the biofilm gets more and more opportunities to contact oxygen and organic matters, which is beneficial to the growth of microorganisms and degrades organic matters that producing turbidity in the water. Thus, the turbidity of effluent water can be reduced. In the meantime, aging biofilms exfoliate because of turbulent current and become large flocs through being combined with tiny suspended particles, which also reduces the turbidity. Here, it can be concluded that turbidity removal of the system relies mainly on the synergism of adsorptive degradation of biofilm in the early stage and sedimentation in the later stage.

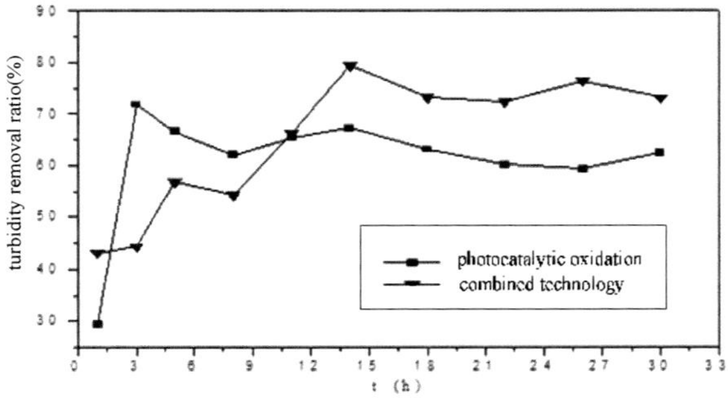

Fig. 2. Effect of time on turbidity removal with combined process

3.2 CODMn Removal Effect of Combined Technology

In the biofilm reactor, organic substances and dissolved oxygen should be transferred to the biofilm surface after going through the liquid phase and then diffuse into the internal part of biofilm to be decomposed and transformed. And ultimately various metabolites are formed (CO_2, H_2O, etc).

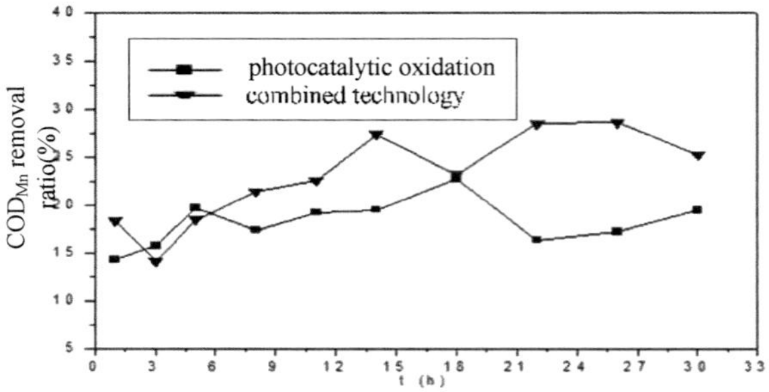

Fig. 3. Effect of time on COD_{Mn} removal with combined process

Figure 3 shows that, after the formulation of biofilm, the polypropylene media also has a preferable removal effect of organic substances by means of incepting, degrading contaminants in the water and mechanical entrapment of pollutants. The average COD_{Mn} removal ratio reaches 22.8%.

3.3 NH3-N Removal Effect of Combined Technology

Figure4 demonstrates that it is the biofilm that plays the major role in removing NH_3-N during the process of combined technology. The average removal ratio of biofilm reaches 29.5%, whereas the NH_3-N removal effect of photocatalytic oxidation is relatively bad which is only 11.9%. The removal effect of NH_3-N tends to fluctuate if photocatalytic oxidation of compound media is adopted only. Especially under the circumstance of a long HRT, the quantity of NH_3-N would rebound, being influenced by anaerobic bacteria. The insufficiency of single technology in processing contaminated raw water can be avoided effectively by combining biological contact oxidation with photocatalytic oxidation.

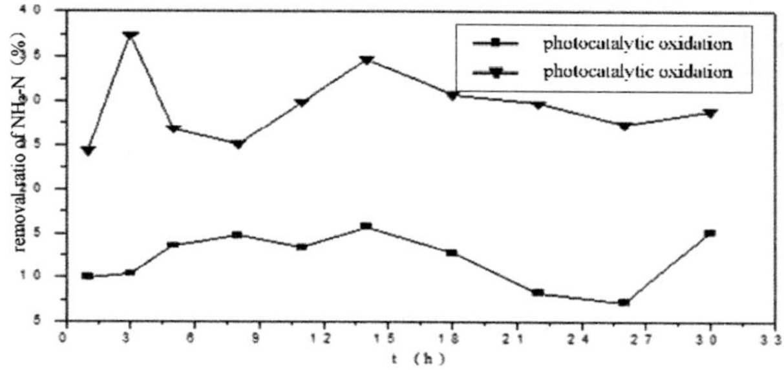

Fig. 4. Effect of time on NH3-N removal with combined process

Biofilm accomplishes the degradation of organic substances through heterotrophic bacteria, while it finishes the NH_3-N removal by autotrophic bacteria. The coexistence of heterotrophic bacteria and autotrophic bacteria leads to a scramble for dissolved oxygen and causes competition[3]. However, during the process of feed water treatment, the concentration of organic substances is relatively low and oxygen tends to be ample so that nitrobacteria and heterotrophic bacteria can be satisfied to the full extend. Consequently, there would not be any obvious competition which can be seen from the general stability of microorganism's removal of organic substances and NH_3-N[4]. The growth and reproducing rate of heterotrophic bacteria increases with the concentration of organic substances in the water. Heterotrophic bacteria take advantage of dissolved oxygen to reproduce in large quantities under the condition of abundant nutrients. Yet nitrobacteria is a kind of strict aerobic bacteria which has less capability of absorbing oxygen than heterotrophic bacteria when the dissolved oxygen is insufficient or the transferring rate of oxygen slows down. All these would impose restrictions on growth and reproduction of nitrobacteria and impact on the nitration reaction so that the NH_3-N removal ratio declines. Therefore, the NH_3-N removal would not be affected largely when the organic matters concentration of inflow water remains in a certain range, while the ratio would possibly decrease when the concentration exceeds a certain limited value.

On the other hand, the concentration of NH_3-N in the water also has some impacts on the removal of organic substances. There are probably two reasons. The first one comes from simultaneous processes of both nitration reaction and degradation of heterotrophic bacteria. Nitrobacteria and heterotrophic bacteria take advantage of each other and compete against each other. Heterotrophic bacteria are not able to grow and reproduce quickly when the concentration of organic substances is low. So their oxidative degradation ratio of organic substances in the water is also low. When the NH_3-N concentration increases, organic cellular substances produced by nitrobacteria can be substrates of heterotrophic bacteria partially so that the growth of heterotrophic bacteria would be accelerated and organic substances would be decomposed and oxidized in the water by heterotrophic bacteria. Hence, to a certain extent, NH_3-N is beneficial to promoting growth of heterotrophic bacteria and oxidative decomposing of trace organics during the process of raw water. Organic cellular substances produced by nitrobacteria can be substrates of heterotrophic bacteria partially. Second, in metabolic process, microorganisms acquire energy from surroundings and synthesize new cellular substances. During the synthesizing process, nutrients should be compounded according to a certain proportion, so as to be material foundations of cellular metabolism which conducts material exchange. And finally form a biological structure with all kinds of physiological functions.

Therefore, when concentrations of organic substances and NH_3-N are in proper ranges, removals of the two supplement and complement with each other so that both organic substances and NH_3-N removal ratios can be increased by biofilm.

3.4 Total Phosphorus Removal Effect of Combined Technology

It can be seen from figure5 that the average removal ratio of total phosphorus is 28.5% during the process of combined technology. Throughout the whole period of the system, the biofilm reactor functions in the way of aerobic-anaerobic so as to eliminate phosphorus. Facultative bacteria of the anaerobic period convert dissolved BOD into organic substances with low molecular weight through fermentation. Phosphate accumulating organisms (PAO) decompose polyphosphates inside cells, produce ATP and expel phosphates produced in decomposition from cells. Taking advantage of ATP, PAO absorb substances with low molecular weight in waste water into cells organic and store them in cells in forms of PHA and glycogen. In the subsequent aerobic period, PAO absorb phosphorus excessively from wastewater by making use of the energy released from oxidative decomposition of PHA and store them up in the form of phosphate. Biofilm falls off and sinks to the bottom. Then the phosphorus removal can be realized through separating sludge and water[5].

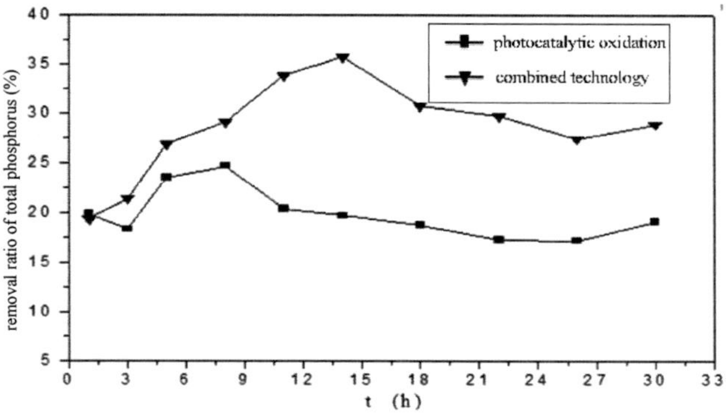

Fig. 5. Effect of time on TP removal with combined process

4 Conclusion

The result of the pilot-scale study on the pretreatment effect of micro-polluted raw water by taking advantage of the combined technology of photocatalysis-biological contact oxidation shows that: turbidity removal ratio of the combined technology reaches more than 70% and the average removal ratio of COD_{Mn} reaches 22.8%; the average removal ratio of ammonia and nitrogen of photocatalytic oxidation is only 11.9%, while biological contact oxidation reaches 29.5%, and of which the average removal ratio of the total phosphorus is 28.5%. The combined technology has a better effect of processing micro-polluted raw water, which not only effectively avoids the inefficiency of single technology, but also can be used for reference in future practical engineering.

References

1. Liu, C.: Nano-photocatalysis and photocatalytic environmental purifying materials. Chemical Industry Press, Beijing (2008)
2. Zhou, Y., He, Y.: Micro-polluted water purifying technology and engineering examples. Chemical Industry Press, Beijing (2003)
3. Xu, B., Xia, S., Hu, C., et al.: Kinetics of nitrification in biological pretreatment of micro-polluted raw water. China Water & Wastewater 19(4), 15–18 (2003)
4. Liu, H.: The whole process of biological treatment of micro-polluted raw water. Chemical Industry Press, Beijing (2003)
5. Zhou, Q., Gao, T.: Environmental engineering microbiology. Higher Education Press, Beijing (2003)

A Singular Integral Calculation of Inverse Vening-Meinesz Formula Based on Simpson Formula

Huang Xiao-ying[1], Li Hou-pu[1], Xiang Cai-bing[1] and Bian Shao-feng[1,2]

[1] Department of Navigation Engineering
Naval University of Engineering
Wuhan, China
[2] Institute of Geodesy and Geophysics
Chinese Academy of Sciences
Wuhan, China
hjgchxy@163.com, sfbian@sina.com

Abstract. A new investigation of singular integral of inverse Vening-Meinesz formula based on Simpson formula is done in this paper. A set of integral calculation formula is given and the precision of the formula is examined, the contrast results before and after non-singularity transformation based on theoretical model of geoidal height are given. Results indicate that the calculation precision of the formula gained after non-singularity transformation is better than 1%, which can absolutely fulfill practical needs.

Keywords: geoidal height; gravity anomaly inversion, inverse Vening-Meinesz formula, non-singular transformation.

1 Introduction

According to WANG [1], geoidal height and deflection of the vertical can be expressed as linear polynomial, biquadratic polynomial, and bicubic polynomial respectively, and a formula to calculate the gravity anomaly of the innermost area is deduced. The formula is analytical yet complicated in form, and has a large number of terms which is inconvenient to use; the coefficients of geoidal height and deflection of the vertical polynomial on known knots will be utilized in calculating the gravity anomaly of the innermost area, which will increase the amount of computing to some extent [2-7]. Based on the knowledge above, an investigation of singular integral of inverse Vening-Meinesz formula based on Simpson formula is done and a set of integral calculation formula is given, which can be utilized to calculate the geoidal height and deflection of the vertical of the known knots directly.

2 The Singular Integral of Inverse Vening-Meinesz Formula Based on Simpson Formula

According to WANG[1], the calculating formula of gravity anomaly of the innermost area is

D. Zeng (Ed.): Advances in Control and Communication, LNEE 137, pp. 365–371.
springerlink.com

$$\Delta g_P = \frac{\gamma}{4\pi}\iint_\sigma (-\frac{2R^2}{l^2})\cdot\left((-\frac{x}{l})\cdot\xi_Q + (-\frac{y}{l})\cdot\eta_Q\right)\cdot\frac{1}{R^2}dxdy$$
$$= \frac{\gamma}{2\pi}\iint_\sigma \frac{\xi_Q\cdot x + \eta_Q\cdot y}{(x^2+y^2)^{3/2}}dxdy \tag{1}$$

Where ξ_Q, η_Q are the components of vertical deflection, γ is the earth normal gravity, $d\sigma$ is the current surface primitive, l is distance between point P and the current surface primitive, $R = 6371km$ is the earth mean radius.

Suppose the unit length of x and y directions are a respectively, if $a = 1$, the innermost area is a square area with $\sigma \in [-1 < x < 1, -1 < y < 1]$, shown as figure 1.

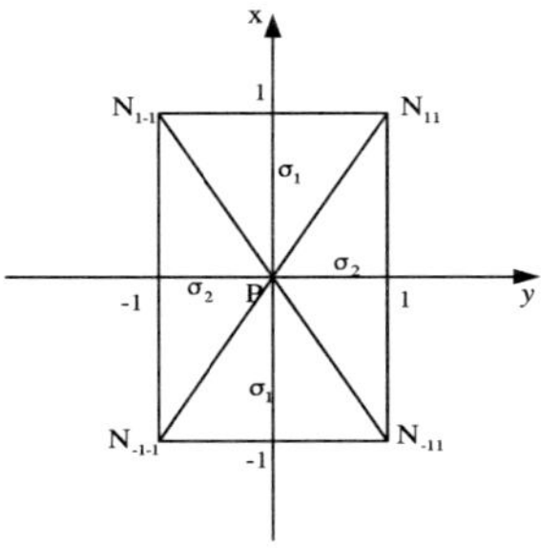

Fig. 1. The sketch map of the innermost area

Suppose $\Delta g_\xi = \frac{\gamma}{2\pi}\iint_\sigma \frac{\xi_Q\cdot x}{(x^2+y^2)^{3/2}}dxdy$, $\Delta g_\eta = \frac{\gamma}{2\pi}\iint_\sigma \frac{\eta_Q\cdot y}{(x^2+y^2)^{3/2}}dxdy$, the gravity anomaly calculating formula (1) with inverse Vening-Meinesz method in the innermost area will be the form of

$$\Delta g_P = \Delta g_\xi + \Delta g_\eta \tag{2}$$

The components ξ_Q, η_Q of the vertical deflection can be expanded to Taylor series [8]

$$\xi(x,y) = \xi_P + \xi_x x + \xi_y y + \frac{1}{2!}(\xi_{xx}x^2 + 2\xi_{xy}xy + \xi_{yy}y^2) + \cdots \tag{3}$$

$$\eta(x,y) = \eta_P + \eta_x x + \eta_y y + \frac{1}{2!}(\eta_{xx}x^2 + 2\eta_{xy}xy + \eta_{yy}y^2) + \cdots \tag{4}$$

Where:

$$\xi_x = \frac{\partial\xi}{\partial x},\ \xi_y = \frac{\partial\xi}{\partial y},\ \xi_{xx} = \frac{\partial^2\xi}{\partial x^2},\ \xi_{yy} = \frac{\partial^2\xi}{\partial y^2},\ \xi_{xy} = \frac{\partial^2\xi}{\partial x\partial y},$$

$$\eta_x = \frac{\partial\eta}{\partial x},\ \eta_y = \frac{\partial\eta}{\partial y},\ \eta_{xx} = \frac{\partial^2\eta}{\partial x^2},\ \eta_{yy} = \frac{\partial^2\eta}{\partial y^2},\ \eta_{xy} = \frac{\partial^2\eta}{\partial x\partial y}$$

Considering the symmetry of the integral area, Δg_ξ, Δg_η can be written as

$$\Delta g_\xi = \frac{\gamma}{2\pi}\iint_\sigma \frac{\xi_Q - \xi_P}{(x^2+y^2)^{3/2}}\cdot xdxdy \tag{5}$$

$$\Delta g_\eta = \frac{\gamma}{2\pi}\iint_\sigma \frac{\eta_Q - \eta_P}{(x^2+y^2)^{3/2}} \cdot ydxdy \tag{6}$$

Divide innermost area into two parts $\sigma_1 \in [-1<x<1, |y|<|x|]$ and $\sigma_2 \in [|x|<|y|, -1<y<1]$, considering the symmetry of the integral area, we have

$$\iint_\sigma \frac{x}{(x^2+y^2)^{3/2}} dxdy = \iint_\sigma \frac{y}{(x^2+y^2)^{3/2}} dxdy = 0 \tag{7}$$

Introducing new integral variable to σ_1

$$\begin{cases} x = x \\ y = kx \end{cases} \tag{8}$$

And σ_2

$$\begin{cases} x = \lambda y \\ y = y \end{cases} \tag{9}$$

Then (5) and (6) can be rewritten as

$$\begin{aligned} \Delta g_\xi = &\frac{\gamma}{2\pi}\int_{-1}^{1}\int_{-1}^{1} \frac{[\xi(x,kx)-\xi_P]/x}{(1+k^2)^{3/2}} \cdot dxdk + \\ &\frac{\gamma}{2\pi}\int_{-1}^{1}\int_{-1}^{1} \frac{[\xi(\lambda y,y)-\xi_P]/y}{(1+\lambda^2)^{3/2}} \cdot \lambda dyd\lambda \end{aligned} \tag{10}$$

$$\begin{aligned} \Delta g_\eta = &\frac{\gamma}{2\pi}\int_{-1}^{1}\int_{-1}^{1} \frac{[\eta(x,kx)-\eta_P]/x}{(1+k^2)^{3/2}} \cdot kdxdk + \\ &\frac{\gamma}{2\pi}\int_{-1}^{1}\int_{-1}^{1} \frac{[\eta(\lambda y,y)-\eta_P]/y}{(1+\lambda^2)^{3/2}} \cdot dyd\lambda \end{aligned} \tag{11}$$

From (3) and (4) we can derive

$$\lim_{x\to 0} \frac{\xi(x,kx)-\xi_P}{x} = \xi_x + k\xi_y < \infty \tag{12}$$

$$\lim_{y\to 0} \frac{\eta(\lambda y,x)-\eta_P}{y} = \lambda\eta_x + \eta_y < \infty \tag{13}$$

Utilizing Simpson formula, the numerical integral of x and y direction will be

$$\begin{aligned} \Delta g_\xi = &\frac{\gamma}{6\pi}\int_{-1}^{1} \frac{-\xi(-1,-k)+4(\xi_x+\xi_y k)+\xi(1,k)}{(1+k^2)^{3/2}} dk \\ &+\frac{\gamma}{6\pi}\int_{-1}^{1} \frac{-\xi(-\lambda,-1)+4(\xi_x\lambda+\xi_y)+\xi(\lambda,1)}{(1+\lambda^2)^{3/2}} \lambda d\lambda \end{aligned} \tag{14}$$

$$\begin{aligned} \Delta g_\eta = &\frac{\gamma}{6\pi}\int_{-1}^{1} \frac{-\eta(-1,-k)+4(\eta_x+\eta_y k)+\eta(1,k)}{(1+k^2)^{3/2}} kdk \\ &+\frac{\gamma}{6\pi}\int_{-1}^{1} \frac{-\eta(-\lambda,-1)+4(\eta_x\lambda+\eta_y)+\eta(\lambda,1)}{(1+\lambda^2)^{3/2}} d\lambda \end{aligned} \tag{15}$$

Note the value of $\xi(x,y)$, $\eta(x,y)$ at $x, y = 0,\pm 1$ as ξ_{ij}, η_{ij} $(i, j = 0,\pm 1)$, utilizing Simpson formula to (14)and (15)at k and λ direction again, considering

$$\int_{-1}^{1}\frac{1}{(1+k^2)^{3/2}}dk=\int_{-1}^{1}\frac{1}{(1+\lambda^2)^{3/2}}d\lambda=\sqrt{2} \tag{16}$$

$$\int_{-1}^{1}\frac{k^2}{(1+k^2)^{3/2}}dk=\int_{-1}^{1}\frac{\lambda^2}{(1+\lambda^2)^{3/2}}d\lambda=-\sqrt{2}+2\ln(1+\sqrt{2}) \tag{17}$$

we can derive

$$\Delta g_{\xi}=\frac{\gamma}{2\pi}\left(\frac{8}{3}\xi_x\ln(1+\sqrt{2})+\frac{4}{9}(\xi_{10}-\xi_{-10})+\frac{\sqrt{2}}{18}(\xi_{11}-\xi_{1-1}+\xi_{-11}-\xi_{-1-1})\right) \tag{18}$$

$$\Delta g_{\eta}=\frac{\gamma}{2\pi}\left(\frac{8}{3}\eta_y\ln(1+\sqrt{2})+\frac{4}{9}(\eta_{01}-\eta_{0-1})+\frac{\sqrt{2}}{18}(\eta_{11}-\eta_{1-1}+\eta_{-11}-\eta_{-1-1})\right) \tag{19}$$

Where ξ_x, η_y can be derived from deflection of vertical at knots according to numerical differential formula

$$\xi_x=\frac{\xi_{10}-\xi_{-10}}{2} \tag{20}$$

$$\eta_y=\frac{\eta_{01}-\eta_{0-1}}{2} \tag{21}$$

Inserting (20) and (21) into (18) and (19), we have

$$\Delta g_{\xi}=\frac{\gamma}{2\pi}\left(\frac{4}{3}[\ln(1+\sqrt{2})+\frac{1}{3}](\xi_{10}-\xi_{-10})+\frac{\sqrt{2}}{18}(\xi_{11}-\xi_{-11}+\xi_{1-1}-\xi_{-1-1})\right) \tag{22}$$

$$\Delta g_{\eta}=\frac{\gamma}{2\pi}\left(\frac{4}{3}[\ln(1+\sqrt{2})+\frac{1}{3}](\eta_{01}-\eta_{0-1})+\frac{\sqrt{2}}{18}(\eta_{11}-\eta_{1-1}+\eta_{-11}-\eta_{-1-1})\right) \tag{23}$$

Usually integral area is not a unit square and a is not unit length though, Δg_{ξ} and Δg_{η} don't have dimension, so (2), (22) and (23) are the numerical integral formulae of inverse gravity anomaly with inverse Vening-Meinesz formula at innermost area.

3 The Accuracy Examination of Singular Integral Formula

Introducing non-singular transformation, non-singularize the singular integral in gravity anomaly inversion of satellite altimetry has very important actual meaning with respect to improving the accuracy of gravity anomaly inversion. The accuracy examination of the formula after non-singular transformation will be carried out by the numerical results before and after non-singular transformation of inverse Vening-Meinesz formula with geoidal height as theoretical model value. The contribution of innermost area to gravity anomaly inversion will be carried out by numerical integral formula (22) and (23) based on inverse Vening-Meinesz formula and real deflection of vertical data.

The mathematical model of geoidal height is

$$N_Q=\sqrt{x^2+y^2+\delta^2}\,(\delta>0)$$

And components of deflection of vertical are

$$\xi_Q = -\frac{\partial N}{\partial x} = -\frac{x}{\sqrt{x^2 + y^2 + \delta^2}} \tag{24}$$

$$\eta_Q = -\frac{\partial N}{\partial y} = -\frac{y}{\sqrt{x^2 + y^2 + \delta^2}} \tag{25}$$

For calculation convenience, the innermost area will be unit square: $\sigma \in [-1 < x < 1, -1 < y < 1]$, the integral before non-singular transformation is

$$\begin{aligned} IV_1 &= \iint_\sigma \frac{\xi_Q \cdot x + \eta_Q \cdot y}{(x^2 + y^2)^{3/2}} dxdy \\ &= -\iint_\sigma \frac{1}{\sqrt{x^2 + y^2}\sqrt{x^2 + y^2 + \delta^2}} dxdy \\ &= -4\int_0^1\int_0^1 \frac{1}{\sqrt{x^2 + y^2}\sqrt{x^2 + y^2 + \delta^2}} dxdy \end{aligned} \tag{26}$$

Introducing new integration variables (8) and (9), integral after non-singular transformation is

$$\begin{aligned} IV_2 &= -4\int_0^1\int_0^1 \frac{1}{\sqrt{x^2 + y^2}\sqrt{x^2 + y^2 + \delta^2}} dxdy \\ &= -4\int_0^1\int_0^1 \frac{1}{\sqrt{1+k^2}\sqrt{x^2 + k^2x^2 + \delta^2}} dxdk \\ &\quad -4\int_0^1\int_0^1 \frac{1}{\sqrt{1+\lambda^2}\sqrt{\lambda^2 y^2 + y^2 + \delta^2}} d\lambda dy \\ &= -8\int_0^1\int_0^1 \frac{1}{\sqrt{1+k^2}\sqrt{x^2 + k^2x^2 + \delta^2}} dxdk \end{aligned} \tag{27}$$

Computed results of IV_1 and IV_2 with different intervals when $\delta = 10$ and $\delta = 100$ are as Table 1.

Table 1. Results of singular integral IV before and after the non-singular transformation

δ	N × N	4 × 4	10 × 10	200 × 200	600 × 600	800 × 800	1000 × 1000
10	IV_1	-0.4869	-0.5828	-0.6916	-0.6989	-0.6999	-0.7005
	IV_2	-0.6721	-0.6914	-0.7030	-0.7034	-0.7034	-0.7035
100	IV_1	-0.0489	-0.0584	-0.0693	-0.0700	-0.0701	-0.0702
	IV_2	-0.0674	-0.0693	-0.0704	-0.0705	-0.0705	-0.0705

From the table we can see the accuracy of IV_2 after non-singular transform with 10×10 intervals almost equal to that of IV_1 before transform with 200×200 intervals, which indicates the computation efficiency has been improved greatly.

With deflection of vertical expressed as biquadratic polynomial interpolation, the calculating formula of gravity anomaly of the innermost area is as follow [1]:

$$\Delta g_P = -\frac{2\gamma}{3\pi}\left(\sqrt{2}(\alpha_{12}+\beta_{21})-(3\alpha_{10}+2\alpha_{12}+2\beta_{21}+3\beta_{01})\ln(1+\sqrt{2})\right)$$

Where α_{ij}, β_{ij} is the undetermined coefficient

$$\begin{pmatrix}\alpha_{00} & \alpha_{01} & \alpha_{02}\\ \alpha_{10} & \alpha_{11} & \alpha_{12}\\ \alpha_{20} & \alpha_{21} & \alpha_{22}\end{pmatrix} = \begin{pmatrix}0 & 1 & 0\\ -\frac{1}{2} & 0 & \frac{1}{2}\\ \frac{1}{2} & -1 & \frac{1}{2}\end{pmatrix}\begin{pmatrix}\xi_{-1-1} & \xi_{-10} & \xi_{-11}\\ \xi_{0-1} & \xi_{00} & \xi_{01}\\ \xi_{1-1} & \xi_{10} & \xi_{11}\end{pmatrix}\begin{pmatrix}0 & -\frac{1}{2} & \frac{1}{2}\\ 1 & 0 & -1\\ 0 & \frac{1}{2} & \frac{1}{2}\end{pmatrix}$$

$$\begin{pmatrix}\beta_{00} & \beta_{01} & \beta_{02}\\ \beta_{10} & \beta_{11} & \beta_{12}\\ \beta_{20} & \beta_{21} & \beta_{22}\end{pmatrix} = \begin{pmatrix}0 & 1 & 0\\ -\frac{1}{2} & 0 & \frac{1}{2}\\ \frac{1}{2} & -1 & \frac{1}{2}\end{pmatrix}\begin{pmatrix}\eta_{-1-1} & \eta_{-10} & \eta_{-11}\\ \eta_{0-1} & \eta_{00} & \eta_{01}\\ \eta_{1-1} & \eta_{10} & \eta_{11}\end{pmatrix}\begin{pmatrix}0 & -\frac{1}{2} & \frac{1}{2}\\ 1 & 0 & -1\\ 0 & \frac{1}{2} & \frac{1}{2}\end{pmatrix}$$

Then, the analytic expression of singular integral IV with deflection of vertical expressed as biquadratic polynomial interpolation is

$$IV = \frac{4}{3}\left(3(\alpha_{10}+\beta_{01})\ln(1+\sqrt{2}) + (2\ln(1+\sqrt{2})-\sqrt{2})(\alpha_{12}+\beta_{21})\right) \tag{28}$$

Note the result of (28) as IV_3, and the results of (22) and (23) as IV_4, take the 1000×1000 interval result of (27) as exact value, values of IV_3, IV_4 and their corresponding errors are shown as Table 2.

Table 2. Comparisons between results of the singular integral IV after the non-singular transformation and exact values

δ	**Exact Value**	IV_3	**Absolute Error**	**Relative Error**	IV_4	**Absolute Error**	**Relative Error**
10	-0.7035	-0.7011	0.0024	0.34%	-0.7069	0.0024	0.34%
100	-0.0705	-0.0705	0.0000	0.00%	-0.0710	0.0005	0.71%

From the table we know that the accuracy of analytic expression IV after non-singular transform is better than that of integral formula, the relative errors of both are smaller than 1%, which indicate both of them can completely fulfill practical application.

4 Conclusions

For complementing the arithmetic theory of the altimetry gravity calculated by the inverse Vening-Meinesz formula, the non-singular transform is introduced to solve singular integral problem in the inverse Vening-Meinesz formula, a formula to calculate gravity anomaly of innermost area is derived whose result can be directly obtained. Based on theoretical model, the accuracy of the deduced formula is analyzed, and the effect of the innermost area on gravity anomaly is discussed. The validity of non-singular transform is verified through comparing the results of before and after transformation, the efficiency is improved remarkably.

Acknowledgment. This work was supported by National Natural Science Foundation of China (Grant Nos. 40125013, 40774002, and 40904018).

References

1. Wang, R.: The Research on the Recovery of Background Field for Gravity Matching Aided Navigation and the Sea Floor Topography from Satellite Altimeter Data. Naval University of Engineering, Wuhan (2009)
2. Rapp, R.H.: Gravity Anomaly Recovery from Satellite Altimeter Data Using Least Square Collocation Techniques. Ohio State University, Columbus (1974)
3. Huang, M.-T., Zhai, G.-J., Ouyang, Y.-Z., et al.: The Recovery of Bathymetry from Altimeter Data. Geomatics and Information Science of Wuhan University 27, 133–137 (2002)
4. Li, J.-C., Chen, J.-Y., Ning, J.-S., et al.: Earth's Gravity Field Approximation Theory and the China 2000 Geoid Determination. Press of Wuhan University, Wuhan (2003)
5. Moritz, H.: Advanced Physical Geodesy. Herbet Wichmann Verlag, Germany (1980)
6. Hwang, C.: Inverse Vening Meinesz Formula and Deflection-Geoid Formula: Applications to the Predictions of Gravity and Geoid over the South China Sea. Journal of Geodesy 72, 304–312 (1998)
7. Bian, S.-F.: Numerical Solution for Geodetic Boundary Value Problem and the Earth's Gravity Field Approximation. Wuhan surveying and mapping University of Science, Wuhan (1992)
8. Ye, Q.-X., Shen, Y.-H.: Practical Math Manual. Press of Science, Beijing (2007)

An Approach in Greenway Analysis and Evaluation: A Case Study of Guangzhou Zengcheng Greenway

Wu Juanyu[1] and Xiao Yi[2]

[1] School of Architecture & State Key Lab of Subtropical Building Science South China University of Technology Guangzhou, China
[2] Chief Engineer Office Guangzhou Residential Architectural Design Institute Guangzhou, China
wujuanyu@scut.edu.cn, earlxy@sina.com

Abstract. Greenway analysis and evaluation is designed to distinguish and measure the suitability of potential sites for greenway development. In this study, we try to present an approach to greenway analysis and evaluation of the equipment system and the recreational suitability of the Licheng section of the Guangzhou Zengcheng Greenway. This study aims to improve the quality of design and to make the greenway network system planning more scientific. The purpose of this paper is to outline the specific features of the Zengcheng greenway (Licheng section) and the strategy for its development as a key part of the greenway system for the Pearl River Delta landscape.

Keywords: Zengcheng greenway, Licheng section, Analysis and evaluation, Equipment system, Recreational suitability.

1 Introduction

The use of the concept of greenway can be identified in China at the beginning of the 21st century as a planning and design tool. In response to "energy-efficient and low-carbon emission reduction" -the target of urban development, in recent years, the Guangdong Province Government has built up 1690 km greenway in the whole Pearl River Delta, which has won initial success to explore a characteristic greenway in China. According to *Pearl River Delta Greenway Network Master Plan Outline*: "From 2010, within 3 years, the Guangdong Province Government should first built in the Pearl River Delta Region total length 2372 km of six regional greenways, series with more than 200 nature protected areas, national parks and historical and cultural sites, to connect three Pearl River metropolitan areas in the direct service a population of 25,650,000 people, to achieve in the region city to city, city and suburb, suburban and rural areas and the mountain, waterfront and other ecological resources and historical and cultural resources connected to the inter-city green network, which plays an important role for improving the quality of living environment." These greenways have become favorite leisure spots among local people and tourists and have also provided new sources of income for local farmers.

D. Zeng (Ed.): Advances in Control and Communication, LNEE 137, pp. 373–380.
springerlink.com

2 Study Context: The Zengcheng Greenway

Zengcheng Greenway is first-built bicycle recreational greenway in Guangdong Province. As a pioneer of the greenway construction, the successful exploration of Guangzhou Zengcheng Greenway provided an experience for greenway network construction in the Pearl River Delta. It is a typical water village of Lingnan area, within the ring of water around the river. It has all the natural resources, such as riverside, canals, scenic roads, natural reserves to build this green infrastructure. Zengcheng Greenway is 100 km length, known as longest cyclist leisure road in present Guangzhou. It is a scenic, predominately rural area, which provides the visitors a beautiful environment and varied activities with ecological, recreational, cultural and aesthetic features of suburb greenway open space. It has been completed in September 2008. Since that, the greenway has attracted a lot of people around the Pearl River Delta come to visit and to practice. Within the last two years the area has experienced rapid growth, and significant changes in land-use patterns. Therefore, Zengcheng Greenway as a component of urban open space plays an important role in public life, urban construction and social development. And the greenway users' viewpoints are the best way to identify and measure the equipment system and recreational suitability of the Zengcheng Greenway. In this paper, we try to present an approach to greenway analysis and evaluation to the equipment system and the recreational suitability of the Licheng section. From the users' point of view, we will set up an investigation report of Zengcheng Greenway through observations, interviews, questionnaires, etc. This study aims to improve the quality of design and to make the greenway network system planning more scientific.

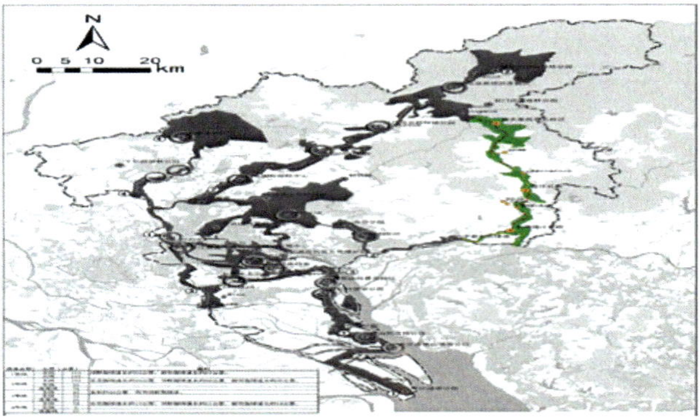

Fig. 1. Situation of Zengcheng Greenway in Guangzhou

3 Method

This research was based on a two-phase process involving equipment system analysis and diagnosis of recreational suitability.

3.1 Analysis of the Equipment System

We organize 20 students in School of Architecture of South China University of Technology to investigate the status of equipment system in Licheng section from 1 April to 30 April 2010. Equipment system analysis includes slow-lane system, green space system, traffic interface system, serve facilities system, signs and maps system, lighting system. These factors were used to identify the quality of the facility within the Zengcheng landscape. They were assessed in terms of safety, comfort, recreation, etc.

3.2 Diagnosis of Recreational Suitability

Specific plan: From1 May 2010 to 20 May 2010, two hundred questionnaires were issued to the user of Zengcheng Greenway (Licheng Section), of which 154 valid questionnaires, the effective rate was 77%. In order to investigate more extensively, the investigation time continued from Monday to the weekend. To avoid a single survey, we intended to investigate the users in the different ages, residences, preferences and activities, hoping the results more true and reliable.

4 Survey of Equipment System

Licheng section of Zengcheng Greenway includes an area of about 700,000 m^2 with a length of 25 km and a width of 2.0-3.6m, providing a diverse range of activities. It's the most important suburban greenway in Zengcheng to ensure the transition from urban to rural areas.

Fig. 2. General plan of Licheng section

4.1 Slow-Lane System

According to the survey, we found that the slow-lane system of Licheng section begins to take shape by combining the existing village roads, plowing roads, embankments and orchards sidewalk. Some belvederes and water platforms are set for tourists in the beautiful landscape sites and open spaces for perspective view. The path of this section is designed for comprehensive slow-lane, that is to say that bicycle and pedestrian paths are mixed to use. During the research, we found that visitors mainly ride a bicycle, but few people walk. Therefore, the safety of pedestrians and cyclists can be assured basically.

4.2 Green Space System

The green space system at the site protects firstly species and ecosystem of the local original forests and vegetations, and secondly combines local native species and ornamental trees to create a good ecological green corridor. For example, on both sides of slow lane, there are *ficus virens*, *ficus microcarpa*, *bauhinia blakeana*, *lychee* trees and other species with Lingnan characteristics as the main trees. So the lanes were fringed with trees and vegetations with a rich and varied landscape, and also have a good visual result and shade effect. Different themes of vegetation, such as the Lotus pool, the Bamboo forest, were planted at distance intervals to attract the visitors.

4.3 Traffic Interface System

The interface between greenways and roads is most important in ensuring a safe and smooth transition. The public parking are set at distance intervals in Zengcheng Greenway, which vehicles parking are located in the edge of regional greenway, away from the ecological sensitive areas; on the other hand, the bicycle parking are located at intervals of 6-10 km.

4.4 Serve Facilities System

In the site, the serve facilities system concludes management center services, consulting point, bicycle rental, public toilets, litter bins, selling points, rest points, telephone booths and other facilities.

• Bicycle rental points: There are totally five bicycle rental points in Licheng Section: Each bicycle rental point is located at the distance intervals about 4-5 km, a vehicle parking beside the point to let drivers to change easily a bike.

• Public toilets: Current public sanitation facilities do not meet the satisfaction in Licheng Section. They are located just in each bicycle rental point, so the numbers of public toilets are not enough in areas.

• Waste bin: the waste bins are not set in some sections, while the waste bins are located at distance intervals in the other sections. But most of the waste bins have not set up a garbage sorting guide.

4.5 Signs and Maps System

The awareness of where one is in relation to one's surroundings is an important component in feeling secure. The signs and maps system is well equipped in most of areas in Licheng section, such as signboards, safety warning signs, road signs have been set up in the greenway. Basically, the directional signs are marked at each intersection, and the junction of each attraction has a large map, marked the location of people at that time, to tell everyone your exact real-time location.

4.6 Lighting Facility System

Where pathways are programmed for night use, lighting should be provided to a level which will allow a use to recognize another person's face at a distance of 25 m. In Licheng section, some part has an even, consistent level of light, but the other part has higher average light levels with greater contrast between bright spots and pools of shadow. The survey show that fewer visitors came at night. So in some forest areas of greenway, it may become the evil black spots.

5 Diagnosis of Recreational Suitability

The diagnosis of recreational suitability in Licheng section includes user's attribution, assessment of the recreational suitability of landscapes, assessment of the visual aesthetic quality of the riverside landscape

5.1 User's Attribution

The greenway user's attribution including gender, age, the way to reach and the visiting hour, is shown in the following Figure 3.

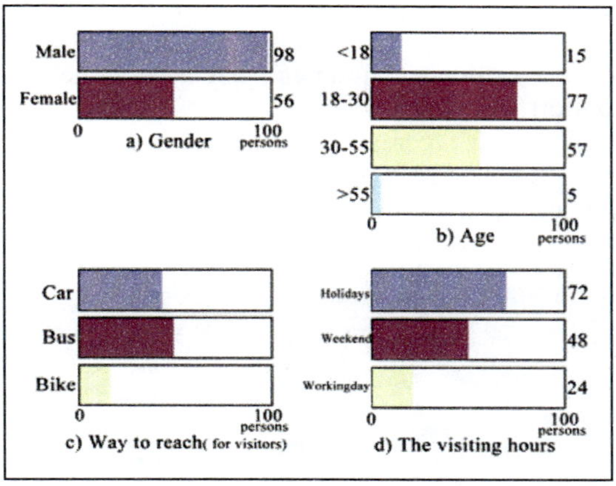

Fig. 3. User's attribution

5.2 User's Behaviors

User's behaviors in the site can be described their flow into two major types:

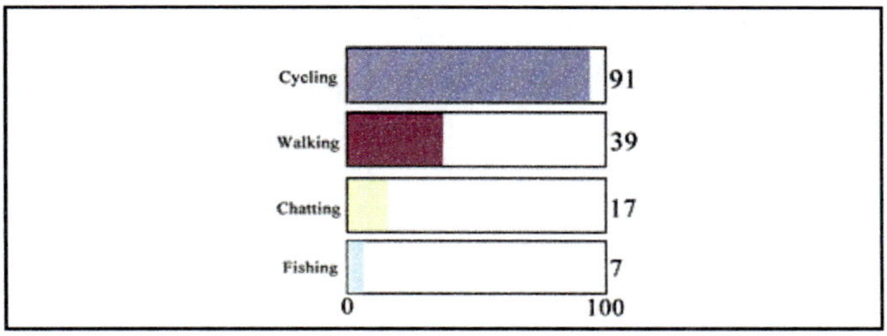

Fig. 4. User's behaviors

1)Mobility, strong activity. Such as cycling, running, walking, walking the dog and so on, with the clear liquidity purpose activities;

2) Stay in state in the process of moving. This type of flow characteristics is relatively static, such as viewing, fishing, resting, chatting and so on.

Stay in state in the process of moving. This type of flow characteristics is relatively static, such as viewing, fishing, resting, chatting and so on.

The data of Figure 3 show that most of the users who come to visit the Zengcheng Greenway are purposed for exercise and fresh air.

5.3 Assessment of the Recreational Suitability of Landscape

On the basis of assessment results, afforested zones with relatively equal recreational value were outlined Figure5. Their classification and areal distribution is as follows:

1) Landscapes with high recreational suitability-180,600 m2 (25.8%). The forests of Licheng section are in good condition, such as the Lychee forest in Liao village and the Bamboo forest in Tansha village.
2) Landscapes with moderate recreational suitability-423,500m2(60.5%). Some spots had a good natural environment, but need to achieve good access and recreation comfort.
3) Landscapes with low recreational suitability-95,900 m2 (13.7%). Some riversides along the greenway need to do the hydrophilic landscape design to create the recreational space for visitors.

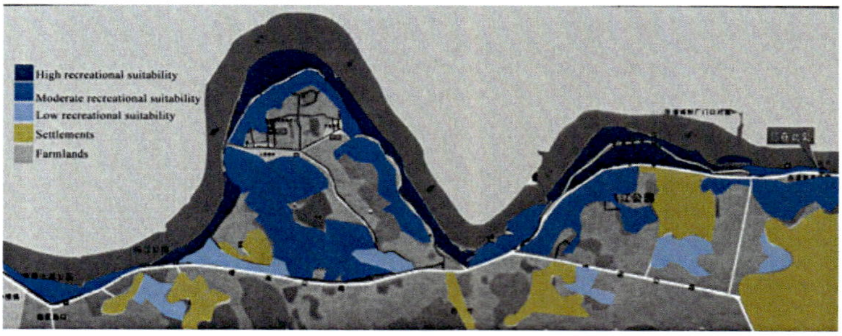

Fig. 5. Recreational suitability of landscape in Licheng section

5.4 Assessment of the Visual Aesthetic Quality of Theriverside Landscapeses

The results of the assessment of the visual aesthetic quality are shown in Figure6. Those landscapes with high aesthetic quality involve woodlands, wetlands, river and sites with slightly developed infrastructure. The forests are with picturesque fringes and rivers flow through scenic valleys. The landscapes with medium aesthetic quality have thick forests (few open spaces), farmlands, vineyards, shelter-belts, and other infrastructure lines. Zones with low aesthetic quality and with strong presence of anthropogenic activity are the existing settlements and their industrial zones.

Fig. 6. Recreational suitability of landscape in Licheng sectio

6 Discussion

On the basis of analysis and assessment on the above, some discussion can be summarized as follows:

6.1 Aspects of Saftey

The purpose of suburban greenway systems to provide for human use and natural ecology illustrates the need for a diversity of spaces within a comprehensive greenway system, balancing the need for safety and the need for nature. In Licheng section, the interface transportation system has not been perfectly improved. Currently, the parking lot is just set at each entrance of greenway, but there is no shuttle bus to transfer the public transportation. We propose that the public shuttle for the greenway should be added to improve public transport accessibility.

6.2 Aspects of Human's Comfort

A system of a well-used, clean serve facility has been identified as an important component of healthy and human's comfort. But in the site, we found that some of service facilities are lack of human's care. Firstly, the number of public toilets is not sufficient; Secondly, the distance between the resting points is far away, and also some resting points have no shade to protect the users. It is proposed to set more pavilions and other structures for users to stay. Currently, the Zengcheng Greenway is in first-phrase of construction, it is newly developed, but usually it needs to be redeveloped and to be improved within the next few years, such as providing bicycle parking in a one-stop facility with mechanics, medical and emergency center, etc. to achieve normal state.

6.3 Aspects of Recreational Suitability

On the basis of analysis and landscape assessment results, natural resources and sites best suited for the leisure activities were identified. The outstanding landscape quality of Licheng section is very much concerned with protecting existing natural resources which determine the existing quality and character. The design of restored natural environments in greenways provides an opportunity to integrate public safety and ecological sustainability. Therefore, the aim of the Zengcheng Greenway the strategy was to link patches and sites along the river of natural and cultural value with a new network of high quality public spaces.

6.4 Aspects of Maintenance

Greenway maintenance is not only related to ultimate realization of project, but also have a direct impact on the use of effects. Currently Zengcheng Greenway is newly complete, and there is no corresponding maintenance system for an effective maintenance of the greenway. Therefore, a scientific greenway maintenance system should be established to promote a healthy and sustainable greenway development. For example, it is proposed to formulate the Greenway Guide, including the opening time, the areas allowed to enter, the caution of the use, etc. that let people to know the safe use of greenway. Otherwise, making a regular schedule for the inspection of service facility, is effective to avoid the potential risk and injure for visitors.

This case study provides us a feedback for the improvement of facilities and recreational suitability of Zengchen Greenway(Licheng section). This study aims to improve the quality of design and to make the greenway network system planning more scientific, and to outline the strategy for its development as a key part of the greenway system for the Pearl River Delta landscape.

Acknowledgment. The research for this paper was funded by the Central University Research Project of S.C.U.T.(No: 20092M0203).

The authors wish to express their gratitude to their students Lu Yun, Tan Tiansheng, Xu Jianxin, Liang Minyan, Zhu Yunxin, etc. at School of Architecture, South China University of Technology in Guangzhou (China), who participate the preliminary investigation in Licheng section of Zengcheng greenway.

References

1. Preiser, W.F.E., Rabinowitz, H.Z., White, E.T.: Post Occupancy Evaluation. Van Nostrand Reinhold Company, New York (1998)
2. Pearl River Delta Greenway Network Master Plan Outline. Construction Supervision and Inspection and Cost Commission, pp. 10–70 (2010)
3. Can, S., Francis, M., Rivlin, L., Stone, A.: Public Space:Environment and Behavior Series. Cambridge University Press, New York (1992)
4. Litton, R.B.: Landscape and aesthetic quality. In: America's Changing Environment. Beacon Press, Boston (1970)
5. Stauskas, V.: Recreation and landscape protection: the planning of outskirt zones at seaside and continental towns in the USSR-report theses. In: Schmid, A.S. (ed.) The Urban Fringe. XXIIth IFLA World Congress, IFLA Yearbook, 1985-1986, Siofok, Lake Balaton (September 1984)
6. Fábos, J.: Introduction and overviewthe greenway movement, use and potentials of greenways. Landscape and Urban Planning 33(1-3), 1–13 (1995)
7. Ahern, J.: Greenways as a planning strategy. Landscape and Urban Planning 33(1-3), 131–155 (1995)

A New Combination Pressure Drop Model Based on the Gas-Liquid Two-Phase Flow Pressure Drop Model

Li Bin[1] and Ma Yu[2]

[1] Institute of Mathematics and Physics
Chongqing University of Science and Technology
Chongqing, China
[2] School of Arts and Humanities
Chongqing University of Science and Technology
Chongqing, China
{libinstudy,swpu-cqzs}@163.com

Abstract. Due to the variability of flow pattern of gas-liquid two-phase flow and complexity of flow mechanism, it is very difficult to seek a single model which is able to predict pressure drop and fit for any flow conditions. Currently, the conventional models of two-phase flow pressure drop which are widely used in engineering practice are mainly based on the experimental data of circular pipe flow, so their applicable conditions are limited, mostly close to oil well producing conditions. As far as producing gas well with formation water is concerned, gas-water ratio is usually much higher than that of producing oil wel; the physical property of water in some aspects is different from that of oil and velocity slip between gas phase and liquid phase is serious. when the existing model of two-phase flow pressure drop is used to predict pressure drop of the conditions of producing gas well,a large error occurs. Therefore, it is necessary, based on the experiment data of gas-water two-phase flow, to research the flow mechanism and discover the regular existing in the process of fluid property changing. On the basis of the current two-phase flow pressure drop model, it is necessary and important to explore modified pressure loss model applicable for producing gas well with water, to improve predictability of the pressure drop of gas wells,and to provide the theory and technology guidance for development of gas reservoir with water.

Keywords: gas-liquid two-phase flow, simulation experiment, pressure drop model, model updating, data processing.

1 Introduction

This research is funded by Research Foundation of Chongqing University of Science & Technology, the project No. is CK2010Z11 and CK2010Z15, and is Supported by Chongqing City Board of Education Natural Science Foundation Project(KJ101402 and KJ101408).

Natural gas is an important economic energy. Sichuan, Xinjiang, Changqing gas field and other big gas field are now in the middle or late stage of development, with the production gradually reduced, water production is rising, causing water flooding in many gas production, the production situation is becoming increasingly difficult.

D. Zeng (Ed.): Advances in Control and Communication, LNEE 137, pp. 381–389.
springerlink.com

Therefore, it is necessary to study the gas-liquid two-phase flow dynamics and promote gas production technology, thereby enhancing the level of development of the gas fields, and advancing economic and social benefits.

Producing water wells gas-liquid two-phase flow dynamics research is to analyze the characteristics of gas well production systems, The core issue is identification of flow pattern of pipe flow, and to explore along the way the pressure loss and its impact factors. The study for a correct analysis of dynamic gas production and liquid gas production process design is of great significance. Flow pattern (distribution interface) greatly affects gas-liquid two-phase flow properties. It will affect the accuracy of flow measurement. Gas-Liquid Flow Pattern Recognition has been the important research direction of two-phase study.

Gas-water two-phase flow patterns and calculation on the pressure gradient have entered the petroleum engineering design and calculation of the field.As the gas-liquid two-phase flow patterns is of variability and complexity of the mechanism of flow, it is difficult to seek to apply to any flow of gas-liquid two-phase flow conditions of the pressure drop calculation. Currently pressure drop calculation method has no viable strict mathematical solution. So far the development of the many different conditions are applicable to the gas-liquid two-phase pressure drop calculation method These approaches are based on the pipe flow experiment data, and the applicable conditions have certain limitations. These are often closer to the oil production wells. For gas- water wells, water-gas ratio is much higher than that of water -oil and physical parameters vary, gas-water two-phase liquid slippage is serious, the existing two-phase pressure drop model for the gas pressure drop has errors in forecasting.

To solve the above problems, it is necessary to study gas water two-phase flow mechanism and parameters change in depth from the gas-water experiments, which is useful to increase the accuracy of production of gas pressure drop distribution and provide important theoretical base of the producing water gas reservoir.

2 Gas-Water Two-Phase Flow Experiments

Experiment has played a pivotal role in research on two-phase flow pressure drop mode. In order to study the gas-water two-phase pipe flow pressure gradient prediction model, simulation tests are carried out in 930m simulation experiment well. The 930m simulation well is one of the "full-scale" wells, with advanced computer monitoring and control systems, that can achieve the experimental data in a short period of time and significantly shorten the research cycle, effectively avoid the new technology blindly putting into use at the scene of the producing wells. Experimental system which consists of wells and underground system parts, mainly includes an experimental 930m wells, a 200m gas wells, the power equipment, ground monitoring and control system technology and automation components[1,2].

Experimental study of gas-water two-phase flow mainly aims at different lifting heights, gas-liquid flows, wellhead back pressures, air-water two-phase flow pressures and parameters such as pressure gradient.Experiments use multi-factor orthogonal experiment design method, design the scope of gas-water two-phase flow experiment involved in producing gas wells currently produce a variety of possible flow. By regulating the gas and water flows, two-phase flow pattern can be observed: bubble flow, slug flow and mist flow [3,4,5]. Based on the actual production, the experimental study of gas-water two-phase flow is divided into three sections: 10~60 m^3 / d; 60 ~ 120m^3 / d; 120m^3 / d or more.

3 Flow Pattern Identification Criteria

Gas-water two-phase pipe flow pressure drop models generally have two categories: one is to neglect the flow pattern; the other is to consider the two-phase flow pattern. In this paper, experiments are carried out to identify flow patterns.

3.1 Direct Measurements to Identify Flow Patterns

(1) Characteristics of differential pressure fluctuations

According to the experimental conditions, pressure fluctuations have relation with flow pattern. The differential pressure fluctuations curves of the bubbly flow, slug flow and mist flow can be seen from figure 1, figures 2 and figures 3.

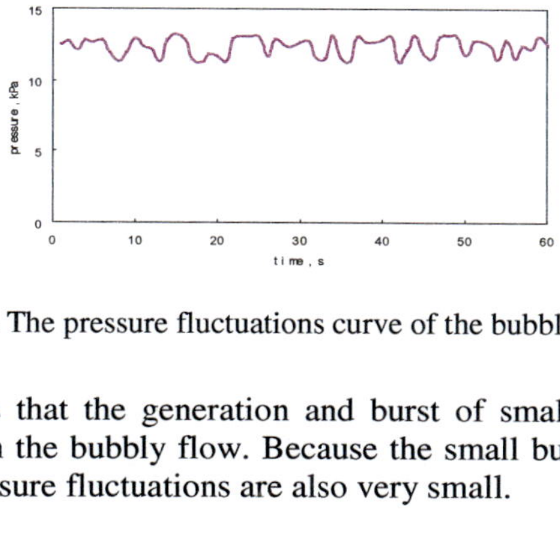

Fig. 1. The pressure fluctuations curve of the bubble flow

The figure 1 shows that the generation and burst of small air bubbles generate pressure fluctuations in the bubbly flow. Because the small bubble is very small, the amplitude of these pressure fluctuations are also very small.

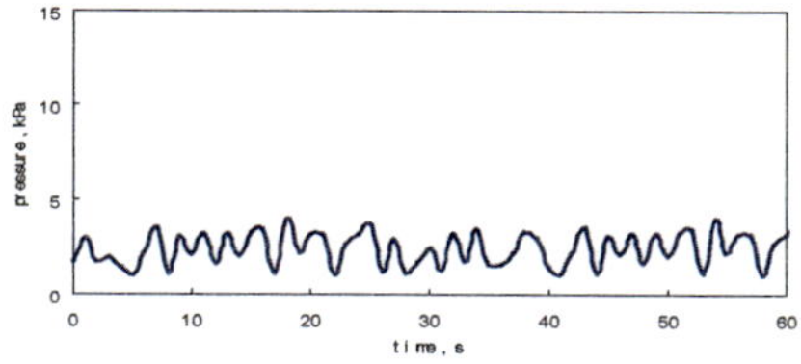

Fig. 2. The pressure fluctuations curve of the mist flow

In figure 2, it can be seen in the fog flow that the liquid film moves upward along the tube wall under the dragging of the gas core, the smaller the amplitude differential pressure, differential pressure fluctuations and more gentle change.

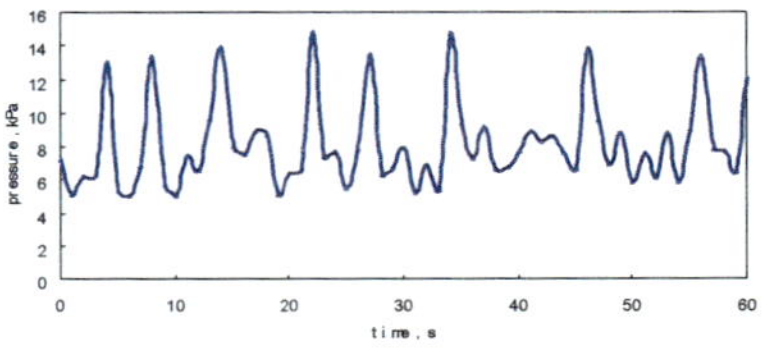

Fig. 3. The pressure fluctuations curve of the slug flow

Figure 3 shows that there are many liquid bombs with small air bubbles between the two gas shells in slug flow conditions. The pressure flow measurement will have a significant segment of the differential pressure fluctuations, so the differential slug flow fluctuations are larger amplitude.

From figure 1, figure 2 and figure 3, we can see that the pressure fluctuation characteristics in bubbly flow and mist flow are different from that in slug flow. Some of the pressure fluctuations of bubbly flow, mist flow are so similar that can not be directly used to identify the flow pattern, still looking for other characteristics.

(2) characteristics of liquid holdup

We can see from figure 4 that the range of holding liquid bubbly flow rate is about 0.70 ~ 0.90 and the bubbly flow rate of the average liquid holdup is about 0.82. Because the flow of gas is very small and uniform, the rate of fluctuations in liquid holdup is small.

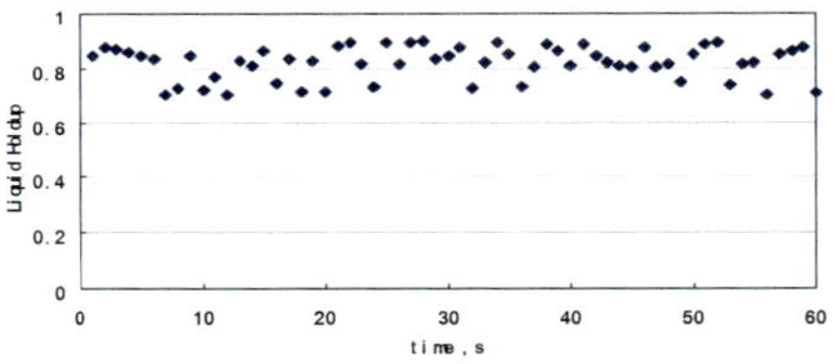

Fig. 4. Liquid Holdup of the bubbly flow

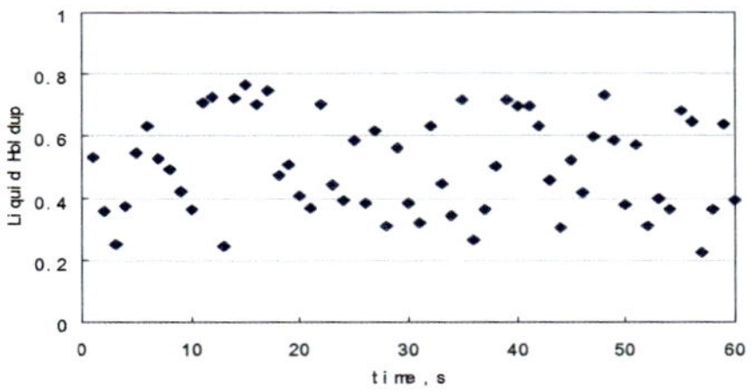

Fig. 5. Liquid Holdup of the slug flow

As what figure 5 shows, hold liquid slug flow rates range from 0.20 to 0.80, a stable slug flow when the average is about 0.5, both the border, one at about 0.2, and the other at about 0.8. Since a period of slug flow with liquid gas for some structure, so the rate of liquid holdup changes a lot.

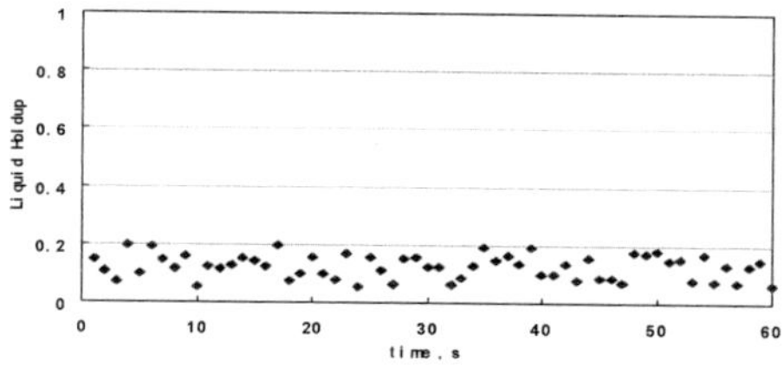

Fig. 6. Liquid Holdup of the slug flow

As Figure 6 shows, fog liquid flow rate fluctuations had a range of 0 ~ 0.20 between the average liquid holdup at the rate of about 0.16.

From Figure 4, Figure 5 and Figure 6 we can see that the bubbly flow, slug flow fluctuations in the rate of the scope of liquid holdup and liquid holdup fog flow fluctuations in the rate of coverage is different. Because there is an overlapping part between the ranges of liquid holdup rates of bubbly flow, and slug flow, which makes they can not be directly used to identify the flow pattern, and we are still needed to look for other characteristics.

3.2 Wavelet Singularity Analysis

Figure 1, Figure 2 and Figure 3 separately show the differential pressure fluctuating curves which genenerate under the conditions of bubbly flow, slug flow and mist flow. But they are not sufficient to quantify the signal using singular value theory of convection-based quantitative analysis of signals.

From the differential pressure fluctuations curves, we can see that different flow patterns have different pressure fluctuation signal, and pressure fluctuation signals are affected by water flow, liquid flow rate, density, surface tension and other parameters. Index α is the quantitative description parameters of physical characteristics of differential pressure signal fluctuations. A large number of experiments and simulations show that the index α of different flow pattern can mold into a dimensionless quantity that uses liquid velocity, gas volume rate, liquid viscosity, etc.

1) Slug flow

Slug flow α-index can be used by the formula (1)

$$\alpha_s = 2.718 N_{lv}^{\ 0.125} \frac{Q_g}{Q} \tag{1}$$

Where as:

α_s --the index of slug flow; Nlv-- the velocity of liquid-phase; Qg—the volume of gas flow, m^3 / s; Q--the volume of gas-water mixture flow, m^3 / s.

For the pressure Δp signal of slug flow, then there is

$$\left|\Delta p(x) - \Delta p(x_0)\right| \le k(t - t_0)^{2.718 N_{lv}^{\ 0.125} \frac{Q_g}{Q}} \tag{2}$$

Find the corresponding wavelet transform to determine the peak a1, a2, a3 in the following equation:

$$\begin{cases} a_1 = 2k(4+\sigma^2)^{\left(2.718 N_{lv}^{0.125}\frac{Q_g}{Q}-1\right)} \\ a_2 = 4k(16+\sigma^2)^{\left(2.718 N_{lv}^{0.125}\frac{Q_g}{Q}-1\right)} \\ a_3 = 8k(64+\sigma^2)^{\left(2.718 N_{lv}^{0.125}\frac{Q_g}{Q}-1\right)} \end{cases} \tag{3}$$

2) Mist flow

$$\alpha_m = 1.782 \times 10^{\left(-0.599-2.694 N_l + 0.521 N_{lv}^{0.0525}\right)} \frac{Q_g}{Q} \tag{4}$$

3) Bubble flow

$$\alpha_b = 2.718 N_{lv}^{0.125} \frac{Q_g}{Q} \tag{5}$$

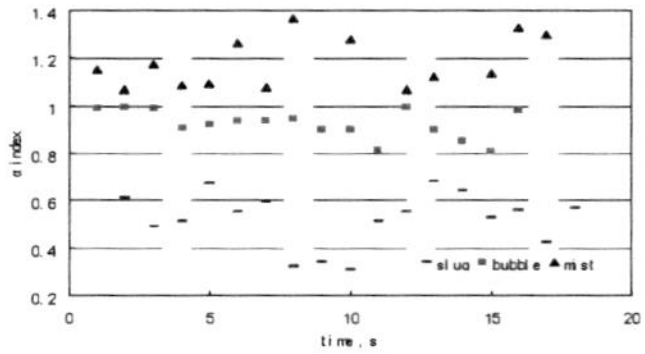

Fig. 7. α Exponential Distribution

From the above chart we can see that slug flow index is between 0 to 0.70, the bubble flow index between 0.70 to 1.0, the fog flow index between 1.0 to 1.38.

3.3 The Judging Criteria of Flow Pattern

By analyzing the relationship of flow pattern between pressure drop, using the wavelet method to identify flow pattern, decomposing the wavelet signals of pressure fluctuation and calculating α index we can draw a conclusion that the α index of different flow patterns are quite different. The α index can identify the slug flow, bubble flow and mist flow and can be used as a new quantitative index to identify flow patterns.

The α-index of slug flow and bubble flow can be expressed by the formula (6)

$$\alpha_s = \alpha_b = 2.718 N_{lv}^{0.125} \frac{Q_g}{Q} \tag{6}$$

Where as, $0 < \alpha_s < 0.70$, $0.7 \le \alpha_b \le 1$;

The α-index mist flow can be expressed by the formula (7)

$$\alpha_m = 1.782 \times 10^{\left(-0.599-2.694 N_l + 0.521 N_{lv}^{0.0525}\right)} \frac{Q_g}{Q} \tag{7}$$

Where as, $\alpha_m \ge 1$.

Through many experiments and theoretical calculations, the initial flow-based criteria have been shown As follows: Bubble Flow($0.7 < \alpha_b \leq 1$, $\alpha_m < 1$),Slug flow ($\alpha_m < 1$, $0 \leq \alpha_s \leq 0.7$),Mist flow ($\alpha_m \geq 1$).

4 A New Combination Pressure Drop Model Based on the Two-Phase Flow Pressure Drop Model

4.1 Optimization of Two-Phase Flow Pressure Drop Model

Based on the four types of two-phase flow pressure drop models commonly used in engineering (Ros & Duns, Orkiszewski, Beggs & Brill and Hasan & Kaber) and statistical analysis of evaluation methods used in different flow patterns, we can apply many experimental testing data which is from experimental wells to optimize two phase flow pressure drop model.

The pipe flow pressure error can be separately calculated according to the conditions of the underground pipe 750m ~ 800m and 350m ~ 400m of experimental Wells. Mainly based on E2 which is the average of flow pressure percentile absolute error, and other error parameters, compared four types of pressure drop calculation model above, the pressure drop model with the smallest average percentile flow pressure absolute error E2 is selected as the optimal model.

(1) Optimization of bubble flow pressure drop model

According to the flow pattern identification criteria, the identification of flow patterns on experimental wells can be obtained by calculating testing experimental data. In the bubble flow, the two-phase flow pressure drop model is used to calculate pressure of downhole measuring points, count statistical error of the flowing pressure. The average error, average absolute error and Standard Error of the two-phase flow pressure drop model. The result shows that the average absolute flow pressure error of Hasan&Kaber model is 5.82%, the best model in all compared model, followed by Beggs & Brill model, Duns&Ros and Orkiszewski model. Therefore, Hasan&Kaber model can be selected as pressure drop model of mist flow.

(2) Optimization of slug flow pressure drop model

According to the flow pattern identification criteria in table 2, the identification of flow patterns on experimental wells can be obtained by calculating testing experimental data. In the slug flow, the two-phase flow pressure drop model is used to calculate pressure of downhole measuring points, count statistical error of the flowing pressure. The result shows that the average absolute flow pressure error of Duns&Ros model is 5.60%, the best model in all compared model, followed by Hasan & Kaber, Beggs & Brill model and Orkiszewski model. Therefore, Duns&Ros model can be selected as pressure drop model of mist flow.

(3) Optimization of mist flow pressure drop model

According to the flow pattern identification criteria in table 2, the identification of flow patterns on experimental wells can be obtained by calculating testing experimental data. In the fog flow, the two-phase flow pressure drop model is used to calculate pressure of downhole measuring points, count statistical error of the flowing pressure. The result shows that the average absolute flow pressure error of

Ros & Duns model is 7.73%, the best model in all compared model, followed by Orkiszewski model, Hasan & Kaber and Beggs & Brill model. Therefore, Ros & Duns model can be selected as pressure drop model of mist flow.

4.2 New Combination Pressure Drop Model

On the basis of the analysis of a number of gas-liquid two-phase flow model by the experimental data of experimental wells in this paper, the identification of flow patterns, the selection of pressure drop calculation method in the different flow patterns, a new pressure drop model finally can be obtained by the comprehensive study (Table 1).

Table 1. The new pressure drop model

Flow Pattern	Selection method
Bubble Flow	Hasan&Kaber
Slug flow	Duns&Ros
Mist flow	Duns&Ros

5 Conclusions

(1) The method using singular exponential α based on wavelet transform to identify flow pattern is developed, and the ranges of singular exponential α of different flow pattern are also preliminarily proposed according to experiment tested and observed result. α can be quantitative parameter which are gas-water flow, liquid flow rate, density and surface tension experimental parameters. The experimental test data determine the α index value of the slug flow, which is between 0-0.70, α index value of the bubble flow between 0.70-1.0, and mist flow between 1.0-1.38.

(2)Characteristics of differential pressure fluctuations and liquid holdup can be used as a partial recognition of two-phase flow patterns. Bubbly flow and mist flow have differential characteristics compared with slug flow. The average value of liquid holdup of bubbly flow is about 0.82. Slug flow has greater change and the average is about 0.5, one boundary at 0.2, and the other at 0.8. The range of the mist flow is 0 ~ 0.20.

(3)Based on the proposed principles of singular exponential α, the flow pattern of flowing fluid in simulation experiment producing well are identified, and then comprehensive evaluations on Ros&Duns, Orkiszewski, Beggs&Brill and Hasan&Kaber correlations widely used in fields are given, a new combined correlation is formulated .The new gas-water two-phase flow pressure drop combined correlation is comprised of Hasan&Kaber model limited to calculate bubble flow condition and Duns&Ros model limited to calculate slug flow and mist flow conditions respectively.

References

1. Liao, K., Li, Y., Liu, Y., et al.: Monitoring system study on gas production process simulating experiment well. Drilling & Production Technology 30(3), 46–48 (2007)
2. Liao, K., Li, Y., Liu, Y., et al.: Development of test device for new gas lift. Oil Field Equipment 36(7), 68–70 (2007)
3. Liao, K., Li, Y., Liu, Y., et al.: Quantitative study on the capability of liquid carrying flow rate for gas wells. Oil Drilling & Production Technology 29(4), 97–99 (2007)
4. Li, Y., Du, Z.: An optimal model for two phase flow in tubing and annulus of gas wells. Acta Petrolei Sinica 20(2), 87–92 (1999)
5. Li, Y., Du, Z.: A new numerical model for two-phase flow in liquid-cut gas wells. SPE 35612, 349–360 (1996)
6. Li, Y.: Production Engineering, pp. 31–49. Petroleum Industry Press (2005)
7. Borthne, Gunnar, Johannessen: Modeling and Implementation of a Database for Multiphase Flow Experiments. In: SPE 35996-MS, pp. 117–129 (1996)
8. Hagedorn, A.R., Brown, K.E.: Experimental study of pressure gradients occurring during continuous two phase flow in small-diameter vertical conduits. J. Pet. Tech. (4), 475–484 (1965)
9. Chokshi, R.N.: Experimental Study and the Development of a Mechanistic Model for Two-Phase Flow Through Vertical Tubing. SPE 35676-MS, 349–360 (1996)
10. Guo, B., Ghalambor, A.: A Systematic Approach to Predicting Liquid Loading in Gas Wells. SPE 94081-PA, 81–88 (2006)

Study on Safety Assessment of Gas Station Based on Fuzzy Theory

Liu Jing-cheng[1,2], Wang Hong-tu[1], Li Wen-hua[2], Zeng Shun-peng[2], and Ma Yu[2]

[1] Chongqing University
Key Laboratory for Exploitation of Southwestern Resources & Environmental Disaster Control Engineering, Ministry of Education
Chongqing, China
[2] Chongqing University of Science and Technology
Institute of Oil & Gas Engineering
Chongqing, China
Liujingcheng1980@126.com

Abstract. Gas Station safety is non-linear systems that refer to many factors and multilayer. There are much flammable and explosive compressed natural gas in gas station, and it is considered as a kind of major hazard source. Therefore, the safety assessment of gas station has their special important and realistic significance. Based on the types of accident and risk factors, a new index system of safety assessment was set up. The weight of the evaluation index would be analyzed by analytic hierarchy process, and then the matrix of evaluating results could be calculated through fuzzy comprehensive evaluation. At last, the grade of safety for gas station could be obtained. This Safety Assessment Model can be applied for every gas station.

Keywords: gas station, analytic hierarchy process, fuzzy comprehensive evaluation, safety assessment.

1 Introduction

With the rapidly increasing of oil price, as well as the growing number of car, conflicts among environmental protection, energy and economy become very serious[1]. Having outstanding price quality, natural gas which is a clean energy now catch public's attention. There are much flammable and explosive compressed natural gas in gas station, and it is considered as a kind of major hazard source. So the safety assessment of gas station is necessary. Presently analytic hierarchy process and fuzzy comprehensive evaluation are the most commonly used index system of safety assessment. This paper selects a gas station as an instance, combined with the advantage of these two approaches, to perform safety assessment using multi-level fuzzy comprehensive evaluation with the help of AHP which gives sub goals and weight of the evaluation index.

D. Zeng (Ed.): Advances in Control and Communication, LNEE 137, pp. 391–398.
springerlink.com

2 Established the Index System of Safety Assessment of Gas Station

The establishment of the index system of safety assessment is a process of dialectical thinking which can be expressed as "concrete-abstract-concrete" after comprehensive consideration of the properties of gas station[2]-[4]. The following principles should be followed when the index system of safety assessment of gas station is established.

2.1 Systemic Principle

Gas station as a system, its hazard exists in all aspects of the system. Without going into detail, without studing the relevant and constraint relation between system and subsystems, all risks of the system can't be maximal recognized. The index system of safety assessment of gas station should be established form Person, Facility, Safety management and Environment, which agree with safe ergonomics principles.

2.2 Concise and Scientific Principle

The index system should be simple, in order to collect information and analysis data, and the truth of safety assessment of gas station will be reflected. The scale of Index system should be appropriate. If the index System is too small and too few levels, the actual safety will not be reflected. If the index System is too large, too many levels and the indexs is too small, which will spoil the ship for half penny worth of tar.

2.3 Make Good Use of Index Information Principle

Some indexs are quantitative indexs, which is stated as some numerical value in order to handle conveniently,the people are accustomed to quantitative indexs. however ,many indexs of the index system of safety assessment of gas station are qualitative, such as Safety education, Safety organization, Safety regulations and Contingency plan etc. So we must establish an appropriate quantitative methods to convert this information into value, thus we can make good use of index information.

2.4 Steady and Comparability Principle

For the convenience of comparison, All indexs should do the quantized treatment. The index system of safety assessment of gas station is fit for different gas station, and evaluation standard is unanimous, in order to safety assessment of gas station is authenticity, objectivity, rationality and unanimous.

2.5 Pointed Principle

Usually, the index system of safety assessment are different for different systems. Different of their risk evaluation index system. There are many differences in details, although they may be much similar, and some sub-index system may be the same. Therefore, the index system of safety assessment of gas station should meet pointed principle.

2.6 Flexible and Operational Principle

The index system of safety assessment of gas station should have a clear hierarchy, each sub-index system is relatively independent, So that users can establish corresponding sub-index system according to their different purposes.

As a consequence of the above, The structure of the established index system of safety assessment is listed in the Table 1.

Table 1. The index system of safety assessment of gas station

Target layer	**Gas station safety condition**			
Rule layer	**Person (B_1)**	**Facility (B_2)**	**Safety management (B_3)**	**Environment (B_4)**
Index layer	Political thought (C_1)	Explosion-proof equipment (C_5)	Safety education (C_{11})	Working environment (C_{15})
	Safety consciou-sness (C_2)	Anti-static equipment (C_6)	Safety organization (C_{12})	Workplace (C_{16})
	Physical condition (C_3)	Thunder electric equipment (C_7)	Safety regulations (C_{13})	surrounding environment (C_{17})
	Homework skill levels (C_4)	Fire protection system (C_8)	Contingency plan (C_{14})	
		Aerated operating system (C_9)		
		gas storage system (C_1)		

3 The Establishment of the Weight of Evaluation Index

3.1 Constructing Paired Comparison Matrix

Based on the hierarchical structure established in the index system of safety assessment elements are compared to each other when A stands for the target which domain $b_1,b_2,...,b_n$ constituting lower level and a_{ij} stands for the ratio scale measuring the importance of b_i or b_j relative to the target. Using multivariate scaling from 1 to 9 the corresponding judgment matrix will be attained.

Accordingly, the judgment matrix of each level of evaluation index is listed in Table 2-6.

Table 2. The judgment matrix of first level and the weight calculation

Factors	**B_1**	**B_2**	**B_3**	**B_4**	**weight**
B_1	1	3	3	5	0.51
B_2	1/3	1	2	4	0.25
B_3	1/3	1/2	1	3	0.17
B_4	1/5	1/4	1/3	1	0.07

Table 3. The judgment matrix of B_1-C and the weight calculation

Factors	C_1	C_2	C_3	C_4	weight
C_1	1	1/3	1/2	1/4	0.09
C_2	3	1	2	1/3	0.23
C_3	2	1/2	1	1/5	0.13
C_4	4	3	5	1	0.55

Table 4. The judgment matrix of B2-C and the weight calculation

Factors	C_5	C_6	C_7	C_8	C_9	C_{10}	Weight
C_5	1	1/3	1/3	1	1	1/2	0.10
C_6	3	1	1/2	2	1	2	0.20
C_7	3	2	1	2	2	2	0.29
C_8	1	1/2	1/2	1	1/2	1/2	0.11
C_9	1	1	1/2	2	1	1	0.15
C_{10}	2	1/2	1/2	2	1	1	0.15

Table 5. The judgment matrix of B_3-C and the weight calculation

Factors	C_{11}	C_{12}	C_{13}	C_{14}	weight
C_{11}	1	2	3	2	0.40
C_{12}	1/2	1	5	1/2	0.23
C_{13}	1/3	1/5	1	1/3	0.09
C_{14}	1/2	2	3	1	0.28

Table 6. The judgment matrix of B4-C and the weight calculation

Factors	C_{15}	C_{16}	C_{17}	Weight
C_{15}	1	2	3	0.54
C_{16}	1/2	1	2	0.30
C_{17}	1/3	1/2	1	0.16

3.2 Consistency Test

We calculate the consistency index (CI= $\lambda_{max}-1/n-1$), and find the corresponding average random consistency index(RI), then calculate consistency ratio($CR = CI/RI$). When CR<0.10, the consistency test is acceptable, otherwise it will be revised.

4 Fuzzy Comprehensive Evaluation of Gas Station

4.1 Research on the Fuzzy Comprehensive Evaluation of Gas Station

Fuzzy comprehensive evaluation is based on the principle of composition of fuzzy relations where the degree of membership of evaluated things are evaluated by measuring multi-indices. The basic approach and steps are as follows:

(1) Determining factor sets U

Factor sets U is a collection of various factors,which impact assessment object, can be expressed as $U=\{u_1,u_2,.....u_n\}$.

(2) Determining evaluation sets V

Evaluation sets V is a collection of may be made the evaluation results by the evaluation object, can be expressed as $V=\{v_1,v_2,.....v_m\}$.

(3) Single factor fuzzy evaluation

The single factor fuzzy evaluation is starting from a single evaluation factors to determine the membership level of evaluation object for evaluation sets element.

Evaluation object carry out evaluating from the i element u_i in factor sets U,the membership level of the j element v_j in evaluation sets V is r_{ij} ,so the evaluation results of the i element u_i in factor sets U can be expressed as $R=\{r_{i1},r_{i2},\dots,r_{im}\}$.

R is Single factor fuzzy evaluation sets, which is fuzzy subset of evaluation sets V.

(4) Fuzzy comprehensive evaluation

Factor sets, evaluation sets and Single factor fuzzy evaluation sets can be set to fuzzy comprehensive evaluation model.

$$B=A\cdot R=(a_1,a_2,..a_n)\begin{pmatrix} r_{11} & r_{12} & \cdots & r_{1m} \\ r_{21} & r_{22} & \cdots & r_{2m} \\ \cdots & \cdots & \cdots & \cdots \\ r_{n1} & r_{n2} & \cdots & r_{nm} \end{pmatrix}=(b_1,b_2,.b_n)$$

b_i is the index of fuzzy comprehensive evaluation.

(5) Evaluation index treatment

Maximum membership degree method can be used to determine the final evaluation results,when obtain evaluation index B.

4.2 Application of the Fuzzy Comprehensive Evaluation of Gas Station

In this paper we select a gas station in Chongqing as an example to verify the fuzzy comprehensive evaluation. This gas station was put into production in December 2005 supplying 1400 m^3/d of gas. Presently maximum sales volume is 12000m^3 and there is a big traffic flow for this gas station.

Based on the weight of each element the following will be obtained:

Should weight : A_0=(0.51 0.25 0.17 0.07)
Person : A_1=(0.09 0.23 0.13 0.55)
Facility : A_2=(0.10 0.20 0.29 0.11 0.15 0.15)
Management : A_3=(0.40 0.23 0.09 0.28)
Environment : A_4=(0.54 0.30 0.16)

Secondly we perform primary assessment and the assessment matrix is listed as follows after expert scoring:

Personnel quality :

$$R_1 = \begin{pmatrix} 0.3 & 0.4 & 0.2 & 0.1 & 0 \\ 0.1 & 0.4 & 0.4 & 0.1 & 0 \\ 0.1 & 0.3 & 0.2 & 0.3 & 0.1 \\ 0.1 & 0.2 & 0.4 & 0.3 & 0 \end{pmatrix}$$

Equipment and facilities:

$$R_2 = \begin{pmatrix} 0.1 & 0.3 & 0.3 & 0.2 & 0.1 \\ 0 & 0.3 & 0.5 & 0.2 & 0 \\ 0.2 & 0.3 & 0.4 & 0.1 & 0 \\ 0.2 & 0.5 & 0.2 & 0.1 & 0 \\ 0.3 & 0.2 & 0.3 & 0.2 & 0 \\ 0.1 & 0.3 & 0.5 & 0.1 & 0 \end{pmatrix}$$

Management factors:

$$R_3 = \begin{pmatrix} 0.4 & 0.3 & 0.2 & 0.1 & 0 \\ 0.2 & 0.5 & 0.2 & 0.1 & 0 \\ 0.1 & 0.5 & 0.3 & 0.1 & 0 \\ 0.5 & 0.3 & 0.2 & 0 & 0 \end{pmatrix}$$

Environment factors:

$$R_4 = \begin{pmatrix} 0.5 & 0.3 & 0.2 & 0 & 0 \\ 0.1 & 0.2 & 0.4 & 0.3 & 0 \\ 0.3 & 0.2 & 0.4 & 0.1 & 0 \end{pmatrix}$$

Then we would get the following assessment results:

$$B_1 = A_1 \times R_1 = (0.118 \quad 0.277 \quad 0.366 \quad 0.226 \quad 0.013)$$

$$B_2 = A_2 \times R_2 = (0.15 \quad 0.307 \quad 0.388 \quad 0.145 \quad 0.01)$$

$$B_3 = A_3 \times R_3 = (0.355 \quad 0.364 \quad 0.209 \quad 0.072 \quad 0)$$

$$B_4 = A_4 \times R_4 = (0.348 \quad 0.254 \quad 0.292 \quad 0.106 \quad 0)$$

We can get the following assessment matrix of two-level assessment based on the assessment performed above:

$$R = \begin{pmatrix} 0.118 & 0.277 & 0.366 & 0.226 & 0.013 \\ 0.15 & 0.307 & 0.388 & 0.145 & 0.01 \\ 0.355 & 0.364 & 0.209 & 0.072 & 0 \\ 0.348 & 0.254 & 0.292 & 0.106 & 0 \end{pmatrix}$$

The membership of two level assessment:

$$V = A_0 \times R = (0.182 \quad 0.217 \quad 0.34 \quad 0.251 \quad 0.01)$$

Based on the principle of maximum membership, the safety level of gas station ranks V_3 namely moderate safety. The deciders will take corresponding measures to make corrections to control the concurrency of accident effectively on the basis of the influence of each element on system safety.

5 Conclusions

(1) Compared to other methods, fuzzy mathematics method can reflect the safe condition of gas station more actually and objectively when we consider multi-factor impact and uncertainty of safety assessment of gas station

(2) The weight will be more exact and reasonable when We get weight using "Analytic hierarchy process" and perform consistency check on the judgment matrix.

(3) Adopt two level fuzzy evaluation, and the level of security of various elements will be determined

Acknowledgment. This work is a result of partial support under the Natural Science Foundation of China(Grant NO.504106),the National Key Fundamental Research and Development Program of China(Grant NO.2005CB221502),the Natural Science Innovation Group Foundation of China(Grant NO. 50921063),the technical plan project of land housing administration(2009-01),Scientific and technical project of

Sichuan coal mine group(2009-08),Natural Science Foundation of Chongqing educational committee(KJ101402 and KJ101408), and Research Foundation of Chongqing University of Science & Technology(CK2010Z11 and CK2010Z15).

References

1. Hou, X.-B., Zhang, P.: Based on fuzzy synthesis of LPG station risk assessment. Natural Gas and Oil 5(27), 14–17 (2009)
2. Zhu, X.-Z., Yang, H.-G.: Study on safety assessment of gas station based on fuzzy theory. Journal of Safety Science and Technology 1(16), 51–53 (2005)
3. Song, G.-X., Yang, D.-L.: Fuzzy judgment matrix consistency inspection and consistency improving methods. System Engineering 21(1), 110–116 (2003)
4. Chu, T.-X., Yang, S.-Q.: Safety assessment of gas station based on SCL-FEC model and its application. Journal of Safety Science and Technology 4(6), 147–151 (2008)
5. Lin, Z., Xin, S.: LPG station security fuzzy comprehensive evaluation model. Natural Gas Technology 1(6), 84–87 (2007)

The Automatic Control System of Large Cotton Picker

Yang Xu-dong, Sun Dong, and Jin Liang-liang

Guizhou University, Guiyang, China, 550000
yxd_gzu@163.com

Abstract. The intelligent control system of 4MZ-5 intelligent self-propelled cotton picker is developed, which is applied with fieldbus technology the first time in China to complete digital transmission of system data. The data acquisition and control of every monitoring control points of the cotton picker is achieved and the CAN network transmission using embedded devices is realized. The system supports the standard of international farm machinery bus protocol, provides general cotton picker control platform, and has equipment diagnostics, maintenance technology.

Keywords: Cotton picker, Vehicle-mounted computer system, Performance monitoring, Automaticly align, Production measurement.

1 Prefaces

China is the world's largest cotton producer in which cotton production has been the first in the world, while cotton picking, which is the bottleneck restricting the development of cotton industry, still rests on using imported equipments or manual picking extensively. Manual picking has some problems like lower efficiency while the imported cotton picker has the problems of high prices, high costs of parts, and long period of after-sales service etc. So, the development of large intelligent self-propelled cotton picker suitable for China is of great urgency.

Around the world, American "Deer" and "Case" of the cotton picker represents the international advanced level, which combines advanced hydraulic technology, control technology, modern processing techniques. In domestic, through years of development and digestion, absorption, innovation of foreign technology, Pingshui company of Guihang Group has developed large self-propelled cotton picker (4MZ-5) which is suitable for cropping patterns in China. At present, the main technical indicators are close to the international mainstream models.

However, the electrical control system for cotton picker (4MZ-5) is backward with higher failure rate. A new generation of automatic, intelligent monitoring and control platform for upgrading electrical system of Domestic cotton picker is needed.

2 UMMARY of Intelligent Control System of Large Intelligent Self-propelled Cotton Picker

The intelligent control system of this machine is applied with fieldbus technology the first time in China to complete digital transmission of system data, complete data

D. Zeng (Ed.): Advances in Control and Communication, LNEE 137, pp. 399–405.
springerlink.com

acquisition and control of every monitoring and control points of the cotton picker and achieve CAN network transmission using embedded devices. The system supports international farm machinery standard bus protocol, provide general cotton picker control platform, and has equipment diagnostics, maintenance technology.

Fig. 1. 4MZ-5 cotton picker

2.1 Key Technologies

Cotton picker key technologies of automatic control system are as follows:

a) structure for body monitoring and diagnosis based on CAN bus.
b) Building system's hardware / software platform based on embedded technology.
c) The systems application architecture based on international standards of open systems.
d) Achieve vehicle system control, optimization and display.
e) Digital information transmission in whole vehicle.

2.2 System's Sub-module

a) Vehicle-mounted computer system with following realization

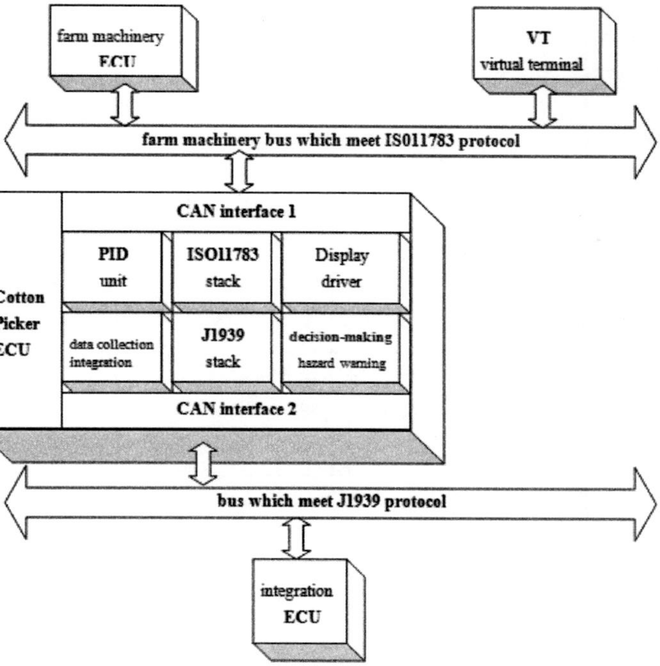

Fig. 2. Vehicle-mounted computer system

b) Performance monitoring, display and data acquisition and processing system which is consist of cotton picker vehicle instrument assembly and body CAN bus modules. Achievement is in the form below:

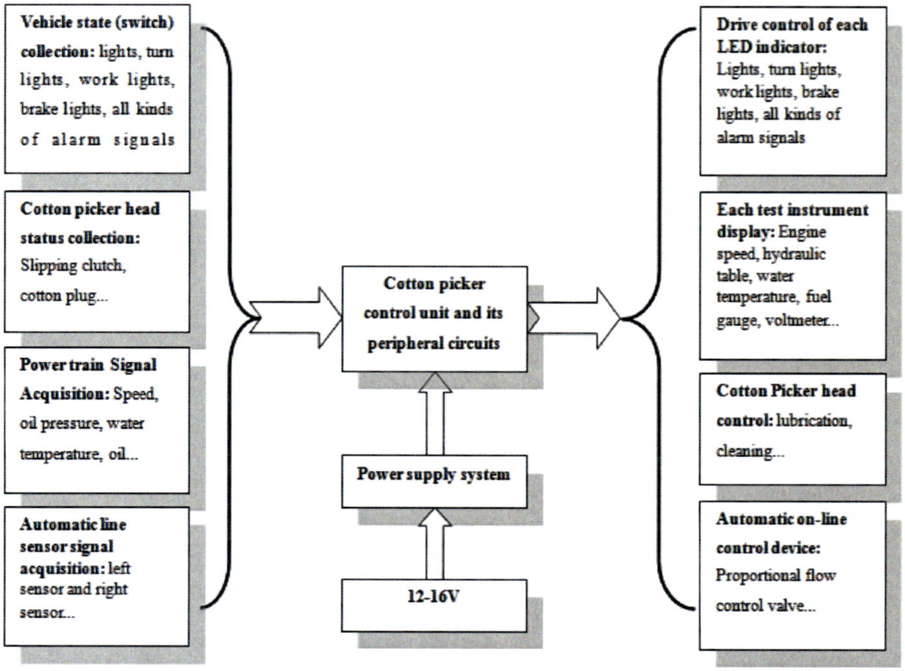

Fig. 3. Vehicle instrument assembly and body CAN bus modules

3 Automaticly Align Technology

When cotton picker working in the field, the more precisely the heads align cotton ridge, the better the picking effect is. Cotton growing for the single-line spacing 76,90 cm plant model is used abroad while in Xinjiang, Spaced 66 +10 or 68 +8 cm single dual-line mode is used, in which the density is about 17000 per acre. The density is so high, that it is difficult to harvest. The harvest mainly depends on the driver's skills, and it increases the driver's labor intensity. Make the cotton picker head in a more accurate alignment of cotton and achieve cruise operations when working through developing automaticly align system. Use hand-operated control when transporting and turning by switching operations and transport model.

The automaticly align system is shown below:

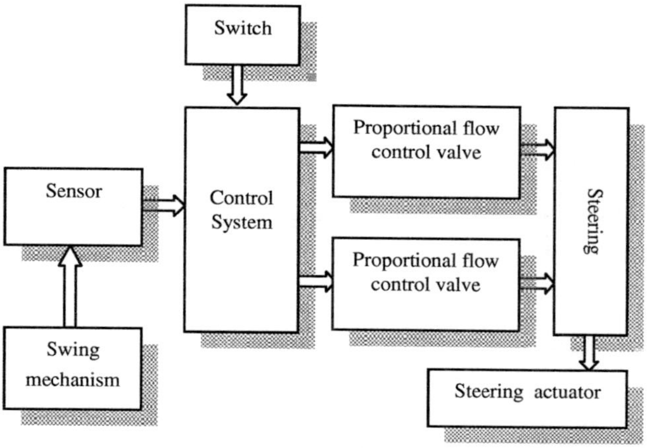

Fig. 4. The automaticly align system

There is a swing mechanism installed at the entrance of cotton stalks; when working in the field, switch to "cruise" position, start the automatic control system. The swing mechanism swings as the body contact with the roots of cotton stalk and sensors are installed to transmit collected signal to the PLC control system which sent the received signal to the proportional flow valves after processing. The valves provide the steering with hydraulic oil with a certain pressure under the control of electric signal to adjust the direction automatically. When transporting and turning, switch to transport model to close automatic control system and implement manual steering.

4 On-Line Production Measurement Techniques

Conversion out of the average cotton flow rate per unit time to obtain cotton production through measuring the near infrared light absorption rate of cotton in cotton tube. Set the flow sensors in the upper part of each cotton transmission pipeline; each set of flow sensor has a set of optical signal transmitter and receiver which is installed in the output of cotton relative to the side wall of the tube. The transmitter emits near-infrared beam while the receiver tests the intensity of this beam and the measured value has the correlation between the total volume of cotton which go through the transmitter and receiver in certain time interval. The data processing algorithms in the monitor process all the sensor signals and estimate the average cotton flow within a certain time interval. Through the weight calibration, conversion the weight of average output of cotton in cotton tube in a certain time interval, and the weight is divided by the picking area (acreage) in that time interval to get the yield per unit area where picking area equals to time×velocity×cut width.

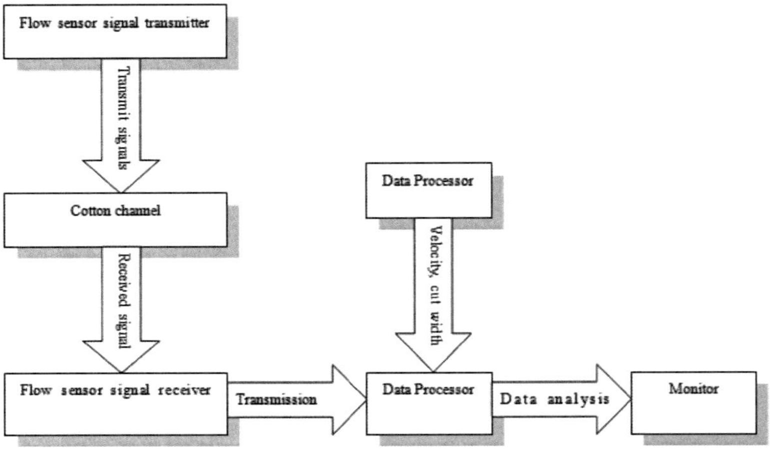

Fig. 5. On-line production measurement

5 Conclusion

As one of the main economic crops in our country, only if use mechanized production, can the costs reduce as much as possible, and make it competitive in price in the domestic and international market. To promote large self-propelled cotton picker and large-scale machine picking technology can reduce the labor intensity and is benefit to scale operation. It also can accomplish the purpose of reducing the cost of cotton, improving labor productivity, promoting the reform of the cotton production, and the development of China's cotton industry. With the mechanization of modern agriculture continues to improve, the development of large self-propelled cotton picker is the general trend of C)hina's cotton industry.

Acknowledgment. The project is supported by Foundation of Guizhou Science and Technology Department (No. [2008]6205, No. [2008]2210 and No. [2009]3009) and Foundation of Guizhou University (No. [2007]25).

References

1. Key, H.A.: Cotton picker development and field performance. Paper - American Society of Agricultural Engineers (1985)
2. Wu, X.-H., Sun, H.-P., Chen, L.-N.: The application research of CAN bus in the city-bus control system. In: 2010 International Conference on Mechanic Automation and Control Engineering, MACE 2010, pp. 3428–3430 (2010)
3. Dahan, M.: On-board data acquisition and processing system. In: Proceedings of International Telemetering Conference, vol. 22, pp. 673–685 (1986)

4. Shi, C., Huang, S., Yan, X.: The Design of an Evolvable on-Board Computer. In: Kang, L., Liu, Y., Zeng, S. (eds.) ICES 2007. LNCS, vol. 4684, pp. 292–296. Springer, Heidelberg (2007)
5. Syed, M.A., Schueler, E.: Reconfigurable parallel computing architecture for on-board data processing. In: Proceedings - First NASA/ESA Conference on Adaptive Hardware and Systems (AHS 2006), vol. 2006, pp. 229–236 (2006)

The Development and Application Prospect of Natural Food Preservative

Wu Jie[1], Sun Bei[2], and Zhu Fei[1]

[1] Bengbu College, Biotechnology and Food Engineering, Anhui, China
[2] Nanjing Agricultural University, College of Food Science and Technology, Nanjing, China

Abstract. This paper mainly introduces types, characteristics, principles, use scope of some newly developed natural food preservative in recent years, as well as the existing problems in the development of natural food preservative and its rational use. This paper also looks into the application prospect of natural food preservative.

Keywords: natural food preservative, development, application, prospect.

1 Introduction

According to the different sources, food preservative can be divided into two types of synthetic and natural preservative. For a long time in food industry the synthetic food preservative mainly consists of potassium sorbate and sodium benzoate has been the dominant, the natural food preservative has not been widely used because of its high cost. As people's living standard and technological advance, synthetic food preservative has non-negligible toxic side effect, there is a growing emphasis on development and research of natural preservative. Further developing safe, efficient, economical natural food preservative has become a hot issue in food additives industry; it is not only the mainstream of overseas development, but also the main research direction of China.

2 Natural Food Preservative

According to different sources, natural food preservative generally has the following categories: animal-derived, plant-derived, microorganisms, natural organic compounds, mineral extraction.

2.1 Preservative of Animal-Derived

1) Protamine (Protamine): Protamine is a poly-cationic peptides present in spermary tissue of various animals, with characteristics of high efficiency, safety, functionality and so on. With broad-spectrum inhibitory activity, protamine can inhibit the bacillus subtilis, bacillus megaterium, licheniform bacillus from growing; It also has a significant inhibitory effect on Gram-positive bacteria, yeast, mold; In addition, protamine can inhibit the spores of aerobic bacteria growing after germination; In

D. Zeng (Ed.): Advances in Control and Communication, LNEE 137, pp. 407–412.
springerlink.com

addition, protamine can effectively inhibit clostridium botulinum of type A, B and E from developing. According to reports, add 0.05% to 0.1% protamine in milk, eggs, pudding, they can be stored below 15 ℃ for 5 to 6 days, while a control group without protamine can only stored for less than 4 days [1].

2) Propolis (Porpolis): Propolis is a resin collected by bees from plant bud and trunk, then mixed with secretions from mandibular gland, beeswax and others to process a opaque colloidal solid with aromatic odor. Its ingredients are complex, due to a large number of active reduction factor, propolis is of strong antioxidant activity, which can be used as natural antioxidants for grease and other food products; It contains polyphenolic compounds which can inhibit and kill bacteria, antisepticize, skin care, and promote immune function, it is widely used in fields of food, medicine, light industry and chemical industry and so on[2].

3) Chitosan (Chitoson): Chitosan is deacetylation chitin, also known as chitosan, which is extracted from crab shell and shrimp shell. It is a polysaccharide hydrolyzed from deacetylation of chitin (Chitin). Its preservative effect mainly behaves in: ① Chitosan can form semi-permeable membrane in the surface of food to effectively control the invasion of microorganisms; ② It can inhibit the growth and propagation of certain microorganisms. The higher the deacetylation degree of chitosan, the stronger the antibacterial activity. When the concentration is 0.4%, chitosan has antibacterial properties on E. coli, pseudomonas fluorescens, proteus vulgaris, staphylococcus aureus and bacillus subtilis. Chitosan has a good application in food, and 0.1% or 1.0% chitosan acetic acid solution can completely inhibit the growth of staphylococcus aureus, pseudomonas aeruginosa and Candida tropicalis; Add some chitosan which can dissolve in vinegar in the seasoning can inhibit bacterial from growing for 7 days, and when used with sodium acetate the days can be increased to 12 days[3], it can been seen from that chitosan is a highly effective natural preservative.

4) Lysozyme (Lysozme): Lysozyme is a basic protein, also known as muramidase or Nacetylmuramideglycanohydrlase, which is widely exist in sap and latex of mammalian, protein of some poultry and parts of plants and micro-organisms. After the function of lysozyme, osmotic pressure is destructed thus causing balance breakdown in microbial cells, thus leading the dissolved material in microbial cells leak to cause cells death. Lysozyme is a non-toxic protein, which is harmless to human body with high safety, and has antibacterial action on Gram-positive bacteria, aerobic spore-forming bacteria, bacillus subtilis, licheniform bacillus and others. As a natural preservative, lysozyme is widely used in low-alcohol wine, sausage, cream, cakes, noodles, beverages, cheese and dairy products and other foods.

2.2 Botanical Preservative

1) Extraction of Perilla Leaf: Distillation of perilla leaves has a function of broad-spectrum antimicrobial activity, the main active ingredients are perillaldehyde and citric acid aldehyde, these two compounds are terpene aldehydes with the similar acting site and a coordinating function. In vitro, Perilla has a significantly inhibition on staphylococcus aureus, beta streptococcus, diphtheria bacillus, bacillus anthracis, pseudomonas aeruginosa, bacillus subtilis. Perilla leaf oil a also has a significantly inhibition on vaccinal and natural contaminated mold and yeast.

2) Pectin Decomposition: The main antibacterial substances of pectinase decomposition of citrus peel and pomace are poly galacturonic acid and galacturonic acid. A Japanese company successfully developed pectin decomposition as a natural preservative[1], which is mainly applied in hamburgers, noodles, cream cakes, pickles and other foods. Such as applying experiment of noodles in soup, the total plate count of samples without pectin decomposition stored at 20 ℃ is more than 103 in second day with a large number of turbidity. While samples with pectin decomposition stored for 5 days, the total plate count is only less than 103 without turbidity.

3) Antibacterial Substances in Garlic: Garlic contains alliin, allicin and allitridin, of which allicin and allitridin are the major antibacterial components of garlic. Garlic has a significant inhibition and bactericidal effect on a variety of coccus and mold.

4) Licorice: Licorice is derived from a variety of roots and rhizomes in plants of leguminous glycyrrhiza, it is not only a very important medicine, but also a raw material and flavor additive commonly used in medicine, food, tobacco and other industries. People selected out the antimicrobial compositions from glycyrrhiza residues after extracting glycyrrhizic acid, and found that extraction of glycyrrhiza residues have a good inhibition on aspergillus niger, aspergillus oryzae, black rhizopus, mucor, paecilomyces. And they also adopted the method of the thin layer chromatography to determine that the key component of anti-mildew is liquorice flavonoids.

5) Tea Polyphenol: Large number of experiments showed that tea polyphenol has a good preservation effect, it has a inhibition on bacillus subtilis, staphylococcus aureus, escherichia coli, tomato ulcer sores, caries streptococcus and mucor, penicillium, gibberella, colletotrichum gloeosporioides and saccharomyces cereviseae.

6) Agar Oligosaccharides: Agar extracted from seaweed, the main components are agarose, the emzyme decomposition is agar oligosaccharide, which has strong function of antibacterial and preventing the starch from aging and retrogradation. When the concentration is up to 3.11%, it can effectively reduce the bacteria production. Now agar oligosaccharides are widely used for noodles, bread and cakes and other foods[1].

2.3 Microbial Natural Preservative

Microbial natural preservative is safe, efficient, and is recently the most promising new class of natural preservative.

1) Nisin (Nisin): Nisin a small peptide generated by some lactococcus lactis of N serological type, which is of antibacterial activity for many G^+ bacteria, especially has a inhibition on activity of clostridium and bacillus[4] . These bacterium produced endospores are the main spoilage organisms of food. Nisin is first discovered by Rosers in 1928, and then Nattick and Hirsch proved the inhibitory effect of such kind of substance in 1944 , thus this active substance is named as Nisin. Nisin is recognized as natural food preservative in 1969 by the British Food Preservatives Union Commission and experts of Federation of the World Health Organization, which is also the first bacteriocin approved to use in food. Currently, Nisin is widely used in almost 50 countries and regions. Our studies on Nisin began in 1989, Institute of Microbiology in Chinese Academy of Sciences completed a basic study on Nisin, which is currently in mass production to put on the market. The Ministry of Health of China implements the GB2760-86 regulations, Nisin can be used in milk and dairy products, canned foods, plant protein foods, meat and other foods, which can effectively inhibit microbial growth, reducing heat treatment time and reducing the impact on flavor and nutritional value, it both improves the quality and saves energy. Nisin has a strong activity and is safe and non-toxic to human, so its application prospect in food industry is optimistic, many countries such as Britain, France and Australia don't have any restrictions on its adding amount. Recent studies show that, Nisin in combination with disodium EDTA has an inhibition on Salmonella and other Gram-negative bacteria, which better extending the applying arrange of Nisin in food preservation[5].

2) Polylysine (Polylysine): Lysine is an essential amino acid for human, while polylysine formed by lysine is a kind of biosynthesis bacteriostat which has a broad spectrum effect on Gram-positive and-negative bacteria, fungi and other microorganisms. Polylysine is not formed with lysine as raw materials but starchiness, and then cultured and fermented with aerobic gas by streptomyces actinomycetes. After the human body takes in polylysine, it can be broken down into lysine nutrients to be absorbed and used without any side effects. Polylysine can be dissolved in water, it does not affect food flavor with good thermal stability, when heated with 100 ℃ for 100min or 120 ℃ for 20min, it still can inhibit microbial from growing. In a wide range of pH value, polylysine still has a bacteriostatic action.

Polylysine was first put into produce by Japanese in 1989, and then entered to the market and used in food preservatives. Currently, Japan's annual consumption of Polylysine is about 800 ~ 900t.

3) Natamycin (Natamycin): Natamycin produced by natalensis streptomyces is a cyclic tetraene compound with activity, the stability of its activity is impacted by pH value, temperature, light intensity, oxidant and metals, so it should avoid exposure to oxide and sulfhydryl compound, at the same time it should be kept in container of glass, plastic or stainless steel. Natamycin can effectively inhibit and kill mold, yeast, filamentous fungi, its mechanism is the reaction between cell walls and cellmembrane of ergosterol and fungal, which resulting in cell rupture. For some microorganisms without similar steroid compounds existing in cells, Natamycin is invalid, such as bacteria and viruses[6]. As a result, natamycin can be directly add to the yogurt and other fermented products to inhibit mold and yeast without killing beneficial bacteria, while other preservatives do not yet have this kind of feature.

4) Tylosin (Tylosin): Tylosin produced by streptomyces is a macrolide compound, its bacteriostatic action is similar to Nisin. The mechanism is: It can inhibit protein synthesis of bacterial, which binds to 50S subunit of ribosome to inhibit the combination of aminoacyl end of aminoacyl tRNA- , and the formation of mRNA-aminoacyl-tRNAribose complexus.

5) Yeast killer toxins: Bactericidal substances existing in yeast was first discovered from saccharomyces cerevisiae in 1983, and then people found that candida and CreweVickers yeast also can secrete bactericidal substances, which are called as killer toxins. They are with the following characteristics: With a protein component and lower best pH value, it can be passivated above 35 ℃[7]. So far, more research is focus on the K 1 toxin of saccharomyces cerevisiae, and its mechanism on cells are: K1 toxin can first adsorb on the ï-1 ,6-D-glucan receptor of yeast's cell wall, after that an energy-requiring process can transfer K 1 toxin to the cell membrane to form cation channel, so that cytoplasm acidulated and K^+ leaked, and finally causing cells death.

Killer toxin can inhibit or kill the black yeast, yeast with the same family and related yeast in the fermentation process[6]. Xie Junjie and some other people found that tea fungus liquid has a significantly inhibition on Gram positive and negative bacteria, and then extracts protein substance with small molecular from broth, this kind of substance can inhibit bacterial from growing under acidic conditions.

2.3 Natural Organic Compounds

There are a lot of natural organic compounds can be applied in food preservation techniques, mainly including organic acids: acetic acid, lactic acid, glycine, etc.; ethanol, polysaccharides: konjac mannose, chitin, etc.; enzymes: protease, oxidase, reductase etc.; vitamin series: VC, VE, VP, fatty acid ester, as well as sugar, oil, starch in general food ingredients and so on. Such as ethanol has antiseptic and antibacterial effect, in addition, ethanol has a strong effect of increasing osmotic pressure, thus improving food preservation. Organic acids play an important role in the fermentation, the organic acids of fermented products such as lactic acid, acetic acid, propionic acid, 3 – hydroxyl propionic acid etc. should be applied to adjust the pH to improve the preservation [8]. Some mechanisms of organic acids are aimed at damage cell membranes and chemosmosis related to membranes.

3 The Existing Problems and Rational Use in the Development of Natural Food Preservative

Some businesses set the additives against security and mislead consumers, which greatly hinders the application of natural food preservative. The developing process of natural food preservative lacks of systematic theoretical guidance, the development of new natural food preservative needs based on theory of food quantum biochemistry, microbiology, biochemistry and other subjects, but experts in these areas are rarely involved in co-operation. Since research and development need to invest too much money, companies are not enthusiastic about participating.

In using process we should strictly control the use scope and usage amount, appropriately increasing the acidity of food, lowering the pH value and reducing the amount of preservative. When adding preservative before heating, pay attention to the heating temperature should not be too high to prevent evaporation along with water vapor, thus losing corrosion. Use coordinately with two or more natural preservatives to reach the complementary synergy.

Natural preservatives still have other problems in practical application, such as losing preservative effect with a little amount, impacting food flavor and quality with a large amount, and even generating toxic side effects. Such as tea polyphenol, when used as a preservative, too high concentration will make people feel bitter taste, and also leading food discolored due to oxidation.

4 Outlook and Prospects

There are many problems as above existing in research and development process, but the characteristics of safety and health of natural preservative decide it will be vigorously promoted, researched and developed; countries have already invested a large number of human and material resources in the study. China has a good resource advantage in this area with a broad prospect for development. To solve the existing problems and promote the development of natural food preservative, it needs the cooperation of every walk of life. Only in this way, safe and efficient natural food preservative can be used in a correct and reasonable way.

Acknowledgment. Fund: Science and Technology Project for 2010 in Anhui Province, (No.: 10020303014).

References

1. Zhang, H., Lin, K., Yan, C.: Research Progress of Domestic and Foreign Natural Food Preservative. Food Processing (3), 57–60 (2004)
2. Zhang, W., Cui, T.: Research of Propolis's Bacteriostatic Action on Pathogen in Food. Food Science 19(3), 40–43 (1998)
3. Chen, C., Cheng, W.: Research and Application of Natural Preservative Chitosan. Journal of Zhengzhou University of Light Industry 13(1), 1–4 (1998)
4. Delves-Broughton, J.: Nisin and its uses as a food preservative. Food Technol. (44), 100–117 (1990)
5. Tang, F., Cai, H.: Research Progress of Microbial Preservative Nisin. Food Science and Technology (11), 46–48 (2002)
6. Holzapfel, W.H., et al.: Biological preservant on foods with reference to protective cultures, bacteiocins and food-grade enzymes. Internation Journal of Food Microbiology (24), 343–362 (1995)
7. Liu, Y., Ning, Z.: Natural Food Preservative – Antimicrobial Peptides. Food and Machinery (3), 10–12 (1999)
8. Hao, S.: Preparation and Application Technology of Food Ddditives, 1st edn., pp. 109–123. Chemical Industry Press, Beijing (2003)

Penalty Function Element-Free Method for Solving Seepage Problem of Complex Earth Dam

Wang Zhi, Shen Zhen-zhong, Xu Li-qun, and Gan Lei

College of Water Conservancy and Hydropower Engineering
Hohai University
Nanjing, Jiangsu Province, China
wangzhiok@126.com, shen6627@yahoo.com.cn

Abstract. The element-free method (EFM) simplifies the pre-processing and has high calculation efficiency since it requires only the information of nodes and boundaries instead of elements. By taking the adaptive influential radius and arranging nodes flexibly, homogenization could be reduced significantly. Such method was applied in this paper to calculate the steady seepage field of a core wall dam with anti-seepage curtain and concrete cushion. The comparison of the calculation results with that calculated by the finite element method shows they match well and both the seepage field and free surface distribution accord with the general rules, which mean the element-free method can be applied to solve the seepage problem of complex earth dam.

Keywords: Element-free method, penalty function, steady seepage field, adaptive influential radius.

1 Introduction

Element-free method (EFM) is a new kind of numerical calculation method which started late in the field of solving seepage problem and has yet to be perfected to calculate the complex seepage field [1]. It was applied to solve the stability seepage field of homogeneous earth dam by Li guang-xin, Ge jin-hong[2]; It was also applied to solve the problem of fracture seepage by Li xiao-chun who considered the penalty function to deal with the boundary conditions while no derivation was given[3];Chen xiao-hu, Shen zhen-zhong used penalty function to treat the boundary and put forward specific expression [4].

It's very important for element-free method to calculate the complex seepage field since it is the base of research on hydraulic fracturing and coupling model of stress field and seepage field in which the element-free method is superior to the finite element method (FEM).

However, when node density or material permeability coefficient between adjacent regions differs significantly, homogenization effect is produced which has a remarkable effect on results. According to the actual situation of the core wall dam with anti-seepage curtain and concrete cushion, homogenization is reduced significantly by treating the boundary conditions with penalty function method and taking the adaptive influential radius and arranging nodes flexibly. The steady seepage field of the core

D. Zeng (Ed.): Advances in Control and Communication, LNEE 137, pp. 413–420.
springerlink.com

wall dam under normal water level was calculated and then the comparison of the calculation results with that calculated by the FEM indicates that they match well which means the EFM can be applied to solve the seepage problem of complex earth dam.

2 Penalty Function Element-Free Method for Solving Seepage Problem

The element-free Galerkin method is based on the moving least squares method [5], many factors are critical to the calculating precision of the method such as selection of base function, determination of influential radius and weight function, discontinuity treatment method, processing of essential boundary conditions, integration and nodal arranging scheme [6-9].

Without considering internal source, the seepage differential equation for two-dimensional, homogeneous and incompressible fluid is [10]:

$$k_x \frac{\partial^2 H}{\partial x^2} + k_z \frac{\partial^2 H}{\partial z^2} = 0 \tag{1}$$

In which, H is the water head field function, k_x and k_z are the permeability coefficients of X direction and Z direction.

In accordance with the moving least squares method that the head field function can be expressed as follows:

$$H = \mathbf{N}^{\mathrm{T}}\mathbf{H} = n_1 H_1 + \ldots + n_j H_j \tag{2}$$

Then the total energy of the system can be expressed as:

$$\boldsymbol{\pi} = \frac{1}{2}\mathbf{H}^{\mathrm{T}}\mathbf{K}\mathbf{H} - \mathbf{H}^{\mathrm{T}}\mathbf{F} + \mathbf{C} \tag{3}$$

Based on the variation principle, the integral equilibrium equation is:

$$\mathbf{KH} + \mathbf{F} = \mathbf{0} \tag{4}$$

In which, $\mathbf{H}$ is the water head array, j is the amount of the nodes, n_j is the shape function of i , $\mathbf{K}$ is the permeability tensor and $\mathbf{F}$ is the right end item of the equation.

2.1 Differential Equation

Energy function of (3) can be expressed as:

$$\boldsymbol{\pi} = \iint_{\Omega} \frac{1}{2}\left[k_x \left(\frac{\partial H}{\partial x} \right)^2 + k_y \left(\frac{\partial H}{\partial z} \right)^2 \right] d\Omega \tag{5}$$

And its extreme value is:

$$\mathbf{K}^{(1)} = \iint_{\Omega} \mathbf{B}(\mathbf{x})^{\mathrm{T}} \mathbf{K} \mathbf{B}(\mathbf{x}) \mathbf{d}\Omega \tag{6}$$

$$\mathbf{F}^{(1)} = (0, 0, \ldots, 0) \tag{7}$$

In which $\mathbf{B}(\mathbf{x}) = [\mathbf{B}_1(\mathbf{x}), \mathbf{B}_2(\mathbf{x}), \ldots, \mathbf{B}_n(\mathbf{x})]$, $\mathbf{K} = \begin{bmatrix} k_x & 0 \\ 0 & k_z \end{bmatrix}$ and

$\mathbf{B}_i(\mathbf{x}) = \begin{bmatrix} \frac{\partial n_i}{\partial \mathrm{x}} & \frac{\partial n_i}{\partial \mathrm{z}} \end{bmatrix}^{\mathrm{T}}$, Ω is the calculated domain.

2.2 Known Water Head Boundary

Generally, there are four boundary conditions in the seepage problem with free surface: Known head boundary, impermeable boundary, permeability surface boundary and the free surface boundary. Penalty function is used to deal with the known head boundary and assuming that there lies a strong permeable layer in the boundary with its thickness ΔS and permeability coefficient k_s, then

$$\mathbf{K}^{(2)} = \int_{\Gamma_1} \frac{k_s}{\Delta S} \mathbf{N}^{\mathrm{T}} \mathbf{N} d\Gamma_1 \tag{8}$$

$$\mathbf{F}^{(2)} = \int_{\Gamma_1} \frac{k_s}{\Delta S} \mathbf{N} H_B d\Gamma_1 \tag{9}$$

In which, $\mathbf{N}$ is the shape function matrix, Γ_1 is the known head boundary, H_B is the value of the head, $\frac{k_s}{\Delta S}$ is penalty factor and $\frac{k_s}{\Delta S} = \frac{c \cdot k}{\Delta b}$, $\Delta b = min(d_n)$, $c = 10$, d_n is the distance between nodes and k is the normal permeability coefficient.

2.3 Impermeable Boundary

Penalty function is used to deal with the impermeable boundary and assuming that there lies an aquitard in the boundary with its thickness Δd, outer boundary Γ_2', inner Γ_2 and zero normal discharge. The normal permeability coefficients of the aquitard and seepage area are k_s and k, then

$$\mathbf{K}^{(3)} = \int_{\Gamma_2} \frac{\Delta d}{k_d} \mathbf{B}(\mathbf{x})^{\mathrm{T}} \mathbf{K}^{\mathrm{T}} \mathbf{R} \mathbf{K} \mathbf{B}(\mathbf{x}) d\Gamma_2 \tag{10}$$

$$\mathbf{F}^{(3)} = (0, 0, \ldots, 0) \tag{11}$$

In which, $\mathbf{B(x)} = [\mathbf{B}_1(\mathbf{x}), \mathbf{B}_2(\mathbf{x}), \ldots, \mathbf{B}_n(\mathbf{x})]$, $\mathbf{K} = \begin{bmatrix} k_x & 0 \\ 0 & k_z \end{bmatrix}$,

$\mathbf{B}_i(\mathbf{x}) = \begin{bmatrix} \frac{\partial n_i}{\partial x} & \frac{\partial n_i}{\partial z} \end{bmatrix}^T$, $\frac{\Delta d}{k_d}$ is penalty factor. $\mathbf{R}$ is the rotation matrix and $\mathbf{R} = \begin{bmatrix} sin^2\alpha & \text{-}sinacos\alpha \\ \text{-}sinacos\alpha & cos^2\alpha \end{bmatrix}$ in which α is the angle of X axis and impervious boundary.

Similar to known water head boundary, $\frac{\Delta d}{k_d} = \frac{c \cdot \Delta b}{k}$, $c = 10$, $\Delta b = R_{\max}$ and $R_{\max}$ is the largest influential radius , k is the normal permeability coefficient.

To sum up, $\mathbf{K}$ and $\mathbf{F}$ in (4) can be regarded as $\mathbf{K}=\mathbf{K}^{(1)}+\mathbf{K}^{(2)}+\mathbf{K}^{(3)}$ and $\mathbf{F}=\mathbf{F}^{(1)}+\mathbf{F}^{(2)}+\mathbf{F}^{(3)}$, that is,

$$K_{ij} = \iint_\Omega \left(k_x \frac{\partial n_i}{\partial x}\frac{\partial n_j}{\partial x} + k_z \frac{\partial n_i}{\partial z}\frac{\partial n_j}{\partial z} \right) d\Omega + \int_{\Gamma_1} \frac{k_s}{\Delta S} n_i n_j d\Gamma_1$$
$$+ \frac{\Delta d}{k_d} \int_{\Gamma_2} (k_x \sin\alpha)^2 \frac{\partial n_i}{\partial x}\frac{\partial n_j}{\partial x} + (k_y \cos\alpha)^2 \frac{\partial n_i}{\partial z}\frac{\partial n_j}{\partial z}$$
$$-k_x k_y \sin\alpha\cos\alpha \left(\frac{\partial n_i}{\partial x}\frac{\partial n_j}{\partial z} + \frac{\partial n_i}{\partial z}\frac{\partial n_j}{\partial x} \right) d\Gamma \tag{12}$$

$$F_i = \int_{\Gamma_1} \frac{k_s}{\Delta S} n_i H_B d\Gamma_1 \tag{13}$$

2.4 The Permeability Surface Boundary

The permeability points are the special points of the free surface, they can be regarded as the known water head points located by the neighboring ones in the free surface and also can be treated as the unknown water head points iterated like other ones belong to the free surface. Through a great amount of calculating, that permeability points taken as known water head ones would accord well with the results derived from that of FEM, so the permeability points will be applied as known water head points in this paper.

2.5 Determination of the Free Surface and the Permeability Point

The position of unknown seepage free surface should be determined through iterative calculation. Since element-free method only needs information at nodes and boundaries and the integral grid is irrelative to the nodes, the finite element grids won't appear distortion and the nodes can be adjusted in the processing of iteration. So in this paper, the variable mesh method of finite element method is referenced to calculate the free surface [11].

First, nodes should be produced in the free surface especially more in the borders of different material to ensure accuracy. Calculate the seepage field and get the water head of the nodes in the free surface after the initial free surface determined, then determine whether it is equal to its Z direction coordinates, if not, adjust the nodes' location in the free surface and recalculate the seepage field until the difference between the water head and the Z direction coordinates less than the control value. Concrete steps as follows: Assuming that the water head of node i in the free surface is $H_i^{(n)}$ with its coordinates $(x_i^{(n)}, x_j^{(n)})$ after having iterated n times, if $\left|H_i^{(n)} - z_i^{(n)}\right| < \varepsilon$ (ε is the error control value), then stop the iteration, if not, change the Z direction coordinate according to the following formula:

$$z_i^{(n+1)} = z_i^{(n)} + m_i (H_i^{(n)} - z_i^{(n)}) \tag{14}$$

In which, m_i is the correction coefficient and $0.5 < m_i < 1$.

3 The Adaptive Influential Radius and the Nodal Arranging Scheme

3.1 The Adaptive Influential Radius

The influential radius is closely related to the weight function. The appropriate influential radius should make the nodes in the influence domain "enough and as few as possible" to not only ensure the accuracy of calculation but also make the Gaussian point not to be affected by the neighboring nodes which should not be considered.

Thus, when treating thin structures like anti-seepage curtain and concrete cushion, the influential radius of the Gaussian points in those areas should be took smaller to make sure the nodes surrounded by the influence domain are from the same area. At the same time, the node refining degree should be controlled to make the number of nodes in the influence domain at about 20. In other areas, influential radius should be larger and so the number remains 20. A large number of numerical examples show that by taking the adaptive influential radius, homogenization could be reduced significantly and a relatively good result can be reached.

3.2 Nodal Arranging Scheme

Nodal arranging scheme has great effect on the calculation results of element-free method and rational arrangement of nodes is the key to ensure accuracy. Advice is given according to the author's experience as follows: background grid and integral grid coincide with each other; nodes are evenly arranged, that is the distance between nodes in horizontal direction is approximately equal to that of vertical direction; trying to make the number of nodes in each influence domain at 20; thin layer structure should be locally appropriate refined. And the calculation experience shows that only when the node density increases more than two times can the calculation accuracy requirements be met when permeability coefficients of adjacent area differ by more than one order of magnitude.

4 Engineering Application

The crest elevation of one core wall dam is 2902.00m with its foundation surface elevation 2587.00m, maximum height 315m, the crest width 18m, maximum bottom width 1193m. The upstream dam slope is 1:2.0 and downstream dam slope is 1:1.9. The elevation of center top of the core wall is 2898.00m and 2587.00m at the bottom, the top width is 5.0m and 127m at the bottom. The core wall is located on concrete cushion. The upstream normal water level is 2985.00m and the downstream normal tail water level is 2618.54m [12]. Necessary simplification is made when calculating and the calculated permeability coefficients adopt design value as shown in Table I.

Due to the existence of such thin structures as core wall, anti-seepage curtain and concrete cushion, the nodes are refined and the influential radiuses of the three areas are reduced according to the requirements above. Free surface calculating uses the base form, error control value is 0.01m.

The calculated free surface and potential lines are shown in Fig. 1, the comparison of the results calculation by EFM and FEM is shown in Table II, the seepage field conditions of the core wall in different influential radius and nodal arranging scheme is shown in Table 3.

It is can be known from Fig.1 that the free surface of the seepage field and the distribution of such potential lines accord with average seepage field distribution rule and also known from Table 2 that the calculated results by EFM and FEM fit well. At the same time, it is can be known from Table 3 that the results by taking the adaptive influential radius and arranging nodes flexibly are much accord than that of regular nodal arranging scheme with the FEM result and homogenization is reduced significantly. Hence the calculation and analysis results by taking penalty function element-free method can basically reflect the actual situation of the seepage field of the dam.

Table 1. Permeability Coefficient of Each Material

Soil layer	Permeability coefficient (cm/s)
Upstream and downstream rock	1.0×10^{-1}
Gravel soil of core wall	1.0×10^{-5}
Concrete cushion	1.0×10^{-7}
Anti-seepage curtain	5.0×10^{-6}

Table 2. Comparison of The Calculation Results by EFM and FEM

X coordinate of the point in the free surface (m)	Element-free method result (m)	Finite element method result (m)	Relative error[a] (%)
989.00	2850.00	2847.91	-0.76
1025.00	2750.00	2748.28	-0.62
1394.00	2645.00	2646.57	0.57

[a]The relative error = (finite element method result - element-free method result)/head difference between upstream and downstream..

Table 3. Seepage Field of The Core Wall by Different Method

Method	Maximum gradient in the core wall	Reduction of the head by the core wall (m)	Reduction percentage[b] (%)
Regular nodal arranging scheme	4.03	225.87	81.70
Adaptive influential radius	4.20	229.73	83.09
Finite element method	4.18	229.61	83.05

[b]The reduction percentage = reduction of the head by the core wall /head difference between upstream and downstream..

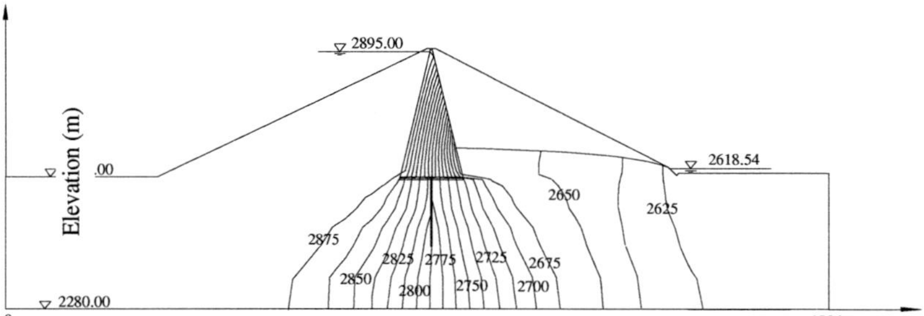

Fig. 1. Free surface and potential line distribution

5 Conclusion

A complex earth dam with core wall, anti-seepage curtain and concrete cushion is studied and the method of taking adaptive influential radius and arranging nodes flexibly is proposed. According to the actual situation of the core wall dam, the method is applied to calculate the steady seepage field of the core wall dam under normal water level and the calculated free surface and potential lines show that the seepage field and free surface distribution accord with the general rules which mean the calculation and analysis results by taking penalty function element-free method can basically reflect the actual situation of the seepage field of the dam and homogenization could be reduced significantly by taking the adaptive influential radius and arranging nodes flexibly. In conclusion, the element-free method has further application to research on hydraulic fracturing and coupling model of stress field and seepage field where the element-free method is superior to the finite element method (FEM) after having the ability to calculate the complex seepage field.

References

1. Belytschko, T., Lu, Y.Y., Gu, L.: Element free Galerkin methods. International Journal for Numerical Methods in Engineering 37(2), 229–256 (1994)
2. Cao, Z., Zhou, B., Chen, D.: Objiect-Oriented element-free method. Chinese Journal of Computational Mechanics 23(5), 483–486 (2006)
3. Zhou, X., Zhou, R.: Current study and development of element-free Galerkin method. Engineering Mechanics 22(11), 12–20 (2005)
4. Li, G., Ge, J., Jie, Y.: Free surface seepage analysis based on the element-free method. Tsinghua University (Sci.&.Tech.) 42(11), 35–37 (2002)
5. Li, X.: Meshless method on seepage in fractured rock mass in Xiaowan dam site with 3D network computational simulating technique for random discontinuous, pp. 25–27. Jilin University Press, Changchun (2005)
6. Shen, Z., Chen, X., Wu, Y.: Penalty function element-free method for solving seepage field of embankments. Hohai University (Nature Sciences) 36(11), 44–48 (2008)
7. Xu, W., Deng, J., Yu, Q.: Two methods to decrease the numerical error in calculating of EFM. Engineering Mechanics 24(5), 72–77 (2007)
8. Raiesh, K.N., Rao, B.N.: Coupled element free Galerkin and fractal finite element method for unbounded problems. Jouranl of Structural Engineering 37(3), 167–179 (2010)
9. Zhuang, X., Auqarde, C.: Aspects of the use of orthogonal basis functions in the element-free Galerkin method. International Journal for Numerical Methods in Engineering 81(3), 366–380 (2010)
10. Mao, C., Duan, X., Li, Z.: Seepage numerical calculation and applications, pp. 117–118. Hohai University Press, Nanjing (1999)
11. Li, L., Liao, H., Liu, G.: A method of water free surface adapting grid system to simulate phreatic surface in seepage flow. Journal of Sichuan University(Engineering Science) 38(5), 76–81 (2006)
12. Shen, Z.: Seepage analysis report, pp. 71–82. Hohai University Press, Nanjing (2010)

The Natural Gas Pools Characteristics in Sulige Gas Field, Ordos Basin, China

Lin Xiaoying[1,2,*], Zeng Jianhui[1], and Zhang Shuichang[3]

[1] State Key Laboratory for Petroleum Resources and Prospecting, China University of Petroleum, Beijing, China
[2] School of Energy Science and Engineering, Henan Polytechnic University, Jiaozuo, Henan, China
[3] Research Institute of Petroleum Exploration and Development, PetroChina, Beijing, China
lxy2002199@126.com

Abstract. There are abundant natural gas resources in Sulige gas field, Ordos Basin. The ascertained resources are relatively low, because of complex geologic setting and reservoir heterogeneity. The characteristics of natural gas pools were analyzed from gas composition, gas saturation, permeability, pressure, and gas-water distribution in this paper. Methane is the main composition of natural gas in Sulige gas field, but C_2+ hydrocarbon content is relatively high, gas dryness is less than 0.95. The gas density is generally 0.6kg/m^3. The carbon isotope of methane is heavier. The original of gas is coal. The reservoir permeability and porosity is low. The gas reservoirs are defined as low porosity, low permeability or very low permeability reservoirs. The low or very low permeability reservoirs are controlled by the radius of pore throat that is thin, generally 0.25-1.5 um according to mercury inject data. The reservoirs are mainly of subnormal pressure reservoirs, and formation pressures are affected by gas-water characteristics. The gas saturations are low, and generally less than 70%. There are some relations between gas saturation and the porosity, permeability. The gas-water distributions are complex. Vertically, multi-gas layer are mutual doublications; laterally, there are some complex characteristics, e.g., the entire layer has gas, there are not unified gas-water contact, the distributions of gas and water are normal, and the distribution are inverted.

Keywords: Sulige gas field, natural gas pools, reservoir permeability, reservoir pressure, gas saturation, gas-water distribution.

1 Introduction

Sulige gas field lies in Suligemiao area of Ordos city, Inner Mongolia autonomous region, and tectonically belongs to the northwest Yishan slope of Ordos Basin. The exploration ranges of Sulige gas field is bounded by the Etuokeqianqi of Inner Mongolia autonomous region in the west, Taolimiao area of Wushenqi in the east, Aobaojiahan of Etuokehouqi in the north, and Dingbian country of Shanxi Province in the south [1-4]. The exploration area covers 2×10^4km^2. The main gas reservoirs are the Shan1 member of Shanxi formation and He8 member of Xiashihezi formation in the

* Corresponding author.

D. Zeng (Ed.): Advances in Control and Communication, LNEE 137, pp. 421–427.
springerlink.com

lower Permian. By the end of 2005, the proved reserve of He8 member is 418.977 billion m^3, the Shan1 is 1 114.675 billion m^3, and the total is 533.652 billion m^3. The massive gas exploration and development began in 2000. In 2009, the daily gas production has reached 20 million m^3. It is predicted that there would have 20 billion m^3 annual gas productions in 2013 [3].

The main hydrocarbon source rocks are coal and dark mudstone developed in carboniferous-Permian, whose distribution area is wide, thickness is great and gas generation potential is big [5]. There are the characteristics of wide hydrocarbon generation. The tectonic thermal event during the Yanshanian orogeny caused an abnormal paleogeothermal field. The source rocks in high temperature conditions reached mature or over mature stage, and generated large amounts of gas. It is provided favorable reservoir-cap combination for natural gas preservation by the sand body north-south stripped spread, widely regional cover layer and lateral mudstone formed in delta plain distributaries channel [1]. It is of great significance to ascertain gas pool distribution and promote natural gas exploration and development through the comprehensive research of gas reservoir characteristics in Sulige gas field.

2 Gas Compositions Characteristics

The natural gas in Upper Paleozoic is mainly made up of hydrocarbons, and less non-hydrocarbon that is CO_2 and N_2. The CO_2 and N_2 content generally is less than 3%. CH_4 is the main components of hydrocarbons; average content is 89% or so. C_{2+} content is relatively greater. C_1 index ($C_1 / (C_2\text{-}C_5)$) is less than 0.95, and give priority to wet gas. The natural gas stable carbon isotopic analysis shows that most of the sample of the methane has the characteristics of coal bed natural gas with relatively rich stable isotopes of 13 C[5]. The $\delta^{13}C_1$ ranges from -37‰ to -30‰, the$\delta^{13}C_2$ from -26‰ to -20‰, and the $\delta^{13}C_1$ from -30‰ to -20‰. The statistics to gas testing results indicated that gas density mainly distributed from 0.55 kg/m^3 to 0.70 kg/m^3, and peak in 0.6 kg/m^3. Gas compression factor mostly is less than 1, which show more easily compressed than ideal gas.

3 Gas-Water Distribution Characteristics

3.1 Gas-Water Distribution Characteristics in the Single Well

The thickness of single gas layer is generally less than 10m, average about 4m in developed wells. There exist interlayers in gas layers. According to logging, the water saturations of gas layers generally are less than 50%, mainly from 30% to 40%; the water saturation of gas layers containing water generally from 40%-60%, averagely 53% or so. The reservoir porosity and permeability has greatly changes vertically. The reservoirs with lower permeability become interlayer. Therefore, the thickness of continuous gas layers commonly is not great. There is close relation between water saturation and the reservoir permeability. The better is the permeability, the lower is water saturation. As is shown in Tab.1, Sue 6 Wells can be divided into six gas layers from bottom to up, it are gas-water layers when permeability is less than 0.4 and porosity is less than 8%, and it are gas layers with 70% gas saturation when permeability is more than 0.4 and porosity is more than 8%.

3.2 Regional Gas-Water Distribution Characteristics

Regionally, there is not only gas production but water production in the Upper Paleozoic reservoir. The gas continuous column heights are usually not more than 50 m because the reservoir permeability horizontally greatly change and relatively high permeability sandstone scatteredly distributed in low permeability sandstone. It is difficult to form the big gas reservoir. As is shown in the east-west section, the east part is high and west is low, there is no unified gas-water interface, and there is no obvious gas-water transitional zone. In the section, the unconnected-distribution of gas layer is obvious. In the west part of section, the reservoir permeability horizontally is good and not big change from well A to well I, the structure is the main factors affecting gas water distribution, and normal gas-water distribution that gas is over water formed. The gas-water distribution is abnormal in well I-well K and well L-well N (Fig.1). The conventional gas reservoir is mainly distributed in the west, and the deep basin gas reservoir is distributed in the east. This kind of distribution probably is related to low permeability in deep structure. Many gas layers are vertically overlapped because of interlayers. In short, the gas-water distribution has the characteristics as follows: 1) the entire reservoir containing gas, 2) gas above water, 3) gas-water inversion, and 4) no unified gas-water interface (Fig.1).

Table 1. The gas layer distribution of Su6 well in Sulige gas field

Depth of top (m)	Depth of bottom (m)	Thickness of sand layer (m)	Porosity (%)	Permeability ($\times10^{-3}\mu m^2$)	Water saturation (%)	Logging explanation
3311	3313	2.2	7.21	0.36	57.6	Gas-water layer
3316	3318	2.5	7.77	0.27	55.14	Gas-water layer
3318	3325	6.1	9.85	0.84	32.56	Gas layer
3325	3329	3.9	11.82	1.26	32.35	Gas layer
3330	3332	2.1	6.47	0.4	52.32	Gas-water layer
3337	3339	2.7	7.32	0.41	48.63	Gas-water layer

4 Reservoir Permeability Characteristics

The sandstone reservoir is low permeability and low porosity, and local relatively higher porosity and permeability. The porosity is ranging from 2% to 8% (Fig. 2). The permeability is ranging from $0.1\times10^{-3}\mu m^2$ to $0.5\times10^{-3}\mu m^2$ (Fig.3). The change of permeability and porosity is not obvious with the increasing of the depth. In the area, the change of porosity and permeability is great in different areas. The permeability of He8 member is relatively good. The threshold pressure of sandstone reservoir ranges from < 0.1 Mpa to 4.97Mpa. The maximum pore throat radius distributes in 0.25 um～181um, mainly 0.25 um -1.5 um, and averagely 14.14 um. The threshold pressure is greater. The median capillary pressure is 0.1096MPa-197MPa, mainly 3MPa-50MPa. Some wells such as shan 291 can't even get the median pressure. The median pore throat radius distribution is ranging from 0.03 um to 2.24 um, averagely 0.4 um. The median capillary pressure is larger too. Those fully explain that the pore throat is thinner in Sulige gas field. The reserve capability of low permeability sandstone reservoir in Sulige gas field is controlled by the pore throats radiuses, especially the minimum throats.

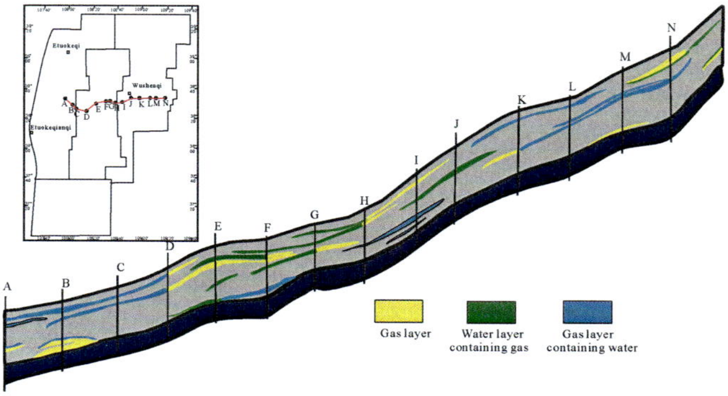

Fig. 1. The gas-water distribution in cross section from west to east, Sulige gas field

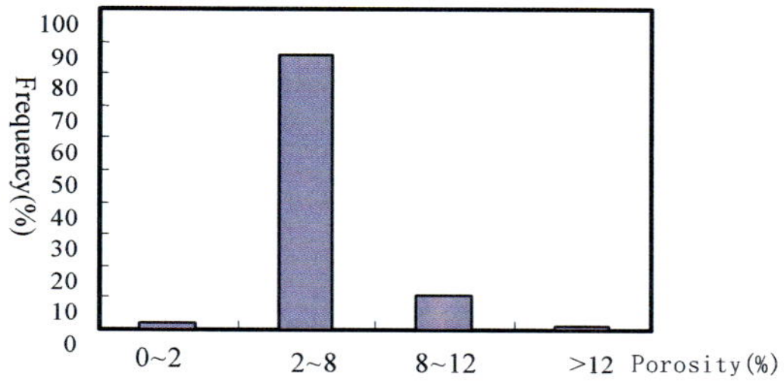

Fig. 2. The porosity distribution of the Upper Paleozoic sand

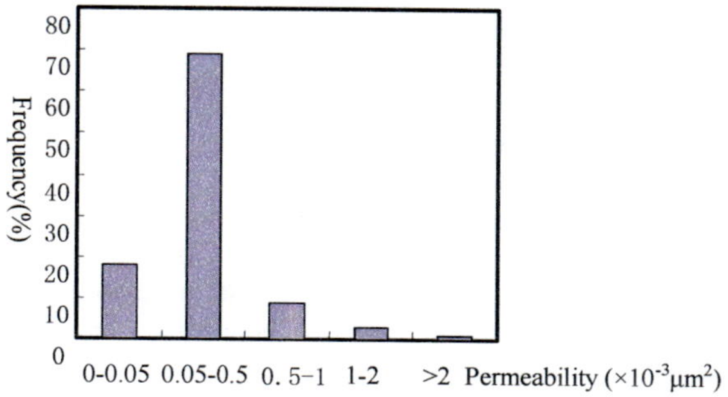

Fig. 3. The permeability distribution of the Upper Paleozoic sand

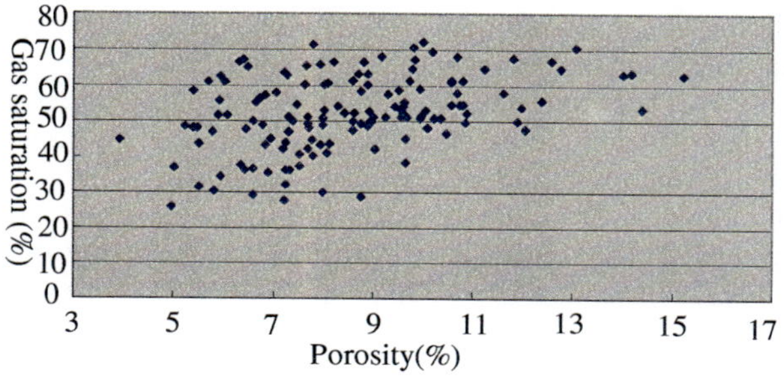

Fig. 4. The relationship between porosity and gas saturation

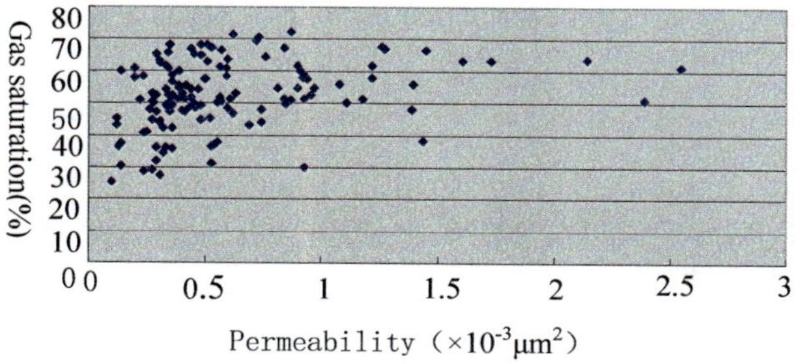

Fig. 5. The relationship between permeability and gas saturation

5 Reservoir Pressure Characteristics

The reservoir pressure is low in the Upper Paleozoic reservoir of Sulige gas field, and is underpressure and ultra underpressure in most areas. The pressure distribution is "low-high-low" from southwest to northeast in the whole of the abnormal low pressure background. The vertical distribution of formation pressure is close relations to the fluid properties. The fluid properties are different; the formation pressure variation characteristics are different. The pressures in water layer trend to the hydrostatic pressure, and increase with the increasing of depth. The pressure coefficients are 0.9 ~ 1.0, which show the normal pressure characteristics. The pressures in gas layers are different in different depth. Above 2800 m, the pressure trends are along the hydrostatic pressure line, and the pressure coefficients are 0.8 ~ 1.1. Below 2800 m, the pressure trends are deviated from the hydrostatic pressure line, and the deviation degree increase with the increasing of depth. The pressure coefficients are less than 1.0, and decreases with the increasing of depth. The pressure coefficients in gas-water layers are between the gas layers and water layers, and are 0.8-1.0. Those show that the gas saturation has important influence on formation pressure.

6 Gas Saturation Characteristics

The gas saturation is generally less than 70% in the Upper Paleozoic reservoir of Sulige gas field, Ordos Basin, mainly concentrated in between 40% and 60%. There is of certain correlation between gas saturation and the porosity, permeability. When the porosity is less than 9% and the permeability is less than 1 x 10-3μm^2, gas saturation has certain increase trend with the increasing of the porosity and permeability. When the porosity is more than 9% and the permeability is more than 1 x 10-3μm^2, gas saturation does not change significantly with the the increasing of the porosity and permeability (Fig. 4, Fig.5).

7 Conclusions

The main gas production layers are the Shanxi and Xiashihezi formation of the lower Permian in study area. Gas reservoirs distribute in the Shan1 and He8 member closing to source rocks. The gas pool distribution is complex because of many factors such as reservoir permeability, and structure.

(1) The reservoir permeability is low; the reservoir is low permeability or ultra low permeability reservoir

The porosity of sandstone in the gas field is low and ranges from 2% to 8%. The permeability is poor, and ranges from $0.1\times10^{-3}\mu m^2$ to $0.5\times10^{-3}\mu m^2$. The vertical change of permeability and porosity is big, and is not obvious relationship with depth.

(2) Reservoir pressure is lower; threshold pressure is higher

There exist ultra low pressure, low pressure and normal pressure in Sulige gas field, mainly ultra low and low pressure. The pressure coefficient generally is less than 1. The vertical distribution of the pressure is mainly controlled by the fluid properties. The pressure coefficient in water layer is higher, and that in gas layer and gas-water layer is lower and generally less than 1. The pore is thin, displacing pressure is great and capillary pressure is big in the reservoir. The gas migration way probably performs non-darcy flow.

(3) Gas-water distribution is complex; gas saturation is low

There is no unified gas-water interface and no obvious gas-water transition zone. The gas-water is inversed. It appears in the same gas production area that gas is above water and water is above gas. Gas saturation is ranging from 30% to 70%. The relationship between the gas saturation and permeability is complex.

References

1. Li, H., Wu, T., Ma, Z., et al.: Research on the basic characteristics and contro factors of high-quality reservoir in Sulige gas Field. Natural Gas Industry 24(8), 12–13 (2004) (in Chinese)
2. Yang, H., Zhang, W., Li, J., et al.: Geochemical study of the Upper Paleozoic gas in the Northern Ordos Basin. Acta Sedimentologica Sinica 22 Suppl., 39–44 (2004)
3. Dai, J., Zhong, N., Liu, D., et al.: Geological Foundation and main controlling factors of coal-formed large and medium gas fields in China. Petroleum Industry Press, Beijing (2000) (in Chinese)
4. Zhang, W., Guo, Y., Tang, D., et al.: Characteristics of fluid inclusions and determination of gas accumulation period in the Upper Paleozoic reservoirs of Sulige Gas Field. Acta Petroleum Sinica 30(5), 685–691 (2005) (in Chinese)
5. Li, J., Luo, X., Shan, X., et al.: Natural gas accumulation in the Upper Paleozoic of Ordos Basin, China. Petroleum Exploration and Development 32(4), 54–59 (2005)

Evaluation of Excavation Geological Conditions of Wawaigou Quarry

Yongcheng Guo[1,2,*], Jianlin Li[1], and Xing Chen[1]

[1] College of Civil Engineering and Architecture,
China Three Gorges University,
Yichang, 443002, P.R. China
[2] School of Water Resource and Hydropower,
Wuhan University,
Wuhan, 430072, P.R. China
ych_guo@126.com

Abstract. Overview of Sanliping power station and material-yard slope's influence on the power station is briefly introduced. The analysis of geology, lithology, the scope of mining area, excavation depth, slope gradient and aggregate reserves of the material slope is done. The environmental geological evaluation is made according to the local position.Some problem and related issues requiring attention in the construction are dicussed in this paper. Finally,optimal excavation design plan of reasonable construction and reinforcement is proposed.

Keywords: Material-yard, geology, excavation, evaluation.

1 Introduction

With the implementation of the Grand Development Strategy of Western China, massive hydropower projects are either planned or under construction. The Construction involves the excavation of large yard. How to effectively control the excavation slope stability is the current hot issue. The security control of the construction process has a good prospect. A lot of progress has been made on the research of excavation slope stability [1~7]. But more for the dam abutment, shiplock, plant, spillway and other high slope of permanent buildings, and the high slope of non-permanent buildings like material-yard class is often overlooked. Material-yard is the production base of filling material or concrete aggregate for hydropower dam. Excavation progress of materials-yard slope has a controlling role on duration and progress of the whole project, therefore, A comprehensive evaluation on the field should be made before excavation. how to ensure the stability of slope and construction works is an important safety technical factor, must be scientific and rational treatment of disposal.

2 Project Overview

Sanliping Hydropower Station is located in Fangxian County, Hubei Province, located in the Nan River's middle reaches, right bank first-level branch of Hanjiang River

* Corresponding author.

D. Zeng (Ed.): Advances in Control and Communication, LNEE 137, pp. 429–435.
springerlink.com

middle reaches.The dam site is in the downstream of Sanliping Village about 1.0 km, at the upper Lianziya 490 m, to Fang County 50 km.

Sanliping station is of normal water level of 416 m, dead water level of 392m, the total capacity of 499 million m3, adjust the capacity of 211 million m3, flood control capacity of 121 million m3, installed capacity 70MW, guaranteed output 12.4MW, annual average generating capacity of 183.4 million kw·h. The completion of power station will also increase the generating capacity and guarantee output of hydropower stations,such as Shiping, Guoduwan, Baishuiyu, Hujiadu, Miaozitou, Jinpengou at the downstream of Nanhe River. The project construction is not only beneficial to the Hanjiang River and Nanhe River flood control, but also conducive to Hubei Power Grid and Central China Power Grid load of electric, is helpful to Fang County poverty.

Sanliping hydropower station is a RCC arch dam. The adopts of aggregate are used on-site excavation of rock from crushing, quarry slope shown in Fig.1. At the yard near the dam, the external transport is closely linked Lizigou Bridge, shown in Fig.2.

Fig. 1. Excavated slope

Fig. 2. Excavated slope and the Lizigou bridge

3 Terrain Geology

Quarry is located in the dam site area on the cores of Lizigou anticline, stratigraphic trend NW-SE direction, core flat formation, to the near north-south; rock tendencies, North Wing is NW355 °~ NE25 °, dip angle, 50°~70°; South Wing is SE135°~155°, dip angle, 60°~80°. Main Stratum is the upper Tianheban Formation ($€_1t^2$), the lower Tianheban Formation ($€_1t^1$) and the lower Shilongdong Formation ($€_1sl^1$), lithofacies stability.

Yard is located in the left bank of the dam slope of about 200 m , the top ridge of yard is sharp, ridge elevation of nearly 700m, slope gradient is about 50°, slope and formation direction is nearly perpendicular, its topography is basic integrity, surface high runoff, valley development, no karst pipe systems, exposed bedrock, thin layer of weathered layer and peel, no obvious phenomenon of geological disasters. Therefore, the yard by topography and geology condition is identified as class I .quarry layout shown in Fig.3, the relevant section shown in Fig.4 and Fig.5.

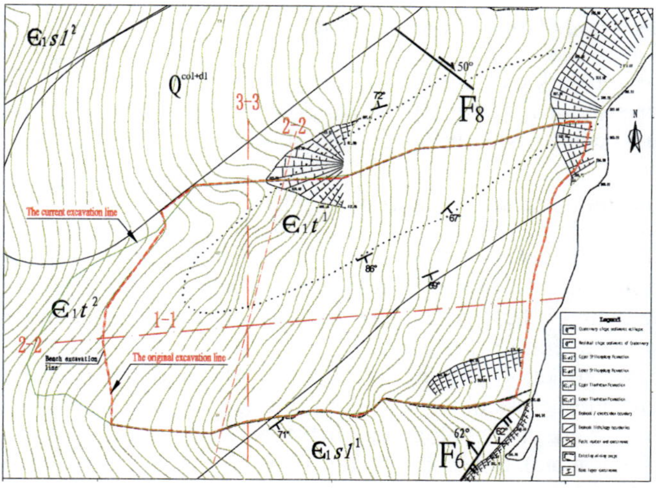

Fig. 3. Comprehensive engineering geological map of quarry

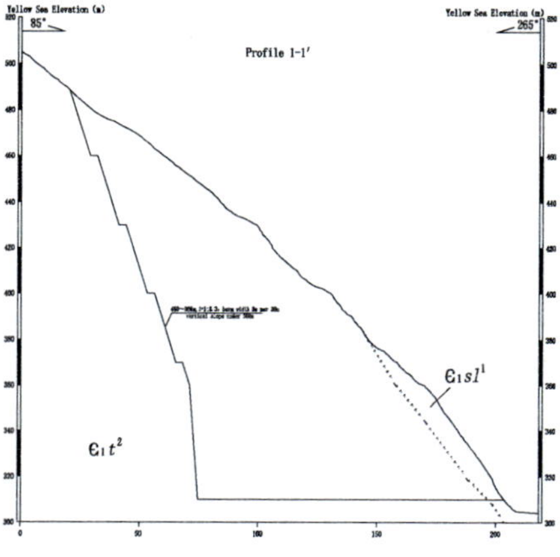

Fig. 4. Profile 1-1

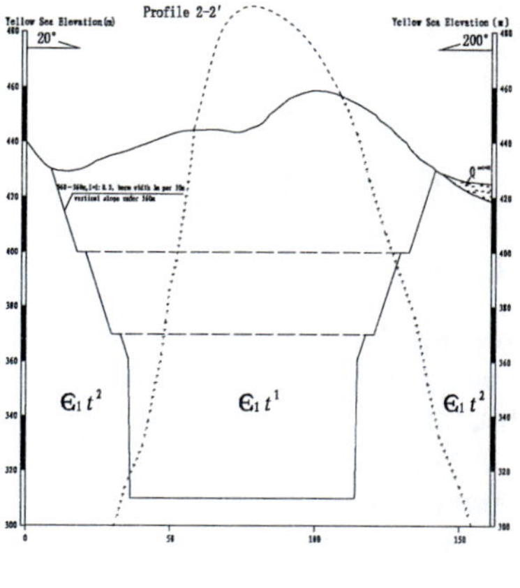

Fig. 5. Profile 2-2

4 Lithology

There are 3 material layers, lower Shilongdong Formation ($\epsilon_1 sl^1$), upper Tianheban Formation ($\epsilon_1 t^2$) and lower Tianheban Formation ($\epsilon_1 t^1$). Most of them are thick structure. They are as follows:

4.1 Lower Tianheban Formation ($\epsilon_1 t^1$)

The top is the bean-like limestone.The upper is dark gray, gray, thick with the thick bands of argillaceous limestone or micrite. The mid-lower part is the oolitic limestone or lime beaded body shale.The thickness is 52.7m.

4.2 Upper Tianheban Formation ($€_1t^2$)

There is gray and dark gray thick layer limestone. The bottom is 0.7m thick gray matter of two folders beaded. The upper is clip dolomitic limestone shale and a little gray dolomite, and contacts with the upper suture line. The thickness is 23.5 m.

4.3 Lower Shilongdong Formation ($€_1sl^1$)

There is gray, light gray crystallization dolomite limestone or gray dolomite. The upper part is dark gray mottled and wavy dolomitic limestone or limestone, constructed with sutures. The central is light gray folder argillaceous dolomite, and more with bedding. The thickness is 24.9m.

The exposed rocks of yard are thick layers. The rock intensity is high, and clay content is low, mostly useful for layers. According to analysis and thin section mineral composition, There are dolomitic limestone with mud microcrystalline ($€_1t$ argillaceous limestone), false sparry oolitic limestone and beans ($€_1t^1$ beans limestone), limestone like the gravel ($€_1t$ oolitic limestone), calcium powder dolomite ($€_1sl^1$ crystalline dolomite), coarse powder dolomitized limestone ($€_1sl^1$ wavy dolomitic limestone) and so on. Stripping layer material of yard are mainly broken rock with surface weathered and argillaceous rock with crack open filled. The thickness is about 2m.

5 Mining Area, Reserves and Quality Assessment

5.1 Determination of Mining Area

The main building of Sanliping hydropower station is RCC arch dam and underground powerhouse. According to design requirements, the main volume of concrete is 509,200 m^3, included in the temporary works, construction losses and geological defects treatment ,total concrete is about 540,100 m^3, total need concrete aggregate is approximately 831,000 m^3, of which coarse aggregate is about 574,000 m^3, fine aggregate (sand) is about 251,000 m^3, altogether to 646,800 m^3. in addition to the already existing reserve with parts and material of about 100,000 m^3, the current still need to aggregate about 500,000 m^3.

According to design requirements and consider the mining area around the terrain and geologic environment, an initial priority mining area is shown in the program one. The whole area of quarry is about 46000m^2, the vertical length is 220m, the maximum length down the river is 200m; priority mining areas, surface area of about 23000m^2, the vertical length of 200m, a maximum length down the river flow is 150m, according to the current status of mining and the owner comments, the upper excavation elevation limit is 493m. The lower limit establishes provisionally 310m. According to the actual amount used, it may be also suitable downward mining.

5.2 Reserves of Yard

Stratigraphic section based on field measurements revealed that the main peeling layer of yard is weathered fracture zone. The average thickness is 2.0m; The useful layers are limestone, microcrystalline limestone of Tianheban Formation and dolomite rock, gray dolomite or dolomitic limestone of Shilongdong Formation.

According to geological report wrote by Changjiang Institute of Survey, Planning, Design and Research, the pre-estimated total reserves of yard is 1.6 million m^3. Mining areas of priority reserves is calculated in the method of parallel sections. Elevation of the top mining sector is respectively 510m and 493m, while the bottom boundary elevation of mining is 320m and 310m. The final results of reserves is shown in Table1. Reserve calculation results show that the maximum aggregate recoverable reserves is 830,000m^3. The amount is basic to meet the design requirements. Other reasons Ruoyin more aggregate, may be appropriate to extend mining to the next bottom. Because if other reason need more aggregates, then the lower mine bound may be suitable to continue to extend.

Table 1. Quarry Reserves Calculation Sheet Priority Mining Areas

Top boundary elevation (m)	510		493		Remarks
Bottom elevation (m)	320	310	320	310	Each 30m wide and 3m leave a berm, 360-460m slope less than 1:0.3, excavation by nearly vertical slope under 360m
Reserves($10^4 m^3$)	74.0	83.0	50.0	55.0	

Table 2. Main Physical and Mechanical Rock Recommended Value Table

Stratum Code	Rock name	Proportion	Density(g/cm^3)		Compressive strength (MPa)		Softened coefficient	Water-absorptivity (%)
			Dry	Wet	Dry	Wet		
$\epsilon_1 t^2$	Argillaceous limestone	2.78	2.75	2.76	100	90	0.80	0.30
$\epsilon_1 t^1$	Lenticular limestone	2.78	2.76	2.77	100	85	0.75	0.24
$\epsilon_1 sl^1$	Dolomite /calcite dolomite	2.77	2.72	2.74	90	65	0.50	0.49

5.3 Quality Evaluation

At the investigation stage of the dam, a great deal of the original rock physical and mechanical tests is done. The major rock proposed physical and mechanical values of quarry are shown in table 2, based on their test results.

Through testing, sand (fine aggregate) quality of stone powder content compared with the code requirement larger, other indicators are in line with regulatory requirements; at coarse aggregate quality testing, the packing density is too small in the specification, the other indicators are in line regulatory requirements; at artificial sand particle size distribution testing, fineness modulus μ_x=3.0, the level belongs to zone 1, in line with regulatory requirements.

6 Environmental Geological Assessment of Mining Areas

The terrain of yard is integrated, landforms simple, no adverse geological phenomenon, the overall structural is stable. Exposed rock strata near the side of the mountain is bedrock of the lower Shilongdong Formation($\epsilon_1 sl^1$), the upper part is thin light gray argillaceous dolomite, the formation is oblique dip slope. If the beginning of mining is at the elevation 500m of the terrain. There will possibly present the side slope along the level slipping phenomenon, leading to larger difficulty of exploitation, increasing mining costs and affecting the progress of mining materials.

There is a collapse accumulation body at the northwest side of the priority mining areas. The horizontal width is 60m, the vertical length is 200m. Spatial form is like drop-shaped. The valley landscape is in the middle of the ridge on both sides of the rock formation. Consolidated excavation yard road on the terrain and geological data related to the yard, the thickness preliminary estimate is up to 18m, the minimum for the "drop" tip, only 1m or so. Accumulation body is mainly composed by the gravelly soil, rock composition of limestone and dolomite. The diameter is generally 2-5cm, and the large is up to 15-20cm. The upper part of 1m is more loosely, while the lower is more dense. The accumulation of whole body is currently in stable condition.

According to plans known, the upstream side of the accumulation body is the downstream exploitation of material-yard boundary. When the top of the extraction line is 510m, there will be a free surface of the accumulated after rock mined. It will slip to mining deformation zone by the weight and rainfall, which will have greater impact on the security of workers and the purity of aggregates. Therefore it is recommended that the 480-520m elevation of accumulation body near the mining areas should be removed, the 520m above is by cutting slopes and other parts is with the reinforcement.

The existence of blasting of artificial aggregate mining quarry may have some impact on the bridge. The slope after the excavation is also very steep, and the maximum elevation difference achieves 180m, there may cause local mining slope instability problems.

Exploitation of material-yard will form high slope, in order to ensure the smooth progress of construction safety, the excavation should be carried out with reference to proposed excavation ratio. After each first-level side slope forms, you must first relieve the dangerous work and carry on the next carriage excavation. If the mining approaches the side of the mountain of the quarry, more attention should be payed to the stability problem of slope. The slope of each ladder should carry on reinforcement before the next ladder excavation.

7 Conclusion

(1)The depression and useless layer of material-yard is small, mostly $\epsilon_1 t^1$, $\epsilon_1 t^2$ and $\epsilon_1 sl^1$ layer are as useful layer, the aggregate amount of reserves is to meet the design requirements, physical and mechanical properties meet the basic requirements of the dam used as aggregate.

(2)The existence blasting of artificial aggregate mining may have some impact on adjacent buildings (especially Lizigou Bridge) and local exploitation of slope stability problems.

Acknowledgments. The research has been supported in part by National Natural Science Foundation of P.R. China (No.50379023, 50679097, 90610029), Innovation Troups of Natural Science Foundation of Hubei Provice of P.R. China (No. 2005ABC009), National Key Technology R&D Program during the 11th Five-Year Plan Period (No.2008BAb03-4), Science Foundation of China Three Gorges University.

References

1. Li, J.L., Wang, L.H., Liu, J., et al.: Rockslope engineering. China Water Power Press, Beijing (2006)
2. Sheng, Q., Yue, Z.Q., Lee, C.F., et al.: Estimating the excavation disturbed zone in the permanent shiplock slopes of the Three Gorges Project. China. International Journal of Rock Mechanics and Mining Sciences 39(2), 165–184 (2002)
3. Griffiths, D.V., Lane, P.A.: Slope stability analysis by finite elements. Geotechnique 49(3), 387–403 (1999)
4. Yang, G.L., Huang, R.Q., Lin, F., et al.: Numerical analysis of deformation damage in hydropower station dam foundation due to excavation in Southwest China. Journal of Engineering Geology 15(3), 356–361 (2007)
5. Zhou, H.M., Sheng, Q., Li, W.S., et al.: Excavation disturbed zone and weaken degree of mechanical properties for rock mass of TGP shiplock slope. Chinese Journal of Rock Mechanics and Engineering 23(7), 1078–1081 (2004)
6. Malmgren, L., Saiang, D., Toyra, J.: The excavation disturbed zone (EDZ) at Kiirunavaara mine, Sweden - by seismic measurements. Journal of Applied Geophysics 61(1), 1–15 (2007)
7. Wu, F.Q., Liu, T., Liu, J.Y., Tang, X.L.: Excavation unloading destruction phenomena in rock dam foundations. Bulletin of Engineering Geology and the Environment 68(2), 257–262 (2009)

The Cause and Characteristics of Land Subsidence in Xi'an, China

Weifeng Wan[1], Juanjuan Zhang[2], and Yunfeng Li[3]

[1] Yellow River Engineering Consulting Co., Ltd.
Zhengzhou, Hennan, China
[2] No.2 Hydrogeological and Engineering Geological Brigade,
Henan Bureau of Geoexploration and MineralDevelopment
Zhengzhou, Hennan, China
[3] Chang'an University
Xi'an, Shaanxi, China
{Wwfmt,zhjj0822,lyfphd}@163.com

Abstract. Land subsidence is one of the most prominent geological disasters in Xi'an, China. Base on the review of the occurrence and development of land subsidence in Xi'an, the causes of it was analysed. The research results indicate that: over-exploitation of confined water is the main cause of land subsidence in Xi'an; regional tectonic activity is one but not the leading cause; the engineering activities have a certain influence on the land subsidence while the loess collapsibility has no general effect on it. Then the characteristics of the land subsidence were summarized. The land subsidence in Xi'an has the characters of the being consistent with confined water cones of depression, being inheritable of subsidence centers and unevenness in space and in time.

Keywords: Xi'an, land subsidence, Groundwater.

1 Introduction

Land subsidence refers to an environmental geological phenomenon of the reduced regional surface elevations due to the natural and (or) artificial soil mass compression below the ground surface. According to Bertoldi G.L., land subsidence is considered to be the loss of surface elevation caused by the movement of underground supporters [1, 2]. It is a geological disaster posing threats to resource utilization, environmental protection, economic development, urban construction, and residents living [3].

Xi'an is the largest city and an important industrial base in Northwest China. With the rapid development of construction and the continuous acceleration of urbanization process, the irrational exploitation and application of groundwater in the urban region have caused many geologically and environmentally negative effects, such as land subsidence and ground fissure. Of them, land subsidence is one of the most prominent geological disasters in Xi'an. Since the 1970s, land subsidence has been a major geological disaster influencing urban construction planning, land utilization, housing construction, and the safe use of underground pipeline engineering in Xi'an, having caused heavy losses to urban construction and planning. The land subsidence in Xi'an is primarily hazardous in damaging the surface buildings with foundation cracks,

D. Zeng (Ed.): Advances in Control and Communication, LNEE 137, pp. 437–443.
springerlink.com

breaking the water supply, sewage, and underground gas pipelines, raising well pipes, deforming road surfaces, etc. Additionally, land subsidence is threatening many cultural relics. The famous Big Wild Goose Pagoda of Tang Dynasty, for example, has titled northeast by 1,004 mm, and the Clock Tower has subsided by as many as 1,000mm [4] .

2 Development Process of Land Subsidence in Xi'an

The land subsidence in Xi'an mainly occurs in the urban and suburban areas. From 1959 to 1996, the area with the accumulated volume of land subsidence of over 100mm reached 150km^2 in Xi'an, involving an area of 200 km^2, and the maximum rate of land subsidence was 300mm/a. In the meantime, several subsidence centers were formed, each with an accumulated volume of land subsidence of over 2,000mm and the maximum reaching 2,600mm.

According to the rate of land subsidence, the formation and development history of the land subsidence in Xi'an can be divided into five phases [5], see table 1.

Table 1. Development Phases of Land Subsidence in Xi'an

Development Phases	slow subsidence phase	accelerated subsidence phase	abnormally accelerated subsidence phase	continuous subsidence phase	decelerated subsidence phase
Time	1959-1971	1972-1978	1979-1983	1984-2001	2002-2006
Subsidence rate (mm/a)	3.5-9.4	13.86-46.7	36.0-110.2	60.0-130	18.0-40

3 Causes of Land Subsidence in Xi'an

3.1 Over-Exploitation of Confined Water Is the Main Cause of Land Subsidence in Xi'an

The acceleration in the rate of land subsidence in Xi'an is consistent with the rapid increase in exploited quantities of self-drilled wells and the sharp fall of confined water level in time. Before 1970, there were few self-drilled wells in Xi'an, the confined water level was basically stable, and the rate of land subsidence was generally less than 3mm. Mainly controlled by the regional geological structure, the phreatic water level was relatively stable, primarily in the precipitation-type dynamics. After 1970, the number of self-drilled wells increased rapidly. With the accelerated falling rate of the confined water level, the rate of land subsidence was correspondingly increasing: the general trend of changes in land subsidence was closely related to the falling rate of the confined water level. Viewed from the dynamic trends of the confined water level after 2001, due to the diversion of Heihe River to Xi'an, the exploited quantities of self-drilled wells have fallen, and thus the water levels are stabilized in most parts of Xi'an; they are even rising in some areas. Therefore, the land subsidence is slowing down significantly in most areas.

Calculations with large amounts of data show that the volume of land subsidence is positively correlated to the decrease in the confined water head with a coefficient of over 0.85 (Fig.1), and the volume of land subsidence in the confined water cones of depression is increasing with the decrease of water levels.

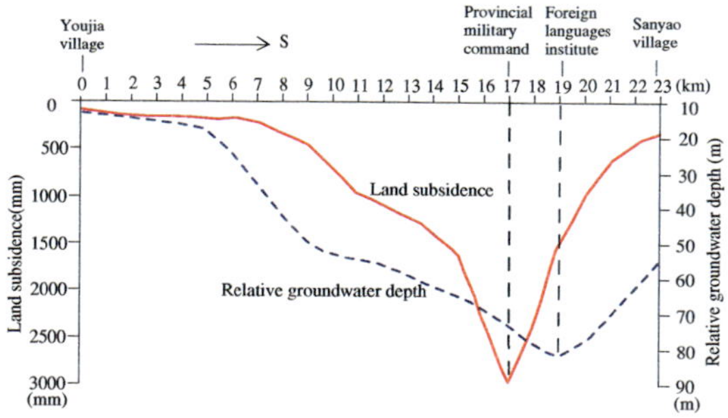

Fig. 1. Profile chart of land subsidence and relative groundwater depth [5]

The distribution of land subsidence in Xi'an is basically consistent with the distribution of depression cones of the regional water level, and the center of land subsidence is the center of the cone of depression. The areas where the confined water level rises a lot are in the eastern and southern suburbs, and several land subsidence cones with relatively large volume of land subsidence are all concentrated in the same areas.

3.2 Regional Tectonic Activity Is One But Not the Leading Cause of Land Subsidence in Xi'an

The area of land subsidence in Xi'an is located at the eastern edge of Xi'an fault, which has been in a state of relative subsidence for a long time.

The subsiding effect of Xi'an fault has continued till now. The repeated leveling results of Wei River Basin from 1970 to 1980 show that a triangular area of subsidence exists at the eastern edge of Xi'an fault zone (Fig.2), where the northern boundary of the subsidence area is the Wei River fault, and the southeastern one is the Chang'an - Lintong fault.

The loess plateaus in the south and southeast of the study area are relatively slowly rising, while the central and northern valley plain areas are the relatively slowly subsiding. The subsidence of the eastern edge of the Xi'an fault zone is related to the activities of the two above-mentioned faults. The multi-year fault leveling results reflect that the Wei River fault is undergoing the fault activities with the northern part rising and the southern subsiding with an average rate of 3.37mm/a and the Chang'an - Lintong fault is undergoing the fault activities with the northern part subsiding and the southern one rising with an rate of 3.98mm/a (Table 2).

Related studies show that the regional volume of land subsidence caused by tectonic activities accounts for 3% of the total[6].

The slow regional rates of land subsidence of the above-mentioned subsidence areas are inconsistent with the considerably constant subsidence in the suburbs of Xi'an since 1971. Thus, the relative subsidence of the Xi'an fault controlled by new tectonic movements is one but not the leading cause of land subsidence in Xi'an.

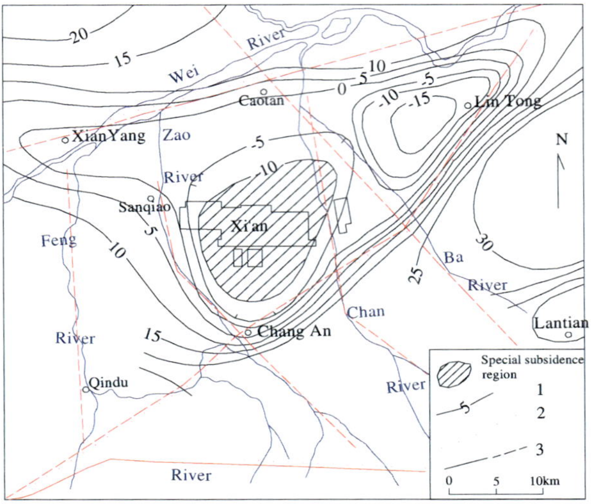

Note: 1- Isoclines of vertical deformation (positive values represent rising and negative values represent descending, and the unit is mm); 2 - geological structure boundary; 3-Faults and insidious faults

Fig. 2. Aggregate vertical deformation of Xi'an during 1970 and 1980[6]

Table 2. The Faults Activity Rate of in Xi'an Area [6]

Position	Wei River fault			Chang'an - Lintong fault				
Monitoring point number	3-9	9-18	1-2	4-3	28-1	27-2	7-6	MJ
Year	1970 -1980	1973 -1980	1970 -1988	1970 -1980	1971 -1980	1971 -1980	1970 -1980	1977 -1987
Rate(mm/a)	3.0	3.0	4.1	3.6	5.8	3.0	2.7	4.8

3.3 Engineering Activities Have a Certain Influence on Land Subsidence

In recent years, constructions of large-scale high-rise and ultra-high-rise buildings in residential areas have also caused land subsidence in some areas. As time goes on, the volume of land subsidence caused by constructions will account for a greater proportion of the total.

3.4 Loess Collapsibility Has No General Effect on Land Subsidence of Xi'an

Malan loess layers of 10 ~ 15 meters thick are widely distributed in the surface of subsidence areas. With the relative collapsibility coefficient of 0.022 ~ 0.035 and the compression coefficient of 0.02-0.46, it is classified as medium compression soil. Since the occurrence of land subsidence in the urban and suburban areas in 1971, some scholars maintain that the land subsidence might be caused by loess collapsibility considering loess distribution and characteristics in the region. However, Xi'an is often dry with few rains and a relatively small precipitation in recent decades. This fact, coupled with an increase in quantities of agricultural exploitation and other causes, the phreatic water level in this region has been stable or slightly descending. Therefore, the loess collapsibility can be concluded to have no universal effect on land subsidence in the entire city. Land subsidence occurring in recent years can not from the result of loess collapsibility caused by the ascending phreatic water level.

4 The Characteristics of Land Subsidence in Xi'an

In the subsidence areas of Xi'an, 14 ground fissures are distributed in the NNE direction, separating the subsidence areas into belts of the same direction and restricting the development of land subsidence in the horizontal direction. The land subsidence areas are oval-shaped as a whole, and most horizontal expansions of all formed depression cones are limited between the two ground fissures, thus forming a series of oval settlers in the NNE direction with the direction of the long axis basically coincident with that of ground fissures. All settlers are generally deep in the north and shallow in the south, and the volume of land subsidence is large to the south of ground fissures, thus forming the belts of terrain mutation in the form of scarps or steep slopes [7].

The main characteristics of land subsidence in Xi'an are as follows:

4.1 Subsidence Centers Are Basically Consistent with Confined Water Cones of Depression

Influenced by the hydrogeological conditions and the distribution of well groups, land subsidence centers in Xi'an are basically consistent with confined water cones of depression, and both are distributed oval-shaped in the NE direction in the overall plane. In the areas where confined water levels descend a lot, the volume of land subsidence is correspondingly greater.

4.2 Settling Velocity Fluctuations within a Year

The multi-layer monitoring data show that since the quantity of exploited confined water varies with quarters in a year, the water level falls at varying rates, thus resulting in fluctuating rates of land subsidence within the year. Generally, the volume of land subsidence is large in the second and third quarters (accounting for 63% of the annual total), followed by the first and fourth quarters (Fig.3).

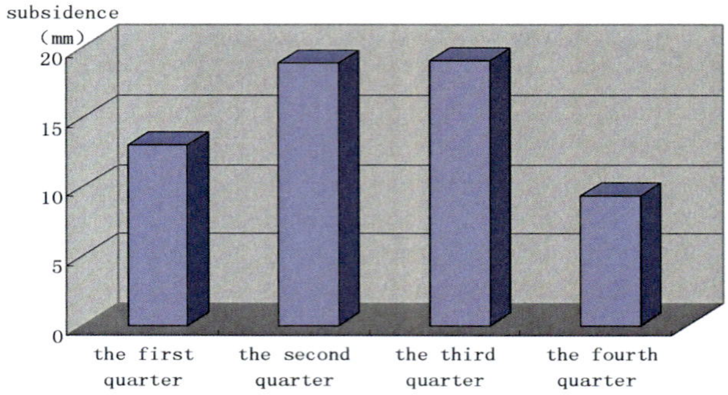

Fig. 3. The Histogram of volume of land subsidence during the year [5]

4.3 Inheritable Development of Subsidence Centers

Since the occurrence of land subsidence in Xi'an in 1959, subsidence centers have formed around Xiaozhai, Shapo, NPU and Hujiamiao, and are still developing. Their distribution and development are inheritable spatially and temporally.

4.4 Subsidence Activities' Spatial Unevenness

1) The land subsidence has rapid vertical development yet slow horizontal development

With the relatively unchanged number of exploited well and exploited quantities, land subsidence cones are horizontally expanding slowly. Since 1985, the subsidence area has been basically kept at around 200 km^2, but the vertical volume and rate of land subsidence in all subsidence centers are growing faster. For example, the rate of land subsidence in Xiaozhai area was 26mm/a in 1970s, but that was up to 125mm/a at the end of 1980s, almost a 5-fold increase in the rate of land subsidence in a decade. The main reason for this phenomenon is that the exploitation of groundwater is not significantly expanded to the surrounding areas but still focused on urban and suburban areas, but the groundwater level continues to fall with the exploitation of groundwater, and the water level cone of depression is developing vertically.

2) Land Subsidence's Regional Differences

Influenced by layers, distribution of well groups, and exploited quantities, etc., the land subsidence in Xi'an is shown uneven spatially. The overall pattern is that land subsidence in the eastern and southern suburbs is greater than that in the northern suburbs, and the general trend is that the volume of land subsidence decreases from the southeast to the northwest. In 1989, the rate of land subsidence has a difference of 3.8-5.5 times between the southern and northern suburbs. By the end of 1995, the monitoring point in northeast suburbs had an accumulated volume of land subsidence of 1,000mm, while that was 2,322mm in one monitoring point of the southern suburbs, resulting in a difference of 1.3 times between the two locations.

3)Land Subsidence's Vertical Differences

The wells for production in Xi'an suburbs mainly exploit the confined aquifers at the buried depth of 100-300m, and most exploited wells are at the buried depth of 100-250m. The multi-layer monitoring data of subsidence in the southern suburbs shows that volume of land subsidence is different vertically. The volume of land subsidence at the layer with a buried depth of 0-100m accounts for about 15% of the total; the volume of land subsidence at the buried depth of 100-300m accounts for approximately 79% ; and that at the buried depth of 300-367m accounts for only 10%.

5 Conclusions

From what has been discussed above, the following conclusions can be drawn.

1) According to the rate of land subsidence, the formation and development history of the land subsidence in Xi'an can be divided into five phases: slow subsidence phase, accelerated subsidence phase, abnormally accelerated subsidence phase, continuous subsidence phase and decelerated subsidence phase. Each phase is closely related to the exploitation of groundwater.

2) Over-exploitation of confined water is the main cause of land subsidence in Xi'an; regional tectonic activity is one but not the leading cause; the engineering activities have a certain influence on the land subsidence while the loess collapsibility has no general effect on it. So shutting down the private wells and reducing groundwater exploitation have very apparent effect on decelerating subsidence development.

3) Being consistent with groundwater depression cones, being inheritable of subsidence centers and unevenness in space and in time are the main characteristics of land subsidence in Xi'an.

References

1. Liu, Y.: Land Subsidence Research Approaches and Advent Problems. Earth Science Frontier 8(2), 273–278 (2001)
2. Bertoldi, G.L.: Where do we go from here: a new initiative for subsidence research in the United States. In: GRA 1995 Annual Meeting., AEG, California (1995)
3. Zheng, X., Wu, Q., Hou, Y., et al.: Advances and Trends in Research on Urban Land Subsidence. Geological Review 48(6), 612–618 (2002)
4. Yin, Y., Zhang, Z., Zhang, K.: Land Subsidence and Countermeasures for Its Prevention in China. The Chinese Journal of Geological Hazard and Control 16(2), 1–7 (2005)
5. Geo-Environmental Monitoring Central Station of Shaanxi Province, "Investigation report of Land Subsidence and Land Fissures in Xi'an", unpublished (2007)
6. No.1 Hydrogeological and Engineering Geological Brigade of Shaanxi Province, "The Comprehensive Report of investigation and Research on Xi'an ground Fissures," unpublished (1990)
7. Wan, W., Li, Y., Zhang, J.: Development of Land Subsidence in Xi'an in Past 50 Years., Advanced Materials Research, Vols. 113-114, pp. 1199-1207 (2010)

Origin Analysis on Breccia of the Lower Mantou Fromation of Cambrian System in Reservoir Area in Hekou Village

Weifeng Wan, Qingjun Liu, Yaojun Wang, and Jichang Gong

Yellow River Engineering Consulting Co., Ltd.
Zhengzhou, Henan, P.R. China
wwfmt@163.com

Abstract. Mantou Fromation of Cambrian System is one of the major strata developed in the reservoir and dam area. It belongs to Mantou Fromation of Cambrian System, which is closely related to the engineering geology and leakage feature of the reservoir, and the breccia is found to distribute in many places in the lower part Mantou Fromation. Geological body and lithological features of breccias are described in significant detail in this paper. The breccia of the lower Mantou Fromation can be classified into 2 types. Origin of the two types of the breccia was analyzed on basis of the result of the sedimentary environment study and mineral chemical composition testing. One type of the breccia was formed in the lower incompetent rock layers of Mantou Fromation, formed by weathering and unloading at the offshore slopes; the other type was developed in the brittle rocks in the lower portion of Mantou Fromation as a result of tectonic movements. This origin analysis offers a basis for the assessment on engineering geological issues and reservoir leakage in the reservoir areas.

Keywords: Reservoir of Hekou Village, Mantou Fromation, Breccia, Dissolution Collapse.

1 Introduction

Reservoir at the estuary of Qinhe River is located in Jiyuan City of Henan Province, the south end of Taihang Mountain valley in the middle reaches of Qinhe River and about 9.0Km away from the outlet of the valley-Wulongkou. Outcrops of Cambrian System in the reservoir and dam area include multiple alternating permeable and waterproof layers of limestone, mud limestone, shale and sandstone interbeddings with gentle strata occurrence. In the process of the thematic exploration work of leakage in this reservoir and dam areas, the breccias were discovered in the 3rd rock layer of Mantou Fromation in the lower Cambrian systerm, mostly outcroping in bank and slope area and audits in the dam area. Revealing the origin of breccia and the distribution code are very important for the understanding of leaking medium's property. Similar breccias were also discovered in the strata of the Mantou Formation of Cambrian system in other areas, such as Xishan area and Huashan area in Beijing

D. Zeng (Ed.): Advances in Control and Communication, LNEE 137, pp. 445–451.
springerlink.com

and Luxi area in Shandong. Through preliminary research of mineralogy and sedimentary face sequence, Fan Kaiqiang [1] analyzed the origin of the breccia in Mantou Fromation of the lower Cambrian system in Xishan area in Beijing, concluding that the breccia of Mantou Fromation in Xishan area belongs to the evaporate dissolving and slump breccias formed as a result of dissolution of gypsum layers, the collapsing of gypseous micritic dolomite under gravity and the dissolving of gypsum contained in gypseous micritic dolomite. Yan Zhen and etc. [2, 3] analyzed the turbidite and slump stack combination discovered in Luxi area, and thought that the once existed paleoslope and tectonic uplifting in North China platform during the Cambrian period was the trigging factor for slump accumulation during diagenesis.

As for the breecia of Mantou Fromation in the reservoir dam area in Hekou Village, some experts and scholars doubted that it belonged to the dissolving and slump breccia of gypsum-dissolving type. Through research on sedimentary environment and mineral chemical composition, the breccia was classified and its origin was analyzed, which is of great significance for getting to know the leakage features of rock layers in the lower portion of Mantou Fromation and for the analysis of karst development.

2 Geological Conditions in the Reservoir Area

The reservoir in Hekou Village is a typical valley river reservoir, and north branch of the regional Pangusi fault crosscuts Qinhe River at the lower reache of the dam site (the nearest distance is about 620m away from the dam axis) and serves as the landform boundary of the medium-height valley area and the low-height valley and basin area.

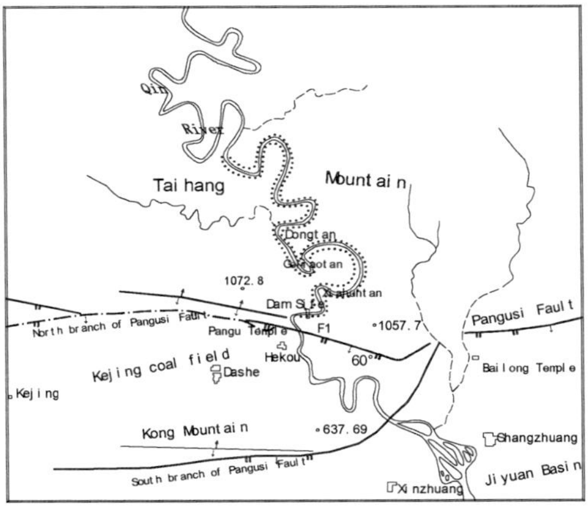

Fig. 1. Schematic diagram of geographical location of reservoir at Hekou Village in Qinhe River

Outcroping strata in the reservoir area upwards are as follows: the Presinian Archean Dengfeng Group, Sinian Mesoproterozoic Ruyang Group, Paleozoic Cambrian System, Cenozoic Ordovician and Quaternary System [4].

Bed rock of reservoir and dam disc is a rock body alternating by pervious and waterproof multi-layers: the lower is metamorphic and clastic rock of the Archean Dengfeng Group and Mesoproterozoic Ruyang Group, which is waterproof rock body; the intermediate and lower portions are rock layers of the Mantou Fromation of Cambrian System including lower parts of $\in_1m^1$, $\in_1m^2$, $\in_1m^3$ and $\in_1m^4$, with thickness from 32m~34m, lithology includes dolomite, marlstone and shale. Under the effect of tectonic movements, dragged folds were well developed and rock bodies have been crushed , accompanied by karst caves and dissolved holes, belonging to permeable layers; the upper portion is the upper Mantou Fromation of the Cambrian System (upper $\in_1m^4$, $\in_2m^5$, $\in_1m^6$), including interbedding limestone, shale and sandstone of Maozhuang and Xuzhuang Formation, which belongs to relatively waterproof rock body; the reservoir disc top is Zhangxia Formation $\in_2z$ (including of Xuzhuang Formation $\in_2X^4$) mainly consisting of karstified limestone with developed karst caves. It has strong water permeability and distributes 275m higher than the reservoir area.

Tectonic structures in the area are predominantly SW normal steep faults, locally accompanied by (near the dam site Guitou Mountain) reverse faults, over-thrust faults and small scale folds and the trend of these faults is 270°~300° in dominancy.

3 Summarization of Breccia Bevelopment Phenomenon in Mantou Fromation €1m3

The strata of Mantou Fromation $€_1m^3$ can be classified into 4 layers from lower to upper, lithology of this rock layer $\in_1m^{3-2}$ is mainly dolomite. Due to the occurrence of calcite veins, which taking in the form of wall, the lithology is brittle and hard. Over and below this layer are $\in_1m^{3-1}$ and $\in_1m^{3-3}$ respectively, composed of incompetent grayish yellow marlstone. Affected by the tectonic movements, dragged folds are frequently seen in inter-layer of $€_1m^3$. At places where there are inter-layer rugosities, deformation of $\in_1m^{3-1}$ and $\in_1m^{3-3}$ is weaker than that of $\in_1m^{3-2}$. "Small fault" is shown in $\in_1m^{3-2}$, yet no obvious displacement was discovered in the marlstones of $\in_1m^{3-1}$ and $\in_1m^{3-3}$ along direction vertical to rock layers, yet the deformation here is mainly reflected by "small rugosities". This phenomenon is not obvious or disappears at the bottom of the upper $\in_1m^4$ close to $\in_1m^3$.

The currently discovered breccia in strata of $€_1m^3$ has been classified into 2 types, the one outcrops in the greyish yellow marlstone of $€_1m^{3-1}$ and $€_1m^{3-3}$(Fig.2), mainly distributed near the slope, and the bedding of this strata gradually becomes more and more obvious farther away from the slope while the breccia becomes weak and disappears finally; the other one is the locally developed breccia in wall-shape dolomite in $€_1m^{3-2}$(Fig.3), and the outcrops are distributed at the right bank cliff of Xiahui Beach and the left bank of Xiezhuang-Xiahuitan Qinhe River.

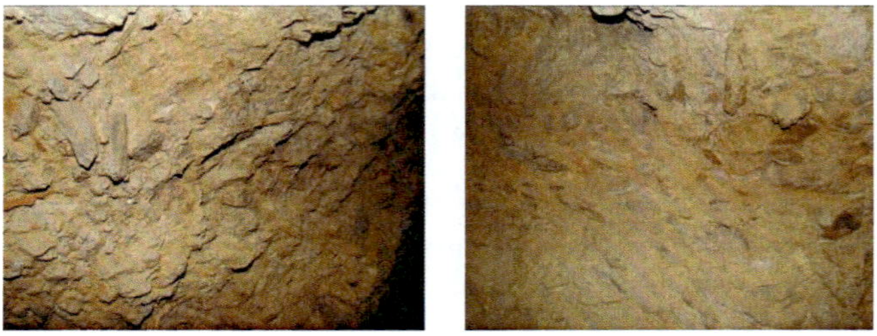

Fig. 2. Breccia in $Є_1m^{3-1}$ and $\in_1m^{3-3}$ greyish yellow marlstone

Fig. 3. Breccia in $Є_1m^{3-2}$ plate-shape dolomite

4 Analysis on Origin of Breccia

4.1 Origin for the 1st Type of Dissolving and Slump Breccia

The 1st type of "breccia" is featured by "dissolving and slump breccias of sypseum dissolving type" in appearance.

According to paleogeography evolution of the Cambrian System, Northern China saw the paleogeography transformation from transgression to regression during the early Cambrian Epoch to the late Cambrian Epoch, and the main sedimentary environments include tidal flat, open platform, platform tidal flat, and the platform tidal flat representing the tidal flat sedimentation in the carbonate platform, which can be further classified into intertidal and supralittoral zone faces [5]. Gypseum rock layer of the Cambrian system was mainly developed on the 3rd member of Xinji Formation in the supralittoral zone. In accordance with the geological investigation and drilling data, Strata of Xinji Formation are missing in most sections of engineering area. There is no strata condition of extensively distributing gypseum karst in the perspective of sedimentary environments.

To further analyze and find out the origins, the mineral compositions of breccia were analyzed. In accordance with the assay results of 12 samples from the rock layers of $Є_1m^{3-1}$ and $Є_1m^{3-3}$, compositions of breccia were calcite, dolomite and chlorite with relatively constant amount of quartz, bronze mica. There were no plasters detected other than in one group samples (Table 1). On basis of the thin-section appraisal results of the rock samples from $Є_1m^{3-1}$ and $Є_1m^{3-3}$, plasters were not discovered. Therefore, the origin of gypsum "dissolving and slump breccia" of gypsum dissolving type is not tenable.

Table 1. Results of Rock Sample Mineral Testing

Experimenting conditions: Lixue DMAX-3C diffractometer, CuKa, Ni filtering												
Sample No.	Sampling location	Stratum code	Testing results (%)									
			Montmorillonite	Illite	Chlorite	Quartz	Feldspar	Calcite	Dolomite	Bronze mica	Talcum	Anhydrous gypsum
1	Horizontal tunnels	$Є_1m^{3-1}$			38	18	1	30		12		
2	Drill Hole	$Є_1m^{3-1}$			14	13		20	22	13	18	
3	Horizontal tunnels	$Є_1m^{3-1}$			22	20	1	47	1	9		
4	Drill Hole	$Є_1m^{3-1}$			24	20		17	20	15	4	
5	Horizontal tunnels	$Є_1m^{3-1}$			45	17	1	26	1	10		
6	Bank Slop	$Є_1m^{3-1}$			31	24	2	35		8		
7	Drill Hole	$Є_1m^{3-3}$			34	15		25	13	13		
8	Horizontal tunnels	$Є_1m^{3-3}$			9	11		27	39	14		
9	Horizontal tunnels	$Є_1m^{3-3}$			10	14		21	43	12		
10	Horizontal tunnels	$Є_1m^{3-3}$			9	18		7	56	11		
11	Horizontal tunnels	$Є_1m^{3-3}$			12	9	1	26	43	9		
12	Drill Hole	$Є_1m^{3-3}$	2	17	6	12		10	51			2

4.2 Origin of the 2nd Type of Dissolving and Slump Breccia

The 2nd type of "breccia" outcrops at Xiahui Beach and at the cliff of the opposite bank. The feature of the breccia is that it occurs in strata of $\in_1m^{3-2}$, namely the wall-shape dolomite. The breccia is irregular, with feature of local dissolving and slump breccias. At the exposure location, there are large amounts of small inter-layer faults and small rugosities that can be seen in the strata of $\in_1m^3$, particularly to the left bank of Xiahui Beach, and $\in_1m^{3-2}$ wall-shape dolomite exposes on the ground surface, which is more obvious.

According to exposure location of breccia and its accompanying location relationship with small faults and rugosity in $\in_1m^3$ strata, it can be concluded that the origin of breccia is controlled by tectonic movements. Under tectonic stress, differentiate deformation happened between the upper and lower incompetent layers and the inter brittle and hard layers, the brittle and hard rock bodies were faulted and deformed along the tectonic function line and bedding direction. The weight of rock body decreased, the structure of original rock body was damaged, even overground phenomenon happened (Fig.4). Located at the slope area on both banks of Qinhe River, activity of the underground water in these strata is strong. After a relatively long geological time period, while the broken rock body is under comprehensive alteration by weathering, unloading, leaching, corrosion and etc., it was also re-cemented (calcium re-cementation), and the dissolving and slump "breccia" was thus formed, namely, the forming process of the breccia undergoes deformation and faulting, corrosion and collapsing, re-cementation.

Fig. 4. Rugosities in Plate Dolomite of $\in_1 m^{3-2}$ in Xiahuntan

5 Conclusions

1) Breccia in $Є_1m^3$ stratum can be classified into 2 types, the 1st type exposes in the greyish yellow marlstone of $Є_1m^{3-1}$ and $Є_1m^{3-3}$; the 2nd type is the locally developed breccia in wall-shape dolomite of $Є_1m^{3-2}$, these two type of breccias have conspicuous differences;

2) Based on analysis of the Paleogeography and paleosedimentary environment, there is no stratum condition of plaster karst widely distributing in the project area. The mineral testing results of rock samples indicate that the breccia developed in this area is not "dissolving and slump breccias" of the gypsum-dissolving type.

3) The formation of the 1st type of the breccia is synergetic result of tectonic movements and weathering, under which the rock layers of $Є_1m^{3-1}$ and $Є_1m^{3-3}$ were softened and deformed; the controlling factor of the 2nd type of breccia is mainly the tectonic movements. Under the tectonic stress, the upper and lower incompetent layers and the inter brittle and competent layers deformed differently, the structure of the later is destroyed and then suffered from a comprehensive transformation by weathering, unloading, leaching, corroding and etc., namely, the forming process of the breccia undergoes deformation and faulting, corrosion and collapsing, and re-cementation.

References

1. Fan, K., Yin, Z., Wang, C.: Origin of the Breccia of the Lower Cambrian Mantou Formation in the Western Hills, Beijing. Bulletin of Institute of Geology, Chinese Academy of Geological Sciences 22, 23–38 (1999) (in Chinese)
2. Yan, Z., Yang, C., Li, J., et al.: Geological Features and Significance of Olistostromes from Lower Part of the Mantou Formation in the Luxi Areas, Shandong Province. Acta Sedimentologica Sinica 23(1), 29–33 (2005) (in Chinese)
3. Yan, Z., Yang, C., Li, J., et al.: Features of Turbidites and Olistostrome in the Mantou Formation in Jinan and Xintai, Shandong, and Its Tectonic Significance. Chinese Journal of Geology 40(4), 579–584 (2005) (in Chinese)
4. Liu, Q.J., et al.: Feasibility Research Repot of Hekoucun Reservoir Project of Qin River. Yellow River Engineering Consulting Co., Ltd., Zhengzhou (2009) (unpublished) (in Chinese)
5. Li, Y., Yuan, J.L., Lin, T.R.: Lower Cambrian trace fossils from the Mantou Formation of Huainan, Anhui. Acta Paleontologica Sinica 38, 114–124 (1999)

The Existence of Analytic Solutions of a Functional Equation for Invariant Curves

Lingxia Liu

Department of Mathematics, Weifang University, Weifang, Shandong 261061, P.R. China
llxmath@126.com

Abstract. This paper is concerned with an iterative functional equation from the problem of invariant curves. By reducing the equation with the *Schröder* transformation to another functional equation without iteration of the unknown function *f*. For technical reasons, in previous work the constant λ given in the *Schröder* transformation is required to fulfil that λ is off the unit circle or lies on the circle with the Diophantine condition. In this paper, we obtain analytic solutions in the case of λ at resonance, i.e., at a root of the unity and the case of λ near resonance under the Brjuno condition.

Keywords: Iterative functional equation, Convergence, Brjuno condition, Analytic solution, Majorant series.

1 Introduction

The iterative functional equations

$$f(f(z)) = 2f(z) - z - \frac{1}{2}(g(f(z)) + g(z)), \quad z \in C \tag{1}$$

is studied on existence, periodicity oscillation and analytic solutions ([1-7]). Eq. (1) is deduced from the problem of invariant curves ([6]) for a delay differential equation with a piecewise constant argument

$$\frac{d^2}{dt^2}x(t) + g(x(t)) = 0, t \in R, \ x \in R, \tag{2}$$

where $[\,t\,]$ denotes the greatest integer less than or equal to t and $g: \mathbf{R} \to \mathbf{R}$ is continuous. In [7], the existence of analytic solutions is discussed in complex field **C**. In Eq. (1), f is unknown and g is a given function which is analytic in a neighbourhood of the origin, such that $g(0) = 0$ and $g'(0) = \xi \neq 0$. As in [7], we reduce the equation to another functional equation without iteration

$$\phi(\lambda^2 z) = 2\phi(\lambda z) - \phi(z) - \frac{1}{2}(g(\phi(\lambda z)) + g(\phi(z))), \ z \in C, \tag{3}$$

called the auxiliary equation of (1), where $\lambda \neq 0$ satisfies the algebraic equation

$$2\lambda^2 - (4-\xi)\lambda + 2 + \xi = 0. \tag{4}$$

D. Zeng (Ed.): Advances in Control and Communication, LNEE 137, pp. 453–461.
springerlink.com

For technical reason, in [7], the constant λ given in the *Schröder* **transformation, is required to fulfil that** α is of the unit circle ($0<|\lambda|\neq 1$) or λ is on the circle i.e., $|\lambda|=1$ with the Diophantine condition: $|\mu|=1$, μ *is not a root of unity, and* $\log\frac{1}{|\mu^n-1|}\leq T\log n,\quad n=2,3,\cdots$ *for some positive constant* T. Roughly speaking, the Diophantine condition requires λ to be far from all roots of unity where the fixed point 0 is irrationally neutral. In this paper, we break the restriction of Diophantine condition and discuss not only those λ near resonance under the Brjuno condition. The complex λ in (3) satisfies (4) and the following hypotheses:

(H1) $\mu=e^{2\pi i\theta}$, where $\theta\in \mathbf{R}\backslash\mathbf{Q}$ is a Brjuno number [11,12], i.e., $B(\theta)=\sum_{k=0}^{\infty}\frac{\log q_{k+1}}{q_k}<\infty$, where $\{p_k/q_k\}$ denotes the sequence of partial fraction of the continued fraction expansion of θ, and is said to satisfy the Brjuno condition.

(H_2) $\alpha=e^{2\pi iq/p}$ for some integers $p\in\mathbf{N}$ with $p\geq 2$ and $q\in\mathbf{Z}\backslash\{0\}$, and $\alpha\neq e^{2\pi il/k}$ for all $1\leq k\leq p-1$ and $l\in \mathbf{Z}\backslash\{0\}$.

The number λ is on the unit circle in (H_1) and (H_2). More difficulties are encountered for λ on the unit circle, as mentioned in the so-called "small-divisor problem". In this paper, considering the Brjuno condition instead of the Diophantine one, we discuss not only (H_1) but also (H_2) for analytic invariant curve (1).

2 Analytic Solution of the Auxiliary Equation (3)

As stated in [10], for a real number θ we let θ denote its integer part and let $\{\theta\}=\theta-[\theta]$. Then every national number θ has a unique expression of the Gauss' continued fraction

$$\theta=a_0+\theta_0=a_0+\frac{1}{a_1+\theta_1}=\ldots,$$

denoted simply by $\theta=[a_0,a_1,\ldots,a_n,\ldots]$, where a_j's and θ_j's are calculated by the algorithm:

(a) $a_0=[\theta],\ \theta_0=\{\theta\}$ and

(b) $a_n=\left[\frac{1}{\theta_{n-1}}\right],\theta_n=\left\{\frac{1}{\theta_{n-1}}\right\}$ for all $n\geq 1$.

Define the sequences $(p_n)_{n\in\mathbf{N}}$ and $(q_n)_{n\in\mathbf{N}}$ as follows

$$q_{-2}=1,\ q_{-1}=0,\ q_n=a_nq_{n-1}+q_{n-2};\qquad p_{-2}=0,\ p_{-1}=1,\ p_n=a_np_{n-1}+p_{n-2}.$$

It is easy to show that

$$p_n/q_n=[a_0,a_1,\ldots,a_n].$$

Thus, for every $\theta\in\mathbf{R}\backslash\mathbf{Q}$ we associate, using its convergence, an arithmetical function

$$B(\theta)=\sum_{n\geq 0}\frac{\log q_{n+1}}{q_n}.$$

We say that θ is a Brjuno number or that it satisfies Brjuno condition if $B(\theta)<+\infty$. The Brjuno condition is weaker than the Diophantine condition. For example, if $a_{n+1}\leq ce^{a_n}$ for all $n\geq 0$, where $c>0$ is a constant, then

$\theta=[a_0,a_1,\ldots,a_n,\ldots]$

is a Brjuno number but is not a Diophantine number. So, the case (H1) contains both Diophantine condition and a part of μ ``near'' resonance.

In order to discuss the existence of the auxiliary equation (4) under (H1), we need to introduce Davie's Lemma. First, we recall some facts in [11] briefly. Let $\theta\in\mathbf{R}\backslash\mathbf{Q}$ ··· and $(q_n)_{n\in\mathbf{N}}$ be the sequence of partial denominators of the Gauss's continued fraction for θ as in the Introduction. As in [10], let

$$A_k=\{n\geq 0\,|\,\|n\theta\|\leq\frac{1}{8q_k}\},$$

$$E_k=\max(q_k,\frac{q_{k+1}}{4}),\ \eta_k=\frac{q_k}{E_k}.$$

Let A_k^* be the set of integers $j\geq 0$ such that either $j\in A_k$ or for some j_1 and j_2 in A_k, with $j_2-j_1<E_k$, one has $j_1<j<j_2$ and q_k divide $j-j_1$. For any integer $n\geq 0$, define

$$l_k(n)=\max\left((1+\eta_k)\frac{n}{q_k}-2,\quad (m_n\eta_k+n)\frac{1}{q_k}-1\right),$$

Where $m_n=\max\{j\,|\,0\leq j\leq n,\ j\in A_k^*\}$. We then define function $h_k:\mathbf{N}\to\mathbf{R}_+$ as follows:

$$\begin{cases}\dfrac{m_n+\eta_k n}{q_k}-1, & \text{if}\quad m_n+q_k\in A_k^*,\\ l_k(n), & \text{if}\quad m_n+q_k\notin A_k^*.\end{cases}$$

Let

$$g_k(n):=\max\left(h_k(n),\left[\frac{n}{q_k}\right]\right),$$

and define $k(n)$ by the condition $q_{k(n)}\leq n\leq q_{k(n)+1}$. Clearly, $k(n)$ is non-decreasing. Then we are able to state the following result:

Lemma 1. (Davie's Lemma [11]) Let

$$K(n)=n\log 2+\sum_{k=0}^{k(n)}g_k(n)\log(2q_{k+1}).$$

Then,

(a) There is a universal constant $\gamma>0$ (independent of n and θ) such that $K(n)\leq n\left(\sum_{k=0}^{k(n)}\frac{\log q_{k+1}}{q_k}+\gamma\right)$,

(b) $K(n_1)+K(n_2)\le K(n_1+n_2)$ for all n_1 and n_2, and

(c) $-\log|\alpha^n-1|\le K(n)-K(n-1)$.

Now we state and prove the following theorem under Brjuno condition. The idea of our proof is acquired from [11].

Theorem 1. Suppose (H1) holds, then for any $\tau\in \mathbf{C}$, the auxiliary equation (3) has an analytic solution $\phi(z)$ in a neighborhood of the origin such that $\phi(0)=0,\ \phi(0)=\tau$.

Proof: If $\tau=0$, (3) has a trivial solution $\phi(z)=0$. Assume $\tau\ne 0$, let

$$g(z)=\sum_{n=1}^{\infty}a_n z^n,\ a_1=\xi. \tag{5}$$

As in [7], we assume that

$$|a_n|\le 1,\ n=2,3\cdots. \tag{6}$$

Furthermore, let

$$\phi(z)=\sum_{n=1}^{\infty}b_n z^n \tag{7}$$

be the expansion of a formal solution $\phi(z)$ of (3). Substituting to (5) and (7) in (3) we have

$$\sum_{n=1}^{\infty}b_n\lambda^{2n}z^n=\sum_{n=1}^{\infty}(2\lambda^n-1)b_nz^n-\frac{1}{2}\lambda^n\sum_{n=1}^{\infty}\Big(\sum_{\substack{l_1+l_2+\cdots+l_k=n\\ k=1,2,\cdots n}}a_kb_{l_1}b_{l_2}\cdots b_{l_k}\Big)z^n-\frac{1}{2}\sum_{n=1}^{\infty}\Big(\sum_{\substack{l_1+l_2+\cdots+l_k=n\\ k=1,2,\cdots n}}a_kb_{l_1}b_{l_2}\cdots b_{l_k}\Big)z^n.$$

Comparing coefficients we obtain

$$(\lambda^{2n}-2\lambda^n+1)b_n=-\frac{1}{2}(\lambda^n+1)\sum_{\substack{l_1+l_2+\cdots+l_k=n\\ k=1,2,\cdots n}}a_kb_{l_1}b_{l_2}\cdots b_{l_k},\quad n=1,2,\cdots.$$

This implies that

$$\left[\lambda^2-2\lambda+1+\frac{1}{2}(\lambda+1)\xi\right]b_1=0, \tag{8}$$

$$\left[\lambda^{2n}-2\lambda^n+1+\frac{1}{2}(\lambda^n+1)\xi\right]b_n=-\frac{1}{2}(\lambda^n+1)\sum_{\substack{l_1+l_2+\cdots+l_k=n\\ k=2,3\cdots n}}a_kb_{l_1}b_{l_2}\cdots b_{l_k},\qquad n=2,3,\cdots. \tag{9}$$

From (4) we have $\xi=-2(\lambda-1)^2/(\lambda+1)$, so (9) is reduced to

$$b_n=-\frac{(\lambda+1)(\lambda^n+1)}{2(\lambda^n-\lambda)(\lambda^{n+1}+\lambda^n+\lambda-3)}\sum_{\substack{l_1+l_2+\cdots+l_k=n\\ k=2,3\cdots n}}a_kb_{l_1}b_{l_2}\cdots b_{l_k},\ n=2,3,\cdots \tag{10}$$

In particular, (4) implies the coefficient of b_1 is zero, then we choose $b_1 = \tau \neq 0$, we can uniquely determine the sequence $\{b_n\}_{n=2}^{\infty}$ by (10) recursively.

Note that $\lambda = e^{2\pi i\theta}, \theta \in \mathbb{R}/\mathbb{Q}$, since (H$_1$) implies $|\lambda| = 1$ and $\lambda^n \neq 1, \forall n = 1, 2, \cdots$. Thus $|\lambda - 3| = |\cos 2\pi\theta + i\sin 2\pi\theta - 3| \geq 3 - \cos 2\pi\theta > 2$.

That is, $N = |\lambda - 3| - 2 > 0$. From (10),

$$b_n \leq \frac{(|\lambda|+1)(|\lambda|^n+1)}{2|\lambda|\cdot|\lambda^{n-1}-1|(|\lambda-3|-|\lambda|^{n+1}-|\lambda|^n)} \sum_{\substack{l_1+l_2+\cdots+l_k=n \\ k=2,3\cdots n}} |b_{l_1}||b_{l_2}|\cdots|b_{l_k}|$$

$$\leq \frac{2}{|\lambda^{n-1}-1|(|\lambda-3|-2)} \sum_{\substack{l_1+l_2+\cdots+l_k=n \\ k=2,3\cdots n}} |b_{l_1}|\cdot|b_{l_2}|\cdots|b_{l_k}|$$

$$\leq \frac{2}{N|\lambda^{n-1}-1|} \sum_{\substack{l_1+l_2+\cdots+l_k=n \\ k=2,3\cdots n}} |b_{l_1}||b_{l_2}|\cdots|b_{l_k}|, \; n \geq 2. \tag{11}$$

To construct a governing series of (7) we consider the implicit function equation

$$V(z) = |\tau| z + \frac{2}{N} \cdot \frac{(V(z))^2}{1 - V(z)} \tag{12}$$

Define the function

$$F(z, V; \tau, N) := V - |\tau| z - \frac{2}{N} \cdot \frac{V^2}{1 - V} \tag{13}$$

for (z, V) in a neighborhood of $(0,0)$. Then $F(0,0;\tau,N) := 0$, $F_V'(0,0;\tau,N) := 1 \neq 0$. Thus, there exists a unique function $V(z;\tau,N)$, analytic in a neighborhood of zero, such that $V(0;\tau,N) = 0$, $V_z'(0;\tau,N) = |\tau|$ and $F(z, V(z;\tau,N);\tau,N) = 0$. So $V(z;\tau,N)$ can be expanded into a convergent power series

$$V(z;\tau,N) = \sum_{n=1}^{\infty} B_n z^n, \; B_1 = |\tau|. \tag{14}$$

Replacing (14) into (12) and comparing coefficients; we obtain that $B_1 = |\tau|$ and

$$B_n = \frac{2}{N} \sum_{\substack{l_1+l_2+\cdots+l_k=n \\ k=2,3\cdots n}} B_{l_1} B_{l_2} \cdots B_{l_k}, \; n \geq 2. \tag{15}$$

Note that the series (14) converges in a neighborhood of the origin, so there is a constant $T > 0$ such that $B_n \leq T^n, n = 1, 2, \cdots$.

Now, we can deduce, by induction, that $|b_n| \leq B_n e^{k(n-1)}$ for $n \geq 1$, where $K : N \to R$ is defined in Lemma 1. In fact, $|b_1| = |\tau| = B_1$, for inductive proof, we assume that $|b_j| \leq B_j e^{K(j-1)}$, $j \leq n-1$. From (11) (15) and lemma 1 we know

$$|b_n| \le \frac{2}{N|\lambda^{n-1}-1|} \sum_{\substack{l_1+l_2+\cdots+l_k=n \\ k=2,3,\cdots n}} B_{l_1} B_{l_2} \cdots B_{l_k} e^{k(l_1-1)+k(l_2-1)+\cdots+k(l_k-1)}$$
$$\le \frac{e^{k(n-2)}}{|\lambda^n-1|} B_n.$$

Note that

$$\mathrm{K}(l_1-1)+\mathrm{K}(l_2-1)+\cdots+\mathrm{K}(l_t-1) \le \mathrm{K}(n-2) \le \mathrm{K}(n-1)+\log|\alpha^{n-1}-1|.$$

Then

$$|b_n| \le B_n e^{\mathrm{K}(n-1)}$$

as required. Note that $k(n) \le n(B(\theta)+\gamma)$ for some universal constant $\gamma > 0$. Then $|b_n| \le T^n e^{(n-1)(B(\theta)+\gamma)}$, that is,

$$\limsup_{n\to\infty}(|b_n|)^{\frac{1}{n}} \le \limsup_{n\to\infty}(T^n e^{(n-1)(B(\theta)+\gamma)})^{\frac{1}{n}} = Te^{B(\theta)+\gamma}.$$

This implies that the convergence radius of (7) is at least $(Te^{B(\theta)+\gamma})^{-1}$. This completes the proof.

In case (H_2), the constant λ is not only on the unit circle in **C**, but also a root of unity. In such a case, the resonant case, both Diophantine condition and Brjuno condition are not satisfied. Let $\{C_n\}_{n=0}^{\infty}$ be a sequence defined by

$$\begin{cases} C_1 = |\tau|, \\ C_n = \dfrac{2\Gamma}{N} \displaystyle\sum_{\substack{l_1+l_2+\cdots+l_k=n \\ k=2,3,\cdots n}} C_{l_1} C_{l_2} \cdots C_{l_k},\ n \ge 2, \end{cases} \tag{16}$$

where

$$\Gamma = \max\{1, |\alpha^i - 1|^{-1} : i = 1, 2, \cdots, p-1\} > 0,$$

and N is defined in Theorem 1.

Theorem 2. Suppose that (H_2) holds, and p is given as above. Let $\{b_n\}_{n=1}^{\infty}$ be determined recursively by $b_1 = \tau$ and

$$-\frac{2\lambda(\lambda^{n+1}+\lambda^n+\lambda-3)(\lambda^{n-1}-1)}{(\lambda+1)(\lambda^n+1)} b_n = \Delta(n,\lambda),\ n \ge 2, \tag{17}$$

where

$$\Delta(n,\lambda) = \sum_{\substack{l_1+l_2+\cdots+l_k=n \\ k=2,3,\cdots n}} a_k b_{l_1} b_{l_2} \cdots b_{l_k}.$$

If $\Delta(vp+1,\lambda)=0$ for all $v=1,2,\cdots$, then Eq. (3) has an analytic solution $\phi(z)$ in a neighborhood of the origin such that $\phi(0)=0$, $\phi'(0)=\tau$ and $\phi^{(vp+1)}(0)=(vp+1)!\tau_{vp+1}$, where all $\tau'_{vp+1}s$ are arbitrary constants satisfying the inequality $|\tau_{vp+1}| \le C_{vp+1}$ and the sequence $\{C_n\}_{n=1}^{\infty}$ is defined in (16). Otherwise, if $\Delta(vp+1, \lambda) \neq 0$ for some $v = 1, 2, \cdots$, then Eq. (3) has no analytic solutions in any neighborhood of the origin.

Proof: We seek a power series solution of (3) of the form (7) as in the proof of Theorem 1, where the equality in (17) is indispensable. If $\Delta(vp+1, \lambda) \neq 0$ for some natural number v, then the equality in (17) does not hold for $n = vp+1$, since $\lambda^{vp} - 1 = 0$. In such a circumstance Eq.(3) has no formal solutions.

When $\Delta(vp+1, \lambda) = 0$ for all natural number v, then for each v, b_{vp+1} in (17) has infinitely many choices in **C**, that is, the formal series solution (7) defines a family of solutions with infinitely many parameters. Choose $b_{vp+1} = \tau_{vp+1}$ arbitrary such that

$$|\tau_{vp+1}| \le D_{vp+1}, v = 1, 2, \cdots, \tag{18}$$

where D_{vp+1} is defined by (16). In what follows, we prove that the formal series solution (7) converges in a neighborhood of the origin. Observe that $|\lambda^n - 1|^{-1} \le \Gamma$ for $n \neq vp$, then

$$b_n \le \frac{2\Gamma}{N} \sum_{\substack{l_1+l_2+\cdots+l_k=n \\ k=2,3,\cdots n}} |b_{l_1}| \cdot |b_{l_2}| \cdots |b_{l_k}|, n \neq vp, v = 1, 2, \cdots,$$

where N is defined in Theorem 1.

Let

$$U(z; \tau, \frac{N}{\Gamma}) = \sum_{n=0}^{\infty} C_n z^n, C_1 = |\tau|. \tag{19}$$

It is easy to check that (19) satisfies the implicit functional equation

$$F(z, U; \tau, \frac{N}{\Gamma}) = 0, \tag{20}$$

where F is defined in (13). Moreover, similarly to the proof of Theorem 1, we can prove that (20) has a unique analytic solution $U(z; \tau, \frac{N}{\Gamma})$ in a neighborhood of the origin such that $U(0; \tau, \frac{N}{\Gamma}) = 0$ and $U'_z(0; \tau, \frac{N}{\Gamma}) = |\tau|$. Thus (19) converges in a neighborhood of the origin. By induction, we can have $|b_n| \le C_n, n = 1, 2, \cdots$.

Therefore, the series (7) converges in a neighborhood of the origin. This completes the proof.

3 Analyticity of Invariant Curves

In this section, we will state and prove our main result.

Theorem 3. Suppose one of the conditions in Theorems 1-2 is fulfilled. Then Eq. (1) has an analytic solution of the form

$$f(z) = \phi(\lambda\phi^{-1}(z))$$

in a neighborhood of the origin, such that $f(0) = 0, f'(0) = \lambda$, where $\phi(z)$ is an analytic solution of the auxiliary equation (3) .

Proof. By Theorems 1-2, we can find an invertible analytic solution $\phi(z)$ of the auxiliary equation (3) in the form of (7) such that $\phi(0) = 0, \phi'(0) = \tau \neq 0$. Let

$$f(z) = \phi(\lambda\phi^{-1}(z)),$$

which is also analytic in a neighborhood of the origin. From (3), it is easy to see

$$\begin{aligned} f(f(z)) = \phi(\lambda^2\phi^{-1}(z)) &= 2\phi(\lambda\phi^{-1}(z)) - z - \frac{1}{2}(g(\phi(\lambda\phi^{-1}(z))) + g(z)) \\ &= 2f(z) - z - \frac{1}{2}(g(f(z)) + g(z)), \end{aligned}$$

and

$$f'(z) = \frac{\lambda\phi'(\lambda\phi^{-1}(z))}{\phi'(\phi^{-1}(z))}.$$

Thus, we have $f(0) = g(\lambda\phi^{-1}(0)) = g(0) = 0$, and

$$f'(0) = \frac{\lambda\phi'(0)}{\phi'(0)} = \lambda.$$

The proof is complete.

References

1. Aftabizadeh, R., Wiener, J.: Oscillatory and periodic solutions of systems of two first order linear differential operations with piecewise constant arguments. Appl. Anal. 26, 327–333 (1988)
2. Aftabizadeh, R., Wiener, J., Xu, J.M.: Oscillatory and periodic properties of delay differential equations with piecewise constant arguments. Amer. Math. Soc. 99, 673–679 (1987)
3. Cooke, K.L., Wiener, J.: Retarded differential equations with piecewise constant arguments. Math. Anal. Appl. 99, 265–297 (1984)
4. Gyori, Ladas, G.: Linearized oscillations for equations with piecewise constant arguments. Differ. Integ. Equa. 2, 123–131 (1989)
5. Kuczma, M.: Functional Equations in a Single Variable. PWN, Warsaw (1968)
6. Ng, T., Zhang, W.: Invariant curves for planar mappings. Differ. Equa. Appl. 3, 147–168 (1997)
7. Si, J.G., Zhang, W.: Analytic Solutions of a Functional Equation for Invariant Curves. Math. Anal. Appl. 256, 83–93 (2001)

8. Bjuno, D.: Analytic form of differential equations. Trans. Moscow Math. Soc. 25, 131–288 (1971)
9. Marmi, S., Moussa, P., Yoccoz, J.-C.: The Brjuno functions and their regularity properties. Comm. Math. Phys. 186(2), 265–293 (1997)
10. Carletti, T., Marmi, S.: Linearization of Analytic and Non-Analytic Germs of Diffeomorphisms of $(\mathbb{C},0)$. Bull. Soc. Math., France 128, 69–85 (2000)
11. Davie, M.: The critical function for the semistandard map. Nonlinearity 7, 219–229 (1994)

Supply Chain Collaboration Based on the Wholesale-Price and Buy-Back Contracts

Jinyu Ren[1], Yongping Hao[1], and Yongxian Liu[2]

[1] School of Mechanical Engineering Shenyang Ligong University, Shenyang, China
[2] School of Mechanical Engineering & Automation Northeastern University, Shenyang, China
rjy27@163.com, yxliu@mail.neu.edu.cn

Abstract. Collaboration between two different business entities is an important way to gain competitive advantage and improve supply chain profit. This paper studies a decentralized supply chain consisting of one manufacturer and one retailer in random demand setting. An incentive function on the wholesale price and buy-back cost is introduced and an incentive scheme based on the wholesale-price contract and buy-back contract has been developed to encourage the retailer to participate in the collaboration ways. Then a numerical example is done and the result shows the proposed collaboration mechanism not only allows the decentralized supply chain to achieve the same performance as the centralized system but also allows both parties in the supply chain to achieve profit sharing by tuning the contract parameters.

Keywords: Supply chain, Collaboration, Random demand, Wholesale-price contract, Buy-back contract.

1 Introduction

Supply chain collaboration, which means to design an effectively incentive scheme so as to allow the decentralized supply chain to realize the overall optimal performance, has been extensively studied in the recent operations management literature[1~4]. Especially it has been widely paid attention to how to set up the effective coordination mechanism by all kinds of contracts. Giannoccaro and Pontrandolfo propose a model of an supply chain contract based on the revenue sharing mechanism and this model allows the system efficiency to be achieved as well as it could improve the profits of all the supply chain actors by tuning the contract parameters[5]. Haoya et al. study a coordination contract for a supplier-retailer channel producing and selling a fashionable product exhibiting a stochastic price-dependent demand [6]. Ozen et al. focus on coordination of the manufacturer and the retailers through buy-back contracts and prove buy–back contracts, in general, cannot make the distribution system achieve the same performance as the centralized system [7]. Chaharsooghi and Heydari develop an incentive scheme based on credit option contracts to coordinate the reorder point and order quantity and increase the overall chain profitability as well as each member's profitability [8].

Different from the conventional studies, this paper considers collaboration of a two-echelon supply chain consisting of one manufacturer and one retailer under

D. Zeng (Ed.): Advances in Control and Communication, LNEE 137, pp. 463–470.
springerlink.com

random demand and finds a collaboration mechanism based on the wholesale-price and buy-back contracts that not only allows the decentralized system to perform just as well as a centralized one, but also provides a flexibility of cost allocation between the two parties of the supply chain.

The paper is organized as follows. In section 2, the basic assumptions of the model are presented and the profit functions of the manufacturer and the retailer are discussed. Section 3 analyzes two extreme cases of the decentralized and centralized systems to evaluate the collaboration mechanisms in the following sections. In section 4, the collaboration mechanism based on the wholesale-price and buy-back contracts with uncertainty demand is developed. In section 5, a numerical example is given. The paper is concluded in Section 6.

2 Model Description

We consider a supply chain with a manufacturer and a retailer in a one-period setting. The retailer faces a stochastic customer demand r and has to determine his stocking quantity, which he orders from the manufacturer, at the beginning of a selling period. When placing his order q, the retailer does not know exact demand realization but knows the distribution of the demand. After a certain period of time, the goods are produced and shipped by the manufacturer to the retailer. Then the demand are realized and satisfied as much as possible. If the ordered amount is more than the market demand so that the surplus products can not be sold at the end of the period, the retailer would salvage any leftover inventory. If the ordered amount is less than the external demand, the retailer needs to pay for the shortage cost.

Now, for a given value of the order quantity, the profit for the manufacturer is given by

$$R^m = (w - c - t)q \tag{1}$$

Notations

R^m Profit for the manufacturer

q Order quantity for the retailer

c Production cost for the manufacturer

w Wholesale price

t Transportation cost from the manufacturer to the retailer, not including purchasing cost

For a given value of the order quantity, the profit for the retailer is given by

$$R^s = p\min\{q,r\} + S[q-r]^+ - \pi[r-q]^+ - wq \tag{2}$$

where $[q-r]^+ = \max\{0, q-r\}, [r-q]^+ = \max\{0, r-q\}$.

Notations

R^s Profit for the retailer

p Selling price for the retailer

S Salvage cost for the retailer

π Shortage value for the retailer

When the retailer places his order, only the distribution of the demand $F(r)$ and the density function $f(r)$ are known. By Eq.(2), the expected profit for the retailer is given by

$$E(R^s) = p\int_0^q rf(r)dr + p\int_q^\infty qf(r)dr + S\int_0^q (q-r)f(r)dr - \pi\int_q^\infty (r-q)f(r)dr - wq \tag{3}$$

3 Supply Chain Decision Analysis

3.1 Supply Chain Decentralized Decision

With supply chain decentralized decision situation the manufacturer and retailer optimize their objective functions independently and the retailer determines the optimal order quantity that achieves maximal expected profit according to the wholesale price of the manufacturer.

Firstly, the first derivative of the function $E(R^s)$ on the order quantity q is solved by

$$\frac{dE(R^s)}{dq} = p\int_q^\infty f(r)dr + S\int_0^q f(r)dr + \pi\int_q^\infty f(r)dr - w \tag{4}$$

Let the right of the Eq.(4) equals zero and the optimal order quantity with decentralized decision situation q_{opt}^{de} can be calculated solving the following equation:

$$F(q_{opt}^{de}) = \frac{\pi + p - w}{\pi + p - S} \tag{5}$$

Where $F(r)$ denotes a continuous probability distribution of the random variable r.

3.2 Supply Chain Centralized Decision

With supply chain centralized decision situation the profit decision problem at this time can be considered as a joint optimal problem. So the actual total profit of the supply chain is given by

$$R^{Tce} = p\min\{q,r\} + \max\{h-t,S\}[q-r]^+ - \pi[r-q]^+ - (c-t)q \tag{6}$$

Where R^{Tce} stands for the actual total profit of the supply chain and h stands for the salvage value for the manufacturer.

The expected total profit is then given by

$$E(R^{Tce}) = p\int_0^q rf(r)dr + p\int_q^\infty qf(r)dr + \max\{h-t,S\}\int_0^q (q-r)f(r)dr - \pi\int_q^\infty (r-q)f(r)dr - (c-t)q \tag{7}$$

By solving the first derivative of the function $E(R^{Tce})$ on the order quantity q and letting it equal zero the manufacturer and the retailer determine the optimal order quantity q_{opt}^{ce} that achieves maximal expected profit for the entire supply chain, viz.

$$F(q_{opt}^{ce}) = \frac{\pi + p - c - t}{\pi + p - \max\{h - t, S\}} \tag{8}$$

Comparing Eq.(5) with Eq.(8), generally, $w \geq c + t$, $S \leq \max\{S, h - t\}$, so it is straightforward to verify that the centralized decision scenario more order quantity will be placed by the retailer than the decentralized decision scenario, which leads to a higher total profit for the entire supply chain. Therefore it is necessary that a collaboration mechanism would be set up to increase the total profit of the supply chain.

4 Collaboration Mechanism

The analysis for supply chain decentralized decision and centralized decision shows that in a decentralized system without collaboration the expected total profit of the entire supply chain is usually lower than in a centralized one. In the following, we will determine appropriate collaboration mechanism that allows the decentralized system to achieve the same performance as a centralized supply chain.

In general, salvaging at the manufacturer is more beneficial than salvaging at the retailer because the manufacturer might redirect the unsold or not-needed units to this market and gain the positive revenue. This opportunity becomes significant if the manufacturer offers buy-back options to the retailers[7]. Therefore here consider a variable buy-back cost which the manufacturer promises to offer to the retailer when the ordered amount of the retailer is more than the market demand at the end of the period. Let R denote the buy-back cost the manufacturer has to pay for every unit.

The actual total profit of the manufacturer and the retailer and the total profit of the entire supply chain after introducing buy-back cost are then given by

$$R^{mc} = (w - c - t)q + (h - R - t)[q - r]^{+} \tag{9}$$

$$R^{sc} = p\min\{q, r\} - \pi[r - q]^{+} - wq + R[q - r]^{+} \tag{10}$$

$$R^{Tc} = p\min\{q, r\} + (h - t)[q - r]^{+} - \pi[r - q]^{+} - (c - t)q \tag{11}$$

By comparing Eq.(6) with Eq.(11), it is obvious that the total profit of the supply chain with supply chain centralized decision and the total profit of the entire supply chain after introducing the incentive function are different depending on the order quantity of the retailer as assuming salvaging at the manufacturer is more beneficial than salvaging at the retailer, that is $h - t \geq S$.

By Eq.(10) the expected profit of the retailer is given by

$$E(R^{sc}) = p\int_0^q rf(r)dr + p\int_q^\infty qf(r)dr - \pi\int_q^\infty (r-q)f(r)dr - wq + R\int_0^q (q-r)f(r)dr \tag{12}$$

Solving the first derivative of the function $E(R^{sc})$ on the order quantity q and letting it equal zero, the retailer determines the optimal order quantity that achieves maximal expected profit. The optimal order quantity q_{opt}^{co} can be calculated solving the following equation:

$$F(q_{opt}^{co}) = \frac{\pi + p - w}{\pi + p - R} \tag{13}$$

Note that only buy-back contract, in general, cannot make the distribution system achieve the same performance as the centralized system [7]. Therefore, it is necessary to introduce the wholesale-price contracts apart from the buy-back contract to realize the perfect collaboration of the supply chain.

Assume that the buy-back price R and the wholesale price w decisions of the manufacturer aim to achieve the higher total profit of the supply chain and the manufacturer directs the retailer to choose the overall optimal decision. To ensure that the retailer chooses this value of order quantity that incurs overall maximal expected total profit, the manufacturer must fix the buy-back cost R and wholesale price w so that the following equation holds:

$$F(q_{opt}^{co}) = F(q_{opt}^{ce}) \Leftrightarrow \frac{\pi + p - w}{\pi + p - R} = \frac{\pi + p - c - t}{\pi + p - \max\{h-t, S\}} \tag{14}$$

By solving Eq.(14) the optimal wholesale price w as a function of buy-back price R is attained:

$$w_{opt} = \pi + p - \frac{\pi + p - c - t}{\pi + p - \max\{h-t, S\}}(\pi + p - R) \tag{15}$$

Each combination of the buy-back cost R and the wholesale price w that satisfies the above equation ensures that the retailer chooses the overall optimal order policy that leads to maximal expected total profit of the entire supply chain.

In order to make the parties of the supply chain agree to such a collaboration model, after introducing the incentive methods the expected profits of the manufacturer and retailer should be more than that with supply chain decentralized decision situation, which require to fulfill the constraints as follow,

$$s.t.\begin{cases} E(R^{sc}) \geq E(R_{de}^{s}) \\ E(R^{mc}) \geq E(R_{de}^{m}) \end{cases} \tag{16}$$

So far, it have been developed that collaboration mechanisms that gain an overall optimal performance of the entire supply chain.

5 Numerical Example

In the following, a numerical example is considered in order to get a better understanding of the underlying collaboration model. The parameters are specified as follows: The wholesale price charged by the manufacturer to the retailer for each unit of the selling product is w=600 yuan. The production cost by one unit for the manufacturer is c =340 yuan and the transportation cost by one unit from the manufacturer to the retailer is t =16 yuan. As to the retailer, the selling price per unit for the external market is p =800 yuan and a shortage cost by one unit, which is π =500 yuan, will be paid when the customer's need cannot be meet. While the order quantity for the retailer exceeds to the external demand, the retailer does not store the surplus products but salvage them at the price per unit S =200 yuan. If the surplus products are done with by the manufacturer, the salvage price is h =300 yuan. Assuming the random variable r on the market demand follows a uniform distribution over the range [100, 200] to simplify numerical calculation. Thus the density function $f(\alpha)$ is given by

$$f(r)=\begin{cases}\dfrac{1}{200-100} & 100\le r\le 200\\ 0 & otherwise\end{cases} \tag{17}$$

Then the continuous probability distribution $F(\alpha)$ is attained by

$$F(r)=\begin{cases}\dfrac{r-100}{200-100} & 100\le r\le 200\\ 0 & otherwise\end{cases} \tag{18}$$

5.1 Supply Chain Decision Analysis

Substituting the density function (18) in (5), the optimal order quantity of the retailer in supply chain decentralized decision scenario can be determined to be

$$\frac{q_{opt}^{de}-100}{200-100}=\frac{\pi+p-w}{\pi+p-S} \tag{19}$$

After that, the optimal order quantity q_{opt}^{de} is 164 units. As a result of both decisions the expected total profit for the entire supply chain is $E(R^T)$ =57288 yuan, in which the expected profit of the manufacturer is $E(R^m)$ =40016 yuan

(see Eq.(1)), whereas the expected profit of the retailer is $E(R^s)$ =17272 yuan(see Eq.(3) and Eq.(17)). On the other hand, the optimal order quantity of the retailer in supply chain is q_{opt}^{ce} =193 units by Eq.[8] and Eq.[18] which achieves maximal expected profit for the entire supply chain, that is $E(R^{Tce})$=63255.08 yuan(see Eq.(7)

and Eq.(17)). The results are summarized in Tab. 1, which shows the expected total profit of the entire supply chain in supply chain centralized decision scenario is about 10% more than in supply chain decentralized decision scenario. Therefore, it is necessary to set up collaboration mechanism to increase the total profit of the entire supply chain.

Table 1. Supply Chain Decentralized And Centralized Desicion

	q_{opt}	$E(R^m)$	$E(R^s)$	$E(R^T)$
Decentralized decision	164	40016	17272	57288
Centralized decision	193	—	—	63255.08

5.2 Collaboration Mechanism Application

As mentioned in Section Ⅳ the optimal wholesale price as a function of the buy-back cost (see Eq. (15)) is given by

$$w_{opt} = 91 + 0.93R \tag{20}$$

Where the constraints on the buy-back cost is attained by Eq. (16), that is

$$s.t. \begin{cases} E(R^{sc}) \geq 17272 \\ E(R^{mc}) \geq 40016 \end{cases} \tag{21}$$

The range of the buy-back cost value is then calculated by Eqs. (9), (12) and (17) as

$$578 \leq R \leq 622 \tag{22}$$

Numerical calculation results based on the constraint (22) are summarized in Tab. 2, which shows the mutual dependency of wholesale price and buy-back cost for some specific discrete values. It can be observed that by increasing the buy-back cost the wholesale price as well as the expected profit of the manufacturer increase, and

Table 2. Optimal Wholesale Price In Dependency Of The Buy-back Cost

R	w	$E(R^m)$	$E(R^s)$	$E(R^{Tc})$
579	629.5	40025.33	23229.75	63255.08
585	635.1	40846.63	22408.45	63255.08
590	639.7	41518.18	21736.90	63255.08
600	649.0	42880.58	20374.50	63255.08
610	658.3	44242.98	19012.10	63255.08
620	667.6	45605.38	17649.70	63255.08

the expected profit of the retailer decreases. Recall that it is by adjusting the wholesale price and buy-back contract parameters well accepted by the actors that the collaboration mechanism introduced above not only ensures that the expected total profit of the entire supply chain correspond to those of supply chain centralized decision but also allows both parties in supply chain to benefit from applying the mechanism, when compared to the case with supply chain decentralized decision.

6 Conclusion

This paper has addressed the problem of supply chain collaboration for a decentralized supply chain consisting of a manufacturer and a retailer with the uncertainty of external market demand. A contract model based on the buy-back cost and wholesale cost mechanism has been proposed. Through an example application, it has been verified that a proper contract design by adjusting the wholesale price and buy-back contract parameters can not only lead the manufacturer and the retailer to improve their profits compared with supply chain decentralized decision setting to achieve a win–win condition but also allow the decentralized supply chain to achieve the optimal overall performance.

Acknowledgment. Supported by the National High-Tech. R&D Program of China (863) under grant No.2009AA04Z167. And the Natural Important Special Program, ID76204043.

References

1. Zimmer, K.: Supply chain coordination with uncertain just-in-time delivery. Production Economics 77, 1–15 (2002)
2. Zhang, C.H., Ren, J.Y., Yu, H.B.: Supply Chain Collaboration Mechanism Based on Penalty and Bonus under Asymmetric Information. Chinese Journal of Management science 14, 32–37 (2006)
3. Ruoning, X., Xiaoyan, Z.: Analysis of supply chain coordination under fuzzy demand in a two-stage supply chain. Applied Mathematical Modelling 34, 129–139 (2010)
4. Tiaojun, X., Xiangtong, Q.: Price competition, cost and demand disruptions and coordination of a supply chain with one manufacturer and two competing retailers. The International Journal of Managemant Science 36, 741–753 (2008)
5. Giannoccaro, I., Pontrandolfo, P.: Supply chain coordination by revenue sharing contracts. Production Economics 89, 131–139 (2004)
6. Haoya, C., Youhua, C., Chunhung, C., Tsanming, C., Suresh, S.: Coordination mechanism for the supply chain with leadtime consideration and price-dependent demand. European Journal of Operational Research 203, 70–80 (2010)
7. Ozen, U., Sosic, G.: A multi-retailer decentralized distribution system with updated demand information [DB/OL] (August 2006), http://fp.tm.tue.nl/beta/publications/working%20papers/Beta_WP193.pdf
8. Chaharsooghi, S.K., Heydari, J.: Supply chain coordination for the joint determination of order quantity and reorder point using credit option. European Journal of Operational Research 204, 86–95 (2010)

Research and Development on a Three-Tier C/S Structure-Based Quality Control System for Nano Plastics Production

Cunrong Li and Chongna Sun

School of Mechanical and Electronic Engineering Wuhan University of Technology Wuhan, China
{Cunrong_li,ziyunying18}@163.com

Abstract. For many nano plastics production companies, there is many problems with the quality data collecting and data analysis during the production process such as poor management and fuzzy records. Based on the three-tier C/S (Client/Server) software structure and the principles and methods of Management Information System Development, a digital input, storage and query system for multiple information of the plastic production is designed and realized. This system provides an effective method to cut down the losses caused by the workshop producing chaos by normalizing operation behaviors, and to some extent, it also can offer the managers a way to monitoring the whole production through sharing variable integrated information delivered by the Intranet. With the implementation of this system, a reliable guarantee is acquired by the quality improvement of the company's products and services.

Keywords: Three-tier C/S structure, Qulity Control, Normalization, Real-time monitoring.

1 Introduction

Nano Plastics, which is also named by Nanocomposite, is the emphasis of researches in new material field all over the world. As a new high-tech material, nowadays Nano Plastics is considered as the successor of the existing plastics. In order to respond the customers' diverse requirements, many companies devote themselves to strengthen hardware construction by bringing in high quality measuring &analysis equipments and developing large-scale production lines for mass production. With the growing production capacity and scale, many problems about management quality and information transmission have come into being from the initial materials purchasing to the final product sales. The following items existed in the process control and manufacturing test should be pay particular attention to.

Records are kept in Excel Spreadsheet. The implementation of this recording method in the monitor and test procedure provides heavier task and less efficiency to the workers, meanwhile, it results in lower data accuracy caused by the large amount of information.

D. Zeng (Ed.): Advances in Control and Communication, LNEE 137, pp. 471–481.
springerlink.com

The "Capacity First, Record Second" principle is followed in the workshop. It results in the poor performance of real-time data acquisition and the lower data reliability because of workers paying lip service in the process.

The information utilization is also at a low level. It is impossible to perform data mining of the potential information, since the existing manual record system can't make full use of data, text, report, chart and many other forms. Furthermore, long-term data analysis and forecasting are not easy to be carried out due to that the paper documents couldn't be kept for a long time as we need.

Based on the above and the three-tier C/S model, in this paper with the Borland Delphi7 software development platform and SQL Server 2000 database management system, a quality control system implemented in the whole enterprise's intranet was developed, which has made the information sharing seamlessly and transparently possible, and it was also proved to be an effective method for the supervisors to find and correct the unreasonable operations by obtaining the time code in the records.

2 System Development Model

With the rapid development and deeper application of the computer technology, the scale and complexity of the software systems are increasing gradually. In the software design process, the problems faced by the people are not only the functioning of the software system, but also the system has to have better robustness, scalability, easier upgrade and maintenance and can quickly adapt to changing business rules. In recent years, under the network environment information management system architecture is mainly in the following two ways: C/S (Client/Server) and B/S (Browser/Server) model.

Built based on the middleware product, C/S model has fast response and high efficiency because both the client and server system can handle the task. The operation mode of two-tier C/S model is the direct communication between the front-end client operation and the back-end database server: the server is running its own data processing mechanism to deal with data requests, which are sent by the client, and then sends processed-data back to the client application. As Fig. 1 shows, three-tier C/S model is mainly composed of the client application (Client), middleware (Middleware) and Server Manager (Server) [1]. As a component for the system users, Client is responsible for connection management, data capture, data representation and interface display; Middleware takes charge of connecting Client and Server and is specialized in processing business rules, transforming the users' queries into the language which can be identified by the Server, transferring the query results and executing specific functional operations; Server presides over effectively managing system resources, which means it could offer optimization management when multiple concurrent clients request the same resource on the server, in the meantime, it undertakes the resource, security, data, query and database system management and many other management responsibilities [2].

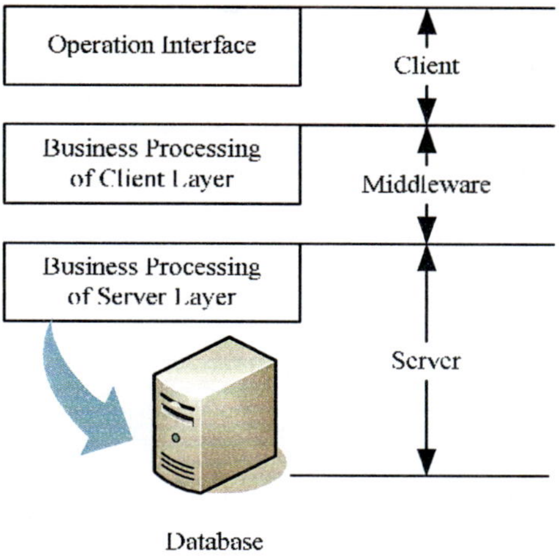

Fig. 1. Structure of general three-tier C/S model

B/S model is a new information management system platform one based on web technologies, which also has three layers: the first layer is common browser software, though which the user interacts with the second layer of web server. With its new online feature, this model has widely spread through business systems and enterprise portals, since it has simplified the client development and maintenance [3]. However, due to the exposure to the Internet in the environment of this open model, its safety is also greatly reduced. On the contrary, the C/S model has a higher security cause of its relatively point-to-point structure mode and generally following a local area network protocol. Compare with the B/S logical three-tier structure, three-layer C/S mode can reduce network traffic and process large amounts of information.

As stated previously, taking into account this quality control system is mainly used in the internal local area network, the development model chooses three-tier C/S mode. Fig. 2 shows the structure of three-tier C/S when the actual software system is constructed. As is shown in Fig. 2, when dealing with the data requests from the monitoring and testing main-station, the model makes the part taking charge of the data communication between the client and database server completely independent as the middle layer, whose data connection with the server is using the "connection pooling" by dynamic allocation and release of data connection to control the number of connections [4]. With the help of the middle layer, the number of data connections and the burden on the database server is significantly alleviated compared with two C/S model, at the same time, database maintenance and its affairs handling are more flexible.

Socket is the basic building block in network communication, as the research adopts Borland Delphi7 as the application development platform this paper uses Borland Socket Server as the server-side Socket. Firstly set “211” as the port number of the local service socket, and then the server invokes method “Listen” to enter a blocked state, waiting connection requests from the clients. The corresponding clients not only have to set the server name (in this system which is designed to be dynamically set in

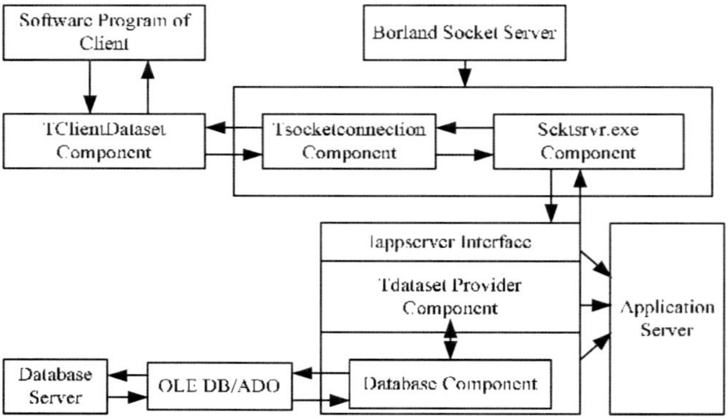

Fig. 2. Structure of C/S model in this system

the program of the server's IP address) as the "Address" of the TSocketConnection component, but also set the corresponding server listening port number (211) as the port number (port), and then set the preceding TSocketConnection's name as RemoteServer of the clients' data query tools TClientDataset, thus the client's operating requirements data can be passed to the TSocketConnection components, which will invoke command "Connect" to call a request to the server. After the server receives a client request, the event "ConnectionRequest" will be triggered to call the "Accept" method to accept the connection if it is willing to provide services, and then the "SendData" or "GetData" can be used by both ends to send or receive data once the connection is established [5].

3 System Architecture

The earlier development stage of the system follows the basic principles of developing management information system. The first one is the systematic principle of building the entire system structure, which means to achieve optimal economic effect from the viewpoint of system theory and the purpose of achieving overall function of the system. The second is the practical principl, and it means that to realize fast and simple operation based on not changing the available computer management system. Lastly is the reliable and secure principle, which emphasizes the system should have sound permissions, data backup and recovery functions, moreover, system equipments could run for a long period reliably and stably [6].

In comply with the above principles and through the overall system analysis, the system architecture will be built up as is shown in Fig. 3. System consists of a production process monitoring subsystem, a product testing subsystem, a managers viewing subsystem, etc. On-site monitoring staff of the monitoring subsystem mainly used in the production process production site in a certain period of time will check the standard parameters of the production machines and manually input them to the client system as data A, including mainframe speed, given feed and load current. R & D(research and development) personnel in the production test subsystem primarily used in new product development lab using a variety of high-precision instruments to test the parameters under certain testing standards and environment and input them into the client system as data B, including impact strength, melt index and heat distortion temperature. And the managers viewing subsystem mostly services for internal managers of different sectors.

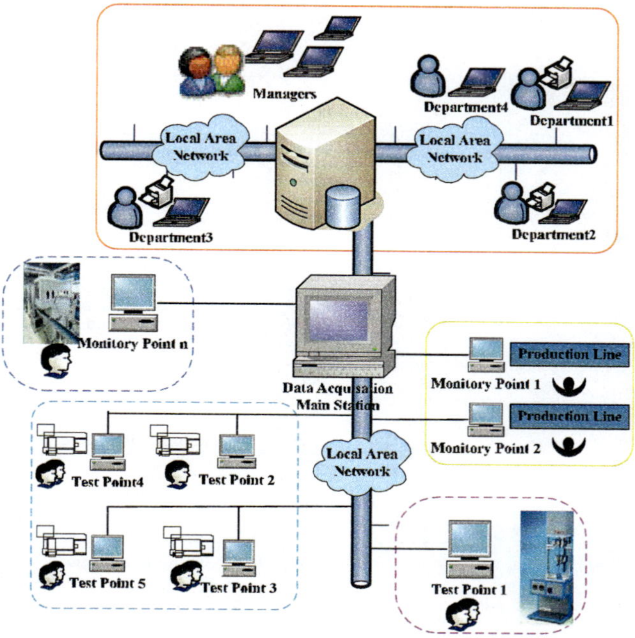

Fig. 3. System Overall Structure

The Server linking the monitoring and testing clients through the Intranet is mainly applied to storing and managing integrated data A and B, and after some format conversions different types of data are stored here following certain rules. All of the operational users constitute the client of three C/S mode, whose requirements contact with the server through the interpretation of the middle tier, which also feeds back the corresponding results to the client [7].

4 System Design

After the conceptual model was proposed, the system entered into the specific design period.

4.1 System Module Division

The system owns two terminals (client and server) since it uses C/S software structure, of which the service server will be installed to an individual computer and the server programs are set to run as the system applications, that is to say, the server programs will kip the Windows login screen to work once the computer has started, just like the anti-virus software. This proved to be an effective way to ensure the security and stability of the server system. The client terminal could be separated to two categories as management client and application client according to different system-using objects. Fig. 4 shows the module structure of the management client.

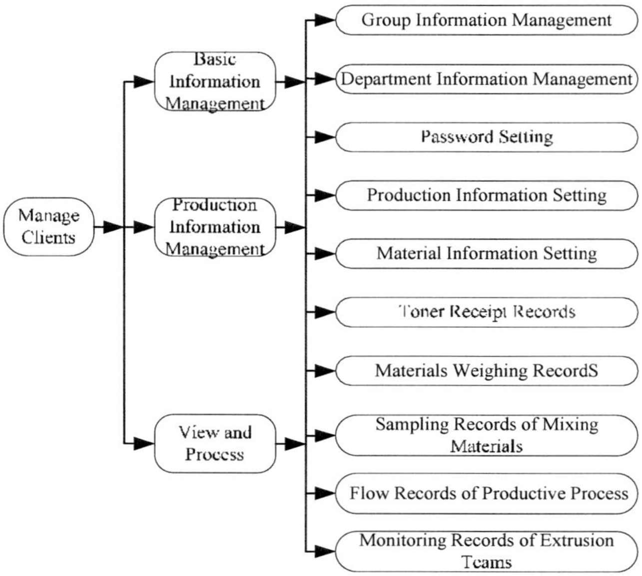

Fig. 4. Module Structure of the Management Client

4.1.1 Basic Information Management

This module is in charge of setting users' permission, and modifying or deleting their basic information.

4.1.2 Product Information Management

This module's responsibility is to establish basic information's initial accounts for the development products, including their names, colors and formula-ratio card numbers, and the names, colors and types of various raw materials required by the products production.

4.1.3 Check and Processing

This is the major function of the management client. The managers or leaders of different departments could query and view recording information in this module which is originally write down in the Excel table to carry out statistical analysis and monitoring treatment comprehensivly.

4.2 Database Design

Database design means to construct an optimal database model under a given application environment, and then for the purpose of meeting various user's demands to establish a database and application systems to store data efficiently [8]. According to the above described relevant contents about database design, this paper mainly create a total of 25 tables, for instance, products basic information table (product), proportioning and mixing ingredients table (plhl), extrusion operating table (jc), the basic information of material weighing Table (clcz_basic), material weighing data sheet (clcz_data), etc. After mapped out those data tables needed to be created, the next procedure is to determine the fields and primary keys in each table, and to establish relationships between all of the tables, which are shown in Fig. 5 [9].

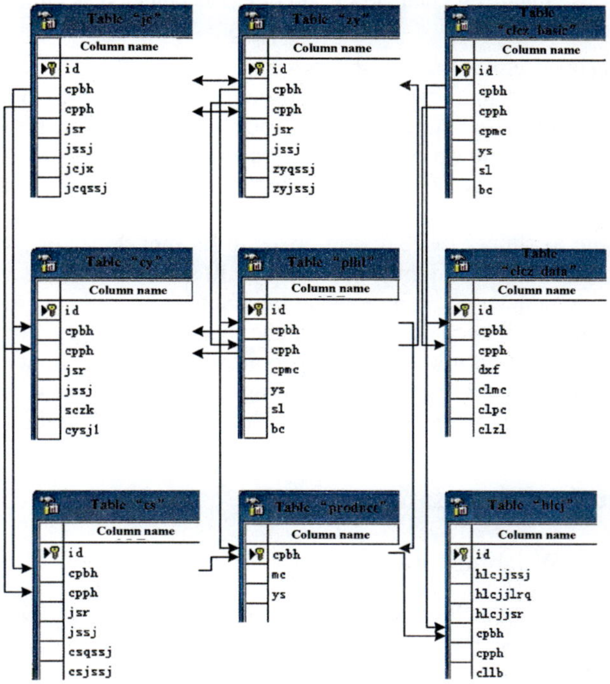

Fig. 5. Relation Diagram of Database Tables

4.3 Main Program Design Process

4.3.1 The Production Flow Process Design

The process goes ahead in the producing workshop, whose main operations are stated in Fig. 6. To resolve the problem of those records of the lag, each of the sessions marked by precise time node and responsible person records will transfer the production information in the first time to the next section, which could then have a crystal-clear understanding of their own assignments and at the same time check on the shop floor whether the work is in place there at the right time according to records. So that through this simple design production information and responsibility will have a high traceability. Utilizing True or False symbol in the programs to identify the process is realized or not has been proved to be feasible.

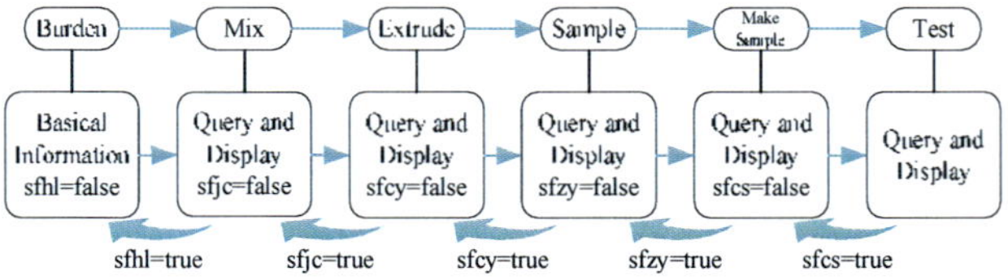

Fig. 6. Program Flow Chart of Production Process Flow

4.3.2 Material Weighing Process Design

Since the weight and ingredient of every raw material as the component of Nano Plastics products has strict standards and controls, so sampling of materials-weighing is a key step in production processes. At present, the process records differ from the realistic situation tremendously and are not convenient for the managers to view, control and modify because they are not standard writing documents. So it is essential for the program to supply more dialog boxes to prompt the errors while the workers run the faulty operations filling the records, and in addition, the input interface should have a stronger characteristic of practical effectiveness.

By means of this flowchart designing, the system could regulate the employee's behaviors when they submit the data to avoid missing important information, and moreover, could effectively prevent inputting unreasonable data or confusing data formats by restricting users to fill in the right location with the right data.

5 System Implementation

5.1 Materials Weighing Interface

If the users didn't follow the steps shown in Fig. 7 in practical operations, the system would popup a dialog box saying there is an error. By rigidly limit the matters needing attention that the administrators have to emphasize orally in each meeting, this system remarkably prevents the occurrence of non-standard records phenomenon.

5.2 The Record View Interface

In this module all of the original Excel tables will be changed into the records automatically generated in this system, whose formats are still retain the existing ones' and enable the users to browse the data information familiarly. Besides, the system supports diverse query requirements, such as a user needs to scan the flow sheet of some product, he/she could inquire it by a single record or in criterion of date and time, and the managers could master the work situation of all employees' and easily control the progress of the projects through by viewing the time node in the records, and

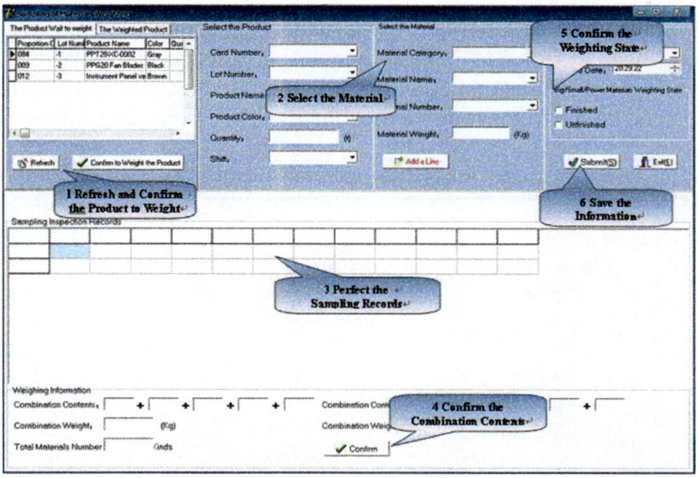

Fig. 7. Materials Weighing Interface and Operational Steps

more important is that the system can meet the managers' needs to storage the records for a long time by providing the function of exchanging the records into Excel documents. Fig. 8 is showing below.

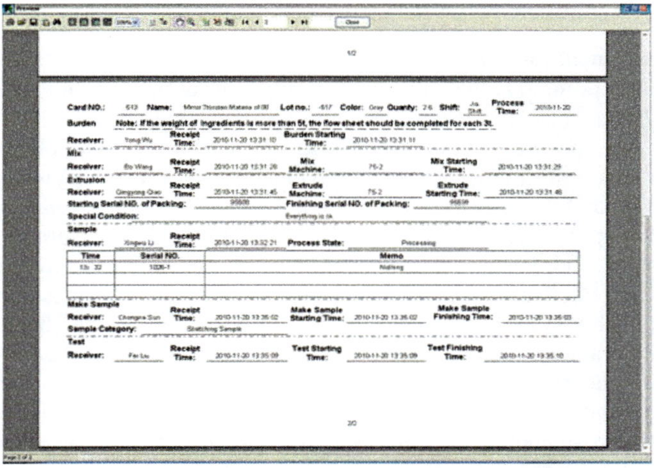

Fig. 8. Records View Interface of The Management Client

System provides for the users as well as the function of viewing the records in forms of chart. The system will filtering data about the performance of a product selected beforehand by the user, and if he/she has set the upper and lower limits of the performance previously in the limits-setting module, those data that is over the maximum or under the minimum will be revealed in the graph form signing in red font, which is conducive to strikingly attract managers' attention to carrying out all-round statistical analysis in order to set an accurate performance index and bring forward corresponding improvement suggestions on the production process, ultimately, to ensure the steady improvement in product quality. The interface is exhibited in Fig. 9.

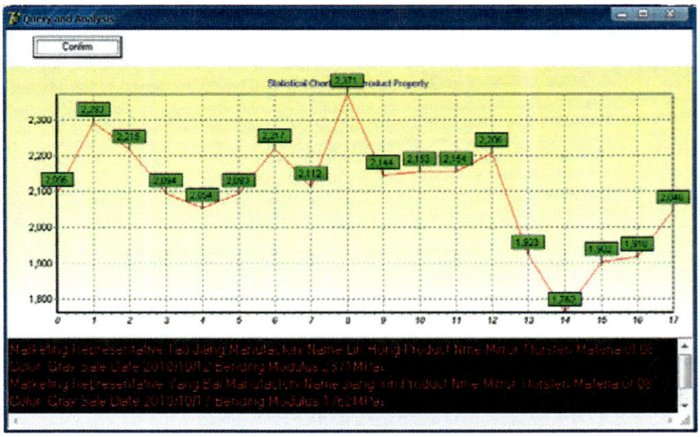

Fig. 9. Curve Graph View of The Management Client

6 Conclusion

Quality management or control system is widely applied in today's enterprises, however, the key point to realize all-round quality control in the company is to develop a quality system that can meet its own specific requirements and reflect the characteristics of the production, which also plays an important role to enable the enterprise to stand out in industry peers implementing quality systems [10]. Based on the investigation and analysis of system requirements during the overall operation in a company, this paper designed and realized this quality control system combined with advanced systems and database development techniques, which is proved to be easy to operate and can effectively meet the managers' and on-site personnel's requirements. Benefited from this system, the nano plastics information can be shared among the whole company rapidly and conveniently and have a stronger traceability than before. In addition, the successful application of three-tier C/S model has greatly enhanced the server security and data transmission rate. What's more, this model provides a vast advantage to improve and upgrade this system while the enterprise's scale expanses.

References

1. Wang, W.: Research and design of chain books management information system based on 3-tier strucrture. Wuhan University of Technology, Wuhan (2006)
2. Xia, K.: Design of enterprise customer service system based on three-layer framework. Computer Knowledge and Technology 6(27), 7470–7471 (2010)

3. Huang, J.: Design and Implementation of university library management system. University of Electronic Science and Technology, Chengdu (2007)
4. Tang, Y., Song, Y.: Architecture research of three-layer C/S mode. Science & Technology Information 09, 5–6 (2009)
5. Zhao, H.: Implemenation of heartbeat program adopting Socket for the programmingi n Delphi and VB, CNKI
6. Wang, Y.: Management Information System. Electronics Industry Press, Beijing (2005)
7. Wu, H., Li, Z., Lu, Y.: The distributed simulation system design of aircraft display control based on C/S structure. Computer System Application 7, 19–22 (2010)
8. Li, C., Song, C.: Research and Development on Worker-oriented Supporting Module in Statistical Process Control System. IEEE 9-10, 516–519 (2010), doi:10.1109/ICLSIM.2010.5461368
9. Lin, F.: The security aplication research of SQL Server database system. Journal of Anhui University of Science and Technology (Natural Science) 29(4), 51–54 (2009)
10. Zhou, Q.: The application of nano materials and industrialization. Journal of Jiangsu Teachers University of Technology 8(4), 83–87 (2002)

A Particle Swarm Optimization Algorithm for Grain Emergency Scheduling Problem

Bao Zhan Biao[1] and Wu Jianjun[2]

[1] Henan University of Economics and Law Zhengzhou, China
[2] Henan University of Technology, Zhengzhou, China
baozhanbiao@126.Com, popp_ly@163.com

Abstract. Grain Emergency Scheduling is a very important and practical research subject, Scheduling problems should regard not only its time effectiveness but also its economical effectiveness. According to the characteristics of grain emergency , establishing a more constraints grain emergency scheduling model, to consider constraints of emergency resource requirements, emergency time constraints and cost constraints. to optimize multi-objective of the earliest start time of emergency and the least number of rescue point. In the experiments, a number of numerical examples are carried out for testing and verification. The Computational results confirm the efficiency of the proposed methodology.

Keywords: Particle Swarm Optimization (PSO), Emergency scheduling, Efficiency and reliability.

1 Introduction

With the development of economy and society, the sudden public events are increasing. The characteristics of emergency logistics objectively requires to draw up an appropriate scheduling plan for transporting the resources required to emergency sites with the greatest probability of rapid, timely and accurately, to make emergency location for emergency activities as soon as possible, so as to reduce losses of emergencies. Grain as an important emergency resources, its emergency logistics are especially important. To improve the capability of the government, enterprises and social organizations to respond large-scale natural disasters and sudden public events triggered grain emergency logistics has become an important issue.

Grain emergency scheduling is main function of grain emergency logistics, which in the achievement of grain emergency logistics to maximize the effectiveness of time plays a vital role. Taking grain as the research object, proceeding from the actual application environment, considering initial traffic limit amount of each distribution center of grain emergency logistics, to construct the model of grain emergency scheduling system that is constrained by emergency resource requirements, emergency time constraints, cost constraints and so on. Using Using PSO to solve and giving out the steps for solving as well, so as to explore the scientific methods of grain emergency scheduling.

D. Zeng (Ed.): Advances in Control and Communication, LNEE 137, pp. 483–488.
springerlink.com

2 Particle Swarm Optimization

2.1 Fundamental Principle of PSO

The Particle Swarm Optimization (PSO) algorithm is an adaptive algorithm based on a social-psychological metaphor. It was originally proposed by J.Kennedy [5].A population of individuals adapts by returning stochastically toward previously successful regions in the search space, which is influenced by the successes of their topological neighbors. PSO is related with Artificial Life, and specifically to swarming theories, and also with Genetic Algorithms (GA). PSO can be easily implemented and it is computationally inexpensive. Moreover, it does not require gradient information of the objective function under consideration, but only its values, and it uses only primitive mathematical operators. PSO has been proved to be an efficient method for many optimization problems, such as Design Combinational Logic Circuits, Evolving Artificial Neural Networks, Multiple Object Problems, and TSP.

Two versions of the PSO algorithm have been developed, one with a global neighborhood, and the other with a local neighborhood. The global version was used in the paper. Each particle moves towards its best previous position and towards the best particle in the whole swarm. On the other hand, according to the local version, each particle moves towards its best previous position and towards the best particle in its restricted neighborhood. The global version PSO algorithm can be described as follows: Suppose that the search space is *D*-dimensional, and then the *i*-th particle of the swarm in the t-th iteration can be represented by a *D*-dimensional vector. The velocity of this particle can be represented by another *D*-dimensional vector. The best previously visited position of the particle in t-th iteration is denoted as Pi,t. The global best particle in t-th iteration denoted as Pg,t. Then the swarm is manipulated according to the following two equations:

$$Vi_{t+1} = c_1 Vi_t + c_2 * r_1 (p_{i,t} - Xi_t) + c_3 * r_2 * (P_{g,t} - Xi_t) \tag{1}$$

$$Xi_{t+1} = Xi_t + Vi_{i+1} \tag{2}$$

Where i=1, 2...P, and P is the total number of particles in the swarm, i.e. the population size; t=1, 2...T, and T is the iteration limited; c1 is an inertia weight which is employed to control the impact of the previous history of velocities on the current one. Accordingly, the parameter c1 regulates the trade-off between the global (wide-ranging) and local (nearby) exploration abilities of the swarm. r1, r2 are random numbers, uniformly distributed in [0, 1]; c2, c3 are two positive constants, called cognitive and social parameter respectively. That proper fine-tuning these two parameters may result in faster convergence and alleviation of local minima. The details of tuning the parameters of PSO were discussed in [6]. Formula (1) is used to calculate a particle's new velocity according to its previous velocity and the distance from its current position to its local best and the global best. Formula (2) is used to calculate a particle's new position by utilizing its experience (i.e., local best) and the best experience of all particles (i.e., global best). Formulas (1) and (2) also reflect the information-sharing mechanism of PSO.

2.2 The Solving Steps of PSO

The steps of using PSO to solve the upper grain emergency scheduling Mathematical Model are as follows:

Step1 first, set the scale of particle swarm, the precision of solution and the maximal iteration number.

Step2 second, randomly initialize the particle swarm: according to constraints (3), initialize the speed and position of all particle in the particle swarm;

Step3 use the aim function of optimization problem (2) as fitness function to calculate the fitness of each particle;

Step4 for the ith particle, compare its fitness with that of the best passed position P_{besti} . if better, then set X_i as the current best position P_{besti} ;

Step5 for the ith particle, compare its fitness with that of the best passed position G_{best} . if better, then set x_i as the best position G_{best} of all current particles;

Step6 update the particle's speed and position, according to formula (8) and (9);

Step7 determine whether reaches the prescriptive iteration number or satisfies the prescriptive standard error. If reaches, then terminates iteration and we get the optimal solution. Otherwise, we turn to the step (3) to do it again.

3 The Model of Grain Emergency Scheduling

Using the grain emergency resource of continuous consumption as the study object, to consider the cost of the rescue in terms of both emergency response system cost and the loss caused by untimely rescue.To research the model of grain emergency scheduling that resource requirements constraints, emergency time constraints, cost constraints and multiple constraints. The following notation is required to formulate the model:

A_i —The No.i grain emergency suppliers or the center of grain emergency scheduling;

A—The emergency disaster place;

v— grain resource consumption rate after the commencement of emergency activities;

T—The termination time of emergency response activities;

x_i' —The quantity of the emergency supplier place A_i (i=1,2,…,n)(the amount can be scheduled),

$x_i' \geq 0$, and $\sum_{i=1}^{n} x_i' \geq v(T-t_1)$;

x_i —The quantity of the scheduled resources from A_i *(i=1,2,⋯,n)* to *A* , $0 \leq x_i \leq x_i'$;

t_i —the time from the supply place A_i *(i=1,2,⋯,n)* to the demand place *A*,supposed $t_{i+1} \geq t_i$ *(i=1,…,n-1)* , and $t_n \leq T$;

$I(t)$ — The remaining amount of grain at A in t time;

C_i —The unit cost from A_i to A, $i=1,2,\cdots N$;

D — The grain loss (penalty) fees per unit of time in a unit price;

B_i —The missing cost of emergency grain loss in the $[t_{i,} t_{i+1}]$ period.(i = 1,2, ..., n), supposed $t_{i+1} = T$;

x_0 — grain initial traffic limit of every grain supplier place A_i (i=1,2,...,n);

Any program can be expressed as Φ :

$$\Phi = \{(A_{i1}, x_{i1,})(A_{i2}, x_{i2}), \cdots, (A_{im} x_{im}\} \tag{3}$$

Questions is required to give the best grain emergency scheduling plans so as to minimize the total cost of the emergency activities under the conditions of minimize emergency time and the supplier place to meet continuous consumption of grain at A. Taking into account the background of the use of the real economy, to demand that if grain storage be asked to join in emergency activities, their number must not be less than the minimum traffic limit. Suppose the time of the accident t =0, time from accident to,the loss of not timely rescue can not be avoided, therefore, the losses in this time can not be considered in the objective function.

According to the characteristics of the above problems, To establish grain emergency scheduling model is as follows:

$$MinZ = \sum_{i=1}^{n} C_i x_i + \sum_{i=1}^{n} B_i \tag{4}$$

$$S.\ t. \begin{cases} x_0 \le x_i \le x_i' or x_i = 0, i = 1,2\cdots n \\ \sum_{i=1}^{n} x_i' \ge v(T - t_1), i = 1,2\cdots n \\ t_i + 1 \ge t_i, and t_{n+1} = T, i = 1,2\cdots n \end{cases} \tag{5}$$

For the calculation of B_i, sub-three cases to consider:

1) When $\exists I(\mathrm{t}_i)<0, i = 1,2\cdots n$,That is at i-time grain has been missing, the lack of capacity at this time:

$$I(\mathrm{t}_i) = \sum_{i=1}^{n} x_i - v(t_i - t_1) \tag{6}$$

In the cycle of $[t_{i,} t_{i+1}]$, the loss cost of grain deficiency:

$$\begin{aligned} B_i &= -DC_{i+1}\int_i^{i+1} [I(\mathrm{t}_i) - vt]dt \\ &= -DC_{i+1}\left[I(\mathrm{t}_i)(t_{i+1} - t_i) - \frac{1}{2} v(t_{i+1}{}^2 - t_i^2) \right] \end{aligned} \tag{7}$$

To (5) type of substitution was:

$$B_i = DC_{i+1}\left[(v(t_i - t_1) - \sum_{i=1}^{n} x_i)(t_{i+1} - t_i) + \frac{1}{2}v(t_{i+1}{}^2 - t_i{}^2)\right] \tag{8}$$

2) When there $t_i^{'} \in \left(t_{i,}t_{i+1}\right)$, $I(t_i) - v(t_i^{'} - t_i) = 0$. That is grain has been missing at $t_i^{'} \in \left(t_{i,}t_{i+1}\right)$. when $1 \le i \le n$, $t_i^{'} = \frac{1}{v}\sum_{i=1}^{n} x_i + t_1$, In the cycle of $[t_{i,}t_{i+1}]$, the loss cost of grain deficiency:

$$B_i = DC_{i+1}\int_i^{i+1} vtdt = \frac{1}{2}vDC_{i+1}\left(t_{i+1}{}^2 - t_i^{'2}\right)$$

$$= \frac{1}{2}vDC_{i+1}\left[t_{i+1}{}^2 - \left(\frac{1}{v}\sum_{i=1}^{n} x_i + t_1\right)^2\right] \tag{9}$$

(3) Other cases, $B_i = 0, 1 \le i \le n.$

4 Running Result and Analysis Running Result and Analysis

It is assumed that there is an emergency in a certain city,which urgent needs to supply 200,000 tons of grain, assuming that grain consumption rate is 5,000 tons per day, the loss cost of grainmissing D is 5 per unit time, the minimum traffic limit x0 is 5,000 tons. In this experiment, the algorithms presented in article were implemented in JAVA language on a Toshiba notebook, as Pentium M 2.0G (T2500), 1024 MB machine running in Windows xp environment. We firstly choose some suitable grain depots, Choice of conditions are: firstly, in the main producing areas; secondly, it is transport facilities; lastly,having the high degree of mechanization loading and unloading, According to the above conditions, we select the eight granary qualified supply points, using road transportation, the time to reach disaster place unit costs, the availability quantity of various grain storage such as shown in Table1. Required to give a reasonable scheduling program, so as to meet the earliest emergency start time, the save point at least, and the minimum total cost.

To solve the model, coding PSO program prepared in Matlab7.0, running experiment on computer P4 2.8G/Windows Server 2003 operating system. The examples select swarm size l = 50, maximum number of iterations 1000, ω = 0.7, observing that the effect under different values of learning factor, determine the selected c1 = 1.5, c2 = 1.6 is better, the number of iterations to 169 times to find the optimal solution. At this time the best scheduling scheme for the ((A1, 0.9), (A2, 1.5), (A4, 3.6)), the first response time is 30h, only 3 grain storage supplier points, then the fitness value of the minimum objective function Z (x) = 20.7 million; but if used in [5,6] the solution out the need to 5 points, the cost will be higher.

Table 1. Grain Storage Supplier Information

A_i	A_1	A_2	A_3	A_4	A_5	A_6	A_7	A_8
t_i (h)	30	36	36	42	48	52	56	60
$x_i^{'}$ (10^4t)	0.9	1.5	3.2	6	0.7	2	2	2.8
C_i(million/ton)	2	3	7	4	6	12	8	10

5 Conclusion

This paper considers the multi-constrained problem of grain emergency scheduling and puts the emergency dispatch costs as the main optimization objective, it has a strong practical background. We also consider the initial dispatch emergency amount to construct a more realistic optimization model. However, this analytical optimization model for this structure is difficult to solve. Using particle swarm algorithm is better for the Characteristics of model, fully able to meet the practical demand, it will provides a solution in the grain emergency scheduling problem for our country.

References

1. Schrage, L.: Formulation and Structure of More Complex/realistic Routing and Scheduling Problems. Networks 11, 229–232 (1981)
2. Sariklis, D., Powell, S.: A Heuristic Method for the Open Vehicle Routing Problem. Journal of the Operational Research Society 51(5), 564–573 (2000)
3. Brandao, J.: A Tabu Search Algorithm for the Open Vehicle Routing Problem. European Journal of Operational Research 157(8), 552–564 (2004)
4. Zhuo, F.: Research for Open Vehicle Routing Problem and its Application. Doctor Dissertation of Central South University 11 (2003)
5. Kennedy, J., Eberhart, R.C.: ParticleSwarm Optimization. In: Proceedings of IEEE International Conference on Neural Networks, Piscataway, NJ, pp. 1942–1948 (1995)
6. Shi, Y., Eberhart, R.C.: Parameter Selection in Particle Swarm Optimization. In: Evolutionary Programming VII: Proceedings of the Seventh Annual Conference on Evolutionary Programming, New York, pp. 591–600 (1998)
7. Ayed, S., Imtiza, A., Sabah, A.-M.: Particle Swarm Optimization for Task Assignment Problem. Microprocessor and Microsystems 26(8), 363–376 (2002)
8. Li, N., Zou, T., De-bao, S.: Particle Swarm Optimization for Vehicle Routing Problem. Journal of Systems Engineering 19(6), 596–600 (2004)
9. Gendreau, M., Hertz, A., Laporte, G.: New Insertion and Postoptimization Procedures for the Traveling Salesman problem. Operations Research 40(8), 1086–1093 (1992)

Fast Sparse Recovery in Underdetermined Blind Source Separation Based on Smoothed ℓ^0 Norm

Tianbao Dong and Jingshu Yang

Teaching and Research Section 702
Electronic Engineering Institute
Hefei, China
dtb_1@163.com

Abstract. In this paper, we propose a fast method to resolve the sparse recovery problem in underdetermined blind source separation. Our method is a modified version of recently proposed method (SL0), which is faster than original SL0 method while acquiring the same estimation quality. Comparison between our method and orthogonal matching pursuit (OMP) is given from viewpoints of efficiency and estimation quality. Experiments using synthetic data show that our method is faster than OMP especially when noise in the source vector exists and can acquire better estimation quality than OMP when signal is sufficiently sparse. We also apply the method to underdetermined blind source separation using real world data. It is experimentally shown that our method is much faster than OMP, while providing almost the same accuracy.

Keywords: Sparse component analysis, Sparse recovery, Underdetermined blind source separation, Orthogonal matching pursuit, Smoothed l0 norm.

1 Introduction

Consider the following noisy instantaneous mixing system:

$$\mathbf{x}(t)=\mathbf{A}\mathbf{s}(t)+\mathbf{n}(t) \tag{1}$$

where $\mathbf{x}(t)=[x_1(t), x_2(t)......x_m(t)]^T$ is the known $m\times 1$ mixing vector at time t, $\mathbf{A}$ is the unknown $m\times n$ mixing matrix, $\mathbf{s}(t)=[s_1(t), s_2(t)......s_n(t)]^T$ is the unknown $n\times 1$ source vector at time t, $\mathbf{n}(t)$ is the unknown $n\times 1$ additive noise vector. Estimating the source signals $\mathbf{s}(t)$ from the mixtures $\mathbf{x}(t)$ without knowing other information is the objective of underdetermined source separation(BSS). When the number of sensors is less than that of sources, i.e., m<n, BSS is said to be underdetermined blind source separation (UBSS).

UBSS has received a great deal of attention in recent years. Many approaches are presented to resolve the UBSS problem, and most of which are discussed in the context of Sparse Component Analysis (SCA). SCA can be viewed as the problem of separating sparse sources. It usually consists of two steps: mixing matrix is estimated at the first step, and then source signals are estimated at the second step. Mixing matrix is usually estimated by means of clustering [1], [2], [3]. Estimating source signals is to find the sparsest solution of (1), which can be formulated as the following optimization problem (regardless of noise):

D. Zeng (Ed.): Advances in Control and Communication, LNEE 137, pp. 489–495.
springerlink.com

$$(P_0)\text{: min } \| \mathbf{s}(t) \|_0 \quad \text{s.t.} \quad \mathbf{x}(t)=\mathbf{A}\mathbf{s}(t), \tag{2}$$

Where the $\| s(t) \|_0$ is defined by the number of nonzero entries of s(t) , which is usually called ℓ^0 norm.

However, it is well known that solving (2) is NP-hard since the computational time for solving this optimization problem increases exponentially as the dimension of **s**(t) increases. Hence, many alternative methods have been proposed, and Basis Pursuit (BP) and Orthogonal Matching Pursuit (OMP) [9] are the most frequently used methods. BP was presented by solving the following optimization problem:

$$(P_1)\text{: min } \| \mathbf{s}(t) \|_1 \quad \text{s.t.} \quad \mathbf{x}(t)=\mathbf{A}\mathbf{s}(t), \tag{3}$$

Many studies have shown that in many cases the solution of ℓ^1 norm is equivalent to that of ℓ^0 norm. BP can be resolved by means of linear programming method, but for many large scale applications the computational complexity is often too high. In recent years, many fast ℓ^1 minimization algorithms have been proposed to solve (P_1) [4], [5], [6], [7], [8].OMP is a greedy algorithm, which is one of the fastest algorithms for sparse recovery.

Recently, an algorithm called Smoothed ℓ^0 (SL0) [10] was proposed to solve (P_0), which provides a fast solution of the sparsest solution. The main idea of SL0 is to approximate the ℓ^0 norm by a smooth function, and then the objective of SL0 is to maximize the smooth function.

In this paper, we propose a modified SL0 algorithm. By means of a parameter selection strategy we provide, our method is faster than SL0 algorithm, while providing almost the same accuracy.

2 Fast Sparse Recovery Algorithm

2.1 Main Idea of SL0

As stated in previous section, SL0 replaces ℓ^0 norm by continuous function which approximates the Kronecker delta function, and then maximizes it by means of a maximization algorithm for continuous functions. [10] provides the following function approximating the kronecker delta function:

$$f_\sigma(s) = \exp(-s^2 / 2\sigma^2) \tag{3}$$

By defining:

$$F_\sigma(\mathbf{s}) = \sum_{i=1}^{n} f_\sigma(s_i), \tag{4}$$

Then $\|s_0\| \approx n - F_\sigma(\mathbf{s})$ for small values of σ, when σ→0 the approximation tends to equality. Hence, the solution of (P_0) can be obtained by maximizing $F_\sigma(\mathbf{s})$ (subject to **x**=**As**). for convenience, we neglect time t here. σ is a critical parameter, which determines how smooth the function $F_\sigma(\mathbf{s})$ is and how the function $F_\sigma(\mathbf{s})$

approximates to ℓ^0 norm: for large value of σ, $F_\sigma(\mathbf{s})$ is smooth and contains less local maxima, and its maximization is easy, but its approximation to ℓ^0 norm is not well; on the other hand, for smaller value of σ, $F_\sigma(\mathbf{s})$ is highly nonsmooth and contains more local maxima, and its maximization is harder, but its approximation to ℓ^0 norm is better. To escape from local minima in maximization algorithm, [12] adopted a decreasing set of values for σ in each iteration. For each value of σ the maximization of $F_\sigma(\mathbf{s})$ is initiated with the maximizer of the $F_\sigma(\mathbf{s})$ for the previous value of σ.

2.2 MSL0 Algorithm

SL0 algorithm uses a decreasing set of values for σ to escape from local minima. When generating decreasing set of values, it uses following formula:

$$\sigma_j = c\sigma_{j-1} \tag{5}$$

where c is a constant, usually chosen between 0.5 and 1. As described before, the larger the value of σ, the smoother the $F_\sigma(\mathbf{s})$ function; the smaller the value of σ, the more nonsmooth the $F_\sigma(\mathbf{s})$ function. Therefore, we can decrease the value of σ as follows: the smaller the value of σ, the smaller the decreasing step size. By doing so, we can balance between computational cost and accuracy. The final algorithm is shown in Fig.1. We call our algorithm Modified SL0 (MSL0).

3 Experiments

In this section, the performance of our method is compared with OMP and SL0 experimentally. In the first two experiments, sparse sources are artificially created using a Bernoulli-Gaussian model and in Experiment 3, we use real world data. In Bernoulli-Gaussian model, each source is distributed as:

$$s_i \sim p\cdot\mathcal{N}(0,\sigma_{on}) + (1-p)\cdot\mathcal{N}(0,\sigma_{off}) \tag{6}$$

where each source is 'active' with probability p, and is 'inactive' with probability 1-p. If it is active, its value is modeled by a zero-mean Gaussian random variable with variance σ_{on}^2; if it is not active, its value is modeled by a zero-mean Gaussian random variable with variance σ_{off}^2, where $\sigma_{off}^2 \ll \sigma_{on}^2$. Sparsity implies that $p \ll 1$. We consider $\sigma_{on} = 1$. Noise in this paper has two meanings. The noise in the source means that the inactive elements of source are not exactly equal to zero, which is modeled by σ_{off}. The noise in the sensors means the additive noise in (1), which is modeled by σ_n (standard deviation).

Signal-to-Noise Ratio (SNR) is used to evaluate the estimation quality. It is defined as $20\log(\|\mathbf{s}\|/\|\mathbf{s}-\hat{\mathbf{s}}\|)$, where $\mathbf{s}$ and $\hat{\mathbf{s}}$ denote the actual source and its estimation, respectively.

- Initialization:
  1. Let $\hat{\mathbf{s}}_0 = \mathbf{A}^T(\mathbf{A}\mathbf{A}^T)^{-1}\mathbf{x}$.
  2. Choose suitable values σ_1, $\sigma_{\min}$, c, $c_{\max}$, ρ, L, μ.
  3. Let $\sigma = \sigma_1$. j=1.
- While $\sigma > \sigma_{\min}$
  1. Maximize the function $F_\sigma(\mathbf{s})$ on the feasible set $S = \{\mathbf{s} \mid \mathbf{x} = \mathbf{A}\mathbf{s}\}$ using L iterations of the steepest ascent algorithm (followed by projection onto the feasible set).
     —Initialization: $\mathbf{s} = \hat{\mathbf{s}}_{j-1}$.
     —For k=1...L(loop L times)
       1) Let $\boldsymbol{\delta} \triangleq [s_1 \exp(-s_1^2/2\sigma^2), \ldots, s_n \exp(-s_n^2/2\sigma^2)]^T$
       2) Let $\mathbf{s} = \mathbf{s} - \mu\boldsymbol{\delta}$.
       3) Project **s** back onto the feasible set S: $\mathbf{s} = \mathbf{s} - \mathbf{A}^T(\mathbf{A}\mathbf{A}^T)^{-1}(\mathbf{A}\mathbf{s} - \mathbf{x})$.
  2. $\sigma = c\sigma$.
  3. If $c > c_{\max}$ then $c = c_{\max}$ else $c = \rho c$.
  4. Set $\hat{\mathbf{s}}_j = \mathbf{s}$. j=j+1.
- Final answer is $\hat{\mathbf{s}} = \hat{\mathbf{s}}_j$.

Fig. 1. The MSL0 algorithm

Our experiments are performed in MATLAB7 environment using Pentium 4 2.8GHz dual processor with 1.5GB of memory, and under Microsoft Windows XP operating system.

Experiment 1. Computational Cost
In this experiment, we study the computational cost of the proposed method versus the number of sources, and compare the method with OMP in two cases: no noise in the sources and noise in the sources.

The values used are m=40, $\sigma_{on} = 1$, $\sigma_n = 0$, $\sigma_1 = 2\max|s_i|$, $\sigma_{\min} = 0.001$,c=0.5, $c_{\max} = 0.8$, $\rho = 1.2$,L=3, $\mu = 1$. In noise case, $\sigma_n = 0.01$, and sparse level k=10. The experiment is repeated 100 times, and then the values of computational time are averaged. In Fig.1, averaged computation time is plotted versus the number of sources. Fig. 1(a) is the noiseless case and Fig. 1(b) is the noisy case. It is clear that the proposed method is faster than OMP and is much faster than OMP when there is noise in the sources, which means that our method is less sensitive to the noise in the sources than OMP.

Experiment 2. Performance comparison
In this experiment, we compare the performance of our method with OMP from different aspects. In the first part of the experiment, we consider the noise in the sources. Then the noise in the sensors is considered in the second part.

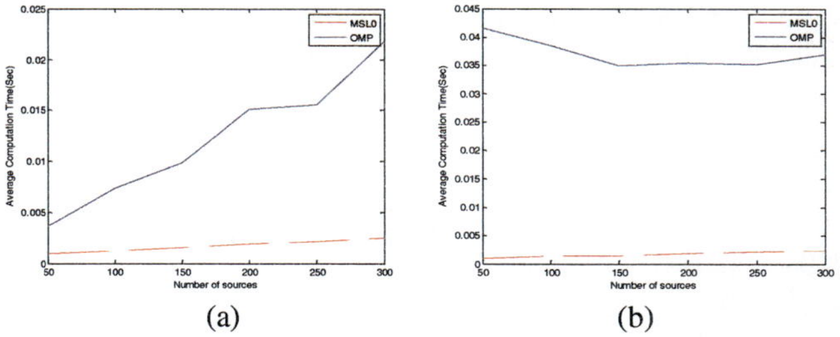

Fig. 1. Averaged computation time versus number of sources. (a) Noiseless case, $\sigma_n = 0$. (b) Noisy case, $\sigma_n = 0.01$

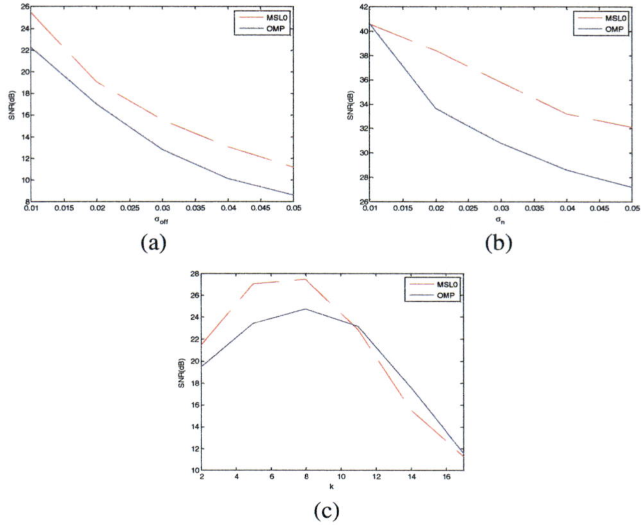

Fig. 2. Performance comparison for the case m=40, n=100. (a) SNR versus the noise in the sources. (b) SNR versus the noise in the sensors. (c) SNR versus the sparsity.

Finally, we compare the effect of sparsity on the performance of the proposed method and OMP. The experiment is also repeated 100 times. Here, m=40, n=100.

Figures 2 give the final results of the comparison. SNR versus the noise in the sources is plotted in Fig. 2(a). The presented method provides better accuracy than OMP when considering the noise in the sources. Fig. 2(b) depicts SNR versus the additive noise. From the comparison, we can see that with the additive noise increasing, the proposed method obtains better estimation quality than OMP. In Fig. 2(c), we compare the performance of our method with OMP for different sparsity. When sources are sufficiently sparse, MSL0 provides better accuracy.

Experiment 3. Sparse recovery in UBSS
To evaluate the performance of the proposed method in UBSS, we use real world data in the simulation. The sources are four flutes, sampled at 44.1 kHz, 30000 samples for each source. The wave forms are shown in Fig. 3. The mixing matrix is set as:

$$A = \begin{bmatrix} 0.9985 & -0.1142 & 0.2152 & 0.4027 \\ 0.0555 & -0.9935 & -0.9766 & 0.9152 \end{bmatrix}$$

We first estimate the mixing matrix using clustering, and then recover the sources using the proposed method, SL0 and OMP respectively.

Figures 4 give the recovered sources using SL0. The decreasing step size c is 0.5 in Fig. 4(a) and is 0.8 in Fig. 4(b). It is clear that the estimation quality of Fig. 4(b) is better than that of Fig. 4(a), but the computation time of the former is longer than that of the latter. The computation time is 5.562382 seconds for Fig. 4(a) and is 6.684151 for Fig. 4 (b). Fig. 5 is the wave forms recovered by the proposed method, where c=0.5, $c_{\max} = 0.8$, $\rho = 1.2$. The estimation quality of Fig. 5 is almost the same as that of Fig 4(b), but Fig. 5 requires less time, 5.678274 seconds. Although OMP can recover the sources well (Fig. 6), it requires much time which is 414.865728 seconds.

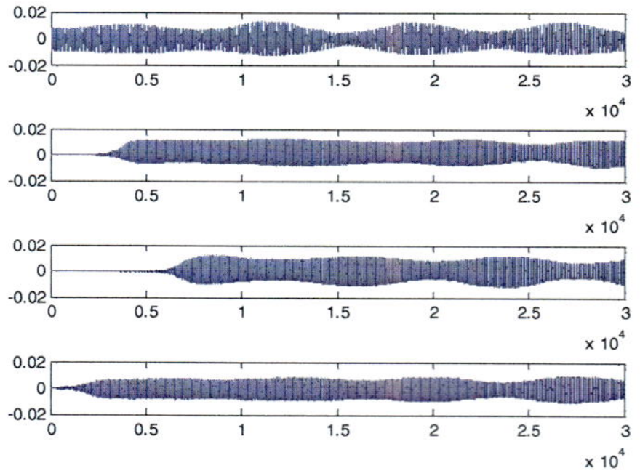

Fig. 3. The original sources wave form

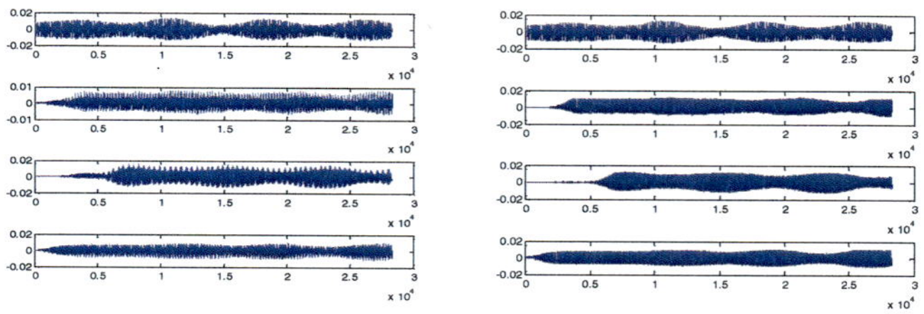

(a) (b)

Fig. 4. The recovered sources using SL0. (a) c=0.5. (b) c=0.8.

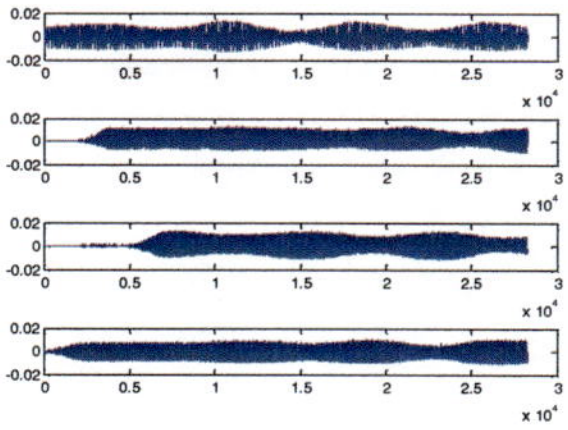

Fig. 5. The recovered sources using MSL0

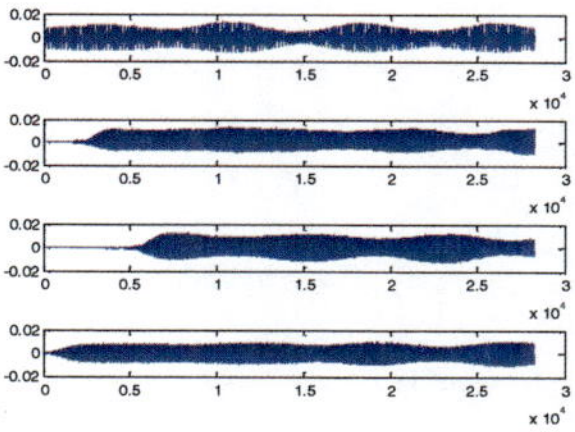

Fig. 6. The recovered sources using OMP

4 Conclusion

We have presented a fast sparse recovery method which is a modified SL0. Compared with OMP, the proposed method runs faster and is less sensitive to the noise in the source. Experiments show that when used in UBSS, the presented method is faster than SL0 and OMP, while providing almost the same estimation quality.

References

1. Bofill, P., Zibulevsky, M.: Underdetermined blind source separation using sparse representations. Signal Processing 81(11), 2353–2362 (2001)
2. Li, Y.Q., Cichocki, A., Amari, S.: Analysis of sparse representation and blind source separation. Neural Computation 16(6), 1193–1234 (2004)
3. Zibulevsky, M., Kisilev, P., Zeevi, Y.Y., Pearlmutter, B.A.: Blind source separation via multinode sparse representation. In: Advances in Neural Information Processing Systems, pp. 1049–1056. MIT, Cambridge (2002)
4. Figueiredo, M., Nowak, R., Wright, S.: Gradient projection forsparse reconstruction: Application to compressed sensing and other inverse problems. IEEE Journal of Selected Topics in Signal Processing 1(4), 586–597 (2007)
5. Osborne, M., Presnell, B., Turlach, B.: A new approach to variable selection in least squares problems. IMA Journal of Numerical Analysis 20, 389–404 (2000)
6. Daubechies, Defrise, M., Mol, C.: An iterative thresholding algorithm for linear inverse problems with a sparsity constraint. Communications on Pure and Applied Math. 57, 1413–1457 (2004)
7. Beck, Teboulle, M.: A fast iterative shrinkage-thresholding algorithm for linear inverse problems. SIAM Journal on Imaging Sciences 2(1), 183–202 (2009)
8. Yang, J., Zhang, Y.: Alternating direction algorithms for l1-problem in compressive sensing, (preprint) arxiv:0912.1185 (2009)
9. Davis, G., Mallat, S., Avellaneda, M.: Adaptive greedy approximations. Journal of Constructive Approximation 13, 57–98 (1997)
10. Mohimani, H., Babaie-Zadeh, M., Jutten, C.: A fast approach for overcomplete sparse decomposition based on smoothed l0-norm. IEEE Trans. Signal Processing 57, 289–301 (2009)

Chinese Text Classification Based on VC-Dimension and BP Neural Network

Wang Yaya[1] and Ji Xueyun[2]

[1] Department of Information Engineering
Binzhou Polytechnic
Binzhou, Shandong, China
[2] Mathematics and Computer Department
Baoding University
Baoding, China
wangyaya@bzpt.edu.cn, glee-1218@163.com

Abstract. Chinese text classification based on BP neural network is an important research topic in information retrieval. In this paper, by analyzing the method of the classification based on BP, we propose that used VC-dimension to certain the number of neurons in hidden layers, describe and Analysis of the algorithm. Then, evaluating results by precision rate, recall rate and F1. Experimental results show that the proposed method has high accuracy and efficiency.

Keywords: BP neural network, Vapnik-Chervonenkis-dimension, Chinese text classification, Gaussian function.

1 Introduction

With the high development of Internet technology, Chinese text classification [1-2] plays an important role in research field of information processing. The task of text classification is processing the unknown document to judging which class it belongs to by automatically. The common methods for Chinese text classification are bayes method [3-7], k-NN, decision tree, rough sets, genetic algorithm, fuzzy set, SVM and neural network. But all these method have their limits, such as lacks precision and high complexity. The common problem in current research is how to improve the recognition effects and generalization ability of the text classifier.

At same time BP[8] has been fully applied in data mining. Lippmann points [9] out that BP neural network with two hidden layers can solve all kinds of the problem in classification. But the structure of BP neural network will directly affect their generalization ability and convergence rate. For complicated problem, we need to add the number of the hidden layers to improve degree of precision. In addition, if there is a few neurons of setting in the .the hidden layer, BP can not training, then the network has a low generalization; else, If the number of neurons in the hidden layers to the excessive, the network will too difficult, waste time and has the possibility of over fitting. Then, The most important things in BP neural network is setting the number of the hidden layers and the number of neurons in hidden layers properly. In this paper, we used VC -dimension to set the number of neurons in hidden layers.

D. Zeng (Ed.): Advances in Control and Communication, LNEE 137, pp. 497–502.
springerlink.com

2 Used Vc-Dimension to Setting the Number of Neurons in Hidden Layers

The VC dimension [10] (for Vapnik-Chervonenkis-dimension) is a measure of the capacity of a classification algorithm. It is one of the core concepts in statistical learning theory. Intuitively, a sample x of length m is shattered by H, or that H shatters x, if H gives all possible classifications of x. H is the maximum length of a sample shattered by H; if there is no such maximum, we say that the VC-dimension of H is infinite. The VC-dimension of bounded real transfer functions can use a threshold transform it to an indicator function.

In this paper, there are two hidden layers and the number of neurons in hidden layers evaluated by VC-dimension [11]. As a learning system whose VC-dimension is h, if the size of training dataset is m, the generalized error will be smaller than e, e satisfies

$$e \leq o\left(\frac{d}{m}\ln\left(\frac{m}{d}\right)\right) \tag{1}$$

For a single hiding layer of neural network, If N is the number of neurons (node) of input layer, P is the number of neurons hidden layer, C is number of neurons of output layer and its VC-dimension is $NP + PC$ .Because N , C and m which is the number of example ,we are already know. Thus, we can estimate P by setting up error upper bound e .It can be seen that P is that lower bound on the number of hidden layer in the worst situation. It makes a possible that need not change network structure while increasing the training dataset.

3 Activation Function and Realization of BP Network

Since neural network to be asked gradient, thus, the function must be met to be has derivative. To make this idea precise, we introduce activation function $f(x)$ based the Gaussian function [12].

$$f(x)=\frac{\sum_{l=1}^{M} y_l[\prod_{i=1}^{n}\exp(-(x_i - \mu_{il})^2)]}{\sum_{l=1}^{M}[\prod_{i=1}^{n}\exp(-(x_i - \mu_{li})^2)]} \tag{2}$$

Above, M means the number of neurons of hidden layer, y can learn as a weight for we can learn.

μ_i After we known y , n means the number of neurons of input layer, μ_i means the weight that we need to learn. For discussion easily, the formula (3) can be decomposition as follow:

$$Z^l = \prod_{i=1}^{n}\left[\exp(-(x_i - \mu_{il})^2)\right] \tag{3}$$

By the sum and weighted sum operation to Zl , assuming that:

$$a = \sum_{l=1}^{M} y_l Z^l, b = \sum_{l=1}^{M} Z^l$$

Then we can obtain the activation function $f(x)$ is: $f(x) = \frac{a}{b}$. The structure of neural network in this paper is illustrated in figure1:

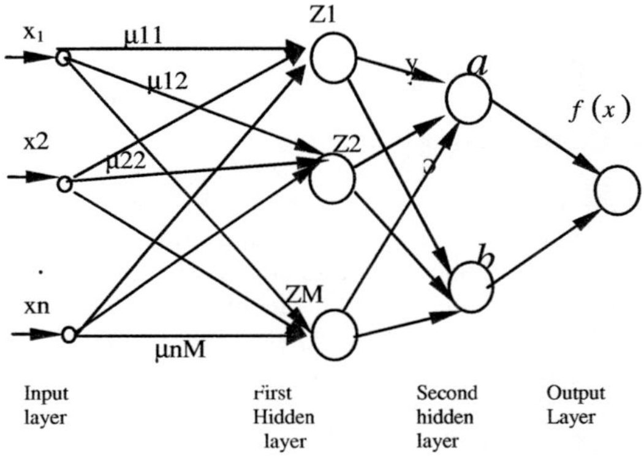

Fig. 1. The Structure of Neural Network in This Paper

For given the dual of input-output $(x_{01}^p, \ldots, x_{0n}^p; y_0^p)$, we should by the activation function which is defined in formula (3) makes the fitting at least:

$$E = \frac{1}{2}\sum_i [f(x_i) - y_i]^2 \tag{4}.$$

This paper used the common BP network to learn the parameter μ_i, we can certain μ_i by formula (5) :

$$\mu_{il}(q+1) = -\alpha \tag{5}$$

By the Derivation for Composite Function:

$$\mu_i(q) \qquad \frac{\partial E}{\partial \mu_{il}}\Big|_q$$

$$\frac{\partial E}{\partial \mu_{il}} = \frac{\partial E}{\partial f}\frac{\partial f}{\partial Z^l}\frac{\partial Z^l}{\partial \mu_{il}} = (f - y)\frac{y - f}{b}Z^l 2(x_i - \mu_{il}) \tag{6}$$

Above, f represent $f(x_i)$, y represent y_i take formula (6) into formula (5), can get the learning process of μ_i , shown in (7):

$$\mu_{il}(q+1) = \mu_{il}(q) - \alpha \ (f - y)\frac{(y(q) - f)}{b} Z' 2(x_i - \mu_{il}(q)) \tag{7}$$

Above, α is learning rate, 0<α<1. Generally speaking α as learning rate is fixed in the BP algorithm. Because of the fixed rate in learning network have a certain stability, but the problem is that is the reducing the adaptive capacity. The important things in learning process of classification are setting the learning rate. Because of α is great then network convergence turn fast and lead to stability decline.

4 Experiment and Analysis

4.1 Dataset and Preprocessing

The dataset or corpus in this paper is based on SoGou corpus and the part of the open-source corpus of Fudan University and added some of Chinese news texts in Sina. Relatively speaking, we established a whole corpus .There are 9 classes in this corpus, financial, military, education, entertainment, technology, sports, computers, social and medicine. We get 2926 documents in this corpus, the documents in training dataset are 1949 and in test dataset are 977.There is no ambiguous category in training dataset and document distribution is illustrated in Figure2.

In the process of pretreatment this experiment, we need to construct a feature item sets.

统计信息 | 财经 | 军事 | 电脑 | 教育 | 科技 | 体育 | 卫生 | 社会

序号	类别名称	训练语料库	测试语料库
1	财经	222	123
2	军事	237	100
3	电脑	225	107
4	教育	116	116
5	科技	239	103
6	体育	200	127
7	卫生	217	119
8	社会	278	82
9	娱乐	215	100
--	----	----	----
0	总计：	1949	977

Fig. 2. The document distribution of corpus in this paper

A reasonably and effectively features item sets must meet the following two conditions, perfectibility and resolution. First, We used the ICTCLAS which is developed by Zhang Huaping in Institute of Computing Technology Chinese Academy of Sciences to putting the word segmentation in this paper. Second, we used the $TF-IDF$ for feature representation and used the developed CHI [13] for feature selection. Then we get a features datasets meet the above two conditions.

4.2 Training Module and Test Module

Chinese text classifier by training modules and testing modules, the corpus of testing module is provided testing dataset and the result of the classification is illustrated in Figure3.

Table 1. The Class and Number of Corpus in this experiment

Class	The Number of Train Dataset	The Number of Test Dataset
Financial	222	123
Military	237	100
Entertainment	226	100
Education	115	116
Technology	239	103
Sports	200	127
Computers	215	107
Medicine	217	119
Social	278	82
Total	1949	977

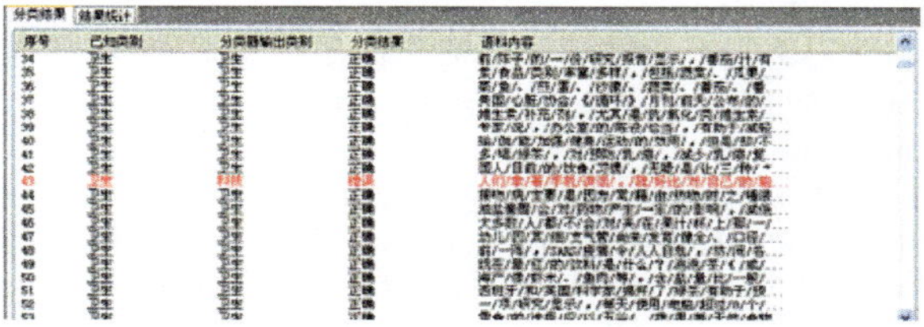

Fig. 3. Classification results of the testing module

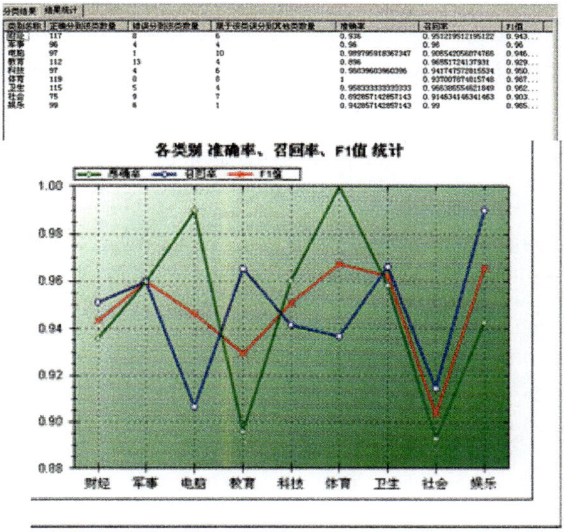

Fig. 4. The Recall Rate, Precision Rate and Fl

There are two kinds of methods for analysis the result, these are micro- average and macro-average. This paper uses the Recall Rate, Precision Rate and Fl to showing the result of experiment.

5 Conclusion

In this paper, by shown the characteristics of neural network we using VC-dimensional method to determine the number of hidden neurons. Given a method for structuring a Chinese text classifier based on improved BP neural network. And improved this method is feasible and effective by some experiments. Furthermore, the learning factor in BP network is a new point for researching in the future.

References

1. Yang, Y.: Sampling Strategies and Learning Efficiency in Text Categorization. In: AAAI Spring Symposium on Machine Leariing in Information Access, pp. 88–95 (1996)
2. Sebastiani, F.: Machine learining in automated text categorization. ACM Computing Surveys 31(1), 1–47 (2002)
3. Shi, Z.: Neural Networks, 1st edn. Publishing House of Electronics Industry, BeiJing (1993)
4. Lv, X., Shi, H.: Multi-label Text Classification Algorithm Based on Frequent Item Sets. Computer Engineering 15(5), 83–85 (2010)
5. Zhao, W.-Q., Zhu, Y.-L., Jiang, B.: A classification model based on SVM and rough set theory. Journal of Communication and Computer (2), 42–45 (2008)
6. Cheng, L.-L., He, P.-L., Sun, Y.-H.: Study on Chinese keyword extraction algorithm based on Naive Bayes model. Computer Applications (12), 2780–2783 (2005)
7. Thong, Z.-Y., Qin, K.-Y.: Text Classification Approach Based On Fuzzy Soft Set Theory. Computer Engineering (13), 90–92 (2010)
8. Simon, H.: Neural Networks:A Comprehensive Foundation, 2nd edn. Prentice Hall/Pearson (January 2004)
9. Lippmann, R.: An Introduction to Computer with Neural Nets. IEEE ASSP Magazine (April 1987)
10. Zhao-Hui: Study on Support Vector Machines Classification methods and their Application in Text Categorization. Dalian University of Technology (2005)
11. Research of the VC-Dimension Computation of Neural Networks. Computer Science (7), 59–63 (2000)
12. Anthony, M.: Probabilistic Analysis of Learning in Artificial Neural Networks: The PAC Model and Its Variants. Neural Computing Surveys, 1–47 (1997)
13. Cheng, C.-P.: Research and Implementation on Intelligent Information Retrieval Based on Classification. Yunnan Normal University (2005)

Research of Credit Risk Based on Credit Transition Matrix

Zheng Zhi-hong[1] and Lu Shi-bao[2]

[1] North China Institute of Water Conservancy and Hydroelectric Power Zhengzhou, 450011, China
[2] Shenzhen Government Investment Project Evaluation Center Shenzhen, 518036, China
346402642@qq.com, Lu5111284@yahoo.com.cn

Abstract. Risk assessment methodologies seek to assess the maximum potential change in the value of a portfolio of financial instruments with a given probability over a pre-set horizon resulting from changes in market factors, credit risk, This paper provides a methodology to assess credit risk. credit transition matrix has a great impact. It will affect the credit rating changes. Credit Change this index ranged from zero when the critical state has a different default status, when the credit quality index greater than zero, the rise in the value of its credit; if the credit quality index is less than zero, the credit quality of the index on the lower Level; If zero, credit rating is to maintain the normal status.

Keywords: Credit, Credit risk, Credit rating, Risk evaluation.

1 Introduction

Risk management is a core activity in asset allocation conducted by banks, insurance and investment companies, or any financial institution that evaluates risks. Credit risk is one of the most important topics in risk management and finance today. The last decade has seen the development of models for pricing credit risky instruments and derivatives, for assessing the credit worthiness of obligors, for managing exposures of derivatives and for computing portfolio credit losses for bonds and loan portfolios. In light of these financial innovations advances the Basle Committee on Banking Supervision has taken the first steps to amend current regulation and is reviewing the applicability of internal

2 Default

Credit risk analysis is a methodology to assess the effect of stochastic changes in credit quality on the value of a fixed income security or a portfolio of fixed income securities. To do this, credit risk methodologies estimate the probability of financial assets migrating to different risk categories over a pre-set horizon. Default occurs, technically, at the same time as delinquency; that is, a loan is in default as soon as the borrower misses a scheduled payment. In this article, however, we reserve the term ''default'' for any of the following four situations [1]:

D. Zeng (Ed.): Advances in Control and Communication, LNEE 137, pp. 503–509.
springerlink.com

• A lender has been forced to foreclose on a mortgage to gain title to the property securing the loan.

• The borrower chooses to give the lender title to the property ''in lieu of foreclosure.''

• The borrower sells the home and makes less than full payment on the mortgage obligation.

• The lender agrees to renegotiate or modify the terms of the loan and forgives some or all of the delinquent principal and interest payments. Loan modifications may take many forms including a change in the interest rate on the loan, an extension of the length of the loan, and an adjustment of the principal balance due.

Because practices differ in the lending industry, not all of the above situations are consistently recorded as defaults by lenders. Moreover, the length of the foreclosure process may vary considerably, affecting the measured default rate.

Default will spend a lot of cost. To the extent that the causes of default are not well understood, lenders may charge a higher average price for mortgage. credit to reflect this uncertainty. Alternatively, lenders may respond to this uncertainty by restricting credit to only the most creditworthy borrowers. By better distinguishing between applicants that are likely to perform well on their loans from those that are less likely to do so, lenders can ensure wider availability of mortgages to borrowers at prices that better reflect underlying risks. Default also imposes great costs both on the borrowers involved in the process and on society in general. For borrowers, default ordinarily results in a lower credit rating and reduced access to credit in the future, a loss of assets, and the costs of finding and moving to a new home[2]. When geographically concentrated, defaults can also have a pronounced social effect because they lower local property values, reduce the incentives to invest in and maintain the homes in the affected neighborhoods, increase the risk of lending in those neighborhoods, and thus reduce the availability of credit there.

3 Transition Probabilities

To illustrate migration rates, consider a simplified loan rating system with two credit risk classes, A and B, signifying low and high risk[3], respectively. An A-rated loan can either remain in class A or migrate to class B in the next period. Similarly, a B-rated loan can either remain in class B or migrate to class A. These possibilities are

Period1 \ Period2	A	B
A	90%	10%
B	15%	85%

Fig. 1. Miguration Alternatives and Probabilities

portrayed in Figure 1.The transition probabilities (Pi values) are estimated from historic migration rates. For example, historic rates of movements among classes might indicate class A loans remain in class A 90% of the time and migrate to class B 10% of the time. Class B loans remain in class B 85% of the time. The combined migration patterns then are expressed by the transition probability matrix in Figure 2. In this case, potential upgrading (downgrading) of a loan is represented by arrow BA (AB), and the stationary class ratings are represented by arrows AA and BB from Figure 1. The aggregate rate of high risk (i.e., being in Class B) would be the weighted average of AB and BB using the relative numbers or volumes of loans A and B as the weights. Continuing the numerical example, if classes A and B comprise 75% and 25%,respectively, of the portfolio, then the weighted average of class B is 28.75%:0.2875 = (0.10)(0.75) + (0.85)(0.25). These rating migrations may be measured for any length of period, although one year is the most common (Saunders; Crouhy, Galai, and Mark). Generally, probabilities for interclass migrations tend to increase as the length of period increases, reflecting the potential for greater fluctuations in economic conditions over longer periods (Carty and Fons). The migration rates may also differ across industries [4-7].

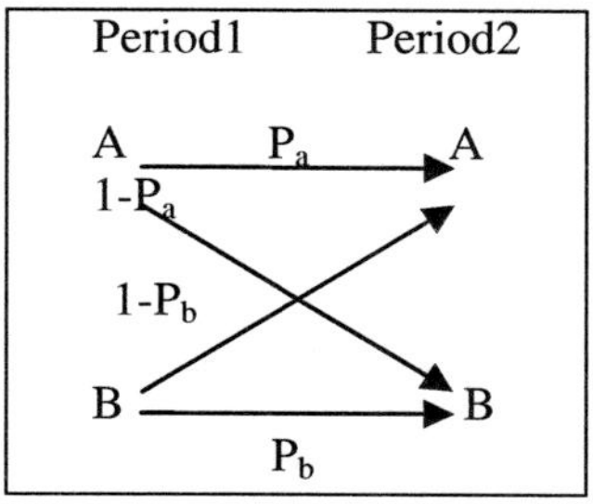

Fig. 2. Trandtion Probability Matrix

4 Conditional Credit Ratingtransition Matrix

The reference[8,9]shows the credit transformation is a dynamic process, which could be expressed by linear relation. Here, Y stands for the continuous credit transformation. And Z stands for a credit quality index at one moment. β stands for regression coefficient whose value is 0.0244. ε_t for a standard free error whose value is from 0 to 1. And its formula is:

$$Y_t = \beta Z_t + \sqrt{1-\beta^2}\varepsilon_t \tag{1}$$

Because Yt is a standard normal distribution, in the finance, ri is the initial credit rating and rm is the final credit rating, we can divide the value of Yt into many intervals. Then, the whole credit rating is:

$$\begin{cases} r_1, \text{ if } Y_t \leq y_1 \\ r_2, \text{ if } y_1 \leq Y_t \leq y_2 \\ r_3, \text{ if } y_2 \leq Y_t \leq y_3 \\ \cdots\cdots \\ r_{m-1}, \text{ if } y_{m-2} \leq Y_t \leq y_{m-1} \\ r_m, \text{ if } y_{m-1} \leq Y_t \end{cases} \tag{2}$$

Meanwhile, we could obtain:

$$\begin{aligned} \theta_{i,j} &= \text{Prob}\{y_{j-1} < Y_t \leq y_j | Z\} \\ &= \text{Prob}\{y_{j-1} < \beta Z_t + \sqrt{1-\beta^2}\varepsilon_t \leq y_j\} \\ &= \text{Prob}\left\{\frac{y_{j-1} - \beta Z_t}{\sqrt{1-\beta^2}} < \varepsilon_t \leq \frac{y_j - \beta Z_t}{\sqrt{1-\beta^2}}\right\} \end{aligned} \tag{3}$$

When we fix j in the scope of j=1, j=m-1 and j=m, we also could deduce:

$$(\theta_{i,j})_{m\times m} = \begin{cases} \Phi\left(\frac{y_j-\beta Z_1}{\sqrt{1-\beta^2}}\right) \ldots\ldots\ldots\ldots j=1 \\ \Phi\left(\frac{y_j-\beta Z_1}{\sqrt{1-\beta^2}}\right) - \Phi\left(\frac{y_{j-1}-\beta Z_1}{\sqrt{1-\beta^2}}\right) \ldots\ldots\ldots\ldots 1<j=m-1 \\ 1-\Phi\left(\frac{y_{m-1}-\beta Z_1}{\sqrt{1-\beta^2}}\right) \ldots\ldots\ldots\ldots j=m \end{cases} \tag{4}$$

The credit rating probability is just built up on the base of conditional credit rating transition matrix and credit quality index. Generally speaking, the credit change index of one construction enterprise depends on its credit quality index [10].

Now, assuming the credit quality index of finance is -1.097, by using of formula (2), we could deduce the credit change index is -0.027; by using of formula (3), we could conclude the conditional credit rating transition matrix of EOY, which is shown as Table 2. (See Table 1) In the same way, now assuming that the credit quality index of finance is 1.003, by using of formula (2), we could deduce the credit change index is 0.024. So Table 3 is the conditional credit rating transition matrix in the circumstance of 0.024 as the credit change index. Also, assuming that the credit quality index is 1.053, then its credit change index is 0.027.(See Table 2)

Table 1. Conditional Credit Rating Transition Probability Matrix (Credit Change Index=-0.027)

Rating at the be Ginning of the year	Rating at end of year(%)							
	Aaa	Aa	A	Baa	Ba	B	Caa-c	Default
Aaa	91.782	6.816	1.110	0.271	0.022	0.000	0.000	0.000
Aa	1.204	91.336	6.408	0.748	0.194	0.033	0.000	0.077
A	0.073	2.351	91.189	5.368	0.738	0.119	0.022	0.142
Baa	0.036	0.249	3.995	89.07	5.511	0.726	0.075	0.036
Ba	0.018	0.082	0.409	4.853	87.001	5.831	0.477	1.329
B	0.000	0.036	0.129	0.644	6.214	85.185	3.693	4.100
Caa-c	0.000	0.018	0.037	0.343	1.363	5.734	78.08	14.419
Default	0.000	0.000	0.000	0.000	0.000	0.000	0.000	100.000

Table 2. Conditional Credit Rating Transition Probability Matrix (Credit Change Index=-0.024)

Rating at the Beginning of the year	Rating at end of year (%)							
	Aaa	Aa	A	Baa	Ba	B	Caa-c	Default
Aaa	92.531	6.239	0.979	0.232	0.018	0.000	0.000	0.000
Aa	1.374	91.862	5.847	0.658	0.168	0.028	0.000	0.064
A	0.087	2.643	91.498	4.883	0.648	0.103	0.019	0.120
Baa	0.044	0.290	4.434	89.219	5.019	0.640	0.066	0.288
Ba	0.022	0.097	0.470	5.354	87.128	5.338	0.427	1.164
B	0.000	0.044	0.151	0.735	6.809	85.189	3.404	3.669
Caa-c	0.000	0.022	0.043	0.0397	1.533	6.250	78.467	13.288
Default	0.000	0.000	0.000	0.000	0.000	0.000	0.000	100.000

From Table 1 and Table 2, we could find that the credit change index just influences greatly on transition matrix. It will influence the changes of credit rating. When the credit change index is -0.027 or 0.024, the corresponding default status is different, i.e., the credit rating is becoming high or becoming low. From Table 1, we find that when the credit quality index is more than 0, the credit value of finance will rise up. If its credit quality index is less than 0, the rating of credit quality index of this finance will decrease. If this index is 0, the credit rating will be kept as normal status.

5 Case Analysis

In this paper, we choose one project of finance as a sample to prove hypothesis. The total capital outlay for this project of finance is about US$ 1.2 billion. Herein, we just choose the every investment index from 1998 to 2001. And we could find some changes from these data. The relevant index in Table 3, we could also calculate from formula (1) and (2). This project was operated by means of BOT. Please refer to Table 3. (See Table 3)

Table 3. HK-Canton Highway: Financial Rations and Credit Changes Indicator (Base Case)

Year	Debt service ratio	Debt/total asset	EBRRDA/asset	Revenue /asset	Credit Quality Index Z	Credit Change Indicator Y
1998	1.348	0.472	0.245	0.267	0.667	0.016
1999	1.391	0.377	0.264	0.291	1.003	0.024
2000	1.587	0.265	0.311	0.347	1.590	0.039
2001	1.678	0.136	0.335	0.375	2.038	0.050

Source: References(9).

From Table 4, we could find out the default probability. In the view of total asset-liability ratio, this ratio is 47.2% in 1998 and till 2001, 13.6%. With time passing, the asset-liability just decreases gradually. In the view of total asset-profit ratio, it is 26.7% in 1998 and till 2001, 34.5%. And the credit quality index also increases from0.667 to 2.038 and credit change index from 0.016 to 0.050. Based on these data, we can say: for a bank, the credit risk to load for this project is low, i.e., its credit value is high.

6 Conclusion

The effect of default correlation can be very large, sometimes as large as the default probabilities themselves. Our goal in this article was to demonstrate the applicability of credit concepts to the evaluation of credit risks. The concept is highlighted by the broad scope of information it provides about the risk stability of loan portfolios, and includes but does not rely solely on measuring stress or default rates.Deriving transition probabilities is a natural use of the output of institutional risk-rating systems. These systems are widely used by lenders, and in the future will form the basis for credit risk measurement in determining regulatory requirements for economic capital held by many banks and other types of financial institutions (Basel Committee on Banking Supervision).The uncertainty of finance just determines the risk in the progress of financing. So, for a bank, it is very important to apprehend the risks of loans and to evaluate by means of scientific method. This thesis is to apprehend the risk and to affirm the default risk responsibility, through the credit rating transition matrix so as to evaluate the credit rating of risk of finacial credit.

References

1. Pennington-Cross, A.: Credit History and the Performance of Prime and Nonprime Mortgages. The Journal of Real Estate Finance and Economics 27(3), 279–301 (2003)
2. Credit Risk, Credit Scoring, and the Performance of Home Mortgages. Federal Reserve Bulletin - federalreserve. gov. (1996)
3. Barry, P.J., Escalante, C.L., Ellinger, P.N.: Credit Risk Migration Analysis of Farm Businesses. Agricultural Finance Review (Spring 2002)
4. Altman, E.I., Saunders, A.: Credit Risk Measurement. Journal of Banking and Finance 21(1), 1721–1742 (1998)
5. Crouhy, M., Galai, D., Mark, R.: Risk Management. McGraw-Hill, New York (2001)
6. Carty, L., Fons, J.S.: Measuring Changes in Credit Quality. Fixed Income 27(5), 27–41 (1996)
7. Moody's Investor Service. Default and Recovery Rates of Corporate Bond Issuers, New York (February 2001)
8. Mausser, H., Rosen, D.: Applying Scenario Optimization to Portfolio Credit Risk. ALGO Res. Quart. 50(2), 19–33 (1999)
9. Dequan, K., Tiong, R.L.K.: Journal of Construction Engineering and Management. Assessment of Credit Risk in Project Finance 134(11), 876–884 (2008)
10. Belkin, B., Forest Jr. L.R., Suchower, S.: A one-parameter representation of credit risk and transition matrices. Credit Metrics Monitor, Third Quarter (3), 46–56 (1998)

A Cluster-Based Load Balancing for Multi-NAT*

Linjian Song, Jianping Wu, and Yong Cui

Department of Computer Science and Technology Tsinghua University, Beijing 100084, China
songlinjian@gmail.com, cy@csnet.cs.tsinghua.edu.cn,
jianping@cernet.edu.cn

Abstract. The Internet currently bears both rapid exhaustion of IPv4 address pool and the slow development of IPv6. IPv4-IPv6 coexistence is proved to be a long period. Multi-NAT, which is desirable in large scaled IPv4-IPv6 coexistent network, has some inherent difficulties in the traffic balancing and failure recovery. In this paper we design a novel Cluster-based Load Balancing scheme for Multi-NAT which is composed of load balancer, extended NAT box and cache server. Additionally in order to load-balance the traffic adaptively according to the state of each NAT box, an Adaptive Reassigning Algorithm (ARA) is proposed to enabling our scheme to accommodate dynamic change of traffic and recover rapidly from failures. In the simulation the cluster-based load balancing outperforms other existing DNS-based Multi-NAT schemes and meets the requirements of large scaled NAT.

Keywords: IPv4-v6 coexistence, Multi-NAT, Load balancing.

1 Introduction

The Internet currently bears both rapid exhaustion of IPv4 address pool and the slow development of IPv6. The IPv4-IPv6 coexistence is proved to be a long period and draws a wide concern and extensive discussions in both industry and academia [1-3]. As key technology in IPv4-IPv6 coexistence, Network address translation (NAT) [4] provides an efficient way to share the same IPv4 address among several hosts. Although proposed as a short-term solution for IP address depletion, NAT is widely used nowadays and extended in many schemes [5-6] to accommodate the future Internet.

However most of the NAT-like functions have limitations [7] for network of large size. A standalone NAT box is not able to support large traffic volume. In addition most of NAT devices are stateful maintaining address binding for network translation which makes it hard to simply split the traffic among multiple NAT devices without standardized state synchronization and coordination. These stateful devices, especially deployed in the CGN architecture [8], will bring high risks on reliability when the network becomes huge. As a result it is a necessity to design more scalable mechanism for Multi-NAT scenario.

* This work is supported by National Major Basic Research Program of China (2009CB320501, 2009CB320503), NSFC (no. 60911130511, 60873252), CNGI Project (no. 2008-102), the National High Technology Development Program of China (No. 2008AA01A324).

D. Zeng (Ed.): Advances in Control and Communication, LNEE 137, pp. 511–519.
springerlink.com

In the context of dynamic traffic load balancing there are lots of works in traffic splitting and multi-path routing which are hot topics in Traffic Engineering. As to traffic splitting Table-based Hashing [9] is static enduring the expense of less optimal load balancing. Algorithms 10] split the traffic dynamically according to the desired proportion, but they introduce packets reordering. The dynamic protocol COPE [11] is designed to avoid hot-spots and cope with failures. Though the existing mechanisms have good performance in load balancing, they do not concern the specific scenario of Multi-NAT which consists of many stateful devices. .

To solve the specific problems on large scaled NAT some DNS-based Multi-NAT drafts advanced recently in IETF BEHAVE WG such as multi-box in NAT64 [12] and Redundancy Framework [13]. These proposals are based on the simple idea that the DNS server replies requests for a single domain name using different IP addresses in load-balance way. There are two drawbacks: firstly it can not react to the dynamic changes of traffic. Secondly once failure occurred, it is hard to split the traffic more properly on the rest of NATs. As far as we know little academic work is done to study the Multi-NAT mechanism and analyze the performance of large scaled NAT.

In this paper we propose a novel cluster-based load balancing for Multi-NAT which is composed of load balancer, extended NAT box and cache server. Additionally in order to split the traffic adaptively according to the state of each NAT box, an Adaptive Reassigning Algorithm (ARA) is advanced so that our scheme is able to accommodate dynamic change of traffic and recover rapidly from failures. To the best of our knowledge it is the first academic work to study the modeling, design and performance of Multi-NAT in IPv4-v6 coexistence.

This paper is organized as follows. In Section 2 the cluster-based Scalable Multi-NAT (SMNAT) system will be introduced. In section 3, SMNAT is compared with other DNS-based algorithm via simulations. Finally we conclude the paper in section 4.

2 Scalable Multi-NAT IN IPv4-v6 Coexistence

2.1 Overview of SMNAT

In Fig. 2 the Multi-NAT Cluster integrates load balancer, extended NAT boxes and cache server as a whole. Software protocol and signaling is running on several NAT boxes that concentrate on taking and sharing the traffic load. In many respects they form a NAT cluster providing high availability and scalability.

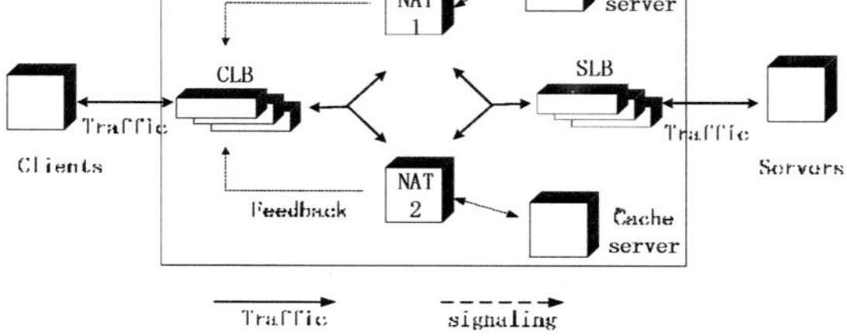

Fig. 1. The Scalable Multi-NAT system

The incoming flow from clients is directed to a dedicated load balancer called Client Load Balancer(CLB). The CLB decides which NAT box should handle the flow and directs the flow to the selected box according to a adaptive reassigning algorithm desicribed in next section. As to traffic from servers it is scheduled by another load balancer called Server Load Balancer (SLB). However SLB is act just as a slave device and follow the flow scheduling made by CLB. The synchronization is done by CLB in the adaptive reassigning algorithm. Note that both CLB and SLB should work with backup devices for redundancy to avoid being single point of failure.

Besides the original functions such as address translation and ALG, the extended NAT box should feedback its state to load balancer indicating its traffic burden. In addtion the extended NAT box should back up its binding information to cache server prepared for flow reassigning in dynamic adaptive algorithm and redundancy requirements.

Cash server in SMNAT design serves as redundancy device in which the address binding information is stored. The address binding can be stored in <key, value> pair which can be rapidly found in a well designed structure such as hash table. If the SMNAT deployed in a distributed manner that the NAT boxes and cache servers may located in different places, memcached is a good choice which is a distributed cache solution that provides a large cache across multiple machines.

2.2 Adaptive Reassigning Algorithm (ARA)

Fig. 3 outlines the inner data structure and logic flow of our proposed Adaptive Reassigning Algorithm (ARA) running in the CLB. There is a set of bins (Set_{bin}) and an Assigning Table (*AT*) which mapping the bins to NAT boxes. Information in IP header is put into the hash function to produce a hash value, which then fall into one of bins. By setting the *AT* properly in a time interval the traffic volume is split dynamically according to the status of each NAT box. If any changes happened at the *AT* binding signaling should be done previously to setup the address binding for the traffic in advance.

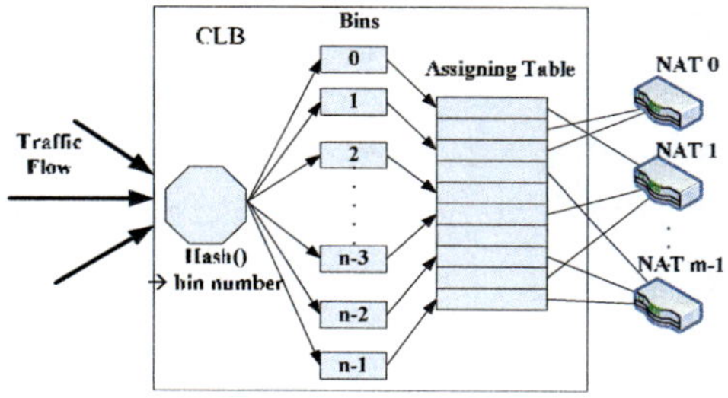

Fig. 2. The data structure of CLB

For hash bin i there is a byte counter C_i which records the volume of traffic and the counter is refresh to zero in a time interval τ . Let the moving average of byte counter passing through hash bin i be mC_i which can be expressed using an exponentially weighted moving average function as follows: $mC_i = (1-\alpha)\times mC_i' + \alpha \times C_i$ where

mC_i' is the moving average value in the previous and α is the smoothing parameter ($0 \le \alpha \le 1$). Without loss of generality, let $Binset_j$ be the set of bins point to NAT *j* and the total traffic volume assigned to NAT *j* at time *t* is defined as $Tv_j = \sum_{i \in Binset_j} mC_i$.

Assume that the CLB get the capacity of each NAT box through signaling or configuration which indicate the proper traffic splitting ration among the NAT boxes : $x_1, x_2, \ldots, x_{m-1}$ where $x_1+x_2+ \ldots,+x_{m-1}=1$. Also the balancer keeps a set of NAT boxes (Set_{NAT}) which are still alive. Note that the load difference between two NAT boxes is expressed as $Diff\,(i, j)$ which is defined in (5):

$$Diff\,(i,j) = \left(\frac{Tv_i}{x_i} - \frac{Tv_j}{x_j}\right) / \sum_{k \in Set_{NAT}} Tv_k \tag{1}$$

The algorithm for ALB is summarized in Table 1.

Table 1. The Adaptive Reassigning Algorithm

```
Algorithm ARA (triggered every time interval τ)
Input: mC, Output: AT; Parameter: n and δ
1:Calculate traffic volume Tv for every NAT box;
2: Find i, j ∈ Set_NAT that Diff(i, j) is maximal;
3: while Diff(i, j) > δ and Binset_i ≠ φ do
4:     Select b_k ∈ Binset_i that mC_k is maximal, Set AT'(k) = j
5.     Update the Tv and Binset; //delete b_k form Binset_i
6.     Calculate Diff(i, j) with the updated Tv;
8:     Do step 2 again find another maximal Diff(i, j);
9.end while
10.Trigger the binding signaling process
11.if ACK message form NAT boxes is received then
12     AT=AT'
13.end if
14.Reset the byte counter to zero for each hash bin
```

In table 1 the traffic volumes for every NAT box are calculated in the first step. Then tow boxes with largest load difference is selected. There is a threshold to decide whether the difference is bigger enough to reassign flows between the two NAT boxes. This helps to split the traffic more balanced among these NAT boxes. The while-do loop selects the proper bins to adjust repeatedly until any $Diff\,(i, j)$ is under the threshold value where all bins will be reassigned to another NAT box in the worst case. Afterwards binding signaling process is triggered.

The binding signaling process can be described in Fig. 4. Firstly load balancer tells the NAT *i* and NAT *j* to exchange address binding for the flows in the bins which should be reassigned to NAT *j*. Then NAT *i* just delete the address binding for the flows while NAT *j* will query the caches sever to fetch the address binding information. NAT *j* will send ACK to traffic balancer once it binds the addresses. It is named as a binding signaling process because it paves the path for the reassigned and incoming flows. When binding signaling process for address binding is established, the *AT* will be updated to let the traffic follow the new assignment. In the end the byte counter will be reset to zero.

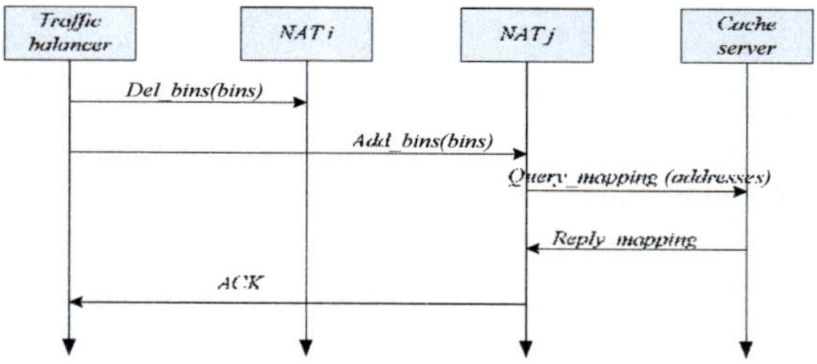

Fig. 3. The Binding Signaling Process

2.3 Redundancy

Redundancy can be easily achieved based on ARA in SMNAT architecture. If one NAT box is failed, the invalid entries in *AT* will be directed to other box with less burden. In table 2 the algorithms of Redundancy is depicted in which the Set of Failed NAT (SFN) is known as an input and a reassigned *AT* is a result of the algorithm. *AT* will be traversed to reassign the invalid entries. The algorithm of Redundancy is viewed as an initialization for ARA. The redundancy algorithm is triggered by the failure of NAT boxes (or links to them)

Table 2. The Algorithm of Redundancy

Algorithm Redundancy

Input: *SFN*; **Output:** *AT*; **Parameter:** *n*

---

1. $Set_{NAT} = Set_{NAT} - SFN$
2. **for each** $b_i \in Set_{bin}$ **do**
3.   **if** $AT(i) \in SFN$ **do**
4.   Select $j \in Set_{NAT}$ that Tv_j is minimal. Set $AT'(k) = j$;
5. $Tv_j = Tv_j + mC_i$;
6.   **end if**
7. **end for**

2.4 Complexitis and Cost

Although the traffic load balancing in SMNAT can react to dynamically traffic and failures, it endures the cost in both reassigning and binding signaling process in every time interval τ indeed. In reassigning process the cost is directly proportional to the number of reassigned bins. In TABLE I it is easily to find that the worst-case time complexity in reassigning process of ARA is $O(n)$, where n is the total number of bins. But the cost in the binding signaling process is not so obvious. It may contain the expenses in the process of each device and the cost of signaling among them, which depends on the exact design of binding signaling process. It is roughly estimated that more reassigned bins will introduce more cost. In addition the τ also should be set properly to handle the tradeoff between the performance and the cost of Multi-NAT.

3 Simulation and Result

In order to simulate and evaluate the performance of SMNAT a simulation test bed is built. Based on the test bed various schedule strategies and load balance mechanisms are simulated and evaluated using the tools of c++ and MATLAB.

3.1 Simulation Methodology

Suppose that there are 10000 clients in 10 groups and 10 servers in service network. Between the clients and servers there are 5 NAT boxes with capacity ca_i for each box i. In order to simulate the traffic volume between of each group of clients, we use a traditional traffic generation model for circuit-switched data as well as packet data, that is the poisson process, where the number of incoming packets follows the poisson distribution with λ as its parameter. In addition we assume that the traffic volume of each group conforms to normal distribution with μ as its mean and σ^2 as its variation which means the traffic parameter λ of any group conforms to the normal distribution. In order to simulate the destination popularities which follows the Zip'f law we simply classify servers to 5 degrees of popularities that more popular sites will attract more traffics.

3.2 Impact of Parameters on the ARA

A test on the sensitivity of parameter δ and n is done by changing δ in the scope of [0.001, 0.02] and n=*100~500* with the time interval τ =1 sec. Fig. 4 shows the changes of reassigning cost and load balancing performance according to the alteration of δ and the number of bins. We simply assume that one bin reassigned in the ARA will bring 1 unit cost and the sum of load deviation of each NAT from the average load is considered as the metrics to evaluate the performance in the load balancing. The result confirms that large δ will cause bad efficiency in load balance while small δ will introduce extra cost in binding signaling process. In addition the test demonstrates that large number of bins offers fine-grained reassigning in ARA with little cost. Note that there is certain point in each line in Fig. 4.a where the cost of reassigning will get worse dramatically if δ is smaller. In the contrast the change of deviation is more smoothly.

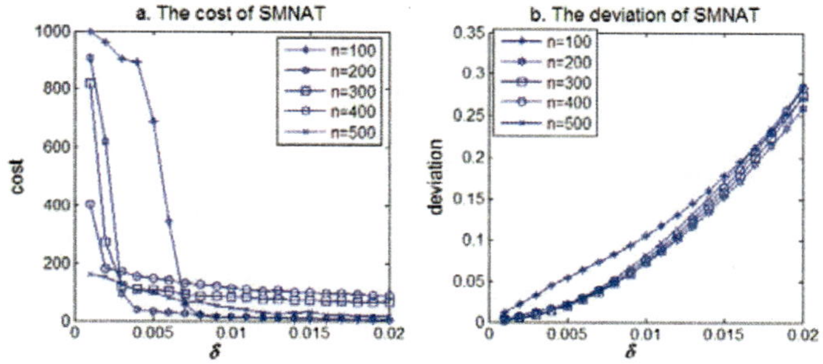

Fig. 4. Sensitivity Analysis of δ and n

Next we change the update interval τ of our ARA algorithm with δ=0.01 and n=100. The average deviations over all 5 NAT boxes of each time interval are summarized by the first row "Ave-Dev" in Table 3. This row shows the overall performance. The "Max-Dev" row presents the max deviation in a certain time interval which shows the volatility of the traffic load on these NAT boxes. Though the load balancing performance is poorer with longer interval, compare with table IV in next section, it is obvious that ARA still outperforms the other static algorithm in DNS-based load balancing.

Table 3. Effect of Update Interval τ

τ (sec)	0.5	1	2	5
Ave-Dev	0.083%	0.091%	0.101%	0.23%
Max-Dev	0.59%	0.713%	0.813%	2.532%

3.3 DNS-Based vs. SMNAT

In this section we will compare the load balancing performance between DNS-based schemes and our SMNAT. The three DNS-based schemes are source-based hash (SAH), destination-based hash (DAH) and round-robin policy (RR) which is suggested in [13]. In SAH the traffic is split by source IP address, while in DAH the destination IP address is taken into account. RR just uses a round-robin way to split traffic flow.

Fig. 5 shows load shares of NAT boxes on different schemes under fixed number of users with δ=0.01 and n=100 . In source-based schemes traffic load lines on each NAT box are separated due to the different behavior of user group in that users are divided into 3 kinds with different traffic volume. The situation is much worse in destination-based schemes shown in Fig. 6.b because the traffic load is split by destinations with different popularities. Fig. 6.c illustrates the RR splitting in which the lines are twisted together but vary in the scope of [8k~10k] bytes per time interval. In Fig. 6.d all lines are confined in a small area due to that the ARA algorithm adjusts the traffic splitting dynamically by reassigning the *AT*.

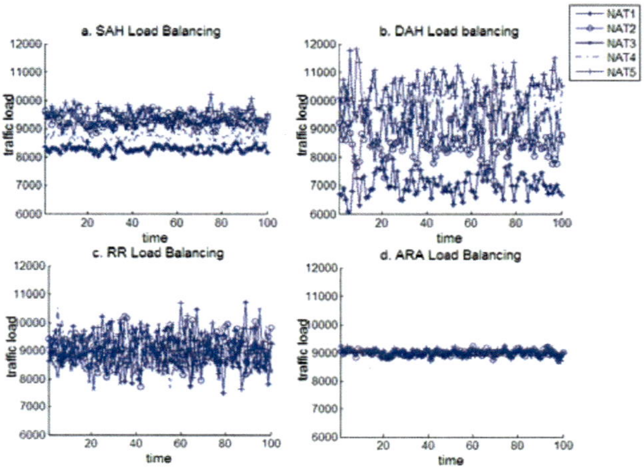

Fig. 5. Traffic load on NAT boxes

Table 4. Traffic splitting proportion in multi-nat system

	SAH	DAH	RR	ARA
Ave-Dev	4.443%	11.107%	0.633%	0.091%
Max-Dev	6.58%	30.133%	7.56%	0.713%

The detail statistic data of these tests is shown in table 4. Ave-Dev and Max-Dev have the same meaning with that in Table 4. The Ave-Pen and Max-Pen rows represent the average penalty and the maximum penalty respectively during the test. All these statistics prove that ARA performs better.

4 Conclusion

In this paper a novel scalable cluster-based load balancing schemes for Multi-NAT, called SMNAT, is proposed for the Multi-NAT scenario in IPv4-IPv6 coexistence. Using the ARA algorithm SMNAT reassign some active flows form the over-utilized NAT boxes to under-utilized ones based on the dynamic load share on each NAT box. The binding signaling process in ARA is delicately designed for Multi-NAT scenarios in order to keep good performance in load balance as well as the availability of translation address binding. The effectiveness of ARA is studied in the simulation test. Compared with the existing DNS-based Multi-NAT schemes, ARA obviously outperforms and shows its good availability and scalability. Next after this work we will focus on the protocol design and the implementation in binding signaling process of our SMNAT.

References

1. Wu, J., Cui, Y., Li, X., Metz, C.: The Transition to IPv6, Part 1: 4over6 for the China Education and Research Network. IEEE Internet Computing 10(3), 80–85 (2006)
2. Cui, Y., Wu, J., Li, X., Xu, M., Metz, C.: The Transition to IPv6, Part II: The Softwire Mesh Framework Solution. IEEE Internet Computing 10(5), 76–80 (2006)
3. Wu, P., Cui, Y., Xu, M., Wu, J., Li, X., Metz, C.: PET: Prefixing, Encapsulation and Translation for IPv4-IPv6 coexistence. In: GLOBECOM (2010)
4. Egevang, K., Francis, P.: The IP Network Address Translation (NAT). IETF RFC 1631 (May 1994)
5. Tsirtsis, G., Srisuresh, P.: Network Address Translation Protocol Translation (NAT-PT). IETF RFC 2766 (2000)
6. Guha, S., Francis, P.: An End-Middle-End Approach to Connection Establishment. In: ACM SIGCOMM (2007)
7. Aoun, C., Davies, E.: Reasons to Move the Network Address Translator - Protocol Translator (NAT-PT) to Historic Status. IETF RFC 4966 (July 2007)
8. Durand, A.: Dual-Stack Lite Broadband Deployments Following IPv4 Exhaustion. draft-ietf-softwire-dual-stack-lite-04 (work in progress) (March 8, 2010)
9. Cao, Z., Wang, Z., Zegura, E.: Performance of Hashing-Based Schemes for Interet Load Balancing. In: Proceedings of IEEE INFOCOM (March 2000)
10. Gojmerac, Ziegler, T., Ricciato, F., Reichl, P.: Adaptive Multipath Routing for Dynamic Traffic Engineering. In: IEEE Globecom, San Francisco (November 2003) (February 24, 1999I)
11. Wang, H., Xie, H., Qiu, L., Yang, Y.R., Zhang, Y., Greenberg, A.: Cope:Traffic engineering in dynamic networks. In: ACM SIGCOMM (2006)
12. Bagnulo, M., Matthews, P., van Beijnum, I.: Stateful NAT64: Network Address and Protocol Translation from IPv6 Clients to IPv4 Servers. draft-ietf-behave-v6v4-xlate-stateful-11 (work in progress) (March 30, 2010)
13. Xu, X., Boucadair, M., Lee, Y., Chen, G.: Redundancy and Load Balancing Framework for Stateful Network Address Translators (NAT). draft-xu-behave-stateful-nat-standby-03 (work in progress) (March 5, 2010)

The Evaluation Theory on Credit Levels of Retailers Based on Analytic Network Process

Ruan Lili[1] and Ma Ke[2]

[1] Glorious Sun School of Business and Management Donghua University Shanghai, China
[2] Bond Issuance Department Donghai Securities Co., Ltd., Shanghai, China
rlili123@hotmail.com, make0098@126.com

Abstract. When the permissible delay in payments is applied in a supply chain, the supplier would face a huge potential credit risk as the retailers need not pay for the items as soon as they are received, thus, the problem of breach of contract should be considered firstly. This paper develops a model by using analytic network process (ANP), an extension of analytic hierarchy process (AHP), which incorporates the relationship of interdependence and feedback in a multilevel decision network. The proposed model is then implemented in a diesel engine manufacturing firm.

Keywords: ANP; Supply chain, Credit level, The permissible delay in payment.

1 Introduction

The traditional economic order quantity (EOQ) model is widely used by practitioners as a decision-making tool for the control of inventory. The EOQ model assumes that the retailer's capitals are unrestricting and must be paid for the items as soon as the items are received. However, this may not be true. In practice, the supplier will offer the retailer a delay period, which is the trade credit period, in paying for the amount of purchasing cost. Before the end of the trade credit period, the retailer can sell the goods and accumulate revenue and earn interest. A higher interest is charged if the payment is not settled by the end of the trade credit period. Recently, many researchers have done researches on the permissible delay in payment. Goyal [1] developed an economic order quantity (EOQ) model under the conditions of permissible delay in payments. Chung [2] then developed an alternative approach to the problem. Chand and Ward [3] analyzed Goyal's problem under assumptions of the classical economic order quantity model, obtaining different results. Next, Aggarwal and Jaggi [4] extended Goyal's model to allow for deteriorating items. Jamal et al. [5] extended Aggarwal and Jaggi's [4] model to shortages. There were several interesting and relevant papers related to the delay of payments such as Chu et al. [6], Chung [7], Hwang and Shinn [8], Jamal et al. [9], Sarker et al. [10-11], Shah [12], Shinn [13], Khouja and Mehrez [14] and their references.

However, these researchers have not considered the breach of contract for the retailers. As the retailers need not pay for the items as soon as they are received, the problem of breach of contract should be considered firstly, otherwise, a huge potential

D. Zeng (Ed.): Advances in Control and Communication, LNEE 137, pp. 521–528.
springerlink.com

credit risk is facing to the supplier. Considering the relationship of financial criteria of the retailers, this paper proposed an evaluation theory on credit levels of the retailers based on the analytic network process (ANP), an extension of analytic hierarchy process (AHP) [15].

This paper is divided into four sections. The following section presents a brief review of the ANP approach. Section 3 proposes a model and then it is implemented in a diesel engine manufacturing firm. Short conclusions are given in Section 4.

2 Analytic Network Process

ANP, first presented by T.L. Saaty in 1996, is a general form of AHP, it allows for more complex interrelationships among the decision levels and attributes [16-18]. ANP provides a general framework to deal with decisions without making assumptions about the independence of the elements between and within levels. It allows both interaction and feedback within clusters (inner dependence) and between clusters (outer dependence). In fact, ANP uses a network to make decisions without the need to specify levels as in a hierarchy [19].

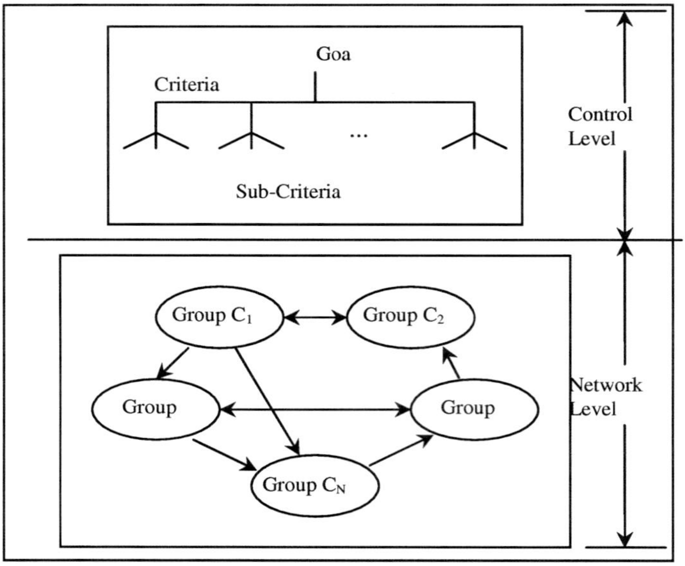

Fig. 1. Structure relationship in ANP method

Figure 1 illustrates the structural relationship in ANP method [20]. The system is divided into two parts. The first part is control hierarchy, consisting of network relationships between goal, criteria and sub-criteria. It affects the internal relationships of the system. The second part is network hierarchy, consisting of network relationships between elements and clusters.

In ANP analysis, the control hierarchy provides overriding criteria for comparing each type of interaction.

There are two types of control criteria [21].

1) One type of control criteria may be directly connected to the structure as the goal of a hierarchy. In this case, the control criteria are called comparison-"linking" criteria.

2) The other type of control criteria do not connect directly to the structure, but "induces" comparisons in a network. In this case the control criteria are called comparison-"inducing" criteria.

ANP comprises four major steps [22-23].

Step 1: Problem structuring and model construction

In the first step, problems should be stated clearly. The goals, criteria and sub-criteria related with the problems should also be selected via brainstorming or other methods. And then, the network with the interactions between and within clusters and elements should be constructed based on all these above.

Step 2: Making pair-wise comparisons

According to the network proposed in the first step, pair-wise comparisons should be made by a number of experts. And then experts' preferences should be integrated to form comparison matrices. Consistency test should also be applied to verify the consistency of each comparison matrix.

Step 3: Supermatrix formation

In Step 3, a supermatrix should be formed, which lists down all the sub-matrices consisting of all the clusters and necessary elements in order on the left and upper sides of the matrix. The supermatrix, whose concept resembles the Markov chain process [22], is a partitioned matrix, it is formed by grouping the local priority vectors obtained in Step 2 and then locating them in appropriate positions based on the flow of effect from one component to another, or from a component to itself as a loop in order to determine the global priorities [24]. Let the components of a decision system denote C_k ($k=1,2,\cdots,n$), let each component k contain m_k elements, denoted by ($e_{k1}, e_{k2}, \cdots, e_{km_k}$), and let W_{ij} , ($i=1,2,\cdots,n$, $j=1,2,\cdots,n$) denotes the sub-matrix. Equation (1) shows a standard form of a supermatrix [22].

$$
\begin{array}{ll}
 & \begin{array}{ccccccccc} & & C_1 & & & C_k & & & C_n \end{array} \\
 & \begin{array}{ccccccccc} e_{11} & \cdots & e_{1m_1} & e_{k1} & \cdots & e_{km_k} & e_{n1} & \cdots & e_{nm_n} \end{array} \\
\begin{array}{cc} & e_{11} \\ & \vdots \\ C_1 & e_{1m_1} \\ \vdots & e_{k1} \\ C_k & \vdots \\ \vdots & e_{km_k} \\ C_n & e_{n1} \\ & \vdots \\ & e_{nm_n} \end{array} &
\begin{bmatrix} W_{11} & \cdots & W_{1k} & \cdots & W_{1n} \\ \vdots & \vdots & \vdots & \vdots & \vdots \\ W_{k1} & \cdots & W_{kk} & \cdots & W_{kn} \\ \vdots & \vdots & \vdots & \vdots & \vdots \\ W_{n1} & \cdots & W_{nk} & \cdots & W_{nn} \end{bmatrix}
\end{array}
\qquad (1)
$$

Step 4: Selection of the best alternative
The supermatrix will be steady state by multiplying the weighted supermatrix by itself until the row values converge to the same value for each column of the matrix. The alternative with the highest overall priority should be selected.

3 Case Implementation

W Company is one of the largest diesel engine manufacturers in China. The products it produced include bus diesel engines, truck diesel engines, generating diesel engines, marine diesel engines, engineering diesel engines and agricultural diesel engines, most of them are very famous in the diesel engine manufacturing industry. During its more than 60 years' development process, W Company gradually organized a diesel engine supply chain system, and became its core firm.

With the market competition more and more fiercely, W Company needs to compress the total cost of the whole supply chain urgently, thus, it intends to apply the permissible delay in payments to its supply chain. However, W Company needs to evaluate its retailers firstly based on their credit levels as the retailers will not pay for the products immediately under the permissible delay in payments. Based on this reason, a team consisting of 15 people has been formed to help W Company with its problem. Take this case for example, W Company needs to evaluate the credit levels of its four retailers named S1 to S3.

In the following part, a model based on ANP approach will be proposed, and it will then be applied in W Company to assess its retailers.

Step 1: Problem structuring and model construction
In Figure 2, the proposed ANP criteria systems on retailer selection are given. As displayed in Figure 2, the goal of the retailer selection problem is selecting the supplier with the highest credit level. To evaluate the credit levels of the retailers, 17 different criteria and their relationships are defined by the team combined with the experiences of the persons who come from the financial department of W Company.

Step 2: Making pair-wise comparisons

After the first step, pair-wise comparisons are performed by the persons from the financial department, the comparison results are combined by geometric mean. Here, the elements in a cluster are compared by applying Saaty's 1-9 scales with respect to an element in another cluster which they are connected to (or on elements in their own cluster). Consistency test also should be applied to verify the consistency of each comparison matrix.

Step 3: Supermatrix formation

The software "Super Decisions" is a simple easy-to-use package for constructing decision models with dependence and feedback. With the helps of Super Decisions, the unweighted supermatrix, weighted supermatrix and the limit supermatrix could be obtained easily. The limit supermatrix is shown in Table 1.

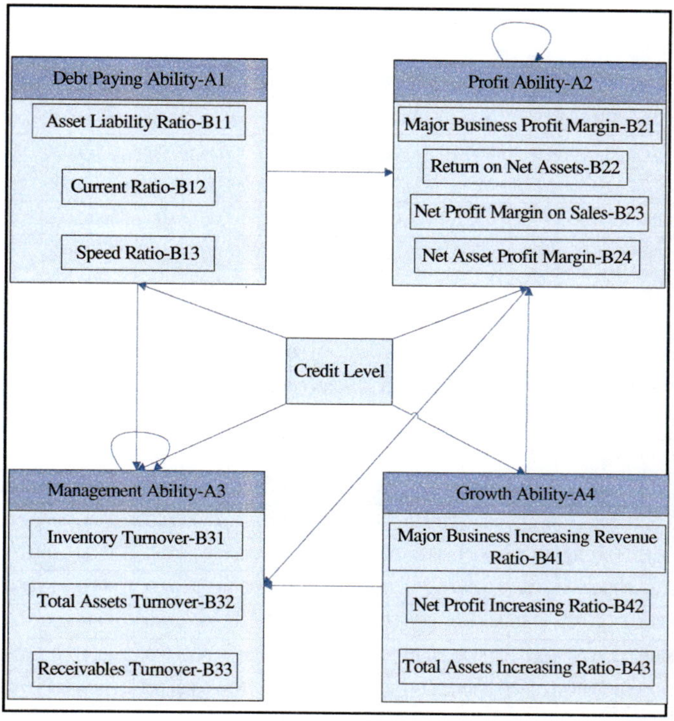

Fig. 2. The proposed ANP criteria systems and their relationship

Step 4: Selection of the best alternatives

After rating results, the weight of the criteria from B11 to B43 is shown in table 2.

The overall scores of the four suppliers could be acquired by using equation (2). In equation (2), w_{Bij} denotes the weight of the criteria from B11 to B43 which is shown in Table 4. v_{ij} denotes the criteria values after normalization in each retailer.

$$Credit \quad Level = \sum_{i=1,\,j=1} w_{Bij} v_{ij} \tag{2}$$

After calculation, the credit levels of each retailer could be acquired, that is, S1=0.863, S2=0.746, S3=0.698. From the result, it could be seen that the retailer S1 has the highest credit level , however, the retailer S3 has the lowest credit level, while the retailer S2 is in the middle position.

Table 1. The Limit Supermatrix

	A1	**A2**	**A3**	**A4**	**B11**	**B12**	**B13**	**B21**	**B22**	**B23**	**B24**	**B31**	**B32**	**B33**	**B41**	**B42**	**B43**	**Goal**
A1	0.000	0.000	0.000	0.000	0.000	0.000	0.000	0.000	0.000	0.000	0.000	0.000	0.000	0.000	0.000	0.000	0.000	0.091
A2	0.114	0.000	0.000	0.068	0.000	0.000	0.000	0.000	0.000	0.000	0.000	0.000	0.000	0.000	0.000	0.000	0.000	0.149
A3	0.171	0.242	0.000	0.238	0.000	0.000	0.000	0.000	0.000	0.000	0.000	0.000	0.000	0.000	0.000	0.000	0.000	0.204
A4	0.000	0.000	0.000	0.000	0.000	0.000	0.000	0.000	0.000	0.000	0.000	0.000	0.000	0.000	0.000	0.000	0.000	0.035
B11	0.094	0.000	0.000	0.000	0.000	0.000	0.000	0.000	0.000	0.000	0.000	0.000	0.000	0.000	0.000	0.000	0.000	0.019
B12	0.059	0.000	0.000	0.000	0.000	0.000	0.000	0.000	0.000	0.000	0.000	0.000	0.000	0.000	0.000	0.000	0.000	0.012
B13	0.075	0.000	0.000	0.000	0.000	0.000	0.000	0.000	0.000	0.000	0.000	0.000	0.000	0.000	0.000	0.000	0.000	0.015
B21	0.006	0.026	0.000	0.004	0.000	0.000	0.000	0.000	0.000	0.000	0.000	0.000	0.000	0.000	0.000	0.000	0.000	0.008
B22	0.020	0.085	0.000	0.012	0.000	0.000	0.000	0.000	0.000	0.000	0.000	0.000	0.000	0.000	0.000	0.000	0.000	0.026
B23	0.017	0.074	0.000	0.010	0.000	0.000	0.000	0.000	0.000	0.000	0.200	0.000	0.000	0.000	0.000	0.000	0.000	0.023
B24	0.020	0.085	0.000	0.012	0.000	0.000	0.000	0.000	0.000	0.000	0.000	0.000	0.000	0.000	0.000	0.000	0.000	0.026
B31	0.120	0.133	0.281	0.107	0.000	0.667	0.000	0.000	0.000	0.000	0.200	0.000	0.500	0.000	0.000	0.000	0.000	0.105
B32	0.106	0.188	0.350	0.137	0.000	0.000	0.000	0.000	0.000	0.000	0.400	0.000	0.000	0.000	0.000	0.000	0.000	0.127
B33	0.198	0.166	0.368	0.139	0.000	0.333	1.000	0.000	0.000	0.000	0.200	0.000	0.500	0.000	0.000	0.000	0.000	0.143
B41	0.000	0.000	0.000	0.057	0.000	0.000	0.000	0.000	0.000	0.000	0.000	0.000	0.000	0.000	0.000	0.000	0.000	0.004
B42	0.000	0.000	0.000	0.150	0.000	0.000	0.000	0.000	0.000	0.000	0.000	0.000	0.000	0.000	0.000	0.000	0.000	0.010
B43	0.000	0.000	0.000	0.065	0.000	0.000	0.000	0.000	0.000	0.000	0.000	0.000	0.000	0.000	0.000	0.000	0.000	0.004
Goal	0.000	0.000	0.000	0.000	0.000	0.000	0.000	0.000	0.000	0.000	0.000	0.000	0.000	0.000	0.000	0.000	0.000	0.000

Table 2. Criteria weight

Criteria	**Nonlayer weight**	**Layer weight**
A1	0.091009	0.19013
A2	0.149134	0.31156
A3	0.203584	0.42531
A4	0.034946	0.07301
B11	0.018775	0.03601
B12	0.011828	0.02269
B13	0.014902	0.02858
B21	0.008136	0.01561
B22	0.026167	0.05019
B23	0.022820	0.04377
B24	0.026166	0.05019
B31	0.104806	0.20104
B32	0.127303	0.24419
B33	0.142952	0.27421
B41	0.003666	0.00703
B42	0.009609	0.01843
B43	0.004198	0.00805
Goal	0.000000	0.00000

4 Conclusions

In this paper, a model by using ANP approach is proposed, it could reflect the criteria relationships, interdependence and feedback more clearly, and this is very close to the real word. In this model proposed, 17 criteria have been introduced in order to evaluate the credit levels of the retailers which belong to a diesel engine manufacturing firm. Pair-wise comparison matrices and consistency tests are acquired easily with the help of the software "Super Decisions". At the end of this paper, the three retailers of the diesel engine manufacturing firm have been evaluated successfully.

References

1. Goyal, S.K.: Economic order quantity under conditions of permissible delay in payments. The Journal of the Operational Research Society 36, 335–338 (1985)
2. Chung, K.J.: A theorem on the determination of economic order quantity under conditions of permissible delay in payments. Comput. Oper. Res. 25, 49–52 (1998)
3. Chand, S., Ward, J.: A note on economic order quantity under conditions of permissible delay in payments. J. Oper. Res. Soc. 38, 83–84 (1987)
4. Aggarwal, S.P., Jaggi, C.K.: Ordering policies of deteriorating items under permissible delay in payments. J. Oper. Res. Soc. 46, 658–662 (1995)
5. Jamal, A.M.M., Sarker, B.R., Wang, S.: An ordering policy for deteriorating items with allowable shortages and permissible delay in payment. J. Oper. Res. Soc. 48, 826–833 (1997)
6. Chu, P., Chung, K.J., Lan, S.P.: Economic order quantity of deteriorating items under permissible delay in payments. Comput. Oper. Res. 25, 817–824 (1998)
7. Chung, K.J.: The inventory replenishment policy for deteriorating items under permissible delay in payments. Oper. Res. 37, 267–281 (2000)
8. Hwang, H., Shinn, S.W.: Retailer's pricing and lot sizing policy for exponentially deteriorating products under the condition of permissible delay in payments. Comput. Oper. Res. 24, 539–547 (1997)
9. Jamal, A.M.M., Sarker, B.R., Wang, S.: Optimal payment time for a retailer under permitted delay of payment by the wholesaler. Int. J. Prod. Econ. 66, 59–66 (2000)
10. Sarker, B.R., Jamal, A.M.M., Wang, S.: Supply chain model for perishable products under inflation and permissible delay in payment. Comput. Oper. Res. 27, 59–75 (2000)
11. Sarker, B.R., Jamal, A.M.M., Wang, S.: Optimal payment time under permissible delay in payment for products with deterioration. Prod. Plan. Control 11, 380–390 (2000)
12. Shah, N.H.: A lot-size model for exponentially decaying inventory when delay in payments is permissible. Cah. du CERO 35, 115–123 (1993)
13. Shinn, S.W.: Determining optimal retail price and lot size under day-terms supplier credit. Comput. Ind. Eng. 33, 717–720 (1997)
14. Khouja, M., Mehrez, A.: Optimal inventory policy under different supplier credits. J. Manuf. Syst. 15, 334–339 (1996)
15. Saaty, T.L.: The Analytic Hierarchy Process. McGraw-Hill, New York (1980)
16. Kameya, N., Miyagi, H., Taira, N., Yamashita, K.: Multiple-criteria decision-making using sectional supermatrix. In: The Proceedings of the International Technical Conference on Circuits/System, Computers and Communications, Phuket, Thailand, pp. 838–840 (2002)

17. Meade, L.M.: Strategic analysis of logistics and supply chain management systems using the analytical network process. Transportation Research Part E-Logistics and Transportation Review 34(3), 201–215 (1998)
18. Raisinghani, M.A.: Balanced Analytic Approach to Strategic Electronic Commerce Decisions: A Framework of the Evaluation Method, Information Technology Evaluation Methods and Management, pp. 185–197. Idea Group Publishing, Hershey (2001)
19. Saaty, T.L.: The Analytic Hierarchy Process, pp. 184–192. RWS Publications, Pittsburg (1990)
20. Wang, L.F.: The theory and algorithm of ANP. System Engineering Theories and Practices 3, 44–50 (2001)
21. Saaty, L.T.: Fundamentals of the analytic network process. In: ISAHP, Japan, pp. 12–14 (1999)
22. Saaty, T.L.: Decision Making with Dependence and Feedback: the Analytic Network Process. RWS, Pittsburgh (1996)
23. Meade, L.M., Sarkis, J.: Analyzing organizational project alternatives for agile manufacturing process-an analytical network approach. International Journal of Production Research 37(2), 241–261 (1999)
24. Chang, C.W., Wu, C.R., Lin, C.T., Lin, H.L.: Evaluating digital video recorder systems using analytic hierarchy and analytic network process. Information Sciences 177(16) (2007)

The Research on the Permissible Delay in Payments in Oligopoly Market

Ruan Lili[1], Ma Ke[2], and Liang Guangyan[1]

[1] Glorious Sun School of Business and Management Donghua University, Shanghai, China
[2] Bond Issuing Department Donghai Securities Co., Ltd. Shanghai, China
rlili123@hotmail.com, make0098@126.com, lianggy985@yahoo.com.cn

Abstract. This paper did some researches on the optimal replenishment and sales policies of the manufacturer and the retailers in oligopoly market under Economic Production Quantity (EPQ) mode when the permissible delay in payments was applied in the supply chain. From the perspective of the manufacturer and the retailers, mathematical models had been established, respectively. The method how to make the optimal replenishment and sales policies was illustrated, and the basic condition whether trigger strategy was deviated was also discussed in this paper by using the infinitely repeated game theory. Finally, this paper illustrated whether the system composed of two retailers could get the maximum profit through a numerical example.

Keywords: Supply chain, Permissible delay in payments, EPQ model, Oligopoly Market, Repeated game.

1 Introduction

The traditional economic order quantity (EOQ) model is widely used as a decision-making tool for the control of inventory. The EOQ model assumes that the retailer's capitals are unrestricting and must be paid for the items as soon as the items are received. However, in practice, the supplier will offer the retailer a delay period, which is the trade credit period, in paying for the amount of purchasing cost. Before the end of the trade credit period, the retailer can sell the goods and accumulate revenue and earn interest. A higher interest is charged if the payment is not settled by the end of the trade credit period. Permissible delay in payments is generally considered to be the equivalent form of price discount contract [1]. Recently, many researchers have done researches on the permissible delay in payment. Goyal [2] firstly developed an economic order quantity (EOQ) model under the conditions of permissible delay in payments. Chand and Ward [1] analyzed Goyal's problem under assumptions of the classical EOQ model, obtaining different results. Next, Aggarwal and Jaggi [3] extended Goyal's model to allow for deteriorating items. Jamal et al. [4] extended Aggarwal and Jaggi's [3] model to shortages. Shin and Hwang [5] studied the optimal pricing and optimal order quantity problem under the permissible delay in payments linked to order quantity. Liao et al [6] and Sarker et al [7] researched permissible delay in payments under inflation.

D. Zeng (Ed.): Advances in Control and Communication, LNEE 137, pp. 529–538.
springerlink.com

John and Henke et al [8] divided supplier into outside supplier and in-house supplier in accordance with the relationship between the supplier and the retailer. According to this division, the researches that the scholars above did all assumed that the product was obtained from the outside supplier, they assumed the bulk arrival, that is, the replenishment rate was infinite. However, for the in-house supplier, the retailer's replenishment rate is equal to its supplier's production rate, that is, the replenishment rate is finite. Recently, some scholars did researches on how to use permissible delay in payments to the in-house supplier, Chung and Huang [9] firstly studied the optimal replenishment policy under the permissible delay in payments with finite replenishment rate. Huang [10] modified Chung and Huang [9] model and proposed the optimal replenishment policy in the condition that the sales price was not equal to the purchase price. Liao [11] researched the replenishment strategy for deteriorating items based on EPQ model. Huang [12] established the replenishment strategy model based on EPQ model in a two-stage supply chain.

All of the researches above did not consider the effect of the market structure to the supply chain members. From the economics theory, the market could be divided into four types: perfect competition market, monopolistic competition market, perfect monopoly market and oligopoly market [13, 14]. Among them, the oligopoly market is more common. The oligopoly market is a market characterized by a small number of manufacturers who often act together to control the supply of a particular good and its market price. In the oligopoly market, there are only a few companies, and each company occupies a very important position, their respective decision-making on the price or output will affect the entire market and the behavior of other competitors. Therefore, when making the decision of the price or output, the company will consider not only the costs or benefits of their own, but also the impact of the decision on the market, and the possible responses of other companies. The purpose of this article is to do some researches on the optimal replenishment and sales policies for the supply chain members under permissible delay in payments based on the EPQ model in the oligopolistic market.

2 Notation and Assumptions

2.1 Notation

s_i : unit selling price per item of good quality for retailer i;

c : unit purchasing price per item for retailer i;

q : unit production cost for the manufacturer;

P : annual replenishment rate;

D_i : demand rate per year for retailer i;

ρ_i : $\rho_i = 1 - \frac{D_i}{\mathrm{P}}$;

h_i : unit stock holding cost per item per year excluding interest charges for retailer i;

A_i : cost of placing an order for retailer i;

$\mathrm{I_k}$: interest charges per \$ investment in inventory per year;

I_e : interest that can be earned per \$ per year;

M_i : the retailer *i*'s trade credit period offered by manufacturer;

Q_i : the replenishment quantity for retailer *i*, $Q_i = D_i T_i$;

$I_1(t)$: the inventory level that changes with time t during production period;

$I_2(t)$: the inventory level that changes with time t during non-production period;

t_i : the maximum replenishment time for retailer *i* ;

T_i : the replenishment time for retailer *i*;

δ : discount coefficient, $0 < \delta < 1$;

Π_B^i : average profit for retailer *i*;

Π_S : average profit for manufacturer.

2.2 Assumptions

1) The demand rate P is a known constant;

2) $P \geq D_i$, $s_i \geq c \geq q$ and $I_k \geq I_e$;

3) An infinite planning horizon is assumed;

4) Supply chain contains one manufacturer and two retailers;

5) Demand Rate $D_i = \alpha - \beta s_i + \gamma s_{-i}$;

6) Shortages are not allowed.

3 Mathematical Model and Analysis

During the production period, the inventory level of the system for the retailer can be described by the following equations: $$\frac{dI_1(t)}{dt} = P - D_i$$

By using boundary condition $I_1(0) = 0$, we have

$I_1(t) = (P - D_i)t$

During the non-production period, production stops and the inventory level decreases due to demand, so the inventory level of the system for the retailer can be described by the following equations: $$\frac{dI_2(t)}{dt} = -D_i$$

Boundary condition is $I_2(T_i) = 0$. We have

$$I_2(t) = D_i(T_i - t)$$

In addition, from the boundary condition $I_1(t_i)=I_2(t_i)$, we have $t_i = \frac{D_i}{P} T_i$

For the moment, the individual costs are now evaluated before they are grouped together.

1) Annual ordering cost for retailer $i = \frac{A_i}{T_i}$

2) Annual stockholding cost for retailer $i = \frac{1}{2} h_i D_i \rho_i T_i$

3) There are three cases to occur in earning of interest and cost of interest charges for the retailer i per year:

Case 1. $T_i \geq \frac{P}{D_i} M_i$

The interest earned per year for retailer i is given by $\frac{s_i I_e}{T_i} \int_0^{M_i} D_i t \mathrm{d}t = \frac{s_i I_e}{T_i}\left[\frac{1}{2} D_i M_i^2\right]$, the interest charged per year is $\frac{cI_k}{T_i}\left(\int_{M_i}^{t_i} I_1(t)\mathrm{d}t + \int_{t_i}^{T_i} I_2(t)\mathrm{d}t\right) = \frac{cI_k \rho_i}{2T_i}\left(D_i T_i^2 - PM_i^2\right)$.

Case 2. $M_i \leq T_i < \frac{P}{D_i} M_i$

The interest earned per year for retailer i is given by $\frac{s_i I_e}{T_i} \int_0^{M_i} D_i t \mathrm{d}t = \frac{s_i I_e}{T_i}\left[\frac{1}{2} D_i M_i^2\right]$, the interest charged per year for retailer i is given by

$$\frac{cI_k}{T_i} \int_{M_i}^{T_i} I_2(t)\mathrm{d}t = \frac{cI_k D_i (T_i - M_i)^2}{2T_i}.$$

Case 3. $0 < T_i < M_i$

The interest earned per year for retailer i is given by $\frac{s_i I_e}{T_i}\left[\int_0^{T_i} D_i t \mathrm{d}t + D_i T_i (M_i - T_i)\right] = s_i I_e D_i \left(M_i - \frac{1}{2} T_i\right)$, the interest charged per year for retailer i is 0.

Therefore, the total variable profit function for retailer i is

$$\Pi_B^i = \begin{cases} \Pi_{B1}^i & \text{if } T_i \geq PM_i/D_i & \text{(a)} \\ \Pi_{B2}^i & \text{if } M_i \leq T_i < PM_i/D_i & \text{(b)} \\ \Pi_{B3}^i & \text{if } 0 < T_i < M_i & \text{(c)} \end{cases} \quad (1)$$

$$\Pi_{B1}^i = \frac{(s_i - c)Q_i}{T_i} - \frac{A_i}{T_i} - \frac{1}{2} h_i D_i \rho_i T_i - \frac{cI_k \rho_i}{2T_i}\left(D_i T_i^2 - PM_i^2\right) + \frac{s_i I_e}{T_i}\left[\frac{1}{2} D_i M_i^2\right]$$

$$\Pi_{B2}^i = \frac{(s_i - c)Q_i}{T_i} - \frac{A_i}{T_i} - \frac{1}{2} h_i D_i \rho_i T_i - \frac{cI_k D_i (T_i - M_i)^2}{2T_i} + \frac{s_i I_e}{T_i}\left[\frac{1}{2} D_i M_i^2\right]$$

$$\Pi_{B3}^i = \frac{(s_i - c)Q_i}{T_i} - \frac{A_i}{T_i} - \frac{1}{2} h_i D_i \rho_i T_i + s_i I_e D_i \left(M_i - \frac{1}{2} T_i\right)$$

The total profit function for manufacturer per unit is

$$\Pi_S = \begin{cases} \Pi_{S1} & \text{if } T_i \geq PM_i/D_i \quad \text{(a)} \\ \Pi_{S2} & \text{if } T_i \leq PM_i/D_i \quad \text{(b)} \end{cases} \tag{2}$$

$$\Pi_{S1} = \frac{(c-q)Q_i}{T_i} - \frac{cI_k}{T_i}\int_0^{M_i} Ptdt = \frac{(c-q)Q_i}{T_i} - \frac{cI_k PM_i^2}{2T_i}$$

$$\begin{aligned}\Pi_{S2} &= \frac{(c-q)Q_i}{T_i} - \frac{cI_k}{T_i}\left(\int_0^{t_i} Ptdt + Pt_i(M_i - t_i)\right) \\ &= \frac{(c-q)Q_i}{T_i} - \frac{cI_k M_i Q_i}{T_i} + \frac{cI_k Q_i^2}{2PT_i}\end{aligned}$$

4 Game Analysis

4.1 Cournot Equilibrium

After the manufacturer offering the trade cournot credit period, the retailer needs to make a decision on the unit selling price and the replenishment quantity of its own. From the cournot equilibrium, we have

$$\begin{cases} s_i^* = \arg\max_{S_i} \Pi_B^i(s_i, s_{-i}^*, M_i) \\ s_{-i}^* = \arg\max_{S_{-i}} \Pi_B^{-i}(s_{-i}, s_i^*, M_{-i}) \end{cases} \tag{3}$$

By solving the above equations, we obtain the unit selling price relationship between the two retailers at the cournot equilibrium $s_i^*(M_i, M_{-i})$, $s_{-i}^*(M_i, M_{-i})$, and then, the replenishment quantity relationship could be expressed as

$$\begin{cases} Q_i^*(M_i, M_{-i}) = (\alpha - \beta s_i^*(M_i, M_{-i}) + \gamma s_{-i}^*(M_i, M_{-i}))T_i \\ Q_{-i}^*(M_i, M_{-i}) = (\alpha - \beta s_{-i}^*(M_i, M_{-i}) + \gamma s_i^*(M_i, M_{-i}))T_{-i} \end{cases}$$

In the supply chain system consisting of a manufacturer and retailers, as a leader of the supply chain, the manufacturer will offer the retailer a certain trade credit period according to the replenishment quantity of the retailer, and as the follower of the supply chain, the retailer will determine the optimal replenishment quantity and the optimal unit selling price in accordance with the trade credit period offered by the manufacturer, this process could be view as the Stackelberg game model with the manufacturer first.

We put $Q_i^*(M_i, M_{-i})$ and $Q_{-i}^*(M_i, M_{-i})$ into equation (2), the optimal trade credit period M_i^* and M_{-i}^* for the two retailers offered by the manufacturer could be acquired

$$(M_i^*, M_{-i}^*) = \arg\max_{(M_i, M_{-i})} \{\Pi_s(Q_i^*, M_i) + \Pi_s(Q_{-i}^*, M_{-i})\} \tag{4}$$

And then, we put M_i and M_{-i} into equation (3), we will have the optimal unit selling price s_i* and $s_{-i}*$ for the two retailers. Furthermore, the optimal replenishment quantity of the two retailers could be obtained by

$$\begin{cases} Q_i* = (\alpha - \beta s_i * + \gamma s_{-i} *) \mathrm{T}_i \\ Q_{-i}* = (\alpha - \beta s_{-i} * + \gamma s_i *) \mathrm{T}_{-i} \end{cases}$$

4.2 Centralized Decision

From the game theory, we learn that the system could not achieve the maximum profit at the cournot equilibrium. $s_0*(M_0)$ is assumed to be the unit selling price of the retailers when the system achieve the maximum profit.

$$s_0*(M_0) = \arg\max_{s_0} \left\{ \Pi_B^i(s_0, M_0) + \Pi_B^{-i}(s_0, M_0) \right\} \tag{5}$$

The maximum profit of the whole system at centralized is $\left\{ \Pi_B^i(s_0*, M_0) + \Pi_B^{-i}(s_0*, M_0) \right\}$, thus, there is the relationship

$\Pi_B^i(s_i*, s_{-i}*, M_i*) + \Pi_B^{-i}(s_{-i}*, s_i*, M_{-i}*) < \Pi_B^i(s_0*, M_0) + \Pi_B^{-i}(s_0*, M_0)$ By using $s_0*(M_0)$, the total replenishment quantity of the system at centralized decision could be expressed as

$$Q_0*(M_0) = (\alpha - \beta s_0*(M_0) + \gamma s_0*(M_0))(\mathrm{T}_i + \mathrm{T}_{-i}) \tag{6}$$

We put $Q_0*(M_0)$ into equation (2), the optimal trade credit at centralized decision can be obtained, that is, $M_0* = \arg\max_{M_0} \Pi_s(Q_0*, M_0)$. From equation (5) and (6), we may get s_0* and Q_0*.

4.3 Infinitely Repeated Game

The unit selling price at cournot equilibrium above is obtained from one-time static game. However, the oligopoly market tend to be stable in real life, every oligopoly has no expectation for the market change, the competition among oligopoly could be considered as infinitely repeated game. Here, we introduce the infinitely repeated Cournot game model to describe the problem.

We propose a trigger strategy, assuming that every game player is trying to cooperate first, it should choose the replenishment quantity and unit selling price which could benefit the both sides. If the uncooperative attitude of the game player is found, the other game player will choose the replenishment quantity and the unit selling price at the cournot equilibrium for revenge.

(1) If both of the game players obey the trigger strategy, the retailer i could have the profit $\Pi_B^i(s_0*, M_0*)$ per phase, and the present value of the total profits for retailer i could be expressed as $\Pi_B^i(s_0*, M_0*)(1 + \delta + \delta^2 + \cdots) = \dfrac{\Pi_B^i(s_0*, M_0*)}{(1-\delta)}$

(2) If one of the game player deviate from the trigger strategy, it would get the short-term excess profit. Here, we assume retailer i deviate from the trigger strategy, the unit selling price for the other game player and the trade credit period offered by the manufacturer will not change in a short term, thus , the unite selling price for retailer i could be expressed as $s_i'*(M_0*) = \arg\max_{s_i} \Pi_B^i(s_i', s_0*, M_0*)$. And the replenishment quantity is $Q_i'* = \left(\alpha - \beta s_i'* + \gamma s_0*\right)\mathrm{T}_i$, the profit of retailer i is $\Pi_B^i(s_i'*, s_0*, M_0*)$. From the game theory, we have $\Pi_B^i(s_i'*, s_0*, M_0*) > \Pi_B^i(s_0*, M_0*)$. While, at this time, the other game player will revenge the retailer i by the trigger strategy, using the replenishment quantity and unit selling price at cournot equilibrium. After this, retailer i will only acquire the profit $\Pi_B^i(s_i*, s_{-i}*, M_i*)$ at cournot equilibrium. And the present value of the total profits is

$$\Pi_B^i(s_i'*, s_0*, M_0*) + \Pi_B^i(s_i*, s_{-i}*, M_i*)\left(1+\delta+\delta^2+\cdots\right)$$
$$= \Pi_B^i(s_i'*, s_0*, M_0*) + \frac{\Pi_B^i(s_i*, s_{-i}*, M_i*)}{1-\delta}$$

Therefore, we could have the following theory:

Theory 1: When δ meet the following condition $\frac{\Pi_B^i(s_0*, M_0*)}{(1-\delta)} \geq \Pi_B^i(s_i'*, s_0*, M_0*) + \frac{\Pi_B^i(s_i*, s_{-i}*, M_i*)}{1-\delta}$, both of the game players will not deviate from the trigger strategy, and the whole system could reach the maximum profit; Otherwise, the game player will deviate from the trigger strategy, and the maximum profit of the system will not be achieved.

5 Numerical Example

As the calculation of the model above is more complicated, it is very difficult for the conventional method to obtain the optimal solution, therefore, we use random experiment method to find the approximate optimal solution.

Given α =2000, β =35, γ =20, c =60, q =30, P =1800, h_1 =30, h_2 =40, A_1 =60, A_2 =80, I_k =0.08, I_e =0.06, T_1=T_2=0.5.

With matlab simulation software, we select experimental points randomly and do simulation for about 10^6 times. Then we could obtain the parameters at the cournot equilibrium as

By contrast the data in table 2 with the one in table 1, we could learn that the profit of the whole system made up of the two retailers at centralized decision is higher than the one at cournot equilibrium.

Table 1. The Parameters At The Cournot Equilibrium

	s^*	M^*	Q^*	Π
Retailer 1	96.6184	0.6574	312.7349	9715.5
Retailer 2	98.3521	0.4176	291.3981	11781.1
Manufacturer	--------	--------	604.1330	47057.0

Similarly, table II shows the parameter conditions at centralized decision.

Table 2. The Parameters At The Centralized Decision

	s^*	M^*	Q^*	Π
Retailer 1	95.3862	0.4085	308.5767	14021.0
Retailer 2	95.3862	0.4085	308.5767	13271.0
Manufacturer	--------	--------	617.1534	52877.0

We set trigger strategy in the system. If both of the two retailers do not deviate from the trigger strategy, the whole system could get the maximum profits.

(1) If retailer 1 deviates from the trigger strategy, the parameters of the supply chain are

Table 3. The Parameters When Retailer 1deviates

	s^*	M^*	Q^*	Π
Retailer 1	85.9899	0.4086	480.6152	21391.2
Retailer 2	96.4192	0.4086	215.7556	10670.8
Manufacturer	--------	--------	696.3708	44006.0

From table 3 we could found that in the short-term after retailer 1 deviate from the trigger strategy, retailer 1 will get more profit, while the profit of retailer 2 becomes lower, then, retailer 2 must revenge retailer 1 by using the trigger strategy.

According to theory 1, if $\delta \geq 0.76$, retailer 1 will not deviate from the trigger strategy; if $\delta < 0.76$, retailer 1 will deviate from the trigger strategy.

(2) If retailer 2 deviates from the trigger strategy, the parameters of the supply chain are

Table 6. The Parameters When Retailer 2 Deviates

	s^*	M^*	Q^*	Π
Retailer 1	97.8372	0.4099	223.0233	9842.1
Retailer 2	89.0091	0.4099	445.1939	18611.0
Manufacturer	--------	--------	739.5779	45354.0

With the same method above, if $\delta \geq 0.92$, retailer 2 will not deviate from the trigger strategy; if $\delta < 0.92$, retailer 2 will deviate from the trigger strategy.

Considering the two conditions, we could get conclusion: If $\delta \geq 0.92$, neither of the two retailers will deviate from the trigger strategy, the system consist of the two retailers will reach the maximum profit; If $0.76 \leq \delta < 0.92$, retailer 2 will deviate from the trigger strategy, the system will not achieve the maximum profits; If $\delta < 0.76$, both of the two retailers will deviate from the trigger strategy, the system will not achieve the maximum profits.

6 Conclusions

This paper does some researches on the supply chain made up of one manufacturer and two retailers in the oligopoly market, and then discusses the replenishment strategy of the supply chain when it applies the permissible delay in payments based on EPQ model. Finally, this paper proposes the basic condition whether the retailer deviates from the trigger strategy by introducing infinitely repeated game theory.

References

1. Chand, S., Ward, J.: A note on economic order quantity under conditions of permissible delay in payments. J. Oper. Res. Soc. 38(1), 83–84 (1987)
2. Goyal, S.K.: Economic order quantity under conditions of permissible delay in payments. The Journal of the Operational Research Society 36(4), 335–338 (1985)
3. Aggarwal, S.P., Jaggi, C.K.: Ordering policies of deteriorating items under permissible delay in payments. J. Oper. Res. Soc. 46(5), 658–662 (1995)
4. Jamal, A.M.M., Sarker, B.R., Wang, S.: An ordering policy for deteriorating items with allowable shortages and permissible delay in payment. J. Oper. Res. Soc. 48, 826–833 (1997)
5. Shinn, S.W., Hwang, H.: Optimal pricing and ordering policies for retailers under order-size-dependent delay in payments. Computers and Operations Research 30(1), 35–50 (2003)
6. Liao, H.C., Tsai, C.H., Su, C.T.: An inventory model with deteriorating items under inflation when a delay in payment is permissible. International Journal of Production Economics 63(2), 207–214 (2000)
7. Sarker, B.R., Jamal, A.M.M., Wang, S.: Supply chain model for perishable products under inflation and permissible delay in payment. Comput. Oper. Res. 27(1), 59–75 (2000)
8. Henke Jr., J.W.: Competing against an in-house supplier. Industrial Marketing Management, 147–154 (1989)
9. Chung, K.J., Huang, Y.F.: The optimal cycle time for EPQ inventory model under permissible delay in payments. International Journal of Production Ecnomics 84(3), 307–318 (2003)
10. Huang, Y.F.: Optimal retailer's replenishment polic for the EPQ model under supplier's trade credit policy. Production Planning and Control 15, 27–33 (2004)

11. Liao, J.J.: On an EPQ model for deteriorating items under permissible delay in payments. Applied Mathematical Modeling 31, 393–403 (2007)
12. Huang, Y.F.: Optimal retailer's replenishment decisions in the EPQ model under two levels of trade credit policy. European Journal of Operational Research 176, 1577–1591 (2007)
13. Liang, X.M.: Western Economics Course. China Statistics Press (1995)
14. Katz, M., Rosen, H.: Microeconomics. China Machine Press (1999)

A Hybrid Method for the Dynamics Response of the Complicated Structure

Jie Gao and Bing Li

Northwestern Polytechnical University, Xi'an, China
gaojie@snnu.edu.cn, libing@mail.nwpu.edu.cn

Abstract. This article based on a hybrid analysis method that couples numerical methods and statistical methods predicting dynamics response of the complicated structures. This research technique may change the numerical method to the model building when the massive node degree-of-freedom's need, only needs the few degree-of-freedom numbers, reduced the computation load greatly. At the same time, has solved the question that the low precision of statistical energy method in the low frequency. Like this we may obtain the computation highly effective, fast, accurate, and conforms to the actual structure dynamics response analysis method.

Keywords: Hybrid FE-SEA, Complicated structure, Dynamic response.

1 Introduction

Because all structure's process of manufacture the materials behavior, the fabrication technology and the assembly process always have some uncertainty. Regardless of how strict manufacture limit, in the hypothesis estimate parameter and the project actual parameter always has certain difference. Has many elements of certainty to exist in the manufacture process. The structure performance to these uncertainty is very sensitive, structure each kind of uncertainty factor to the structure response's influence, has become the restriction structure optimization design and performance evaluation development important attributes one, needs the accurate structural design process to guarantee the project the security and the success.

This article to has the hybrid Finite Element - Statistical Energy Analysis (FE-SEA) method promotion the uncertainty complicated structure, changed the numerical method to the model massive node degree-of-freedom need, only needs the few degree-of-freedom number, reduced the computation load greatly; Meanwhile did to the method of average theory part revises, has solved the method of average in the low frequency precision not high limitation. Thus might arrive at the computation highly effective, fast, accurate, and conformed to the actual structure response analysis method.

2 The Hybrid Finite Element-Statistical Energy Analysis Method

When generally regarding structure dynamics response computation, in low frequency is suitable for the Finite Element(FE) method, but in high frequency suits uses the statistical energy(SEA) method. The hybrid method, unified the FE method and the

D. Zeng (Ed.): Advances in Control and Communication, LNEE 137, pp. 539–545.
springerlink.com

SEA method, disseminates the wave basis the structure in or not to divide into the direct field and the reverberation field after the reflection, drove produces the wave, namely the incident ray, is called "the direct field"; Wave superimposition which and reflects many times after a reflection, is called "the reverberation field". Extracts the direct field using FEM the response, the reverberation field the additional strength which produces to the direct field influence obtains by SEA.

Here, definite part by the FE simulation, but the stochastic part simulates by the SEA subsystem. Must first determine that the system response these parts may use the SEA subsystem to describe, the part which is left over ("determinism" part) simulate with FE. The dynamic equation is:

$$D_{tot}q = f + \sum_k f^{(k)}{}_{rev} \tag{1}$$

$$D_{tot} = D_d + \sum_k D^{(k)}{}_{dir} \tag{2}$$

Where $D_{dir}{}^{(k)}$ is subsystem k direct field power rigidity. Moreover; D_d, system determinism partial finite element model power rigidity. f, exerts system this part of external forces. $f_{rev}{}^{(k)}$, on subsystem k the reverberation field produces strength. D_{tot}, after increasing each sub-system direct field power rigidity FE model power rigidity.

The following formula is about the reverberation strength cross spectrum matrix:

$$S^{(k)}{}_{ff,rev} \equiv E[f_{rev}{}^{(k)} f_{rev}{}^{(k)*T}] = (\frac{4E_k}{\omega\pi n_k})\,\mathrm{Im}\{D_{dir}{}^{(k)}\} \tag{3}$$

Where E_k, n_k respectively is the ensemble average energy and the kth subsystem's modality density. The formula (3) has connected subsystem's energy and the boundary elastic force, it is coupling SEA and the FE essential formula.

By the formula (1), responds q to be possible to expand for the below form:

$$\begin{aligned} q &= q_d + \sum_k q^{(k)} \\ q_d &= D^{-1}{}_{tot} f \\ q^{(k)} &= D^{-1}{}_{tot} f^{(k)}{}_{rev} \end{aligned} \tag{4}$$

The subsystem j direct field average power may write:

$$\begin{aligned} P_{in,j} &= (\omega/2)\,\mathrm{Im}\{E[q^{*T} D^{(j)}{}_{dir} q]\} \\ &= (\omega/2)\sum_{rs} \mathrm{Im}\{D^{(j)}{}_{dir,rs}\} S_{qq,rs} \end{aligned} \tag{5}$$

By the formula (3), (4), (5) may be get:

$$P_{in,j} = P^{ext}{}_{in,j} + \sum_k \omega\eta_{jk} n_j (E_k / n_k) \tag{6}$$

Where:

$$P^{ext}{}_{in,j} = (\omega / 2) \sum_{rs} \mathrm{Im}\{D^{(j)}{}_{dir,rs}\} (D^{-1}{}_{tot} S_{ff} D^{-1*T}{}_{tot})_{rs} \tag{7}$$

$$\begin{aligned} \omega\eta_{jk} n_j = (2/\pi) \sum_{rs} \mathrm{Im}\{D^{(j)}{}_{dir,rs}\} (D^{-1}{}_{tot} \\ \mathrm{Im}\{D_{dir}{}^{(k)}\} D^{-1*T}{}_{tot})_{rs} \end{aligned} \tag{8}$$

We can get subsystem j the energy balance equation:

$$\begin{aligned} \omega(\eta_j + \eta_{d,j}) E_j + \sum_k \omega\eta_{jk} n_j (E_j / n_j - E_k / n_k) \\ = P^{ext}{}_{in,j} \end{aligned} \tag{9}$$

And the cross spectrum matrix of responds q to be possible by the formula (3), (4) can be get:

$$S_{qq} = D^{-1}{}_{tot} [S_{ff} + \sum_k (\frac{4E_k}{\omega\pi n_k}) \mathrm{Im}\{D_{dir}{}^{(k)}\}] D^{-1*T}{}_{tot} \tag{10}$$

Where the formula (9) (10) these two equation very good union FE and SEA method, the formula (9) is the SEA form, but the coupling loss factor and the loss factor actually must obtain from the FE model computation; Moreover, the type male (10) is the standard stochastic FE analysis form, but increased the strength which the SEA subsystem's reverberation energy produces.

3 Establishes the Panel-Beam Complicated Structure the Hybrid Model

The panel-beam built-up section is in the project the common one kind of complicated structures. Carries on dynamics response analysis to this kind of panel-beam structure, establishes the mixed method model, has the very vital practical significance.

3.1 Establishes the Panel-Beam Structural Model

Uses the finite element method the panel-beam structure model building which uses frequently to a project, the structure is composed of a Liang frame and four thin steel plates, as shown in figure 1. The length of each beam is 0.7m, the section is the hollow, the most outer layer width is 25.4mm, wall thickness 3.175mm. The design parameter is $E = 71GPa$, $v = 0.33$, $\rho = 2700kg/m^3$, The material quality is an aluminum.

The panel size is 0.6m*0.6m, thickness is 1mm. The connection of panel and beam is point connection. The damping loss factor is 1%. The force apply on a beam of the frame, for details sees figure 1:

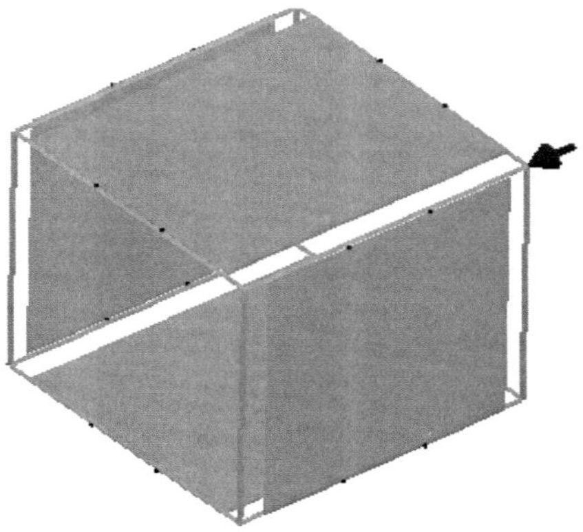

Fig. 1. Panel-beam built-up section model

3.2 Divides the System into the Definite Part and the Statistical Part

According to hybrid the FE-SEA, first, we should divides the system into the definite part and the statistical part. This kind of division may the sound wave length decide based on the frequency band. In the plate the shear wave and the longitudinal wave length in the 300Hz time wave length respectively is 18m and 16m. The movement may divide with these wave related board in is the FE part. In board flexural wave length when 100Hz is 0.31m, when 300Hz is 0.18m. This is equal to the shortest size has 2~4 wave lengths in panels, outside the notice board movement should describe by SEA. When is lower than 50Hz, the wriggle wavelength is bigger than the board the size. By now SEA was not suitable, should use the complete FE law. In Liang, the first four dissemination's wave is, a reverse wave, a longitudinal wave, two flexural waves (other higher order wave to 9650Hz). Corresponding wave length when 300Hz is bigger than 1m, therefore should use FE to the beam

In four panels probably include some 250 curving modalities, do not have the plane modality. In the frame probably have some 30 modalities, through the use board meandering movement's statistical description, compared to purely uses FE to describe, the efficiency enhanced greatly

3.3 Contrasts Pure Fe Method and Hybrid Method to the Structure Grid Division Result

Establishes the good system definite part the FE model. Except is located at located at subsystem boundary these parts of degrees-of-freedom, will possess about the SEA subsystem's degree-of-freedom henceforth in the model obliterates. because the mixing method is divided in the net not including to subsystem's shortwave response, therefore compares with the traditional FE law, only needs the minority degree-of-freedom then.

This may reduce the computation load greatly. Figure 2 and Figure 3 respectively be uses the FE method and the hybrid method carries on the modelling for the panel-beam built-up section.

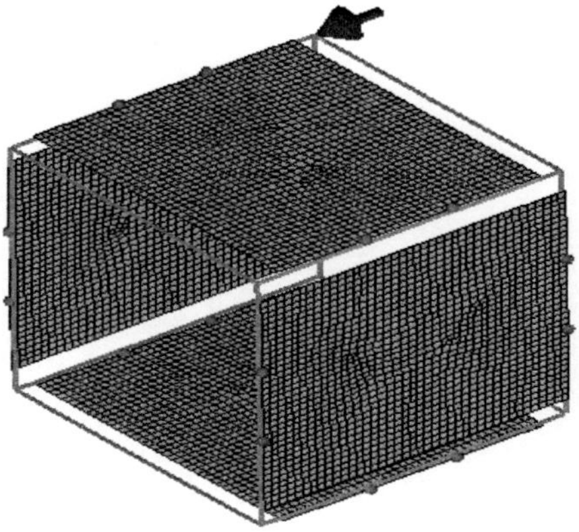

Fig. 2. Pure finite element method division grid

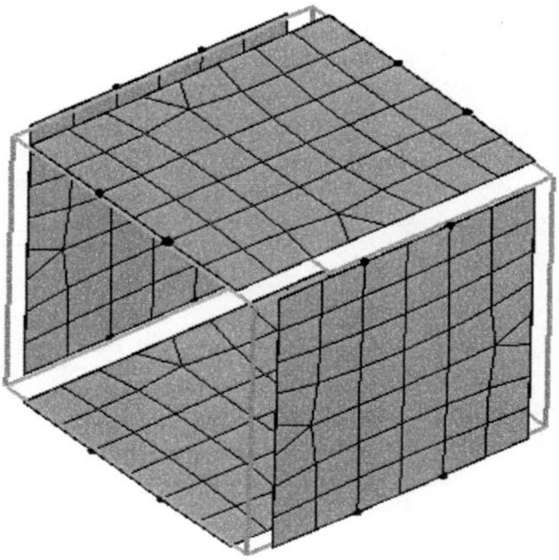

Fig. 3. Hybrid method division grid

Figure 2 is the pure FE model analysis, uses the Ansys software, the model has 6021 elements; Figure 3 is the hybrid method FE model analysis, the frame and a connection divides the grid by pure FE. But the board by the sketchier grid description, is composed of 520 triangle units, because only considers the plane motion, coupling FE and the SEA subsystem, described through a series of 16 spot connection. The model has 559 elements.

Because time use hybrid method division grid not including to subsystem's shortwave response, therefore compares with the traditional FE law, only needs the minority degree-of-freedom then, greatly reduces the computation load.

4 Panel-Beam Complicated Structure Vibration Response Result Contrast and Analysis Example

The distinction use pure finite element method and hybrid the FE-SEA method to do to on example panel-beam structure result speed of response calculates, may obtain the following speed response contrastive analysis result:

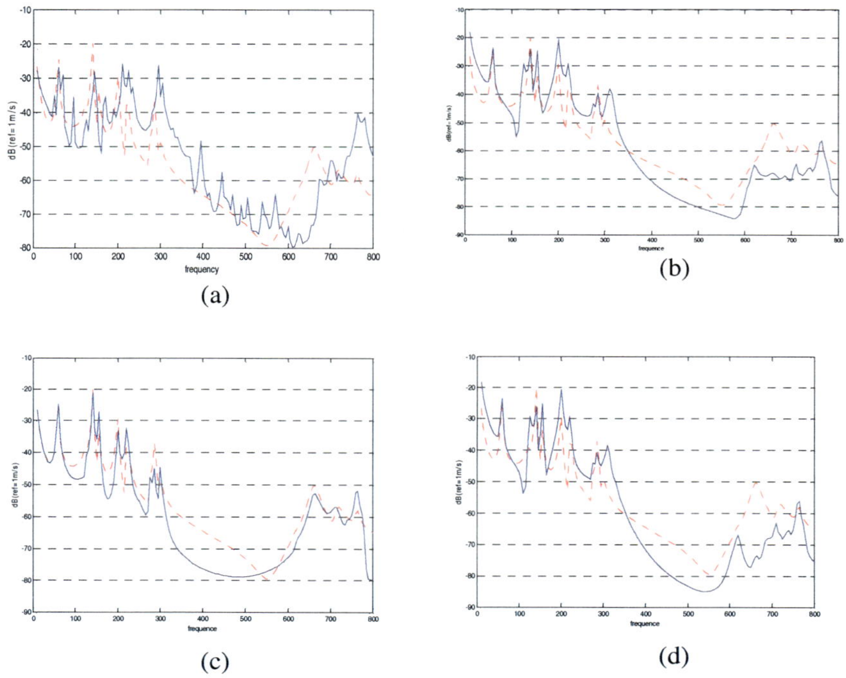

Fig. 4. Pure finite element method and hybrid method dynamics response result contrast. Dotted line, hybrid method; solid linethod, pure FE method. (a),(b),(c),(d)respectively is four panels of the structure computation contrast result.

We can see that uses the FE method and the hybrid method modelling, obtains the speed response spectrum finally. Two methods obtain result basic consistent. Except has approximately the 10dB difference about high frequency 800Hz. This is because the finite element method creates in the high frequency limitation. The entire computational process use hybrid method needs the time compared to the use finite element method to reduce greatly.

5 Summary

This article to has the FE-SEA hybrid method promotion the uncertainty built-up section. Changed the numerical method to the model massive node degree-of-freedom need, only needs the few degree-of-freedom number. Therefore reduced the computation load greatly. Meanwhile did to the method of average theory part revises, has solved the method of average in the low frequency precision not high limitation. Thus might arrive at the computation highly effective, fast, accurate, and conformed to the actual structure response analysis method. May see that the mixing method has provided a ideal analysis method to built-up section's response analysis. It has satisfied, the low expenditure, the low risk request highly effective. And can obviously raise the SEA forecast level.

References

1. Langley, R.S., Bremner, P.: A hybrid method for the vibration analysis of complex structural-acoustic systems. Journal of the Acoustical Society of America 105, 1657–1671 (1999)
2. Zhao, J.: With the statistical energy analysis forecast propeller-driven aircraft cabin internal noise. Northwestern Polytechnical University master paper, Xi'an (1990)
3. Langley, R.S.: A general derivation of the Statistical Energy Analysis equations for coupled dynamic systems. Journal of Sound and Vibration 135(3), 499–508 (1989)
4. Langley, R.S., Cordioli, J.A.: Hybrid determ inistic-statistical analysis of vibro-acoustic systems with domain couplings on statistical components. Journal of Sound and Vibration 321, 893–912 (2009)
5. Wang, Y.: The airplane typical cabin structure transaudient statistics energy analyzes the. Tongji University doctoral dissertation, Shanghai (2005)
6. Yao, D., Wang, Q.: The statistical energy analysis principle and applies. Beijing Institute of Technology Publishing house (1995)
7. Langley, R.S.: Mid and high-frequency vibration analysis of structures with uncertain properties. In: Proceedings of the Eleventh International Congress on Sound and Vibration, St. Petersburg, Russia (2004)
8. Cordioli, J.A., Triches Jr., M., Gerges, S.N.Y.: Applications of the Statistical Energy Analysis to Vibro-Acoustic Modelling of Vehicles. SAE

Optimization of Coatings for Vascular Stent with Ultrasonic Spray Technology

Gu Xingzhong[1,2], Huang Jie[1,3], and Ni Zhonghua[1,3]

[1] School of Mechanical Engineering, Southeast University, Nanjing, China
[2] Jiangsu Key Laboratory for Design and Manufacture of Micro-Nano Biomedical Instruments, Nanjing, China
[3] Suzhou Key Laboratory for Design and Manufacture of Micro-Nano Biomedical Instruments, Suzhou, China
aresgu@seu.edu.cn, huangjie_seu@163.com, nzh@seu.edu.cn

Abstract. Ultrasonic spray technology is used to prepare the coating of stent. Based on the mechanism of coating surface and the parametric model of ultrasonic spray process, an orthogonal experiment using THF and 30% PEO-PSF is carried out to optimize the processing parameters. It shows that the flow rate, gas pressure and feeding speed are the main parameters influencing the quality of coating surface. Besides, experiments have verified the rationality of optimized parameters, and the optimized process eliminates the common surface defects such as webbing, bridge, etc. AFM image indicates that within a 2μm×2μm scope, the coating roughness is reduced from 6.35nm to 1.72nm which improves the surface quality remarkably. The weighted results of coating mass tells the average mass deviation is about 7.5μg which shows the ultrasonic spray process has good repeatability.

Keywords: Stent, Coating, Optimization, Ultrasonic atomization.

1 Introduction

Since the birth of the vascular stent, it plays an important role in interventional therapy of heart disease, and with the development of technology, the stent starts containing various functional coatings [1][2]. The technical difficulty of coating arises from precisely controlling the coating mass and fine surface quality on the stent which has micron and complicated structure. In general, stent's coating can be achieved by soaking [3], spraying [4], or chemical deposition [5] process. However, experiment has shown that soaking technology be only applicable to the stents with simple structure at the early stage, when facing complicated stents, it causes envelope and bridge on parts of the stent, which lowers down the stent's performance. Besides, chemical deposition has special requirements in stent's material and coating material, making it can't be widely used. As a new kind of spray technology, ultrasonic spray has low flow rate with precisely control, fine spray with controllable droplet size, flexible coating ways, and low energy consumption. It is a potential technology for stent coating preparation.

D. Zeng (Ed.): Advances in Control and Communication, LNEE 137, pp. 547–556.
springerlink.com

There have been few reports systematically introduce coating preparation technology. The FDA rules focus on testing guides for medical and mechanical performance of stent, but have no specific requirements on coating surface quality [6]. This paper will discuss the character and mechanism of the coating surface, as well as the process of ultrasonic spray on stent coating preparation.

2 The Analysis of Stent Coating Surface

In order to make sure the stent has good performance after implantation, the coating should be uniform and smooth as far as possible, and the coating quality can be analyzed from two categories: the macroscopic surface quality and microcosmic surface quality, the macroscopic surface quality can be surveyed from ordinary magnifying glass while the microcosmic surface quality needs microscope for observation.

2.1 The Analysis of Macroscopic Surface Quality

According to experimental observation and statistics, the defects of stent macroscopic surface include bridge, webbing, envelope and wiredrawing, as shown in Fig.1. During implantation, the stent will expand and the defects may cause the coating be broken or rent, which reduce the degree of adhesion force between the coating and the substrate, and shorten the working life of stent.

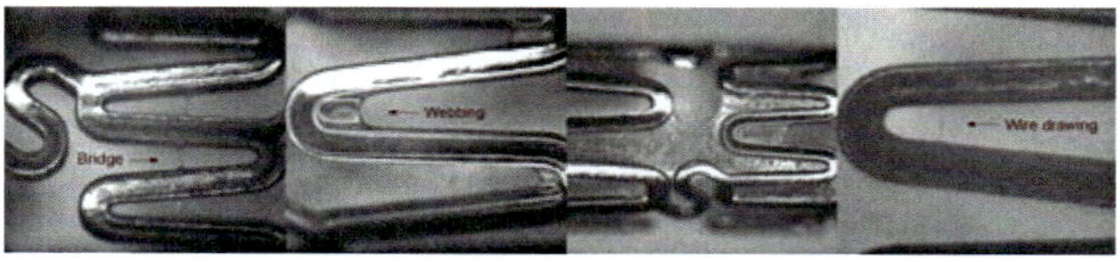

Fig. 1. The three defects of macro surface quality: bridge, webbing, envelope and wire drawing (shot by Zoom645 stereo microscope)

2.2 The Analysis of Microcosmic Quality

From experiments we have concluded that for stent coating, the micro defects of its surface include cratering, wrinkle, and unevenness (see Fig.2) which need to be avoided in spray process. These defects originate from the surface tension gradient which is an expression to the leveling property of liquid film on stent surface.

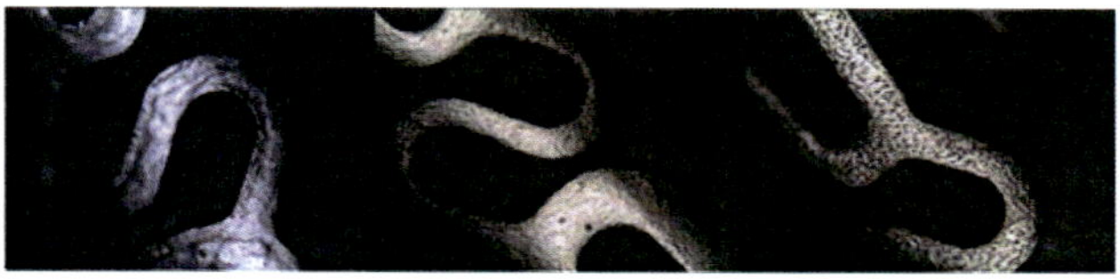

Fig. 2. Micro defects of stent coating: unevenness, cratering and wrinkle from left to right (shot by Olympus BX51 optical microscope)

3 The Analysis of Ultrasonic Spray Process for Stent Coating

To avoid above-mentioned defects of stent coating, a rational spray process is required. We have done more research about ultrasonic spray process; here we will introduce the development of ultrasonic spray technology in stent coating preparation.

3.1 Ultrasonic Spray System

In the lab, we use MediCoat spray system (see Fig.3) for coating preparation, including spray chamber, syringe pump, assistant gas, ultrasonic generator, etc. The system uses ultrasonic generator to produce ultrahigh-frequency signal and stimulate the piezoelectric ceramics which make the nozzle tip vibrate and transfer the liquid carried by syringe pump into aerosol, then by the motivation of assistant nitrogen, the aerosol is changed into spray. With the help of two servo motors, the stent has motion in two dimensions. The working way of the system is controlled by programming.

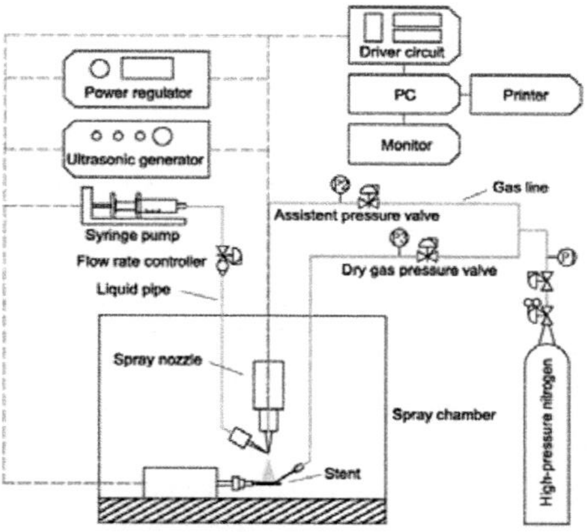

Fig. 3. The ultrasonic spray system structure

3.2 Parametric Model of Ultrasonic Spray Process

The ultrasonic spray process parameters for stent coating preparation include ultrasonic frequency *f*, liquid flow rate *Q*, input power *P*, assistant gas pressure *p*, stent rotational speed *r* and stent feeding speed *v*.

The spray nozzle is driven by piezoelectric transducer, according to Lang's [7] theory, under the excitation of ultrasonic frequency, the numerical median diameter d_{10} is described by following equation:

$$d_{10} = 0.34(\frac{8\pi\sigma}{\rho f^2})^{1/3} \tag{1}$$

It can be seen that the numerical median diameter is proportional to the surface tension and inversely proportional to the density and ultrasonic frequency. If we define distribution scope: $\sigma_d = d_{30}/d_{10}$, here d_{30} is the droplet mass median diameter. When the frequency uses 25kHz, 35kHz, 48kHz, 120kHz respectively, σ_d almost remains unchanged [8], which indicates that the frequency variation has no effect on droplet size distribution. For a certain spray liquid, the ultrasonic frequency is the main way to control spray droplet size. In order to get fine spray, high frequency should be chosen as far as possible.

For a liquid film, according to the Orchard levelling property equations [9], the time needed for levelling is described by following equation:

$$t = \frac{3\lambda^4 \eta}{16\pi^4 \sigma h^3} \ln\left(\frac{a_0}{a_t}\right) \tag{2}$$

Where t is the leveling time, a_0 is the initial amplitude, a_t is the amplitude at time t, λ is the wavelength, h is the average film thickness, σ is the liquid surface tension, and η is the liquid viscosity.

The liquid flow rate determines the feeding volume for spray per unit time, for ultrasonic spraying system, the maximum flow rate per unit area is inversely proportional to the ultrasonic frequency [8], when the nozzle tip area is S, the maximum flow can be written by following:

$$Q_{\max} = kf^{-2/3} \cdot S \tag{3}$$

Here k is a constant relates to the nozzle structure. Experiments have shown that flow rate is a key factor to spray process and its influence on surface quality is shown in Fig.4(1).

As the aerosol after ultrasonic atomization is dense, with the help of gas, it is formed into spray. Different gas pressure makes different spray cone angle and density. The probability of webbing and bridge together with the spray efficiency is inversely proportional to the pressure value while the loss of spray liquid increase with pressure rising. From Experiments, the influence of the pressure to the probability of wire drawing is shown in Fig.4(2), when the rotation speed is 200 *rpm* and the gas pressure is about 2 *psi*, the curve has a peak. Besides, high pressure may accelerate the evaporating rate of liquid film, which may lower down the levelling property and cause wrinkles or other micro defects on stent surface.

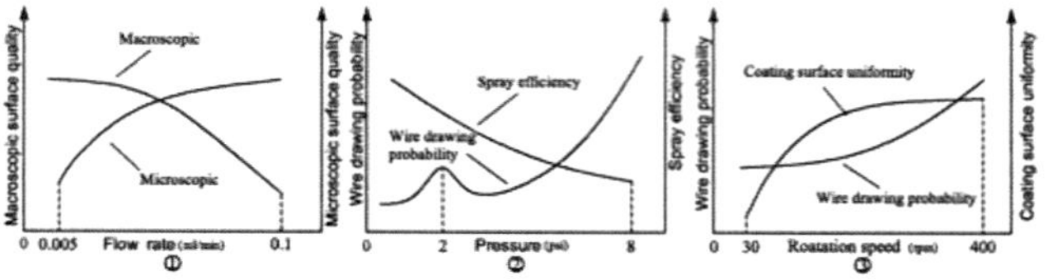

Fig. 4. Influence of flow rate, gas pressure and rotation speed

Power is a key factor of successful atomization in ultrasonic spray process for coating preparation. Rajan's study [10] shows that the energy density per unit area can be described by following equation:

$$\mathrm{I} = 0.5\rho\zeta(2\pi f A_m)^2 \tag{4}$$

Here I is the energy density, ζ is the ultrasonic wave speed transferring in spray liquid, A_m is the amplitude of capillary wave on nozzle tip. According to the theory of interfacial tension [11], the critical amplitude A_{mcrit} of the ultrasonic atomization is:

$$A_{mcrit} = \left(\frac{2\eta}{\rho}\right)\left(\frac{\rho}{\pi\sigma f}\right)^{1/3} \tag{5}$$

So the critical input power P_{crit} is:

$$P_{crit} = 0.5\rho\zeta\left[2\pi f\left(\frac{2\eta}{\rho}\right)\left(\frac{\rho}{\pi\sigma f}\right)^{1/3}\right]^2 \times \frac{Q}{fA_{mcrit}} \tag{6}$$

Take the energy loss into consideration, the actual power of ultrasonic spray is $P=\alpha \cdot P_{crit}$, α is a constant relevant to the spray system. Unlike Mohan's [12] study, precise experiments show that the droplet diameter decreases a little when power increase. For coating preparation process, the input power should be larger than the critical power, only if the power is not large enough to interrupt spray stability.

The spraying time is the actual time of stent in ultrasonic spray process, which controls the thickness of stent coating. Experiments have shown that the coating thickness is linearly proportional to the spraying time. For a predetermined coating amount, the spray time can be derived from following equation:

$$t = \frac{\pi m s^2}{Qldc\varepsilon\left(2\xi - \xi^2\right)} \cdot \tan^2\frac{\theta}{2} \tag{7}$$

Here θ is the spray cone angle, s is the distance between the stent and the nozzle tip, Q is the flow rate, l is the stent length and d is diameter of the stent, c is the liquid concentration, ξ is the effective ratio of available stent area, and ε is the adhesion efficiency on stent surface. Generally, ε equals to 1, and when taking droplet size into consideration, the approximation of ξ is replaced by following equation:

$$\xi = \frac{\left(w + 2d_0\right)}{w}\xi_0 \tag{8}$$

During spray process, the stent motion is cyclical, it includes three elements: rotation speed r, feeding speed v and delay time t_d. The feeding speed should not be too low, otherwise it may cause wire drawing and low spraying efficiency. The rotation speed has effect on the efficiency of the particle adhesion to stent's surface, as shown in Fig. 4(3), when the rotation speed is high, the tangential speed of the spray particles to the stent increases, which increases the probability of tangential wire drawing. Besides, there should be a proper delay time for evaporation of liquid film on stent surface each circle; otherwise, high liquid accumulation will appear at narrow parts of stent and create webbings or bridges.

3.3 The Optimization Experiment of the Stent Coating

As for many factors, we use orthogonal design. Based on parametric model of ultrasonic spray process, the input power is chosen as 3W, the ultrasonic frequency is 120 kHz, and the mass concentration of PEO-PSF to THF is 3%, other factors are shown in Table 1. As there are no cross actions among factors, the L_{18} (2×3^7) orthogonal table is chosen to design the testing program.

Table 1. The level of the stent's coating experiment factors

Factor / Level	DSN (mm)	Flow rate (ml/min)	Gas pressure (psi)	Feeding speed (cm/s)	Rotation speed (rpm)	Delay time (s)
1	10	0.02	1	0.15	100	2
2	5	0.03	2	0.30	150	5
3		0.04	3	0.45	200	10

3.3.1 The Result and Analysis of the Orthogonal Experiment

After stent coating preparation, we use Zoom645 stereo microscope to observe and gather macroscopic defects, and use BX51 optical microscope of Olympus Corporation to observe and gather statistics for microscopic defects, the coating surface quality of the eighteen stents is divided into five levels and marked. Spray time is recorded and the stent is weighted before and after spray. For the factor weights, we design the whole score is 100, and QS is 90 when TE is 10. According to the statistics, the results are converted into relevant scores (See Table 2). The experiment images show that there are craterings and wrinkles on the stent coating (see Fig.2), but no orange peel or cracking defects seen in conventional spray process.

Table 2. The results of orthogonal experiment

No	Macroscopic SQ	Microcosmic SQ	Time (s)	Coating mass (mg)	Spray time (s)	SQ	TE	Synthesis
1	45	28	22min31s	0.637	3.80	73	3.8	76.8
2	50	12	15min46s	0.307	2.5	62	2.5	64.5
3	40	4	16min19s	0.152	1	44	1	45
4	50	28	24min34s	0.854	4.76	78	4.76	82.76
5	40	28	18min30s	0.712	5.31	68	5.31	73.31
6	50	12	10min30s	0.242	3.02	62	3.02	65.02
7	50	36	19min02s	0.586	4.17	86	4.17	90.17
8	50	12	10min52s	0.304	3.76	62	3.76	65.76
9	40	36	24min43s	1.085	6.10	76	6.10	82.10
10	48	4	13min1s	0.284	2.84	52	2.84	54.84
11	33	28	27min50s	0.820	3.97	61	3.97	64.97
12	33	4	13min54s	0.393	3.8	37	3.8	40.8
13	50	20	13min48s	0.576	5.78	70	5.78	75.78
14	50	4	12min50s	0.382	4.02	54	4.02	58.02
15	50	36	27min52s	1.192	5.94	86	5.94	91.94
16	50	20	16min9s	0.438	3.63	70	3.63	73.63
17	50	36	22min44s	1.598	9	86	9	95
18	50	28	15min44s	0.652	5.74	78	5.74	83.74

In order to do further research about the influence of parameters, the diagram of SQ and TE versus different factors is show in Fig.5. The SQ chart shows flow rate, gas pressure and feeding speed are the main factors to coating surface quality. When the flow rate increases, the SQ is improved and with no webbing or bridge appearing, that may because the current flow rate is still within a small scope, and more sensitive to microscopic SQ than macroscopic SQ. When gas pressure and feeding speed increases, the liquid film leveling property is decreased, and the coating quality is lowered down.

From the TE chart, it can be found that the DSN, flow rate, feeding and rotation speed are the main factors to time efficiency. Besides, when the rotation speed is high, the time efficiency is increased more or less, that may because high rotation speed may lead to a bigger effective surface area, however the rotation speed should not be too high, that may cause more wire drawings among stent branches.

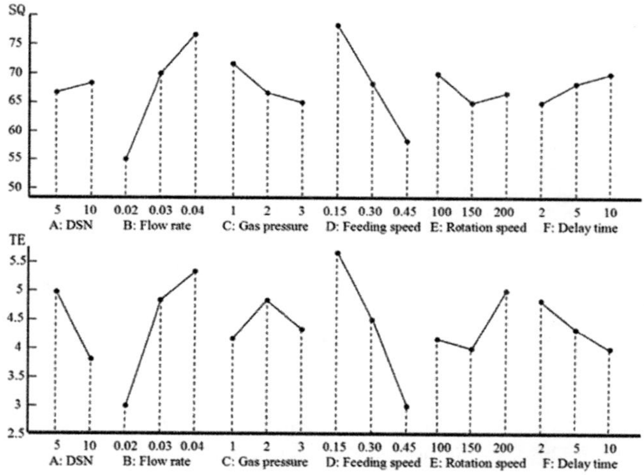

Fig. 5. Analysis of SQ and TE diagrams versus different factors

From the comparison between SQ and TE diagram, we can see that the flow rate, feeding speed and rotation speed have the same trend when the DSN, gas pressure and delay time are different. Since delay time is not the main factor, and to reduce spray time, we choose 2s finally. Including the factors outside the parameters orthogonal, the optimized process parameters are shown in Table 3.

Table 3. Optimized process parameters

Ultrasonic Frequency (KHz)	Mass Concentration (%)		Input power (W)	DSN (mm)
120	3		3	5
Gas pressure (psi)	Flow rate (ml/min)	Feeding speed (cm/s)	Rotation speed (rpm)	Delay time (s)
1	0.04	0.15	100	2

3.3.2 Verification of Optimized Process Parameters

To verify the optimized process parameters in Table 3, we use two kinds of stents with L605 alloy and 316L stainless steel as material. The coating is observed using the Olympus BX51 optical microscope and FEI Quanta 200 environmental scanning electron microscope (ESEM), and use MFP-3D Atomic force microscopy (AFM) of Asylum Research Corporation for nanoscopic observation. Besides, we gather the amount of each defect for evaluation of surface quality, all stents are weighted by UMX2 balance before and after spray to get the coating mass, and the spray time is recorded, all results are shown in Table 4.

Table 4. Experiment results for verification of optimized process

No	Stent type	Macroscopic defects Quantities	Microscopic SQ	Spray time	Δm (mg)
1	L605 9#	0	Fine	11min01s	0.700
2	316L 1#	0	Fine	11min01s	0.710
3	L605 9#	0	Fine	11min01s	0.668
4	316L 10#	0	Fine	11min01s	0.718
5	L605 9#	0	Fine	11min01s	0.685
6	316L 2#	5	Fine	11min02s	0.720
7	L605 9#	0	Fine	11min02s	0.675
8	L605 9#	0	Fine	11min02s	0.680
9	316L 3#	0	Fine	11min02s	0.670
10	L605 9#	0	Fine	11min02s	0.686

By Comparing Fig.6 with Fig.2, we can find that the coating surface after optimization is much better than before. We may see that in test No.6, there are bridges on stent surface, that mainly because the stent structure is much denser than others, which increase the probability of bridge appearance. When the spray time is about 11 minutes, the average coating mass is about 0.682*mg*, and the spray efficiency is 0.062mg/min. According to the displacement of camera shot of microscope to bare stent and stent with coating, the coating thickness is about 12 μm. For the same stent sprayed in six times, the coating mass deviation is 7.5μg, indicating that the optimized process have good repeatability.

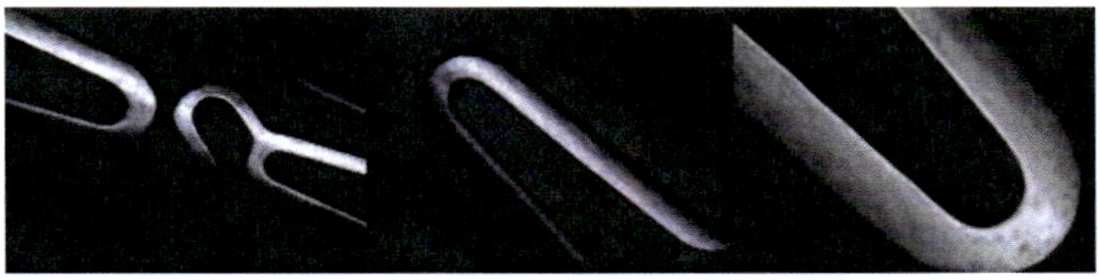

Fig. 6. Coating surface of test No.1, No.4 and No.8 (Shot by Olympus BX51 optical microscope)

For the restriction of DOF of the BX51 microscope, we use ESEM for observation in higher power scope, as shown in Fig.7, when magnifying power is 200×, there are not webbing, bridge and other defects on stent surface, and the coating is uniform and smooth; When the magnifying power is 2000×, there are no cratering, wrinkle or concavo convex topography on surface with L605or Stainless Steel, which indicates that the optimized process could eliminate the common microscopic defects and produce fine coating.

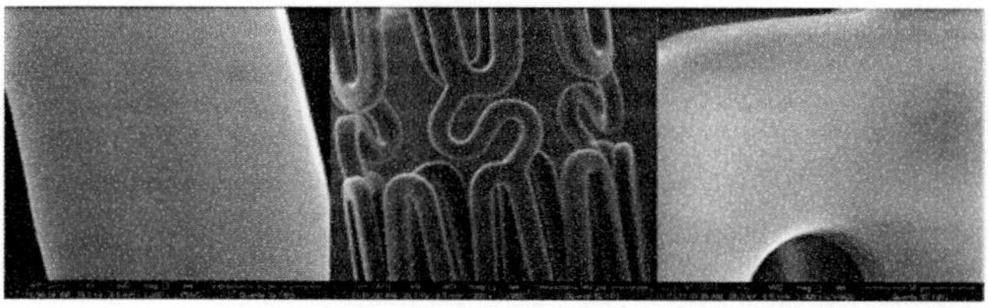

Fig. 7. Coating surface of test No.1, No.4 and No.9 (Shot by Quanta 200 ESEM)

In order to observe the nanoscopic features of the stent coating, ESEM is used to scan the spray coating surface, the 3D model images are shown in Fig.8, all images are shot within $2\mu m \times 2\mu m$ scope, The left image is the bare stent surface, the middle image is the coating surface before optimization, and the right image is after optimization. According to measurements, the surface roughness of the bare stent is about 12.4 *nm*, and the coating roughness is about 6.35 *nm* but decrease to about 1.72 *nm* before and after optimization. However, both coating roughness are less than bare stent roughness which help to explain why the biocompatibility of stent is increased after coated. And we can make a conclusion that the roughness of stent surface can be reduced by coating, and the uniformity of optimized coating surface is much better than before.

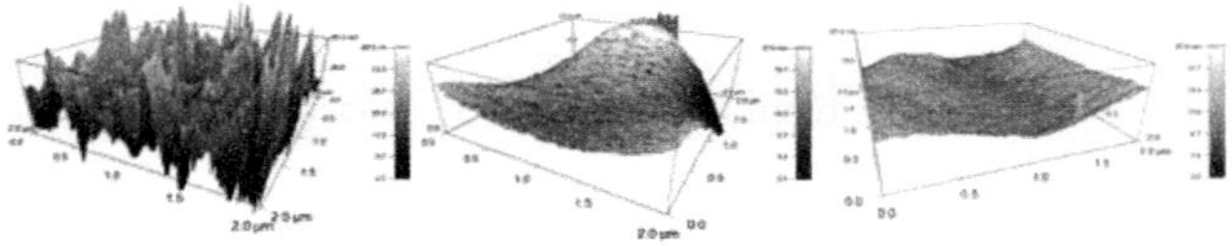

Fig. 8. From left to right: Bear stent surface and coating surface before and after optimization(Shot by MFP-3D AFM)

4 Summary

Based on ultrasonic atomization theory, the parametric model of ultrasonic spray technology on stent coating is established. Then, an orthogonal experiment is carried out to verify the influence of each processing parameter. The result shows that the ultrasonic frequency, liquid flow rate, gas pressure, rotation speeding and feeding speed are the key factors to stent coating process. And from the analysis of SQ and TE charts versus different factors, the optimized process parameters are obtained. Then, experiments have been done to verify the rationality of optimized spray process, the results show that the optimized process has good reliability and repeatability, and the common defects have been well eliminated. Besides, the coating has low roughness which is even superior to the polished metal surface, and the coating is more uniform than ever before, which have positive effect on improving stent biomedical performance.

Acknowledgment. This research is supported by Doctoral Foundation of Ministry of Education of China (No. 20090092120007) and Jiangsu Province Natural Science Foundation (BK2010398 and BE 2009054).

References

1. Pan, C.J., Tang, J.J., et al.: Preparation, characterization and anticoagulation of curcumin-eluting controlled biodegradable coating stents. Journal of Controlled Release 116, 42–49 (2006)
2. Hanefeld, P., Agarwal, S., et al.: In Vitro Study of Dexamethasone Release From Poly (p-xylylene) Films. Macromol. Chem. Phys. 211, 265–269 (2010)
3. Hossainy, S.F.A., Sanders-Millare, D., Guruwaiya, J., et al.: Bio compatible coating. PCT Publication: No WO 01/45763 (June 28, 2001)
4. Jan, H., Eftichia-Vassiliki, Z.: Heparin stent. PCT Publication: No WO 03/011250 (February 13, 2003)
5. Dubois-Rande, J., Doan, T.L., Pham, M.C., et al.: Implantable device covered with polymer Capable of releasing biologically active substances. USA: No 6 468 304 (October 22, 2002)
6. Guidance for Industry:Coronary Drug-Eluting Stents—Nonclinical and Clinical Studies. FDA & CDRH&CDER, USA (March 26, 2008)
7. Lang, R.J.: Ultrasonic atomization of liquids. Acoust. Soc. Am. 34, 6 (1962)
8. Berger, H.L.: Ultrasonic liquid atomization: Theory and Application. Partridge Hill Publishers, USA (2006)
9. Orchard, S.E.: On surface leveling in viscous liquids and gels. Appl. Sci. Res. A(11), 451–464 (1961)
10. Rajan, R., Pandit, A.B.: Correlations to predict droplet size in ultrasonic atomization. Ultrasonics 39, 235–255 (2001)
11. Pohlman, R., Heisler, K., Cichos, M.: Powdering aluminium and aluminium alloys by ultrasound. Ultraconics 12(1), 11–15 (1974)
12. Avvaru, B., Patil, M.N., Gogate, P.R., Pandit, A.B.: Ultrasonic atomization: Effect of liquid phase properties. Ultrasonics 44, 146–158 (2006)

Research on Photoelectric-Precise Monitoring System of Thin-Film Deposition

Xu Shijun[1], Zhang Chunmin[1], and Ren Xiaoling[2]

[1] Dept. of Physics, School of Science,
Xi'an Jiaotong University, Xi'an Technological University,
Xi'an, China
[2] School of Computer, Xi'an Polytechnic University,
Xi'an, China
xushijun000@sina.com, zcm@mail.xjtu.edu.cn,
renxiaoling2@163.com

Abstract. Existing coating equipments have low precision and automation level about monitoring thin-film thickness. A new photoelectric-control-analysis system had been researched. The proposed system included a double-frequency modulation equipment and a photoelectric dual-light path system. A comprehensive digital processing-controlling system and a dual-lock-phase circuit system were used to process monitoring signals. In the test, all automation-monitoring functions were achieved with a control-analysis system. The experimental results show that the static and dynamic stabilities and the thin-film thickness control precision are extremely increased. The standard deviation of thin-film thickness, which indicates duplication of monitoring thin-film thickness, is less than 0.50%. The display resolving limit on reflectivity is 0.02%. The signal-noise ratio approaches 1300. in the proposed system, the linearity of drift is very high, and the static drift ratio approaches 1.63%/h. It is concluded that the photoelectric-precise monitoring system has advantages of the static and dynamic stabilities, the signal-noise ratio, and the control precision for thin-film thickness.

Keywords: photoelectric engineering, optical film, monitoring thin-film thickness, dual-light path, precision.

1 Introduction

Because optical thin-film thickness directly affects the transmitted spectrum or spectrum of reflection and the polarization characteristic of thin-film[1,2], the accurate monitoring-controlling of the thickness is a key technique for deposition of optical thin-film[3,4]. In many fields on coating optical thin-film and monitoring spectrum characteristic or optical constants, the exploration and development in abroad groups are in the lead. For example, the coating equipments of Leybold corporation in Germany and Carl Zeiss corporation in Switzerland have a higher level on automatic control, the thin-film thickness instruments of Honglun in Japan have good monitoring and control functions[5,6]. By comparison, the monitoring coating

D. Zeng (Ed.): Advances in Control and Communication, LNEE 137, pp. 557–564.
springerlink.com

process of domestic equipments almost depends upon fixing human eyes on display number or instrument needle, and upon analyzing their changes. Consequently, the manual operations with lots of experience inevitably and subjectivity lead to inferior duplication and low qualified product ratio. Also, the poor automation level means lower efficiency and higher labor intensity in coatings.

To the domestic coating equipment with low precision and automation level on monitoring thin-film thickness, a new photoelectric control and analysis system was investigated. The new system included a dual-optical path photoelectric system and a dual-lock-phase circuit system and a comprehensive digital processing-controlling-analysis system. The test results show that the static and dynamic stabilities are extremely increased, along with controlling precision in monitoring optical thin-film thickness.

2 Basic Principles and New Thin-Film Thickness Monitoring System Construction

The main methods of monitoring thin-film thickness primarily contain PEEMM (photoelectric extrema monitoring method), quartz crystal oscillation method, scanning ellipsometer by synchronous rotation of polarizer and analyzer, and wide spectrum monitoring method [6,10],etc. In monitoring optical thin-film thickness, especially, the application of PEEMM is more extensive than other methods, which directly detects reflectance or transmittance of thin-film to monitor its thickness by photoelectric detection technology.

PEEMM principle of is the film-interference theory in wave optics. While optical thickness of film, *nd*, changes by $\lambda/4$, *R* (reflectivity) or *T* (transmittivity) of vertical incidence light appears with an extremum. By means of monitoring the extrema in a depositing process, the monitoring thickness of $\lambda/4$ film series is realized. In the regions nearby the extremum point, the change defined as $\frac{\partial R}{\partial (nd)}$ or $\frac{\partial T}{\partial (nd)}$ is very small, and the unsteady factors (e.g. noise etc.) are relatively strong, as a result, it is very difficult to improve the duplication and precision of monitoring thin-film thickness. To this defect of principle, although many remedial measures have been proposed, such as the extremum-point-later control technique (belonging to a hysteretic controlling) and the pre-deposited layer technique and the multistage control technique[4,5], the quality of monitoring thickness is dependent on static and dynamic stability and sensitivity of the monitoring system. This stability and sensitivity are just the purpose of this paper.

2.1 The Original System Construction

In the original system of monitoring thin-film thickness, a light beam from a light source is modulated with single-frequency, and divided into two bunches by a beam splitter. One light beam is received and concerted into a lock-phase demodulation signal by the optoelectronic sensor. After other bunch penetrates the control slice, the light, which has some information of thin-film thickness variety, passes through a monochrometer and a photomultiplier, therefore a measurement electric signal is acquired. Firstly, the two signals are processed to a low-noise transmitted signal in filter circuit and lock-phase amplifier circuit. Secondly, the signal is displayed on a numeral monitor. Finally, the monitoring thickness is accomplished by watching and determining the display number.

2.2 The New System Construction

As shown in Fig.1, a light beam from a light source is modulated with double-frequency. Two different lock-phase demodulated signals of reference path and measuring path are gained with two silicon-photo battery, at the same time, a reference light beam and a measuring light beam are also obtained. The reference light beam transmitted with a optical fiber is directly received by a monochrometer and a photomultiplier. As well as, the measuring light beam, which is reflected on the control slice and contains some information of thin-film thickness variety, is received by the same optical fiber and monochrometer and photomultiplier. The superimposition of reference signal and measuring signal is separated, and processed in dual-lock-phase amplifier circuit for digital acquisition, filtering procedure and division operation. A high quality signal including information of thin-film thickness variety is then acquired and displayed on numeral phototube. Moreover, the function of judgment thin-film thickness is realized with starting the judgment procedure and a servo system.

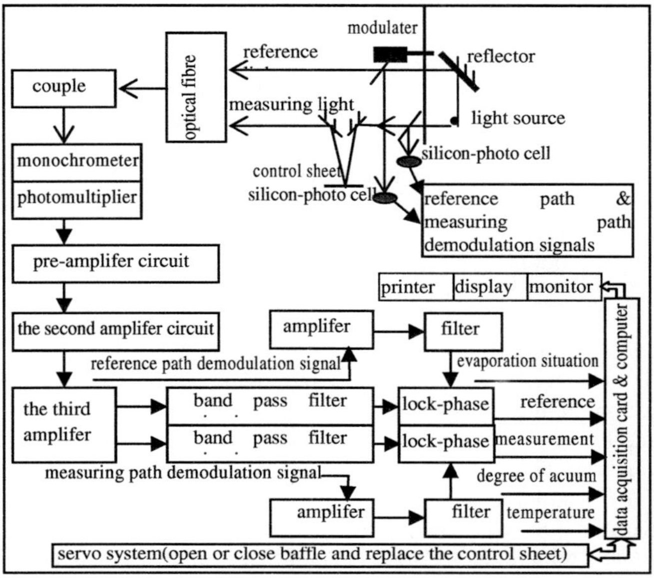

Fig. 1. The sketch map of proposed system

2.3 The Defects of Original System

In the original system, anti-interference ability and stability of monitoring thin-film thickness are poor, and these main reasons are as follows.

The construction in which the measurement beam directly projects upon the incidence slit of monochrometer makes the abilities of anti-vibration and excluding the stray light fairly weak. At last, the stability of the monitoring signal is affected.

After the signal is processed only in analog filter circuit without further digital filter, it is showed on numeral phototube. Therefore, the numeral display of monitoring signal has bad stability.

The stabilities of light source and photomultiplier and their power supplies affect remarkably the stability of monitoring signal. For example, an infinitesimal fluctuation of high voltage of photomultiplier will cause an enormous output fluctuation that exceeds a input stage one order of magnitude, the luminous intensity of light source directly effects measuring beam intensity. In the monitoring system, consequently, very high demands of these stabilities and anti-interference abilities have been put forward.

3 Main Techniques of New Thin-Film Thickness Monitoring System

Compared with the original system, the new technical characteristics are as follows.

3.1 Double-Frequency Modulator and Dual-Light Path

Optical modulation has many effects as follows. To measurement result, influence of the stray light is reduced, the bad effect of photomultiplier dark current is eliminated. The modulation signal provides the common work signal to lock-phase amplifier and select-frequency amplifier.

For the domestic deposition equipment (Peking Instrument Factory DMD-700J), the proposed system changes single-frequency modulator into double-frequency modulator, 33Hz measuring light path and 383Hz reference light path thus come into being. Besides, the double-frequency modulator is a basis of realizing dual-optical path system, dual-lock-phase circuit system and digital division (measuring data by reference data). Thus, some common interference of light source, modulation motor, monochrometer, photomultiplier, etc., is extremely repressed.

The dual-light path system includes 33Hz modulated measuring light path taking some information of thin-film thickness variety, 383Hz modulated reference light path with some information of light source modulated, 33Hz and 383Hz two modem light paths which are used for a dual-lock-phase amplifier circuit of cross correlation detection. In the original system, measuring light path and demodulation light path modulated with single-frequency, and this construction makes the stability of measuring signal remarkably hinging upon the stabilities of light source and photomultiplier and all power supplies. To resolve the problem, the measurement path and reference path are employed in the proposed system. The signals from two paths have the common fluctuation and interference of light source and photomultiplier. If the signals are divided, all this common fluctuation and interference should be cleared up. Although the dual-optical path is more complicated than single optical path of the original system, the work stability demands of these key equipments are greatly reduced, furthermore, the static stability of monitoring control system and the duplication of coatings are extremely improved.

3.2 The Same Monochrometer and Photomultiplier and Optical Fiber

The optical fiber is connected to the monochrometer for transmitting the light in the proposed system so that the utilization of optical energy is increased, while the measuring beam is directly projected upon the monochrometer slit in the original system. To the transmittal optical signal, the interference of the stray light can be eliminated. Particularly, the connection of optical fiber with monochrometer can eliminate the influence of coating equipment vibration to the measuring signal.

We transmit the measuring light and the reference light by a same set of optical fiber, monochrometer and photomultiplier. Opposed to two sets of monochrometers and photomultipliers, this design can greatly reduce the difference mode interference, and is of benefit to the repression of the common mode interference in later division operation.

3.3 Dual-Lock-Phase Amplifier

The dual-lock-phase amplifier circuit, without the single-lock-phase amplifier, is corresponding to above dual-optical path system. The pre-amplifier and lock-phase circuit are designed with a symmetric arrangement, and chooses high and similar performance components. The above measures create some favorable conditions for later repressing the circuit drift with "digital division operation".

3.4 Comprehensive Digital Processing-Analysis Technique (Compound Digital Filter and Digital Division and Numeral Display)

For further reduction of differential-mode interferences from stray light in vacuum chamber and noise jamming in circuit etc, a compound digital filter technique is primarily adopted. In comparison with analog filter circuit, the technique has many advantages: only one filter program for many data channels, no hardware, reliable stability, lower frequency signal can be filtered, and the filtering parameters can be changed conveniently.

The measuring signal is digitally divided by the reference signal, which has higher accuracy and stability than the analog division. Especially, the digital division can eliminate the common mode interference from light source and photomultiplier etc. The ultimate purpose of the dual-optical path system and the dual-lock-phase amplifier circuit will be realized, and the stability of monitoring system should be raised.

After the film thickness signal is handled by the dual-lock-phase amplifier circuit and the comprehensive digital processing-analysis technique (compound digital filter, digital division, linearization of numeral display), it is finally showed in numeral phototube. This technological stability is very high by contrast to the numeral display only carried in analog circuit in the original system.

4 System Tests

Our experiment scheme is testing, contrasting, analyzing proposed system and original system, i.e., the static stability and the dynamic monitoring characteristics of the two systems are respectively tested, analyzed, compared on the DMD-700J deposition equipment.

The data of monitoring system are recorded and respectively fit with a linearity expression and a quadric polynomial expression. After the two expressions are evaluated with RSE (residual standard error) and CCC (composite correlation coefficient) and amended CCC, the accurate static drift law is got. While the numeral show of proposed system is linearly demarcated with quadric polynomial expression, the demarcated result is analyzed by variance and correlation coefficient, and the show sensitivity of the proposed system is determined. For four kinds of film series, which are ZnS and MgF2 single-layer and poly-layer optical thin-film, we have proceeded duplicate-coating experiments for 10 times in the two systems, respectively, guaranteeing the completely same technological conditions to the same film series in each experiment. Using the variance-analysis method and uncertainty-analysis method, we have studied all monitoring data and spectrum-test data of each kind of film series, and got lots of quantitative film thickness indexes.

5 Results

The fundamental monitoring functions of the whole system have realized, which include the dual-light path lens imaging system, the dual-lock-phase amplifier circuit system, the digital data acquisition-filter, the numeral display of analog signal, and the final automatic control. Thus the research lays a solid foundation for the next step product developments. The specific research conclusions are as follows.

The static drift characteristic of DMD-700J has been found

$$\hat{y} = 5.02322\times10^{-8}x^2 - 0.000055x + 0.95836 \tag{1}$$

The expression provides a mathematical amended model for automatic monitoring thin-film thickness, as well as gives a basis for analyzing influences of film spectrum from extremum-point-later control value.

The revised variance analysis method by the author, is applied to the analysis of coating stability or deposition duplication level and control precision. Some detailed quantitative parameters are given. All this results provide a basis for design and experimental operation of coating film.

In the proposed system, the numeral display properties have four aspects. First, the experimental results show that the signal-noise ratio approaches 1300. Second, the linearity of numeral display is very high ($dS/dR \approx 1$) and the residual deviation is very minor. This means that it can certainly be used as the photometer. Thirdly, the display-resolving limit of low reflectivity film, ΔR, is equal to or more than 0.02%. Lastly, in the all reflectivity scope, the display sensitivity of responding to reflectivity is higher, while the lower reflectivity corresponds to the higher show sensitivity.

Many static properties of the two systems are listed in Table 1, considering low reflectivity. Firstly, it has been found from S_R (regression deviation sum) that the linearity of monitoring drift is greatly higher in the proposed system than in the original system, particularly, for the low reflectivity, and this conclusion provides some conveniences for experimental control and operation. Secondly, based on $\hat{\beta}$ (drift coefficient), the drift ratio of low reflectivity is very tiny and approaches zero or -1.63%. In general, drift ratio in the proposed system is smaller than the old one. Thirdly, it can be seen from the comparison of S_r of the two residual deviation sums that the static stability in the proposed system is better than the old. To the high reflectivity film, the static stability is usually not only about 2 times higher than the original, but also further higher to the low reflectivity film. At last, the lowest digit-flutter-scope, l, means an uncertainty scope of the numeral display. The uncertainty of new system is obviously tiny, that is approximately ±1,±2, ±3.5 (in the original system it is ±4.5).

Table 1. Static properties of two systems

category	S_R	$\hat{\beta}$ [%/h]	S_r	l
original system	0.00034017	-14.7	0.00025774	±4.5
proposed system	0.00000011	-1.63	0.00000191	±1

Dynamic properties of proposed system have been tested and various film series have been successfully monitored. The system's anti-interference ability has been greatly enhanced. For the single layer film of ZnS, the comparison of monitoring stability or duplication of the proposed system and old system is in Table 2.

The degree of remarkable difference is defined consistency analysis of variance, which means the more obvious difference between the calculated value and the critical value and the poorer consistency. S_t is defined standard deviation of transmissivity and the standard deviation of thin-film thickness (S_f) is an object relativity index of duplication. It is obvious that the duplication of proposed system is perfect, S_f and S_t, respectively, are equal to 0.50% and 0.0562.

Table 2. Stability or duplication of two systems

category	degree of remarkable difference	S_t	S_f
original system	$191 > F_{0.01}(9,10) = 4.94$	0.0874	1.25%
proposed system	$102 > F_{0.01}(6,7) = 7.19$	0.0562	0.50%

6 Conclusion

The thin-film thickness standard deviation of the proposed system, which indicates duplication of monitoring thin-film thickness, is less than 0.50%. The display resolving limit on reflectivity is 0.02%. The signal-noise ratio approaches 1300. The linearity of drift is very high, and the static drift ratio approaches 1.63%/h. In a word, based on higher sensitivity and stability of the proposed system, the static and dynamic stabilities or duplication are extremely raised, along with controlling precision in monitoring optical thin-film thickness.

Acknowledgment. Many techniques in the paper were proposed with a help from HAN Jun professor, so author expresses heartfelt thanks to him ! This paper is sponsored by the Scientific Research Fund from Shaanxi Province Education Committee (06JK283).

References

1. Hong, D.M., Zhu, Z., Yue, W.: Optical coatings for fiber lasers. Opt. Precision Eng. 13, 480–482 (2005)
2. Kaiser, N., Stenfeigl, T., Stenzel, O.: Optical coatings: trends and challenges. Opt. Precision Eng. 13, 389–396 (2005)
3. Macleod, H.A.: Monitoring of optical coatings. Appl. Opt. 20, 95–100 (1981)
4. Tang, J.F., Gu, P.F.: Thin-film Optics and Technology, 1st edn., pp. 258–274. Machine Press Publications, Beijing (1989) (in Chinese)
5. Tami, K., Kiyoshi, Y.: Simultaneous measurement of refractive index and thickness of thin film by polarized reflectances. Appl. Opt. 29, 5069–5070 (1990)
6. Xu, S.J.: The development of real-time monitoring system of optical thin-film thickness, pp. 36–45. Dissertation of Xi'an Institute of Technology, Xi'an, China (2002) (in Chinese)
7. Benedic, F., Bruno, P., Pigeat, P.: Real-time optical monitoring of thin film growth by in situ pyrometry through multiple layers and effective media approximation modeling. Applied Physics Letters 90, 1341–1344 (2007)
8. Ehlers, H., Grob, T., Lappschies, M.: Optical broadband monitoring of thin film growth. Opt. Precision Eng. 13, 403–412 (2007)
9. Navarro, M., Fandino, J.: Virtual substrate model for in-situ thin film thickness determination. Surface Engineering 16, 70–72 (2000)
10. Bulkin, P., Daineka, D., Kouznetsov, D.: Deposition of optical coatings with real time control by the spectroellipsometry. Applied Physics 28, 235–242 (2004)

Research on J2ME-Based Mobile Random Chat System

Gang Qiang, Yanqi Liu, Chong Zhang, Chao Mu, Bin Tan, and Sha Ding*

Sichuan University, Jinjiang College, Pengshan, Sichuan, China
465119148@qq.com, dingsha7320@gmail.com, tanb_ice@163.com

Abstract. First, there is a brief introduction about the development of mobile instant messaging, and JAVA architecture and various levels, including the Java virtual machine layer, the configuration layer, Jane surface and mobile information device Jane surface. Then it focus on the introduction of mobile random chat system, including the design of sever-side interface and the function of each function.

Keywords: JAVA, Instant Messaging, Mobile random chat system, Server-side, User-side.

1 Prolegomenon

Mobile instant messaging is to achieve instant messaging by cell phone. In recent years, universal mobile messaging industry develops fast, especially in developing countries which the achievement is more obvious, resulting in the quick increase of total cell phone users. The increasing speed has already surpassed the users increasing velocity of telephone and Internet the two big industries. People also become more and more unsatisfied with communication according to personal computer and the work station of office. It has been the urgent demand to accord to the potable and wireless devices.

Mobile chat system is a chat software which has been exploited based on mobile platform. The available software's owns QQ, MSN, UC, Fetion and so on at present market.

2 Brief Introductiin of JAVA

Java is a development made by the company's next-generation programming language. It can be used in a variety of different kinds of machines, different kinds of network operating platform software development environment. Whether you are using a WWW browser, which, what kind of computer, what kind of operating system, as long as the WWW browser specified above, the "support Java", you can see the vivid home page. Java is becoming the main Internet application development language. It completely changed the application software development model, has been brought

* Corresponding author.

D. Zeng (Ed.): Advances in Control and Communication, LNEE 137, pp. 565–570.
springerlink.com

from the PC, another technological revolution, for the rapidly developing world of information has added new vitality.

2.1 The Architecture of JAVA

Java platform by the Java Virtual Machine (Java Virtual Machine) and Java application programming interfaces (Application Programming Interface, referred to as API) composition. Java application programming interface for Java applications independent of the operating system of a standard interface, can be divided into the basic part and extension. In the hardware or operating system platform, after installing a Java platform, Java applications can run. Embedded Java platform now has almost all the operating systems. This Java program can only compile it once, you can run a variety of systems.

3 Mobile Random Chat System

3.1 The Overall Design Essentials of Chat System

The design of chat system has many differences from comment websites. The factors that the common websites design consider ,for example ,common websites demand to greatly beautify the layout and animation design, but the chat room just needs to provide with the direct instant chat that can satisfy visitors both sides. So, make sure to consider the following design essentials in the process of design chat system:

1)Realize thought

The chat program in Internet usually provides with server-side connective response by server. Clients login to server according to the user-side, then they can communicate with the users in the same server. This is a communication process that face to connection. So, with the environment of TCP/IP, program realizes the server-side and user-side two parts programs.

2)The work flow of server

After server-side creates a Socket number-group according to socket() system call(settles the maximum account of accepting connection users),binds with the appointed native port, then can listen listen() in port .If there is a request from user-side, then choose an empty Socket in number-group, assigning the user-side address to the socket. Then the users who login successfully can chat in server.

3)The work flow of user-side

The user-side program is simple relatively. It only demands to establish a socket to connect with server-side, then it can send and receive dates according to the socket.

3.2 The Design Steps and Function Module Division of Chat System

The operating principle graph of chat system

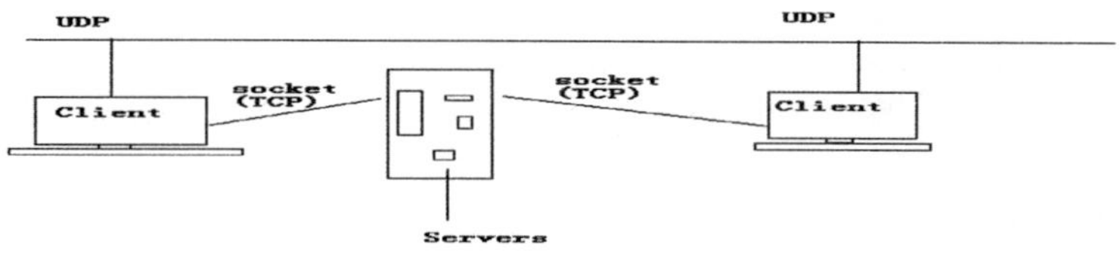

Fig. 1.

3.3 The Each Function Module of Chat System

1)The program module of server
The server and the accounts connect with each other by Socket(TCP).It's quite easy to use socket in Java. The Java API offers a class Java. net. socket for dealing with the messaging of socket, which makes it easier relatively to compile applied program. The server adopts multithreading to meet the request from the accounts by creating a server socket object, which the original port is 8080, then by the means of calling accept ()endlessly to accept the connection of account program..

2)The module of account program
Clients create the connection with server by Socket(Internet Address, pot).The server and clients both create in and out stream by establishing Buffered Reader ,Print Writer ,then both sides pass information mutually by in and out stream. Once accepts the request from account side, the server returns to a new creating socket object. Then the account side send information to server, for example document transportation. After receiving the requests from account side, it deals with the requests directing to different information. The specific original code as followings

3.4 The Generation of Random Process

Random process is meant to let the system generate a random number appears, and this random number as a client. At this point the customer service end of the connection. The corresponding connection is successful. At this point they form a random process. The cycle of this process, you can ensure that the process is not interrupted. Random number generation process: In Java, the concept of random numbers from a broad sense, there are two.

1. To create the Random object with no seeds Random random = new Random ();
2. Create the Random object with no seeds in two ways:1) Random random = new Random (555 L);2) Random random = new Random (); random. Set Seed (555 L); and random chat is not the same need to generate numbers. So it is necessary without needs.

For example, as we have a random number 1-100. Does not appear 1 and 8, int num = new Random (). Next Int (100);

```
While (num == 1 || num == 8) {
   Num = new Random ( ).nextInt (100);
```

```
}
System. out. Println (num);
```

System, there will be unrestricted random number. Each output of one. Not duplicate each generate a random number. The system will connect the appropriate randomized values. Once the connection is finished. When we disconnect, the system will connect again. That line into a loop.

4 Experiment

4.1 Hardware Environment

The demanded environments that the system operates environment are:

Hardware configuration:

CPU: 2.0 G CPU
Hard disk: over 500 GB hard disk
EMS memory: 2 GB DDR EMS memory
Display device; 17 inch display device
Display card: 512 MB WINDOWS XP SERVER version, tomcat 5.0, All-In-One – Eclipse-311, j2sdk -1.42.

4.2 The Test of Website Function Module

The compiled environment and platform in this course design is Java. The mainly be used is the class socket of class Java. net.

The server can set the port number that the server is listening, which must be between 1024 and 65535, the middle is a Text Field which be used to show the dialogue records. While the contents are greater than length an width of the Text Field, it appears horizontal scroll bar or vertical scroll bar spontaneously, dragging them to view all the contents. The following is a textbox which mainly is to send information from server-side to user-side. The length of information is not restricted, that is to say Chinese or letter both are ok. The lowest line is control button. After clicking "send information" would send the contents in the textbox to user-side, and clear the contents in the textbox up at the same time ,and the text field of the server would show ''The server says…""cut" The server send" bye" to user-side after clicking and then the server is to discontinue listening, closing applied program.

The text which the "Host computer IP" correspond must write your connective target host computer IP, the textbox that the port number correspond write the port number which the applied program monitors in the connective target host computer, and the port number should be between 1024 and 65535 all the same. The following text field and textbox are the same with the function of server: The following order button, after clicking "connect", client sends a host computer IP that for connecting request parameter and a port number that the applied program listen, clicking "send information" and "cut" is same with the function of servers 's accomplish.

1)The function of expand

The client can send a whole document "*。 txt" server. Specific as follows: clicking "send text" In user-side interface shows an open dialogue box, as pic. 3, then choosing the text(*。 txt)that you would send upon then to click ok. The user-side send a" your document is kept in c;\ a. txt" to server . If there is a a. txt document, then cover it. If there is not, then create a. txt document .Thus the user-side can pass a document in native disk to the content in C disk of server, accomplishing the network transmission of the whole document.

2)The working process

The IP of the host computer that lies in the server is "127.0.0.1". The port number that is listened is "8000". Starting the server and user-side applied program successively. After inputting "8000" in the port of server, the server commences to listen 8000 port number .Input "127.0.0.1" in the user-side target host computer IP, the port number is "8000". Click connective button, when the server receives the connective request form user-side, if the connection is allowed, then send a "****the information that connect successfully *****" to user-side, and display in the text field of user-side is to remind clients. After establishing connection both sides can communicate mutually: The client inputs the information that he would like to send in the textbox. After clicking send, the send information(the client says...)would display in the text field of server and user-side. The same, the server inputs the information that he would like to send in the textbox. After clicking information button, the send information(the server says...) would display in the text field of server and user-side. If you would like to send text, click "send text" button and then the opening dialogue box would occur, choosing what you want to send in it is ok. Then the text would be kept in the content of C disk in server, the text field in the server would shows" your text is kept in c:/a. txt" at the same time. If both sides ,no matter which side ,would like to finish connection ,it only need to click' cut" button, then send a information" bye" to the other side ,and then terminate connection, closing applied program.

3)The final effects

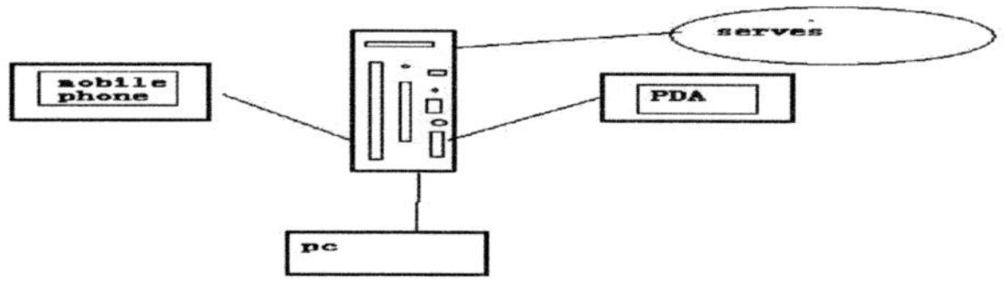

Fig. 2.

On both sides of the above for data transmission process server and client runtime interface, both sides can send information mutually no matter Chinese or characters .It can establish freely and dismantle connection.

5 Summary

The fine characters of J2ME brings great converse for the mobile exploit of cellphone. But due to the limit ness of current internet, mobile instant messaging can not be available to every place in the world like internet. So, the mobile instant messaging that crosses the network, crosses the border has become what people expect. This also brings new chance and challenge for universal mobile operators.

References

1. Geary, D.M.: Java 2 Graphic Design. Swing (United States), vol. II
2. Eckel, B.: Java2 programming ideas (U.S.)
3. Richard Stevens, W.: JAVA Network Programming (U.S.), vol. I
4. Wen, I. (ed.): Java Language and Object-Oriented Programming. Tsinghua University Publishing
5. Shi, J.J.: Chifeng College_Natural Science

Optimization of Bilinear Interpolation Based on Ant Colony Algorithm

Olivier Rukundo[1,*], Hanqiang Cao[1], and Minghu Huang[1]

Department of Electronics and Information Engineering,
Huazhong University of Science and Technology, Wuhan 430074, China
caohq@mail.hust.edu.cn, orukundo@gmail.com

Abstract. This paper proposes a novel method optimizing the bilinear interpolation using Ant Colony Algorithm (ACA). The isotropous-assignment of the old-pixels' average-value to an unknown point (or pixel), in bilinear interpolation, is responsible for many artifacts which corrupt the resulting image features, edges in particular. To correct, we apply the ant colony algorithm to find a reasonable value from all the directions in the neighborhoods of the unknown point. The experimental results show better immanent image features than bilinear interpolation.

Keywords: bilinear interpolation, ant colony algorithm, anisotropic value, optimization.

1 Introduction

The image interpolation algorithms have been widely studied and developed. In our novelty, we propose an optimization method for image interpolation based on ant colony algorithm [1]. The simplest algorithm assigns the nearest neighborhood pixel value to an unknown location or point. The conventional bilinear interpolation algorithm isotropically assigns the average value of the pixels surrounding the randomly chosen unknown location within that neighborhood. Such pixel value assignment is not a good one for an overall output image since that there results in some visible blurring artifacts which are crucial challenges for today's perceived-approximation [2-13]. The optimization of the existing image interpolation algorithms is one of the simplest solutions. In this paper we applied the ant colony algorithm to achieve that. The ant colony algorithm is a novel nature-inspired heuristic applicable to solve problems requiring optimal solutions [14-18]. Ant colony algorithm functions made possible an anisotropic pixels' value assignment. The overall algorithm output better and impressing results than those generated by the conventional bilinear interpolation algorithm.

This paper is organized as follows: Section-II introduces the ant colony algorithm. Section-III presents the proposed optimization procedure. Section-IV shows experimental results and Section-V concludes the proposed method.

---

[*] Mr. Rukundo Olivier, PhD student, Department of Electronics and Information Engineering, Huazhong University of Science and Technology, Wuhan, China.

D. Zeng (Ed.): Advances in Control and Communication, LNEE 137, pp. 571–580.
springerlink.com

2 Ant Colony Algorithm

The development of ant colony algorithm was inspired by the observation of ant colonies. Many researchers are still interested in their foraging behaviors (which results in finding the shortest path between two distant locations). Initially, real ants forage independently from one another. However, when one of the foraging-ants discovers food, it's more likely to return to the nest through the same path leaving chemical pheromones on the ground thus giving that path more chances to be detected and followed by the other foraging-ants.

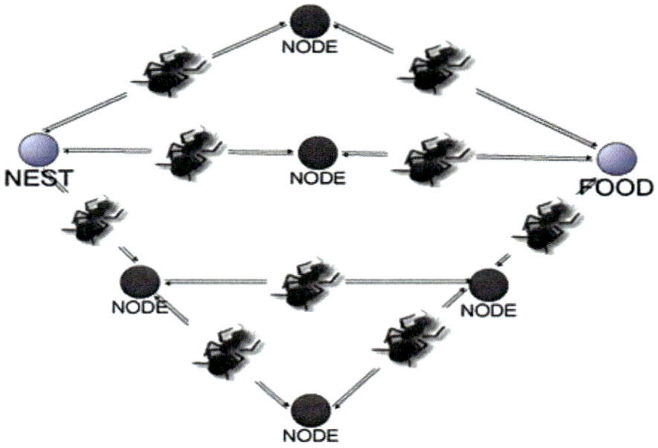

Fig. 1. The process of finding the shortest-path between the ants' nest and food source. An ant starting in the nest can choose the shortest path between the upper, middle and lower parts of the graph.

In general, any of the moving ants moves from nest (x) to food source (y) with the probability

$$p_{x,y}^{(k)} = \frac{(\tau_{x,y})^{\alpha}(\eta_{x,y})^{\beta}}{\sum(\tau_{x,y})^{\alpha}(\eta_{x,y})^{\beta}} \tag{1}$$

where $\tau_{x,y}$ is the amount of pheromone deposited for transition from state x to y, 0 $\leq \alpha$ is a parameter to control the influence of $\tau_{x,y}$, $\eta_{x,y}$ is the desirability of state transition *x,y* (a priori knowledge, typically 1 / $d_{x,y}$, where d is the distance) and $\beta \leq 1$ is a parameter to control the influence of $\eta_{x,y}$. When all ants have completed a solution, the pheromone are updated by

$$\tau_{x,y}^{k} = (1-\rho)\tau_{x,y}^{k} + \Delta\tau_{x,y}^{k} \tag{2}$$

where $\tau^{k}_{x,y}$ is the amount of pheromone deposited for a state transition *x,y*, ρ is the pheromone evaporation coefficient and $\Delta\tau^{k}_{x,y}$ is the amount of pheromone deposited, typically given by

$$\Delta\tau^{k}_{x,y} = \begin{cases} Q/L_k & \textit{if ant } k \textit{ uses curve } x,y \textit{ in its tour} \\ 0 & \textit{other} \end{cases} \quad (3)$$

where L_k is the cost of the kth ant's tour (typically length) and Q is a constant.

3 Optimization Procedure

Assume, we have a gray scale image of the size $M_1 x M_2$ as depicted in the Fig. 2-[a], a group of known value (or main) pixels (N_1, N_2, N_3 and N_4) and an unknown value location P as depicted in the Fig. 2-[b].

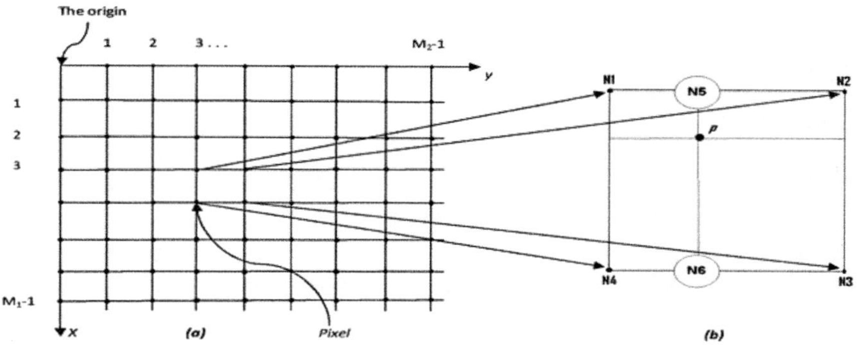

Fig. 2. A gray Scale image (a) and 4 pixels values surrounding an unknown pixel value P in (b)

Now, let's consider each pixel's original-value as a luminous-ray with a certain intensity value. Assume that, the four pixels inside the square are all off simultaneously. In this case, only the closest will omnidirectionally keep emitting the luminous-rays at the location P . Probably, this location will be reached by rays with weaker intensities and an intensity received at the point P is proportional to all the intensity values reaching that point. Now, let's consider each 90 degree angle like receiving the so-called emitted intensities as shown in Fig.3-[L]. For each angle, we used the ant colony algorithm probabilistic function to select properly the closest known-value location, out of the four main pixels neighborhood, containing the most influential pixel value and added it to the main pixel belonging to that angle.

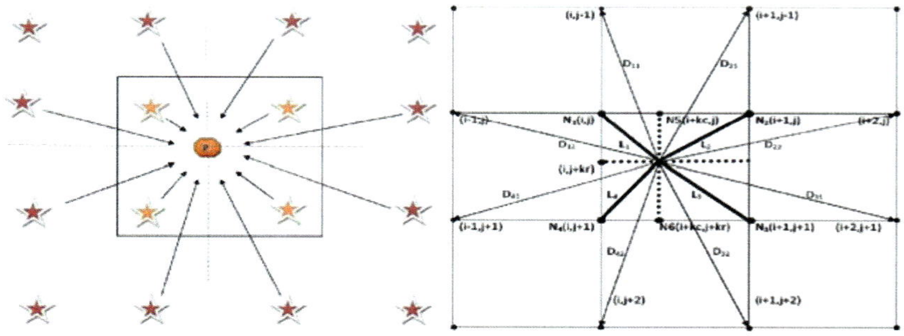

Fig. 3. [L] - The star symbols replaced the pixels locations. Each star emits rays which reach at the unknown value point *P*. [R] - The pixels locations are represented by black dots. The search starts at an unknown point or pixel location and it is performed according to Eq. (7).

The ant colony algorithm availed that smallest potent pixels values V_{aca} overlooked by the conventional bilinear interpolation algorithm. The overall value at an unknown point location is represented by P_{aca} , which, from Fig.3-[R], has coordinates (*kc*[x], *kr*[y]) and can be obtained from the following steps:

3.1 Step One

In this step, we considered (as hint) the conventional bilinear interpolation formulas for $f(N_6)$ and $f(N_5)$ and modified them as shown below

$$f(N_6) \approx \frac{x_2 - x}{x_2 - x_1} f(N_4 + V_{aca4}) + \frac{x - x_1}{x_2 - x_1} f(N_3 + V_{aca3}) \tag{4}$$

where N_6 has $(i + k_c, j + k_r)$ coordinates and;

$$f(N_5) \approx \frac{x_2 - x}{x_2 - x_1} f(N_1 + V_{aca1}) + \frac{x - x_1}{x_2 - x_1} f(N_2 + V_{aca2}) \tag{5}$$

where N_5 has $(i + k_c, j)$ coordinates. Next, we consider the y-direction and obtain the unknown point value formula as shown below

$$f(P_{aca}) \approx \frac{y_2 - y}{y_2 - y_1} f(N_6) + \frac{y - y_1}{y_2 - y_1} f(N_5) \tag{6}$$

At this stage, P_{aca} has (k_c, k_r) coordinates.

3.2 Step Two

In this step, we found the smallest potent value V_{aca} by randomly distributing on a gray scale image of the size M_1 x M_2 a certain number of K ants (Here, $K = [\sqrt{M_1 \times M_2}]$). After the distribution, the ants start searching in each angle direction towards the outside of the 4 connectivity neighborhood the potent pixels values according to

$$p^{(n)}_{(kc,kr),(i\pm(r-1),j\pm(s-1))} = \frac{(\tau^{(n-1)}_{(i\pm(r-1),j\pm(s-1))})^{\alpha}(\eta_{(i\pm(r-1),j\pm(s-1))})^{\beta}}{\sum_{(kc,kr)\in\Omega_{(i\pm(r-1),j\pm(s-1))}}(\tau^{(n-1)}_{(i\pm(r-1),j\pm(s-1))})^{\alpha}(\eta_{(i\pm(r-1),j\pm(s-1))})^{\beta}} \quad (7)$$

where, $\tau^{(n-1)}_{i\pm(r-1),j\pm(s-1)}$ is the pheromone information on the node $(i\pm(r-1),j\pm(s-1))$, $\Omega_{(k_c,k_r)}$ represents the node neighborhood of the ant's nest node (k_c,k_r), $\eta_{(i\pm(r-1),j\pm(s-1))}$ is the heuristic information at $(i\pm(r-1),j\pm(s-1))$. And, α and β are pheromone and heuristic control coefficients. The factors r and s have potential influence on the right search direction and they range between 0 and 1. The heuristic information $\eta_{(i\pm(r-1),j\pm(s-1))}$ is determined at the pixel position $(i\pm(r-1),j\pm(s-1))$ and is equal to

$$\eta_{(i\pm(r-1),j\pm(s-1))} = \frac{I_{(i\pm(r-1),j\pm(s-1))}}{d_{((k_c,k_r),(i\pm(r-1),j\pm(s-1)))}} \quad (8)$$

where, $I_{i\pm(r-1),j\pm(s-1)}$ is the value of pixel intensity at the position $(i\pm(r-1),j\pm(s-1))$. The empirical permissible range of ants' movements Ω could reach 8-connectivity neighborhood and could be controlled by r and s factors.

3.3 Update Procedure

**Update one* is performed after each ant moves out of the square according to

$$\tau^{(n-1)}_{i\pm(r-1),j\pm(s-1)} = \begin{cases} (1-\rho).\tau^{(n-1)}_{i\pm(r-1),j\pm(s-1)}+\rho.\Delta^{(k)}_{i\pm(r-1),j\pm(s-1)} & if\,(i\pm(r-1),j\pm(s-1))\,belongs\ to\ the\ best\ tour \\ \tau^{(n-1)}_{i\pm(r-1),j\pm(s-1)}, & Otherwise \end{cases} \quad (9)$$

where, ρ is the evaporation rate and $\Delta^{(k)}_{i,j}$ is equivalent to the heuristic information.

**Update two* is performed after all unblocked ants complete their tours according to

$$\tau^{(n)} = (1 - \psi) \cdot \tau^{(n-1)} + \psi \cdot \tau^{(0)} \tag{10}$$

where ψ is defined as the decay coefficient.

3.4 Step Three

In this step, the smallest potent value conventionally belonged to each angle direction. In each direction, there were potent pixels (i.e. $N_x^{'}$ $N_x^{''}$ $N_x^{'''}$, etc.). The smallest potent value needed was located within the proximity of the main pixels. The product of the probabilistic function and the potent pixel minus the main pixel value yielded the V_{aca} value as shown by the Eq. (11) .

$$V_{aca} = \begin{cases} [N_x^{'} . P^{(n)}_{((k_c, k_r),(i \pm (r-1),(j \pm (s-1))}] - N_x & \text{if } P^{(n)}_{((k_c, k_r),(i \pm (r-1),(j \pm (s-1)) \neq 0} \\ 0 & \text{otherwise} \end{cases} \tag{11}$$

Here, $V_{aca} > 0$ (always positive)

Note, that, there can be found one or more different potent pixel values ($N_x^{'}$, $N_x^{''}$ etc.) for all x-directions. Here N_x is the main pixel value. The overall value at an unknown location P_{aca} was obtained according to

$$P_{aca} = \frac{f(N_4 + V_{aca4})}{(x_2 - x_1)(y_2 - y_1)}(x_2 - x)(y_2 - y) + \frac{f(N_3 + V_{aca3})}{(x_2 - x_1)(y_2 - y_1)}(x_2 - x)(y_2 - y) + \frac{f(N_1 + V_{aca1})}{(x_2 - x_1)(y_2 - y_1)}(x_2 - x)(y - y_1) + \frac{f(N_2 + V_{aca2})}{(x_2 - x_1)(y_2 - y_1)}(x - x_1)(y - y_1) \tag{12}$$

where $V_{aca(x)} =$ represents the smallest potent value in x-directions.

4 Experimental Results

We did two experiments using Matlab. First, we cropped a gray scale image into four local images (Fig.4-[L]) and interpolated at the ratio n=4. In the second experiment, we interpolated at n=2 four full images (Fig.4-[R]). For that case, we calculated the Peak Signal to Noise Ratio (PSNR) and Signal to Noise Ratio (SNR). The results were shown in Table 1.

4.1 Experiment 1: Local Image Analysis

The aim is to analyze the visual effects and efficiency of our proposed method using small cropped parts of a full Lenna image. The output images were clearer than those interpolated using the conventional bilinear interpolation algorithm.

Fig. 4. [L] - Full gray scale Lenna image of the size 512*512 cropped into four local images of the size 128*128. [R] - (a) Lenna, (b) Lake, (c) Pepper and (d) Cameraman are full gray scale images of the size 256*256 used in this second experiment.

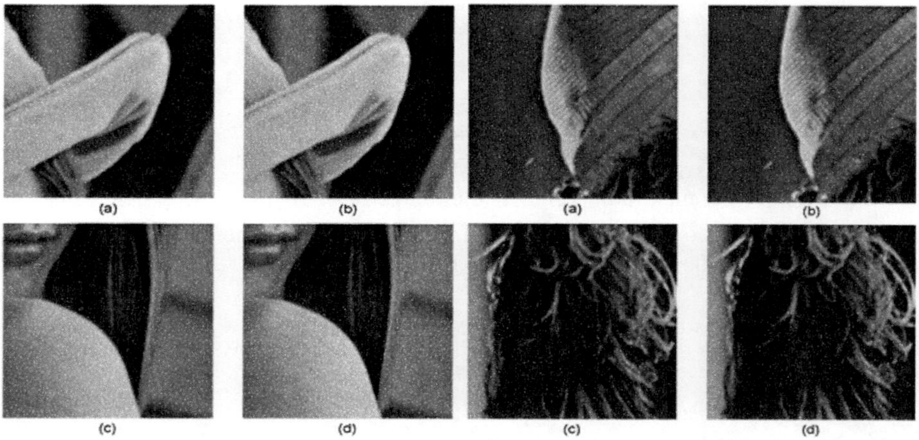

Fig. 5. [L - R] - From the above figures (a), (c) are output images generated by the conventional bilinear algorithm. (b), (d) are output images generated by the proposed method.

4.2 Experiment 2: Full Image Analysis

Unlike the previous experiment, we used n=2 and full images of 256*256. A numerical analysis was also used to show more clearly the difference between the proposed method and conventional bilinear interpolation.

A higher Peak Signal to Noisy Ratio (PSNR) or Signal to Noise Ratio (SNR) would normally indicate the higher quality. It is most easily defined via the Mean Squared Error where one of the monochrome images I and K is considered as a noisy approximation of the other.

Fig. 6. [L - R] - From the above figures (a), (c) are output images generated by the conventional bilinear algorithm. (b), (d) are output images generated by the proposed method.

$$MSE = \frac{1}{mn}\sum_{i=0}^{m-1}\sum_{j=0}^{n-1}\left[I(i,j) - K(i,j)\right]^2 \tag{13}$$

The PSNR is defined as:

$$PSNR = 10 \cdot \log_{10}\left(\frac{MAX_I^2}{MSE}\right) = 20 \cdot \log_{10}\left(\frac{MAX_I}{\sqrt{MSE}}\right) \tag{14}$$

where, MAX_I is the maximum value of the image pixel. Typical the PSNR values in lossy image and video compression range from 30 to 50 dB. When the two images are identical, the MSE=0. In that case, the PSNR is undefined.

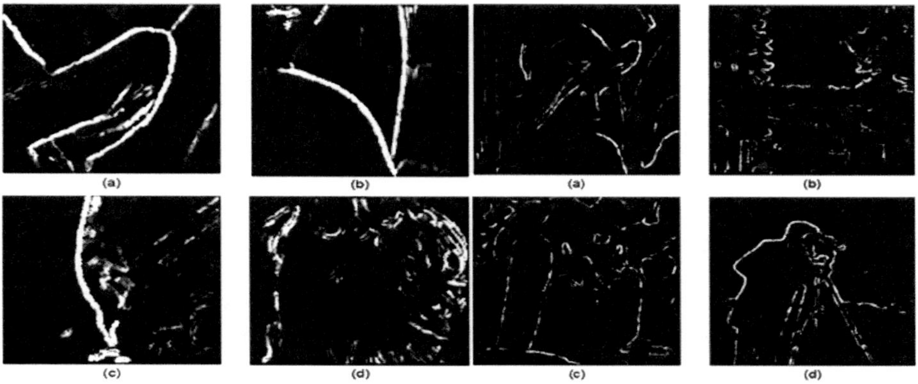

Fig. 7. [L - R] - (a), (b), (c) and (d) images show again the differences between original images and images interpolated using the conventional bilinear interpolation and the proposed method

Table 1. We used an error free image to calculate the PSNR and SNR between the both methods

Images	Bilinear Interpolation		Proposed method	
	PSNR	**SNR**	**PSNR**	**SNR**
Lenna	23.8437	10.1784	23.8561	10.1907
Lake	22.1675	10.3704	22.1684	10.3714
Pepper	24.9382	11.4067	24.9443	11.4128
Cameraman	23.5519	11.2683	23.5662	11.2826

As evidenced in the above Table 1, the proposed optimization method shows better results than bilinear interpolation.

5 Conclusion

In this paper, the method optimizing the conventional bilinear interpolation based on ant colony algorithm was presented and demonstrated. Since, the conventional bilinear interpolation isotropically yields a big number of blurring edge artifacts; we applied the ant colony algorithm functions to find the smallest potent neighborhood-wide pixels values overlooked by the conventional bilinear interpolation algorithm with aim to obtain an optimal value at an unknown location. The proposed method performs better than conventional bilinear interpolation. The results showed in Table 1.evidenced our conclusion.

Acknowledgments. This work was supported by the National Anti-counterfeit Engineering Research Centre and Laboratory for Information Security and Identity Research Team of Huazhong University of Science and Technology, Wuhan, China. Authors would like to use this opportunity to thank the reviewers for their helpful comments.

References

1. Dorigo, M., Stutzle, T.: Ant Colony Optimization. MIT Press, MA (2004) ISBN: 0262042193
2. Kim, Jin, Yang, Jeong: Enhanced Edge-Weighted Image Interpolation Algorithm. In: Eighth International Conference on Hybrid Intelligent Systems (2008)
3. Wittman, T.: Mathematical Techniques for Image Interpolation. Department of Mathematics, University of Minessota
4. Luong, H.Q., De Semet, P., Philips, W.: Image Interpolation Using Constrained Adaptive Contrast Enhancement Techniques (2005)
5. Shi, Z., Yao, S., Li, B., Cao, Q.: A novel Image Interpolation Technique Based on Fractal Theory. IEEE Computer Society (2008)
6. Ledda, A., Luong, H.Q., Philips, W., De Witte, V., Kerre, E.E.: Image Interpolation Using Mathematical Morphology. IEEE Computer Society (2006)
7. Ni, K.S., Nguyen, T.Q.: An Adaptable k-Nearest Neighbors Algorithms for MMSE Image Interpolation. IEEE Transactions on Image Processing 18(9) (September 2009)

8. Aftab, H., Mansoor, A.B., Asim, M.: A New Single Image Interpolation Technique for Super Resolution. In: Proceedings of the 12th IEEE International Multitopic Conference (2008)
9. Kim, H., Park, S., Wang, J., Kim, Y., Jeong, J.: Advanced Bilinear Image Interpolation Based On Edge Features. In: First International Conference on Advances in Multimedia. IEEE (2009)
10. Hsu, C.-T., Chen, M.J., Huang, C.-H.: An Edge-Based Interpolation Algorithm for Resolution Enhancement at Low Bit-Rate. In: The 2004 IEEE Asia-Pacific Conference on Circuits and Systems, December 6-9 (2004)
11. Li, X., Orchard, M.T.: New Edge-Directed Interpolation. IEEE Transaction on Image Processing 10(10) (October 2001)
12. Chen, M.-J., Huang, C.-H., Lee, W.-L.: A Fast edge oriented Algorithms for Image Interpolation. Elsevier Image and Visual Computing 23 (2005)
13. Miklos, P.: Comparison of Convolutional Based Interpolation Technique in Digital Image Processing. In: 5th International Symposium on Intelligent Systems and Informatics, SISY (2007)
14. Li, L., Ren, Y., Gong, X.: Medical Image Segmentation Based on Modified Ant Colony Algorithm with GVF Snake Model. IEEE Computer Society (2008)
15. Laptik, R., Navakauskas, D.: Application of Ant Colony Optimization for Image Segmentation. Department of Electronics Systems, Vilnius Gediminas Technical University 8(80) (2007) ISSN 1392-1215
16. Yan, Z., Gu, H.-M.: Research of the Image Segmentation based on Ant Colony Algorithms. IEEE Computer Society (2009)
17. Tian, J., Yu, W., Xie, S.: An Ant Colony Optimization Algorithms for Image Edge Detection. In: 2008 IEEE Congress on Evolutinary Computation, CEC 2008 (2008)
18. Otsu, N.: A Threshold Selection Method from Gray-Level Histograms. IEEE Trans. Systems, Man and Cybernetics SMS-9(1) (January 1979)

Parallel Artificial Fish Swarm Algorithm

Mingyan Jiang and Dongfeng Yuan

School of Information Science and Engineering, Shandong University,
250100 Jinan, China
{jiangmingyan,dfyuan}@sdu.edu.cn

Abstract. Artificial fish swarm algorithm (AFSA) is one of the novel optimization algorithms based on swarm intelligence which is developed in recent years. As a nature inspired algorithm, AFSA has obtained prominence in real world optimization problems in many fields. We first show the theory research and applications of AFSA in different subjects, such as quantum theory, chaotic theory and simulated annealing, etc. Also many other improvements are also revealed, such as designing new behaviors and investigating the global information of AFSA. Considered the modifications in recent years to increase the search quality of AFSA, this computational intensive algorithm could be realized in the way of parallelization. We discuss the possibility of parallelization of AFSA and make the analysis of parallel AFSA, and the detailed implementation methods for parallel AFSA are given in the form of pseudo code with complete explanation.

Keywords: parallel computing, improved AFSA, standard parallel AFSA, decomposed parallel AFSA.

1 Introduction

Artificial fish swarm algorithm (AFSA) is developed by Li [1] in 2002, it is a form of swarm intelligence in which the behavior of a biological social system like a school of fish is simulated [2][3]. AFSA is also one of the advanced evolutionary computation methods and like other evolutionary algorithms AFSA need not to consider the mathematical conditions, such as the convexity of the feasible set, the continuity and the differentiability of the objective functions. The standard AFSA is iterated by evaluating the basic fish behaviors according to the objective function value which decides the moving direction of the whole population in next step.

AFSA has many advantages, such as good robustness, global search ability, tolerance of parameter setting, and it is also proved to be insensitive to initial values.

AFSA has appeared in many applications in the fields of traveling salesman problem, production scheduling, graphic division, function optimization, machine learning and so on, in which good performances are obtained respectively. In paper [4], a novel hybrid evolutionary algorithm based on AFSA and particle swarm optimization algorithm (PSO) is proposed to realize feed forward neural network training. A new PID parameters optimization scheme based on AFSA is applied to single loop control system and main steam temperature cascade control system in

D. Zeng (Ed.): Advances in Control and Communication, LNEE 137, pp. 581–589.
springerlink.com

paper [5]. In order to speed up the segmentation procedure and solve the problem of noise-sensibility in image segmentation, the paper [6] suggests a fast synthetic aperture radar (SAR) image segmentation method, which integrates stationary wavelet transform (SWT) and AFSA. Based on AFSA and combined with genetic algorithm (GA) and simulated annealing (SA) algorithm, a hybrid algorithm is developed in the paper [7] to solve aircraft landing sequence problem. The paper [8] proposes some improvements for AFSA aiming to manage robot tasks to minimize the deviation between the resource requirements and stated desirable levels. To overcome the problems of parameter selecting for support vector machine, the paper [9] proposes the AFSA to choose the parameters of least squares support vector machine (LS-SVM) automatically in time series prediction.

Though the standard AFSA has been achieving more and more successful engineering applications, there still are many problems to be solved to the new intelligence algorithm. Many researchers have proposed their improvements for standard AFSA aiming to cut down the computation load, avoid falling into the local solution set and have enhanced the performance by coupling AFSA with other methods or algorithms.

The leaping behavior [3] is proposed to increase the probability to leap out local extremes. If the objective function is almost the same or the difference of the objective function values is smaller than a proportion during the given iterations, choose some fish randomly in the whole population and set parameters randomly to the selected individuals. The swallowing behavior [10] is executed if the fitness function value is bigger (for minimum optimization) than a given threshold in computing process of AFSA. The fish with worse objective function value will be swallowed to save the memory space so as to speed up iteration, in which the complexity will be cut down obviously especially for the large scale problem.

However, the above improvements mainly focus on how to use the local and individual information to search the better positions while ignoring the global information which may affects the search performance dramatically. Since AFSA is a new developed optimization method, it is still in its developmental infancy and further studies are necessary. For example, the optimizing precision and convergence speed of the standard AFSA is not satisfactory in the later search stage. In order to overcome the drawback of AFSA, the paper [11] proposes a new AFSA based on global information which could converge to the global optimum with fast speed and high precision.

With the fast development of different fields, the merging of different subjects has been focused on more and more especially in the background of economic globalization. Many new concepts and emerging theories are introduced into standard AFSA continually which improve the performance of AFSA greatly.

Since chaotic [12][13] mapping enjoys certainty, ergodicity and stochastic property, chaotic search can serve as a kind of method for global optimization. Feedback can also act as a strategy to lead the movement of artificial fish. A new AFSA with chaotic search and feedback strategy is proposed in the paper [14] which can overcome the drawback of AFSA that falling into local extreme values.

In order to improve the global search ability and the convergence speed of AFSA, a novel quantum AFSA based on the concepts and principles of quantum computing, such as the quantum bit and quantum gate, is proposed in the paper [15]. The position of the artificial fish is encoded by the angle in $[0,2\pi]$ based on the qubit's polar coordinate representation in the 2-dimension Hilbert space [16][17][18].

Besides of the quantum theory, another method is also introduced into AFSA to maintain the local search ability and global search ability. The paper [19] presents a novel stochastic approach called the simulated annealing-artificial fish swarm algorithm for solving some multimodal problems. The proposed algorithm incorporates the simulated annealing [20] into AFSA to improve the performance of standard AFSA. Not like the paper [21] in which mutation operator is added to AFSA to escape from the local area which also leads to more computation load, the paper [19] makes some improvements by using new behaviors in the paper [3][10] and other technicals to cut down the computing complexity.

However, there are many other practical engineering problems that not only the global optimum is needed but also the other sub-optimum or local extremes are needed, which is called niche problem [22]. In the paper [23], the quantum mechanism is introduced into the AFSA to increase the diversity of species. The whole population is divided into several sub-populations to form the niche and the restricted competition selection strategy (RCS) [24] is used to maintain the niche. The results in [23] show that the quantum AFSA is effective to solve niche problem.

Similar to niche problem, the multiobjective optimization problems [25] are also very complicated in many fields. The paper [26] proposes a new AFSA which converges to the Pareto front with fast speed and high precision by investigating the global information of AFSA in the paper [11]. The proposed method could maintain the diversity and wide distribution of the feasible set in the Pareto front, which provides a new method for multiobjective optimization problems.

The above discussion shows that the AFSA has achieved better performances in many fields; however, most of the current researches have been limited in the serial computing field. In the case of single processor, some complicated problems can not be computed instantly and effectively using AFSA which can not meet the practical demand, for example the weather forecast computing needs to deal with a great amount of data. To solve the problem, parallel computing may be a good idea. As parallel algorithms are distributed computing, many practical problems that require strict executing time can be solved effectively. As AFSA is computationally intensive, complex system can be alleviated through parallel processing using parallel AFSA. In this paper, we extend the standard AFSA and modify it to provide both a synchronous and asynchronous parallel AFSA, the detailed implementation methods are also given in the following parts.

2 The Aims of AFSA's Parallelization

Parallel machines are divided into six categories, they are single instruction multipe data machine (SIMD), parallel vector processing machine (PVP), symmetric multiprocessor (SMP), massively parallel processing machine (MPP), cluster of workstations (COW), and distribution shared storage multiprocessor (DSM) respectively. Actually, the performance of the parallel and serial algorithms or the comparisons of several parallel algorithms are evaluated by using their speedup ratio, efficiency and utility. Among various parallel programming environment, Massege

Passing Interface (MPI) standard and Parallel Virtual Machine (PVM) environment are widely applied. To learn the programming technique of distributed parallel calculation using PVM and MPI environment, you can refer to site [27].

AFSA is a new swarm intelligence optimization algorithm using simultaneously optimization method with multiple parameters and multiple individuals, which is developed by imitating the behaviors of fish swarm. Considering the theory research and real word applications in current years, AFSA has demonstrated great ability in solving the optimization problems in complex system, especially for the non-deterministic polynomial (NP) combinatorial optimization problems and large scale, non-linear complicated optimization problems.

Since the standard AFSA makes every individual fish as an operand and the operations on each individual are independent with each other to some degree, the parallel structure is inherent in AFSA. Though the individual processing model implies the parallelization, the actual process and the realization of AFSA still have the serialization in essence. In solving some practical problems, the process of AFSA may last a long duration and even fails to meet the convergence speed requirement due to the operations on each individual fish, especially in the calculations of the fitness function values for each fish with a large population size. Considering this issue, the parallelization for AFSA arises great interest in theory research and applications. So, it is very necessary to develop some new parallel AFSAs based on various parallel machines or local area network in order to improve the computing speed of AFSA and to meet the requirements in practical engineering applications.

3 Analysis and Implementation for Parallel AFSA

3.1 Parallelization Analysis of AFSA

AFSA is an iterative optimization process in which new candidate solutions are generated through evaluating the basic behaviors of fish and the loop is terminated when the ending conditions are meet. Aims to realize the algorithm parallelization, some directions and improvements can be achieved in the following parts according to the standard AFSA process.

(1). Parallelization of fitness function evaluations

Though the fitness function evaluations for the fish swarm often last for a long time, the methods of parallel fitness function evaluations for individual can be further developed to improve the computing efficiency through some research work. The possibility of realizing parallelization and its computing efficiency rely on not only the expression form of fitness function but also the research and development of different numerical calculation parallel algorithms. Since the process of fitness function evaluation for every individual is independent with each other, it provides the possibility that the evaluation operations or other computing process for different individuals can be distributed in different machines and be computed simultaneously.

(2). Parallelization of subpopulations.

The basic fish behaviors can be operated independently for each individual and a large scale population can be divided into many subpopulations which will be distributed in different machines or processors to improve the computing efficiency.

Sometimes, the information exchange strategy will be adopted between different machines (or different subpopulations) to implement the search process together, in which the search process of different individuals and different subpopulations are also parallel. The best performance can be achieved by coordinating the information exchange strategies among the subpopulations on different machines. This method receives more and more attention for that it can be implemented simply in the parallel machines and local area network.

3.2 The Classification of Implementation Methods

The parallel AFSAs based on parallel machines or local area network includes: standard parallel AFSA and decomposed parallel AFSA.

(1). The standard parallel AFSA searches in a whole environment and needs to utilize an united population. In the search process, besides of the necessary global memory, a special functional agent is designed to coordinate the whole population and control the communication process in the population. Even though we have a good coordinating strategy, this kind of implementation method can not increase the computing speed too much due to the time consuming of coordinating strategy for the whole population.

(2). The main idea of decomposed parallel AFSA is that the original fish population is divided into many subpopulations which will be distributed to different machines. Each subpopulation executes the standard AFSA and exchanges the information with other subpopulations from different machines. This method can be implemented simply for that it takes full advantage of the inherent parallelization and the swarm attribute of AFSA. There are three grouping models according to the simulation of different living environments and the animals' survival characteristics. They are respectively the stepping stone model, island model and neighborhood model, consequently there are three parallel AFSAs based on the three grouping models.

4 Implementation Methods

4.1 Implementation Methods for Standard Parallel AFSA

This method doesn't change the structure of basic AFSA which means that each individual fish still searches the best solutions in the uniform environment. The basic starting point is to develop the parallelization of each individual's search process. In the whole search using AFSA, the basic fish behaviors and fitness function evaluation are executed independently for each individual fish. Thus, each individual fish's search process can be realized, by adopting the parallel processing machines with shared memory, to increase the computation speed of AFSA. Commonly, this method is fit to solve the problems with large amount of evaluations for fitness function.

```
The pseudo code:
Begin
    Initialize the population.
    While running do
      for =1 to N do
         Calculate the fitness function value for fish i (parallel part).
      Execute the basic fish behaviors.
      Generate the new individuals.
    End while
End
```

where N denotes the population size, for =1 to N do is the parallel executable statement. Character i presents the index number of the processing machine while character N also means the number of machines. The master-slave parallelization method is used where the basic fish behaviors are executed by the host processor using a serial method. Considering the fitness function value of each individual is computed parallelly, the computing speed can be greatly increased if the communication cost among the processors is low. However, the efficiency of standard parallel AFSA is still very low due to the demand of synchronous mechanism among the processors. In order to avoid the host processor waiting for other subprocessors under the synchronous mechanism, asynchronous parallel method may be a good choice for that the host processor needn't to wait for the subprocessors with slow computing speed. Hence the new individuals on different subprocessors are not generated synchronously.

In the implementation process, the host processor must obtain every individual's fitness function values and the comparison operations are carried out according to the full information of the fitness function values, partly rely on the communication requirements.

4.2 Implementation Methods for Decomposed Parallel AFSA

The idea of decomposed parallel AFSA is to divide the whole population into several subpopulations so that it can work well in the environment with not enough processors. All the subpopulations are distributed to different processing machines so that each subpopulation can run the AFSA independently in the way of parallelization. The characteristic of this method is to investigate the parallelization of the population search from the global angle by information sharing mechanism, in which the best individuals among the subpopulations emigrate to their neighbors after some iterations. This method has changed the basic structure of AFSA for that the search process is not executed by the whole population using the same search mechanism but the groups of subpopulations using their own search mechanisms independently. Allowing for that the decomposed parallel AFSA distributes the subpopulations to the local memories on each processor to carry out the search process independently; it can get the approximate linear speedup ratio and increase the computing speed effectively. This method also can successfully avoid the convergence to the local extremum too early for that it maintains the local characteristics of each subpopulation distributed on each subprocessor. The ways of grouping the whole population and exchanging information should be considered carefully when designing the decomposed parallel AFSA.

The pseudo code:

```
Begin
    Initialize one population and divide it into P subpopulations.
    For 1 to P do
      Initiation
      Evaluate the fitness function values for the first generation.
      While condition do
        for j=1 to G(generations) do
          Execute the basic fish behaviors.
            Calculate the fitness function value for each individual fish.
        Select the individuals to be migrated.
        do step (1) and (2) in parallel
          (1) Send the migrated individuals.
          (2) Receive the immigrated individuals.
       End while
    End for
End
```

where P presents the number of processors and also denotes the population size. Sending and receiving the migrated individuals are executed simultaneously, and how to design the migration model is a key factor that affects efficiency of the algorithm. The migration mode includes synchronous mode and asynchronous mode: in synchronous mode, the migration is happened periodically; while in asynchronous mode, the migration happens when the condition meets the threshold value, hence it is very important to execute the migration operation at the right time. Besides, the performance of decomposed parallel AFSA is related with the topographic connection ways (static, dynamic, connection radius) of the parallel processing machines.

When there are enough processors in the whole environment, the parallel AFSA will achieve the biggest parallelization if each individual is allocated one processor to implement the search process independently with each other.

The pseudo code:

```
Begin
   Initialize one population and distribute the population to P=N processors.
   For 1 to N do
     While running do
       do step (1) and (2) in parallel
         (1) Send the migrated individuals.
         (2) Receive the immigrated individuals.
       Evaluate the fitness function value for immigrated individuals.
       Execute the basic fish behaviors.
       Calculate and compare the fitness function values for each individual fish;
         then generate the new individuals.
     End while
   End for
End
```

where the fish population size N equals to the number of processors P.

5 Conclusion

The current research and development of AFSA have been previewed; both the improvements on the parameter setting, introducing new behaviors, investigating global information and the new hybrid algorithms with other methods are presented to give researchers new hints. The main contribution of this paper is the detailed investigation of the parallel AFSA based on the previous modification on the algorithm, including the implementation methods and the classification of parallel AFSA. Optimizing quality of AFSA partly depends on population size, large number of fish leads to better results. But more fish also means more objective function evaluations and longer computing time. Using effective communication scheme, the distributed processors can reduce the computing time greatly when applying the parallel AFSA. Parallel computing also enhances the local area search ability, improves the search precision in later period and avoids the premature of the whole population. What can be forecast is that parallel AFSA will be applied to more complex systems successfully in real word due to its advanced advantages over traditional methods. As a future work, we are exploring the dynamically changing AFSA parameters to track the real statement of some complex dynamic systems. Some experiments using parallel AFSA (for example, synchronous AFSA, asynchronous AFSA) will be tested in several architectures, such as shared memory machine and distributed memory machine using MPI.

Acknowledgments. This work is supported by the Natural Science Foundation of Shandong Province (No.ZR2010FM040), key provincial project (No. 2009ZHZX1A1401, No. 2010ZHZX1A1001).

References

1. Li, X.L.: A New Intelligent Optimization-Artificial Fish Swarm Algorithm. In: PhD thesis, Zhejiang University, China (June 2003)
2. Zhu, K.C., Jiang, M.Y.: The Optimization of Job shop Scheduling Problem based on Artificial Fish Swarm Algorithm with tabu search strategy. In: Third International Workshop on Advanced Computational Intelligence, pp. 323–327 (2010)
3. Jiang, M.Y., Wang, Y.: Spread Spectrum Code Estimation by Artificial Fish Swarm Algorithm. In: IEEE International Symposium on Intelligent Signal Processing, pp. 1–6 (2007)
4. Chen, X.J., Wang, J.Z., Sun, D.H., Liang, J.Z.: A Novel Hybrid Evolutionary Algorithm Based on PSO and AFSA for Feedforward Neural Network Training. In: Wireless Communications, Networking and Mobile Computing, pp. 1–5 (2008)
5. Luo, Y., Zhang, J.T., Li, X.X.: The Optimization of PID Controller Parameters Based on Artificial Fish Swarm Algorithm. In: Automation and Logistics, pp. 1058–1062 (2007)
6. Ma, M., Liang, J.H., Sun, L., Wang, M.: SAR Image Segmentation Based on SWT and Improved AFSA. In: Intelligent Information Technology and Security Informatics (IITSI), pp. 146–149 (2010)
7. Wang, F., Xu, X.H., Zhang, J.: Strategy for aircraft sequencing based on artificial fish school algorithm. In: Control and Decision Conference (CCDC 2008), pp. 861–864 (2008)

8. Tian, W.J., Liu, J.C.: An Improved Artificial Fish Swarm Algorithm for Multi Robot Task Scheduling. In: Natural Computation (ICNC 2009), vol. 4, pp. 127–130 (2009)
9. Chen, X.J., Wang, J.Z., Sun, D.H., Liang, J.Z.: Time Series Forecasting Based on Novel Support Vector Machine Using Artificial Fish Swarm Algorithm. In: Natural Computation (ICNC 2008), vol. 2, pp. 206–211 (2008)
10. Cheng, Y.M., Jiang, M.Y., Yuan, D.F.: Novel Clustering Algorithms Based on Improved Artificial Fish Swarm Algorithm. In: Proceedings of the 6th International Conference on Fuzzy Systems and Knowledge Discovery (FSKD 2009), pp. 141–145 (2009)
11. Jiang, M.Y., Yuan, D.F., Cheng, Y.M.: Improved Artificial Fish Swarm Algorithm. In: 5th International Conference on Natural Computation (ICNC 2009), pp. 281–285 (2009)
12. Wang, L., Zheng, D.Z., Lin, Q.S.: Survey on chaotic optimization methods. Comput. Technol. Automat., 1–5 (2001)
13. Bucolo, M., Caponetto, R., Fortuna, L., Frasca, M., Rizzo, A.: Does chaos work better than noise? Circuits and Systems Magazine, 4–19 (2002)
14. Zhu, K.C., Jiang, M.Y.: An Improved Artificial Fish Swarm Algorithm Based On Chaotic Search And Feedback Strategy. In: The International Conference on Computational Intelligence and Software Engineering, pp. 1–4 (2009)
15. Zhu, K.C., Jiang, M.Y.: Quantum Artificial Fish Swarm Algorithm. In: The 8th World Congress on Intelligent Control and Automation, pp. 1–5 (2010)
16. Sun, J., Xu, W.B., Feng, B.: A global search strategy of quantum behaved particle swarm optimization. In: Proceedings of IEEE Conference on Cybernetics and Intelligent Systems, pp. 111–116 (2004)
17. Li, P.C., Li, S.Y.: Quantum ant colony algorithm for continuous space optimization. Control Theory and Applications 25, 237–241 (2008) (in Chinese)
18. Hey, T.: Quantum computing: An introduction. Comput. Control. Eng. J. 10(3), 105–112 (1999)
19. Jiang, M.Y., Cheng, Y.M.: Simulated annealing artificial fish swarm algorithm. In: 8th IEEE World Congress on Intelligent Control and Automation, pp. 1590–1593 (2010)
20. Singh, H.K., Ray, I.A.: A simulated annealing algorithm for constrained Multi-Objective Optimization. In: IEEE congress on Evolutionary Computation, pp. 1655–1662 (2008)
21. Zhang, M.F., Shao, C., Gan, Y.: Hybrid Artificial Fish Swarm Optimization Algorithm Based on Mutation Operator and Simulated Annealing. Acta Electronica Sinica 34, 1381–1385 (2006)
22. Gu, J.H., Li, N.N., Tan, Q., Wei, W.: A Novel Niche Genetic Algorithm with Local Search Ability. In: IEEE Congress on Evolutionary Computation (2007)
23. Zhu, K.C., Jiang, M.Y.: Niche Artificial Fish Swarm Algorithm Based on Quantum Theory. In: 10th IEEE International Conference on Signal Processing, pp. 1425–1428 (2010)
24. Lee, C.G., Cho, D.H., Jung, H.K.: Niching Genetic Algorithm with Restricted Competition Selection for Multimodal Function Optimization. IEEE Transactions on Magnetics 35(3) (1999)
25. Deb, K., Pratap, A., Agarwal, S., Meyarivan, T.: A fast and elitist multiobjective genetic algorithm: NSGA-II. IEEE Transactions on Evolutionary Computation, 182–197 (2002)
26. Jiang, M.Y., Zhu, K.C.: Multiobjective Optimization by Artificial Fish Swarm Algorithm. In: 2011 IEEE International Conference on Computer Science and Automation Engineering, pp. 506–511 (2011)
27. MPI documents, `http://www.mpi-forum.org`

A Modified Discrete Shuffled Flog Leaping Algorithm for RNA Secondary Structure Prediction

Juan Lin, Yiwen Zhong, and Jun Zhang

College of Computer and Information Science,
Fujian Agriculture and Forestry University, Fuzhou, China
sannuo@126.com

Abstract. A modified discrete shuffled flog leaping algorithm is designed for the RNA secondary prediction problem. According to the characteristics of RNA folding, new search space and individual location updating rules are redefined to search the RNA secondary structure with minimal free energy in the combinatorial space of stems. The algorithm introduces an iteration improvement strategy to get the real-time update for global optimal solution. The simulation results comparing with two typical algorithms and its original version show that the proposed algorithm has good performance.

Keywords: RNA secondary structure prediction, Shuffled flog leaping algorithm, Iterative Improvement, Minimal free energy, Combinatorial optimization of stems.

1 Introduction

RNA is one of the most important elements in the biological system. A large amount of structural information is available, while it's not easy for experiment methods such as X-ray crystallography and NMR to determine RNA 3D-structures, both of them are expensive and time consuming. Research indicates it is hard to get RNA tertiary structure directly from the primary structure. So determining the secondary structure of RNA often constitutes an essential step towards predicting the tertiary structure [1].

Different algorithms approaches exist for RNA secondary prediction. Among them, the most accurate ones are comparative sequence analysis methods which require a collection of homologuous sequences and hence they are not applicable to the prediction of many novel sequence structures. Thermodynamic folding algorithms is an ab initio prediction method, starting from a single sequence, calculating its generated by base pairing to predict the minimal free energy(MFE) RNA secondary structure, which are more practical. Moreover, dynamic programming(DP) algorithms are the most widely used methods to compute the optimal MFE secondary structure, which are computation-intensive and hence expensive in probing RNA folding dynamics over long time scales.

In contrast to DP, combinatorial optimization algorithms can simulate the actual folding process of an RNA sequence and therefore achieve higher prediction rates of base pairs. Among those several procedures using intelligent optimization algorithm

D. Zeng (Ed.): Advances in Control and Communication, LNEE 137, pp. 591–599.
springerlink.com

based on population for RNA secondary structure prediction have been proposed. These methods find the optimal stem combination whose corresponding secondary structure with MFE.

The typical algorithm is Genetic Algorithm (GA)[2-4]. Some proposed particle swarm optimization (PSO). Marais Neethling et al. introduced set-based PSO[5]. Michael Geis et al. combined with a GA that found a good starting population for new PSO approach called HelixPSO[6]. Jun Yu et al. used the modified ant colony algorithm[7] and He Jingyuan et al. applied a self-adaptive artificial fish swarm optimization algorithm for RNA secondary structure prediction [8].

The shuffled frog-leaping algorithm (SFLA) was proposed by Muzaffar Eusuff et al. for solving combinatorial optimization problems[9]. Since SFLA is simple in concept, easy to implement, which has fewer control parameters, it has been widely used in many fields. SFLA has applied successfully to numerical optimization problems[10-11].The experiments show that the extended version of SFLA has a good performance. Xia Li et al proposed a hybrid scheme that combines the merits of extremal optimisation, exhibited strong robustness and fast convergence for high-dimensional continuous function optimisation[12].Tacher Niknam et al. combined the self-adaptive particle swarm optimization to find optimal configuration of distribution network[13].

This paper proposes a new SFLA approach called DSFLA for finding RNA Structures with MFE. Based on DSFLA, a modified approach called MDSFLA is also present. The free energy is calculated with the Individual Nearest Neighbour-Hydrogen Bond (INN-HB) model. DSFLA and MDSFLA are compared to two algorithms RNAStructure[14] and pknotsRG[15], which are considered to be the benchmark in RNA folding algorithms.

The rest of paper is organized as follow. A short overview over RNA secondary structure is given in Section 2. A description of SFLA algorithm is given in Section 3. A new DSFLA algorithm and its modified algorithm MDSFLA are introduced in Section 4. The simulation results obtained are presented and discussed in section 5. Finally, the conclusion is in Section 6.

2 RNA Secondary Structures

A secondary structure, S, on an RNA sequence, $R=r_1, r_2, r_3...r_n$, is a set of base pairs (BPs) including GC, CG, AU, UA, GU or UG. These BPs are called canonical BPs or Waston-Crick BPs. Although other BPs are possible and do exist, they are not as common and are not usually considered. Then a BP between nucleotide r_i and r_j ($i<j$) is denoted by $i.\ j$. A few constraints are imposed: i) for each BP $i.\ j$, j-$i\geq 3$ and ii) two BPs, $i.\ j$ and $i'.\ j'$, assuming $i<i'$, are either $i<i'<j'<j$ or $i<j<i'<$j', means pseudoknots are prohibited.

A main element of secondary RNA structures is stems, which are sets of two or more adjacent BPs. Thus, a stem of size $k\geq 2$ consist of k BPs with indices (i, j), (i+1, j-1)..., (i+k, j-k) where $1\leq i<i+k<j-k<j<n$ and n is the length of the RNA sequences. An RNA secondary structure is defined by a set of stems. So the algorithm can start with computing the set Ω of all possible stems of an RNA molecule and then try to

find a subset $\Omega_{min} \subseteq \Omega$ that defines an optimal secondary structure which is the MFE structure.

3 SFLA

Consider a group of frogs leaping in a swamp. The swamp has a number of stones at discrete locations that the frogs can leap onto. The frog's goal is to find the stone with the maximum amount of available food as quickly as possible by improving their memes. The sample of virtual frogs constitutes a population that is partitioned into subset referred to as memeplex. Each frog searches the space in different directions in parallel. Frog leaping improves an individual's meme and enhances its performance towards the goal. The different memeplexes are considered as different cultures of frogs, each performing a local search. Within each memeplex, the individual frogs hold ideas, that can be influenced by the ideas of other frogs, evolve through a process of memetic evolution. After a defined number of memetic evolution steps, ideas are passed among memeplexes in a shuffling process. The local search and the shuffling processes continue until defined convergence criteria are satisfied.

First of all, the SFLA randomly generates initial *SN* solutions of population, each X_i (i=1, 2…*SN*) is a vector of *D* dimension. The flogs are sorted in a descending order according to their fitness. Then, the entire population is divided into *m* memeplexes, each containing *n* frogs (i.e. $SN=m\times n$). In this process, the first frog goes to the first memeplex, the second frog goes to the second memeplex, frog *m* goes to the *m*th memeplex, and frog *m*+1 goes back to the first memeplex, etc.

Within each memeplex, record the frog with best fitness as X_b and worst frog as X_w. Also, the frog with the global best fitness is identified as X_g. The step and new position are adjusted for the frog X_b in the memeplex by

$$\text{Step size } D_i = rand() \times (X_b - X_w)\,. \quad (1)$$

$$\text{New position } X_w = \text{current position } X_w + D_i; \qquad D_{\max} \geq D_i \geq -D_{\max}\,. \quad (2)$$

where *rand()* is a random number in the range [0, 1] and D_{max} is the maximum step size allowed change in a frog's position. If this process produces a better solution, it replaces X_w. Otherwise, the calculations in Eqs. (1) and (2) repeated with X_g replaces X_b. If no improvement becomes possible in this case, then a new solution is randomly generated to replace that frog. The main steps of the SFLA are listed as following:

Step1:Initialize and evaluate each solution X_i of the swarm.
Step2:Rank frogs and partition frogs into memeplexes, update X_g.
Step3:Memetic evolution within each memeplex.
Step4:If the stopping criterion is satisfied, output the best solution; otherwise, go to step 2.

Below are details of step3:

step1: X_w produce and evaluate new solution according to (1) and (2).
step2:Determine the new solution, if not better, replace X_b to X_g, go to step 1.

step3:Determine the solution, if not better still, replace X_w with a randomly compatible stems set.

4 DSFLA and MDSFLA

This section firstly proposes a different version of SFLA, namely DSFLA which operates on mathematical sets in order to solve set-based combinatorial problems. DSFLA therefore applies to discrete search spaces. To achieve this, new operators are defined to work on sets instead of the usual continuous vector space, reasonable individual disturbance are designed to balance exploration and exploitation. On the basis of the DSFLA, X_g is updated in time make sure to guide local search with the high quality solution in MDSFLA.

4.1 Solution Space and Initialize Frogs

The solutions generated with the DSFLA are sets. The universal set U contains all possible stems. Thus the frog's position X is initialized to a random subset of stems from the universal set U, that is, $X_i \subseteq U$, $1 \leq i \leq SN$.

4.2 Individual Vector Update

In the standard SFLA, the update of X_w is an abstract concept, it represents X_w follows the local extreme value or global extreme value in continuous space. *rand()* represents the inherited degree from the local extremum. For the discrete combinatorial optimization problem, a new update strategy must be redesigned. For the RNA secondary structure prediction, each flogs represents a set of compatible stems, the move of position corresponds to add or remove stems from the set. The individual approaches to good solution through add the stems which consist in the X_b. By the same token, add or remove stems must follow the compatible constraint condition. If always add stems to the set, it will not be able to add new stems in a very short time. So before adding, it must remove the stems not existing in X_b, followed adding stems in X_b to X_w. D_{max} corresponds to the follow operation: First remove stems exist in X_w but not exist in X_b with probability P_i, then add stems exist in X_b to X_w with probability P_r. So D_{max} is replaced by two parameters: P_r and P_i, through adding stems exist in X_b to the individual gravitate toward good solution. Then the update of individual may be represented by Eqs. (3) and (4):

$$X_w = X_w - O. \tag{3}$$

$$X_w = X_b + C. \tag{4}$$

where O represents the set of stems which are selected from X_w with P_i. C represents the set of stems which are selected from X_b with P_r. Addition operator represents add operation and subtraction operator represents remove operation. If it can produce better solution, use this solution replace X_w, otherwise replace X_b to X_g, and repeat the operation. If no improvement becomes possible in this case, then a new solution is randomly generated to replace X_g.

4.3 The Modified DSFLA

In the initial SFLA, the update of X_g is done after each memeplex finish the execution of the iteration. That is, the X_g used in the memeplex may not be the current global optimal solution, it will affect subsequent local search process in each memeplex, thus slow the convergence speed. MDSFLA sets the local search in parallel thread for processing. After the update of each memeplex, X_g compares with X_b at once. If X_b in the memeplex is better than X_g, replace it immediately. Continuously update with X_g strengthens the global information, promotes exchange of global information and the combination of internal communication mechanism. Thus guide the algorithm to a better search direction.

5 Simulation Results

5.1 RNA Sequences

The test set includes 10 tRNA sequences randomly selected from Genomic tRNA Database[16]. They are named for three numbers. The first number means serial number, behind two means the sequence's length. The test set also includes 3 5SrRNA sequences selected from the Comparative RNA Website[17] (the accession number and the sequences' length are given in brackets): i) Saccharmoyces cerevisiae 5SrRNA(X67579, 117nt), Haloarcula marismortui 5SrRNA (AF034620, 122nt), Thermus aquaticus 5SrRNA(X01590, 123nt). For these sequences the known secondary structure are available from the two websites.

5.2 Objective Function

The objective function is a thermodynamic energy function which aim to minimize the free energy of the structure. The energy function is calculated with the INN-HB model same with RNAStructure and same thermodynamic parameters are used.

5.3 Parameters Selection

Like other intelligent optimization algorithms, parameter selection is critical to SFLA. SFLA have five parameters: the solution number *SN*, the number m of memeplexes, the number n of frogs in a memeplex, so $SN=m \times n$. The number N of evolution steps in a memeplex and the maximum step size D_{max} allowed during an evolutionary step. There is still no guidance to select proper values for the parameters. According to [10], set m=5, n=10, N=10, Pr=0.6, the total iteration number *IterNum*=200. P_i starts with higher value 0.9 and decreasing 0.1 during the iteration, which similar to the inertia weight in PSO.

5.4 Experimental Results

The accuracy of RNA secondary algorithms is measured by both sensitivity(*SE*) and specificity(*SP*). Let true positive(*TP*) be the number of correctly predicted base pairs

Table 1. Performances Compare of Four Algorithms

sequence	algorithm	*TP*	*FN*	*FP*	*SE*/%	*SP*/%
short172	MDSFLA	16	4	2	80%	88.9%
	DSFLA	16	4	4	80%	80%
	RNAStructure	12	8	15	60%	44%
	pknotsRG	12	8	10	60%	54.5%
short273	MDSFLA	19	1	0	95%	100%
	DSFLA	19	1	0	95%	100%
	RNAStructure	5	15	17	25%	22.7%
	pknotsRG	5	15	15	25%	25%
short373	MDSFLA	20	1	0	95.2%	100%
	DSFLA	21	0	0	100%	100%
	RNAStructure	21	0	0	100%	100%
	pknotsRG	17	4	3	81%	85%
short473	MDSFLA	20	1	0	95.2%	100%
	DSFLA	21	0	0	100%	100%
	RNAStructure	21	0	0	100%	100%
	pknotsRG	21	0	0	100%	100%
short573	MDSFLA	19	1	0	95%	100%
	DSFLA	19	1	0	95%	100%
	RNAStructure	5	15	16	25%	23.8%
	pknotsRG	5	15	15	25%	25%
short676	MDSFLA	21	0	1	100%	95.5%
	DSFLA	21	0	2	100%	91.3%
	RNAStructure	21	0	2	100%	91.3%
	pknotsRG	21	0	2	100%	91.3%
short777	MDSFLA	21	0	0	100%	100%
	DSFLA	21	0	0	100%	100%
	RNAStructure	15	6	8	71.4%	65.2%
	pknotsRG	12	9	8	57.1%	60%
short882	MDSFLA	23	1	0	95.8%	100%
	DSFLA	21	3	4	87.5%	84%
	RNAStructure	12	12	16	50%	42.3%
	pknotsRG	17	7	9	70.8%	65.4%
short984	MDSFLA	22	2	0	91.7%	100%
	DSFLA	20	4	4	83.3%	83.3%
	RNAStructure	18	6	8	75%	69.2%
	pknotsRG	16	8	9	66.7%	64%
short1086	MDSFLA	22	2	4	91.7%	84.6%
	DSFLA	21	3	3	87.5%	87.5%
	RNAStructure	21	3	5	87.5%	80.8%
	pknotsRG	12	12	16	50%	42.9%
S.cerevisiae	MDSFLA	33	4	6	89.2%	84.6%
	DSFLA	33	4	5	89.2%	86.8%
	RNAStructure	32	5	7	86.5%	82.1%
	pknotsRG	28	9	14	75.7%	66.7%
H.marismortui	MDSFLA	30	8	2	78.9%	93.8%
	DSFLA	31	7	7	81.6%	81.6%
	RNAStructure	31	7	7	81.6%	81.6%
	pknotsRG	31	7	5	81.6%	86.1%
T.aquaticus	MDSFLA	23	17	13	57.5%	63.9%
	DSFLA	29	11	7	72.5%	80.6%
	RNAStructure	29	11	14	72.5%	67.4%
	pknotsRG	8	32	28	20%	22.2%

and false negative(*FN*) be the number of base pairs in the correct structure which were not predicted, false positive(*FP*) be the number of incorrectly predicted base pairs that do not exist in the correct structure. Define *SE* as *TP/(TP+FN)*, and *SP* as *TP/(TP+FP)*.

Table 1 is the experiment results of four algorithms on the selected test sequences. From the table we can see that MDSFLA and DSFLA both show relatively higher sensitivity and specificity, In add to the last two sequences, the *SE* and *SP* of other sequences are more than 80%, outperform the RNAStructure and pknotsRG. Except for the third and the fourth sequences, MDSFLA shows the best performance on short sequences. The performance of MDSFLA is not so good on the last two longer sequences. The reason may lie in that MDSFLA loses the diversity of individual in a certain extent while accelerating convergence.

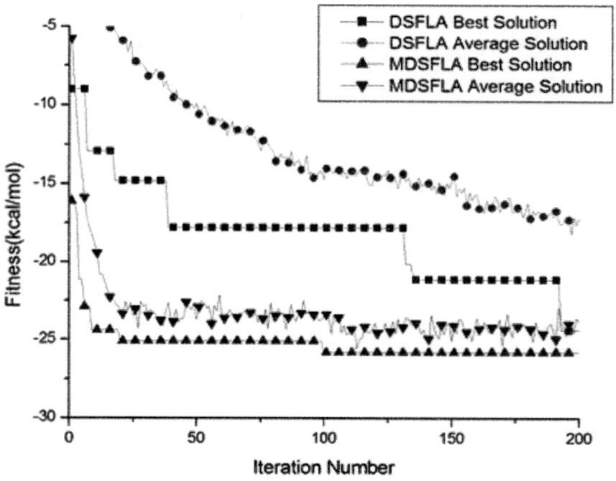

Fig. 1. Iteration process of DSFLA and MDSFLA on short984

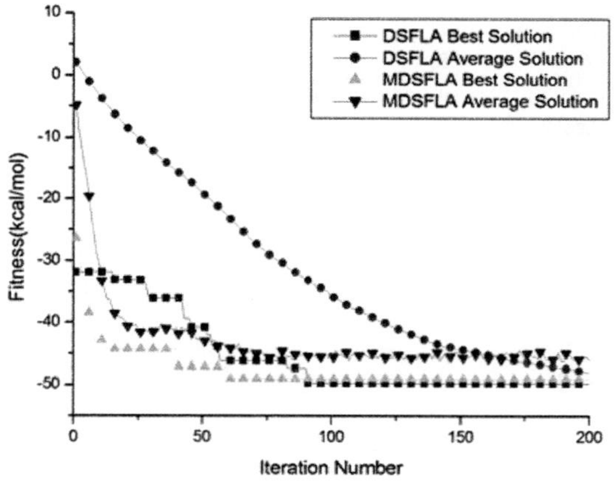

Fig. 2. Iteration process of DSFLA and MDSFLA on S.cerevisiae

In order to obverse the convergence process of DSFLA and MDSFLA, we draw the typical iteration process of both algorithms on two sequences. Fig.1 is the graph of best fitness and average fitness for each generation in one run corresponding to the short984. Fig.2 is the graph of best fitness and average fitness for each generation in one run corresponding to the S.cerevisiae. The iteration and the data of the corresponding rows in table1 show that, MDSFLA and DSFLA both have good convergence speed and better final result, further MDSFLA has faster convergence speed.

6 Conclusions

This paper presents a discrete SFLA to make it adoptable for simulating the folding process of RNA secondary structure and its modified version. In the DSFLA, the update rules of frog's position are redesigned, parameters are properly set to balance the space search ability and local refinement capacity. Based on it, MDSFLA uses new update strategy to make better result and faster convergence speed. The experiment simulations compared with RNAStructure and pknotsRG on several sequences indicate that both DSFLA and MDSFLA have good accuracy of prediction.

Future work will include strategies to increase the diversity of individuals and detailed analysis of influence of parameters for the algorithms. To present an effective thermodynamic model for including calculations for pseudoknots is the next step.

Acknowledgments. This work is supported by Natural Science Foundation of Fujian province of China (2008J0316).

References

1. Jiang, T., Xu, Y., Zhang, M.Q.: Current Topics in Computational Molecular Biology. Tsinghua University Press, Beijing (2002)
2. van Batenburg, F.H.D., Gultyaev, A.P., Pleij, C.W.A.: An APL-programmed Genetic Algorithm For the Prediction of RNA Secondary Structure. Journal of Theoretic Biology 174(3), 269–280 (1995)
3. Shapiro, B.A., Wu, J.C.: An Annealing Mutation Operator in the Genetic Algorithms for RNA Folding. Bioinformatics 12(3), 171–180 (1996)
4. Kay, C.W., Alain, A.D., Andrew, G.H.: RnaPredict-An Evolutionary Algorithm for RNA Secondary Structure Prediction. Transactions on Computational Biology and Bioinformatics 5(15(1)), 25–40 (2008)
5. Neehling, M., Engelbrecht, A.P.: Determining RNA Secondary Structure using Set-based Particle Swarm Optimization: Evolutionary Computation. In: IEEE Evolutionary Computation, Vancouver BC, Canada (2006)
6. Geis, M., Middendorf, M.: A Particle Swarm Optimizer for Finding Minimum Free Energy RNA Secondary Structures. In: IEEE Swarm Intelligence Symposium (2007)
7. Yu, J., Zhang, C., Liu, Y., et al.: Simulating the Folding Pathway of RNA Secondary Structure Using the Modified Ant Colony Algorithm. Bionic Engineering 7, 382–389 (2001)

8. He, J., Zhou, D., He, Z.: Self-adaptive Artificial Fish Swarm Algorithm for RNA Secondary Structure Prediction. Journal of System Simulation 22(6), 1370–1374 (2010)
9. Muzaffar, M.E., Kevin, E.L.: Optimization of Water Distribution Network Design Using the Shuffled Frog Leaping Algorithm. Journal of Water Resources Planning and Management 129(3), 210–225 (2003)
10. Muzaffar, E., Kevin, L., Fayzul, P.: Shuffled Frog-leaping Algorithm: A Memetic Meta-heuristic for Discrete Optimization. Engineering Optimization 8(2), 129–154 (2006)
11. Elbeltagi, E., Hegazy, T., Grierson, D.: Comparison among Five Volutionary_based Optimization Algorithms. Advanced Engineering Informatics 19(1), 43–53 (2005)
12. Li, X., Luo, J., Chen, M.-R., et al.: An Improved Shuffled Frog-leaping Algorithm with Extremal Optimisation for Continuous Optimization. Information Sciences (2010)
13. Niknam, T., Farsani, E.A.: A hybrid Self-adaptive Particle Swarm Optimization and Modified Shuffled Frog Leaping Algorithm for Distribution Feeder Reconfiguration. Engineering Application of Artificial Intelligence 23, 1340–1349 (2010)
14. Zuker, M., Mathews, D.H., Turner, D.H.: Algorithms and Thermodynamics for RNA Secondary Structure Prediction: A Practical Guide In RNA Biochemistry and Biotechnology. 11–43 (1999)
15. Reeder, J., Giegerich, R.: Design, Implementation and Evaluation of a Practical Pseudoknots Folding Algorithm based on Thermodynamics. BMC Bioinformatics 5, 104 (2004)
16. Lowe T M., Eddy S R.: tRNAscan-SE A Program for Improved Detection of Transfer RNA Genes in GenomicSequence. Nucleic Acids Research. Vol. 25, pp. 955–964(1997)
17. Cannone, J.J., et al.: The Comparative RNA Web (crw) Site: an Online Database of Comparative Sequence and Structure Information for Ribosomal, Intron, and other RNAs. BMC Bioinformatics 3 (2002)

On the Equilibrium of Outsourcing Service Competition When Buyer Has Different Preference towards Supplier

Shengli Chen and Xiaodong Liu

School of Management Engineering, Xian University of Finance and Economic
212101csl@163.com, 212101csl@sina.com

Abstract. We formulate respectively the outsourcing service competition models with simultaneous and sequential decision order when buyer has preference toward every supplier and analyze the suppliers' contest equilibrium in outsourcing service competition by applying dynamic game theory and contest theory. Then we explore the effect of sequential decision order on competition equilibrium. It is showed that the sequential decision order not only influences the equilibrium service level of the suppliers, but also has an effect on both total equilibrium service level of the supplier and the average equilibrium service level acquired by the purchaser.

Keywords: equilibrium, outsourcing service, preference, allocation function.

1 Introduction

Outsourcing has emerged as a major trend in many manufacturing and service industries. It has already been a new form of international division of labor and cooperation and industrial transfer. Thus, more organizations focus on how to improve outsourcing service quality by outsourcing service competition. Benjaafar and Elahi explore two approaches the buyer could use to orchestrate this competition, a Supplier-Allocation approach and a Supplier-Selection approach, and compare the service level, expected payoff and service cost of these two approaches[1].Meland and Straume study expost outsourcing of production in an imperfectly discriminating contest, interpreted as a research tournament or a procurement contest for being awarded some production contract[2].Wang and Gerchak use a market attraction function to model competition between two supply chains, where effort is measured by fill rate[3].Li and Wang analyze outsourcing behavior of organizations applying static game model[4].Cao,Yang and Yu construct cooperative game model between core enterprise and outsourcing enterprise[5].They find the cooperation condition of the core enterprise and outsourcing enterprise. These literatures do not consider the effect of service competition with different action timing on outsourcing strategy. Instead, researchers always analyze outsourcing service competition on basis of the same decision order, which is a static game. Whereas, all the suppliers do not choose their service level simultaneously, in fact, suppliers often confirm their strategies with sequential decision order[6].Therefore, the outsourcing service competition should be a dynamic game rather than a static game. We will analyze the equilibrium of

D. Zeng (Ed.): Advances in Control and Communication, LNEE 137, pp. 601–609.
springerlink.com

outsourcing service competition model based on contest model, and explore the effect of sequential decision order on the equilibrium.

2 Supplies' Competition Model

Consider a system with a single buyer that seeks to outsource a provisioning of a product with an expected demand quantity Q and price p to N potential suppliers. The price of product p and the demand quantity Q are fixed and identical across all suppliers. We establish an outsourcing service competition model by Supplier-Allocation approach,

The buyer declares the item of provisioning contract, including the amount of demand Q , the price of the product p and the number of supplies N .The supplier $i(i=1,2,\cdots,N)$) observing the items of provisioning contract offers a service level $x_i \geq 0$ to the buyer. The buyer allocates the amount of demand Q_i to supplier $i(i=1,2,\cdots,N)$ in light of allocation function and its service level x_i , and contracts with suppliers finally. The supplier $i(i=1,2,\cdots,N)$ provides the product to meet the demand allocation Q_i and achieve the service level x_i .

Let $\varphi_i(x_i, x_{-i})$ denote the allocation rule of buyer, which is called allocation function. $x_{-i}=(x_1,\cdots,x_{i-1},x_{i+1},\cdots,x_N)$ denotes the vector of service level that suppliers offer except supplier i . From the view of contest theory, the suppliers' competition vying for provisioning contract is a contest. Therefore, we consider Contest Success Function to define allocation function, which is given by

$$\varphi_i(x_i, x_{-i}) = h_i(x_i) \Big/ \sum_{j=1}^{N} h_j(x_j), i=1,2,\cdots,N \tag{1}$$

Eq.(1) represents that the proportional allocation is increasing with its service level. After defining the allocation function, we present the outsourcing service competition model by Supplier-Selection approach.

However, organization prefers partial outsourcing whilst maintaining internal capacity rather than entire outsourcing without any self-made capability in some important field related to the competitive advantage of organization. Partial outsourcing can allow the sourcing organization to assess its own internal performance and stimulate inter performance improvement. Thus, we investigate a slightly more general situation where the buyer may retain the demand with appositive probability, and the buyer preference is different toward every supplier, and the allocation function with self-made possibility is

$$\varphi_i(x_i, x_{-i}) = \frac{a_i h_i(x_i)}{t + \sum_{j=1}^{N} a_j h_j(x_j)}, i=1,2,\ldots,N \tag{2}$$

Where a_i is constant which denotes the buyer's preference toward every supplier .When $t \ge 0$ there is some positive probability that none of the suppliers will get the demand.

Given demand allocation Q_i, let $C_i(Q_i)$ denote supplier *i*'s production cost and $f_i(x_i, Q_i)$ denote supplier *i*'s cost in providing service level x_i .Given the above allocation function $\varphi_i(x_i, x_{-i})$,the expected payoff of supplier i , $i = 1, 2, \cdots, N$, is

$$R_i(x_i) = Q_i p - C_i(Q_i) - f_i(x_i, Q_i) \tag{3}$$

Where $Q_i = \varphi_i(x_i, x_{-i})Q$.

Under Supplier-Allocation model, this implies supplier i will choose x_i^* to maximize its payoff $R_i(x)$ as follow:

$$x_i^* \in \arg\max_{x_i \ge 0} R_i(x_i, x_{-i}) = \varphi_i(x_i, x_{-i})Q_i p - C_i(Q_i) - f_i(x_i, \varphi_i(x_i, x_{-i})Q) \tag{4}$$

The profile $(x_1^*, x_2^*, \cdots x_N^*)$ of service level satisfying Eq. (4) is Nash equilibrium of the model. We define allocation function as homogeneous linear Tullock contest success function, which is given by

$$\varphi_i(x_i, x_{-i}) = a_i\, x_i \Big/ [t + a_j \textstyle\sum_{j=1}^{N} x_j], i = 1, 2, \cdots, N \tag{5}$$

Especially, when there are only two supplies in the system, we may define the supplies' allocation function $\varphi_i(x_i, x_{-i})$ as follows:

$$\begin{cases} \varphi_1(x_1, x_2) = \dfrac{kx_1}{t + kx_1 + x_2} \\ \varphi_2(x_1, x_2) = \dfrac{x_2}{t + kx_1 + x_2} \end{cases} \tag{6}$$

Where k is a constant which reflects the buyer's preference toward every supplier. When $0 < k < 1$ the buyer has preference toward supplier 2; When $k > 1$ the buyer has preference toward supplier 1; When $k = 1$ the buyer has no preference toward every supplier.

3 Services Equilibrium

When suppliers act simultaneously, suppliers' outsourcing services competition is a static game. To solve the outsourcing service competition model above, we should also specify the cost function. Suppose $C_i(Q_i) = cQ_i$, which represents unit production cost remain the same regardless of the service level offered, and assume $f_i(x_i, Q_i)$ is non-decreasing in both if x_i and Q_i , and $f_i(0, Q_i) = 0$ .We consider the service cost function as a linear function of service level promised. Suppose $f_i(x_i, Q_i) = vx_i$.

Without loss of generality, we suppose the number of potential supplier is two Then, supplier 1 and supplier 2'revenue can be denoted respectively as follows:

$$R_1(x_1) = \frac{Qkx_1}{t + kx_1 + x_2}(p - c) - vx_1 \tag{7}$$

$$R_2(x_2) = \frac{Qx_2}{t + kx_1 + x_2}(p - c) - vx_2 \tag{8}$$

By Eq.(7),(8),we could obtain the Nash equilibrium. The equilibrium service level are given by

$$\begin{cases} x_1^* = \dfrac{Qrk^2 - 2vt(k+1) + kD}{2vk(k+1)^2} \\ x_2^* = \dfrac{Qrk - 2vtk(k+1) + D}{2v(k+1)^2} \end{cases} \tag{9}$$

Where $r = p - c$, $D = \sqrt{Qrk[Qrk + 4vt(k+1)}$.

While the supplies' allocation ratio of orders is given by

$$\begin{cases} \varphi_1 = \dfrac{Qrk^2 - 2vt(k+1) + kD}{(k+1)(Qrk + D)} \\ \varphi_2 = \dfrac{Qrk - 2vtk(k+1) + D}{(k+1)(Qrk + D)} \end{cases} \tag{10}$$

The corresponding total equilibrium service level is given by

$$T_x = x_1^* + x_2^* = \frac{Qrk^2 - vt(k+1)(k^2+1) + kD}{vk(k+1)^2} \tag{11}$$

Furthermore, the average equilibrium service level is given by

$$L = \frac{(Qrk + D)[Qrk^2 - vt(k+1)(k^2+1) + kD]}{Qvk(k+1)^2[Qrk - 2vt(k+1) + D]} \tag{12}$$

Taking the first derivative of x_1^* with respect to k, solving and simplifying, we have:

$$k = k_1^* = \frac{Qr + \sqrt{Qr(Qr + 16vt)}}{2Qr} \tag{13}$$

Therefore, when the buyer's preference satisfies Eq.(13), the supplier 1' service level reaches the maximum Similarly, taking the first derivative of x_2^* with respect to k, solving and simplifying, we have:

$$k = k_2^* = \frac{Qr}{Qr + 4vt} \tag{14}$$

So, when the buyer's preference satisfies Eq.(14), the supplier 2' service level reaches the maximum When $k < \frac{Qr}{Qr + 4vt}$, supplier i's equilibrium service level decrease with the buyer's preference k; When $k > \frac{Qr + \sqrt{Qr(Qr + 16vt)}}{2Qr}$, supplier i's equilibrium service level also decrease with the buyer's preference k. Therefore, only when $k \in [\frac{Qr}{Qr + 4vt}, \frac{Qr + \sqrt{Qr(Qr + 16vt)}}{2Qr}]$, the suppliers' total equilibrium service level can reach the maximum. Particularly, when buyer does not choose self-made, namely $t = 0$, the suppliers' total equilibrium service level can rewrite as follows

$$T_x = \frac{2Qrk}{v(k+1)^2} \tag{15}$$

According to above Eq.(15) , when the buyer prefers some supplier the suppliers' total equilibrium service level can not reach the maximum. However, suppliers may promise service level sequentially in the outsourcing service competition process.

We will explore the effect of sequential decision order on the equilibrium of outsourcing service competition in following cases. The Effect of Sequential Decision Order on the Equilibrium of Outsourcing without Internal Capacity Without losing generalization, we suppose that supplier 1promises an outsourcing service level first and suppliers 2 observing the strategy of supplier 1 offers a service level. Let x_{11}^* and x_{22}^* denote the equilibrium service level of supplier 1and 2 with sequential decision order respectively. Let L_{12} denote the average service level associated with sequential decision order. By applying backwards induction, we analyze the choice of supplier 2 on condition that the buyer could not retain internal capacity ($t = 0$). Given the service level x_1 of supplier 1, the supplier 2's decision problem can be formulated by

$$\max_{x_2} \frac{Qrx_2}{kx_1 + x_2} - vx_2 \tag{16}$$

By Eq.(16), the supplier 2's reaction function is

$$x_{22}^* = S_2(x_1) = \sqrt{\frac{Qr(kx_1 + t)}{v}} - kx_1 - t \tag{17}$$

Where $S_2(x_1)$ is denoted as the reaction function of supplier2 toward supplier 1's service level

x_1 .Since supplier 1 can solve supplier 2's problem as supplier 2 does, supplier 1 should anticipate that the service level choice x_1 is met with the reaction function $S_2(x_1)$ .Thus, supplier l's problem in the first stage of the game amounts to

$$\max_{x_1} \frac{kQrx_1}{kx_1 + x_{22}} - vx_1 \tag{18}$$

where $x_{22}^* = S_2(x_1)$,and yields equilibrium service level

$$x_{11}^* = \frac{Qrk}{4v} \tag{19}$$

$$x_{22}^* = \begin{cases} \dfrac{Qrk(2-k)}{4v}, k \le 2 \\ 0, k > 2 \end{cases} \tag{20}$$

While the corresponding total equilibrium service level is

$$T_x' = x_{11}^* + x_{22}^* = \frac{Qrk(3-k)}{4v} \tag{21}$$

And the supplies' allocation ratio of orders is given by

$$\begin{cases} \varphi_1 = \dfrac{k}{2} \\ \varphi_2 = \dfrac{2-k}{2} \end{cases} \tag{22}$$

And the average equilibrium service level is

$$L_{12} = \frac{rk(3-k)}{4v} \tag{23}$$

According to above Eq.(19), supplier 1's equilibrium service level increase with the buyer's preference k .By Eq.(20), when $0 < k < 1$ supplier 2's equilibrium service level increase with the buyer's preference k ; when $k > 1$ supplier 2's equilibrium service level decrease with the buyer's preference k . According to above Eq.(21) and Eq.(23) , suppliers' total equilibrium service level and average equilibrium service level increase with the buyer's preference k when $0 < k < 1.5$; when $k > 1.5$ suppliers' total equilibrium service level and average equilibrium service level decrease with the buyer's preference k

4 Comparative Analysis

By Eq.(9), Eq.(11) and Eq.(12), we find that if the buyer allocates demand without internal capacity when suppliers act simultaneously , the equilibrium service level are given by

$$\begin{cases} x_1^* = \dfrac{Qrk}{v(k+1)^2} \\ x_2^* = \dfrac{Qrk}{v(k+1)^2} \end{cases} \tag{24}$$

While the suppliers' total equilibrium service level is

$$T_x = x_1^* + x_2^* = \frac{2Qrk}{v(k+1)^2} \tag{25}$$

Furthermore, the average equilibrium service level is given by

$$L = \frac{2rk}{v(k+1)^2} \tag{26}$$

In order to facilitate comparative analysis, if we denote $\Delta x_i = x_i^* - x_{ii}^*$ as supplier i's change of equilibrium service level in two different decision order, then the following Equation can be give by

$$\begin{cases} \Delta x_1 = \dfrac{Qrk(1-k)(k+3)}{4v(k+1)^2} \\ \Delta x_2 = \dfrac{Qrk(1+k)^2(k-2)}{4v(k+1)^2} \end{cases} \tag{27}$$

According to Eq.(27), when $0 < k < 1$, $\Delta x_1 > 0, \Delta x_2 < 0$, this means that if buyer has preference toward supplier 2 ,supplier 1's the equilibrium service level when decision order is static is higher than that when decision order is dynastic, while supplier 2's the equilibrium service level when decision order is static is lower than that when decision order is dynastic. When $k > 2$, $\Delta x_1 < 0, \Delta x_2 > 0$, this means that when the buyer prefers supplier 1, supplier 1's the equilibrium service level when decision order is static is lower than that when decision order is dynastic, while supplier 2's the equilibrium service level when decision order is static is higher than that when decision order is dynastic. Therefore, when the buyer has more preference toward supplier 2, namely $0 < k < 1$, the supplier 2 should choose to act firstly and the supplier 1 should choose to act at the same time; when the buyer has more preference toward supplier 1, namely $k > 2$, the supplier 1 should choose to act firstly and the supplier 2 should choose to act at the same time.

If we denote $\Delta T_x = T_x - T_x'$ as supplier i's change of total equilibrium service level in two different decision order and denote $\Delta L = L - L'$ as supplier i's change of total equilibrium service level in two different decision order, then the following Equation can be give by

$$\Delta T_x = \frac{Qrk(k-1)(k+\sqrt{5})(k-\sqrt{5})}{4v(k+1)^2} \tag{28}$$

$$\Delta L = \frac{rk(k-1)(k+\sqrt{5})(k-\sqrt{5})}{4v(k+1)^2} \tag{29}$$

By Eq.(28) and (29), when $0 < k < 1$ or $k > \sqrt{5}$, $\Delta T_x > 0$, this means that if buyer has more preference toward some supplier, total equilibrium service level and average equilibrium service level when decision order is static is higher than that when decision order is dynastic. When $1 < k < \sqrt{5}$, $\Delta T_x < 0$, this means that if buyer has preference toward supplier 1, total equilibrium service level and average equilibrium service level when decision order is static is lower than that when decision order is dynastic.

5 Conclusions

In this paper, we consider the impact of action timing (also called decision order) on the equilibrium of outsourcing service competition when the buyer has preference toward every supplier. We find that (1) whether the decision orders will influence the equilibrium or not depend on the buyer's attitude towards self-made and the buyer's preference toward every supplier. Once the buyer decides outsourcing without self-made, the decision order will not affect the equilibrium of outsourcing service competition. Whereas the buyer remains certain internal capacity, the decision order will impact the outcome of outsourcing service equilibrium. (2) the equilibrium service level with sequential decision order is different from the one with simultaneous decision order when the buyer chooses partial outsourcing. Comparing the equilibrium service level of outsourcing service competition with simultaneous decision order and the equilibrium service level of outsourcing service competition with sequential decision order, if buyer has preference toward some supplier, the equilibrium service level offered by suppliers who act simultaneously is higher than the one offered by the follower supplier, but lower than the one offered by the leader supplier.(3) if buyer has more preference toward some supplier , total equilibrium service level and average equilibrium service level when decision order is static is higher than that when decision order is dynastic. If the buyer has no preference toward every supplier, different decision order has not impact on total equilibrium service level and average equilibrium service level Therefore, once the buyer decides to remain certain internal capacity, variation of decision order has an impact on outsourcing service competition. The total equilibrium service level and average service level with sequential decision order are much larger, and the buyer prefers sequential decision order to simultaneous decision order. In order to implement outsourcing strategy more efficiently and ensure the demand safety whilst maximizing its payoff, the buyer should intervene suppliers' action timing and require suppliers competing with sequential order.

Acknowledgment. Financial supports from a project with special fund for the construction of key disciplines in Shanxi Province and Shaanxi Nature Science Foundation under Grant No 2010JQ9006 and Shanxi Department of Education

Foundation under Grant No 2010JK552. The helpful comments from anonymous reviewers are also gratefully acknowledged.

References

1. Benjaafar, S., Elahi, E.: Outsourcing via service competition. Management Science 53(2), 241–259 (2007)
2. Meland, F., Straume, O.R.: Outsourcing in contest. PublicChoice 131(7), 315–331 (2007)
3. Wang, Y., Gerchak, Y.: Supply chain cooperation when demand is self-space dependent. Manufacturing&Service Operations Management 3(11), 82–87 (2001)
4. Li, X., Wang, N.: Analysis of the risk game about enterprises business outsourcing. Modern Enterprise 9(9), 43–44 (2004)
5. Cao, Q.K., Yang, Y.L., Yu, R.L.: A strategic analysis of outsourcing game. Modern Management Science (7), 35–36 (2006)
6. Amir, M., Amir, R., Jin, J.: Sequencing R&D decisions in a two-period duopoly with spillovers. Economic Theory 15(3), 297–317 (2000)

On Interval Assignment Games

Xiaodong Liu, Minghai Zhang, and Zhengyu Zang

School of Management Engineering, Xian University of Finance and Economic
xdliu@yeah.net, 212101csl@163.com, 212101csl@sina.com

Abstract. It extends the assignment games defined by Shapley and Shubik to the uncertainty case with interval number. It is proved that the interval core of the interval assignment games is nonempty, and the structure of the interval core in many-to-one matching model is studied.

Keywords: interval assignment games, optimal matching, interval core, partial order.

1 Introduction

In recent years, many scholars made some researches on assignment games, and got some results. Shapley and Shubik[1] proved that the core of the assignment games is nonempty and has a lattice structure. Demange[2] proved that this maximum payoff of a player in the core of the assignment games is his marginal contribution. Quint[3] proved that the class of possible cores of such games is exactly the same as a special class of polytopes called "45°-lattices". Núñez and Rafels[4] proved that the τ-value of the assignment games always lies in the core. Ester Camina[5] proved that the core is nonempty in many-to-one matching model and studies the structure of the core.

In 2008, Tijs[6] introduced the interval numbers to the traditional cooperative games. So the interval cooperative game gets widely attention.

In [7], it discussed the interval assignment problems by using Moore's [8] interval analysis. In this paper, the interval assignment game with two-side market is studied.

2 The Assignment Game

In 1972, Shapley and Shubik[1] studied the assignment game with two-side market. The player set is the finite set $P \cup Q$, where $P=\{1,2,\ldots,n\}$ denotes the set of buyers, and $Q=\{1,2,\ldots,n)$ is the set of sellers, So a matrix, $A=(a_{ij})_{(i,j)\in P\times Q}$, which reflects the profit by each buyer-seller pair is obtained. Where, the jth seller values his goods at c_j, while the ith buyer values the same goods at γ_{ij}. If $\gamma_{ij}>c_j$ then a price favorable to both parties exists. If j sells his good to i for p_j, then j's final profit is exactly p_j-c_j, and i's is exactly γ_{ij}-p_j. That is to say, the profit $a_{ij}=\gamma_{ij}-c_j$ is shared by the buyer-seller pair. We assumed that u_i is distributed to buyer i and v_j is distributed to seller i. So, $a_{ij}=u_i+v_j$.

Definition 1. A cooperative game with transferable utility is a pair (N, v), where $N=\{1,2,\ldots,n\}$ is the player set and $v:2^N\rightarrow\mathbb{R}$ is the characteristic function with $v(\varnothing)=0$.

D. Zeng (Ed.): Advances in Control and Communication, LNEE 137, pp. 611–616.
springerlink.com

Definition 2. An assignment game is a pair ($P\cup Q$, w). Where $P\cup Q$ is the set of players, and w: $2^{P\times Q}\to\mathbb{R}$ is the characteristic function with $w(\emptyset)=0$.

The player set $P\cup Q$ consists of two finite disjoint sets P and Q, while $P=\{1,2,\ldots,n\}$ is the set of buyers and $Q=\{1,2,\ldots,n\}$ is the set of sellers. The characteristic function w is determined by matrix A in the following way. Given $B\subseteq P$ and $S\subseteq Q$, $w(B\cup S)=\max\{\sum_{(i,j)\in\mu} a_{ij} | \mu \in M(B,S)\}$, where $M(B,S)$ denotes the set of matching between B and S. A matching μ between B and S is an element of $B\times S$.

A matching μ^* is optimal for the assignment game if for all $\mu\in M(P,Q)$, we have $\sum_{(i,j)\in\mu^*} c_{ij}\geq\sum_{(i,j)\in\mu} c_{ij}$, and we denote the set of optimal matching by $M^*(P,Q)$.

Shapley and Shubik proved that the core, $c(w)$, of the assignment game ($P\cup Q$, w) is nonempty, and can be represented in terms of any optimal matching μ^* of $P\cup Q$ by

$$c(w)=\left\{(u,v)\in\mathbb{R}_+^P\times\mathbb{R}_+^Q \left| \begin{array}{l} u_i\geq 0, v_j\geq 0, \forall i\in P, j\in Q \\ u_i+v_j=a_{ij}, \forall (i,j)\in\mu^* \\ u_i+v_j\geq a_{ij}, \forall (i,j)\notin\mu^* \end{array}\right.\right\},$$

where R_+ stands for the set of non-negative real numbers.

It still proved that there exists a buyers-optimal core allocation $(\overline{u},\underline{v})$, where each buyer attains his maximum payoff. Similarly, there exists a sellers-optimal core allocation $(\underline{u},\overline{v})$, where each seller attains his maximum payoff.

From Demange[2] and Leonard[9] we know that the maximum payoff of any player coincides with his marginal contribution.

$$\overline{u}_i=w(P\cup Q)-w(P\cup Q\backslash\{i\}) \text{ for all } i\in P, \tag{1}$$

and

$$\overline{v}_j=w(P\cup Q)-w(P\cup Q\backslash\{j\}) \text{ for all } j\in Q. \tag{2}$$

It follows easily that when $(i,j)\in\mu^*$,

$$\underline{u}_i=w(P\cup Q\backslash\{j\})-w(P\cup Q\backslash\{i,j\}), \tag{3}$$

$$\underline{v}_j=w(P\cup Q\backslash\{i\})-w(P\cup Q\backslash\{i,j\}) . \tag{4}$$

But in real life, costs c_j maybe not a constant. Because of uncertainty factors, the cost c_j is showed in the form of interval value. In this case, c_j becomes $[c_j^-,c_j^+]$ for each $j\in P$. But γ_{ij} do not change because it denotes the maximal value, so the profit is also showed in the form of interval form, i.e.,

$$A=([a_{ij}])_{(i,j)\in P\times Q}=([a_{ij}^-,a_{ij}^+])_{(i,j)\in P\times Q}=([\gamma_{ij}-c_j^+ , \gamma_{ij}-c_j^-])_{(i,j)\in P\times Q}. \tag{5}$$

So the interval assignment game is raised.

3 The Interval Assignment Game

In [7], it has studied the assignment problem with interval uncertainty and discussed its properties.

Definition 3. A cooperative interval game in coalitional form is a pair (N, λ), where $N=\{1,2,\ldots,n\}$ is the set of players, and λ: $2^N \to I(\mathbb{R})$is the characteristic function satisfied $\lambda(\varnothing)=[0,0]$, where $I(\mathbb{R})$ is the set of closed intervals in $\mathbb{R}$. For each $S \subseteq N$, the worth set $\lambda(S)$ of the coalition S in the interval game (N, λ) is of the form $[\lambda(S)^-, \lambda(S)^+]$, where $\lambda(S)^-$ and $\lambda(S)^+$ are the minimum profit and the maximum profit of the coalition S, respectively.

Definition 4. An interval assignment game is a pair $(P\cup Q, \omega)$. Where $P\cup Q$ is the set of players, and $w{:}2^{P\times Q}\to\mathbb{R}$ is the characteristic function with $w(\varnothing)=[0,0]$.

The player set $P\cup Q$ consists of two finite disjoint sets P and Q, while $P=\{1,2,\ldots,n\}$ is the set of buyers and $Q=\{1,2,\ldots,n\}$ is the set of sellers. The characteristic function ω is determined by matrix A with interval uncertainty elements in the following way. Given $B\subseteq P$ and $S\subseteq Q$,

$$\omega(B\cup S)=[\omega(B\cup S)^-, \omega(B\cup S)^+], \tag{6}$$

where $\omega(B\cup S)^-=\min\{\sum_{(i,j)\in\mu} a_{ij}^- | \mu\in M(B,S)\}$ and $\omega(B\cup S)^-=\max\{\sum_{(i,j)\in\mu} a_{ij}^- | \mu\in M(B,S)\}$ are the lower bound and the upper bound of $\omega(B\cup S)$, respectively. $M(B,S)$ denotes the set of matching between B and S. A matching μ between B and S is an element of $B\times S$.

From formula (1), we know that the interval length of each column in matrix A is the same. So, according to [7], the optimal matching exists in the interval assignment problem of a bilateral market.

When c_j is a constant, from formulas (1) and (3), it has the following result.

$$\overline{u_i} = w(P\cup Q) - w(P\cup Q\setminus\{i\}) ,\quad \underline{u_i} = w(P\cup Q\setminus\{j\}) - w(P\cup Q\setminus\{i,j\}) .$$

where,

$$w(P\cup Q\setminus\{j\}) = \max\{\textstyle\sum_{(i,j)\in\mu}\gamma_{ij}\,|\mu\in M(P,Q\setminus\{j\})\} - \sum_{\substack{i=1\\ i\neq j}}^{n} c_i ,$$

$$w(P\cup Q\setminus\{i,j\}) = \max\{\textstyle\sum_{(i,j)\in\mu}\gamma_{ij}\,|\mu\in M(P\setminus\{i\},Q\setminus\{j\})\} - \sum_{\substack{i=1\\ i\neq j}}^{n} c_i .$$

So we have

$$\begin{aligned} &w(P\cup Q\setminus\{j\}) - w(P\cup Q\setminus\{i,j\}) \\ &= \max\{\textstyle\sum_{(i,j)\in\mu}\gamma_{ij}\,|\mu\in M(P,Q\setminus\{j\})\} - \max\{\sum_{(i,j)\in\mu}\gamma_{ij}\,|\mu\in M(P\setminus\{i\},Q\setminus\{j\})\} \end{aligned}.$$

That is to say $\underline{u_i}$ is irrelevant with c_j and affected only by the values γ_{ij}.

So, the core $c(\omega)$ of the interval assignment game $(P\cup Q, \omega)$ can be represented by

$$c(\omega) = \left\{ ([u],[v]) \in I(\mathbb{R}_+^P)\times I(\mathbb{R}_+^Q) \,\middle|\, \begin{array}{l} [u_i],[v_j]\succeq[0,0], \forall i\in P, j\in Q; \\ [u_i]+[v_j]=[a_{ij}], \forall (i,j)\in\mu^* \\ [u_i]+[v_j]\succeq[a_{ij}], \forall (i,j)\notin\mu^* \end{array} \right\},$$

where the relation "$\succeq$" is the partial order relation[7].

Theorem 5. The core of the interval assignment game of a bilateral market is nonempty.

Proof: Because the optimal matching of the interval assignment problem exists[8], the upper bound optimal matching is the same as the lower bound optimal matching. We suppose that the optimal matching is μ^*. When we take the lower bound of each element in interval matrix A, we obtain a matrix $(a_{ij}^-)_{(i,j)\in P\times Q}$ with its element is real number. From [1], we can find an $(\underline{u_i}, \overline{v_j})$, which is the core element for according assignment game. Similarly, for the matrix $(a_{ij}^+)_{(i,j)\in P\times Q}$, we can find another core extreme $(\underline{u_i}', \overline{v_j}')$. We can get that if $(i.\ j)\in\mu^*$, then $\underline{u_i}+\overline{v_j}=a_{ij}^-$ and $a_{ij}^+=\underline{u_i}'+\overline{v_j}'$; if $(i,j)\notin\mu^*$, then $\underline{u_i}+\overline{v_j}\ge a_{ij}^-$ and $\underline{u_i}'+\overline{v_j}'\ge a_{ij}^+$. Since $\underline{u_i}$ is not affected by the cost c_j, we get $\underline{u_i}=\underline{u_i}'$. As $\underline{u_i}+\overline{v_j}=a_{ij}^-\le a_{ij}^+=\underline{u_i}'+\overline{v_j}'$, we get $\overline{v_j}\le\overline{v_j}'$. We can at least find an interval core payoff $([u],[v])$ where $[u_i]=[\underline{u_i},\underline{u_i}']$ and $[v_j]=[\overline{v_j},\overline{v_j}']$. It satisfies the condition that

$$[u_i]+[v_j]=[\underline{u_i},\underline{u_i}']+[\overline{v_j},\overline{v_j}']=[\underline{u_i}+\overline{v_j},\underline{u_i}'+\overline{v_j}']=[a_{ij}^-,a_{ij}^+] \quad \text{for} \quad (i,j)\in\mu^*,$$

and

$$[u_i]+[v_j]=[\underline{u_i},\underline{u_i}']+[\overline{v_j},\overline{v_j}']=[\underline{u_i}+\overline{v_j},\underline{u_i}'+\overline{v_j}']\succeq[a_{ij}^-,a_{ij}^+] \quad \text{for} \quad (i,j)\notin\mu^*.$$

Thus $([u_i],[v_j])$ is one of the interval core elements of the interval assignment game. This result shows that the interval core of the bilateral market is nonempty.

4 Interval Partial Order

If these goods are owned by the same seller, the total payoff of the seller can be expressed as $K=\sum_{j=1}^{n}[v_j]$. This is a many-to-one matching model. For this model, the interval core element is $([u_i],[K])$. In the following, we discuss the structure of the set of the interval core.

Definition 6. For any two interval core elements for many-to-one matching model in the interval assignment game, we say that $([u_i],[K])\succeq_P([u_i'],[K'])$ if $\frac{u_i^-+u_i^+}{2}\ge\frac{u_i^-{}'+u_i^+{}'}{2}$ for all $i\in P$. Similarly, $([u_i],[K])\succeq_S([u_i'],[K'])$ means that $\frac{K^-+K^+}{2}\ge\frac{K^-{}'+K^+{}'}{2}$.

Lemma 7. Let $([u_i],[K])$ and $([u_i'],[K'])$ are two interval core elements. If $([u_i],[K])\succeq_P([u_i'],[K'])$, then $([u_i'],[K'])\succeq_S([u_i],[K])$.

Proof: From assumption of the Lemma 7, we know that $([u_i],[K])$ and $([u_i'],[K'])$ are two interval core elements. Because $[u_i]=[u_i^-,u_i^+]$, $[K]=[K^-,K^+]$, $[u_i']=[u_i^-{}',u_i^+{}']$ and $[K']=[K^-{}',K^+{}']$, we have $\sum[u_i]+[K]=[\max\sum a_{ij}^-,\max\sum a_{ij}^+]$. Furthermore, from $\sum u_i^- + K^- = \sum u_i^-{}' + K^-{}' = \max\sum a_{ij}^-$ and $\sum u_i^+ + K^+ = \sum u_i^+{}' + K^+{}' = \max\sum a_{ij}^+$, we have $\sum u_i^- + K^- + \sum u_i^+ + K^+ = \max\sum a_{ij}^- + \max\sum a_{ij}^+$.

Because $([u_i],[K])\succeq_P([u_i'],[K'])$, then $\frac{\sum u_i^- + \sum u_i^+}{2} \geq \frac{\sum u_i^-{}' + \sum u_i^+{}'}{2}$ for all $i\in P$, which implies $\frac{K^- + K^+}{2} \leq \frac{K^-{}' + K^+{}'}{2}$, then $([u_i'],[K'])\succeq_S([u_i],[K])$.

Remark 8. Let $([u_i],[K])$ and $([u_i'],[K'])$ are two interval core elements. The condition $([u_i'],[K'])\succeq_S([u_i],[K])$ do not imply the result $([u_i],[K])\succeq_P([u_i'],[K'])$.

Definition 9. An interval core element $([u_i]^*,[K]_*)$ is called P-optimal, if for any interval core element $([u],[K])$, $([u_i]^*,[K]_*)\succeq_P([u_i],[K])$ is always satisfied. An interval core element $([u_i]_*,[K]^*)$ is called S-optimal, if for any interval core element $([u],[K])$, $([u_i]^*,[K]_*)\succeq_S([u_i],[K])$ is always true.

Remark 10: There exist a unique P-optimal core element $([u_i]^*,[K]_*)$ and a unique S-optimal core element $([u_i]_*,[K]^*)$. Moreover, for any core element $([u_i],[K])$, $([u_i]^*,[K]_*)\succeq_P([u_i],[K])\succeq_P([u_i]_*,[K]^*)$ and $([u_i]_*,[K]^*)\succeq_S([u_i],[K])\succeq_S([u_i]^*,[K]_*)$

5 Numerical Examples

Suppose there is an interval assignment game of the interval bilateral market, the profit matrix with interval value is

$$\begin{pmatrix} [3.3,5.6] & [2,3.1] & [1,3.5] \\ [2.5,4.8] & [4.6,6.3] & [2,4.5] \\ [3.5,5.8] & [3.2,4.9] & [1.5,4] \end{pmatrix}.$$

Then the optimal matching is $\mu^* = \{(1,1'),(2,2'),(3,3')\}$. When we take the lower bound and the upper bound, the corresponding assignment games have the core elements $(\underline{u_1}=0,\underline{u_2}=0.7,\underline{u_3}=0.2,\overline{v_1}=3.3,\overline{v_2}=3.9,\overline{v_3}=1.3)$ and $(\underline{u_1}'=0,\underline{u_2}'=0.7,\underline{u_3}'=0.2,\overline{v_1}'=5.6,\overline{v_2}'=5.6,\overline{v_3}'=3.8)$, respectively. Thus, the interval core element for the interval assignment game is

$$\begin{pmatrix} [u_1]=[0,0],[u_2]=[0.7,0.7],[u_3]=[0.2,0.2], \\ [v_1]=[3.3,5.6],[v_2]=[3.9,5.6],[v_3]=[1.3,3.8] \end{pmatrix}$$

6 Conclusions

Based on the interval assignment games with the profit matrix has an interval uncertainty, we proved the interval core is nonempty. This is very important for the decision makers to understand their situation and make the right decision

Acknowledgment. Financial supports from a project with special fund for the construction of key disciplines in Shanxi Province and Shaanxi Nature Science Foundation under Grant No 2100JM9011.The helpful comments from anonymous reviewers are also gratefully acknowledged.

References

1. Shapley, L.S., Shubik, M.: The assignment game I: The core. International Journal of Game Theory 1, 111–130 (1972)
2. Demange, G.: Strategyproofness in the assignment market game. Laboratorie d'Econometrie de l'Ecole Politechnique, Paris, Mimeo (1982)
3. Quint, T.: Characterization of cores of assignment games. International Journal of Game Theory 19, 413–420 (1991)
4. Nunez, M., Rafels, C.: The assignment game: the T-value. International Journal of Game Theory 31, 411–422 (2002)
5. Camina, E.: A generalized assignment game. Mathematical Social Sciences 52, 152–161 (2006)
6. Branzei, R., Tijs, S., Alparslan Gök, S.Z.: Bankruptcy problems with interval uncertainty, working paper (2003)
7. Liu, X., Zhang, M., Zang, Z.: On Interval Assignment Problem. Journal of Xi an University of Finance and Economics 24(1), 19–22 (2011)
8. Moore, R.: Methods and applications of interval analysis. SIAM, Philadelphia (1979)
9. Leonard, H.B.: Elicitation of honest preferences for the assignment of individuals to positions. The Journal of Political Economy 91(3), 461–479 (1983)

A Trust Computing Method Based on Feedback Credibility

Runlian Zhang, Xiaonian Wu, and Qihong Liu

School of Information and Communication,
Guilin University of Electronic Technology,
Guilin, China
{zhangrl,xnwu}@guet.edu.cn, lancy868@163.com

Abstract. Trust is an important issue in distributed systems. Recommendation trust has very strong subjectivity. Mendacious recommendation will mislead trust evaluation and threat the system security. To address the problem, this paper proposed a trust computing method based on feedback credibility. Feedback credibility is used to evaluate the authenticity of the recommendation trust by computing the difference between the recommendation trust and mean recommendation trust. The time attribute of feedback information was identified by time decay function. Simulation experiments show that this method can reduce effectively the impact of mendacious recommendation trust, and compute accurately the trust of entity according to entity's behaviors. It provides reliable information to correctly make the security decisions for the system.

Keywords: trust, mendacious recommendation, feedback credibility.

1 Introduction

Distributed systems, such as cloud computing, grid computing, P2P, etc., improve greatly the sharing of resource and support large-scale distributed computing through distributed cooperation. One of the most critical issues in distributed systems is security. However, under above distributed systems, user to access by cross-domain and the behaviors uncertainty of user make some security technologies based on traditional software morphology fail to meet their security needs. The trust mechanism provides a new way for addressing security issues in distributed environments. It helps guide to make the security decisions and provides the effective security guarantee for distributed systems.

Trust [1] is a multi-dimensional and multi-level dynamic concept. It has been an efficient security solution for open environments. In past years, a number of research works have been done. PTM [2], Eigen-Trust [3], PowerTrust [4] and PeerTrust [5] are reputation-based trust management. Srivaramangai [6] proposed two new reputation based trust models. It is effective to evaluate the trustworthiness of different entities and address various malicious behaviors. Azzedin [7] proposed a new honesty checking algorithm. This is the first attempt at linking the behavior change to honesty checking.

D. Zeng (Ed.): Advances in Control and Communication, LNEE 137, pp. 617–624.
springerlink.com

We propose a trust computing method based on feedback credibility in the Grid environment based on the last work [8]. Feedback control mechanism was used to adjust dynamically the recommendation trust. There are different weights for different recommendation trusts by the feedback credibility. The weight indicates the degree of prizing good recommendation and punishing mendacious recommendation. The proposed method also uses time frame and time decay function to identify the time attribute of recommendation trust. This method can be against the mendacious recommendation effectively, and for the system security decision.

The remainder of this paper is organized as follows. In section 2, the proposed trust computing method is introduced. In section 3, the experimental results are shown and discussed. Finally, section 4 provides the conclusion and proposes directions for future work.

2 The Trust Computing Method

2.1 Direct Trust

Direct trust is closely related to entities' direct interactions. It satisfies Beta distribution. Based on probability theory, we evaluate direct trust that entity *i* to entity *n* by equation (1).

$$DT_{in} = \frac{\alpha_{in} + 1}{\alpha_{in} + \beta_{in} + 2} \tag{1}$$

Here, α_{in} expresses the frequency of successful interacting between entity *i* and entity *n*. β_{in} expresses the frequency of unsuccessful interacting between entity *i* and entity *n*.

2.2 Time Decay Function

The recommendation trust is variation with time. The importance of the recommendation trust decays over time. The recommendation trust is obtained earlier, its effects on the trust evaluation is smaller. Here, the time decay function shows the time attribute of the recommendation trust. The time is divided into time frames. Time decay function is expressed as equation (2).

$$S_{in}^{t} = \rho^{t-k} \tag{2}$$

Where, $0 \le \rho \le 1$, *t* is the current time frame, and *k* ($1 \le k \le t$) is the time frame when entity *i* shown recommendation trust about entity *n*.

2.3 Feedback Credibility

To resist the possible aggressive behaviors of collusive fraud and imputation in recommendation trust, we introduce feedback credibility. Feedback credibility measures the credible level of the recommendation trust. If feedback credibility is higher, we think the recommendation trust is more reliable. When feedback credibility is low, the server may be provided the mendacious recommendations. We use feedback credibility as a weighting of recommendation trust to compute entities' trust. The illusive recommendation trust has smaller or negative weighting, and it would have a smaller or negative effect on the trustworthiness of entity. So, according to feedback credibility, we can restrain effectively the impact of mendacious recommendation by smaller weighting or punish mendacious recommendation by negative weighting, which contribute to reduce collusive fraud behaviors.

Feedback credibility C_i^t of server i is expressed as follows:

$$C_i^t = \begin{cases} C_i^{t-1} + \alpha \times \dfrac{(1 - C_i^{t-1})}{2} \times f_{in}^t, f_{in}^t = -1 \\ C_i^{t-1} + \beta \times \dfrac{(1 - C_i^{t-1})}{2} \times f_{in}^t, f_{in}^t = 1 \end{cases} \quad (3)$$

Here, $\alpha, \beta (0 < \alpha, \beta < 1)$ as credible weighting factors are used to adjust the changing of feedback credibility according to the principle that fast decrease and slow increase. $C_i^{t-1} \in [0,1]$, is the current feedback credibility of server i at time t-1 and its initialization is 0.5.

f_{in}^t is the mapping function of evaluation differences among recommendation trusts provided different entity. And mapping function f_{in}^t is determined by $diff_{in}^t$ and ε, which can be expressed as equation (4).

$$f_{in}^t = \begin{cases} 1, diff_{in}^t \le \varepsilon \\ -1, diff_{in}^t > \varepsilon \end{cases} \quad (4)$$

Here, ε is the largest evaluation difference which can be tolerated. $diff_{in}^t$ is the evaluation difference of entity n providing by server i, it can be expressed as equation (5). Equation (4) denotes that if $diff_{in}^t \le \varepsilon$, we can consider that the recommended information is reliable, so set f_{in}^t =1; in contrary if $diff_{in}^t > \varepsilon$, set f_{in}^t =-1.

$$diff_{in}^t = \left| e_{in}^t - \sum_{i=0}^{m} \frac{e_{in}^t \times C_i^{t-1} \times S_{in}^t}{\sum_{i=0}^{m} C_i^{t-1} \times S_{in}^t} \right| \quad (5)$$

Here, $i \in [0, m]$, S_{in}^t is the time decay of the recommendation trust about entity n offered by server i at time t. C_i^{t-1} is the current feedback credibility of server i at time *t-1*. $e_{in}^{t} \in [0,1]$ is the recommendation trust of entity *n* offered by server *i*. Recommendation trust e_{in}^{t} is a subjective evaluation provided by entity *i* who had interacted directly with entity *n*. e_{in}^{t} is assigned according to two aspects: whether the resources that the entity *n* has occupied are beyond the specified scope, and whether the behaviors of entity *n* are permitted or not. If abused resources and had malicious behaviors, entity *n* would be assigned a low recommendation trust e_{in}^{t}, otherwise entity *n* would be assigned a high recommendation trust e_{in}^{t}.

2.4 The Combined Recommendation Trust

The combined recommendation trust RT_{n}^{t} is a combined evaluation that all recommenders to entity *n*, it is calculated as follows:

$$RT_{n}^{t} = \sum_{i=0}^{m} \frac{e_{in}^{t} \times C_{i}^{t} \times S_{in}^{t}}{\sum_{i=0}^{m} C_{i}^{t} \times S_{in}^{t}} \tag{6}$$

Here, $e_{in}^{t} \in [0,1]$ is the recommendation trust of entity *n* offered by server *i* at time *t*. S_{in}^{t} is the time decay of the recommendation trust about entity *n* offered by server *i* at time *t*. C_{i}^{t} is the current feedback credibility of server *i* at time *t*.

2.5 The Overall Trust

The overall trust is the combination of direct trust and the combined recommendation trust. It is expressed as Equation (7):

$$T_{in}^{t} = \mu \times T_{in}^{t-1} + \nu \times DT_{in}^{t} + \eta \times RT_{n}^{t} \tag{7}$$

Here, μ, ν, ρ represent the weighting of the last time trust, direct trust and combined recommendation trust. T_{in}^{t-1} denotes the last trust value of the entity *n*, DT_{in}^{t} denotes the direct trust value of entity *i* to entity *n*, RT_{n}^{t} denotes the combined recommendation trust value of the entity *n*.

3 Simulation Experiments

In order to evaluate the effectiveness and anti-attack of our trust computing method, we had made experiments in simulation software omnet++ 4.0. In simulation experiments, there are 40 users and 60 servers in 4 domains. In every domain, there are an agent and a trust server. The agent achieves inter-domain communication, receives the accessing requests and makes the security decisions. The trust server computes feedback credibility, combined recommendation trust and overall trust value.

3.1 The Effectiveness of Trust Computing Method

Firstly, we tested and computed the change of feedback credibility of honest servers and malicious servers under different conditions. There were legal users, malicious users, honest servers and malicious servers in the simulation experiments.

The ratio of malicious entities to legal entities was 1:2. Feedback credibility initializes 0.5. In experiment, we observed 4 malicious recommending entities (m1~m4) and 4 honest recommending entities (h1~h4) with different credible weighting factors (β,α), such as h1(0.2, 0.2) denoted honest recommending entity h1 with credible weighting factors β =0.2 and α =0.2. The change of feedback credibility of recommending entities with different weighting factors (β,α) according to equation (3) has been shown in Fig.1.

The experiment results show that feedback credibility of the honest recommending entity kept increasing and in contrary feedback credibility of the malicious recommending entity kept descending with fluctuating. It can be seen that the change of feedback credibility accords with the rule of the fast decrease and slow increase. The fluctuation was appeared because the system could not absolutely identify collusive beguiling behaviors of malicious recommending entities.

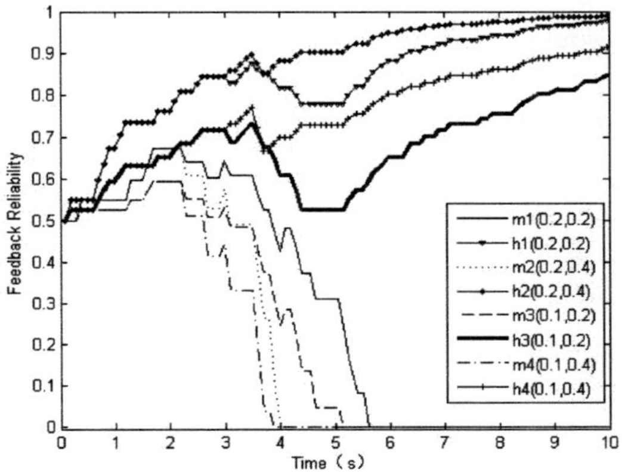

Fig. 1. The change of feedback credibility with different weighting factors

Secondly, based on direct trust, combined recommendation trust and feedback credibility, we tested and calculated the trust value of the entity. In experiment, users were classified as follows: honest user, malicious user, cycle user and random user. Honest user kept legal behaviors and malicious user kept illegal behaviors all the time. Cycle user accessed resources with some legal behaviors and an illegal behavior in turn. Random user had uncertain behaviors. The initiation of user's overall trust is 0.5.

Table 1. Analysis and Simulation Parameters

parameters	α	β	ρ	ε	μ	ν	η
values	0.2	0.1	0.9	0.05	0.2	0.6	0.2

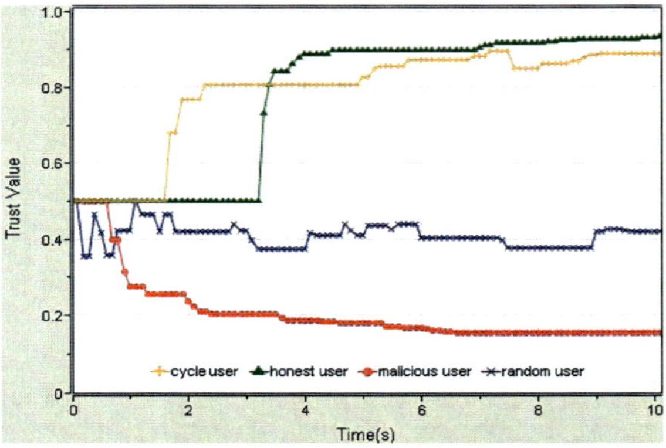

Fig. 2. The change of trust value

Several useful parameters for analyses and simulations are given in Table 1. The change of trust value has been shown in Fig.2. From Fig.2, we can see the honest user's trust value would keep rising with its good behaviors. Because of illegal behaviors, the malicious user's trust value would reduce to a small threshold, and the malicious user couldn't any more get the services, so its trust value would not update again. The cycle user's legal behaviors are more than illegal behaviors; therefore its trust value kept rising on the whole, but was under the honest user's trust value. The random user's trust value fluctuated randomly. The change of user trust value is consistent with user's behaviors on the whole. It indicates our method can identify malicious behaviors and restrain malicious users.

3.2 Anti-attack of the System

Success interaction rate is the ratio of successful interactions to overall interactions in the simulation time. It can reflect the anti-attack of the system in a certain extent. Thus we measure anti-attack of the system by success interaction rate. In experiment, we had done some comparative tests with same conditions, such as same honest and malicious users, etc., under trust computing method and without trust computing method respectively. The experiment result has been shown in Fig.3.

From Fig.3, we can see the success interaction rate of the system with trust computing is similar with it without trust computing at the beginning. After a time, the success interaction of the system with trust computing can maintain a high level and the success interaction of the system without trust computing can keep falling. It is reasonable in that the system with trust computing can identify the malicious entity

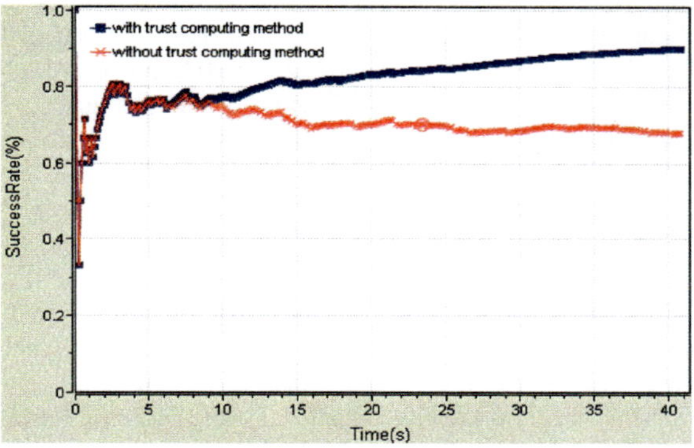

Fig. 3. The change of success interaction rate

and refuse to provide service. So the success interaction rate was enhanced with time increasing. However, the system without trust computing would accept all of requests so some interactions would suffer defeat after defeat. It has shown trust computing can enhance the anti-attack of the system because it can help to the system correctly identifying the malicious users.

4 Conclusions

This paper proposed a trust computing method based on feedback credibility. Feedback credibility can help the system identifying the reliability of recommendation trust and punish malicious recommendation to some extent. And the overall trust based on direct trust and combined recommendation trust according to feedback credibility can correctly identify the honest users and malicious users and compute their trust value. Simulation experiments show that our method can effectively identify the mendacious recommendation and malicious entities, and provide reliable information to correctly make the security decisions for the system. Future, we will look for ways to make the approach more robust against collusive deception behaviors.

Acknowledgments. This work is supported by the National Science Foundation of China (No.60862001) and the Foundation of Guangxi education department (No.201010LX156).

References

1. Blaze, M., Kannan, S., Lee, I., et al.: Dynamic Trust Management. Computer 42(2), 44–52 (2009)
2. Almenarez, F., Marin, A., Diaz, D., et al.: Developing a Model for Trust Management in Pervasive Devices. In: Proc. of the 3rd IEEE Int'l Workshop on Pervasive Computing and Communication Security (PerSec 2006), pp. 267–272 (2006)

3. Kamvar, S.D., Schlosser, M.T.: EigenRep:Reputation Management in P2P Networks. In: Proc.of the 12th Int'l World Wide Web Conf(WWW 2003), pp. 123–134 (2003)
4. Zhou, R., Hwang, K.: PowerTrust: A Robust and Scalable Reputation System for Trusted Peer-to-Peer Computing. IEEE Transactions on Parallel and Distributed Systems 18(4), 460–473 (2007)
5. Xiong, L., Liu, L.: PeerTrust:Sopporting Reputation-based Trust for Peer-to-Peer Electronic Communities. IEEE Transactions on Knowledge and Data Engineering 16(7), 843–857 (2004)
6. Srivaramangai, P., Srinivasan, R.: Enhancements to Reputation based Trust Models for improved reliability in Grid Computing. WSEAS Transactions on Computers 10(7), 220–233 (2011)
7. Azzedin, F., Ridha, A.: Feedback Behavior and its Role in Trust Assessment for Peer-to-Peer Systems. Telecommunication Systems 44(3), 253–266 (2010)
8. Zhang, R.-L., Wu, X.-N., Zhou, S.-Y., Dong, X.-S.: A Trust Mode Based on Behaviors Risk Evaluation. Chinese Journal of Computer 32(4), 689–698 (2009)

The Factors Impact Customer Satisfaction in Online Banking Sector: The Chinese Perspective*

Zhengwei Ma

School of Business Administration,
China University of Petroleum Beijing,
Changping, Beijing, China
mzw8632425@yahoo.com.cn

Abstract. The purpose of this research is to study factors that may influence customer satisfaction in online banking sector in China. Moreover, the paper also aims to understand the relationship between customer satisfaction and website quality, and to find some major factors for keeping high level customer satisfaction in online banking sector. The paper describes the positive effect of website quality about customer satisfaction in online banking sector. After the validation of measurement scales, the hypothesis is contrasted through structural modeling. Finally, the authors validate the hypothesis and a measurement model in the paper. The data showed that website quality have direct and significant effect to customer satisfaction in online banking sector. Besides this, the authors found that some factors are positively related to customer satisfaction. Finally, it is observed that efficiency, interactivity, security, information, ease of use and content are major factors to affect customer satisfaction in the online banking sector. This study proposes a model for analyzing empirically the link between website quality and customer satisfaction in the online banking sector.

Keywords: Website quality, Customer satisfaction, Online banking sector.

1 Introduction

Given the fact that banks invest billions in the internet infrastructure, customer satisfaction and customer retention are increasingly developing into key success factors in e-banking (Bauer et al., 2004). But low customer satisfaction is a major encumbrance to depress development of online banking service in China. Patricio et al.'s (2003) focus-group study found that customers with different patterns of use (e.g., frequency of use and type of operations performed) for an e-banking service tend to value different web-site attributes, several of which are related to website quality. Therefore, website quality has significance relationship with customer satisfaction. Website quality is a key factor to affect customer satisfaction (Arne Floh &Horst Treiblmaier, 2003).

In the present study, the creation of the measurement variables went as follows.

* Sponsored by Research Funds of China University of Petroleum-Beijing.

D. Zeng (Ed.): Advances in Control and Communication, LNEE 137, pp. 625–632.
springerlink.com

Initially, lists of variables from existing academic paper were compiled that would capture the five broad dimensions of website quality identified by Zeithaml et al. (2002). They are information quality, privacy/security, graphic style, ease of use and fulfillment. Where theory is less than well developed, it is beneficial to use both practical and academic perspectives (Hensley, 1999, p. 348). Therefore, using these base variables, several times iterative focus-group discussions were conducted with mid-level managers from several state-owned commercial banks' online banking departments in an attempt to choose one item to adequately represent each of the main quality dimensions in online banking service. During these focus-group discussions, it was considered important, for the topic of website quality. The author breaks the notion of quality into nine dimensions. They are information, efficiency, interactivity, security, technology, content, accuracy, design and ease of use.

2 Literature Review

Security is the freedom from risks, doubts or danger. It involves financial security, physical safety and confidentiality. It is one important fact that may affect customers' intention to adopt online banking services. Encryption technology is the most common feature at all bank websites to secure information privacy, supplemented by a combination of various unique identifiers, such as mother's maiden name, a memorable date or a password. Besides, the "secure socket layer," a widely-used protocol for online credit card payments, is designed to provide a private and reliable channel between two communicating entities. So, security is dimension that should be taken account of when measuring online banking website quality.

Interactivity is concerned with how an online banking website interacts with its customers or visitors. It is defined as the facility for customers and online banking websites to communicate directly with one another. Lowry et al (2006) asserted that the more interactive a website is, the more likely a user is to experience satisfaction. Two-way communication and active control of users are considered critical facts of interactivity, and they both play a key role in determining customer satisfaction (Hoffmanet al., 2000). Hoffman et al. (2000) underlined the importance of interactivity, which also determines website quality.

Furthermore, Zeithaml et al. (2002) reported a significant correlation between efficiency and online banking quality. Downloading speeds and response time, in consumers' perception, are two crucial facts of online banking efficiency. Downloading speed depends on the nature of the website from which information is downloaded, the computing hardwares and method of connection used to download information (Jayawardhena& Foley, 2000).

Website quality is important, because it enhances customer loyalty (Boyer &Hult, 2005), is a key to the success of online services (Reichheld & Schefter, 2000). Zeithaml et al. (2002) and Parasuraman et al. (2005) provide excellent summaries of many of these studies, in which information quality is identified as one the main fact of website quality. Among other results, Novak et al. (2000) found, through structural equation modeling, that information quality has a significant impact on website quality. In view of these findings, information should include in when a researcher measures website quality.

Ease of use is defined as how easy it is for customers to use the website. All websites should be user-friendly, making it easy to navigate, search and use. Ease of use, as an important fact of online banking quality, has been highly rated in website quality measurement and has been noted by some scholars (Zeithaml et al., 2002; Yang, 2001; Fassnacht&Koese, 2006). So, it should be taken account of when measuring online banking website quality.

Pikkarainen et al. (2004), claim that the content on an online banking website will influence the website's acceptance among customers. The factor "content" denotes the design of the service. If it is clearly understood and updated, it creates value if such designs fit customers' needs. Content, needless to say, is a dimension that will affect website quality in online banking services.

If online banking information is accurate, objective, authentic and up-to-date, it will be considered reliable. Madu and Madu (2002), as well as Xiao and Dasgupta (2002), found that accuracy of information played an important role in the formation of customer satisfaction. Accurate information is that can be used effectively for a given purpose. Therefore, many scholars (Waite &Harisson, 2002) believed that it is essential that accurate information be introduced into online banking services and that accuracy is an important benchmark by which online banking customers judge the website quality.

Zeithaml et al. (2000) state that online service is a web service delivered through the Internet. In online services, customers interact with or contact service providers through technologies. Customers have to rely entirely on information technologies in an online service encounter (Zeithaml, 2000). An online banking service's deeds, performance or effort are mediated by IT (information technologies). Zeithaml (2002) stated that some dimensions of SERVQUAL may be applied to online service quality, but in online service there are additional facts, many of which are specifically related to technologies. So, technologies are a dimension that affects website quality.

In the virtual environment of online service for customers, a website is a main access to online banking and to a successful online process. Thus, a deficiency in website design can result in a negative impression of the website quality, and result in customers exiting an online banking transaction. An online banking website is a starting point for customers to gain confidence with the banks (Li, 2009). Website design can influences customers' perceived image of a bank. With good navigation and useful information on its website, a bank may easily attract customers to its online banking services. So, clearly, design is an important aspect of website quality.

Lee et al. (2009) study showed that customer complaints had a direct influence on customer satisfaction. They regarded that as one-dimensional attributes increased, the level of overall customer satisfaction also increased. Ahmed et al. (2003) found that major gains in customer satisfaction were likely to come from an alleviation of complaints. These scholars, overall, concur that the number of complains is an index of overall customer satisfaction. This is why, in the present study, the numbers of complaints were considered to measure customer satisfaction.

Service quality is defined as a long-term cognitive judgment (Gwynne et al., 2004) regarding an organization's "excellence or superiority" (Parasuraman et al., 1988, p. 15). A customer-oriented quality strategy is critical to service companies as it drives customers' behavioral intention with, for instance, highly perceived service quality leading to repeat patronage and customer loyalty (Zeithaml et al., 2000; van Riel et al., 2003). Accordingly, substandard service quality will lead to negative word-of-mouth, which may result in a loss of sales and profits as the consumers migrate to competitors (Zeithaml et al., 2000; Yang, 2004). These factors stress the importance of delivering high-level services, especially within an electronic environment, where consumers can readily compare service firms and where switching costs are low (van Riel et al., 2001).

3 Constructs for the Present Study and Hypothesis

According to the possible connection between customer satisfaction and website quality in handling private data, a direct relationship might be established between the two aspects. And follow the prior study; one construct is addressed in the present study: customer satisfaction and website quality, all of which are elaborated in prior paragraphs. The relationships between these constructs, as embedded in the hypothesis, are now illustrated in the following Figure 1.

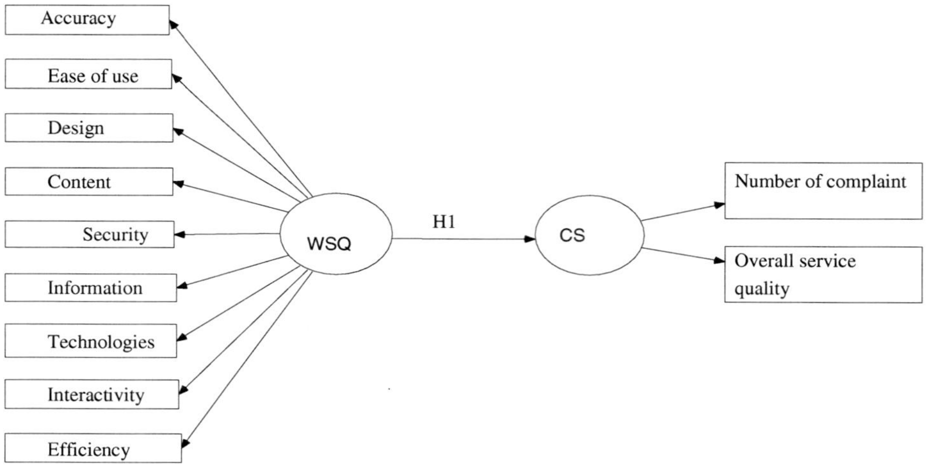

Fig. 1. Research Factors and Hypotheses in the Present Study

Taking into account the previous considerations, the relationship between customer satisfaction website quality and is evident in personal data handling and should be examined in greater detail. With the aim of testing this connection in the online banking customer satisfaction, the following hypothesis is proposed:

H1.There will be a positive relationship between website quality and customer satisfaction.

4 Data Collection and Measures Validation

The generation of the initial questionnaire was ascertained by experts and managers interviews at banks as well as through in-depth discussions with online banking customers. Pre-tests of the initial 24-item questionnaire was carried out with 30 online customers to improve the questionnaire. The resulting modified 11-item pool was presented to customers of online banking in drop in survey. Respondents were asked to refer to their own online banking service (the one they use regularly) when answering the questionnaires. Non-random method of collecting the data (volunteer sampling) generated 198 fully usable questionnaires. The questionnaires of collection are non-random samples. So authors compared some of the survey results with available information about the population. The results are very similar and as a consequence, authors may conclude that our sample represents the profile of the average online banking customers.

The available measuring process started with an initial exploratory analysis of reliability and dimensionality (Anderson and Gerbing, 1988).The Cronbach alpha indicator was used to assess the initial reliability of the scales, considering a minimum value of 0.7 (e.g. Cronbach, 1970). The item-total correlation was used to improve the levels of the Cronbach alpha, considering a minimum value of 0.3 (Nurosis, 1993). All results were adjusted to the required levels (Casalo´ et al., 2007). Secondly, authors proceeded to evaluate the unidimensionality of the proposed scales by carrying out a principal components analysis. Factor extraction was based on the existence of Eigen values higher than 1. And, it was required that factorial loadings were higher than 0.5 points and a significant total explained variance (Hairet al., 1998). All results were adjusted to the required level (Casalo´ et al., 2007).

The Cronbach alpha indicator is the most frequently used test for assessing reliability. Some scholars consider that it underestimates reliability (Smithet al., 1974).Consequently, the use of composite reliability has been suggested (Jo¨reskog, 1971), using a cut-off value of 0.65 (Steenkamp and Geyskens, 2006). The results show the value for website quality is 0.922, and the value for customer satisfaction is 0.656.This is satisfactory.

5 Results and Discussion

Authors developed a structural equations model, which the objective of testing is the proposed hypothesis (Figure 2). Authors observed that the hypothesis was supported at the 0.01 level and, in a similar way. Model fit was acceptable (Chi–square =103.13, 43 df, $p < 0.001$; goodness-of-fit index (GFI) = 0.908; adjusted goodness-of-fit index (AGFI) = 0.859; Comparative Fit Index (CFI) = 0.968; Bollen(IFI)Fit Index= 0.968; Bentler-Bonett Normed Fit Index (NFI) =0.946;the root mean square error of approximation (RMSEA) = 0.084; normed Chi–Square = 2.398).

It was also notable that this model has allowed authors to explain at a very high level the website quality in customer satisfaction of online banking sector. Besides, according to the standardized estimates, authors may say that customer satisfaction is clearly and positively influenced by website quality in handling personal data (β=0.73). And authors found that securities, ease of use, information, efficiency, interactivity and content have larger effect than other three factors for customer satisfaction (β>0.85).

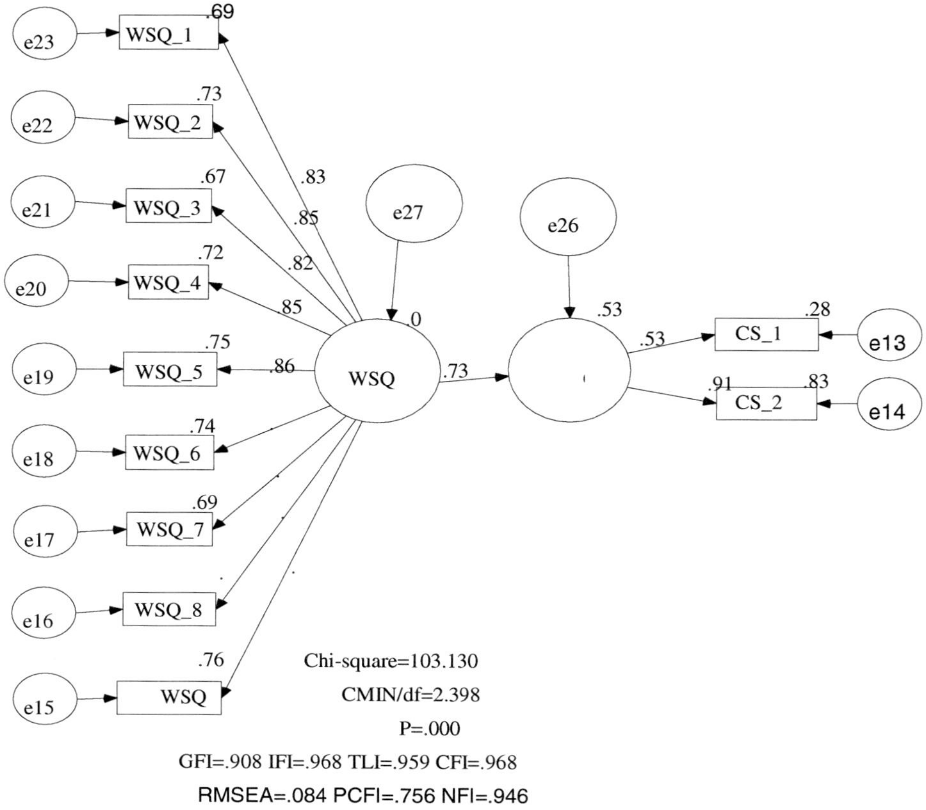

Fig. 2. The structural equation model

Review the results, online banking customer satisfaction cannot be described as one fact construct. Instead, it represents a multi-factor construct that is composed of website quality judgments with regard to the online service categories. This study provides validated measurement scales for each factor. The empirical results strongly support the understanding of website quality as integral solutions. Based on our findings, management can establish a sequential priority to improve website quality in online banking service. The sequential priority depends on the effect to customer satisfaction. For example, when limited resources become the barrier to improve all of nine factors, banks can improve the first efficiency, information, securities, interactivity, ease of use and content to be first step. And the banks could put accuracy, design and technology to be second step.

6 Limitations

There are several limitations to the study. First, the sample was China-focused, with all of the respondents residing in China. The participants in this survey may have possessed attributes and behaviors that differed from other countries in the world. Second, the sample was restricted to the customers of banks and may have possessed attributes and behaviors that differ from those of customers in other business sectors. Next, as mentioned earlier, in the data collection section, since it was impossible to send follow-up surveys, no attempts were made to ascertain the existence of non-response bias by comparing responses to the first-wave surveys with those to a second wave.

References

1. Ahmed, I., Gul, S., Hayat, U., Qasim, M.: Service quality, service features and customer complaint handling as the major determinants of customer satisfaction in banking sector: A case study of National Bank of Pakistan (2001), http://www.wbiconpro.com/5%5B1%5D.ISHFA.pdf (retrieved January 14, 2001); Boyer, K., Hult, G.: Extending the supply chain: Integrating operations and marketing in the online grocery industry. Journal of Operations Management 23(6), 642–61 (2001)
2. Anderson, J., Gerbing, D.: Structural modeling in practice: a review and recommendedtwo-step approach. Psychological Bulletin 103(3), 411–423 (1988)
3. Bauer, H.H., Hammerschmidt, M.: Financial portals in the internet. Paper Presented at the WSEAS Conference on E-Commerce, Janeiro, Brazil (October 2002)
4. Casalo, L.V., Flavian, C., Guinaliu, M.: The role of security, privacy, usability and reputation in the development of online banking. Online Information Review 31(5), 583–603 (2007)
5. Cronbach, L.J.: Essentials of Psychological Testing. Harper and Row, New York (1970)
6. Davis, R.D.: Perceived usefulness, perceived ease of use and user acceptance of information technology. MIS Quarterly 13(1), 319–339 (1989)
7. Doll, W.J., Torkzadeh, G.: The measurement of end-user computing satisfaction. MIS Quarterly 12(2), 259–274 (1988)
8. Fassnacht, M., Koese, I.: Quality of electronic services: Conceptualizing and testing a hierarchical model. Journal of Service Research 9(1), 19–31 (2006)
9. Hair Jr., J.F., Anderson, R.E., Tatham, R.L., Black, W.C.: Multivariate Data Analysis. Prentice-Hall, Englewood Cliffs (1998)
10. Jöreskog, K.G.: Statistical analysis of sets of congeneric tests. Psychometrika 36, 109–133 (1971)
11. Li, H., Suomi, R.: Evaluating electronic service quality: A transaction process based evaluation model. Paper Presented at the European Conference on Information Management and Evaluation (September 2007)
12. Lowry, P.B., Spaulding, T., Wells, T., Moody, G., Moffit, K., Madariaga, S.: A theoretical model and empirical results linking website interactivity and usability satisfaction. Paper Presented at the 39th Hawaii International Conference on System Sciences, Hawaii (September 2007)
13. Madu, C.N., Madu, A.A.: Dimensions of e-quality. International Journal of Quality & Reliability Management 19(3), 246–258 (2002)

14. Moore, G.C., Benbasat, I.: Development of an instrument to measure the perceptions of adopting an information technology innovation. Information Systems Research 2(3), 192–222 (1991)
15. Novak, T., Hoffman, D., Yung, Y.: Measuring the customer experience in online environments: a structural modeling approach. Marketing Science 19(1), 22–42 (2000)
16. Nurosis, M.: Statistical Data Analysis. SPSS Inc., Chicago (1993)
17. Parasuraman, A., Zinkhan, G.: Marketing to and serving customers through the internet: An overview and research agenda. Journal of the Academy of Marketing Science 30(4), 286–295 (2002)
18. Patricio, L., Fisk, R., Cunha, J.: Improving satisfaction with bank service offerings: Measuring the contribution of each delivery channel. Managing Service Quality 13(6), 471–482 (2003)
19. Pikkarainen, T., Pikkarainen, K., Karjaluoto, H., Pahnila, S.: Consumer acceptance of online banking: An extension of the technology acceptance model. Internet Research 14(3), 224–235 (2004)
20. Hensley, R.: A review of operations management studies using scale development techniques. Journal of Operations Management 17(3), 343–358 (1999)
21. Reichheld, F., Schefter, P.: E-loyalty: Your secret weapon on the web. Harvard Business Review 78(4), 105–114 (2000)
22. Smith, K.W.: On estimating the reliability of composite indexes through factor analysis. Sociological Methods & Research 2, 485–510 (1974)
23. Steenkamp, J.B.E.M., Geyskens, I.: How country characteristics affect the perceivedvalue of a website. Journal of Marketing 70(3), 136–150 (2006)
24. Van-Riel, A.C.R., Liljander, V., Jurriens, P.: Exploring consumer evaluations of eservices: A portal site. International Journal of Service Industry Management 12(4), 359–377 (2001)
25. Waite, K., Harrison, T.: Consumer expectations of online information provided by bank websites. Journal of Financial Services Marketing 6(4), 9–322 (2002)
26. Yang, Z.: Consumer perceptions of service quality in Internet based electronic commerce. Paper Presented at the EMAC Conference, Bergen, Norway (August 2001)
27. Zeithaml, V., Parasuraman, A., Malhotra, A.: Service quality delivery through web sites: A critical review of extant knowledge. Journal of the Academy of Marketing Science 30(4), 362–375 (2002)
28. Zeithaml, V.A.: Service quality, profitability and the economic worth of customers: What we know and what we need to learn. Journal of the Academy of Marketing Science 28(1), 67–85 (2000).

Post Environmental-Sustainable Assessment of the Soil and Water Conservation Project of Yanhe River Watershed in China

Chen Li

Department of Management Engineering,
Anhui Institute of Architectural and Industry,
Hefei, China
ChinaliChina@163.com

Abstract. The main purpose of assessment for the soil and water conservation project of Yanhe River Watershed is to summary the success or failure of the project and to improve the World Bank decision-making and implementation of future project. The conclusion of the post evaluation in the Yanhe River Basin environmental assessment of World Bank project is middle. According to the sustainable of Yan River basin six subsystems projects, we can draw that the sustainable development in Yanan City, Ansai County, Yanchang county, Baota is uneven.

Keywords: environmental assessment, sustainable assessment, the soil and water conservation project.

1 Introduction

Soil and water conservation project of Yanhe River Watershed in China is built in 1994, and it is the World Bank loan project. After four years of successful implementation, it has made remarkable achievements. In 1999, the World Bank began to administrate for soil and water conservation project two (That is project II). The main purpose of assessment for the World Bank loan project is to summary the success or failure of the project and to improve the World Bank decision-making and implementation of future project.

Project in loess plateau watershed rehabilitation is one large soil and water conservation project for our government by foreign investment. Total investment in the first phase of the project is 2.1645 billion yuan, and the loan of the World Bank is 150 million U.S. dollars. The main construction contents include land development, vegetation construction, nurseries, construction, small water conservancy facilities and support services. Basic objective of the project is "effective and sustainable manner through the use of land and water resources, increase food production and agricultural output" and " reduce soil erosion for the Yellow River ". They want to break a sustainable development model between poverty and ecological deterioration in the cycle. World Bank loan in Yanhe River Watershed began in 1999.It is completed in 2004, and the implementation period is five years.

D. Zeng (Ed.): Advances in Control and Communication, LNEE 137, pp. 633–639.
springerlink.com

2 Post Environmental Assessment of the Soil and Water Conservation Project

Post environmental assessment of the soil and water conservation project is an important part of the post-evaluation. The purpose of the project assessment is that soil and water resources are fully rational used, and beneficial effects can be fully tapped. At last, the ecological environment can achieve a virtuous circle.

Loess plateau watershed rehabilitation project is a positive benefit project of the environmental improvement. However, it is not only to protect the environment, but also to bring some negative effects. The situation must be implemented on the water and sediment, water, soil, biological and pests, rural energy, engineering safety, climate, environmental impact of construction, run-time environment monitoring and management and protection.

2.1 The Changes of Typical Plots in Soil Properties

Grooving in the soil properties in Yanan, and analysising the monitoring data, we can see that the various measures of soil physical and chemical properties change significantly.

Soil moisture has been significantly increased. The soil of the land has been a fundamental change in the environment by the comprehensive management of small watershed. Increasing the infiltration of surface runoff, we can control the soil erosion. Thus we can also control measures to increase soil moisture. The average annual rate of moisture content in sloping than that of terraces is 1.66 percent. The average annual rate of moisture content than that of wasteland is 1.37 percent. At the same time, soil fertility is enhanced.

The implementation of the project area in loess plateau watershed rehabilitation is to improve the physical and chemical properties of soil, improve soil fertility, and increase soil moisture. Measures in Yanhe River Watershed project area are to improve the soil N, P, K and organic matter .The results show that the nutrient content of the soil slope land and wasteland is not changed. The content of soil nutrients in the six engineering measures showed an increasing trend. After the refurbishment of terraced micro-topography, erosion control, and storing the runoff and sediment, it is to increase soil moisture, maintain soil fertility, and improv soil aggregate structure and micro-organisms living environment.

2.2 Vegetation Changes in the Typical Project Area Land

We selected representative forest of artificial trees, shrub, orchard and pasture as examples. From 1999, we continuous to have a typical block in the project area every year.

Total of reforestation, grass and orchard is 41253hm^2, and degree of vegetation is 17.3 percent in project management early time. In recent years, total of afforestation, grass, and orchard is 85769.7hm^2.Degree of cover area is 78210hm^2, and vegetation degrees increased to 32.7 percent. The project increased vegetation cover, restore and protect biodiversity. Large-scale project engineering measures in soil and water conservation district, especially the large area of forests and pastures (or block) protected effectively the diversity of the ecosystem system, and thus produce a major impact in the genetic diversity of biological species.

2.3 Effectiveness of Water and Soil Conservation

Integrated watershed management monitoring of water and soil conservation benefits shown in Figure 1. The results showed that a small water catchment management and soil conservation benefits is obvious .Degree of treatment decreased from 44.2 percent increased to 69.2 percent in recent year. Water management measures effect was obviously.

The difference between the control measures and actual soil conservation mainly due to floods in recent years, and little rain, soil erosion is not seriously. Project area and soil conservation measures reduced the flood disasters and siltation of river bank, especially in reducing the yellow sediment .It is that control soil erosion effectively.

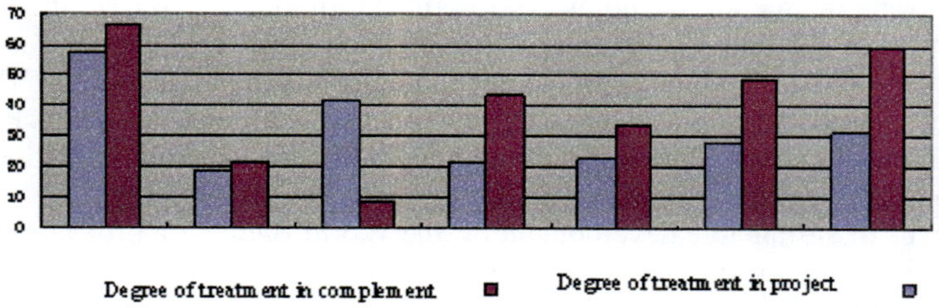

Fig. 1.

Post evaluation in the Yanhe River Basin environmental assessment of World Bank projects. First, we need to establish the environmental evaluation system of the World Bank loan in Yanhe River area. World Bank Loan Project environmental and ecological subsystems: tree forest, shrub forest, economic forest, artificial grass, an area of water conservation measures, control the degree of the total area percentage of the benefits of sediment reduction; resource subsystem: the proportion of agricultural land, orchard land-use ratio, pasture land, per capita basic farmland, the total irrigated area, water-saving irrigation, terracing, dam, water, land.

The conclusion of the Post evaluation in the Yanhe River Basin environmental assessment of World Bank projects is middle.

3 Post Sustainable Assessment of the Soil and Water Conservation Project

3.1 The Structure of the Land Has Been Rationalized, and the Production Conditions Has Been Improved

After the soil and water conservation project has been implement several years. barren hills and slopes have been developed fast. Agricultural economics has been changed from a single phase of food production to agriculture, forestry, fruit, animal husbandry, and laid the foundation for sustainable development.

3.2 Rural Infrastructure Has Been Strengthened

The structure of the land has been rationalized. Production conditions of the soil and water conservation phase II project has been improved in Yan River Basin through the controlling of the soil erosion. A substantial increase in vegetation cover, so the trend of deteriorating ecological environment has been effectively curbed.

The rural roads has been extend in all directions. Drinking water problem is completely resolved. The labor intensity decreased significantly. A large number of workers has been liberated. The secondary and tertiary industries made the agricultural production and rural economic a substantial increase quickly. The sustainable development in Yanan city, Ansai county, Yanchang county, Baota is uneven.

According to the sustainable of Yan River basin six subsystems projects, we can draw the following conclusions: Ansai county economic development rate has increased ,but degree of economic development in other parts of the economic subsystem did not increased. Degree of technological development and social development in Ansai county, Baota technological is relatively good. Degree of ecological subsystem and resource development in Baota is ideal. Population subsystem development degree of Ansai county and Yanchang county is better than other counties.

The focus of sustainable development of the World Bank loan project are follows: (1) A prerequisite to achieve sustainable development projects is optimizing and maintaining the environment. (2) The rational use of World Bank loans is the base of achieving sustainable development. (3) Personnel training and strengthen researching project is the ensurience to achieve sustainable development.

4 Assessment Conclusion

World Bank Loan Project Yanhe River Basin is mainly reflected in the effectiveness of the "three increased, three decreased, three improvement", that is an increasing of basic farmland, increasing of food production, increasing of people's income, the area reducing of sloping land, soil erosion reducing, poverty reducing; improvement of agricultural production conditions , improvement of the ecological environment, improvement of people's living standards. It promoted the using of water resources and ecological environment for sustainable maintenance in China. It played an important economic and social development foundation effect.

Yan River Basin World Bank project evaluation can be drawn: the income of farmers increased year by year,and living standards have improved significantly. Project area of social and ecological environment has been improved, and serious soil erosion has been effectively curbed. People living in the project area, greatly improved,and production conditions and living standards have been improved markedly.The rapid development of cultural and educational health, social status of women improved considerably. Regional weather patterns in the distribution has been apparently changed in control and soil erosion has been controlled effectivly. From the sustainable development project in the Yanhe River Basin, we can conclude that sustainable development of Yan'an City, Ansai County, Yanchang county, Baota is uneven, and sustainable development is good.

4.1 Experience in Project Implementation

4.1.1 The Participation of Farmers Is the Fundamental Guarantee for Success of the Project

Improving agricultural production conditions and increasing their income, farmers in the area eliminate poverty as the fundamental goal of the project. So farmers are both participants in the project and the fundamental guarantee for the smooth implementation of the project.

The design of the project plan is to fully respect the wishes of farmers, and in the implementation process is based on the implementation results and farmers can adjust the content of the project. Completion of the project depends on the content of the active participation of farmers, such as large-scale eco-grass planting, terracing and so can not do without the active participation of farmers. In addition, farmers have local knowledge and project management and protection of the results in long-term benefits, such as eco-forestry, pasture management and protection is entirely dependent on the corresponding land contractors.

4.1.2 Sound Organization and Management, System Security Is the Organization to Ensure Successful Implementation of the Project

In project implementation, the project leading group, the technical committee and project office effectively integrated and coordinated management played a role in ensuring the full implementation of the project objectives. Sound system and the smooth organization and management of high-quality and relatively .Stable team is to ensure the smooth operations of the project. Effectiveness of project is not only the result of a single measure, but also is more dependent on project interventions interwoven system to achieve. The World Bank project related systems on the macro level, as well as the specific circumstances of each project area designated under the rules.

4.1.3 Sound Financial Management System Is to Ensure the Successful Implementation of the Project

According to relevant provisions and requirements of the World Bank, a set of financial management and other aspects of the way to the use of funds and provisions is the guarantee for the project funds. (1) Project funds and foreign-funded infrastructure projects is combined. (2) World Bank pay for the use of project funds and conduct a comprehensive check every year. (3) Establishing and improving a financial audit system can ensure the rational use of project funds.

4.1.4 Deepen the Land Property System and Implementation of Grazing Prohibition Is a Policy to Ensure the Successful Implementation of the Project

During the project preparation, we provide the full implementation of the project area to land and contract policies to ensure the same 30 years of basic farmland, forest and grass for 50 years. Therefore, in the implementation period, the land contract has been quickly implemented in the project area, which greatly aroused the enthusiasm of farmers to control soil erosion and to improve the ecological environment of enthusiasm. They speed up the pace of governance and make the results of the post-treatment management .The protection has been implemented, and efficiency has been improved. On the policy side, the entire project area has issued a number of policies and management methods, and they received good results.

4.1.5 Training of Science and Technology Is the Ensure of the Successful Implementation of Project

Through training, which greatly improved the levels of management and their quality of farmers in the project area, may result in excellent economic, social and ecological benefits. Through studying, the professional quality of project staff and management level not only have been significantly improved, but also have been teach farmers some methods and experience ,such as scientific governance, poverty alleviation.

Yan River Basin actual situation of the project area, widely promote the use of agricultural science and technology of new technologies, and enhance water retention capacity, and improve the survival rate of trees. At the same time, provide technical guidance and services in the production activities of farmers. That is to say, such a comprehensive technology system can not only create significant existing benefits, skills training for farmers in the project area, but also provide better development of human resources.

4.2 Problems in Project Implementation

The standard of Eco-efficient measures supplement is too low, and it is affected the balanced development of the measures. Progress in the implementation of the project is differences between the counties. Some counties exceeded the project tasks and investment, but individual counties are unable to complete the task. Farmers are able to accept the ecological benefits of the project, but they are difficult to accept poor economic shrubs which lead to plant quality, and affecting the rate of seedling survival and preservation.

In addition, the local financial difficulties has led to more or less project implementation progress and the soil and water conservation project of Yanhe River Watershed.

Acknowledgment. CHEN Li (1966.12), Prof. Dr. Department of Management Engineering Anhui Institute of Architectural and Industry Hefei, China.This research has been funded by the Natural Science Project of Anhui Education Department (Project No.: KJ2009A67). Scientific Research Project, Doctor- Master of aiai(2009) NO.94: Study on the theory and method of post-project evaluation of intelligent integrated modeling.

References

1. Jennings, N.R., Faratin, P., Norman, T.J., et al.: Implementing a business process management system using adept: A real-world case study. International Journal of Applied Artificial Intelligence 14(5), 421–465 (2000)

2. Lester, A., Ing, E., Ceng: FICE. FI Meche FI Structure. MAPM. Project Planning and Control, 3rd edn., pp. 326–355. Reed Education and Professional Publishing Ltd., USA (2000)
3. Leu, W.S., Williams, W.P., Bark, A.W.: Development of an environmental impact assessment evaluation model and its application. Taiwan case study. Environmental Impact Assessment Review (16), 115–133 (1996)
4. Broad, K.: Book reviews of the international handbook of social impact assessment. conceptual and methodological advances. Agricultural Systems (83), 101–111 (2005)
5. Chen, L., Zhu, W.-D.: The enviromental quality assessment of neural network algorithm trained by particle swarm optimization. Acta Ecologica Sinica (3), 1072–1079 (2008)

Semi-supervised Laplacian Eigenmaps on Grassmann Manifold

Xianhua Zeng

Institute of Cognitive Computing, College of Computer Science & Technology, Chongqing University of Posts and Telecommunications, Chongqing 400065, China
xianhuazeng@gmail.com

Abstract. Laplacian Eigenmaps (LE) is a classical nonlinear manifold learning algorithm defined on Euclidean space. One contribution of this paper is to extend the LE algorithm on Grassmann manifold which is the set of linear subspaces. Another contribution is to integrate the label information of training subspaces and the geometric distribution information of all subspaces. A Semi-Supervised Laplacian Eigenmaps algorithm on Grassmann Manifold (SSLE-GM) is proposed. Experimental results on Cambridge gesture database showed that SSLE-GM obtained the highest recognition rate by comparing with Mutual Subspace Method (MSM), Constrained Mutual Subspace Method (CMSM), Tensor Canonical Correlation Analysis (TCCA), Canonical Correlation Analysis (CCA), and probabilistic Latent Semantic Analysis (pLSA).

Keywords: Laplacian Eigenmaps, Grassmann manifold, Semi-supervised.

1 Introduction

One of the central problems in machine learning, computer vision and pattern recognition is to develop appropriate representations for complex data. Most of traditional manifold learning algorithms assume that those observed data lie on or close to intrinsic low-dimensional manifolds embedded in the high-dimensional Euclidean space. But in some practical application fields of computer vision and pattern recognition, many tasks are considered as learning problems over vector sets or image sets (included videos)[1]. And the robust performance of object recognition can be achieved by efficiently using set information rather than a single vector or image as input of the model.

Laplacian Eigenmaps (LE) [2,3] is a classical nonlinear manifold learning algorithm defined on Euclidean space. It assumes that neighborhood points in Euclidean space stay as close as possible in low-dimensional space. Both constructing the neighborhood graph and computing the neighborhood similarities are based on Euclidean distance metric, as introduced in section 2. It is necessary that LE should be extended to process the set data which consist of some sets such as image sets, vector sets, videos etc. we use linear subspace to describe this kind of data. And Grassmann manifold is the set of linear subspaces with the fixed dimension [4]. Two points on Grassmann manifold correspond to two subspaces. In this paper, one of the contributions is to extend LE on Grassmann manifold. Another contribution is to use

D. Zeng (Ed.): Advances in Control and Communication, LNEE 137, pp. 641–648.
springerlink.com

the label information of training subspaces and the geometric distribution information of all subspaces included those labeled subspaces and the unlabeled subspaces. So a semi-supervised Laplacian eigenmap algorithm on Grassmann manifold is proposed, named as SSLE-GM algorithm. As for those set recognition tasks, Mutual Subspace Method (MSM) [5] and Constrained Mutual Subspace Method (CMSM) [6] are two classical linear methods for subspace recognition. The proposed SSLE-GM can find a mapping relation from Grassmann Manifold to low-dimensional Euclidean space. Specially, SSLE-GM can find intrinsic low-dimensional structures of subspaces and obtain the higher recognition rate than traditional subspace recognition methods.

The rest of this paper is organized as follows: Section 2 mainly introduces the classical Laplacian Eigenmaps algorithm. In Section 3, the construction of neighborhood graph and the neighborhood similarities are analyzed by exploiting the principle angles between subspaces. Then, a semi-supervised Laplacian eigenmaps algorithm on Grassmann manifold is proposed. Experimental results on Cambridge hand gesture database are shown in Section 4. Finally, we give some concluding remarks and future works in Section 5.

2 Laplacian Eigenmaps (LE)

Laplacian Eigenmaps algorithm (LE) is a classical manifold learning algorithm with the most theoretical foundation [2,3,7]. The research of LE algorithm has become a hot issue in the field of manifold learning. LE can be understood as an algorithm framework based on the Laplace-Beltrami operator for data representation and dimensionality reduction. Its intuitive idea is to make neighborhood points in observed space stay as close as possible in low-dimensional space. One of LE's main advantages is that a natural clustering can emerge in low-dimensional space. LE algorithm includes: building the neighborhood graph, choosing the weights for edges in the neighborhood graph, eigen-decomposition of the graph Laplacian matrix and forming the low-dimensional embedding. The key steps of LE are described as follows:

LE Algorithm:

Input: $x_i \in R^D (i = 1, 2, \cdots, n)$, observed data in Euclidean space

$\boldsymbol{k}$, the number of nearest neighbors based on Euclidean distance

Output: $u_i \in R^d (i = 1, 2, \cdots, n), U = \{u_1, ..., u_n\}$, intrinsic $\boldsymbol{d}$-dimensional coordinates in Euclidean space

Procedures:

Step1, Construct the neighborhood graph $\boldsymbol{G}$ by finding k- nearest neighbors of each data point $x_i \in \boldsymbol{X}$ and connecting these edges.

Step2, Compute the neighborhood similarities W_{ij} by choosing Heat kernel or simple mode.

Heat kernel: $W_{ij} = \exp\{-\|x_i - x_j\|^2 / 2t^2\}$ if x_i, x_j are connected on the graph G, and where $\boldsymbol{t}$ is Heat kernel parameter.

Simple-mode: if x_i, x_j are connected, $W_{ij} = 1$; otherwise $W_{ij} = 0$.

Step3, Compute low dimensional embedding $U = \{u_1,...,u_n\}$ by optimizing the following objection function:

$$\begin{cases} \underset{U}{Min}\, L(U) = \underset{U}{Min}\, tr(ULU^T) \\ Subject \quad to: UU^T = I \end{cases} \tag{1}$$

where Laplacian matrix L=D-W, and diagonal matrix $D_{ii} = \sum_j W_{ij}$.

In Step 3 of LE algorithm, it is equivalent to compute the bottom ***d***+1 eigenvectors $v_0, v_1, ..., v_d$ associated with the ***d***+1 smallest eigenvalues of matrix $L = D - W$ [2]. The first eigenvector (composed of 1's) is excluded. The remaining ***d*** eigenvectors with corresponding to ***d*** smallest nonzero eigenvalues form the embedding $U = \{u_1,...,u_n\}$ where the column vector presentation of the *i*-th low-dimensional coordinate point $u_i = [v_1(i),...,v_d(i)]'$.

3 Semi-supervised LE on Grassmann Manifold (SSLE-GM)

Facing with the visualization and recognition problems over vector sets or image sets (included videos), the observed data consists of sets. It assumes that each set in the observed data can be presented by a linear subspace. Because a Grassmann manifold is the set of linear subspaces with the fixed dimension and any two points on Grassmann manifold correspond to two subspaces, so the extension of LE algorithm on Grassmann manifold mainly is to define the distance metric between subspaces, to construct the neighborhood graph and to computing the neighborhood similarities between linear subspaces.

3.1 Distance Metric between Subspaces

The angles between linear subspaces intuitively become a reasonable measurement for describing the distance between subspaces. So the similarity metric between two subspaces is defined by usually using the basic principal angles. A popular metric method is the cosines of principal angles between any two ***m***-dimensional linear subspaces. The cosines of principal angles are referred to as canonical correlations [1, 8, 9]. The canonical correlations between two subspaces are defines as follows:

Definition 1. Let Y_1 and Y_2 be two ***m***-dimensional subspaces. The cosine of the ***k***-th principal angle (that is, canonical correlation) between two subspaces span(Y_1) and span(Y_2) are defined by:

$$\begin{cases} \cos(\theta_k) = \max\limits_{u_k \in Span(Y_1)} \max\limits_{v_k \in Span(Y_2)} u_k' v_k \\ Subject \quad to: u_k' u_k = v_k' v_k = 1, u_k' u_i = v_k' v_i = 0 \quad (k \neq i) \end{cases} \tag{2}$$

where $0 \le \theta_k \le \pi/2$, $k = 1,...,m$; and u_i *and* v_i respectively are basis vectors in two subspaces.

Obviously, the first principal angle θ_1 is the smallest angle between all pairs of unit basis vectors which are respectively from the first subspace and the second subspace. The rest of the principal angles can be gained by other basis vectors in each subspace. The cosines of all principal angles can be computed by a more numerically stable method, that is, the Singular Value Decomposition (SVD) solution which was proved by Björck & Golub[9]. Let Y_1 and Y_2 be two $D \times m$-dimensional orthogonal matrixes which are respectively consisted of orthogonal bases from two subspaces. Then, the cosines of all principal angles between the two subspaces are computed by the following SVD equation:

Theorem 1. If Y_1 and Y_2 are two $\boldsymbol{m}$-dimensional subspaces, then

$$Y_1 ' Y_2 = U \Lambda V ' \tag{3}$$

where the diagonal matrix $\Lambda =$ diag($\cos\theta_1$, $\cos\theta_2$, ..., $\cos\theta_m$), and $UU' = I_m$, $VV' = I_m$.

3.2 Construct the Neighborhood Graph on Grassmann Manifold

Two points on Grassmann manifold correspond to two subspaces. For constructing the neighborhood graph on Grassmann manifold, a metric of describing neighborhood relation needs to be defined. At present, some valid distance metric methods have been used, such as Projection metric, Binet-Cauchy metric, Max Correlation, Min Correlation, Procrustes metric, etc [1]. In the following algorithm, the similarity $S(Y_1, Y_2)$ between two subspaces Y_1 and Y_2 is defined by the average of all canonical correlations, that is:

$$S(Y_1, Y_2) = \frac{1}{m} \sum_{l=1}^{m} \cos^2 \theta_l \tag{4}$$

Each point on the neighborhood graph, corresponding to a linear subspace, has $\boldsymbol{k}$ edges to link the $\boldsymbol{k}$-nearest neighbors based on Equation (5). The neighborhood graph $\boldsymbol{G}$ is constructed by finding k-nearest neighbors of each subspace Y_i and connecting these edges, with corresponding to the $\boldsymbol{k}$ largest similarities from one subspace to other subspaces. Those edge weights on graph $\boldsymbol{G}$ is computed as follows:

$$G_{ij} = \begin{cases} \frac{1}{m} \sum_{l=1}^{m} \cos^2 \theta_l , & \begin{array}{l} \textit{if subspace } Y_i \textit{ is among k nearest neighbors of subspace } Y_j \textit{ or} \\ \textit{subspace } Y_j \textit{ is among k nearest neighbors of subspace } Y_i . \end{array} \\ 0 & , \textit{ otherwise} \end{cases} \tag{5}$$

where $\cos\theta_1, \cos\theta_2, ..., \cos\theta_m$ is the cosines of $\boldsymbol{m}$ principal angles between two subspaces Y_i and Y_j.

3.3 The Detailed Procedures of SSLE-GM

For improving the effectiveness and robustness of the proposed algorithm, those label information of training subspaces is integrated into the computation of the neighborhood similarity. A semi-supervised neighborhood similarity matrix on graph G is defined as:

$$M_{ij} = \begin{cases} \alpha + (1-\alpha)G_{ij}, & \textit{if subspace } Y_i \textit{ and } Y_j \textit{ belong to the same class} \\ 0, & \textit{if } Y_i \textit{ and } Y_j \textit{ belong to the different class} \\ G_{ij}, & \textit{otherwise} \end{cases} \quad (6)$$

where $\boldsymbol{\alpha}$ is the balance factor between the label information of training subspaces and the local neighborhood geometric characteristics.

Collecting the above analysis and the key steps of Laplacian Eigenmaps, a semi-supervised Laplacian eigenmaps algorithm on Grassmann manifold is proposed, called as SSLE-GM. The SSLE-GM algorithm implements the semi-supervised feature representation from Grassmann manifold to the low-dimensional Euclidean space. And the visualization of the geometric distribution about subspace set, and the combination with classification techniques in Euclidean space are carried out. A simple illustration about SSLE-DM is shown in Fig. 1.

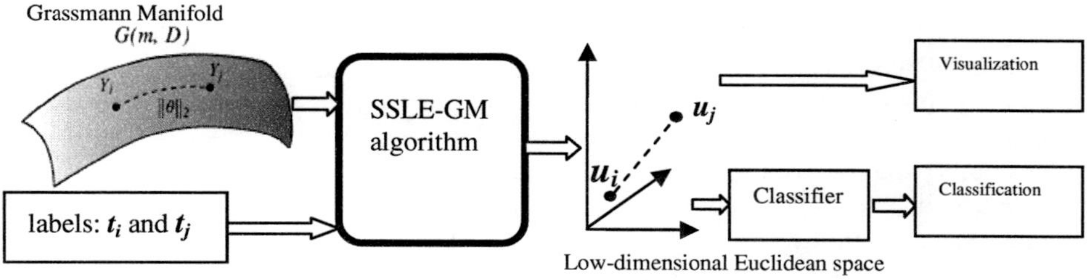

Fig. 1. The working illustration of SSLE-GM

Finally, the key steps of the proposed SSLE-GM algorithm are described as follows:

SSLE-GM Algorithm:

Input: The labeled subspace set $\{Y_i, t_i\}_{i=1}^{l}$, where t_j is the category label of the $\boldsymbol{i}$-th subspace Y_i; let the training subspace set $Y_{train} = \{Y_i\}_{i=1}^{l}$ and the unlabeled testing subspace set $Y_{test} = \{Y_i\}_{i=l+1}^{l+u}$.

Output: (1) Embedding: $u_i \in R^d \ (i = 1, 2, \cdots, l+u), U = \{u_1, ..., u_{l+u}\}$, intrinsic $\boldsymbol{d}$-dimensional coordinates in Euclidean space

(2) Classification: the labels of testing subspace set $Y_{test} = \{Y_i\}_{i=l+1}^{l+u}$

Procedures:

Step1, Compute the unsupervised neighborhood similarity matrix $\boldsymbol{G}$ between the $l+u$ subspaces included training subspaces and testing subspaces. The matrix $\boldsymbol{G}$ is defined in Equation (5).

Step2, Compute the semi-supervised neighborhood similarity matrix $\boldsymbol{M}$.

The semi-supervised neighborhood similarity matrix is defined in Equation (6).

Step3, Compute the embedding coordinates $U = \{U_{train}, U_{test}\} = \{u_1, ..., u_l, ..., u_{l+u}\}$ by mapping all subspaces on Grassmann manifold ***G*** to the Euclidean space R^d

Obtain the embedding coordinates in the Euclidean space by optimizing the objective function which is same as Equation (1).

Step4. Apply the coordinates $U_{train} = \{u_1, ..., u_l\}$ of training subspaces and the corresponding labels $L_{train} = \{t_i\}_{i=1}^{l}$ to train the classifier *T*.

Step5. Use the classifier *T* to recognize the label $L_{test} = \{t_i\}_{i=l+1}^{l+u}$ of the coordinates $U_{test} = \{u_l, ..., u_{l+u}\}$ with corresponding to testing subspaces.

4 Visualization and Classification Experiments

The experimental comparison is completed on the first set of Cambridge gesture database [10]. Each class contains 20 image sequences with the same illumination in the front of the fixed light. In this experiments, each sequence (corresponding to a image set) was uniformly cut into the first 40 frames with size 30*40 pixels with transforming each pixel into grey-value at the interval [0,1]. For experimental repeatability, the first ten sequences in each class are used as training and the rest ten sequences in each class are used as testing. So the training data consists of 90 sequences and the testing data consists of 90 sequences. For obtaining the representation on Grassmann manifold, each image sequence is extracted the 20-dimensional linear subspace (corresponding to the coarse granular representation) by principal component analysis (PCA) method.

The proposed SSLE-GM algorithm can find a mapping from Grassmann manifold to the low-dimensional Euclidean space. Each image sequence is mapped to a point in low-dimensional Euclidean space under the mechanism of LE algorithm. When the dimension in the embedded space is set to 2 or 3, the visualization of sequence set can be observed. Fig. 2 shows the visualization of the first 10 sequences for each gesture class on the first set of Cambridge gesture database. Three primitive shapes from experimental data have distinctive distribution law, respectively located in the left-top, left-down and right-middle in two-dimensional Euclidean space.

As for evaluating the classification performance of SSLE-GM algorithm, the experimental task is to classify an unknown image sequence to one of the training classes. We analyzed the recognition performance of the proposed SSLE-GM algorithm at different values of balance factor and the dimension, and the used classifier is the simple nearest neighbor classifier [11] based on Euclidean distance. Fig. 3 shows the recognition rate surface of image sequences by using SSLE-GM algorithm in different value of balance factor $\boldsymbol{\alpha}$ and different dimension in low-dimensional Euclidean space. When balance factor $\boldsymbol{\alpha}$=0.1 and the dimension d=18 in low-dimensional Euclidean space, SSLE-GM algorithm obtains the highest recognition rate 82.22% on the first set of Cambridge gesture database. The best recognition performance of SSLE-GM for subspace recognition is compared with the reported results by using Mutual Subspace Method (MSM)[5], Constrained Mutual

Subspace Method (CMSM)[6], Tensor Canonical Correlation Analysis(TCCA)[10], Canonical Correlation Analysis (CCA)[12], and probabilistic Latent Semantic Analysis (pLSA)[13], as shown in Fig.4.

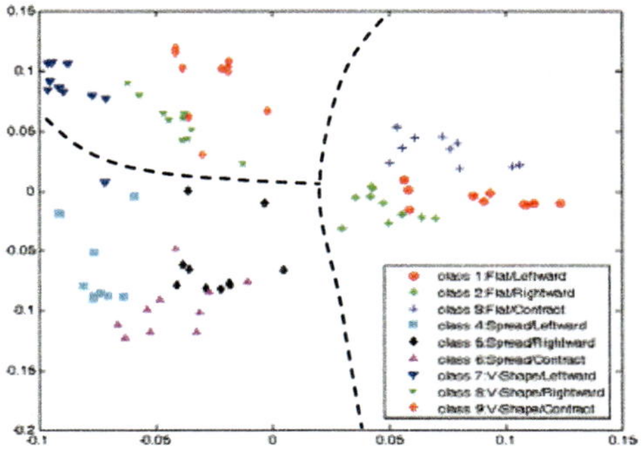

Fig. 2. Visualization of the first ten sequences for each gesture class

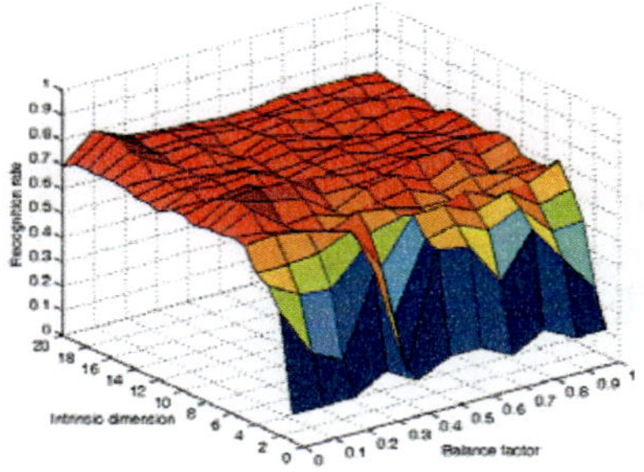

Fig. 3. Recognition rate surface by using SSLE-GM on Cambridge gesture database at different value of balance factor and different dimension

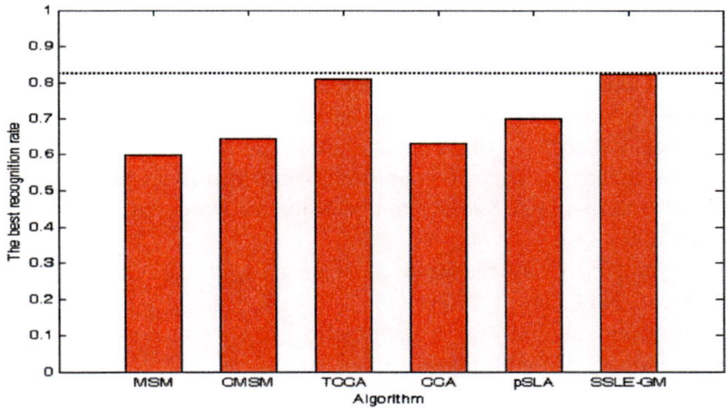

Fig. 4. The best recognition performance by using MSM [5], CMSM [6], TCCA [10], CCA [12], pLSA [13] and the proposed SSLE-GM algorithms

5 Conclusion

In this paper, the LE algorithm is extended to process the set data which consist of some sets such as image sets, and the proposed algorithm integrates the label information of training subspaces and the geometric distribution information of all subspaces included those labeled subspaces and the unlabeled subspaces. A semi-supervised Laplacian eigenmaps algorithm on Grassmann manifold is proposed. Visualization and classification experiments on Cambridge gesture database are completed by using SSLE-GM. Experimental results show that SSLE-GM is effective and obtains the highest recognition rate by comparing with MSM, CMSM, TCCA, CCA and pLSA algorithms. Our future research tasks are to design the multiple-granularity criterion function for SSLE-GM algorithm and to build the relation in different granular space.

Acknowledgments. The authors would like to thank the anonymous reviewers for their help. This work was supported by the National Natural Science Foundation of China (No. 61075019), the Chongqing Natural Science Foundation (No. CSTC-2010BB2406), the Doctor Foundation of Chongqing University of Posts and Telecommunications (No. A2009-24) and Science & Technology Research Program of Chongqing Education Commission (No.KJ110512).

References

1. Hamm, J.: Subspace-based Learning with Grassmann Manifolds. Ph.D thesis, University of Pennsylvania (2008)
2. Belkin, M., Niyogi, P.: Laplacian Eigenmaps for Dimensionality Reduction and Data Representation. Technical Report, University of Chicago, U.S (2001)
3. Belkin, M.: Problems of Learning on Manifolds. Ph.D thesis, University of Chicago, US (2003)
4. Warner, F.: Foundations of Differentiable Manifolds and Lie Groups. Springer, New York (1983)
5. Yamaguchi, O., Fukui, K., Maeda, K.: Face recognition Using Temporal Image sequence. In: The Third Proceedings of the International Conference on Face and Gesture Recognition, Washington, DC, USA, pp. 318–323 (1998)
6. Fukui, K., Yamaguchi, O.: Face Recognition Using Multi-viewpoint Patterns for Robot Vision. In: 11th International Symposium of Robotics Research, pp. 192–201. Springer, Heidelberg (2003)
7. Belkin, M., Niyogi, P.: Convergence of Laplacian Eigenmaps. In: Proceedings of NIPS 2006, pp. 129–136. MIT Press (2007)
8. Kim, T.K.: Discriminant Analysis of Patterns in Images, Image Ensembles, and Videos. Ph.D thesis, University of Cambridge (2007)
9. Björck, A., Golub, G.H.: Numerical Methods for Computing Angles between Linear Subspaces. Math. Computation 123, 579–594 (1973)
10. Kim, T.K., Cipolla, R.: Canonical Correlation Analysis of Video Volume Tensors for Action Categorization and Detection. IEEE Trans. Pattern Analysis and Machine Intelligence 8, 1415–1428 (2009)
11. Cover, T.M., Hart, P.E.: Nearest Neighbor Pattern Classification. IEEE Transactions on Information Theory 1, 21–27 (1967)
12. Kim, T.K., Kittler, J., Cipolla, R.: Discriminative Learning and Recognition of Image Set Classes Using Canonical Correlations. IEEE Trans. Pattern Analysis and Machine Intelligence 6, 1005–1018 (2007)
13. Niebles, J.C., Wang, H., Li, F.F.: Unsupervised Learning of Human Action Categories Using Spatial-Temporal Words. In: Proc. British Machine Vision Conference, pp. 1249–1258. British Machine Vision Association (2006)

Application of Closed Gap-Constrained Sequential Pattern Mining in Web Log Data

Xiuming Yu[1], Meijing Li[1], Dong Gyu Lee[1], Kwang Deuk Kim[2], and Keun Ho Ryu[1]

[1] Database / Bioinformatics Laboratory, Chungbuk National University Korea, Korea
{yuxiuming,mjlee,dglee,khryu}@dblab.chungbuk.ac.kr
[2] Department of Renewable Energy Research, Korea Institute of Energy Research, Korea
kdkim@kier.re.kr

Abstract. Discovery of information in web log data is a very popular research area in the field of data mining. Two of the objectives of favorite applications are to obtain useful information of web users' behavior and to analyze the structure of web sites. In this paper, we suggest a novel approach to generate web sequential patterns using the gap-constrained method in web log data. The process of mining task in the proposed approach is described as follows. First, pre-process of the raw web log data is introduced by removing irrelevant or redundant items, gathering the same users and transforming the web log data into a set of tuples (sequence identifier, sequence) constrained by visiting time. Second, web access patterns, which are closed sequential patterns with gap constraints, are generated using the Gap-BIDE algorithm in web log data with two parameters, minimum support threshold and gap constraint. In the experiment, a data set is derived from http://www.vtsns.edu.rs/maja/, which is proposed in [1]. The result shows that, with the application of sequential pattern mining in the web log data presented in this paper, we can find information about navigational behavior of web users and the structure of the web page can be designed more legitimately by the order of obtained patterns.

Keywords: data mining, web log data, gap-constrained mining, sequential pattern.

1 Introduction

The Web is a popular and interactive medium of information, and it is widely used in our lives. The Web is related to nearly all industries, for example, communication, banking and shopping can be done on line. Over the Web, stress can be relieved, business transactions can be completed easily, and thus creating a key driving force in the rapid growth of electronic commerce. The Web is so closed to our lives, and it helps us overcome lots of problems. Due to those reason, exploring specific information on the Web becomes an essential task today.

The task of Web Mining is to extract the interesting and potentially useful patterns based on activities related to the World Wide Web (WWW). There are roughly three knowledge discovery domains that pertain to web mining: web content mining, web structure mining and web usage mining [2]. In Web Usage Mining (WUM), also

D. Zeng (Ed.): Advances in Control and Communication, LNEE 137, pp. 649–656.
springerlink.com

known as web access, web access pattern tracking can be defined as web page history [3]; the mining task is a process of extracting interesting patterns in web access logs. There are so many techniques of mining usage including statistical analysis [4], association rules [5], sequential patterns [6, 7], classification [8, 9, 10] and clustering [11, 12, 13]. Sequential pattern mining is one of the popular approaches of web usage mining, which extracts frequent subsequences from a sequence data-base [14]. As we know, sequential pattern mining techniques provide knowledge about the behavior of web users in order to extract relationships from the recorded data.

Earlier frequent pattern mining of web data is mentioned by Renata Ivancsy and Istvan Vajk using ItemsetCode algorithm in 2006 [15]. Analyzing the results of the application of frequent mining in web log data, an advertising process can be made more successful and the web page structure can be changed such that the pages contained by the rules are accessible from each other. For the analysis of web structure, orderly patterns should be discovered to determine the sequence of access clearly. Frequent sequential pattern mining can solve the disorderly problem in web log data. The approach was proposed in [16, 17], and the result of Sequential Pattern (SP) mining in web log data shows one achievement of the SP approach in that it can discover web users' navigational patterns.

Gap-constrained sequential pattern mining is a novel approach to be applied in web log data mining, compared with basic sequential pattern mining, this approach can discover more useful patterns, which are not only considered the order of web site items but also no continuous web site items that can find no adjoining patterns. In this paper, we propose mining closed sequential patterns with Gap-BIDE Algorithm in web log data.

The rest of this paper is organized as follows. Section 2 focuses on the process of the mining task: preprocessing, pattern discovery and pattern analysis. We also talk about the algorithm principle. In section 3, we present an extensive performance study. Finally, we conclude the study in section 4.

2 Process of Mining Task

The process of closed gap-constrained sequential pattern mining mainly includes data preprocessing and pattern discovery. Fig. 1 illustrates the process in detail.

2.1 Data Preprocessing

Generally, the raw data in web log files are not fit for the task of mining sequential patterns. The preprocessing of input data becomes very important for closed frequent pattern mining. In order to avoid noisy and outliers data and find the sequence that adapts to the mining task, we need to do the process of data cleaning and data transformation. In this section, we describe preprocessing by introducing the process of removing irrelevant data and transforming the pre-processed data into a set of tuples (sequence identifier, sequence).

2.1.1 Common Log Format

In our experiments, we used the log files of a web server based on the data of web site http://www.vtsns.edu.rs, which had been referenced in paper "Discovering Interesting

Association Rules in the Web Log Usage Data". The experiment log data describes the page visits of users who visited http://www.vtsns.edu.rs on November 16, 2009. Visits are recorded at the level of URL category and are recorded in time order. Each row in the web log file corresponds to the page visits of a user within a twenty-four hour period.

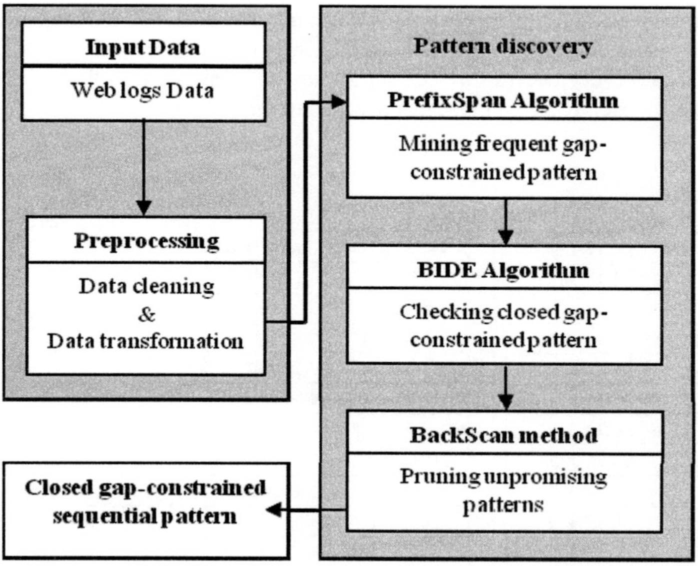

Fig. 1. This shows the process of mining task. After data preprocessing, discovery of pattern is run. The algorithm of this process is composed with the algorithm of PrefixSpan, BIDE and BackScan.

In this paper, web access log data is used for mining task, which records all the kinds of requests that are processed by server [18]. The format of web access log data is specified, different web servers have their own format of log files, and the format of a record in our web access log file is composed of the attributes of IP address, user id time, method, URL, protocol, status, size, referrer and agent.

2.1.2 Data Cleaning

The aim of our task is to determine the behavior of web users to legitimately design the structure of web page. In order to avoid interference, we need to remove the irrelevant request of web page. In our experimental data, the requests of css, jpg, png, ico and so on are defined to be irrelevant requests. We are only interested in the records which are with the URL of web pages. So we need to eliminate the records which contain the irrelevant request and only leave the records who’s URLs are web page addresses.

After completing the above operation, there are 31 different kinds of web page URLs in the left records. In order to be transformed into sequences clearly, these kinds of URLs are related to codes.

In order to clearly determine the struts of the data, we transform the sample data in log file, which deal with removing irrelevant requests into a table data. We choose a sample data from web log file, and there are lots of unimportant attributes of data that are unnecessary in our mining task, so we can remove the irrelevant attributes for clarity. For instance, the attributes of user id, method, protocol, status, size, referrer and agent are not very useful for sequential pattern mining work. These columns can be eliminated. After removing the irrelevant attributes and transforming web URLs into codes, the rest of data are shown in Fig. 2 (a).

#	IP Address	Time	URL
1	82.1.202.158	00:12:15	
2	82.117.202.158	00:12:18	1
3	82.117.202.158	00:12:22	2
4	82.117.202.158	00:12:23	5
5	82.117.202.158	00:12:23	7
6	82.117.202.158	00:12:25	1
7	188.2.176.146	00:13:43	2
8	79.101.52.1	00:17:39	1
9	79.101.52.1	00:17:39	3
10	79.101.52.1	00:17:40	2
11	79.101.52.1	00:17:40	1
12	82.117.202.158	00:17:46	3

(a)

#	IP Address	Time	URL
1	82.117.202.158	00:12:18	1
2	82.117.202.158	00:12:22	2
3	82.117.202.158	00:12:23	5
4	82.117.202.158	00:12:23	7
5	82.117.202.158	00:12:25	1
6	188.2.176.146	00:13:43	2
7	79.101.52.1	00:17:39	1
8	79.101.52.1	00:17:39	3
9	79.101.52.1	00:17:40	2
10	79.101.52.1	00:17:40	1
11	82.117.202.158	00:17:46	3

(b)

Sequence identifier	Sequence
82.117.202.158	<1,2,5,7,1,3>
188.2.176.146	<2>
79.101.52.1	<1,3,2,1>

(c)

Fig. 2. This shows the process of data preprocessing. (a): Remove the irrelevant attributes; (b): Remove the missing value data; (c): Transform the web log data into sequence.

There are also lots of missing value data in the web logs file which are caused by breaking of loading web page, that lead to the URL of the page cannot be obtained. For instance, the first record in Fig. 2 (a), its URL is empty. So the records which contain missing value data should be removed. Removing the missing value data, then we get the cleaning mining data, as shown in Fig. 2 (b).

2.1.3 Data Folding

After completing the above process, we present the process of transforming the web log data into a set of tuples of sequence. The tuples consist of a sequence identifier and sequence. Because we want to find the closed sequential patterns and analyze the relationship between web users and web sites, we make the IP Address that can identify web users as a sequence identifier, and we then group the web sites accessed by same user together in series as sequences.

Here, the time constraint is used to limit the grouping rule in the same sequence identifier. Generally, there is Cookie, also known as web Cookie and HTTP Cookie that may be set by the server with or without an expiration date. Within expiration date, the access of web user is effective. For web access security, web pages are always designed by the time constraint of Cookie. For this mining task, 30 minutes is the period defined as the expiration date.

For instance, in Fig. 2 (b), we assign the IP Address "82.117.202.158" to be the sequence identifier. According to the time constraint, the web sites "1, 2, 5, 7, 1, 3" that the user with IP Address "82.117.202.158" had accessed and the whole access period within the expiration date are identified to be a sequence. We transform the data in the Fig. 2 (b) into sequences, the result as shown in Fig. 2 (c); the final pre-processed data.

2.2 Pattern Discovery Using Gap-Constrained Sequential Mining

In this section, we present the process of pattern discovery in following steps: mining frequent gap-constrained patterns, checking closed gap-constrained patterns and pruning the unpromising patterns. After completing the above steps, the closed gap-constrained sequential patterns will be discovered in the web log data.

Step 1: mining frequent gap-constrained patterns. There are lots of previous algorithms for mining frequent gap-constrained patterns, but many of such algorithms have to store all mined patterns and cost a lot of memory if the databases are dense and huge. We choose an improved PrefixSpan algorithm from these algorithms. This algorithm does not need to store all the mined patterns, and it costs less memory than other algorithms when the databases are dense and huge. With the PrefixSpan algorithm, we can find the patterns via the detail process in [19].

Step 2: checking closed gap-constrained patterns. In previous content, we find all the frequent gap-constrained sequential patterns, and we then need to find their closed gap-constrained sequential patterns. Most of the previous algorithms for finding closed pattern tend to cost a lot of memory because they have to maintain the candidates if the newly discovered patterns are closed. We can overcome the latter problem by using the BIDE algorithm. As we know, if we use the BIDE algorithm, then we should first know the definition of backward space and forward space.

Given a pattern P [M, N] with triple (sequence identifier, beginPos, endPos), the backward space of pattern P is the position from beginPos - N to beginPos - M and from 0 to beginPos. The forward space of pattern P is the position from endPos + M to endPos + N and form endPos to l, where l is the length of sequence.

For a gap-constrained pattern P, if and only if this pattern has no backward extension item nor forward extension item, then pattern P is closed. For instance, we scan the backward and forward space of the sequence in Fig. 2 (c), pattern <3> has no backward extension item nor forward extension item, so pattern <3> is closed.

Step 3: pruning the unpromising patterns. We present the BackScan method for pruning unpromising patterns. As presented in [10], scan each backward space of the appearance of a given gap-constrained pattern; if there exists an item whose total number of appearance is equal to the given pattern's number of appearance, then we can safely prune the given gap-constrained pattern. In the running example shown in Fig. 2 (c), in the forward spaces of pattern <1> we find item 3 with a number of appearance of 2, which is equal to the total number of appearance of pattern <1>, so pattern <1> can be safely pruned.

3 Experimental Result and Discussion

Based on the algorithm presented in this paper, we do an experiment to discuss the application of Gap-BIDE algorithm in web log data.

3.1 Result

We choose the log file of web site http://www.vtsns.edu.rs as the raw data of this experiment. The number of original records in the raw web log file is 5999 after data cleaning, 4777 irrelevant requests are removed, and there are 1222 records left. Further, after data folding, 267 sequences are created for the experiment. As we know, each row of the records in the log file corresponds to a record of a user visiting the intranet of the web site http://www.vtsns.edu.rs. Moreover, each item in the created sequences corresponds to a request of a user for the page of the Intranet.

With the experimental data, closed sequential patterns are discovered by the Gap-BIDE algorithm. We run the open source of Gap-BIDE Algorithm in the web site http://code.google.com/p/pygapbide/.

In the experiment, there are two parameters needed to be set for the algorithm, minimum support threshold and the values of gap constraint. The minimum support threshold means the times of a user visiting the same intranets in a certain period. Here, we define the value of minimum support threshold as 5. For the gap constraint, it contains the value of mingap and maxgap. In our application, when we visit a web page, the next two visited web pages will be seen as the related request, so we define the value of gap constraint as [0, 1].

We run the experiment using the cleaning sequence with above parameters, and 98 sequential patterns are then obtained.

3.2 Discussion

According to the experiment result, we can see that every number in the output sequential patterns represents a web site or a request of web users. For example: [6, 18, 18] is the Nineteenth recode in the result, from this sequential pattern and the categories code table, we know that web users often visit the web page of ispit_raspored_god.php and then visit the web page of oglasna.php, and then visit web page oglasna.php again in a continuous time. According to this discovery, we can change the structure of a web site to offer users a good experience. For instance, for the above sequential pattern [6, 18, 18], we should design a hyperlink in the web page of ispit_raspored_god.php point to the web site of oglasna.php, and in page oglasna.php. We also can build a hyperlink point to itself for easy return. The main idea is shown in Fig. 3.

This approach can be applied to online shopping. We can improve the design of a shopping page by analyzing the web log file of online shopping server. The related and ordered web pages can be discovered, and hyperlinks should be designed into web pages that have the next favorite merchandise.

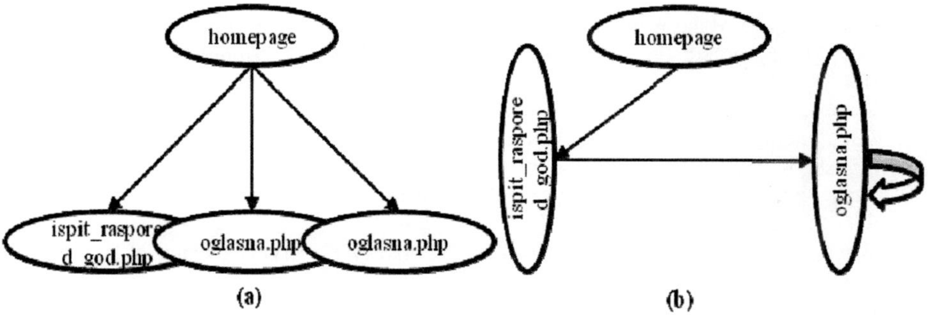

Fig. 3. This shows the main idea of improvement of web design. (a): Original design of web pages; (b): Improved design of web pages.

4 Conclusion

In this paper, we presented the application of mining closed gap-constrained sequential pattern in web log data for getting new web access patterns. In the process of data preprocessing, we removed irrelevant attributes and transformed URLs into code numbers for simplicity, and we removed the missing value data for increasing the quality of data. For getting experimental data for a mining task, we transformed the web log data into sequences based on a time constraint, the value of time is decided by an expiration date of Cookies. As a result, we obtained new web access patterns that expressed the order of accessing the web sites based on the Gap-BIDE algorithm. Compared with frequent pattern mining and sequence pattern mining, our approach considered the patterns, which are not adjacent but ordered patterns using Gap-BIDE algorithm. The Gap-BIDE algorithm is used during the process of task mining, so as to enhance efficiency and accuracy. This approach can help the web site designer to improve the layout of web pages, and make the web pages more comfortable for web users. For future work, we will try to find a more efficient algorithm for mining the closed gap constraint sequential patterns and a more efficient way for transforming web log files into sequence patterns.

Acknowledgments. This work was supported by the grant of the Korean Ministry of Education, Science and Technology (The Regional Core Research Program / Chungbuk BIT Research-Oriented University Consortium) and by the National Research Foundation of Korea (NRF) grant funded by the Korea government (MEST) (No. 2011-0001044) and by Korea Biobank project (4851-307) of the Korea Centers for Disease Control and by Prevention and the Korea Institute of Energy Research (KIER).

References

1. Dimitrijević, M., Bošnjak, Z., Subotica, S.: Discovering Interesting Association Rules in the Web Log Usage Data. Interdisciplinary Journal of Information, Knowledge, and Management 5, 191–207 (2010)

2. Grace, L.K.J., Maheswari, V., Nagamalai, D.: Analysis of Web Logs and Web User in Web Mining. International Journal of Network Security & Its Applications (IJNSA) 3(1) (2011)
3. Saxena, K., Shukla, R.: Significant Interval and Frequent Pattern Discovery in Web Log Data. International Journal of Computer Science Issues, IJCSI 7(1(3)) (2010)
4. Suresh, K., Paul, S.: Distributed Linear Programming for Weblog Data using Mining Techniques in Distributed Environment. International Journal of Computer Applications (0975-8887) 11(7) (2010)
5. Wang, Y., Le, J., Huang, D.: A Method for Privacy Preserving Mining of Association Rules Based on Web Usage Mining. In: 2010 International Conference on Web Information Systems and Mining, WISM, vol. 1, pp. 33–37. IEEE Computer Society, Washington, DC (2010)
6. Wei, C., Sen, W., Yuan, Z., Chang, C.L.: Algorithm of mining sequential patterns for web personalization services. In: ACM SIGMIS Database, vol. 40. ACM. New York (2009)
7. Zhu, J., Wu, H., Gao, G.: An Efficient Method of Web Sequential Pattern Mining Based on Session Filter and Transaction Identification. Journal of Networks (JNW) 5(9), 1017–1024 (2010) ISSN 1796-2056
8. Santini, M.: Cross-Testing a Genre Classification Model for the Web. Genres on the Web, Part 3 42, 87–128 (2011)
9. Rho, J.J., Moon, B.J., Kim, Y.J., Yang, D.H.: Internet Customer Segmentation Using Web Log Data. Journal of Business & Economics Research (JBER) 2(11) (2004)
10. Kejžar, N., Černe, S.K., Batagelj, V.: Network Analysis of Works on Clustering and Classification from Web of Science. In: Proceedings of the 11th IFCS, Part 3, pp. 525–536 (2010)
11. Xu, G., Zong, Y., Dolog, P., Zhang, Y.: Co-clustering Analysis of Weblogs Using Bipartite Spectral Projection Approach. In: Setchi, R., Jordanov, I., Howlett, R.J., Jain, L.C. (eds.) KES 2010. LNCS, vol. 6278, pp. 398–407. Springer, Heidelberg (2010)
12. Makanju, A.A.O., Zincir-Heywood, A.N., Milios, E.E.: Clustering Event Logs Using Iterative Partitioning. In: Proceeding of KDD 2009, pp. 1255–1263 (2009)
13. Wang, J., Mo, Y., Huang, B., Wen, J., He, L.: Web Search Results Clustering Based on a Novel Suffix Tree Structure. In: Rong, C., Jaatun, M.G., Sandnes, F.E., Yang, L.T., Ma, J. (eds.) ATC 2008. LNCS, vol. 5060, pp. 540–554. Springer, Heidelberg (2008)
14. Chen, J., Cook, T.: Mining contiguous sequential patterns from web logs. In: Proceedings of WWW, pp. 1177–1178 (2007)
15. Iváncsy, R., Vajk, I.: Frequent Pattern Mining in Web Log Data. Acta Polytechnica Hungarica 3(1), 77–90 (2006)
16. Mabroukeh, N.R., Ezeife, C.I.: A taxonomy of sequential pattern mining algorithms. ACM Computing Surveys (CSUR) 43(1) (2010)
17. Shivaprasad, G., Subbareddy, N.V., Acharya, U.D.: Knowledge Discovery from Web Usage Data: Research and Development of Web Access Pattern Tree Based Sequential Pattern Mining Techniques: A Survey. In: AIP Conference, vol. 1324, pp. 319–323 (2010)
18. Grace, L.K.J., Maheswari, V., Nagamalai, D.: Web Log Data Analysis and Mining. In: Meghanathan, N., Kaushik, B.K., Nagamalai, D. (eds.) CCSIT 2011, Part III. Communications in Computer and Information Science (CCIS), vol. 133, pp. 459–469. Springer, Heidelberg (2011)
19. Li, C., Wang, J.: Efficiently Mining Closed Subsequences with gap constraints. In: Proceeding of 2008 SIAM International Conference on Data Mining, pp. 313–322 (2008)

Kernelized K-Means Algorithm Based on Gaussian Kernel

Kuo-Lung Wu and You-Jun Lin

Department of Information Management, Kun Shan University,
Yung-Kang, Tainan 71023, Taiwan
klwu@mail.ksu.edu.tw

Abstract. We proposed a kernelized k-means algorithm based on the Gaussian kernel function according to the concepts of support vector clustering and kernel methods. A statistical point of view of robust properties of the proposed method is analyzed. The cluster center estimates obtained by the proposed method can be represented by an M-estimate with a bounded φ function. This provides the theoretical advance to support the robustness of our clustering method. Numerical examples also show the superiority of the proposed method.

Keywords: kernel method, k-means, robust.

1 Introduction

Kernel-based methods are widely applied in many applications [1,2]. There are two ways of implementing kernel-based methods along with supervised and unsupervised learning. One is to transform the data space into a high-dimensional feature space F where the inner products in F can be represented by a Mercer kernel function defined on the data space (see Refs. [1-5]). Another is to find a kernel density estimate on the data space and then search the modes of the estimated density [6]. The mean shift [7], [8] and the mountain method [9] are two simple techniques that can be used to find the modes of a kernel density estimate.

Most partition methods suppose that the data set can be represented by finite cluster prototypes with their own objective functions. Therefore, to define the dissimilarity (or distance) between a point and the cluster prototype is essential for partition methods. Theoretically, if the dissimilarity between a point and a cluster prototype is defined, its corresponding clustering algorithm can be developed. The most popular partition method with cluster prototypes is k-means [10,11] (or called hard c-means). Let $X=\{x_1,\cdots,x_n\}$ be a data set in an s-dimensional Euclidean space R^s and $\{X_1,\cdots,X_c\}$ be a partition of X with corresponding c cluster centers $\{a_1,\cdots,a_c\}$. The objective of k-means is to minimize the following sum of squared within-cluster errors:

$$J_{k-means}=\sum_{i=1}^{c}\sum_{x_j\in X_i} d(x_j,a_i)^2=\sum_{i=1}^{c}\sum_{x_j\in X_i}\left\|x_j-a_i\right\|^2=\sum_{i=1}^{c}\sum_{j=1}^{n} w_{ij}\left\|x_j-a_i\right\|^2, \tag{1}$$

D. Zeng (Ed.): Advances in Control and Communication, LNEE 137, pp. 657–664.
springerlink.com

where $w_{ij} = w_i(x_j) = 1$ if $x_j \in X_i$ and $w_{ij} = 0$ if $x_j \notin X_i$. The *k*-means algorithm (or called hard *c*-means) is a procedure to search the minimizers $\{w_1, \cdots, w_c\}$ and $\{a_1, \cdots, a_c\}$ of $J_{k-means}$ using the following necessary conditions:

$$a_i = \sum_{j=1}^{n} w_{ij} x_j \Big/ \sum_{j=1}^{n} w_{ij} \tag{2}$$

and

$$w_{ij} = w_i(x_j) = \begin{cases} 1, & \text{if } \left\| x_j - a_i \right\|^2 = \min\limits_{1 \le k \le c} \left\| x_j - a_k \right\|^2 \\ 0, & \text{otherwise.} \end{cases} \tag{3}$$

The k-means clustering was early also called a vector quantization [12]. The numerical and stochastic convergence properties of the k-means algorithm had been well studied [13,14]. A central limit theorem was established by Pollard [15]. Recently, a general k-means model has been proposed by Yu [16]. Some applications of k-means can be found in [17-21].

In this paper, we will combine the concept of kernel method and k-means to propose a kernelized k-means based on the Gaussian kernel function in Section 2. We then discussed the robust properties according to the φ function of the M-estimator in Section3. Examples and conclusions are presented in Section 3 and 4.

2 Kernelized K-Means

The kernel methods have been widely studied and applied in pattern recognition and function approximation and are one of the most important subjects in machine learning (see Refs. [22-25]). Girolami [4] first developed a feature space sum-of-squares objective function to extend k-means to feature space and then used the radial basis function (RBF) kernel and stochastic optimization to derive a kernel-based clustering algorithm. Li et al. [26] proposed a cluster labeling method for SVC based on some invariant topological properties of a trained kernel radius function. Leski [27] proposed a kernel fuzzy c-varieties/elliptotypes (KFCV/E) by extending the feature space FCM with linear varieties or elliptotypes. Zhang and Chen [22] first mentioned that these kernel clustering methods with clustering prototypes lied in the feature space may be lack of clear and intuitive descriptions.

The common concepts of the kernel method are to transform the data space into a high-dimensional feature space F with $\Phi : X \to F$ where the inner products in F can be represented by a Mercer kernel function defined on the data space. A kernel $K(x,y)$ in the feature space can then be denoted by $K(x,y)=<\Phi(x)\Phi(y)>$ which is the inner product of $\Phi(x)$ and $\Phi(y)$. We now have

$$\begin{aligned} &\left\| \Phi(x) - \Phi(y) \right\|^2 = (\Phi(x) - \Phi(y))^T (\Phi(x) - \Phi(y)) \\ &= \Phi(x)^T \Phi(x) - \Phi(x)^T \Phi(y) - \Phi(y)^T \Phi(x) + \Phi(y)^T \Phi(y) \\ &= K(x,x) - K(x,y) - K(y,x) + K(y,y). \end{aligned} \tag{4}$$

If we take the Gaussian Radial basis function kernel:

$$K_G(x,y)=\exp(-\| x-y \|^2/\beta) \tag{5}$$

Equation (4) will become

$$\|\Phi(x)-\Phi(y)\|^2 = 2(1-K_G(x,y)) . \tag{6}$$

The advantages of using the kernel methods consist of inducing a robust non-Euclidean distance measure and enhancing robustness of the algorithm to noise and outliers.

Combining the concepts of kernel method, the objective function of the kernelized k-means based on Gaussian kernel (K_G-k-means) can be represented by

$$J_{K_G-k-means} = \sum_{i=1}^{c}\sum_{j=1}^{n} w_{ij} \|\Phi(x_j)-\Phi(a_i)\|^2 = 2\sum_{i=1}^{c}\sum_{j=1}^{n} w_{ij}(1-K_G(x_j,a_i)) . \tag{7}$$

The normalization parameter β in (5) is similar to the dispersion measure (variance) of the data set which is corresponding to the bandwidth of kernel density estimation. We use a dispersion estimator $\beta=(1/n)\sum_{i=1}^{n}\| x_i-\overline{x} \|^2$ to estimate β. The necessary conditions to search the minimizers $\{w_1,\cdots,w_c\}$ and $\{a_1,\cdots,a_c\}$ of $J_{K_G-k-means}$ are as follows:

$$a_i = \sum_{j=1}^{n} w_{ij} K_G(x_j,a_i)x_j \Big/ \sum_{j=1}^{n} w_{ij} K_G(x_j,a_i) \tag{8}$$

and

$$w_{ij} = w_i(x_j) = \begin{cases} 1, & \text{if } \|x_j-a_i\|^2 = \min_{1\le k\le c} \|x_j-a_k\|^2 \\ 0, & \text{otherwise.} \end{cases} . \tag{9}$$

The kernelized k-means algorithm can then be summarized as following steps:

K_G-k-means algorithm

Step 1: Fix $2 \le c \le n$ and $\varepsilon > 0$. Initialize $a^{(0)} = (a_1^{(0)},\cdots,a_c^{(0)})$

Step 2: Update $w^{(s)}$ with $a^{(s-1)}$ using (9).

Step 3: Compute the cluster centers $a^{(s)}$ with $w^{(s)}$ using (8).

Step 4: IF $\|a^{(s)}-a^{(s-1)}\| < \varepsilon$ then STOP; ELSE s=s+1 and return to step 2.

3 Robust Analysis of Kernelized K-Means

A suitable clustering method should have the ability to tolerate noise and detect outliers in the data set. Many criteria such as the breakdown point, local-shift sensitivity, gross error sensitivity and influence functions can be used to measure the level of robustness that is often employed in clustering. Here, we provide a more detailed analysis of the robustness of the mean shift procedures.

The influence curve (IC) can help us to assess the relative influence of an individual observation toward the value of an estimate. The influence function of an M-estimator is proportional to its φ function [28]. If the influence function of an estimator is unbounded, an outlier might lead to problems when the φ function is used for denoting the degree of the influence. Let the loss between the data point x_j and the cluster center a_i be

$$\rho(x_j,a_i)=w_{ij}(1-K_G(x_j,a_i)) \tag{10}$$

and

$$\varphi(x_j,a_i)=\frac{d}{da_i}\rho(x_j,a_i)=(2/\beta)w_{ij}(x_j-a_i)K_G(x_j,a_i) \tag{11}$$

By solving the equation $\sum_{j=1}^{n}\varphi(x_j,a_i)=0$, we have the result shown in Equation (8). Therefore, the cluster center estimate (8) of K_G-k-means algorithm is an M-estimator with the loss function (10) and φ function (11). In the location problem, the influence of adding a data point y on an M-estimate can be denoted by $IC(y;F,\theta)$. The $IC(y;F,\theta)$ of M-estimator has been shown that is proportional to its φ function [25]. If the influence function of an estimator is unbounded, an outlier might cause trouble where the φ function is used to denote the degree of influence. By applying the L'Hospital's rule, we derive $\lim_{x_j\to\pm\infty}\varphi(x_j,a_i)=0$. Thus, we have the influence curve of the clustering center estimate (8) with $IC(x_j;F,x)=0$ when x_j tends to positive or negative infinity. This means that the influence curve is bounded and the influence of an extremely large or small x_j on the clustering center estimate of K_G-k-means algorithm is very small. We can also find the location of x_j which

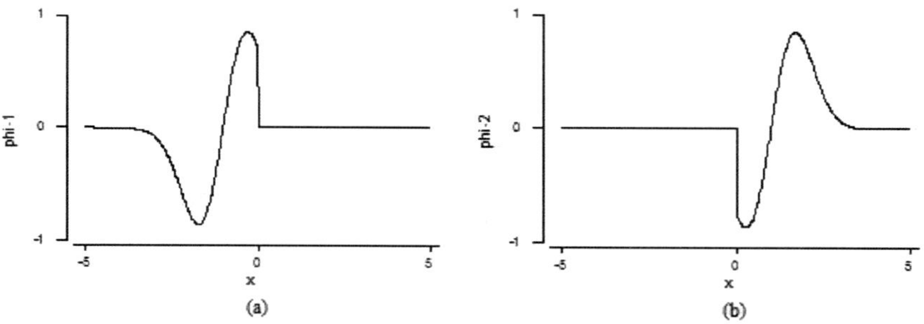

Fig. 1. (a) The influence of an individual x_j on $a_1=-1$. (b) The influence of an individual x_j on $a_2=1$.

has a maximum influence on (8) by solving $d\{\varphi(x_j - x)\}/dx = 0$. The φ function for the cluster center estimators of K_G-k-means algorithm with $\beta = 1$ is shown in Fig. 1. Suppose that there are two cluster centers with $a_1 = -1$ and $a_2 = 1$, the influence of an individual x_j on a_1 and a_2 are denoted by $\varphi_1 = \varphi(x_j, a_1)$ and $\varphi_2 = \varphi(x_j, a_2)$. The φ functions of φ_1 and φ_2 are illustrated in Figs. 1(a) and 1(b), respectively.

4 Examples

Fig. 2(a) shows a 16-group data set. We name this data set the Uniform-16. Data points in each group are generated uniformly from each rectangle. Uniform-16 data set is a two-dimensional data with sample size n = 800 consisting of 50 points in each of 16 clusters. The cluster center estimates generated by k-means and K_G-k-means can well present the data structures as shown in Figs. 2(b) and 2(c), respectively. We now add 100 uniform noise points in the Uniform-16 data set as shown in Fig. 3(a). The clustering result of k-means is heavily influenced by these noise points as shown in Fig. 3(b). However, the result of proposed kernelized k-means algorithm is robust to these noises as shown in Fig. 3(c). We also add a single outlier point in the Uniform-16 data set with coordinate (50,50) and the results of k-means and K_G-k-means are shown in Figs. 4(a) and 4(b). The k-means cluster centers are affected by this outlier point in which the outlier becomes one cluster center alone. Under the same initial values and stopping conditions, the centers obtained by K_G-k-means are not affected by the outlier point.

This is a well-known clustering problem that there are two clusters with great differences of sample numbers. The large separate cluster will be split because this can reduces the sum-of-square error criterion (1) used in k-means and the result is shown in Fig. 5(a). Although there are many clustering methods can improve this weakness in k-means, K_G-k-means can solve this problem simply as shown in Fig. 5(b).

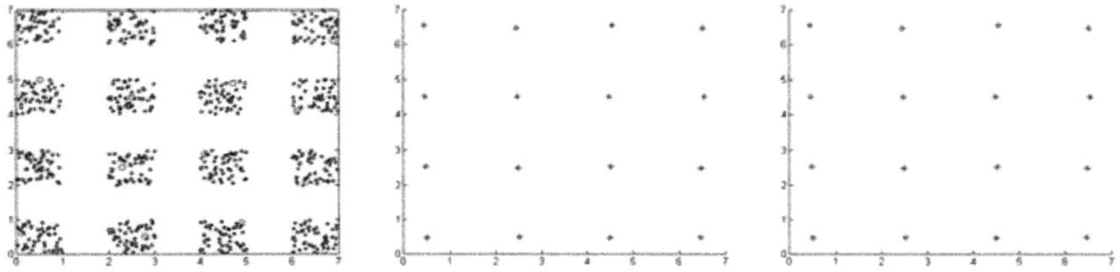

(a) (b) (c)

Fig. 2. (a) The Uniform-16 data set. (b) Cluster centers obtained by k-means. (c) Cluster centers obtained by K_G-k-means.

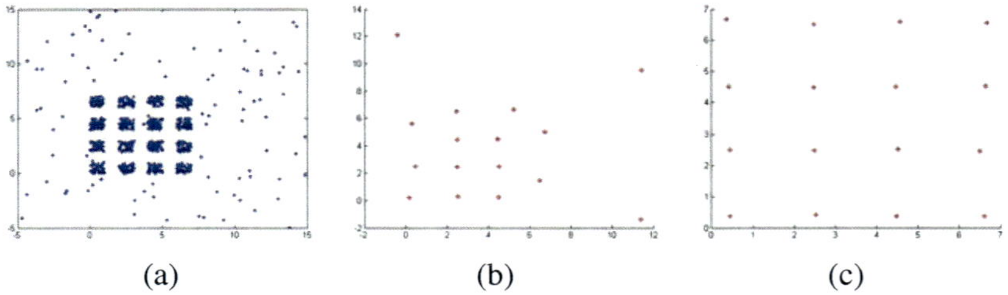

Fig. 3. (a) The Uniform-16 data set with 100 uniform noise points. (b) Cluster centers obtained by k-means. (c) Cluster centers obtained by K_G-k-means.

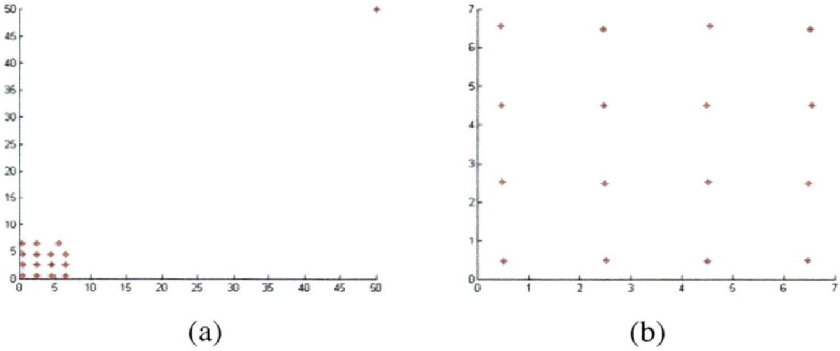

Fig. 4. The results of the Uniform-16 data set with an outlier point (50,50). (a) Cluster centers obtained by k-means. (b) Cluster centers obtained by K_G-k-means.

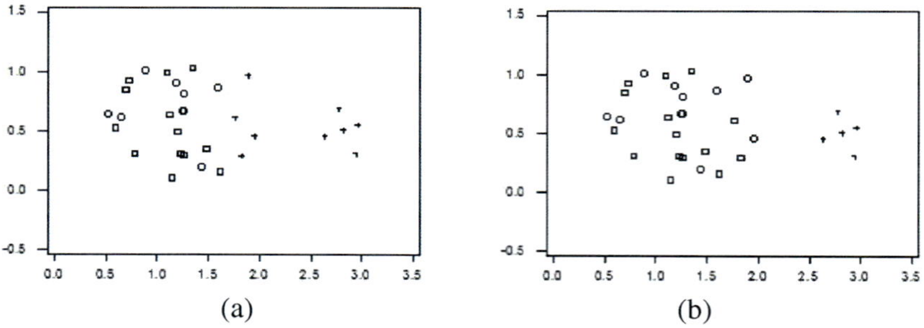

Fig. 5. (a) Partitions obtained by k-means. (b) Partitions obtained by K_G-k-means

Table 1. Error counts for Iris data set

	Setosa	Versicolor	Virginica	total
k-means	0	3	14	17
K_G-k-means	0	3	14	17

In this example, we will implement the Iris data set [29,30], the error counts will be calculated after clustering. Note that the error counts obtained by k-means for this data set are approximately 17 in many literatures. The Iris data set has n=150 points in a four-dimensional space with sepal length, sepal width, petal length, petal width from each of three species of Iris Setosa, Iris Versicolor and Iris Virginica. The clustering error counts of k-means and K_G-k-means are both 17 as Table 1 shows. Although there are two clusters are substantial overlapping in Iris, the data does not contain the noise or outliers. In our simulations, k-means and K_G-k-means will obtain similar results for a data set without noise and outliers. However, if the data contains noise and outliers, the proposed K_G-k-means will give more robust results than the k-means.

5 Conclusions

According to the concepts of support vector clustering and kernel methods, we proposed a kernelized k-means (K_G-k-means) based on the Gaussian kernel function. The robust properties of K_G-k-means are analyzed by a statistical point of view. The cluster center estimates obtained by K_G-k-means can be represented by an M-estimate with a bounded φ function (11). Numerical examples show the superiority of the proposed method.

Acknowledgments. This work was supported in part by the National Science Council of Taiwan, under Kuo-Lung Wu's Grant: NSC-99-2118-M-168-001.

References

1. Vapnik, V.N.: Statistical Learning Theory. Wiley, New York (1998)
2. Cristianini, N., Shawe-Taylor, J.: An Introduction to Support Vector Machines. Cambridge Univ. Press, Cambridge (2000)
3. Camastra, F., Verri, A.: A Novel Kernel Method for Clustering. IEEE Trans. PAMI 27, 801–805 (2005)
4. Girolami, M.: Mercer Kernel Based Clustering in Feature Space. IEEE Trans. Neural Networks 13, 780–784 (2002)
5. Muller, K.R., Mika, S., Rätsch, G., Tsuda, K., Schölkopf, B.: An Introduction to Kernel-Based Learning Algorithm. IEEE Trans. Neural Networks 12, 181–202 (2001)
6. Silverman, B.W.: Density Estimation for Statistics and Data Analysis. Chapman and Hall, New York (1998)
7. Fukunaga, K., Hostetler, L.D.: The Estimation of the Gradient of a Density Function with Applications in Pattern Recognition. IEEE Trans. Information Theory 21, 32–40 (1975)
8. Cheng, Y.: Mean Shift, Mode Seeking, and Clustering. IEEE Trans. PAMI 17, 790–799 (1995)
9. Yager, R.R., Filev, D.P.: Approximate Clustering Via the Mountain Method. IEEE Trans. Systems, Man and Cybernetics 24, 1279–1284 (1994)
10. MacQueen: Some Methods for Classification and Analysis of Multivariate Observations. In: Proc. of 5th Berkeley Symposium, pp. 281–297 (1967)
11. Hartigan, J.A.: Clustering Algorithms. Wiley, New York (1975)

12. Lloyd, S.: Least Squares Quantization in pcm. Bell Telephone Laboratories Papers, Marray Hill (1957)
13. Pollard, D.: Strong Consistency of K-Means Clustering. Annals of Statistics 9, 135–140 (1981)
14. Selim, S.Z., Ismail, M.A.: K-Means-Type Algorithms: A Generalized Convergence Theorem and Characterization of Local Optimality. IEEE Trans. PAMI 6, 81–86 (1999)
15. Pollard, D.: A Central Limit Theorem for K-Means Clustering. Annals of Probability 10, 919–926 (1982)
16. Yu, J.: General C-Means Clustering Model. IEEE Trans. PAMI 27, 1197–1211 (2005)
17. Su, M.C., Chou, C.H.: A Modified Version of the K-Means Algorithm with a Distance Based on Cluster Symmetry. IEEE Trans. PAMI 23, 674–680 (2001)
18. Vrahatis, M.N., Boutsinas, B., Alevizos, P., Pavlides, G.: The New K-Windows Algorithm for Improving the K-Means Clustering Algorithm. Journal of Complexity 18, 375–391 (2002)
19. Charalampidis, D.: A Modified K-Means Algorithm for Circular Invariant Clustering. IEEE Trans. PAMI 27, 1856–1865 (2005)
20. Huang, J.Z., Ng, M.K., Rong, H., Li, Z.: Automated Variable Weighting in K-Means Type Clustering. IEEE Trans. PAMI 27, 657–668 (2005)
21. Laszol, M., Mukherjee, S.: A Genetic Algorithm Using Hyper-Quadtrees for Low-Dimensional K-Means Clustering. IEEE Trans. PAMI 28, 533–543 (2006)
22. Zhang, D.Q., Chen, S.C.: A Novel Kernelized Fuzzy C-Means Algorithm with Application in Medical Image Segmentation. Artificial Intelligence in Medicine 32, 37–50 (2004)
23. Kang, J., Min, L., Luan, Q., Li, X., Liu, J.: Novel Modified Fuzzy C-Means Algorithm with Applications. Digital Signal Processing 19, 309–319 (2009)
24. Kim, D.W., Lee, K.Y., Lee, D., Lee, K.H.: A Kernel-Based Subtractive Clustering Method. Pattern Recognition Letters 26, 879–891 (2005)
25. Zanaty, E.A., Aljahdali, S., Debnath, N.: A Kernelized Fuzzy C-means Algorithm for Automatic Magnetic Resonance Image Segmentation. Journal of Computational Methods in Science and Engineering (JCMES), 123–136 (2009)
26. Li, Z., Tang, S., Xue, J., Jiang, J.: Modified FCM Clustering Based on Kernel Mapping. In: Proc. of SPIE, The Intern. Society for Optical Engineering, vol. 4554, pp. 241–245 (2001)
27. Leski, J.M.: Fuzzy C-Varieties/Elliptotypes Clustering in Reproducing Kernel Hilbert Space. Fuzzy Sets and Systems 141, 259–280 (2004)
28. Huber, P.J.: Robust Statistics. Wiley (1981)
29. Anderson, E.: The Irises of the Gaspe Peninsula. Bull. Am. IRIS Soc. 59, 2–5 (1935)
30. Bezdek, J.C., Keller, J.M., Krishnapuram, R., Kuncheva, L.I., Pal, N.R.: Will the Iris Data Please Stand Up? IEEE Trans. Fuzzy Syst. 7, 368–369 (1999)

A Novel Sparse Learning Method: Compressible Bayesian Elastic Net Model

Cheng Ke-yang[1,2], Mao Qi-rong[1], and Zhan Yong-zhao[1]

[1] School of Computer Science & Telecommunications Engineering, Jiangsu University, Zhenjiang, Jiangsu, China, 212013
[2] School of Information Science & Technology, Nanjing University of aeronautics & astronautics, Nanjing, Jiangsu, China, 210016
kycheng@ujs.edu.cn

Abstract. In this paper, we study the combination of compression and Bayesian elastic net. By including a compression operation into the ℓ1 and ℓ2 regularization, the assumption on model sparsity is relaxed to compressibility. We show that use of a compression operation provides an opportunity to leverage auxiliary information from various sources. We conduct extensive experiments on braincomputer interfacing, handwritten character recognition and text classification. Empirical results show clear improvements in prediction performance by including compression in Bayesian elastic net. We also analyze the learned model coefficients under appropriate compressibility assumptions, which further demonstrate the advantages of learning compressible models instead of sparse models.

Keywords: Sparse Learning, compression operation, Bayesian elastic net.

1 Introduction

Zou and Hastie (2005) [1] emphasized three inherent drawbacks of the lasso estimator. Firstly, due to the nature of the convex optimization problem, the lasso method cannot select more predictors than the sample size. But in practice there are often studies that involve much more predictors than the sample size, e.g. microarray data analysis (Guyon et al. 2002)[2]. Secondly, when there is some group structure among the predictors, the lasso estimator usually selects only one predictor from a group while ignoring others. Thirdly, when the predictors are highly correlated, the lasso estimator performs unsatisfactorily. Zou and Hastie (2005) proposed the elastic net (en) estimator to achieve improved performance in these cases. The en estimator can also be viewed as a penalized least squares method where the penalty term is a convex combination of the lasso penalty and the ridge penalty.

Another shortcoming of Lasso is that the sparsity assumption on model coefficients might be too restrictive and not necessarily appropriate in many application domains. Indeed, many signals in the real world (e.g., images, audio, videos, time series) are found to be compressible (i.e., sparse in certain compressed domain) but not directly sparse in the observed space. Naturally, the assumption of sparsity can be relaxed to

D. Zeng (Ed.): Advances in Control and Communication, LNEE 137, pp. 665–671.
springerlink.com

compressibility. Inspired by the recent development of compressive sampling (or compressed sensing) (Candes, 2006[3]; Donoho, 2006[4]), we study learning compressible models: a compression on model coefficients can be included in the ℓ1 and ℓ2 penalty, and model is assumed to be sparse after compression.

The rest of this paper is organized as follows. In section 2 we will briefly introduce naïve elastic net. In Section 3 we discuss the definition, computation issues and potential benefits of learning compressible Bayesian elastic net model. In Sections 4, experimental as results demonstrate the advantages of compressible Bayesian elastic net (cben) than compressible Bayesian lasso (cbl), elastic net (en) and lasso. Section 5 concludes and mentions some discussions.

2 Compressible Bayesian Elastic Net Model

2.1 Learning Compressible Model

Assuming the model to be sparse and shrinking model coefficients to exactly zero might not be the most appropriate inductive bias in many problems. Interestingly, compressive sampling (Candes, 2006)[3] or compressed sensing (Donoho, 2006) [4]was recently developed in signal acquisition. The target signal is assumed to be compressible and sparse after being compressed. We consider the problem of learning compressible models as follows:

$$\min L(y, 1\alpha + X\beta) + \lambda_1 \| W(\beta) \|_1 + \lambda_2 \| W(\beta) \|_2^2 \tag{1}$$

The loss function L depends on the prediction model, e.g., sum of squares loss for linear regression, log-likelihood loss for logistic regression, hinge loss for SVMs, and so forth. The compression operation W() encodes our assumption on compressibility: model coefficients are compressed by W() before being penalized, and thus tend to follow the compression pattern rather than simply shrink to zero.

For simplicity, we restrict our attention to linear compression. Given that the compression operation is a linear and invertible transform, learning compressible models is represented by eq. (1).

The p×p matrix W denotes the linear and invertible compression transform, where p is the dimensionality of model coefficients. The optimization of eq. (1) can be achieved by applying the inverse compression transform to the feature space and solving the elastic net model in the transformed space. First, transform the training examples by

$$\tilde{X} = XW^{-1} \tag{2}$$

Second, solve the standard elastic net model:

$$\min L(y, 1\alpha + \tilde{X}\tilde{\beta}) + \lambda_1 \| W\tilde{\beta} \|_1 + \lambda_2 \| W\tilde{\beta} \|_2^2 \tag{3}$$

Finally, the solution for eq. (1) is obtained by:

$$\beta =: W^{-1}\tilde{\beta}, \alpha = \alpha . \tag{4}$$

This equivalence is derived from $X\beta = XW^{-1}\tilde{\beta} = \tilde{X}\tilde{\beta}$, $\| W\beta \|_1 = \| WW^{-1}\tilde{\beta} \|_1 = \| \tilde{\beta} \|_1$ and $\| W\beta \|_2^2 = \| WW^{-1}\tilde{\beta} \|_2^2 = \| \tilde{\beta} \|_2^2$.

Why do we want to learn compressible models, which are not necessarily sparse in the original space? Compressible models are useful in several aspects. The first is model fitting and prediction accuracy. The inductive bias of model compressibility might be more appropriate than model sparsity, especially if an informative compression operation is specified based on additional information from domain knowledge, unlabeled data or related problems. The second reason, as claimed for standard sparse models, is interpretability. Model coefficients that are sparse in a compressed domain can still be insightful in the original space. The third reason is that, when the compression operation is known in advance, compressible models are very efficient for storage and transmission in the compressed domain. This advantage has been widely recognized in compressive sensing (Candes, 2006[3]; Donoho, 2006[4]) for general signals and thus also valid when signals are model coefficients.

2.2 Compression Operation

In this section, we discuss how to get the compression operation matrix W. The local smoothness assumption on models is related to compression operation. Smoothness characterizes the properties of derivatives of a function. For example, a constant (or piecewise constant) function has zero first-order derivatives at all (or most) locations, and a quadratic function has zero third-order derivatives at all locations. Here we will show that use of a compression transform is very flexible and can represent various smoothness assumptions on model coefficients.

Suppose we have a natural order over model coefficients $\{\beta_j\}_{j=1}^{p}$, e.g., in temporal domains where each dimension corresponds to a time point, or spectral domains where each dimension corresponds to a frequency. Order-1 smoothness assumes the coefficients "do not change very often" along the natural order. Such an assumption characterizes the first-order derivatives. It has been studied in fused lasso (Tibshirani et al., 2005[5]) where absolute values of the difference of successive coefficients, i.e., $\sum_{j=2}^{p} |\beta_j - \beta_{j-1}|$, are penalized. This idea was also explored in total variation minimization for noise removal and image enhancement (Rudin et al., 1992[6]). As a motivating example, we show that the fused lasso penalty can be approximated by a linear and invertible compression in the ℓ1 penalty.

The p × p matrix W for model compression based on order-1 smoothness can be defined as:

$$W = S_p^1 = \begin{bmatrix} 1/p & 1/p & \cdots & \cdots & \cdots & 1/p \\ 1 & -1 & 0 & \cdots & \cdots & 0 \\ 0 & 1 & -1 & 0 & \cdots & 0 \\ \cdots & \cdots & \cdots & \cdots & \cdots & \cdots \\ 0 & 0 & & & 1 & -1 \end{bmatrix} \tag{5}$$

Model coefficients in the compressed domain $W\beta = [\overline{\beta}, \beta_1 - \beta_2, \ldots, \beta_{p-1} - \beta_p]$ tend to be sparse due to ℓ1 regularization, which achieves the order-1 smoothness. The averaging operation in the first row of W makes the transform invertible. Note that if the first row of W is multiplied by a small constant (e.g., 0.001), $\|W\beta\|_1$ approximates the fused lasso penalty. In our study, we will use the compression in eq. (5) without scaling the averaging operation. Also, we keep the compression operation invertible to make the optimization efficient, as discussed in eq. (2) - eq. (4).

Smoothness of higher orders is also common. For example, a piecewise linear function has piecewise constant first-order derivatives, indicating zero second-order derivatives at most locations. This is defined as order-2 smoothness. In this case, the p × p compression transform W can be:

$$W = S_p^2 = \begin{bmatrix} 1 & 0^T \\ 0 & S_{p-1}^1 \end{bmatrix} S_p^1 \tag{6}$$

where 0 is a (p−1)×1 column vector. By this definition, model coefficients in the compressed domain are $W\beta = [\bar{\beta}, \overline{\Delta\beta}, \Delta\beta_{1,2} - \Delta\beta_{2,3}, \ldots, \Delta\beta_{p-2,p-1} - \Delta\beta_{p-1,p}]$, where $\Delta\beta_{i,i+1} = \beta_i - \beta_{i+1}$. In this sense, sparsity of the compressed model coefficients corresponds to order-2 smoothness assumption in the original space. Also, S_p^2 is invertible since both S_{p-1}^1 and S_p^1 are invertible. Finally, model compression for higher-order smoothness can be defined recursively.

Sometimes features under consideration do not follow a universal order, but can be divided into groups, where each group of features has an order or at least some groups of features are ordered. The compression operation can be defined as a block matrix to handle the use of different groups of features. For example, suppose features can be divided into three groups. We assume p_1 model coefficients on the first group of features satisfy order-1 smoothness, p_2 coefficients on the second group of features satisfy order-2 smoothness, and we have no knowledge about the third group of p_3 features. In this case, model compression is defined as:

$$W = \begin{bmatrix} S_{p_1}^1 & 0 & 0 \\ 0 & S_{p_2}^2 & 0 \\ 0 & 0 & I_{p_3} \end{bmatrix} \tag{7}$$

2.3 Model Hierarchy and Prior Distributions of Compressible Bayesian Elastic Net Model

Consider the linear model $E(y \mid X, \beta) = X\beta$; where we assume that the response variables follow the normal distribution conditionally, i.e. $y \mid X, \beta \sim N(X\beta, \sigma^2 I_n)$ We assume that all analysis hereafter is conditional on X. Zou and Hastie (2005)[1] pointed out that, under these assumptions, solving the cben problem is equivalent to finding the marginal posterior mode of β|y when the prior distribution of β is given by:

$$\pi(\beta) \propto \exp\{-\lambda_1 \|\beta\|_1 - \lambda_2 \|\beta\|_2^2\} \tag{8}$$

a compromise between Gaussian and Laplacian priors. Specifically, the conditional posterior distribution has the probability density function (pdf)

$$f(\beta|\sigma^2,y)\propto \exp\{-\frac{1}{2\sigma^2}(y-X\beta)^T(y-X\beta)-\lambda_1\|\beta\|_1-\lambda_2\|\beta\|_2^2\}=$$
$$\exp\{-\frac{1}{2\sigma^2}(y-X\beta)^T(y-X\beta)+(2\sigma^2\lambda_1)\|\beta\|_1+(2\sigma^2\lambda_2)\|\beta\|_2^2\} \quad (9)$$

However, under (8), neither the conditional posterior mode of $\beta|\sigma^2, y$ nor the marginal posterior mode of $\beta|y$ would be equivalent to the cben estimator $\hat{\beta}_{EN}$ unless the analysis is conditional on σ^2 or σ^2 is given a point-mass prior. Instead we propose a prior for β, which is conditional on σ^2, as

$$\pi(\beta|\sigma^2)\propto \exp\{-\frac{1}{2\sigma^2}(\lambda_1\|\beta\|_1+\lambda_2\|\beta\|_2^2\}\} \quad (10)$$

A noninformative prior is then assigned for σ^2,i.e. $\pi(\sigma^2)\propto 1/\sigma^2$ In this setup, the marginal posterior distribution for $\beta|y$ has the pdf

$$f(\beta|y)=\int_0^\infty \frac{C(\lambda_1,\lambda_1,\sigma^2)}{(2\pi\sigma^2)^{n/2}}\exp\{-\frac{\|y-X\beta\|_2^2+\lambda_1\|\beta\|_1+\lambda_2\|\beta\|_2^2}{2\sigma^2}\}\pi(\sigma^2)d\sigma^2 \quad (11)$$

where $C(\lambda_1,\lambda_2,\sigma^2)$ is the normalizing constant later mentioned in Lemma 1. From (11), we can see that the cben estimator $\hat{\beta}$ maximizes the integrand for each σ^2, and thus equals the marginal posterior mode of $\beta|y$. This conditional prior specification is also used by Park and Casella (2008)[7] for the sake of accelerating the convergence of the Gibbs sampler in that paper. Based on the discussion above, we have the following hierarchial model.

$$\begin{aligned}&y|\beta,\sigma^2 \sim N(X\beta,\sigma^2 I_n),\\&\beta|\sigma^2 \sim \exp\{-\frac{1}{2\sigma^2}(\lambda_1\|\beta\|_1+\lambda_2\|\beta\|_2^2)\}\\&\sigma^2 \sim \frac{1}{\sigma^2}\end{aligned} \quad (12)$$

3 Experimental Results and Analysis

We carry out Monte Carlo simulations and use the median of the prediction mean-squared errors (MMSE) to compare the performance of cben, cbl, en and lasso in prediction accuracy and variable selection. For cben, instead of using the penalty parameters (λ_1; λ_2), it is equivalent to use (s,λ_2), where $s=\|\beta\|_1/\|\beta_{OLS}\|_1$ with β_{OLS} being the ordinary least squares (ols) estimate (Zou and Hastie 2005[1]). Similarly, s instead of λ_1 is used in lasso.

The data are simulated from the linear model,

$$y=X\beta+\varepsilon,\varepsilon \sim N(0,\sigma^2 I) \quad (13)$$

Each simulated sample was partitioned into a training set, a validation set and a testing set. The validation set is used to select s for lasso and (S,λ_2) for en. After the penalty parameter is selected, we combine the training set and the validation set together to estimate β. For Bayesian methods, we directly combine the training and validation sets together for estimation.

For the first two simulation studies, we simulate 50 data sets, each of which has 20 observations for the training set, 20 for the validation set, and 200 for the testing set. In Simulation 1, we set $\beta = (3,1.5,0,0,2,0,0,0)^T$ and $\sigma^2 = 9$. The design matrix X is generated from the multivariate normal distribution with mean 0, variance 1 and pairwise correlations between x_i and x_j equal to $0.5^{|i-j|}$ for all i and j. In Simulation 2, we set $\beta = (0.85,0.85,0.85,0.85,0.85,0.85,0.85,0.85)^T$, and leave other setups the same as in Simulation 1. In Simulation 3, we first generate Z_1, Z_2 and Z_3 independently from $N(0,1)$. Then let $x_i = Z_1 + \varepsilon_i$, $i = 1,\ldots, 5$, $x_i = Z_2 + \varepsilon_i$, $i = 6,\ldots,10$, $x_i = Z_3 + \varepsilon_i$, $i = 11,\ldots,15$ and $x_i \sim N(0, 1)$, $i = 16,\ldots,30$, where $\varepsilon_i \sim N(0,0.01)$, $i = 1,\ldots,15$. We perform 50 simulations, in each of which we have a training set of size 100, a validation set of size 100 and a testing set of size 400. The parameters are set as $\sigma^2 = 225$ and $\beta = (\underbrace{3,\ldots,3}_{5},\underbrace{3,\ldots,3}_{5},\underbrace{3,\ldots,3}_{5},\underbrace{0,\ldots,0}_{15})^T$.

In Simulation 4, we set the sizes of the training set and the validation set both to 200, while leaving other setups the same as in Simulation 3. In Simulation 5, we set the sizes of the training set and the validation set both to 20, and the true parameter value to $\beta = (\underbrace{3,\ldots,3}_{10},\underbrace{0,\ldots,0}_{10},\underbrace{3,\ldots,3}_{10})$, while leaving other setups the same as in Simulation 3.

Table 1. Comparison of the four methods (ben, bl, en, and lasso) on prediction accuracy

Method	Simulation1 MMSE(SE)	Simulation2 MMSE(SE)	Simulation3 MMSE(SE)	Simulation4 MMSE(SE)	Simulation5 MMSE(SE)
cben	11.99(0.28)	11.39(0.29)	232.3(3.83)	215.1(1.95)	335.3(4.17)
en	11.29(0.53)	11.10(0.29)	281.4(6.80)	243.0(4.24)	376.9(9.81)
cbl	10.45(0.24)	10.55(0.21)	227.1(4.20)	211.4(2.36)	352.6(13.1)
lasso	10.99(0.20)	11.41(0.33)	280.1(8.00)	243.0(4.34)	384.9(9.61)

Table 1 shows that cbl has a better prediction accuracy than other methods in most simulation studies. Secondly, when the true model structure is relatively simple (Simulations 1 and 2), the four methods perform comparably in prediction accuracy. But when the true model has a complex structure (Simulations 3, 4 and 5), cben and cbl outperform en and lasso significantly.

4 Conclusion and Discussion

By including a compression operation into ℓ1 and ℓ2 regularized learning, model coefficients are compressed before being penalized and sparsity is achieved in a compressed domain. Real data examples and simulation studies show that cben behaves comparably to en in prediction accuracy en does slightly better than cben for simple models, but cben performs significantly better than en for more complicated models. Simulation studies suggest that cben outperforms en in variable selection when coupled with the scaled neighborhood criterion with a proper probability threshold, and cben gives better prediction accuracy than cbl for small samples from less sparse models.

Acknowledgments. This research is supported by the national science foundation of China (NFSC) No.61003183, natural science foundation of Jiangsu province No.BK2009199, provincial universities natural science foundation of Jiangsu province(2011), funding of Jiangsu Innovation Program for Graduate Education (CXZZ11_0216) and Doctor innovation foundation of Nanjing University of aeronautics & astronautics(2011).

References

1. Zou, H., Hastie, T.: Regularization and variable selection via the elastic net. Journal of the Royal Statistical Society B 67, 301–320 (2005)
2. Guyon, I., Weston, J., Barnhill, S., Vapnik, V.: Gene Selection for Cancer Classification Using Support Vector Machines. Mach. Learn. 46, 389–422 (2002)
3. Candes, E.J.: Compressive Sampling. In: Proceedings of International Congress of Mathematicians (2006)
4. Donoho, D.L.: Compressed Sensing. IEEE Trans. Information Theory 52(4), 1289–1306 (2006)
5. Tibshirani, R.: Regression Shrinkage and Selection via the Lasso. Journal of the Royal Statistical Soceity, Series B 58(1), 267–288 (1996)
6. Rudin, L., Osher, S., Fatemi, E.: Nonlinear total variation based noise removal algorithms. Physica D 60, 259–268 (1992)
7. Park, T., Casella, G.: The Bayesian Lasso. J. Amer. Statist. Assoc. 103, 681–686 (2008)

Evolutional Diagnostic Rules Mining for Heart Disease Classification Using ECG Signal Data

Minghao Piao[1], Yongjun Piao[1], Ho Sun Shon[1], Jang-Whan Bae[2], and Keun Ho Ryu[1,*]

[1] DB/Bioinformatics Lab., Chungbuk National University, Cheongju City, South Korea
[2] Dep. of Internal Medicine, Chungbuk National University, Cheongju City, South Korea
{bluemhp,pyz,shon0621,khryu}@dblab.chungbuk.ac.kr

Abstract. Medical information related data sets are useful for the diagnosis and treatment of disease. With the development of technology and devices in biomedical engineering, it leads data overflow nowadays. Traditional data mining methods like SVM, ANN and decision tree are applied to perform the classification of arrhythmia disease. However, traditional analysis methods are far beyond the capacity and speed to deal with large scale of information. Techniques that have capability to handle the coming data sets in incremental learning phase can solve those problems. Therefore, in this paper, we proposed an incremental decision trees induction method which uses ensemble method for mining evolutional diagnostic rules for cardiac arrhythmia classification. Experimental results show that our proposed method performs better than other algorithms in our study.

Keywords: significant rule, ensemble method, incremental mining, decision tree.

1 Introduction

With the development of database technology and measuring instruments in biomedical engineering, it not only becomes possible to record a huge amount of medical information but also leads to the exploration of the data, nowadays. Those data are including objective physiological values and medical images and signals, such as X-ray CT and ECG, and also detailed patient's background information, such as age, gender, height, weight and medical histories. Those data are rich in classes and useful for the diagnosis and treatment of disease.

ECG is the one of the typical medical signal, it records electronic activities of the heart, and has been widely studied for the diagnosis of heart disease [1, 2, 3, 4, 5], such as cardiac arrhythmia. Early diagnosis of cardiac arrhythmia makes it possible to improve the treatment of arrhythmia. There are many studies about the application of Support Vector Machine [6, 7], Artificial Neural Network [8, 9, 10, 11, 12, 13] and Decision Tree [20, 21]. However, traditional analysis methods are far beyond the capacity and speed to deal with these information, and hard to discover useful and understandable knowledge from incrementally updated data sets.

---

[*] Corresponding author.

D. Zeng (Ed.): Advances in Control and Communication, LNEE 137, pp. 673–680.
springerlink.com

In order to solve those problems, we applied an incremental decision trees mining method for mining evolutional diagnostic rules for heart disease classification using ECG signals.

2 Previous Studies of Decision Tree

A decision tree is a decision support tool that uses a tree-like graph or models of decisions and their possible consequences. Decision trees are commonly used for gaining information for the purpose of decision making. For inductive learning, decision tree is attractive for 3 reasons: (1) Decision tree is a good generalization for unobserved instance, only if the instances are described in terms of features that are correlated with the target concept. (2) The methods are efficient in computation that is proportional to the number of observed training instances. (3) The result of decision tree provides a representation of the concept that explainable to human.

The basis of many existing decision trees is Hunt's algorithm and developed trees are ID3 [14], C4.5 [15], and C5.0 [16]. Tree induction of those non-incremental approaches is based on a consistent set of labeled examples and the process for inducing a decision tree is quite inexpensive because exactly one tree is generated and maps a single batch of examples to a particular tree. ID4 [17], ID5R [18] and ITI [19] are incremental approaches and the results are as correct as non-incremental approaches do. In [22], i^+ learning (Intelligent, Incremental and Interactive learning) theory is proposed which regarding the new attributes with incremental learning ability. Three interesting observations are described in [25] based on the study of many ensemble methods: (1) Many ensembles constructed by the Boosting method were singletons. Due to this constraint, deriving classification rules have a limitation: decision trees are not encouraged to derive many significant rules and they are mutually exclusive and covering the entire of training samples exactly only once. (2) Many top-ranked features possess similar discriminating merits with little difference for classification. This indicates that it is worthwhile to employ different top-ranked features as the root nodes for building multiple decision trees. (3) Fragmentation problem is another problem does those ensemble methods have: as less and less training data are used to search for root nodes of sub-trees.

3 Evolutional Diagnostic Rules Mining

In bio-medical research mining area, the useful diagnostic or prognostic knowledge from the result is very important. Only explainable results (e.g. rule based results) are easy to understand and applicable to the bio-medical and diagnosis of a disease [23], [24]. A rule is a set of conjunctive conditions with a predictive term. The general form of rules is presented as: *IF* $condition_1$ *&* $condition_2$ *& ... &* $condition_m$*, THEN a predictive term*. The predictive term in a rule refers to a single class label. For useful clinical diagnosis purpose, using those rules we can address issues in understanding the mechanism of a disease and improve the discriminating power of the rules. A *Significant Rule* is one with a largest coverage which the coverage satisfies a given threshold. For example, the given threshold is 60%, if one rule's coverage is larger than 60% then it is called significant rule.

3.1 Evolutional Rules Discovery

When using data mining techniques to perform the classification of biomedical datasets, it is important to provide faster and more cost-effective models because of the large size of data set. For this purpose, feature selection techniques are used to find a set of relevant features. These selected features are may be ranked and top-ranked features will be considered as most important one when construct classification models. However, most of time, features are dependent from each other and different combinations between them will result in different models. Specially, for decision trees, different test conditions on different nodes will result in different tree structures. It means if all features have chance to be the root node, then we can get the number of different trees as many as number of features. Furthermore, we can find more classification rules from trees to perform the better classification work.

However, using traditional ensemble method to build and refine the tree committee in incremental phase and derive significant classification rules is still impossible and it is hard. So, we introduce a new incremental decision learning algorithm which accepts the Cascading-and-Sharing ensemble method of CS4 [26] to produce many significant rules from the decision trees and combine those rules discriminating power to accomplish the prediction process. We call this algorithm as **I**ncremental **M**ining of **S**ignificant **R**ules (IMSR) and the main steps of the construction process are shown in below. We will mainly present about details of how to construct tree committee to produce evolutional classification rules here and for more details of how to incrementally construct a decision tree are presented in ITI algorithm [19].

Step 1: Incorporating an example to the tree. When an example is to be incorporated into an empty tree, the tree is replaced by a leaf node that indicates the class of the leaf and the example is saved at the leaf node.

Step 2: Ensure that every decision node has the desired test. This algorithm recursively transposes the tree to install the best test at the node. When incorporating an example or transposing a tree it will cause the changes of examples on a subtree and then the test is marked as stale.

In the original ITI algorithm, after an instance has been incorporated into the tree, the tree is traversed immediately from the root, recursively ensuring that the best test is installed at a node. However, if a feature's rank which is calculated from the whole data set is not changed, we do not have to check the test conditions on nodes immediately, even if a number of examples are added into the tree and their information gains are changed. In our work, the transposition is triggered when the rank of features against whole data set is changed, or there were no more instances incorporated into the tree. Here, we give an example to explain the situation when the transposition is not needed. As shown in Table 1, each instance consists of three attributes: *height*, *hair color* and *color of eyes*. All attribute values are categorical and the class label has two values, positive [+] and negative [-].

Figure1 shows that the 1st to 5th instances are incorporated into the tree. When incorporating the 6th instance *(+, height=tall, hair=red, eyes=blue)*, it does not require changing the test attribute at the root. Because *red* is a new value for *hair*, a new value branch is grown. The 7th and 8th instances only update the various counts. After the 8th instance is added into the tree, the tree is Figure 2.

Table 1. Example data set

Class	Height	Hair	Eyes
-	short	blond	brown
-	tall	dark	brown
+	tall	blond	blue
-	tall	dark	blue
-	short	dark	blue
+	tall	red	blue
-	tall	blond	brown
+	short	blond	blue

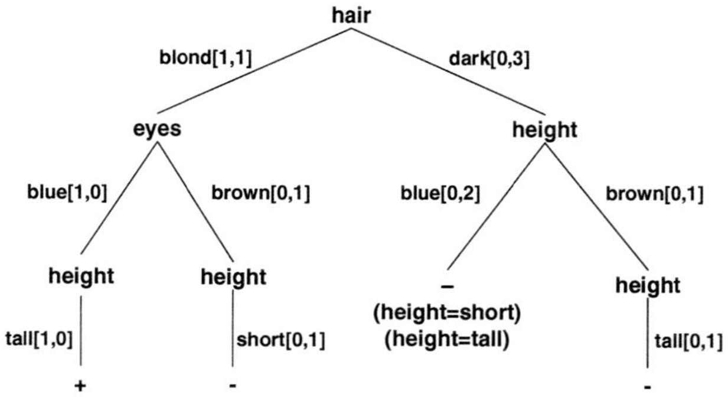

Fig. 1. After 1st to 5th instances are incorporated into the tree

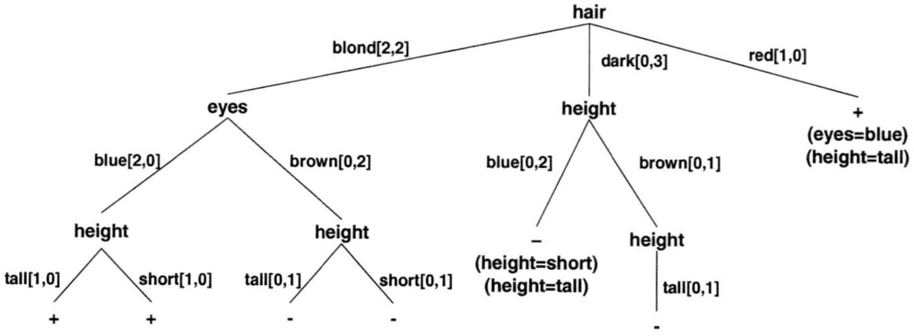

Fig. 2. Adding 6th, 7th and 8th instances into the tree

Step 3: Producing tree committees to derive evolutional rules. It works when there are test examples which need to be assigned a class label. At the moment when the first test example comes, the procedure uses the selected features used in the construction of first decision tree to construct the tree committees and derive evolutional significant rules.

There are two options to construct tree committees: construct at the point when we need to perform classification or start from the beginning of tree induction and use incremental manners to construct all trees. However, there are only a fixed number of selected features. That means that using an incremental manner to construct the tree committees is not suitable. After deriving those rules, we use the aggregate score to perform the prediction task. The classification score [12] for a specific class, say class *C*, is calculated as:

$$Score^{C}(T)=\sum_{i=1}^{K_C} Coverage(rule_i^{C}) \tag{1}$$

Here, K_C denotes the number of rules in class *C*, and $rule^{C}_{i}$ denotes *i*th rules in class *C*. If the score for one class *C* is larger than other classes, then the class label for instance *T* is assigned as *C*.

4 Experimental Results

In this section, we report the empirical behavior of the algorithm discussed above using the Arrhythmia Dataset which is taken from the University of California at Irvine (UCI) machine learning repository [10]. This database contains 452 instances with 279 attributes, 206 of which are linear valued and the rest are nominal. Class 01 refers to "normal" ECG classes 02 to 15 refer to different classes of arrhythmia and class 16 refers to the rest of unclassified ones.

Table 2. Classification results of IMSR on new data

TP Rate	*FP Rate*	*Precision*	*F-Measure*	*Class*
0.789	*0.139*	*0.798*	*0.793*	*Arrhythmia*
0.849	*0.227*	*0.816*	*0.832*	*Normal*
0	*0.033*	*0*	*0*	*Unclassified*

Table 3. Confusion matrix of original & predicted

		Original		
		arrhythmia	*normal*	*unclassified*
Classified	*arrhythmia*	*146*	*36*	*3*
	normal	*26*	*208*	*11*
	unclassified	*11*	*11*	*0*

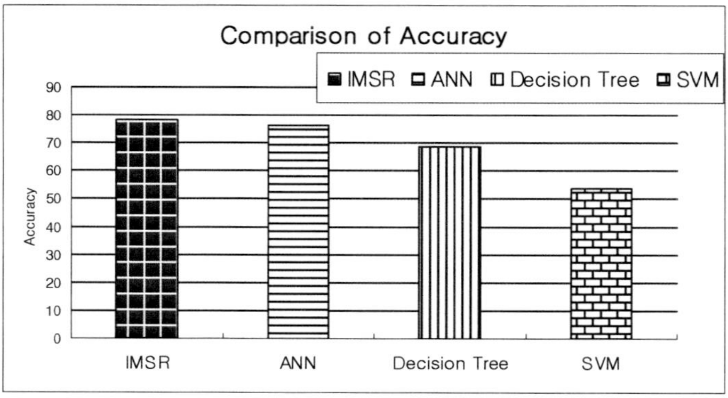

Fig. 3. Comparison of Accuracy with SVM, ANN and Decision tree

We treated class 02 to 15 as one and named as "Arrhythmia" group which means there are three classes such as "normal", "arrhythmia" and "unclassified". Table 5 shows the classification results and table 6 shows the confusion matrix based on the new data set, and figure 1 shows the comparison of accuracy with SVM, ANN and Decision tree. Figure 3 shows that our proposed algorithm has better classification accuracy than others. For the evaluation, Pearson's Chi-Square is used and 199 features whose importance (1 - *p*-value) greater than 0.95 are selected for input.

5 Conclusion

In this paper, base on well-accepted design goals we introduced an approach for discovering evolutional significant classification rules from incrementally induced decision trees by using the Cascading and Sharing method. Tree induction offers a highly practical method for generalizing from examples whose class label is known, and the average cost of constructing the tree committee is much lower than the cost of building a new decision tree committee.

For testing the performance of proposed algorithm, we used the example data which is about diagnosis of Arrhythmia from UCI repository. All 279 features are evaluated by using variable selection method and 199 useful features are given by the threshold of the feature choice. The first tree is constructed in incremental induction manner, and others are constructed when there are instances that need to be assign a class label. Because most of classes do not have enough instances to train the classifier, we reorganized the data into three classes "normal" (class 01), "arrhythmia" (class 02 to 15) and "unclassified" (class 16). Experimental results show that proposed method performs better than others in our study. We also applied the algorithm on our project and tested it on the real data set for diagnosis of ischemia.

However, we found some weakness of the algorithm and there are still some limitations have to be modified. Our proposed algorithm still could not work with new attributes in incremental learning manner and still need users to initialize the number of trees. Our future work will focus on above limitations and try to consider the relationship between features during the tree construction.

Acknowledgments. This research is supported by Basic Science Research Program through the National Research Foundation of Korea (NRF) funded by the Ministry of Education, Science and Technology (2011-0001044) and the Korean Ministry of Education, Science and Technology (The Regional Core Research Program / Chungbuk BIT Research-Oriented University Consortium).

References

1. Yasin, H., Yasin, M., Ardil, C.: Acute Coronary Syndrome Prediction Using Data Mining Techniques- An Application. International Journal of Computational Intelligence 5(4), 295–299 (2009)
2. Ting, S.L., Shum, C.C., Kwok, S.K., Tsang, A.H.C., Lee, W.B.: Data Mining in Biomedicine: Current Applications and Further Directions for Research. Journal of Software Engineering & Applications 2, 150–159 (2009)
3. Li, Y., Li, X.S.: The Data Mining and Knowledge Discovery in Biomedicine, In: Proceedings of 2010 5th International Conference on Computer Science and Education, pp. 1050–1052 (2010)
4. Sitar-Taut, D.-A., Sitar-Taut, A.-V.: Overview on How Data Mining Tools May Support Cardiovascular Disease Prediction. Journal of Applied Computer Science & Mathematics 8(4), 57–62 (2010)
5. Asha, R., Reena, G.S.: Diagnosis of Heart Disease Using Data mining Algorithm. Global Journal of Computer Science and Technology 10(10), 38–43 (2010)
6. Nasiri, J.A., Naghibzadeh, M., Yazdi, H.S., Naghibzadeh, B.: ECG arrhythmia classification with support vector machines and genetic algorithm, In. In: 2009 Third UKSim European Symposium on Computer Modeling and Simulation, pp. 187–192 (2009)
7. Asl, B.M., Setarehdan, S.K., Mohebbi, M.: Support vector machine-based arrhythmia classification using reduced features of heart rate variability signal. Artificial Intelligence in Medicine 44, 51–64 (2008)
8. Shivajirao, M.J., Sanjay, L.N., Ashok, A.G.: Modular Neural Netwrok based Arrhythmia Classification System using ECG Signal Data. International Journal of Information Technology and Knowledge Management 4(1), 205–209 (2011)
9. Jadhav, S.M., Nalbalwar, S.L., Ghatol, A.: Artificial Neural Network Based Cardiac Arrhythmia Classification Using ECG Signal Data. In: Proceedings of 2010 International Conference on Electronics and Information Engineering, pp. 228–231 (2010)
10. Frank, A., Asuncion, A.: UCI Machine Learning Repository. University of California, School of Information and Computer Science, Irvine (2010), `http://archive.ics.uci.edu/ml`
11. Raut, R.D., Dudul, S.V.: Arrhythmia Classification with MLP Neural Network and Statistical Analysis. In: Proceedings of International Journal of Engineering Research and Technology, pp. 553–558 (2008)
12. Yaghouby, F., Ayatollahi, A., Soleimani, R.: Classification of Cardiac Abnormalities Using Reduced Features of Heart Rate Variability Signal. World Applied Sciences Journal 6(11), 1547–1554 (2009)

13. Shantakumar, B.P., Kumaraswamy, Y.S.: Intelligent and Effective Heart Attack Prediction System Using Data mining and Artificial Neural Network. European Journal of Scientific Research 31(4), 642–656 (2009)
14. Quinlan, J.R.: Induction of Decision Trees. Machine Learning, 81–106 (1986)
15. Quinlan, J.R.: C4.5: Programs for Machine Learning. Morgan Kaufmann, San Francisco (1993)
16. RuleQues Research Data Mining Tools, `http://www.rulequest.com/`
17. Schlimmer, J.C., Fisher, D.: A case study of incremental concept induction. In: Proceedings of the Fifth National Conference on Artificial Intelligence, pp. 496–501 (1986)
18. Utgoff, P.E.: Incremental Induction of decision trees. Machine Learning, 161–186 (1989)
19. Utgoff, P.E., Berkman, N.C., Clouse, J.A.: Decision Tree Induction Based on Efficient Tree Restructuring. Machine Learning, 5–44 (1997)
20. Soman, T., Bobbie, P.O.: Classification of Arrhythmia Using Machine Learning Techniques. WSEAS Transactions on Computers 4(6), 548–552 (2005)
21. Marilin, K.M.: Knowledge Discovery and Data Mining of Biomedical Data with Decision Trees. In: Proceedings of CONISAR (2008)
22. Chao, S., Wong, F.: An Incremental Decision Tree Learning Methodology Regarding Attributes in Medical Data Mining. In: Proceedings of Eighth International Conference on Machine Learning and Cybernetics, vol. 3, pp. 1694–1699 (2009)
23. Lee, H.G., Noh, K.Y., Ryu, K.H.: Mining Biosignal Data: Coronary Artery Disease Diagnosis Using Linear and Nonlinear Features of HRV. In: Washio, T., Zhou, Z.-H., Huang, J.Z., Hu, X., Li, J., Xie, C., He, J., Zou, D., Li, K.-C., Freire, M.M. (eds.) PAKDD 2007. LNCS (LNAI), vol. 4819, pp. 218–228. Springer, Heidelberg (2007)
24. Ryu, K.H., Kim, W.S., Lee, H.G.: A Data Mining Approach and Framework of Intelligent Diagnosis System for Coronary Artery Disease Prediction. In: Proceedings of the 4th Korea-Japan International Database Workshop (2008)
25. Li, J.Y., Liu, H.A., Ng, S.-K., Limsoon Wong, L.S.: Discovery of significant rules for classifying cancer diagnosis data. Bioinformatics (2003)
26. Li, J., Liu, H.: Ensembles of cascading trees. In: Proceedings of Third IEEE International Conference on Data Mining, pp. 585–588 (2003)

Extracting Image Semantic Object Based on Artificial Bee Colony Algorithm

Xiao Yong-hao[1,2], Chen Yong-chang[1], Yu Wei-yu[1,2,*], and Tian Jing[3]

[1] School of Electronic and Information Engineering, South China University of Technology, Guangzhou 510641, China
[2] Provincial Key Laboratory for Computer Information Processing Technology, Soochow University, Suzhou, China
[3] School of Electrical and Electronic Engineering, Nanyang Technological University, 639798, Singapore
yuweiyu@scut.edu.cnï

Abstract. In this paper, an efficient approach for extracting semantic object using artificial bee colony algorithm (ABCA) has been proposed. First, we reduce speckle noise in the image. Then fitness function of ABC algorithm is constructed, and image pixels are classified into different regions. Further semantic objects are extracted in terms of color information. The simulation results show that the color clustering via bee colony algorithm gives superior results in enhancing cluster compactness.

Keywords: artificial bee colony algorithm, image semantic, image segmentation.

1 Introduction

Image segmentation plays an important role in image processing and computer vision applications, since it serves as the low-level processing of image(e.g., large collections of images and videos in digital libraries need to be catalogued and stored in order to efficiently browse and retrieve visual information). As a classic problem in the field of computer vision, image segmentation method is deeply studied [1-5].

A more formal definition of image segmentation can be given in the following way: an image f can be segmented into different objects or regions, which are a set of N regions $R_n, n = 1,...,N$, such that:1) $\bigcup_{n=1}^{N} R_n = f$ with $R_n \bigcap R_m \neq 0, n \neq m$ 2) $\gamma(R_n) = true \quad \forall n$; 3) $\gamma(R_n \bigcup R_m) = false \quad \forall R_n \quad and \quad R_m$ adjacent.

All the pixels in an object or a region are similar with respect to some features, such as intensity, color, or texture. Color information permits a more complete representation of the image and a more reliable segmentation of it. Accordingly, there has been a numerous research works in the area of segmentation of color images, such as: threshold segmentation [6], region growing [7], mean shift [8], graph theory [9],

---

* Corresponding author.

D. Zeng (Ed.): Advances in Control and Communication, LNEE 137, pp. 681–688.
springerlink.com

fuzzy theory [10]. In addition to the above methods, other methods are also proposed, such as morphology, Bayesian model, and neural networks.

The rest of this paper is organized as follows. In section 2, color representation is introduced. In section 3, artificial bee colony algorithm is presented. In section 4, the proposed approach is described in detail. In section5, experiments results are listed. Finally, Section 6 concludes this paper.

2 Color Representation

Currently, several color representations are used in color image processing. In the most common used RGB model, color is represented varying intensities of red, green, and blue components in an orthogonal Cartesian space. This is in agreement with the tristimulus theory of color according to which the human visual system acquires color imagery by means of three band pass filters (three different kinds of photoreceptors in the retina called cones) whose spectral responses are tuned to the wavelengths of red, green, and blue. However, the RGB space does not lend itself to mimic the higher level processes which allow the perception of color of the human visual system; in this sense, color is better represented in terms of hue, saturation, and intensity.

Image semantics is a concept in high level of image analysis, so a key problem is how to describe semantic concepts by appropriate structures. It is well known that objects in an image or segments of a scene in video can be represented by tree structure. Semantic concept of *Visual Objects* (VO) in an image or video can be characterized by a tree structure with directed arcs among its nodes at various layers.

Color information is a component of formal tuple expression for semantic of image VO.

$C^{(VOi)}$ expresses as color parameters of VO_i.

$$C^{(VO_i)} = \left[\left(C_p^{(1)}, C_p^{(2)}, \ldots C_p^{(n)}\right), \quad h_c, \quad d_c\right] \tag{1}$$

Where $C_p^{(1)}, C_P^{(2)}, \ldots, C_P^{(n)}$ denote the primary color of subsets of VO_i, color of VO_i is expressed by n color regions, the number n is determined by the precision on request, and h_c is the color histogram of VO_i. d_c is the color spatial distribution of VO_i.

The RGB color space is a popular representation of color image. A color image $I(x,y)$ in a vector form is represented as $I(x,y) = [R(x,y) \quad G(x,y) \quad B(x,y)] \in I^3, x, y \in I^2$, where $R(x,y)$, $G(x,y)$, $B(x,y)$ are the red, green, and blue channels conveniently normalized between 0 and 1 respectively. There are two drawbacks of the RGB color space. First, the RGB space is not perceptually uniform. Second, the RGB space is highly correlated for natural image.

The HSV space is an intuitive model where a specific color is described by its hue, saturation and intensity. But although color space is a very useful construct in the measurement and specification of color. It does not provide a measure of the perceived difference between two colors. The determination of a quantity that describes the difference between stimuli depends on the ability of an observer to judge the relative magnitude of such a difference. A person's judgment depends on the condition of observation and the kind of stimuli presented. To avoid this drawback, the CIE recommended two approximate uniform color spaces that were chosen from several similar merits to promote uniformity of practice. They are $L^*u^*v^*$ and $L^*a^*b^*$ color spaces. The $L^*a^*b^*$ color space is based on Munsell's uniform color space. It is produced by plotting the quantities $L^*a^*b^*$. $L^*u^*v^*$ color space has been used for color segmentation of human flesh. The $L^*a^*b^*$ space has been employed in general color image segmentation.

3 Artificial Bee Colony Algorithm

Bonabeau [11] has first defined the swarm intelligence as "any attempt to design algorithms or distribute problem-solving devices inspired by the collective behavior of social insect colonies and other animal societies". Algorithms are motivated by the collective behavior of social insects (such as ant, bees and fish). Swarm intelligence algorithm features are robustness, scalability, and self-organized. From 1990s, ant colony algorithm [12], particles swarm optimization [13], fish school algorithm [14] were proposed. With bionic swarm intelligence algorithm emerging, it greatly enhanced the ability to solve practical engineering problems.

Seely [19] first proposed self-organization swarm simulation model. Self-organization is a key feature of a swarm system which results in global level response by means of low level interactions. Bonabeau et al. interpreted the self-organization in swarms through four characteristics: (1) Positive feedback is a simple behavioral "rules of thumb" that promotes the creation of convenient structures. (2) Negative feedback can stabilize the collective pattern. In order to avoid the saturation which might occur in terms of available foragers, food source exhaustion, crowding or competition at the food sources, a negative feedback mechanism is needed. (3) Randomness is often crucial for emergent structures since it enables the discovery of new solutions. (4) Multiple interactions occur since agents in the swarm use the information coming from the other agents so that the information and data spread to all network.

Pham [16] described the Bees Algorithm (BA) which performs a neighborhood random search. For neighborhood selection, the highest fitness is chosen as selected bees. For recruitment, bees are assigned based on the fitness associated with the sites "scout" bee search. The elitist bee is selected in order to form the next generation of the colony. Drias [17] introduced a meta-heuristic named "Bees Swarm Optimization", based on the behavior of real bees. An adaptation to the features of the MAX-W-SAT problem was introduced to contribute to its resolution. Sadeg et.al [18] presented a parallel algorithm of the Bees Swarm Optimization (BSO) meta-heuristic and experienced on the performances of the sequential and the parallel algorithms. Chong etal.[20] described a Bee Colony Optimization algorithm based on foraging and waggle dance. This method uses dance durations to select a new path and the algorithm was applied to job shop scheduling.

In model, although different types of bees only complete a single task, the bees can collaborate to complete complex tasks, such as building nest and harvesting pollen and other offspring the job through the swing, smell and other communication methods. After finding a rich food source, the bee performs special movements called "dance" on the area of the comb in order to share her information about the food source such as quantity, its direction and distance and recruits the other bees for exploiting that rich food source. Hamdan [15] discovered that the bee dance have three kinds of dances, round dance, waggle dance, and tremble dance. If the distance of the source to the hive is less than 100 meters, round dance is performed but it does not give direction information. Waggle dance is performed if the source is far away. In case of waggle dance, direction of the source according to the sun is transferred to other bees. The tremble dance is performed when the foraging bee perceives a long delay in unloading its nectar.

In 2005, Baykasoglu et.al.[21] proposed an artificial bee colony algorithm which is applied to no-restricted and restricted numerical functions optimization problems.

Artificial Bee Colony algorithm consists of three basic elements: employer bees, employee bees, scout bees.

- Food sources: The "profitability" of a food source is related to several factors such as its closeness to the nest, richness of food and the ease of extracting the energy from the source.
- Employer bees: Employer bees are associated with a specific food source they exploited. They carry information about the specific source such as its distance, direction from the hive and the profitability of the source. They share this information with the forager bees waiting in the hive by dancing which is an example of multiple interactions.
- Employee bees: onlookers and scouts. The scouts randomly search food sources and this behavior is a kind of fluctuations which is vital for self-organization; and the onlookers waiting in the hive find a food source by means of the information presented by employed foragers. The number of scouts is about 5~10% of the foragers.

4 Proposed Approach

The data are classified according to the degree of similarity automatically. These classified methods include sequence clustering algorithm, hierarchical clustering algorithm, and the optimal cost function based clustering algorithm. In most instances, the data clustered constituted by the real value vector, and Euclidean distance is an effective tool for measures of similarity.

In this paper, optimal cost function-based clustering approach is proposed. The nectar is defined as the set of cluster center. Bees foraging process is to find the optimal value of nectar, i.e., to find the cluster center of an image, which will make the cost function arrive the optimal value.

The whole process can be described as follows:

Step1. Initialize the parameters of bee colony. Bee colony size, employed bee, scout bee, neighbor search radius r , clustering number k, and iteration number;
Step 2. Assign 5-10% of total bees as scout bee. Scout bees conduct random searches for food source, and evaluate the values of food sources;
Step 3.Sort the values of the food sources searched, and determine the number of employed bees;

Step 4. While (Satisfy the cycle conditions);
Step 5. Move the scouts for searching new food sources;
Step 6. Assess the values of food sources searched;
Step 7. Best food source is the optimal cluster center, and calculate the center of each cluster via Euclidean distance, all pixels are classified to the nearest center of class. Image segmentation is completed when all pixels are classified;
Step 8. end.

EU's in algorithm is denoted in Equ. (3)

$$d(x,y)=\left(\sum_{i=1}^{V}|x_i-y_i|^p\right)^{1/p} \tag{3}$$

Here, V is the dimensional of data sample, $p=2$. And cost function is denoted in Eq. (4).

The process of proposed optimal cost function can be expressed via Eq. (5).

$$f(x,c)=\sum_{i=1}^{M}\mathbf{min}\left\{\left\|x_i-c_j\right\|^2,|\, j=1,2,\cdots,K\right\} \tag{4}$$

$$f_{opt}(x,c_{opt})=\mathbf{min}\{f(x,c)\} \tag{5}$$

Here, x is image pixels, c is clustering centers.

5 Experimental Results

In order to evaluate the proposed approach, we conducted experiments using some standard gray-scale images by software of Matlab 2009, which running at a 2.8GHz CPU, 4GHz Memory computer.

In this section, the salt-peppers noise is eliminated first via particle filtering method, which can avoid the impact of noise on the segmentation result. Background extracted from the simple or complex nature objects has the same features. The experiment listed the extraction of several types of image objects, as shown in Figure 1-4. Fig (a) is original image, Fig.1 (b), (c) extract red flower and green leaf respectively. Figure 2(b),(c),(d) extract red flower, green leaf and background, Figure 3(b),(c),(d) extract fish, fish school and background, Figure 4(b),(c),(d),(e) extract Different objects. Experiment results indicated proposed approach is very effective.

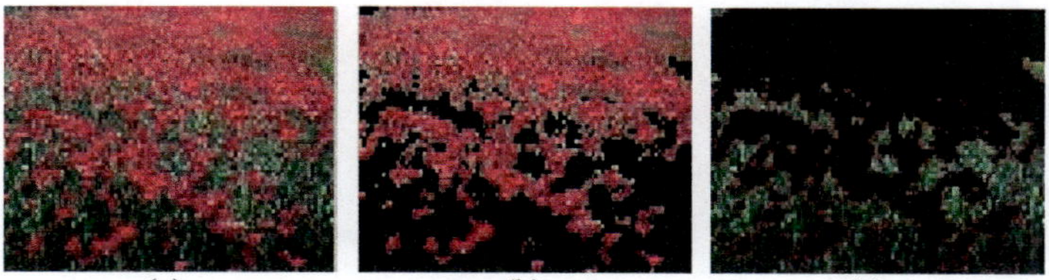

(a) (b) (c)

Fig. 1. Segmentation result, k=2

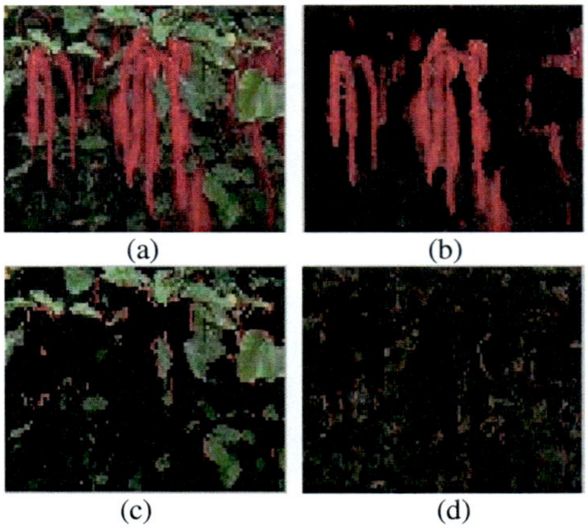

Fig. 2. Segmentation result, k=2

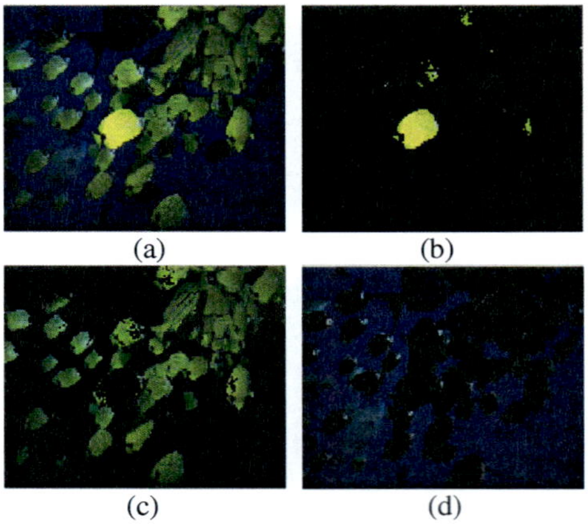

Fig. 3. Segmentation result, k=3

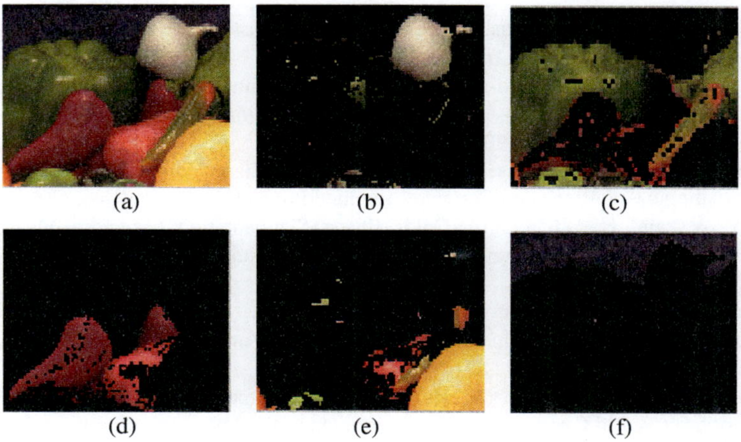

Fig. 4. Segmentation result, k=5

6 Conclusions

We proposed a novel image semantic object extraction using artificial bee colony algorithm in this paper. This approach can extract the semantic objects well, since the HSV colour space obeys a uniform distribution. Experiment results demonstrated the proposed approach is robust and more effective. When the semantic object types are less, the extraction result will be better, and the more types of objects or uneven illumination the pre-extraction image have, the more background the extracted objects have. Future works include avoiding the algorithm falling into local optimization, and how to extract semantics of the objects quick and effective.

Acknowledgements. This paper is supported by the National Natural Science Foundation of China (Grant No. 60872123, 60972133), the Joint Fund of the National Natural Science Foundation and the Guangdong Provincial Natural Science Foundation (Grant No. U0835001), the Fundamental Research Funds for the Central Universities, SCUT, Provincial Key Laboratory for Computer Information Processing Technology, Soochow University, (Grant No. KJS0922).

References

1. Cheng, H.D., Jiang, X.H., Sun, Y., et al.: Color image segmentation: advances and prospects. Pattern Recognition 34(12), 2259–2281 (2001)
2. Noble, J.A., Boukerroui: Ultrasound image segmentation: a survey. IEEE Transactions on Medical Imaging 25(8), 987–1010 (2006)
3. Di Zenzo, S.: Advances in image segmentation. Image and Vision Computing 1(4), 196–210 (1983)
4. Fu, K.S., Mui, J.K.: A survey on image segmentation. Pattern Recognition 13(1), 3–16 (1981)

5. Tobias, O.J., Seara, R.: Image Segmentation by Histogram Thresholding Using Fuzzy Sets. IEEE Transactions on Image Processing 11(12), 1457–1465 (2002)
6. Fu, X., Ding, M.: A multi-threshold image segmentation method with adaptive fuzzy entropy. In: 2nd Interna. Confere. on Information Technology and Management Engineering, pp. 171–174 (2009)
7. Jun, T.: A color image segmentation algorithm based on region growing. In: 2nd International Conference on Computer Engineering and Technology, pp. 634–637 (2010)
8. Tao, W., Jin, H., Zhang, Y.: Color Image Segmentation based on Mean Shift and Normalized Cuts. IEEE Trans. on Systems, Man, and Cybernetics, Part B: Cybernetics 37(5), 1382–1389 (2007)
9. Wu, Z., Leahy, R.: An optimal graph theoretic approach to data clustering: theory and its application to image segmentation. IEEE Trans. on Pattern Analysis and Machine Intelligence 15(11), 1101–1113 (1993)
10. Liu, Y., Zhang, B., Zhang, L., et al.: Research on Image Segmentation Based on Fuzzy Theory. In: World Congress on Computer Science and Information Engineering, vol. 4, pp. 790–794 (2009)
11. Bonabeau, E., Dorigo, M., Theraulaz, G.: Swarm intelligence: from Natural to Artificial Systems. Oxford University Press, New York (1999)
12. Dorigo, M., Socha, K.: An introduction to ant colony optimization. Technical Report series, pp. 3781–3794 (2005)
13. Lu, L., Luo, Q., Liu, J.-Y., et al.: An improved particle swarm optimization algorithm. In: IEEE International Conference on Granular Computing, GrC 2008, pp. 486–490 (2008)
14. Filho, C.J.A.B., de Lima Neto, F.B., Lins, et al.: A novel search algorithm based on fish school behavior. In: IEEE International Conference on Systems, Man and Cybernetics, SMC 2008, pp. 2646–2651 (2008)
15. Hamdan, K.: How do bees make honey. Bee Research Unit, National Center for Agriculture Research and Technology Transfer, bee, `http://www.jordanbru.info/howdoBeesmakehony.htm`
16. Pham, D.T., Ghanbarzadeh, A., Koc, E., et al.: The bees algorithm. Technical report, Manufacturing Engineering Centre, Cardiff University, UK (2005)
17. Drias, H., Sadeg, S., Yahi, S.: Cooperative Bees Swarm for Solving the Maximum Weighted Satisfiability Problem. In: Cabestany, J., Prieto, A.G., Sandoval, F. (eds.) IWANN 2005. LNCS, vol. 3512, pp. 318–325. Springer, Heidelberg (2005)
18. Sadeg, S., Drias, H.: A selective approach to parallelise bees swarm optimization metaheuristic. Int. J. Innov. Comput. Appl. 1(2), 146–158 (2007)
19. Seeley, T.D.: The wisdom of the hive: the social physiology of honey bee colonies. Harvard University Press (1995)
20. Chong, C.S., Malcolm Low, Y.H., Sivakumar, A.I., et al.: Using a bee colony algorithm for neighborhood search in job shop scheduling problems. In: 21st European Conference on Modeling and Simulation, ECMS (2007)
21. Baykasoglu, A., Ozbakir, L., Tapkan, P.: Artificial bee colony algorithm and its application to generalized assignment problem. In: Swarm Intelligence Focus on Ant and Particle Swarm Optimization. I-Tech Education and Publishing, Vienna, pp. 113–144 (2007)

Comparing Content Analysis of Mathematics Websites between Taiwan and China

Hsiu-fei Lee

National Taitung University, Department of Special Education, 684, Sec.1, Chung-hua Rd., Taitung 950, Taiwan
fei@nttu.edu.tw

Abstract. Taiwan is famous for personal computers (PC) and the PC related products. China is the world's biggest PC manufacturing center and consuming market. Students from Taiwan and China are strong in international assessment of mathematics. This study aims to explore the relationship between computer and mathematics worlds with the content analysis method and four virtual spaces of ICDT model to analyze mathematics websites of Taiwan and China. The results showed that most math websites in Taiwan and China remained at the stage of providing information; they were less for communication, and rarely for trade and delivery services.

Keywords: Content Analysis, Mathematics Websites, ICDT.

1 Introduction

Taiwan has a nickname for PC kingdom which is famous for personal computers and PC related products, ranging from notebooks, tablet computers, to smart phones and internet peripherals. China has become the world's biggest PC manufacturing center due to Taiwan's investment and influences. With the advance of internet technology and smart phones, the population of internet users has dramatically increased worldwide recently, including Taiwan and China.

Students of Taiwan and China have impressed the world for recent international assessments on mathematics [1]. Taiwanese 15-year-old students ranked number one in mathematics in 2006 on Program for International Student Assessment (PISA), a test executed by the Organization for Economic Cooperation and Development (OECD). China ranked number one for the 2009 PISA. Given the facts above, it is interesting to learn how these two factors, in terms of math learning and websites are integrated in Taiwan and China. Thus, this study aims to compare math websites between Taiwan and China by adopting the content analysis method and four virtual spaces of ICDT model. In addition to the e-commerce opportunities, website design can serve as a good platform to develop a country's cultural and creative industries. The results of this study can provide the government, educational sectors, and the technology industries some advice for developing the 'soft power'.

D. Zeng (Ed.): Advances in Control and Communication, LNEE 137, pp. 689–697.
springerlink.com

2 Literature Review

2.1 Mathematics Websites

It is effective to utilize website technology to teach math [2]. The positive effects are mainly attributed to the merits of the design and technology of websites in terms of instant feedbacks, learner self-paced process, timeless constraints, vivid and diversified demonstrations and/or representations to boost the learners' motivation, attention, and comprehension.

2.2 Content Analysis

Content analysis (CA) is a methodology which started in the field of social sciences for studying the content of communication. Weber [3] defined content analysis as a methodology providing the procedures towards logical reasoning when analyzing texts or subject matters. Later, CA is defined differently by different theorists. Generally speaking, CA is a summarizing technique that relies on the scientific methods that can provide quantitative and qualitative information for the analyses. Thus, the results of this study can help us understand the types and applications of technology forservices which can provide marketing strategies and decision makings for the development of cultural and creative industries.

2.3 Content Analysis of Websites

Content analysis has become a common method to analyze websites. There are many studies in Taiwan applied this method to study marketing, technology design, and education. However, only one study directly analyzed the mathematics websites in Taiwan[4]. Su [5] compared e-retail websites between Taiwan and China. Lin& Shih [6] studied the leisure farms in Taiwan and China. However, there is no study comparing Taiwan and China for the content analysis of math websites.

2.4 ICDT

The ICDT model was developed by Albert Angehrn [7], which is a systematic approach to the analysis and classification of business-related Internet strategies. This model serves as a basis for identifying how existing products and services can be extended and/or redesigned in order to take advantage of the Internet, as well as suggesting how new goods and services become possible through this new medium.

This model makes the virtual market space into four areas. They are Virtual Information Space (VIS), Virtual Communication Space (VCS), Virtual Distribution Space (VDS), Virtual Transaction Space (VTS).

3 Methodology

This study has reviewed the math websites in Taiwan and China, compared the functions of current business websites, and combined the results of the investigations with the ICDT model. The framework for the analysis of this study has been generated after synthesizing the data collected from the proceduresmentionedabove. As a result, the content framework of math websites for the content analysis of this study is shown in Table 1.

Table 1. Content Framework of Mathematics Websites

Virtual Space	Categorization Variables	
VIS	Math Website Info	News, activities, websites of math department, educational websites
	MathWebsite Services	Teaching materials, games, tests, teaching videos, journal articles/ reports, links, math articles
	Math Teaching Units	Number and quantity, geometry, algebra, statistics and probability, time
	Teaching Levels	K to elementary school, elementary school, junior high school, high school, college
VCS	Members only	Join member, contact us, newsletter, shopping cart info, membership value-added services, membership discount
	Interaction	Discussion board, message board, visitor counts
VTS	Products and ServiceInfo	Cram schools, advertisement info, publishers, commercial online classes
	Type of Payment	Online credit payment, money transfer, TEL/FAX
	Security Agreement	Privacy policy, account and password security, copy rights
	Services	Return and exchange policy, Q&A services, membership services
VDS	Delivery	Logistics distribution, post office, pick-ups at convenience stores
	Return Policy	Pick-ups at home, post office, return by other delivery companies

3.1 Content Framework of Mathematics Websites

The VIS framework for math websites combined the current functions of VIS with the concepts of *search* and *comparison* in thesurfingprocess,and of*problem recognition*in the purchase decision-making process as well as the features of math websites. As a result, this study's VIS included four parts: math website information, math website service, math teaching units, and teaching levels.

The VCS framework was based on the *examination* concept in the internet transactionsprocess. Examination means to examine or inspect the products that were on the shopping list of the customers. It consisted of two sections, members only and interaction. However, it requires the customers to register as members before they can read the basic information or to have any further inquiry of information on most math websites.

The VTS framework was modified based on the theory of three-stage internet transactions from the customers' perspectives, including purchase determination, purchase consumption, and post-purchase interaction [9]. Yet, the procedures or stages were not necessarily linear; they could be coming back and forth. Many math websites did not provide any commercial services, thus only commercial related websites were categorized in this section.

Lastly, the VDS framework included the last stage of the internet transactions. It included two parts: delivery and return policy. The area of delivery was outdated with the advance of delivery business and competitions of global economy.

3.2 Sampling

The source of data came from two biggest internet search engines in Taiwan—Yahoo and Google with keywords of mathematics (teaching) websites. The data collection time was from June 15 to November 15, 2010. There were 62 and 100 pages of results searched from Google and Yahoo, respectively. Each page consisted of 10 websites. As a result, 609 and 997 results were found from Google and Yahoo, respectively. In other words, 1606 results were collected and searched. Then, after deleting the results which were dead, unrelated to math websites, repetitive, or with risk alert from Google, the same screening procedures were applied to the results from Yahoo. Consequently, 900 internet links was gathered, from which consisted of 227 and 42 websites from Taiwan and China, respectively.

3.3 Reliability

Neuendorf [10] suggests that when human coders are used in content analysis, reliability translates to inter-coder/ inter-rater reliability. In other words, it means the amount of agreement or correspondence among two or more coders. Wang [11] suggests the reliability coefficient to be 80% for the standards of the Gerbner's cultural index in the content analysis of communication. Thus, the reliability coefficients between 67-80% are acceptable, of which results should be carefully defined and explained. In general, the reliability coefficient should be greater than 85% to show that the encoding results of the inter-coders are acceptable and reliable.

The formula for content analysis is as below:

Intercoder agreement $= \frac{2M}{N_1+N_2}$, M is the number of total agreement, N_1 is the agreement number of the first encoder, N_2 is the agreement number of the second encoder

$$\text{Reliability} = \frac{\text{n} * \text{Average Intercoder Agreement}}{1 + [(\text{n} - 1)\text{Average Intercoder Agreement}]}$$

n is the number of encoders

Table 2. Inter-rater Agreement and Reliability Coefficient Value of the Math Websites in Taiwan and China

Constructs / Coefficient value / Items	VIS		VCS		VTS		VDS	
	Taiwan	China	Taiwan	China	Taiwan	China	Taiwan	China
Inter-rater agreement	0.93	0.94	0.91	0.92	0.93	0.93	0.92	0.94
Reliability	0.96	0.97	0.95	0.96	0.96	0.96	0.96	0.97

Two researchers coded the websites based on the construct of the ICDT model. The results in Table 2 showed that the inter-rater agreement and reliability coefficient value of the math websites was on average above .90. It means the results of the coding were reliable.

4 Results

4.1 Analysis of VIS

The results of analysis of VIS showed that both Taiwan and China's math websites had the same ranking order in the section of math website information. The educational websites came first, then activities information, math related news, and websites of math department. Providing information for other educational websites was 33.5% and 38.1% for Taiwan and China, respectively. No information was found for websites of math department in China's math websites.

Math website service (MS) was the major function in both Taiwan's and China's math websites. Providing links to other resources came as the first (52.4% vs.50.1%) and teaching materials (42.3% vs.47.6%) as the second frequent function for both areas. For Taiwan, tests (32.2%) and math articles (34.4%) accounted for 1/3 of frequency. Games and teaching video were less frequent but still made up roughly 20 percent. Journal articles/research reports was the least frequent. The reasons could be that journal articles were collected and linked to various databases. Therefore, the designer of the math websites in Taiwan did not provide this service. Such needs could be supplemented by the function of 'links'. For China, the first 4 rankings were the same as Taiwan's. However, China's math websites provided more math articles (42.9% vs.34.4%), journal articles/ reports (28.6% vs.1.3%), tests (38.1% vs.32.2%), and teaching materials than those of Taiwan. No teaching videos and very few games (2.4%) were found in China's math websites.

The rest two sections were related to teaching materials. Since the entrance examinations for high schools and colleges are still in practice and highly valued by the parents in both areas, it is not surprising to find that the content of teaching units matched with the teaching levels since these units were the core topics of math subject in the entrance examinations. The percentages for teaching levels were similar from elementary to high schools making up 1/3 of frequency for Taiwan; yet, the elementary school level was higher for China (50%). The teaching level from kindergarten to elementary (usually below the 2nd grade) made up only 1.3% and 2.4% for Taiwan and China, respectively.

Interestingly, except for the teaching unit 'time', the other four units shared the same percentages in China's math websites. Time is a difficult topic for many children due to its abstract nature. We can measure time, but it is hard to really feel its existence. The trouble thing for children in learning time concept is to remember the different units and to convert them. Taiwan's math websites covered less on time unit than China (11% vs. 16.7%).

Table 3. Statistics of VIS Categorization of Math Websites in Taiwan and China

Categorization	Percentage	Taiwan (%)	China (%)	Average (%)
Math Website Information (MI)	news	4	7.1	5.60%
	activities	10.6	9.5	10.10%
	websites of math department	7.5	0	3.80%
	educational websites	33.5	38.1	35.80%
Math Website Service (MS)	teaching materials	42.3	47.6	45.00%
	games	19.4	2.4	10.90%
	tests	32.2	38.1	35.20%
	teaching videos	19.4	0	9.70%
	journal articles/ reports	1.3	28.6	15.00%
	links	52.4	50	51.20%
	math articles	34.4	42.9	38.70%
Math Teaching Units (MT)	number and quantity	23.8	28.6	26.20%
	geometry	26	28.6	27.30%
	algebra	20.7	28.6	24.70%
	statistics and probability	17.6	28.6	23.10%
	time	11	16.7	13.90%
Teaching Levels (TL)	kindergarten to elementary school	1.3	2.4	1.90%
	elementary school	33.9	50	42.00%
	junior high school	33.5	38.1	35.80%
	high school	33.5	31	32.30%
	college	16.7	9.5	13.10%

4.2 Analysis of VCS

The results of VCS showed that the functions of 'contact us', 'join member' and 'message board' were the three most commonly used functions on the math websites in Taiwan and China, although the percentages were not the same. For the section of members only, the percentages did not exceed 20% for both areas. Shopping cart info, membership value-added services, membership discount were even less than 5%.

The differences between discussion board and message board were the *directions* of communication. Discussion board was more bi-directional; that is, back and forth discussions occurred in this section for the issues or questions posed. The level for the topics was usually more elaborated in discussion board. On the other hand, message board was usually for a quick or short question. The results showed that the message board was used more often than the discussion board for both areas, and the proportions of math-worried parents might be higher in China than in Taiwan.

4.3 Analysis of VTS

Generally speaking, the percentage for the analysis of VTS was below 10% for both areas. The results showed that only 4% of the math websites in Taiwan and 0% in China engaged with e-commerce services.

Table 4. Statistics of VCS Categorization of Math Websites in Taiwan and China

Categorization variables	Percentage	Taiwan (%)	China (%)	Average (%)
Members only	join member	12.8	14.3	13.6%
	contact us	18.1	11.9	15.0%
	newsletter	4.8	2.4	3.6%
	shopping cart info	3.5	2.4	3.0%
	membership value-added services	1.8	4.8	3.3%
	membership discount	1.8	2.4	2.1%
Interaction	discussion board	1.8	4.8	3.3%
	message board	11.5	23.8	17.7%
	visitor counts	11.5	4.8	8.2%

Table 5. Statistics of VTS Categorization of Math Websites in Taiwan and China

Categorization variables	Percentage	Taiwan (%)	China (%)	Average (%)
Products and Service Information	cram schools	3.5	2.4	3.0%
	single ad info	1.3	0.0	0.7%
	publishers	0.4	2.4	1.4%
	commercial online classes	4.4	0.0	2.2%
Type of Payment	online credit payment	9.3	4.8	7.1%
	money transfer	9.3	4.8	7.1%
	TEL/FAX	7.9	4.8	6.4%
Security Agreement	privacy policy	4.4	7.1	5.8%
	account and password security	4.8	9.5	7.2%
	copy rights	5.3	7.1	6.2%
Customer Services	return and exchange policy	1.3	0.0	0.7%
	Q&A services	4.8	2.4	3.6%
	membership services	3.5	2.4	3.0%

Table 6. Statistics of VDS Categorization of Math Websites in Taiwan and China

Categorization variable	Percentage	Taiwan (%)	China (%)	Average (%)
Delivery	Logistics distribution	1.8	0.0	0.9%
	post office	1.3	0.0	0.7%
	pick-ups at convenience stores	1.3	0.0	0.7%
Return Policy	Pick-ups at home	0.9	0.0	0.5%
	post office	0.4	0.0	0.2%
	return by other delivery companies	1.3	0.0	0.7%

4.4 Analysis of VDS

Given the fact that math websites in Taiwan and China engaged in e-commerce was low, it is not surprising to learn that the results of analysis of VDS were very low too. In other words, when there was less trade, then there would be less delivery and return of products. Such VDS concept seemed to be novel in China and worth further development in Taiwan.

5 Conclusions

The comparisons of analysis of math websites between Taiwan and China showed some variations. However, surprisingly, it seemed to be more resemblance between both areas in terms of the math websites based on the ICDT model. The operational functions of math websites in both Taiwan and China were unevenly distributed in the four sections of the ICDT model. The order was VIS, VCS, DTS, and VDS. In other words, the math websites still remained at the stage of providing information and adopted more traditional way of communication (one way rather than two or multiple ways). Information for math teaching materials, tests, and links to other related websites took up the major functions for VIS. Teaching units and teaching levels were interrelated. That is, the major functions of VIS were strongly affected by the national junior high and high school entrance examinations. The very low percentages of math games, kindergarten levels, math teaching videos, and e-commerce functions in the math websites suggested that these areas should be further developed, given the fact that math education is highly valued by the parents in both areas. Therefore, parents are very willing to invest for their children in this regard. On top of that, China is a very big potential market for math education. The findings of this study can provide the government, educational sectors, and the technology industries in Taiwan and China some guidelines for developing the cultural and creative industries, the so called 'soft power'.

References

1. PISA 2006, 2009 results, http://www.oecd.org/document/2/0,3343,en_32252351_32236191_39718850_1_1_1_1,00.html
2. http://www.oecd.org/document/61/0,3343,en_2649_35845621_46567613_1_1_1_1,00.html
3. Sang, M.C.: The Influence of Integrating 'Changba'sMathland'Website into Mathematics Instruction on Learning Achievement and Attitudes toward Mathematics, unpublished thesis, FoGuang University, Yi-lang, Taiwan (2005)
4. Weber, R.P.: Basic content analysis, 2nd edn. Sage Publications, U.S.A (1990)
5. Lee, H.F.: Content Analysis of Mathematics Websites in Taiwan. International Journal of Computer Science & Emerging Technologies (IJCSET) 1(2), 83–88 (2011)
6. Su, C.H.: Analysis of Content Framework of E-Retail Websites by Cluster Analysis and Association Rule Technique, unpublished thesis, KainanUniversity, Tainan, Taiwan (2005)

7. Lin, C.N., Shih, M.L.: Content Analysis on the Website Function Structure of Leisure Farms across the Taiwan and Mainland China. J. International Cooperation 4(1), 16–30 (2009)
8. Angehrn, A.: Designing Mature Internet Business Strategies: The ICDT Model. European Management Journal 15(4), 361–369 (1997)
9. Leung, A.: The ICDT Model: A Framework for E-business, `http://www.mediacircus.net/icdt.html`
10. Kalakota, R.: &Whinston, A.B.: Frontiers of Electronic Commerce. Addison-Wesley, U.S.A (1996)
11. Neuendorf, K.A.: The Content Analysis Guidebook. Sage Publications, U.S.A (2002)
12. Wang, S.F.: Communication Content Analysis: Theory and Demonstration. Youth Books, Taiwan (1992)

Analysis of Online Word of Mouth of Student Teachers in Internship–A View on Mathematics Teaching

Hsiu-fei Lee

National Taitung University, Department of Special Education, 684, Sec.1, Chung-hua Rd., Taitung 950, Taiwan
fei@nttu.edu.tw

Abstract. This study analyzed online word of mouth of student teachers in internship from three groups--general education, special education, and child education at college. The results showed that general subjects, rules and relationship in classroom management and pedagogical content knowledge in teaching methods were the commonly shared concerns for all three groups. The findings can provide the institutes of teacher education some guidelines in designing their internship programs which tailor to the needs of their students.

Keywords: Online Word of Mouth, Student Teachers, Internship, Math Teaching.

1 Introduction

Taiwan is one of the top-performing countries for many recent international mathematics assessments. For instance, Taiwan's 15-year-olds ranked number one in mathematics for the PISA of 2006. However, Taiwan's ranking fell from the number one to five for the PISA of 2009 [1]. This is a warning sign for Taiwan's mathematics education.

In addition, the average birth rate for each woman in Taiwan has been plummeting drastically in recent years. It has become 0.94% in the year of 2010 in Taiwan according to ministry of home affairs [2]. The less-child-phenomenon has caused an impact on various sectors. The institutes of teacher education are facing the challenges of attracting more competent students to be teachers as well as how to train more competent teachers to make Taiwan stay competitive in the world in terms of mathematics education and thus economy. As a member of teacher education programs, I feel the urge to study how student teachers feel and concern about in internship. Since mathematics is the subject that can be taught by any student teacher in Taiwan, what role the teaching of mathematics plays in the internship is my original concern for this study. With advances of technology and accessibility of internet, the results yielded from adopting the online word of mouth method to investigate the student teachers in internship can help the institutes of teacher education better understand their 'clients' and thus improve their internship programs.

D. Zeng (Ed.): Advances in Control and Communication, LNEE 137, pp. 699–705.
springerlink.com

2 Literature Review

2.1 Online Word of Mouth (OWoM)

With cheaper notebooks and accessibility of smart phones and i-pad, the number of internet users in Taiwan reached 14.47 millions in 2010 for the age of above 12 [3]. The growing number of internet-users engage in online activities to communicate, interact, and/or in contact with other users. Via internet, with more diversified information exchange platforms have made the exchange of information flows at a more constant and instant way. For instance, BBS, message boards, forums, chat rooms, blogs, facebook (FB), and twitter are all possible platforms that allow users to collect and exchange information including questions, experiences, knowledge, and comments of other users.

Word of mouth (WoM) is defined as the process of sharing personal experiences, opinions, and ideas regarding products between at least two consumers without commercial intentions [4]. Word of mouth can contain positive or negative comments. Studies show that online word of mouth affect customer's decision-making behaviors more than the traditional WoM for American college students [5].

2.2 Online Discussion of Internship

There was very little research studying the online discussion of internship. Jiang studied 36 student teachers participating in a web-based virtual community for their Interaction behavior and professional development [6]. The results showed that the students from the one year post-college programs were more enthusiastic in interactions with others than their college counterparts. The interactions with the internship advisors mostly were questions regarding classroom management. Teaching methods regarding content knowledge were seldom asked by students. Generally speaking, providing personal experiences, giving advices to problems, and giving quick tips were three most commonly discussed topics, whereas higher levels of comments such as reflections for teaching were very rarely seen in online discussions.

3 Methodology

3.1 Correspondence Analysis

This study investigated the feedbacks from the student teachers in internship with the aim to study what topics and/or areas that the student teachers concerned about and discussed most. Keywords of internship, student teachers, internship of special education, and internship of child education were used for the search. The data were gathered and categorized by the dimensions and items, and then key in on excel. The analysis of frequency and percentage was generated from excel.

Correspondence Analysis (CA) was adopted as the analysis methodology. The cognitive map generated by the CA method can help the management sector recognize the consumer's behaviors of the target markets, such as the factors relating to the consumption behaviors of the consumers [7]. Excel and XLSTAT were adopted for the statistical analysis. A spatial diagram was generated in which the relative positions of the categorized data were used to explain the relationships between the characteristic items and the internship of three groups.

3.2 Sampling

This study classified teacher education programs into three major groups--general education, special education, and child education in that this study aimed to investigate the internship of mathematics teaching in the first place. Student teachers of physical education, music and art education, and/or social studies were treated as the group of general education.

The data came from online content such as forum of NTU PTT, BBS, Yahoo and Wretch Blog. Keywords of internship, student teachers, internship of special education, and internship of child education were searched on Yahoo. The articles dated from January 1, 2008 to December 31, 2010. There were 900 valid samples analyzed in this study after eliminating invalid ones.

4 Results

4.1 Results of Analysis of OWoM of the Student Teachers in Internship

There were 9 dimensions included for the analysis of frequency and percentage for the student teachers in internship (see Table 1). The results showed that general subject was the most frequently discussed online across the three groups, which all consisted of more than 20% of the frequency counts. That is, 29.58%, 20.43%, and 26.92% of the postings were from student teachers of the general education, special education, and child education, respectively. Surprisingly, mathematics seemed to be no concerns or questions to the student teachers of special education and child education. Even for the general education, the percentage was very low, only 3.23%.

In the section of classroom management, rules and relationship were the two most frequently discussed topics for all three groups, although the proportions were slightly different among groups. Student teachers of general education valued questions regarding rules and then student-student and/or student-teacher relationship (15.33% and 14.85%). Student teachers of special education shared the same pattern (15.33% and 14.85%) with their general education peers. Those in the child education group more concerned about the relationship than rules. Whether the reasons resided in the age of the children may need further investigations, such as interviews of the student teachers.

In the section of teaching methods, variations were found in the results. Student teachers of general education valued only pedagogical content knowledge (PCK), whereas their peers of special education considered all three factors in teaching methods, ranging from content knowledge (CK), pedagogical knowledge (PK) to PCK. On the other hand, PK was not a big concern for the student teachers of child education when compared with their peers in special education (1.28% vs. 8.95%).

Table 1. Analysis of Online Word of Mouth of the Student Teachers in Internship

Dimension	Items	General Education		Special Education		Child Education	
		(1)	(2)	(1)	(2)	(1)	(2)
Subjects	General	247	29.58	274	20.43	274	26.92
	Mathematics	27	3.23	0	0.00	0	0.00
Classroom Management	Rules	128	15.33	133	9.92	119	11.69
	Relationship	124	14.85	127	9.47	133	13.06
	Inclusion	11	1.32	15	1.12	0	0.00
	Gender Issues	4	0.48	1	0.07	2	0.20
Teaching Methods	Content Knowledge (CK)	6	0.72	80	5.97	53	5.21
	Pedagogical Knowledge (PK)	5	0.60	120	8.95	13	1.28
	Pedagogical Content Knowledge (PCK)	64	7.66	92	6.86	65	6.39
Resources Sharing	Qualified Exam & Job	27	3.23	21	1.57	16	1.57
	Internship Information	35	4.19	42	3.13	33	3.24
School Administration (SA)	SA-Neutral Comments	30	3.59	31	2.31	6	0.59
	SA-Negative Comments	8	0.96	11	0.82	4	0.39
	SA-Positive Comments	8	0.96	14	1.04	13	1.28
Interactions with Parents (P)	P-Neutral Comments	5	0.60	20	1.49	2	0.20
	P-Negative Comments	4	0.48	4	0.30	1	0.10
	P-Positive Comments	5	0.60	9	0.67	6	0.59
Interactions with Internship Advisors (IA)	IA-Neutral Comments	13	1.56	32	2.39	11	1.08
	IA-Negative Comments	5	0.60	4	0.30	13	1.28
	IA-Positive Comments	9	1.08	87	6.49	78	7.66
Others	Worry Future	1	0.12	29	2.16	29	2.85
	Teaching Competence	25	2.99	74	5.52	5	0.49
	Teaching Place	6	0.72	9	0.67	120	11.79
	Miscellaneous	38	4.55	112	8.35	22	2.16
	Function Margin	835	100.00	1341	100.00	1018	100.00
Evaluations	Positive	22	25.29	110	51.89	97	72.39
	Negative	17	19.54	19	8.96	18	13.43

(1)Frequency (2)Percentage Data Source: this study

In the section of interactions with internship advisors, the positive comments were highest for student teachers of child education than those of special education and general education (7.66%, 6.49%, 1.08%). The reasons why student teachers of general education made so few comments on this section needs further investigations.

In the section of others, the group of special education showed highest anxiety in teaching competence. The percentages were 2.29%, 5.52%, and 0.49% for the groups of general, special, and child education, respectively. Similarly, issues related to teaching place seemed to concern only student teachers of child education (11.79%).

In summary, only three sections (school administration, interactions with parents, and interactions with internship advisors) were coded for positive and negative evaluations. The positive evaluations were highest for child education, then special education, and general education (72.39%, 51.89%, 25.29%). The reasons attributed to the variations need further investigations.

Symmetric Graphs
(Multiple axes F1 and F2: 100.00%)

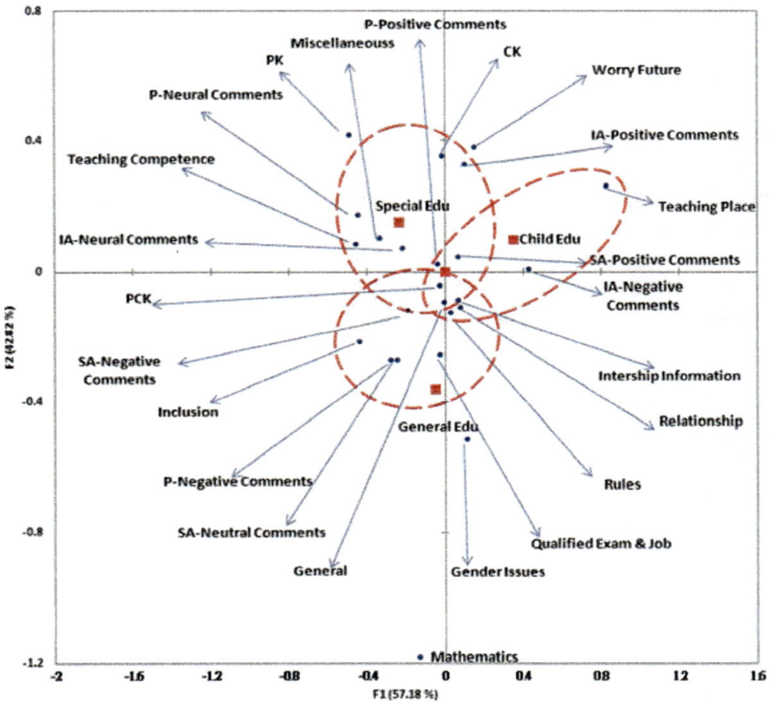

Fig. 1. Position Map of Online Word of Mouth of Student Teachers in Internship

4.2 Results of Correspondence Analysis and Coordinate Diagram

A two-dimensional position map was used to illustrate the relationships between the analyzed items among the three groups of student teachers in internship based on their online word of mouth postings. The corresponding relationships of different strengths were demonstrated by proximity and distance of the dots, which represent the analyzed items.

As figure 1 showed, these 3 groups had their own preferences in internship; however, they also shared some common concerns. These included general subjects, rules and relationship in classroom management, PCK in teaching methods. Moreover, mathematics was another trait commonly shared by the 3 groups; that is, teaching mathematics was not valued in internship by these student teachers, especially for those of special and child education. This can be seen from the dot of mathematics located in the most remote area from the 3 circles.

For general education, negative comments from school administration and parents, questions for the qualified examination and job, as well as inclusion were special concerns of this group. For special education, teaching competence, concerns in PK and CK seemed to perplex this group more. For child education, teaching place appeared to be their peculiar worries.

5 Conclusions

Below are the conclusions of this study. First, very little research has utilized online platforms to study the internship. Second, the information online regarding internship was less than expected. Many centers in charge of internship do not have links or any data for the feedback or reports of internship. Very few institutes of teacher education provide online platforms to monitor, and/or provide necessary and related services to the student teachers. Besides the traditional forms of internship courses, with advances of technology the teacher training programs should be more tech-sensitive, and thus provide more instant and constant advices and services for internship. Third, mathematics teaching in internship should be valued more. Fourth, teaching methods regarding CK, PCK of mathematics should be emphasized more in internship, especially for the student teachers of special education and child education.

Lastly, but not least, only can the institutes of teacher education tailor the teacher training programs to the needs of their students, these institutes will not need to worry to be out of business. As a result, more competent students would be interested in becoming teachers. Finally, the nightmare which the strong mathematical achievement of Taiwanese students becomes the history will not come.

References

1. PISA 2006, 2009 results, http://www.oecd.org/document/2/0,3343,en_32252351_32236191_39718850_1_1_1_1,00.html, http://www.oecd.org/document/61/0,3343,en_2649_35845621_46567613_1_1_1_1,00.html
2. News, http://tw.news.yahoo.com/article/url/d/a/110505/135/2r07x.html
3. Research, Development, and Evaluation Commission, Executive Yuan, http://www.rdec.gov.tw/ct.asp?xItem=4530943&ctNode=12232&mp=100

4. Blackwell, R.D., Miniard, P.W., Engel, J.F.: Consumer Behavior, South Western College, Cincinnati, Ohio (2006)
5. Steffes, E.M., Burgee, L.E.: Social Ties and Online Word of Mouth. Internet Research 19, 42–59 (2009)
6. Jiang, S.Y.: The Study of Pre-service Teachers' Interaction Behavior and Professional Development in Web-Based Virtual Community, Unpublished thesis, Taipei Municipal University of Education, Taipei, Taiwan (2006)
7. Jung, T.S.: SSPS and Research Method, Wu Nan, Taipei, Taiwan (2005)

An Effective Sequence Image Mosaicing Approach towards Auto-parking System

Jinze Song, Bin Dai, and Yuqiang Fang

College of Mechatronic Engineering and Automation,
National University of Defense Technology, 410073 Changsha, China
{songjinze,daibin,fangyuqiang}@nudt.edu.cn

Abstract. For its enormous market demand, Auto-parking system has come to be a hot research topic in Intelligent Vehicle research area recently. To develop the Auto-parking system, this paper presented a new method for parking spaces detection based on the image mosaicing. Considering the characteristics, two pre-processing were presented to improve the results of image mosaicing: firstly, the Adaptive Non-maximal Suppression method (ANMS) was introduced in choosing corners, which improving the distribute characteristic of the corners and the accuracy of the image matching; secondly, the Inverse Perspective Mapping(IMP) is introduced to transform the images, which partially eliminate the perspective effects of the original images and then successfully solve the difficulty of image mosaicing for two images with too different view angles; finally an improved RANSAC method were presented, which effectively improved the accuracy of the registration parameter's estimation from the corrupted matching points. The experiments proved that this proposed method was simple and effective.

Keywords: Autonomous Parking, Image Mosaicing, Inverse Perspective Mapping, RANSAC.

1 Introduction

For the huge of marketing demand, self-parking system has become a hot issue for intelligent vehicle. Currently, a large number of new parking systems continuously available, such as the autonomous parking system City Park of Citroen, Park Assist Vision system of Volkswagen, and LS460L (Intelligent Parking Assist) of Lexus and so on.

Autonomous or auxiliary parking technology mainly uses sensors to accurately perceive parking environment, and then plans an optimal or feasible path in drivable space, finally make the vehicle automatically control the path tracking let it safe and accurate to stop into parking spaces. According to the different parking environment divide into three kinds: *parallel parking*, *perpendicular parking* and *oblique parking*. In this paper, we focus on *parallel parking*. The most difficulty of it is how to use sensors for precise modeling of the parking environment.

Currently, available parking technologies and products often use ultrasonic radar to achieve the detection and approximate location; however, ultrasound has the following disadvantages:

D. Zeng (Ed.): Advances in Control and Communication, LNEE 137, pp. 707–715.
springerlink.com

(1) Detection accuracy is not high enough to meet the demand of parking plan and control for accuracy.

(2) The limit detection range can't recognize the pit of floor and some smaller objects which could impact parking.

Also, Fig.1 shows, the general parking space is larger than then target parking space, so the specific location of target parking space in parking areas also be required to know. From the above analysis, the determination of the target parking areas needs the driver's experience and the actual situation, at this time, the ultrasonic sensor can't provide enough information to the driver to determine the parking area. But, as shown in Fig.1, the end of the vehicle location, which the vision sensor often can't see or see insufficiently of parking space. So we want to getting a wide field of view of the parking space images when the vehicle passing the parking space. Currently, there are two ways to obtain wide field images: splicing the sequence of images and the use of fish-eye camera. The major problem of using fish-eye camera is the lower of image resolution, which make the possibility of the diver use the images confirm the parking spaces more difficult. In additionally, the fish-eye lens obtain images have significant distortion, although which can be restored, it still have more severe distortion so that effect the end of parking spaces measurement accuracy. Based on the above analysis, we choose to use image mosaicing to obtain wide view of images and model the parking environment.

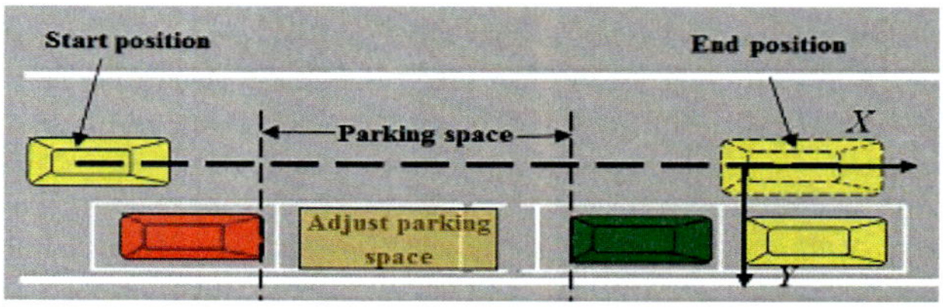

Fig. 1. Parking space for parallel parking

Image mosaicing refers to make a series overlap between any two images synthesis a big angle of panorama, these images can be from different viewpoints or with different perspectives. The aim is to expend the scope of the camera's view. The image mosaicing includes the image matching and image fusion, of which the image matching is to obtain the transform relations between images. And the image fusion is used to smooth overlap, eliminating the effects of light or other factors on the quality of mosaicing. Now, the image mosaicing has been wide used in video monitoring [1], virtual modeling [2], object identification [3] and other fields. However, for the different application background, the application of image mosaicing is different. For autonomous parking system, it is mainly used for modeling parking environment, therefore, the image mosaicing described in this paper will focus on accounting the particularity and targeted of the parking environment. For parking environment modeling tasks, proposed a targeted image mosaicing program.

The paper is organized as follows: firstly, in section 2, analyzing the characteristics of the image in parking environment, according to these characteristics give the basic process of image mosaicing method for the parking environment; secondly, in section 3, introduces all aspects of the mosaic algorithm in detail, including feature extraction, image matching, image fusion and so on. Finally, in section 4, experimental results demonstrate the effectiveness of the proposed method by this paper.

2 Image Mosaicking for Auto-parking

Image mosaicing method can be divided into two kinds: region-based and feature-based method. But no matter which method is used, the general steps are as follows: feature extracting, image matching, image registration (estimate transform parameters) and image fusion. Because of different algorithms for different application background has certain flexibility, so when design the method should consider the special of the parking environment. We can find the image in known parking environment has the following characteristics:

(1) the parking environment modeling is mainly use the information on the ground, the figure shows information on the ground is less, thus using the line features and regional features has great limitations, this paper using the point features to matching.

(2) because of there is certain camera shooting angle, and the camera close to the scene when modeling process in parking environment, making the adjacent two images have a big disparity which let the transformation relation between two image is nonlinear, and can't directly to do image matching, for this the paper introduces the inverse perspective projection transformation to eliminate the parallax effects. So consider these problem, we suggest a new image mosaicing method for auto-parking need, which can describes as follow:

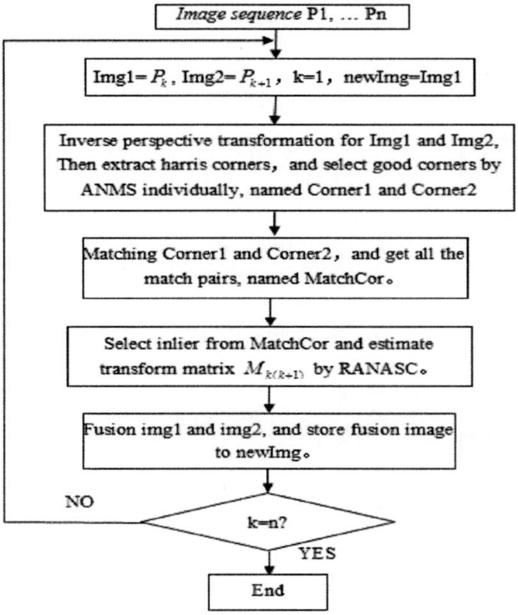

Fig. 2. Proposed method for auto-parking image mosaicing

3 Implementation Details

3.1 Image Matching

Image matching generally include two steps: feature extraction and matching. For the special parking environment, the paper first using Harris corner detection, and then using adaptive non-maximal suppression (ANMS) select the corner, finally introducing inverse perspective projection transformation to improve the corner matching accuracy.

A. Corner extraction. The feature points commonly used in image matching include: SIFT [4] etc. Although similar to the SIFT this kind of high-dimension feature points with high matching accuracy, but low calculating. Here we suggest using a simple Harris corner operator [5] extract matching feature points, which has high calculating speed and suppression for light and noise.

After detected the interest points we can set a certain threshold to extract the matching corner points care about. Direct threshold method makes the corners' distribution is very uneven, as shown in Fig.3 left. We hope to extract most of the corner points in the image on the ground, and according the above method the extracted corner will significantly concentrated in the wall.

To solve this problem, we use adaptive non-maximal suppression (ANMS) [6] to extract corner points. The most important of the ANMS is that it can select corner points according the needed and make them uniform distribute in the image. ANMS is also based on the interest value of the point, which obtained to get the minimum inhibitory radius of each point, in the local area of minimum radius pixels where is the maximum inhibitory radius of the interest point. In practical application, firstly, all the detected interest points are arranged in the sequence to obtain the set of interest points, and then calculate the minimum inhibitory radius of each interest point by the following formula

$$r_i = \min_j \left| x_i - x_j \right|, \qquad s.t. \qquad CRF(x_i) < k * CRF(x_j), \qquad x_j \in SI \qquad (1)$$

Here, k represents adaptive constant, general k taking 0.9, x said pixel coordinate in the image, is a score function defines in [5]. After calculated the minimum inhibitory of each point r, arrange them in sequence, according to the required number of corners N select the appropriate minimum inhibitory radius r, make the number of corners which satisfy r>R is N, then selecting corner is finished. From the above

Fig. 3. Extract corners. Left: thresholding, right: ANMS

analysis, ANMS method can select a smaller threshold value of interest to getting enough interest points shown in Fig.3 right.

B. corner matching and inverse perspective transformation. In the practical application, the requirement for the speed of splicing is necessary. Therefore, the paper use a simple region-based corner matching method: first, using a certain searching strategy search the points from P2 which are maximum similar to the each corner in P1, then search corner from P1 which are biggest similar to the each corner in P2. Obtain two sets of matching points by the two-way matching; finally select the high similarity (more than 0.7) and the corresponding point on the same from the two sets of matching points as the final matching point pairs. That adopted the zero mean normalized cross correlation (ZNCC) as the similarity measure, the formula as the following:

$$C_{ZNCC} = \frac{\sum (f_1(x,y) - \overline{f}_1(x,y))(f_2(x,y) - \overline{f}_2(x,y))}{\sqrt{(\sum f_1^2(x,y) - n\overline{f}_1^2(x,y))(\sum f_2^2(x,y) - n\overline{f}_2^2(x,y))}} \tag{2}$$

Here, $f(x, y)$ is the gray value; $\overline{f}_1(x, y)$ is the average gray value in the region, n is the number of pixels in the region. At the same time, adding the linear constraints in the searching, making the search scope limited in the 15 pixels positive and negative coordinates of original corner y. However, for the close distance of the camera and the camera itself with a certain angle, so that a large disparity between images, resulting in the big difference transformation among different regions in the image. We suggest inverse perspective projection transformation to this problem, which convert the former view to the top view, as shown in Fig.4. Also usually after transform, both sides of the image there is large black void area which can be cut out. As shown in Fig.4. Then corner matching can effectively reduce the impact of the differences of perspective; obtain the high accuracy matching results in Fig. 4.

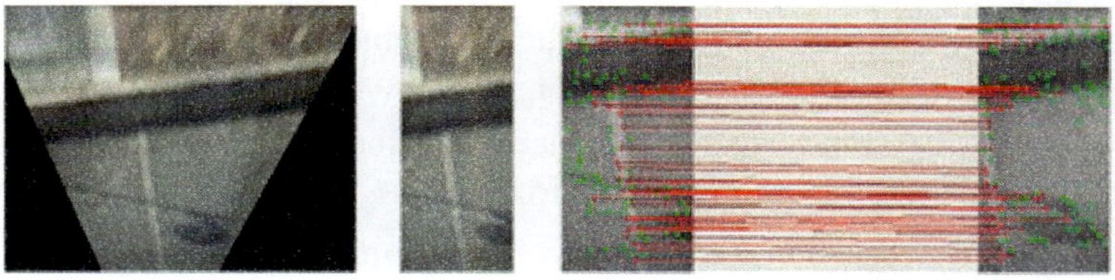

Fig. 4. Inverse Perspective Transformation. Left: transformation image; middle: cutting image; right: matching result

3.2 Image Transformation Model Estimation

After got the matching points between images, we can search the corresponding relationship through these points, while the corresponding relationship between images relate to the transformation model. Image transformation model closely relate to the camera motion, this paper focus on the parking environment modeling. When obtain image the camera is along with the moving of the vehicle, so the vehicle movement decide the camera movement. Consider actual diving, affected by the road conditions and the standard of the diver, vehicle couldn't travel on straight line, so we adopt the general affine transformation model to describe with parameter. So it requires at least 3 matching points to get solution.

Most commonly used methods of estimation of the transform matrix are M-estimation [7], the least median squares [8] and random sample consensus (RANASC) [9]. For those methods, RANSAC has strongest fault tolerance, which can effectively remove foreign points, so this paper adopts this method.

But RANSAC with a quite large calculation, and size of N (iteration limit) and t (distance threshold) impact on the precision of parameter estimation, where t is the basic set, N is the estimated. If N is large enough can improve the estimation accuracy, but the speed will be significantly decreased; if N is too small will fast the speed, while the accuracy is not high. It can be found that N in the constant, random access points are sometimes redundant, which resulted in N is relatively smaller. This paper on the basis of analyzes the reasons of generating extra operations propose the following improvements:

(1) If the three selected points in a straight line, then the equation has no solution, this time the matrix operations are redundant. For this, the paper adding a constraint to determination when selecting interior points, that is before operating matrix determine these points whether in a straight line, or do re-selection.

(2) Selecting 3 points from the points set is a permutation problem; there will be a situation that selected 3 points in the different orders. Therefore, this paper will store each time selected points, then compare the following select points with the stores, if appears the repeat data do the re-selection. At this time, added up to N (N-1)/2 comparisons operations, this is smaller than a RANSAC calculation.

3.3 Multiple Image Fusion

After obtained the transformation matrix between images, it needs to splice multiple images which were collected during the vehicle movements into one image.

Supposed the transformation matrix $M_{(n-1)n}$ used to denote the transformation matrix from first n-1 image to the n image, then for n images, it can obtain n-1 adjacent two images transformation matrix $M_{12}, M_{23}, \cdots, M_{(n-1)n}$. Assuming the reference frame established in the first image, at this time it can get the transformation relation from any image to the reference frame M_{1n}.

$$M_{1n} = M_{12} M_{23} \cdots M_{n(n-1)} \tag{3}$$

However, due to the each matrix itself estimation error and the rounding error of matrix multiplication, trying to use (3) to calculate, it will lead to get greater cumulative error of transformation matrix. This situation is particularly conspicuous when the vehicle trajectory is bending. As shown in Fig.5 left. To address this problem, we suggest the specific method: first, start to splice from the n image, splicing the n and n-1 images will get spliced image P, transformed P then to splice with the n-2 image, so, splicing the rest of images. It will get the renderings as the Fig.5 right shows:

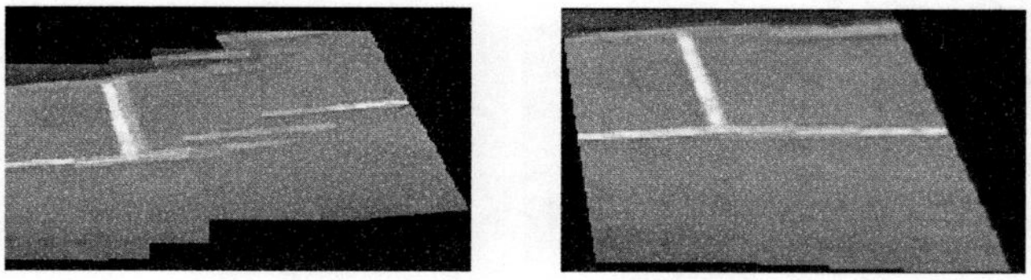

Fig. 5. Multiple image fusion

However, because of the presence of the light and other registration error factors, it will make the images appear more obvious seams. In order to weaken the impact, it needs smooth the overlap region. In this paper, linear transition method was suggested. The specific steps are as follows:

1. Obtain the width L of the overlap between images;
2. Calculate transition factor σ which corresponded to each position in the overlap region.

$$\sigma = \frac{x_{max} - x}{x_{max} - x_{min}} \tag{5}$$

Here, x_{min} (x_{max}) represents the corresponding minimum (maximum) abscissa.

3. The new gray value of each pixel in the overlap region:

$$f(x, y) = \sigma f_1(x, y) + (1-\sigma) f_2(x, y) \tag{6}$$

4 Experiment and Analysis

Above all section describe the specific methods which used in the process of image mosaicing. This section describes the experiment. Two different experiment results are shown, referring to straight and turning route. When the vehicle is driving along the straight line and 24 original images used in the experiment, as shown in Fig.6 (left upper) are the 16 of them which show the image sequences in parking environment. Fig.6 (left bottom) shows the mosaicing results.

In the experiment the vehicle basically driving along a straight line, where the white frame is the standard parking spaces from which we can find when the vehicle driving along a straight line according to the above method could accurately recover the scene of parking environment.

In actual process of driving, the vehicle turning the encoder information can be used, use the proposed method to get the information of when the vehicle is turning. Fig.6 (right upper) shows some of the collected images which contains both turning and straight line driving case images, 21 in total, there are 12 of them. Fig.6 (right bottom) is the results which recover the turning scene accurately.

Fig. 6. Scene images when driving along a straight (left upper); Mosaicing result (left bottom); Scene images when the vehicle is turning (right upper); Mosaicing result (right bottom).

5 Conclusion

In this paper, an effective image mosaicing approach is proposed, which consider the special of the parking environment. From experiment we found this method is simple but effective for application in different situation. But toward auto-parking system, we will try to develop a fusion system with other sensors like radar in the future work. Also human-computer interaction is a promising direction for auto-parking system.

References

1. Joshua, G., Shree, K.N., Keith, J.T.: Real-time omnidirectional and panoramic stereo. In: Proceedings of the DARPA Image Understanding Workshop, pp. 299–303 (1998)
2. Richard, S.: Video Mosaics for Virtual Environments. IEEE Computer Graphics and Applications 16, 22–33 (1996)
3. Fan, Y., Michel, P., Herve, A., Dominique, A.: Fast Image Mosaicing for Panoramic Face recognition. Journal of Multimedia 1, 656–665 (2006)
4. Matthew, B., David, L.: Invariant Features from Interest Point Groups. In: British Machine Vision Conference, pp. 656–665 (2002)
5. Chris, H., Mike, S.: A Combined Corner and Edge Detection. In: Proceedings of The Fourth Alvey Vision Conference, pp. 147–151 (1988)
6. Matthew, B., Richard, S., Simon, W.: Multi-Image Matching using Multi-Scale Oriented Patches, Microsoft Research Technical Report (2004)

7. Maiywan, S., Kashyap, R.L.: A cluster based approach to robust regression and outlier detection. In: IEEE International Conference on Systems, Man and Cybernetics, pp. 2561–2565. IEEE Press, New York (1994)
8. Shmuel, P., Joshua, H.: Panoramic mosaics by manifold projection. In: IEEE Computer Society Conference on CVPR, pp. 338–343. IEEE Press, New York (1997)
9. Martin, A.F., Robert, C.B.: Random sample consensus: a paradigm for model fitting with applications to image analysis and automated cartography. Communications. ACM 24(6), 381–395 (1981)

Pro-detection of Atrial Fibrillation with ECG Parameters Mining Technique

Mohamed Ezzeldin A. Bashir[1], Kwang Sun Ryu[1], Soo Ho Park[1], Dong Gyu Lee[1], Jang-Whan Bae[2], Ho Sun Shon[1], and Keun Ho Ryu[1,*]

[1] Database/Bioinformatics Laboratory, Chungbuk National University, Korea
{mohamed,ksryu,soohopark,dglee,
shon0621,khryu}@dblab.chungbuk.ac.kr
[2] College of Medicine, Chungbuk National University, Cheongju City, South Korea
drcorazon@hanmail.net

Abstract. Reliable detection of atrial fibrillation (AF) in ECG monitoring systems is significant for early treatment and health risks reduction. Various ECG mining and analysis efforts have addressed a wide variety of clinical and technical issues. However, there is still scope for improvement mostly in the number and the types of ECG parameters necessity to detect AF arrhythmia with high quality that encounter a massive number of challenges in relation to computational efforts and time consuming. In this paper, we proposed a technique that caters these limitations. It select features related to the ECG parameters, so as to design a unique feature set that could be employed to describe AF in very sensitive manner. The performance of our proposed technique showed a sensitivity of 95% and a specificity of 99.6%, and overall accuracy of 99.2%.

Keywords: Electrocardiogram parameters, atrial fibrillation, and Classification.

1 Introduction

Atrial fibrillation (AF) is one of the most common cardiac arrhythmias. It is affecting the populations over the age of 75 in most cases, and its prevalence decrease with age degreasing [1]. AF causes the heart to beat irregularly, leading to inefficient pumping of blood and changing the blood flow dynamics. These effects can increase the risk of stroke into 15–20% [2]. When AF occurs, the normal electrical signals provided by sinus node are replaced by rapid circulating waves of irregular electrical signals leading to uncoordinated atrial activation [3]. These multiple fibrillatory waves randomly circulate across the atrial myocardium and result in a frequently rapid ventricular response. Accordingly, the atrial rhythm is out of synchronization with the ventricular rhythm [4]. Accurate detection of AF in real time is demanding to save people lives. The availability of remote monitoring of patients has become clinically important for better controlling of risks and threads [5, 6]. Electrocardiogram (ECG)

---

* Corresponding author.

D. Zeng (Ed.): Advances in Control and Communication, LNEE 137, pp. 717–724.
springerlink.com

provides an essential tool to detect AF. There are some methods developed to enumerate and detect AF regarding different features related to the ECG parameters. But there is massive number of limitations for example, the number and the types of ECG parameters necessity to detect AF arrhythmia with high quality encounter a massive number of challenges in relation to computational efforts and time consuming. Therefore, the current systems cannot detect the AF arrhythmia accurately or detect it but afterward.

In this paper we proposed an ECG parameters mining technique to achieve better AF arrhythmias detection in real-time applications. The features related to the QRS complex plus those, which related to P or T waves through the mining method, are selected. Aiming to design a unique feature set that could be employed to describe AF arrhythmia in very sensitive manner with less computation complexity. The proposed technique is based on our previous work that tunes the ECG parameters to detect fifteen arrhythmias [7]. In this work we are focusing on detecting only the AF arrhythmia. In the rest of this paper we will provide a brief background of AF detection related work, the description of the customization method that follow, then the experimental works, and finally the conclusion.

2 (AF) Detection Techniques

Largely, there are some methods developed to quantify and detect AF regarding features related to one of three main ECG parameters: absence of P wave, atrial activity in fluctuating waveforms, and abnormality of RR intervals. Such algorithms should be accurately able to detect episodes of AF and at the same time have a low computational complexity in order to analyze the ECG signals in real time. Although the P wave-based methods shows outstanding performance among the others [8, 9], but, they have significant limitations. The spectral characteristic of a normal P wave is usually considered to be low frequency, below 10–15 Hz, which is very small and it is widely affected by the noise and interfering signals. So, the identification of its absence or presence in current real-time applications is very challenging task [10]. In the other hand, the repetition rate of the fibrillation waves, which is considered as the AF frequency, plays an important role in detecting the AF arrhythmia using fibrillatory waveform. This technique is quite inaccurate since there are many types of arrhythmias fibrate the ECG waveforms. It can be used in hospital in conjunction with other direct medical investigation methods, but in remote monitoring it is not effectual and useful enough to distinguish AF. In addition to the described methods, the irregularity analysis of the RR intervals has extensively been utilized to detect AF. It is known as heart rate variability (HRV) denotes the variations in the beat-to-beat rotation in the heart beat intervals. Formerly, the irregularity in the RR intervals has been acquired by simple methods that attempt to measure randomness of the RR intervals such as calculating the variance among them [11–13]. More precise approaches for the AF detection commonly build a model to define the RR irregularity. Some of these methods are neural networks [3], Markov model [14], and logistic regression [15]. Mohebbi and Ghassemian [16] used a support vector machine algorithm to detect the AF episodes using the linear and nonlinear features of HRV. RR interval employ to specify many arrhythmias, such as a normal heart beat, premature ventricular contractions, left and right bundled branch blocks, and paced beats. There for, depending only on RR interval to detect AF can Couse such kind of misleading in practical remote real-time applications. Recently, a probability density function method has been proposed by Hong-wei et al [17], which examine the reconstructed phase space of R-R intervals of normal sinus rhythm and AF is studied. Such model to some extent, can relief the distinguishing RR irregularity of AF from that of other cardiac arrhythmias.

Mostly, all these techniques utilize either QRS complex mainly the R, or P waves. The literature never shows the employment of other ECG parameters and their intervals to detect AF arrhythmia. While it displays in the cardiography by altering all the parameters through modifying their shapes and intervals as it shown in Figure 1.

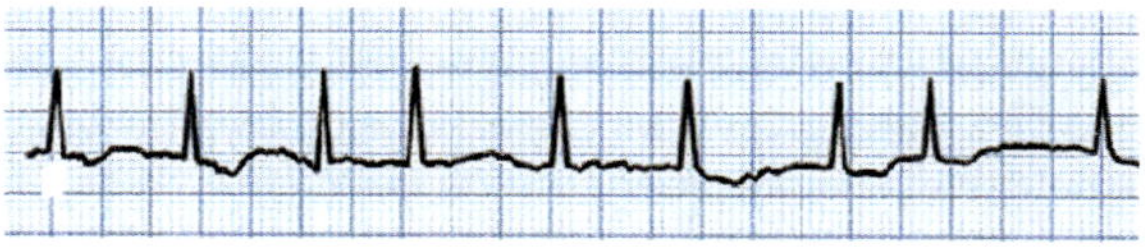

Fig. 1. AF Arrhythmia Shape

In contrast, there are some arrhythmias though they may have different causes, apparent themselves in similar ways on the ECG. Accordingly, analyzing the QRS, P-wave and other elements of the ECG, and measuring the time interval between these elements, are required in real time AF detection systems. Nevertheless, this is technically not feasible in the current systems because of computation considerations.

3 ECG Parameters Mining Technique

The ECG mining technique design a unique feature set that could be employed to describe AF arrhythmia specifically in very sensitive manner.

Similar arrhythmias often share a similar features generated by specific parameters. Therefore, it is useful to predict the required parameters to detect AF. The mining technique uses similar AF arrhythmias collected from the training data, based on general features G^f. The collected cases are used to calculate the parameter scores $P^c(f)$, which utilize to measure the Parameters involvements. The overall parameters lists, which represent the AF arrhythmia class label, are created from the collected similar cases. The parameters with high Parameter score $P^c(f)$ are grouped together generating the overall parameters list, which indicates the possibilities to assign the AF arrhythmia class to the case with a specific feature set *(f)* (distributed through different parameters that included in the overall parameters lists). Accordingly, there will be a unique parameter list to detect AF arrhythmia, which enhances the accuracy and at the same time reduces the computation efforts since we exclude the other unrelated parameters.

First, the most similar ten arrhythmias cases to the AF are collected. Parameters selection process is based on the general features G^f that calculated according to the following formula:

$$G(C^{parameter}, C^{feature}) = \sum_{f=1}^{n} \log \frac{C^{parameter}}{C^{feature}} \tag{1}$$

Where $C^{feature}$ and $C^{parameter}$ are the elements to specify the parameters related to the AF arrhythmia class. *f=1 to n* represent the feature list, which are the general features of the input and training the classifier model to state the arrhythmia's classes. The collected arrhythmias used for manual labeling to generate the binary maps B^c, which indicates the presence '1' or absence '0' of the feature *f* to represent the *AF* arrhythmia class *C*:

$$B^f(C) = \begin{cases} 1 & \text{if handssing(c)} = \text{f} \\ 0 & \text{others} \end{cases} \tag{2}$$

Binary labeled maps B^C are combined to create one general Parameter score P^C. As shown in Figure 2, general parameter score P^C is created through four steps: Gaussian weighted sum B^C, first maximization process O^{1P}, Gaussian weighted average O^{2P}, and final maximization process O^{3P}

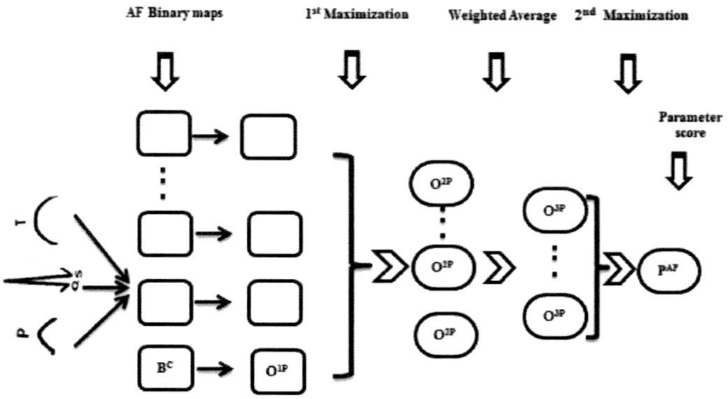

Fig. 2. ECG parameters mining sequence

Phase 1: Weighted sum

Ten maps $B^c p_x$ *(x =1, 2, 3, ... , 10)* for parameter *P* are smoothed out using an isotropic Gaussian function $g\sigma_{sum}$ with a mean of zero and standard deviation of σ_{sum}

$$O_x^{1P}(C) = \sum_{x=1} g^{\sigma} sum(C) B_x^C \tag{3}$$

It will give the highest values to parameter *P* that can be used to specify *AF*.

Phase 2: First maximization process

The maximum value among the ten outputs $O_x^{1P}(C)$ is taken for any arrhythmia:

$$O_x^{2p}(C) = \max O_x^{1p}(C) \tag{4}$$

Phase 3: Gaussian weighted average
The output Ox^{2P} is smoothed using a Gaussian function whose mean is the focused parameter *P*:

$$O_x^{3p}(C) = \frac{1}{S}\sum_{x=1}^{10} g^x \sigma_{avg}(C) O_x^{2p}(C) \tag{5}$$

Where a standard deviation of g^K σ_{avg} is σ_{avg} and *S* is the number of features that represents *AF*. It makes the smooth distribution of scores centering on the focused parameter *P*.

Phase 4: Final maximization process
Finally, the P^C calculated from the maximum value mid Ox^{3P} of the ten cases for *AF*:

$$P^C(f) = \max O_x^{3p}(C) \tag{6}$$

Consequently, the unique features set refer to the different ECG parameters, which mainly specified to describe the AF in very sensitive manner is achieved.

4 The Experimental Works

4.1 Environment

We used a database generated at the University of California, Irvine [18]. It was obtained from Waikato Environment for Knowledge Analysis (WEKA), containing 279 attributes and 452 instances [19]. The classes from 01 to 15 were distributed to describe normal rhythm, Ischemic changes (Coronary Artery Disease), Old Anterior Myocardial Infarction, Old Inferior Myocardial Infarction, Sinus tachycardy, Sinus bradycardy, Ventricular Premature Contraction (PVC), Supraventricular Premature Contraction, Left bundle branch block, Right bundle branch block, degree AtrioVentricular block, degree AV block, degree AV block, Left ventricule hypertrophy, Atrial Fibrillation or Flutter, and Others types of arrhythmias Respectively. The experiments were conducted in WEKA 3.6.1 environment, and carried out by a PC with an Intel Core processor (T M) 2 DUO, speed 2.40 GHz. And RAM 2 GB. We generate a subset of this database that contains only upon Atrial Fibrillation arrhythmia and normal rhythm. We duplicate the AF and generate a total of 249 instances, including 21 cases of Atrial Fibrillation with the rest a normal rhythm, and we applied the mining technique using the J48 algorithm.

4.2 Results

Our experiments prove that, *AF* can be described much more accurate with feature set related to *QRS* and *P*-wave parameters. For a detailed performance analysis, the sensitivity, specificity, and accuracy were obtained. The classification performance is generally presented by a confusion matrix, where TP, TN, FP, and FN stand for true positive, true negative, false positive and false negative, respectively. Consequently, we evaluated: the accuracy, expressed in percentage of the division of the sum of correctly detected AF (TP+TN) by the sum of all parameters (TP+TN+ FP+ FN), resulting in a measure of the precision of the algorithm. Sensibility, expressed in percentage of the division of all true AF (TP) by the sum of TP + FN provides a measure of the capacity of the technique to detect AF. Specificity, expressed in percentage of the division of all non- Atrial Fibrillation AF (TN) by the sum of TN + FP, provides a measure of the capacity of the technique to confirm the non-presence of *AF* episodes in the ECG.

Table 1 shows the parameters figures obtained when applying parameter mining technique using the J48 to detect AF. The results imply that it has good predictive abilities and generalization performance when comparing with *QRS* only, and *P* only. Based on the results, the mining parameter model has provided an outstanding performance. In particular, the sensibility and specificity on the testing data are 95%, 99.6% respectively, and its accuracy is 99.2%.

Table 1. The performance evaluation of mining technique

parameters	*TP*	*FN*	*FP*	*TN*	*sensitivity*	*specificity*	*accuracy*
QRS only	17	3	4	225	85.5%	98.3%	97.2%
P only	18	3	3	225	85.7%	98.7%	97.5%
Mining technique	19	1	1	228	95.0%	99.6%	99.2%

Several researchers have addressed the *AF* arrhythmia detection problem using the ECG signals directly or by analyzing the heart rate variability signal [20–23]. Table 2 summarizes the testing results obtained by different methods. It can be observed from this table that the models derived using parameter mining provide a better accuracy than those obtained by other methods that are reported in the literature.

Table 2. Comparative results of different AF detection methods

Author	*Database*	*Method*	*sensitivity*	*Specificity*	*Accuracy*
Chri. et al. [20]	ECG	P wave	95.7%	-	98.8%
Log. & Heal. [21]	HRV	RR irregularity	96%	89%	-
Rod. & Silv. [22]	HRV	RR interval	91.4%	-	-
Fukun. et al. [23]	ECG	P wave	91%	76%	-
Proposed method	ECG	QRS and P	95%	99.6%	99.2%

5 Conclusion

Detecting the *AF* arrhythmias through ECG monitoring is mature research achievement. Wired ECG monitoring in hospital are very crucial for saving people's lives. However, this kind of monitoring is insufficient for coronary cardiac disease's patients, who need continued follow ups.

Analyzing the *QRS*, *P*-wave and other elements of the ECG, and measuring the time interval between these elements is required in real-time *AF* detection systems. Nevertheless, this is technically not feasible in the current systems because of computation considerations.

In this paper, we presented a parameter mining technique as a proposed solution to solve these problems. The performance of the proposed method has been evaluated using various approaches. Furthermore, the results demonstrate the effectiveness of our proposed method. In the future, we plan to perform more experiments to solve the problem of the interrelated features among different ECG parameters. In order to improve the accuracy much more and to reduce the computational cost as much as possible.

Acknowledgment. This work was supported by the grant of the Korean Ministry of Education, Science and Technology (The Regional Core Research Program / Chungbuk BIT Research-Oriented University Consortium), and the Basic Science Research Program through the National Research Foundation of Korea (NRF) funded by the Ministry of Education, Science and Technology (NRF No. 2011-0001044).

References

1. Wheeldon, N.M.: Atrial fibrillation and anticoagulant therapy. Euro. Heart J. 16, 302–312 (1995)
2. Moody, G., Mark, R.G.: A new method for detecting atrial fibrillation using R–R intervals. Computers in Cardiology, 227–230 (1983)
3. Kara, S., Okandan, M.: Atrial fibrillation classification with artificial neural networks. Pattern Recogn. 40(11), 2967–2973 (2007)
4. Chiarugi, F., Varanini, M., Cantini, F., Conforti, F., Vrouchos, G.: Noninvasive ECG as a tool for predicting termination of paroxysmal atrial fibrillation. Trans. Biomed. Eng. 54(8), 1399–1406 (2007)
5. Ricci, R.P., Russo, M., Santini, M.: Management of atrial fibrillation — what are the possibilities of early detection with home monitoring? Clin. Res. Cardiol. 95(3), 1861–1892 (2006)
6. Bashir, M.E.A., Ryu, K.S., Park, S.H., Lee, D.G., Bae, J.W., Shon, H.S., Ryu, K.H., Bae, E.J., Cho, M., Yoo, C.: Superiority Real-Time Cardiac Arrhythmias Detection using Trigger Learning Method. In: DEXA, Toulouse, France (2011)
7. Bashir, M.E.A., Min Yi, G., Piao, M., Shon, H.S., Ryu, K.H.: Fine-tuning ECG Parameters Technique for Precise Abnormalities Detection. In: ICBBB, Singapore (2011)
8. Stridh, M., Sornmo, L.: Shape characterization of atrial fibrillation using time – frequency analysis. Comput. Cardiol., 17–20 (2002)
9. Guidera, S., Steinberg, J.: The signal-averaged P wave duration: a rapid and noninvasive marker of risk of atrial fibrillation. Am. Coll. Cardiol. 21, 1645–1651 (1993)
10. Bashir, M.E.A., Akasha, M., Lee, D.G., Yi, M., Ryu, K.H., Bae, E.J., Cho, M., Yoo, C.: Highlighting the Current Issues with Pride Suggestions for Improving the Performance of Real Time Cardiac Health Monitoring. In: DEXA, Bilbao, Spain (2010)
11. Tateno, K., Glass, L.: A method for detection of atrial fibrillation using R–R intervals. Comput. Cardiol. 27, 391–394 (2000)

12. Tateno, K., Glass, L.: Automatic detection of atrial fibrillation using the coefficient of variation and density histograms of RR and DRR intervals. Med. Biol. Eng. Comput. 39, 664–671 (2001)
13. Logan, B.H.: Robust detection of atrial fibrillation for a long term telemonitoring system. Computer Cardiol. 32, 619–622 (2005)
14. Young, B., Brodnick, D., Spaulding, R.: A comparative study of a hidden Markov model detector for atrial fibrillation. In: Proceedings of the Neural Networks for Signal Processing IX (IEEE Signal Processing Society Workshop), pp. 468–476 (1999)
15. Kim, D., Seo, Y., Youn, C.H.: Detection of atrial fibrillation episodes using multiple heart rate variability features in different time periods. In: 30th Annual International Conference of the IEEE, IEMBS 2008, pp. 5482–5485 (2008)
16. Mohebbi, M., Ghassemian, H.: Detection of atrial fibrillation episodes using SVM. In: 30th Annual International Conference of the IEEE, IEMBS 2008, pp. 177–180 (2008)
17. Hong-wei, L., Ying, S., Min, L., Pi–ding, L., Zheng, Z.: A probability density function method for detecting atrial fibrillation using R–R intervals. J. Med. Eng. Phys. 31, 116–123 (2009)
18. UCI Machine Learning Repository, `http://www.ics.uci.edu/~mlearn/MLRepository.html`
19. WEKA web site, `http://www.cs.waikato.ac.nz/~ml/weka/index.html`
20. Christov, I., Bortolan, G., Daskalov, I.: Sequential analysis for automatic detection of atrial fibrillation and flutter. Comput. In Cardiol., 293–296 (2001)
21. Logan, B., Healey, J.: Robust detection of atrial fibrillation for a long term telemonitoring system. Computer Cardiol. 32, 619–622 (2005)
22. Rodriguez, C.A.R., Silveira, M.A.H.: Multi-thread implementation of a fuzzy neural network for automatic ECG arrhythmia detection. Comput. Cardiol. 28, 297–300 (2001)
23. Fukunami, M., Yamada, T., Ohmori, M., Kumagai, K., Umemoto, K., Sakai, A., Kondoh, N., Minamino, T., Hoki, N.: Detection of patients at risk for paroxysmal atrial fibrillation during sinus rhythm by P wave-triggered signal-averaged electrocardiogram. Circulation 83, 162–169 (1991)

On Supporting the High-Throughput and Low-Delay Media Access for IEEE 802.11 Enhancements

Shih-Tsung Liang[1] and Jin-Lin Kuan[2]

[1] Taipei Municipal University of Education, Taipei, Taiwan
stliang@ieee.org
[2] National Sun Yat-Sen University, Kaohsiung, Taiwan
kuanjinlin@hotmail.com

Abstract. It has been shown that the Binary Exponential Back-off (BEB) algorithm, which is adopted in the IEEE 802.11 Distributed Coordination Function (DCF), can lead to significant performance degradation due to the high collision rate of transmissions among a large number of competing wireless nodes under the heavy traffic condition. To cope with this problem, a number of researchers proposed the mechanisms known as XIXDs to adjust the increment and decrement of Contention Window (CW) for the failure and success of any frame transmission, respectively. While XIXDs can improve aggregated throughput of the IEEE 802.11 network under high traffic conditions, they may also introduce additional network access delay under dynamic traffic conditions. In this paper, a general enhancement of XIXDs is proposed to provide the high through and the low access delay of the IEEE 802.11 network under unexpected traffic conditions. Numeric results show that the proposed enhancement can be applied to any existing XIXD and the reduced access delay can be achieved.

Keywords: IEEE 802.11, Distributed Coordination Function (DCF), Binary Exponential Back-off (BEB), XIXD, Contention Window (CW), access delay.

1 Introduction

The IEEE 802.11 wireless Local Networks (WLANs) [1] protocol has been adopted worldwide. Growth in competing wireless stations promotes reconsiderations of its original contention resolution mechanism—Carrier Sense Multiple Access/Collision Avoidance (CSMA/CA), in which the binary exponential back-off (BEB) mechanism is applied to resolve possible contention after a busy period of wireless medium.

To stagger the transmissions among multiple wireless stations, BEB schedules the transmission at the time when a random number of idle time-slots elapse. The random number is in turn uniformly distributed on the range [0, CW-1], where CW is the size of the contention window. To correspond with the congestion condition, BEB doubles CW until CW_{max} being reached upon each collision; and reset CW to CWmin upon each successful transmission.

It has been shown that BEB may suffer severely from throughput degradation with the increasing number of wireless terminals because of the high collision rate.

D. Zeng (Ed.): Advances in Control and Communication, LNEE 137, pp. 725–732.
springerlink.com

Therefore, in the past, several innovative back-off schemes [2-12] have been proposed to improve the aggregated throughput of CSMA/CA. In particular, some of these schemes can be classified as XIXDs (X-Increment X-Decrement) in which the strategy for the adjustments of CW are different from that of BEB. In general, instead of resetting CW to CWmin XIXDs decrease CW progressively after each transmission success and thus avoid another collision which may occur if CW is reset and the high contention situation remains unchanged.

Numerical results have shown that XIXDs achieve improved aggregated throughput of the 802.11 network under high traffic conditions. However, XIXDs may incur additional network access delay under dynamic traffic conditions. For instance, for wireless stations sending burst traffic, the CW setting upon completion of current transmission may not apply for the next transmission being not expected to occur in a short time. In this paper, a general enhancement of XIXDs is proposed to provide the high through and the low access delay of the IEEE 802.11 network under dynamic and unexpected traffic conditions.

The rest of the paper is organized as follows: Section 2 reviews the basic operation of IEEE 802.11 DCF mechanism and some XIXD back-off schemes. The general mechanism for the enhancement of XIXDs is proposed in section 3. Section 4 presents some numerical results. Finally, concluding remarks are given in Section 5.

2 IEEE 802.11 DCF Enhancements

The basic access mechanism of IEEE 802.11 DCF is based on Carrier-Sense Multiple Access with Collision Avoidance (CSMA/CA). As shown in Fig. 1, a station experiencing idle medium for a period of DIFS (DCF Inter-Frame Space) time-slots immediately following its transmission attempt starts to transmit after the DIFS. Otherwise, the station waits for another DIFS followed by a random back-off interval determined by BEB in an attempt to avoid colliding among multiple stations contending for the wireless medium.

It is not difficult to see that the performance of DCF is closely related with the randomly selected back-off intervals. Insufficient back-off intervals lead to high probability of collisions while excessive ones result in redundant channel idle time, all cause the degradation of network performance. In order to dynamically adjust back-off intervals to the appropriate setting, IEEE 802.11 adopts the binary exponential

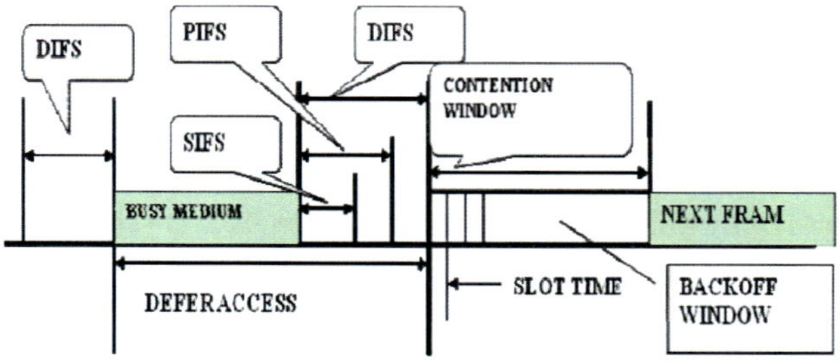

Fig. 1. IEEE 802.11 DCF

back-off (BEB) algorithm. As shown in Fig. 2, BEB schedules the transmission at the time when a random number of idle time-slots elapse. The random number is in turn uniformly distributed on the range [0, CW-1], where CW is the size of the contention window. To correspond with the congestion condition, BEB doubles CW until CW_{max} being reached upon each collision; and reset CW to CW_{min} upon each successful transmission.

A major drawback of BEB has been the high collision rate induced from resetting CW to CW_{min} after each transmission success even under heavy contention. To cope with the drawback, XIXDs adopt a general strategy which decreases CW progressively after each transmission success and thus avoid another collision which may occur if CW is reset and the high contention situation remains unchanged.

In the following, a number of XIXD mechanisms including EIED (Exponential Increase Exponential Decrease) [9], LILD (Linear Increase Linear Decrease) [10], ELBA (Exponential-Linear Backoff Algorithm) [11], MILD (Multiplicative Increase Linear Decrease) [12], etc. are summarized. Without lose of generality, the value of CW_{min} and CW_{max} are assumed to be 32 and 1024, respectively.

2.1 EIED

The major difference between EIED and BEB lies to the set of CW values. As shown in Fig. 3, EIED doubles CW until CW_{max} being reached upon each collision; and halves CW until it reaches CW_{min} upon each successful transmission. When a collision is encountered, EIED follows the same strategy as BEB. When a successful transmission is achieved, however, instead of resetting CW to CW_{min} like BEB, EIED

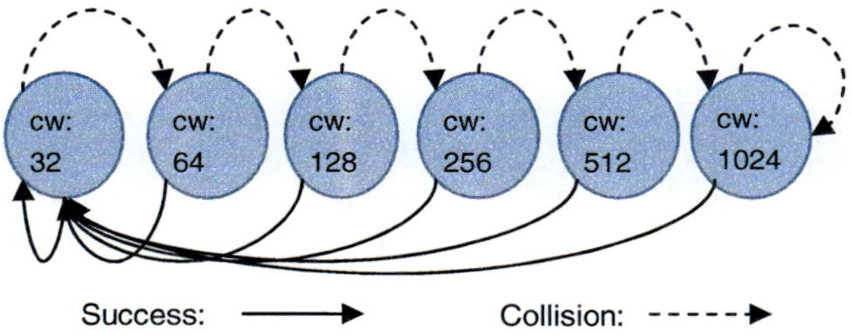

Fig. 2. Markov chain model for BEB

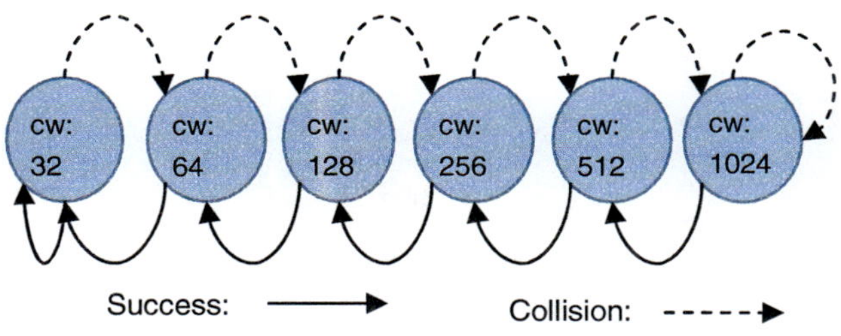

Fig. 3. Markov chain model for EIED

adopts a progressive approach by halving CW in an attempt to avoid another collision which may occur if CW is reset and the high contention situation remains unchanged.

2.2 LILD

As shown in Fig. 4, LILD's approach is to increase and decrease CW in a linear fashion and hence the dynamics of the contention window is reduced when compare to BEB and EIED. In particular, LILD increases and decreases CW by CW_{min} upon each collision and each successful transmission respectively until the CW_{min} or CW_{max} is reached. Because of the change of CW is linear, LILD can achieve good performance in the wireless environment with a large number of competing nodes and low dynamics of traffic loads.

2.3 ELBA

As shown in Fig. 5, ELBA combines concepts from EIED and LILD. In the first few times of collisions when the CW value is still small, ELBA doubles CW like EIED for getting rid of colliding as soon as possible. When a certain number of consecutive collisions and hence a relative large of the CW value are reached, ELBA adjusts CW linearly just as the way LILD does in order to avoid unnecessary delay caused by the excessive setting of CW. It should be noted that what ELBA works depends on a predefined parameter — $CW_{Threshold}$ which acts as a threshold to decide whether the linear or exponential increment/decrement is adopted. In Figure 5, $CW_{Threshold}$ is set to 512.

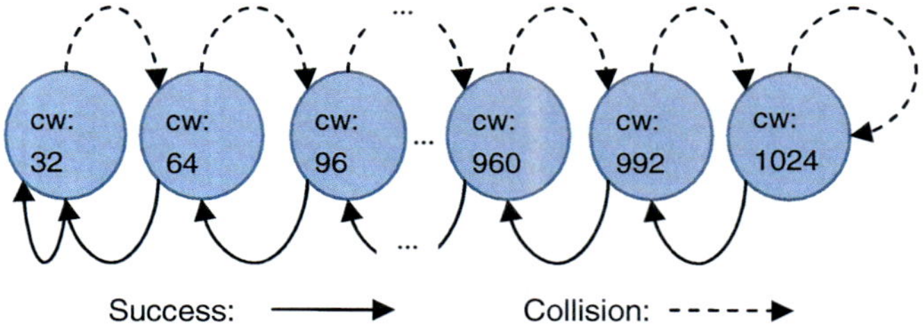

Fig. 4. Markov chain model for LILD

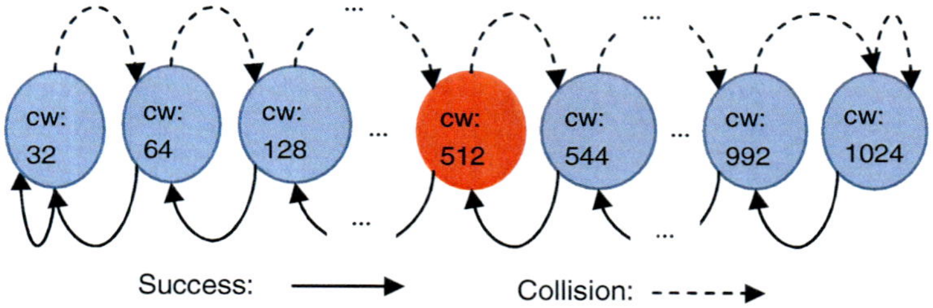

Fig. 5. Markov chain model for ELBA

2.4 MILD

MILD when increasing CW adopts an increasing rate between that of LILD and BEB; and when decreasing CW adopts the minimal decreasing rate among all the XIXD mechanisms mentioned above. As shown in Fig. 6, MILD multiplies CW by 1.5 upon each collision and decreases CW by one upon each successful transmission. Therefore, MILD achieves superior performance when the traffic load and the number of competing stations are very large. If this is not the case, however, the performance of MILD is not prominent.

3 General Enhancement Mechanism for XIXDs

To deal with the hazard of high collision rate induced by BEB which resets CW to CW_{min} after each successful transmission even the traffic contention remains high, XIXDs adopt a general strategy which decreases CW progressively after each transmission success.

Results from existing researches have shown that XIXDs can much improve the saturated throughput of the IEEE 802.11 network under heavy traffic conditions with high contention among a large number of stations each has a greedy traffic source. The progressive decrease of CW upon a successful transmission in XIXDs in this case can effectively correlates to the traffic condition for the next frame transmission. Under bursty traffic conditions, however, certain period of time may have passed from the CW setting until the next transmission. Hence the progressive decrease of CW is not effective and may introduce unnecessary access delay. As shown in Fig. 7, a wireless station with bursty traffic may experience severe congestion and have CW grow rapidly during the transmission of (k-1)-th frame. Some time later, then, to send the k-th frame, a large CW is applied according to XIXDs even when the congestion is alleviated and unnecessary access delay is introduced. In this paper, a general enhancement of XIXDs is proposed to provide the high through and the low access delay of the IEEE 802.11 network under dynamic and unexpected traffic conditions.

The rationale behind the general enhancement of XIXDs is to take into account the time gap between setting CW upon the completion of the current transmission and applying the CW for the next frame when it arriving to the head of transmission queue. In particular, the XIXD timer is introduced and implemented in the MAC

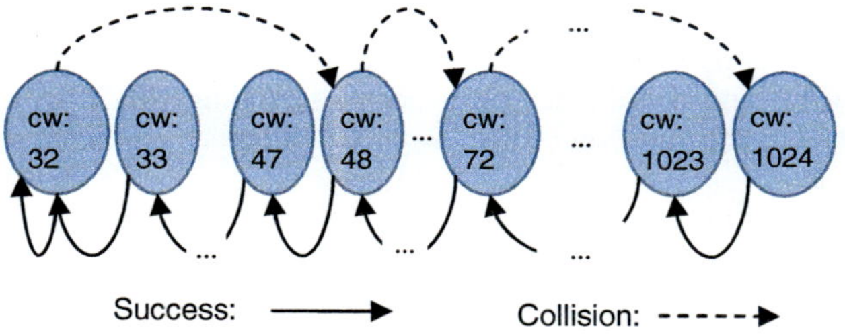

Fig. 6. Markov chain mode for MILD

layer. When a frame is transmitted successfully, XIXD sets CW according to its decreasing algorithm and in the mean while the XIXD timer starts counting down for δ time slots. If time expired prior to the next frame arrival, one of the following two strategies can be adopted for setting CW:

1) Reset CW to CW_{min} ; and
2) Decrease CW according to the XIXD decreasing algorithm and restart the XIXD timer.

Otherwise, the XIXD timer is stopped.

It is worth noting that two strategies are both simple. While the first is beneficial when considering the energy efficient operation of IEEE 802.11 allowing idle stations to enter the power-save mode, the second achieves better correlations of CW settings on traffic conditions changing along with time. As for choosing the appropriate value of δ, δ in the second strategy can be set to the duration of a single transmission in an attempt to capture the gradual alleviation of traffic intensity. While it is difficult to conclude an optimum value of δ, the same value as in the second strategy is adopted.s

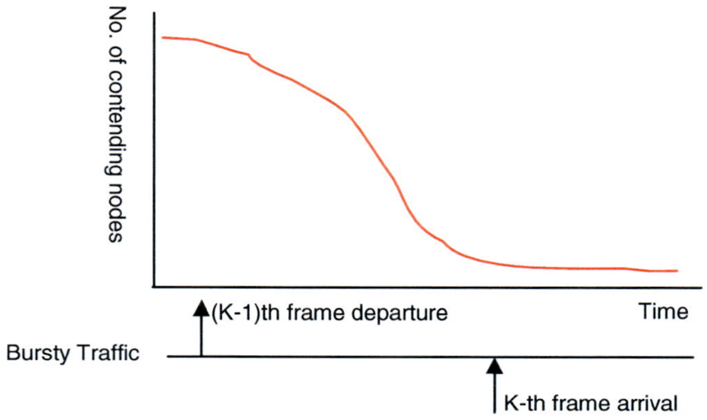

Fig. 7. Illustration of the bursty traffic condition

4 Numeric Results

To conduct the experiment, we use the well-known simulation tool NS-2 [13] as the test-bed platform. The XIXD back-off schemes including EIED, LILD, ELBA, and MILD are embedded into the 802.11 MAC module of NS-2 by C++ code implementation.

The back-off schemes are evaluated in an IEEE 802.11 infrastructure network with a number of wireless stations surrounding the access point. In particular, networks with 32, 48, 64, and 80 wireless stations are considered. Each wireless station sends its data to the access point based on the Pareto traffic model with packet size of 1500 bytes, burst time of 100 ms, idle time of fifty-fifty percent chance of either 100 or 500 ms, and rate of 2/N Mbps where N is the number of wireless stations. The link capacity of the IEEE 802.11 WLAN is assumed to be 2Mbps.

To demonstrate the effectiveness of the proposed general enhancement of XIXDs, comparisons of the mean access delay derived from BEB, a number of XIXDs, and XIXDs each augmented with the XIXD timer mechanism are presented. As shown in Fig. 8, with the XIXD timer mechanism, the mean access delay can be effectively reduced compared to that derived from its original XIXD.

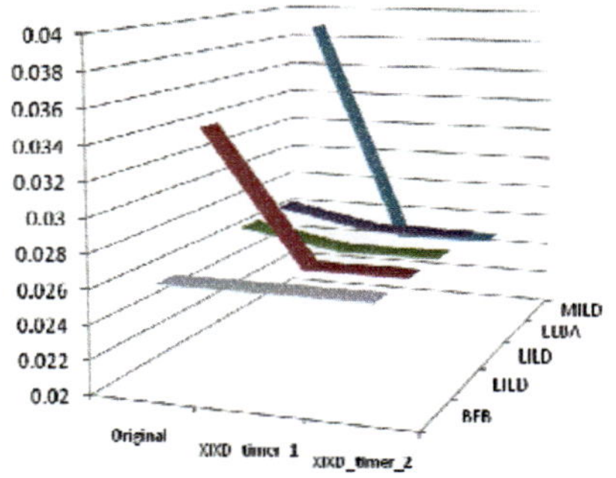

(a) N=32

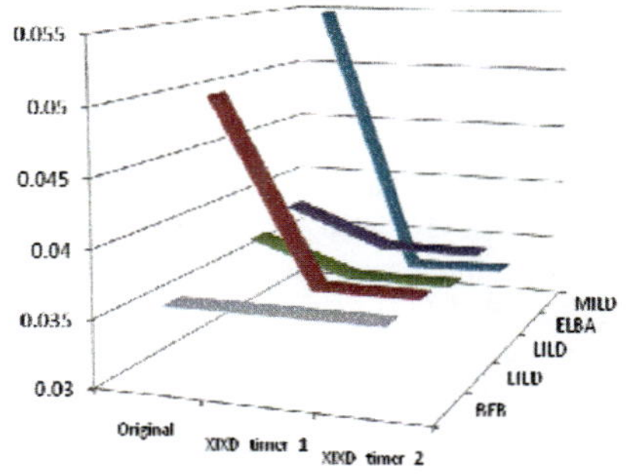

(b) N=48

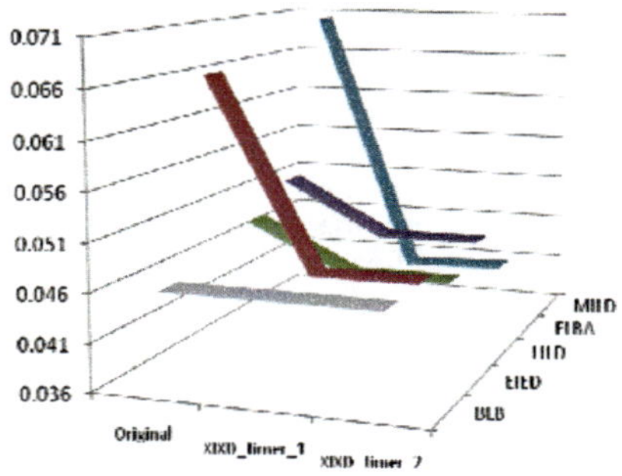

(c) N=64

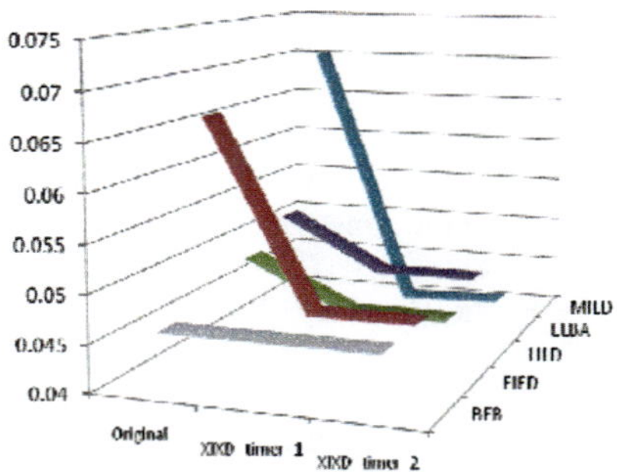

(d) N=80

Fig. 8. Mean access delay

5 Conclusions

In this paper, we have presented a general enhancement of XIXDs which can be applied to any existing XIXDs. In particular, two strategies for the setting of δ, the key parameter in the proposed enhancement of XIXDs, have been proposed and both have shown to be able to achieve the low access delay of the IEEE 802.11 network under unexpected traffic conditions. It is worth noting that two strategies are both simple. While the first is beneficial when considering the energy efficient operation of IEEE 802.11 allowing idle stations to enter the power-save mode, the second achieves better correlations of CW settings on traffic conditions changing along with time. Numeric results have shown that the proposed enhancement achieves the effective reduction of the mean access delay for existing XIXDs.

References

1. IEEE Standard 802.11-1999: Part 11: Wireless LAN Medium Access Control (MAC) and Physical Layer (PHY) Specifications (1999)
2. Balador, A., Movaghar, A.: The Novel Contention Window Control Scheme for IEEE 802.11 Mac Protocol. In: 2010 Second International Conference on Networks Security Wireless Communications and Trusted Computing (NSWCTC), vol. 2, pp. 134–137 (2010)
3. Kang, S.-W., Cha, J.-R., Kim, J.-H.: A Novel Estimation-Based Backoff Algorithm in the IEEE 802.11 Based Wireless Network. In: The 7th IEEE Consumer Communications and Networking Conference (CCNC), pp. 1–5 (2010)
4. Zhou, B., Marshall, A., Lee, T.-H.: A k-round Elimination Contention Scheme for WLANs. IEEE Trans. Mobile Computing 6(11), 1230–1244 (2007)
5. Ye, S.-R., Tseng, Y.-C.: A Multichain Backoff Mechanism for IEEE 802.11 WLANs. IEEE Transactions on Vehicular Technology 55(5), 1613–1620 (2006)
6. Choi, N., Seok, Y., Choi, Y., Lee, G., Kim, S., Jung, H.: P-DCF: Enhanced Backoff Scheme for the IEEE 802.11 DCF. In: IEEE VTC 2005, Stockholm, Sweden (Spring 2005)
7. Ni, Q., Aad, I., Barakat, C., Turletti, T.: Modeling and analysis of slow CW decrease for IEEE 802.11 WLAN. In: Proc. IEEE PIMRC 2003, Beijing, China, pp. 1717–1721 (2003)
8. Wu, H., Cheng, S.: IEEE 802.11 Distributed Coordination Function (DCF): Analysis and Enhancement. In: Proc. IEEE International Conference on Communications, vol. 1, pp. 605–609 (2002)
9. Song, N., Kwak, B., Song, J., Miller, L.E.: Enhancement of IEEE 802.11 distributed coordination function with exponential increase exponential decrease backoff algorithm. In: Proc. IEEE VTC 2003-Spring, vol. 4, pp. 2775–2778 (2003)
10. Deng, J., Varshney, P.K., Haas, Z.J.: A new backoff algorithm for the IEEE 802.11 distributed coordination function. In: Proc. CNDS 2004, Sandiego, CA (January 2004)
11. Ke, C.-H., Wei, C.-C., Wu, T.-Y., Deng, D.-J.: A Smart Exponential-Threshold-Linear Backoff Algorithm to Enhance the Performance of IEEE 802.11 DCF. In: The Fourth International Conference on Communications and Networking in China, Xi'an, China, August 26-28 (2009)
12. Bharghavan, V., Demers, A., Shenker, S., Zhang, L.: MACAW: A Media Access Control Protocol for Wireless LANs. In: Proc. ACM SIGCOMM 1994, pp. 212–225 (1994)
13. The NS-2 Network Simulator, http://www.isi.edu/nsnam/ns/

A Study on the Degree Complement Based on Computational Linguistics

Li Cai

College of Chinese Language and Culture, Jinan University,
510610 Guangzhou, China
caili@hwy.jnu.edu.cn

Abstract. Based on the Corpus of Contemporary Chinese of Peking University and the methods of computational linguistics, this paper had automatically extracted the compositional pairs of degree complements and their adnexes. Based on such compositional pairs ,the combination characteristics and rules of adnexes and 43 degree complements had been studied multi-anglely by using of the computational linguistics methods of word segmentation and part-of-speech tagging, emotional tendency analysis, automatic syntactic analysis technology and chunk analysis. This paper attempts to break through the conventional research paradigm of descriptive studies based on small samples and enumeration, and to introduce computational linguistics methods into conventional grammatical research methods. Linguistic sense experiences have been combined with mathematical statistics, in expectation of providing reference for the study of rules of Chinese syntax and semantic combinations.

Keywords: degree complement, compositional pairs, part of speech tagging, computational linguistics.

1 Introduction

There remains much to explore in the many subtle relationships in the combination of the complement of degree and predicate, rendering it necessary to conduct a comprehensive study of the rules governing the combination of the complement of degree and predicate using statistical analysis in computing linguistics from the perspective of bidirectional semantic selection.

Based on all the relevant corpora retrieved from the Modern Chinese Corpus (hereinafter referred to as "the Corpus") of Peking University, this paper carries out bidirectional and automatic extraction of main pairs of predicates and their complements of degree by using the method of computing linguistics (Liu Hua, 2010). Based on these bidirectional pairs, the paper automatically categorizes the bidirectional pairs of predicates and complements of degree and studies bidirectional semantic selection, based on which the paper studies the grammatical functions of the pairs of predicates and complements of degree by semantic combination and selection category and examines and analyzes the characteristics of 43 phrases where the complement of degree serves as the predicate, summarizing some rules contained therein.

D. Zeng (Ed.): Advances in Control and Communication, LNEE 137, pp. 733–740.
springerlink.com

2 Basic Method and Idea of Corpus Computing

43 search items are set based on the structural characteristics of each complement of degree and entered into the Modern Chinese Corpus of Peking University to extract raw corpora and carry out manual intervention of the relevant corpora.

2.1 Automatic Word Segmentation and Gender Tagging and Manual Proofreading

First, automatic word segmentation and gender tagging are carried out for the strings preceding complements of degree in the 36,235 sample sentences extracted, which will then be subject to manual proofreading. For example:

"祁老人和天佑太太听说瑞丰得了科长，喜欢/v [得什么似的]！【文件名:\现代\文学\老舍 四世同堂.TXT 文章标题:四世同堂 作者:老舍】"

In this paper, the immediate word preceding the complement of degree, such as the "like" in the sentence above, is called "the left collocate".

2.2 Entries of Left Collocates Automatically Extracted and Computing of the Times and Frequency of Their Appearance

Then, for all samples sentences of each complement of degree, the entries of left collocates and the times and frequency of their appearance are determined and listed. For example, a total of 38 entries contain the left collocate of "(得)什么似的", which appears 77 times:

Table 1. Left collocate data (only those appearing more than twice are listed)

Entry	Times	Frequency	Gender
吓	10	12.99	v
高兴	7	9.09	a
喜欢	4	5.19	v
兴奋	4	5.19	v
急	4	5.19	a
乐	3	3.90	a
气	3	3.90	v
恨	3	3.90	v

2.3 Automatic Statistics of Genders of Left Collocates

In the meantime, the genders of left collocates of each complement of degree in all sample sentences are summarized, such as the genders of the left collocates of "什么似的", as shown below:

Table 2. Genders of left collocates

Gender	Total times of appearance	Total frequency of word appearance	Total entries	Total frequency of entry appearance
v	47	61.04	22	57.89
a	30	38.96	16	42.11

2.4 Automatic Statistics of the Emotional Connotations of Left Collocates, Assisted with Manual Proofreading

The emotional connotations of the left collocates of each complement of degree in all sample sentences are summarized based on an established database of emotional connotations of words (20,000 entries in all, including favorable, derogative and neutral), assisted with manual proofreading.

3 Analysis of the Use of Complements of Degree in Discourse

This paper examines a total of 43 complements of degree. We have determined the times of appearance of 43 complements of degree in the corpus and arranged the sequence of these complements of degree by the times of their appearance, with the following results:

Table 3. Times of appearance of complements of degree in the corpus

No.	Entry	Times	No.	Entry	Times	No.	Entry	Times	No.	Entry	Times
1	多	11619	12	死	721	23	不行	300	34	去了	34
2	极	4423	13	要命	652	24	疯	147	35	离谱	33
3	很	3184	14	不得了	632	25	吓人	136	35	过分	33
4	不堪	3025	15	慌	467	26	非常	113	37	蒙	26
5	万分	1318	16	可怜	458	27	傻	109	38	邪乎	8
6	无比	1198	17	之至	431	28	够呛	103	39	够瞧的	7
7	坏	1067	18	绝伦	417	29	可以	102	40	够劲	6
8	透	1035	19	出奇	347	30	了不得	84	41	够受的	2
9	不过	1032	20	惊人	339	31	绝顶	77	42	不成	1
10	异常	800	21	要死	323	32	什么似的	63	42	邪行	1
11	厉害	790	22	透顶	320	33	凶				

The 43 complements of degree appear in 36,235 sample sentences in the corpus, consisting of 16,530 combination complements of degree which account for 45.62% and 19,705 adhesive complements of degree which account for 54.38%. With instances of relative complements of degree excluded, there are 24,616 absolute complements of degree, composed of 8,218 combination complements of degree which account for 33.38% and 16,398 adhesive complements of degree which account for 66.62%. As can be seen, adhesive complements of degree are used more often in the discourse.

From the perspective of semantics, complements of degree are classified by the level of the complement of degree into excessive, top and advanced complements of degree. Excessive complements of degree include “过分,离谱”, which number a mere 66 and account for 0.27%; top complements of degree include “”, which number 19,450 and account for 79.01%, the highest proportion of all; while top complements of degree consist of 1,801 favorable ones which account for 9.26%, 9,896 neutral ones which account for 50.88% and 7,753 derogative ones which account for 39.66%; advanced complements of degree include “”, composed of 5,100 favorable ones which account for 20.72%, 4,522 neutral ones which account for 88.67% and 578 derogative ones which account for 11.33%.

As can be seen from the above data, the use of complements shows the following preferences when judged from the perspective of their levels: (1) In terms of form, the language user tends to use complements of degree without tags. (2) In terms of the level of the complement of degree, the user tends to use polar complements of degree and seldom users excessive complements of degree. (3) In terms of emotional connotations, the user prefers neutral complements of degree most, followed by derogative complements of degree; as favorable complements of degree are the fewest in number (4, accounting for 9.3%), they are generally least used.

4 Analysis of Combination Capabilities of Complements of Degree

4.1 Rules of Inclining Selection by Complements of Degree of the Syllables of Collocates

For some complements of degree, there are certain conditions for the number of syllables of their collocates. The following complements of degree tend to combine with double-syllable words instead of single-syllable words, including:. We examined the syllables of words combined with these nine complements of degree:

Table 4. Analysis of the syllables of words combined with nine complements of degree

Complement of degree	Number of single-syllable words		Number of double-syllable wordsdoul		Number of multi-syllable words		Total
	Number	%	Number	%	Number	%	
无比	4	0.33	1193	99.58	1	0.09	1198
万分	0	0	1318	100	0	0	1318
不堪	7	0.23	3017	99.74	1	0.03	3025
异常	4	0.5	796	99.5	0	0	800
之至	4	0.93	378	87.7	49	11.37	431
绝伦	1	0.24	416	99.76	0	0	417

The above complements of degree inclined to be combined with double-syllable words are all adhesive complements of degree. This is because of the rhythm rules of the Chinese language, under which two syllables are required to form an independent foot (Feng Shengli, 2009), and when the complement of degree contains two syllables, its predicate must tend to have two syllable accordingly to make the rhythm coordinated by allowing the two adjacent words to have an equal number of syllables. In some cases, if the word to be combined has only one syllable, the speaker will seek ways to convert it into a double-syllable word to be combined with these complements of degree.

4.2 Rules of Preferential Selection by Complements of Degree of the Types of Collocates

Some tend to combine with psychological and perceptive verbs or adjectives, such as ""; while some tend to combine with general verbs, such as "凶", as shown in the table below:

Table 5. Combination data of complements of degree with conspicuous inclination in the distribution of types of words

Comp-lement of degree	Psychological, perspective verbs/adjectives		General adjectives		General verbs	
	Number	%	Number	%	Number	%
万分	1271	96.43	47	3.57	0	0
什么似的	72	93.51	3	3.89	2	2.60
了不得	86	84.31	13	12.75	3	2.94
不行	240	80	46	15.33	14	4.67
绝伦	0	0	417	100	0	0
绝顶	0	0	84	100	0	0
去了	0	0	34	100	0	0
可怜	2	0.44	455	99.34	1	0.22
惊人	0	0	334	98.53	5	1.47
可以	2	1.94	100	97.09	1	0.97
出奇	12	3.46	335	96.54	0	0
吓人	1	0.68	140	95.24	6	4.08

4.3 Rules of Preferential Selection by Complements of Degree of Emotional Connotations of Collocates

As complements of degree contain emotional connotations, they are selective toward the emotional connotations of their collocates. This paper comprehensively examines the distribution of emotional connotations of favorable and derogative complements of degree by using the emotional connotation analysis approach in computing linguistics, automatically tagging the emotional connotations of collocates and carrying out manual proofreading.

A. let's take a look at the combinations of favorable complements of degree, namely "无比", "绝伦", "绝顶" and "了不得".

Table 6. Emotional connotations of combinations of favorable complements of degree

Comp-lement of degree	Favorable or positive		Neutral		Derogative or negative	
	Number	%	Number	%	Number	%
无比	727	60.68	229	19.12	242	20.20
绝伦	329	78.90	4	0.96	84	20.14
绝顶	77	91.67	1	1.19	6	7.14
了不得	68	66.67	7	6.86	27	26.47

We can draw the following conclusions based on the data above:

First, in general, favorable complements of degree tend to combine with words with favorable or positive connotations, while it is not impossible for them to express derogative meanings as Zhang Yisheng (2000) claimed. For the 4 favorable complements of degree, combinations with derogative or negative terms account for 19.93%, which suggests these combinations are not incidental or incorrect, but an objective phenomenon.

Second, as complements of degree are functional to different degrees, their respective combinations and flexibility differ. Among the four favorable complements of degree, "绝顶" has the narrowest range of combinations and lowest flexibility in collocate selection, with only one derogative collocate of "荒谬" and a limited number of favorable collocates. "绝伦" has the second narrowest range of combinations and second lowest flexibility, while "无比" and "了不得" have broader ranges of combinations and much higher flexibility.

B. Combinations of derogative complements of degree.

Among the 19 derogative complements of degree, "" seldom appear in the sample sentences of the corpus, numbering 1, 8 and 2, respectively. Due to the scarcity of samples, these three complements of degree are put into a single group for analysis.

From the statistical data of Table 7, the following combination rules can be derived:

First, in general, derogative complements of degree tend to combine with derogative or negative terms, which combinations account for 87.66%; while less often combine with neutral and favorable terms, which combinations account for 9.03% and 3.31% respectively.

Second, there are cases where derogative complements of degree except "慌, 蒙, 凶" combine with favorable terms, suggesting that derogative complements of degree are not fully incapable of combing with favorable terms as Zhang Yisheng (2000) claimed.

Table 7. Analysis of emotional connotations of words combined with derogative complements of degree

Complement of degree	Favorable or positive		Neutral		Derogative or negative	
	Number	%	Number	%	Number	%
死	625	86.69	68	9.43	28	3.88
透	867	83.77	122	11.79	46	4.44
透顶	295	92.19	11	3.44	14	4.38
慌	425	91.01	42	8.99	0	0
要死	267	82.66	44	13.62	12	3.72
要命	432	66.26	143	21.92	77	11.81
疯	178	70.63	60	23.81	14	5.56
傻	85	75.22	24	21.24	4	3.54
蒙	23	100	0	0	0	0
坏	951	89.13	57	5.34	59	5.53
凶	36	57.14	27	42.86	0	0
邪乎、邪行	3	27.27	6	54.55	2	18.18
够呛	91	83.49	15	13.76	3	2.75
不堪	3005	99.34	17	0.56	3	0.10
吓人	36	24.49	106	72.11	5	3.4
过分	8	24.24	15	45.46	10	30.3
离谱	31	93.93	1	3.03	1	3.03
Total	7358	87.66	758	9.03	278	3.31

Third, complements of degree are functional to different degrees. The inclination of complements of degree for their collocates can be used as one of the criteria for judging their degrees of functionality. Among the 19 derogative complements of degree, "" are mainly combined with derogative or negative terms, accounting for over 90% and suggesting they preserve many of the traits of substantial words; while cases of "吓人" and "过分" combing with derogative or negative words account for less than 30%, suggesting they are highly functional.

The intensity of emotional connotations contained in derogative and favorable complements of degree differ, as derogative complements of degree have more intense emotional connotations. This is manifested in two aspects: first, after a favorable complement of degree combines with a derogative word, the entire combination remains derogative, such as "荒谬绝顶" and "阴毒绝伦"; while after a derogative complement of degree combines with a favorable word, the entire combination appears somewhat favorable, such as "好得离谱" and "干净得过分". Second, in terms of the overall frequency of combinations of favorable and derogative complements of degree, favorable complements of degree combine with derogative or negative words significantly more often than derogative complements of degree combine with favorable or positive words, accounting for 19.93% and 3.31%, respectively.

5 Summary

This paper studies the rules of combination of complements of degree with predicates by using the corpus linguistics method, automatic word segmentation and gender tagging, and automatic semantic and syntax analysis in computing linguistics in an attempt to break through the conventional norms of descriptive studies based on small samples and enumeration and to introduce computing linguistics methods based on conventional grammatical research methods, combining linguistic sense experience with mathematical statistics, in expectation of providing reference for the study of rules Chinese syntax and semantic combinations.

Around this study, we have established a resource pool of combinations of complements of degree and predicates and sample sentences graded by difficulty, which can be used to assist the preparation of textbooks for teaching Chinese to non-native speakers and Chinese dictionaries as well as Chinese learning, and to provide references for the teaching of complements. Meanwhile, this resource pool is of certain importance to the processing of Chinese language information, and the study of components, phrase structure grammar and dependent syntax, especially the study of the core verbs of sentences and predicate/complement structure. These will be otherwise covered due to the limit of article size.

References

1. Liu, Y., Pan, W., Gu, W.: Practical Grammar of Modern Chinese (expanded edition). Commercial Press, Beijing (2001)
2. Fang, Y.: Practical Chinese Grammar. Peking University Press, Beijing (2001)
3. Sun, J.: Verbs with Complements of Degree. Journal of Guyuan Normal Vocational School 3 (1994)
4. Ma, Q.: Predicate-Complement Structure with Complements of Degree. In: Grammar Study and Exploration. Commercial Press, Beijing (1988)
5. Zhang, Y.: Multi-dimension Examination of Adverbs of Degree Serving as Complements. World Chinese Teaching (2) (2000)
6. Li, C.: Range and Category of Complements of Degree in Modern Chinese. Journal of Ningxia University 4 (2011)

Stability in Compressed Sensing for Some Sparse Signals

Sheng Zhang and Peixin Ye*

School of Mathematical Sciences and LPMC, Nankai University, Tianjin 300071, China
szhang@nankai.edu.cn, yepx@nankai.edu.cn

Abstract. In this paper, it is proved that every s-sparse signal vector can be recovered stably from the measurement vector $y = Ax$ via l_1 minimization as soon as the $2s$-th restricted isometry constant of the measurement matrix A is smaller than $3/(4+\sqrt{6})$. While for the large values of s, the constant can be improved to $4/(6+\sqrt{6})$. Note that our results contain the case of noisy data, therefore previous known results in the literature are extended and improved.

Keywords: Compressed sensing, Stability, l_1 minimization, Restricted isometry constant.

1 Introduction

Compressed Sensing is a new paradigm in signal and image processing. It seeks to faithfully capture a signal or image with the fewest number of measurements, cf. [1-9]. Rather than model a signal as a bandlimited function or an image as a pixel array, it models both of these as a sparse vector in some representation system. This model fits well real world signals and images. For example, images are well approximated by a sparse wavelet decomposition. One replaces the bandlimited model of signals by the assumption that the signal is sparse or compressible with respect to some basis or dictionary of waveform and enlarges the concept of sample to include the applications of any linear functional. Given this model, how should we design a sensor to capture the signal with the fewest number of measurements? we will focus on the discrete sensing problem where we are given a vector in R^N with N large and we wish to capture it through n measurements given by inner products with fixed vectors. Such a measurement system can be represented by an $n \times N$ matrix A. The vector $y = Ax$ is the vector of n measurements we make of x. The information that y holds about x is extracted through a decoder Δ . So $\Delta(x)$ should be designed to be a faithful approximation to x. The fact that this may be possible is embedded in some old mathematical results in functional analysis, geometry and approximation, cf. [10-14]. We will discuss what are the best matrices to use in sensing and how to extract the information contained in the sensed vector y. We shall focus on the relation between the number of samples we take of a signal and how well we can approximate the signal.

* Corresponding author.

D. Zeng (Ed.): Advances in Control and Communication, LNEE 137, pp. 741–749.
springerlink.com

Let us start by explaining the general setup of compressed sensing. We denote the set of all vectors in which have at most s nonzero coordinates as Σ_s. Defined l_p^N norms as$\|x\|_{l_p^N} = (\sum_{i=1}^{N} |x_i|^p)^{1/p}$. For simplicity we write $\|x\|_p$ for $\|x\|_{l_p^N}$. The restricted isometry constants δ_s of the $m \times N$ easurements matrix A is defined as the smallest positive constant such that

$$(1-\delta_s)\|z\|_2^2 \le \|Az\|_2^2 \le (1+\delta_s)\|z\|_2^2 \tag{1}$$

for all $z \in \Sigma_s$. Let $\sigma_s(x)$ denotes the error of best s-term approximation to x with respect to ℓ_1-quasi-norm, i.e. $\sigma_s(x) := \inf_{z\in\Sigma_s} \|x-z\|_1$. Note that the instance-optimality will automatically recover exactly any vector $x \in \Sigma_s$, i.e., any s-sparse vector.

2 Main Result

Consider the classical problem of Compressive Sensing of recovering a vector $x \in \mathcal{C}^N$ from the mere knowledge of a measurement vector $y \in \mathcal{C}^m$ by solving the minimization problem

$$\min_{\|Az-y\|_2 \le \theta} \|z\|_1 \ , \qquad \|Ax-y\|_2 \le \theta. \tag{2}$$

Lemma 1. [15] A solution of (2) exists for any $\theta \ge 0$.

Lemma 2. [16] Given integers $k, l \ge 1$, for a sequence $a_1 \ge a_2 \ge ... \ge a_{k+l} \ge 0$, we have

$$[\sum_{j=l+1}^{k+l} a_j^2]^{1/2} \le \max\left[\frac{1}{\sqrt{4l}}, \frac{1}{\sqrt{k}}\right]\left[\sum_{j=1}^{k} a_j\right].$$

Theorem 1. if $\delta_{2s} < \frac{3}{4+\sqrt{6}}$,then a solution x^* of (2) approximate the original vector x with errors

$$\|x-x^*\|_1 \le C_1\sigma_s(x) + D_1 s^{1/2}\theta, \qquad \|x-x^*\|_2 \le C_2 s^{-1/2}\sigma_s(x) + D_2\theta \tag{3}$$

where the constants C_1, C_2, D_1, D_2 depend only on δ_{2s}.

Proof: For any vectors u and $v \in \mathcal{R}^n$, one has $\|u\|_1 \le n^{1/2}\|u\|_2$, $\|u+v\|_1 \le \|u\|_1 + \|v\|_1$.

Step1:(Consequence of the assumption on δ_{2s})
Consider an arbitrary index set $s := s_0$ with $|S| \le s$. Let $v \in \mathcal{R}^N$, we may assume without loss of generality that the entries of v are sorted in decreasing order $|v_1| \ge |v_2| \ge \cdots \ge |v_N|$. Set $S = \{1, ..., s\}$.

(I) For the case $s = 5n + r$ with $1 \le r \le 5$.

We partition $\bar{S} = \{s+1, \cdots, N\}$ in two ways as $\bar{S} = S' \bigcup T_1 \bigcup T_2 \bigcup \cdots$ and $\bar{S} = S' \bigcup U_1 \bigcup U_2 \bigcup \cdots$, where $S' := \{s+1, ..., s+s'\}$ is of size s', $T_1 := \{s+s'+1, ..., s+s'+t\}$, $T_2 := \{s+s'+t+1, ..., s+s'+2t\}, \cdots$, are of size t, $U_1 := \{s+s'+1, ..., s+s'+u\}$, $U_2 := \{s+s'+u+1, ..., s+s'+2u\}, \cdots$ are of size u. We impose the size of the sets $S \bigcup S', S \bigcup T_k$, and $S' \bigcup U_k$ to be at most $2s$, i.e. $s' \le s, t \le s, s'+u \le 2s$, with $\delta = \delta_{2s}$, then it follows:

$$\|v_S\|_2^2 + \|v_S'\|_2^2 = \|v_S + v_S'\|_2^2 \le \frac{1}{1-\delta}\Big[\langle Av_S, A(v - \sum_{k\ge 1} v_{T_k})\rangle + \langle Av_{S'}, A(v - \sum_{k \ge 1} v_{U_k})\rangle\Big] \tag{4}$$

From Lemma 2, we have $\|v_{T_k}\|_2 \le \max\left[\frac{1}{\sqrt{4s'}}, \frac{1}{\sqrt{t}}\right]\|v_{\tilde{T}_k}\|_1$, $\|v_{U_k}\|_2 \le \max\left[\frac{1}{\sqrt{4s'}}, \frac{1}{\sqrt{u}}\right]\|v_{\tilde{U}_k}\|_1$, where $\tilde{T_1} := \{s+1, \cdots, s+t\}, \tilde{T_2} := \{s+t+1, \cdots, s+2t\}, \cdots$ and $\tilde{U_1} := \{s+1, \cdots, s+u\}, \tilde{U_2} := \{s+u+1, \cdots, s+2u\}, \cdots$. Also, it is easy to obtain $< A(v_S + v_{S'}), Av > \le \|Av\|_2 \|A(v_S + v_{S'})\|_2 \le \sqrt{1+\delta}\|Av\|_2[\|v_S\|_2 + \|v_{S'}\|_2]$ and $|\langle Av_S, A(-v_{T_k})\rangle| \le \delta \|v_S\|_2 \|v_{T_k}\|_2$. Then we obtain

$$\begin{aligned}\|v_S\|_2^2 + \|v_{S'}\|_2^2 \le{}& \frac{\sqrt{1+\delta}}{1-\delta}\|Av\|_2\left[\|v_S\|_2 + \|v_{S'}\|_2\right] \\ &+ \frac{\delta}{1-\delta}\left\{\|v_S\|_2 \max\left[\frac{1}{\sqrt{4s'}}, \frac{1}{\sqrt{t}}\right]\|v_{\bar{S}}\|_1 + \|v_{S'}\|_2 \max\left[\frac{1}{\sqrt{4s'}}, \frac{1}{\sqrt{u}}\right]\|v_{\bar{S}}\|_1\right\}.\end{aligned} \tag{5}$$

To minimize the first maximum, we take t as large as possible, i.e. $t = s$. For the second maximum, the point (s', u) belongs to the region $\{s' \ge 0, u \ge 0, s' \le s, s' + u \le 2s\}$ which is divided into two parts by the line L of equation $u = 4s'$. Below this line, the maximum equals $\frac{1}{\sqrt{u}}$, which is minimized for a large u. Above this line, the maximum equals $\frac{1}{\sqrt{4s'}}$, which is minimized for large s'. So, the maximum is minimized at the intersection of the line L with the boundary of the region other than the origin which is given by $s^* = \frac{2s}{5}, u^* = \frac{8s}{5}$. If s is a multiple of 5, then we can choose (s', u) to be (s^*, u^*). In this case, (5) becomes

$$\|v_S\|_2^2 + \|v_{S'}\|_2^2 \le \left[\frac{\sqrt{1+\delta}}{1-\delta}\|Av\|_2 + \frac{\delta\|v_{\bar{S}}\|_1}{(1-\delta)\sqrt{s}}\right]\|v_S\|_2 + \left[\frac{\sqrt{1+\delta}}{1-\delta}\|Av\|_2 + \frac{\delta\sqrt{C}\|v_{\bar{S}}\|_1}{(1-\delta)\sqrt{s}}\right]\|v_{S'}\|_2,$$

with $C = \frac{5}{8}$. Let $\tilde{c} = \frac{\sqrt{1+\delta}}{1-\delta}\|Av\|_2$ and $\tilde{d} = \frac{\delta\|v_{\bar{S}}\|_1}{(1-\delta)\sqrt{s}}$. Then it follows

$$\left[\|v_S\|_2 - \frac{\tilde{c}+\tilde{d}}{2}\right]^2 + \left[\|v_{S'}\|_2 - \frac{\tilde{c}+\sqrt{C}\tilde{d}}{2}\right]^2 \le \left(\frac{\tilde{c}+\tilde{d}}{2}\right)^2 + \left(\frac{\tilde{c}+\sqrt{C}\tilde{d}}{2}\right)^2.$$

From $\left(\frac{\tilde{c}+\tilde{d}}{2}\right)^2+\left(\frac{\tilde{c}+\sqrt{C}\tilde{d}}{2}\right)^2 \le \left[\frac{\sqrt{1+C}}{2}\tilde{d}+\frac{1}{\sqrt{1+C}}\tilde{c}\right]^2$, we have

$$\|v_S\|_2 \le \frac{\delta(1+\sqrt{1+C})\|V_{\bar{S}}\|_1}{2\sqrt{s}(1-\delta)}+\left[1+\frac{2}{\sqrt{1+C}}\right]\frac{\sqrt{1+\delta}}{2(1-\delta)}\|Av\|_2 \tag{6}$$

and

$$\|v_S\|_1 \le s^{1/2}\|v_S\|_2 \le \frac{\delta(1+\sqrt{1+C})}{2(1-\delta)}\|v_{\bar{S}}\|_1+\left[1+\frac{2}{\sqrt{1+C}}\right]\frac{\sqrt{1+\delta}\sqrt{s}}{2(1-\delta)}\|Av\|_2 .$$

Let $\mu=\frac{\delta(1+\sqrt{1+C})}{2(1-\delta)}$ and $\lambda=\left[1+\frac{2}{\sqrt{1+C}}\right]\frac{\sqrt{1+\delta}}{2(1-\delta)}$. Then it follows $\|v_S\|_2 \le \frac{\mu}{\sqrt{s}}\|v_{\bar{S}}\|_1+\lambda\|Av\|_2$ and $\|v_S\|_1 \le \mu\|v_{\bar{S}}\|_1+\lambda\sqrt{s}\|Av\|_2$. Therefore, if $\delta<\frac{2}{3+\sqrt{1+C}}$, then $\mu<1$.

(II) For the case s is not a multiple of 5, one can not choose (s',u) to be (s^*,u^*). So we choose it to be a corner of the square $[\lfloor s^*\rfloor,\lceil s^*\rceil]\times[\lfloor u^*\rfloor,\lceil u^*\rceil]$. The corner $(\lceil s^*\rceil,\lfloor u^*\rfloor)$ is inadmissible, and among the three admissible corners, the smallest value of $\max\left(\frac{1}{\sqrt{4s'}},\frac{1}{\sqrt{u}}\right)$ is achieved for $(\lceil s^*\rceil,\lfloor u^*\rceil)$. With this choice, (6) holds with $C=\frac{s}{\lfloor 8s/5\rfloor}$ for this case. Thus, the same arguments as before yield the sufficient condition $\delta<\frac{2}{3+\sqrt{1+s/\lfloor 8s/5\rfloor}}$.

With **(I)** and **(II)**, it follows that $\delta<\frac{2}{3+\sqrt{1+s/(8n+\lfloor 8r/5\rfloor)}}$ when $s=5n+r$ with $1\le r\le 5$ is a sufficient condition for the recovery.

Next, we will prove that (a) $\delta<1/2$, when $s=1$ (b) $\delta<\frac{3}{4+\sqrt{(6s-2r)/(s-1)}}$, when $s=3n+r$ with $1\le r\le 3$ (c) $\delta<\frac{4}{5+\sqrt{(12s-3r)/(s-1)}}$, when $s=4n+r$ with $1\le r\le 4$. The proofs of (a),(b) and (c)are put in the appendix. With (a),(b),(c) and (d) $\delta<\frac{2}{3+\sqrt{1+s/(8n+\lfloor 8r/5\rfloor)}}$, when $s=5n+r$ with $1\le r\le 5$,which we just verified before, we can conclude that the inequality $\delta<\frac{3}{4+\sqrt{6}}\approx 0.4652$ ensures s-sparse recovery following the same method as those of the proof of Theorem 1 of [16].

Step 2 (Consequence of the l_1-minimization)
Let S be specified as the set of indices of the s largest absolute value components of x , $v=x-x^*$, where x^* is a minimizer of (2). Then it follows $\|x\|_1\ge\|x^*\|_1$, *i.e.* $\|x_s\|_1+\|x_{\bar{s}}\|_1\ge\|x^*_s\|_1+\|x^*_{\bar{s}}\|_1$. Also, it is easy to see that $\|x^*_s\|_1=\|x_s-v_s\|_1\ge\|x_s\|_1-\|v_s\|_1$ and $\|x^*_{\bar{s}}\|_1=\|x_{\bar{s}}-v_{\bar{s}}\|_1=\|v_{\bar{s}}-x_{\bar{s}}\|_1\ge\|v_{\bar{s}}\|_1-\|x_{\bar{s}}\|_1$. Thus we obtain that

$$\|v_{\bar{s}}\|_1\le 2\|x_{\bar{s}}\|_1+\|v_s\|_1=2\sigma_s(x)+\|v_s\|_1 . \tag{7}$$

Step 3 (Error estimate)
It is easy to see $\|Av\|_2=\|Ax-Ax^*\|_2\le\|Ax-y\|_2+\|Ax^*-y\|_2\le 2\theta$. Thus, we obtain

$$\|v_{\bar{s}}\|_1 \leq 2\sigma_s(x) + \|v_s\|_1 \leq 2\sigma_s(x) + \mu\|v_{\bar{s}}\|_1 + \lambda\sqrt{s}\|Av\|_2 \leq 2\sigma_s(x) + \mu\|v_{\bar{s}}\|_1 + 2\theta\lambda s^{1/2} ,$$

from which follows $\|v_{\bar{s}}\|_1 \leq \dfrac{2}{1-\mu}\sigma_s(x) + \dfrac{2\theta\lambda s^{1/2}}{1-\mu}$, since $\mu < 1$. With

$$\|v\|_1 = \|v_s\|_1 + \|v_{\bar{s}}\|_1 \leq (1+\mu)\|v_{\bar{s}}\|_1 + \lambda\sqrt{s}\|Av\|_2 \leq (1+\mu)\|v_{\bar{s}}\|_1 + 2\theta\lambda\sqrt{s} ,$$

we have

$$\|v\|_1 \leq 2\theta\lambda\sqrt{s} + (1+\mu)[\frac{2}{1-\mu}\sigma_s(x) + \frac{2\theta\lambda s^{1/2}}{1-\mu}] = \frac{2(1+\mu)}{1-\mu}\sigma_s(x) + \frac{4\theta\lambda\sqrt{s}}{1-\mu} .$$

Let $C_1 = \frac{2(1+\mu)}{1-\mu}$ and $D_1 = \frac{4\lambda}{1-\mu}$, then we obtain

$$\|v\|_1 \leq C_1\sigma_s(x) + D_1\sqrt{s}\theta \tag{8}$$

where C_1 and D_1 only depend on δ_{2s} because μ and λ only depend on δ_{2s}.

For the l_2-error, we note that $\|v_{s'}\|_2 \leq \dfrac{\mu}{\sqrt{s}}\|v_{\bar{s}}\|_1 + \lambda\|Av\|_2$, from which follows

$$\|v\|_2 \leq \|v_s\|_2 + \|v_{s'}\|_2 + \sum_{k\geq 1}\|v_{T_k}\|_2 \leq 2\mu S^{-\frac{1}{2}}\|v_{\bar{s}}\|_1 + 4\lambda\theta + \sum_{k\geq 1}\|v_{T_k}\|_2 .$$

For $i \in T_k$ and $j \in T_{k-1}$, it is easy to see that $\|v_i\| \leq S^{-1}\|v_{T_{k-1}}\|_1$ and $\|v_{T_k}\|_2 \leq S^{-\frac{1}{2}}\|v_{T_{k-1}}\|_1$ since $|v_i| \leq |v_j|$. Also, it follows $\|v_{T_1}\|_2 \leq S^{-\frac{1}{2}}\|v_s\|_1$ because $|v_i| \leq |v_j|$ for $i \in T_i$ and $j \in S$. With these inequalities, we obtain

$$\|v\|_2 \leq (3\mu+1)S^{-\frac{1}{2}}\|v_{\bar{s}}\|_1 + 6\lambda\theta \leq \frac{2(3\mu+1)}{(1-\mu)\sqrt{s}}\sigma_s(x) + [\frac{2(3\mu+1)}{(1-\mu)\sqrt{s}} + 6]\lambda\theta .$$

Let $\widetilde{C_2} = \frac{2(3\mu+1)}{(1-\mu)}$ and $\widetilde{D_2} = [\frac{2(3\mu+1)}{(1-\mu)+4}]\lambda$. Then we have

$$\|v\|_2 \leq \widetilde{C_2}S^{-\frac{1}{2}}\sigma_s(x) + \widetilde{D_2}\theta, \tag{9}$$

where $\widetilde{C_2}$ and $\widetilde{D_2}$ only depend on δ_{2s} because μ and λ only depend on δ_{2s} . This completes the proof.

For the case of large s, we can get the following result:

Theorem 2. For large s, if $\delta_{2s} < \frac{4}{6+\sqrt{6}}$, then a solution x^* of (2) approximate the original vector x with errors

$$\|x - x^*\|_1 \leq \tilde{C_1}\sigma_s(x) + \tilde{D_1}s^{-1/2}\theta, \quad \|x - x^*\|_2 \leq \tilde{C_2}s^{-1/2}\sigma_s(x) + \tilde{D_2}s^{-1}\theta \tag{10}$$

where the constants $\tilde{C_1}, \tilde{C_2}, \tilde{D_1}, \tilde{D_2}$ depend only on δ_{2s}.

Proof. Step1:(Consequence of the assumption on δ_{2s})

We first prove that, for large s, $\delta < \frac{1}{1+\sqrt{s\tilde{s}/[8(\tilde{s}-s)(3s-2\tilde{s})]}}$, where $\tilde{s} = \lfloor\sqrt{\frac{3}{2s}}\rfloor$ is a sufficient condition for the recovery. With this, we can obtain that, for large s, $\delta < \frac{3}{4+\sqrt{6}}$ ensures the unique minimizer of (2) following the same method as [16].

Partition $\bar{S} = \{s+1, s+2, \cdots, N\}$ as $\bar{S} = S' \cup T_1 \cup T_2 \cup \cdots$, where $S' := \{s+1, s+2, \cdots, s+s'\}$ is of size s', $T_1 := \{s+s'+1, s+s'+2, \cdots, s+s'+t\}$, $T_2 := \{s+s'+t+1, s+s'+t+2, \cdots, s+s'+2t\}$, $\cdots$ are of size t. Let v be sorted with $|v_1| \geq |v_2| \geq \cdots \geq |v_N|$ and $S = \{1, 2, 3, \cdots, s\}$. For $r \leq s+s'$, we consider the r -sparse vectors $w_1 := v_{\{1,2,\cdots,r\}}$, $w_2 := v_{\{2,3,\cdots,r+1\}}$, $\cdots$, $w_{s+s'-r+1} := v_{\{s+s'-r+1, s+s'-r+2, \cdots, s+s'\}}$, $w_{s+s'-r+2} := v_{\{s+s'-r+2, s+s'-r+3, \cdots, s+s', 1\}}$, $\cdots$, $w_{s+s'} := v_{\{s+s', 1, 2, \cdots, r-1\}}$, then it follows:

$$\sum_{j=1}^{s+s'} w_j = r(v_S + v_{S'}), \quad m\sum_{j=1}^{s+s'} \|w_j\|_2^2 = r\|v_S + v_{S'}\|_2^2 .$$

It is easy to see $\|Av\|_2 = \|Ax - Ax^*\|_2 \leq \|Ax - y\|_2 + \|Ax^* - y\|_2 \leq 2\theta$. Set $s' \leq s$, $r + t \leq 2s$, $t \leq 4s'$ and $\delta := \delta_{2s}$. Then we have

$$\begin{aligned}
\|v_S + v_{S'}\|_2^2 &\leq \frac{1}{1-\delta}\|A(v_S + v_{S'})\|_2^2 = \frac{1}{r(1-\delta)}\langle A(\sum_{j=1}^{s+s'} w_j), A(v - \sum_{k\geq 1} v_{T_k})\rangle \\
&= \frac{1}{r(1-\delta)}\left[\langle A(v_S + v_{S'}), Av\rangle - \sum_{j=1}^{s+s'}\langle Aw_j, \sum_{k\geq 1} Av_{T_k}\rangle\right] \\
&\leq \frac{1}{r(1-\delta)}\left[\sqrt{1+\delta}\|Av\|_2\|v_S + v_{S'}\|_2 + \delta\sum_{j=1}^{s+s'}\|w_j\|_2\sum_{k\geq 1}\|v_{T_k}\|_2\right] \\
&\leq \frac{1}{r(1-\delta)}\left[\sqrt{1+\delta}\|Av\|_2\|v_S + v_{S'}\|_2 + \delta\sum_{j=1}^{s+s'}\|w_j\|_2\frac{\|v_{\bar{S}}\|_1}{\sqrt{t}}\right] \\
&\leq \frac{1}{r(1-\delta)}\left[\sqrt{1+\delta}\|Av\|_2\|v_S + v_{S'}\|_2 + \delta\sqrt{s+s'}\sqrt{\sum_{j=1}^{s+s'}\|w_j\|_2^2}\frac{\|v_{\bar{S}}\|_1}{\sqrt{t}}\right] \\
&= \frac{1}{r(1-\delta)}\left[\sqrt{1+\delta}\|Av\|_2 + \delta\sqrt{\frac{(s+s')r}{t}}\|v_{\bar{S}}\|_1\right]\|v_S + v_{S'}\|_2 .
\end{aligned}$$

With $\|v_S\|_1 \leq \sqrt{s}\|v_S\|_2 \leq \sqrt{s}\|v_S + v_{S'}\|_2$, we obtain

$$\|v_S\|_1 \leq \sqrt{s}\|v_S + v_{S'}\|_2 \leq \frac{\sqrt{s(1+\delta)}}{r(1-\delta)}\|Av\|_2 + \frac{\sqrt{s}\delta}{1-\delta}\sqrt{\frac{s+s'}{rt}}\|v_{\bar{S}}\|_1 . \tag{11}$$

Let $\sigma := \frac{s'}{s}$, $\rho := \frac{r}{s}$, $\tau := \frac{t}{s}$. Then it follows $\sqrt{\frac{s(s+s')}{rt}} = \sqrt{\frac{1+\sigma}{\rho\tau}}$. We first minimize $\frac{1+\delta}{\rho\tau}$ subject to $\sigma \leq 1$, $\rho + \tau \leq 2$ and $\tau \leq 4\sigma$. The minimum is achieved when ρ is largest possible, i.e. $\rho = 2 - \tau$. Then the minimum of $\frac{1+\sigma}{(2-\tau)\tau}$ subject to

$\sigma \le 1$, $\tau \le 2$ and $\tau \le 4\sigma$ is achieved when σ is largest possible, i.e. $\sigma = \frac{\tau}{4}$. Furthermore, we can find the minimum of $\frac{1+\tau/4}{(2-\tau)\tau}$ subject to $\tau \le 2$ is achieved for $\tau = 2\sqrt{6} - 4$. This corresponds to $\sigma = \sqrt{3/2} - 1$ and $\rho = 6 - 2\sqrt{6}$, i.e. $s' = \lfloor (\sqrt{3/2} - 1)s \rfloor = \tilde{s} - s$, $t = 4s' = 4(\tilde{s} - s)$ and $r = 2s - t = 6s - 4\tilde{s}$. Thus, (11) becomes

$$\|v_S\|_1 \le \frac{\sqrt{s(1+\delta)}}{r(1-\delta)}\|Av\|_2 + \sqrt{\frac{s\tilde{s}}{8(\tilde{s}-s)(3s-2\tilde{s})}\frac{\delta}{1-\delta}}\|v_{\bar{S}}\|_1 \tag{12}$$

Let $\mu = \sqrt{\frac{s\tilde{s}}{8(\tilde{s}-s)(3s-2\tilde{s})}\frac{\delta}{1-\delta}}$, then $\mu < 1$ is satisfied as soon as $\delta < \dfrac{1}{1 + \sqrt{s\tilde{s}/[8(\tilde{s}-s)(3s-2\tilde{s})]}}$, where $\tilde{s} = \lfloor \sqrt{3/2}s \rfloor$.

Step 2 (Consequence of the l_1-minimization)

Let S be specified as the set of indices of the s largest absolute value components of x, $v = x - x^*$, where x^* is a minimizer of (2). Then it follows $\|x\|_1 \ge \|x^*\|_1$, *i.e.* $\|x_s\|_1 + \|x_{\bar{s}}\|_1 \ge \|x_s^*\|_1 + \|x_{\bar{s}}^*\|_1$. Also, it is easy to see that $\|x_s^*\|_1 = \|x_s - v_s\|_1 \ge \|x_s\|_1 - \|v_s\|_1$, and $\|x_{\bar{s}}^*\|_1 = \|x_{\bar{s}} - v_{\bar{s}}\|_1 = \|v_{\bar{s}} - x_{\bar{s}}\|_1 \ge \|v_{\bar{s}}\|_1 - \|x_{\bar{s}}\|_1$. Thus we obtain that $\|x_s\|_1 + \|x_{\bar{s}}\|_1 \ge \|x_s^*\|_1 + \|x_{\bar{s}}^*\|_1 \ge \|x_s\|_1 - \|v_s\|_1 + \|v_{\bar{s}}\|_1 - \|x_{\bar{s}}\|_1$, i.e.,

$$\|v_{\bar{s}}\|_1 \le 2\|x_{\bar{s}}\|_1 + \|v_s\|_1 = 2\sigma_s(x) + \|v_s\|_1 . \tag{13}$$

Step 3 (Error estimate)

Let $\lambda = \dfrac{s\sqrt{1+\delta}}{r(1-\delta)} = \dfrac{1}{6 - 4\tilde{s}/s}\dfrac{\sqrt{1+\delta}}{1-\delta}$ and $\mu = \sqrt{\dfrac{s\tilde{s}}{8(\tilde{s}-s)(3s-2\tilde{s})}\dfrac{\delta}{1-\delta}} = \sqrt{\dfrac{\tilde{s}/s}{8(\tilde{s}/s-1)(3-2\tilde{s}/s)}\dfrac{\delta}{1-\delta}}$, then λ and μ are only dependent on δ and independent on s because $\tilde{s}/s = \lfloor \sqrt{3/2}s \rfloor / s$ is independent on s. So, (12) becomes $\|v_S\|_1 \le \lambda s^{-1/2}\|Av\|_2 + \mu\|v_{\bar{S}}\|_1$ with $\mu < 1$. Thus, it is easy to get

$$\|v\|_1 = \|v_S\|_1 + \|v_{\bar{S}}\|_1 \le 2\theta\lambda s^{-1/2} + (1+\mu)\|v_{\bar{S}}\|_1 .$$

Then it follows $\|v_{\bar{S}}\|_1 \le 2\sigma_s(x) + \|v_S\|_1 \le 2\sigma_s(x) + 2\theta\lambda s^{-1/2} + \mu\|v_{\bar{S}}\|_1$, i.e. ,

$$\|v_{\bar{S}}\|_1 \le \frac{2\sigma_s(x)}{1-\mu} + \frac{2\theta\lambda s^{-1/2}}{1-\mu} \tag{14}$$

Therefore, we obtain that

$$\|v\|_1 \le 2\theta\lambda s^{-1/2} + (1+\mu)\|v_{\bar{S}}\|_1 \le 2\theta\lambda s^{-1/2} + (1+\mu)\left[\frac{2\sigma_s(x)}{1-\mu} + \frac{2\theta\lambda s^{-1/2}}{1-\mu}\right]$$
$$= \frac{2(1+\mu)}{1-\mu}\sigma_s(x) + \frac{4\lambda}{1-\mu}s^{-1/2}\theta = \widetilde{C_1}\sigma_s(x) + \widetilde{D_1}s^{-1/2}\theta \tag{15}$$

where $\widetilde{C_1} = \frac{2(1+\mu)}{1-\mu}$ and $\widetilde{D_1} = \frac{4\lambda}{1-\mu}$ are only dependent on δ because λ and μ are only dependent on δ.

For the ℓ_2-error, we note that $\|v_S\|_2 \le \|v_S + v_{S'}\|_2 \le \lambda s^{-1}\|Av\|_2 + \mu s^{-1/2}\|v_{\bar{S}}\|_1$ and $\|v_{S'}\|_2 \le \|v_S + v_{S'}\|_2 \le \lambda s^{-1}\|Av\|_2 + \mu s^{-1/2}\|v_{\bar{S}}\|_1$. Then it follows

$$\|v\|_2 \le \|v_S\|_2 + \|v_{S'}\|_2 + \sum_{k\ge 1} \|v_{T_k}\|_2 \le (2\mu+1)s^{-1/2}\|v_{\bar{S}}\|_1 + 4\lambda s^{-1}\theta \, .$$

Therefore, we obtain

$$\begin{aligned}\|v\|_2 &\le (2\mu+1)s^{-1/2}\|v_{\bar{S}}\|_1 + 4\lambda s^{-1}\theta \le 4\lambda s^{-1}\theta + (2\mu+1)s^{-1/2}\left[\frac{2\sigma_s(x)}{1-\mu} + \frac{2\theta\lambda s^{-1/2}}{1-\mu}\right] \\ &= \frac{2(2\mu+1)}{1-\mu}s^{-1/2}\sigma_s(x) + \left[\frac{2(2\mu+1)}{1-\mu} + 4\right]\lambda s^{-1}\theta = \widetilde{C_3}s^{-1/2}\sigma_s(x) + \widetilde{D_3}s^{-1}\theta \, ,\end{aligned} \tag{16}$$

where $\widetilde{C_3} = \frac{2(2\mu+1)}{1-\mu}$ and $\widetilde{D_3} = \left[\frac{2(2\mu+1)}{1-\mu} + 4\right]\lambda$ are only dependent on δ because μ and λ are only dependent on δ. This completes the proof.

Acknowledgment. This work was supported by the National Natural Science Foundation of China (Grant No. 10971251), and the work of the first author was also supported by National Natural Science Foundation of China (Grant No. 11071132, 10601045), Projects of International Cooperation and Exchanges NSFC 10811120281 and the Scientific Research Foundation for the Returned Overseas Chinese Scholars, State Education Ministry of China.

References

1. Baraniuk, T., Davenport, M., DeVore, R., Wakin, M.: A simple proof of the restricted isometry property for random matrices. Constr. Approx. 28(3), 253–263 (2008)
2. Cai, T., Wang, L., Xu, G.: Shifting inequality and recovery of sparse signals. IEEE Trans. Signal Process 58(3), 1300–1308 (2010)
3. Candès, E., Romberg, J., Tao, T.: Robust uncertainty principles: exact signal reconstruction from highly incomplete frequency information. IEEE Trans. Inf. Theory 52(2), 489–509 (2006)
4. Candès, E., Romberg, J., Tao, T.: Stable signal recovery from incomplete and inaccurate measurements. Commun. Pure Appl. Math. 59(8), 1207–1223 (2005)
5. Candès, E., Tao, T.: Decoding by linear programming. IEEE Trans. Inf. Theory 51(12), 4203–4215 (2005)
6. Candès, E.: The restricted isometry property and its implications for compressed sensing. Compte Rendus de l'Academie des Sciences, Paris, Series I 346(9-10), 589–592 (2008)
7. Cohen, A., Dahmen, W., DeVore, R.: Compressed sensing and best k - term approximation. J. Amer. Math. Soc. 22(1), 211–231 (2009)
8. DeVore, R., Petrova, G., Wojtaszczyk, P.: Instance-optimality in probability with an l_1-minimization decoder. Appl. Comput. Harmon. Anal. 27(3), 275–288 (2009)
9. Donoho, D.: Compressed sensing. IEEE Trans. Inf. Theory 52(4), 1289–1306 (2006)

10. Garnaev, A., Gluskin, E.: The widths of Euclidean balls. Dokl. An. SSSR 277, 1048–1052 (1984)
11. Kashin, B.: The widths of certain finite dimensional sets and classes of smooth functions. Izvestia 41, 334–351 (1977)
12. Ledoux, M.: The Concentration of Measure Phenomenon. Am. Math. Soc., Providence (2001)
13. Litvak, A., Pajor, A., Rudelson, M., Tomczak-Jaegermann, N.: Smallest singular value of random matrices and geometry of random polytopes. Adv. Math. 195(2), 491–523 (2005)
14. Lorentz, G., von Golitschek, M., Makovoz, Y.: Constructive Approximation: Advanced Problems. Springer, Berlin (1996)
15. Foucart, S., Lai, M.: Sparest solutions of underdetermined linear systems via minimization for $0<q\leq 1$. Appl. Comput. Harmon. Anal. 26, 395–407 (2009)
16. Foucart, S.: A note on guaranteed sparse recovery via l_1 minimization. Appl. Comput. Harmon. Anal. 29, 97–103 (2010)

Research and Comparison on the Algorithms of Sampled Data Preprocess in Power System On-Line Insulation Monitoring

Xuzheng Chai[1], Xishan Wen[2], Yan Li[1], and Yi Liu[1]

[1] Henan Electric Power Dispatching & Communication Center, Zhengzhou, Henan, 450002, China
[2] College of Electrical Engineering, Wuhan University, Wuhan, Hubei, 430072, China
chaixuzheng@tsinghua.org.cn, xswen@whu.edu.cn,
{liyan,liuyi}@ha.sgcc.com.cn

Abstract. Along with the rapid development of computer technique, the precision of on-line insulation monitoring system is requested to be higher. Using statistical theory, an algorithm of preprocessing the sampling data in power system on-line insulation monitoring is presented in this paper. The wavelet transformation process used to preprocess the sampling data in on-line insulation monitoring is analyzed. The results of two methods used to preprocess the practical sampling data are given. Their effect to decrease the impulse interferences, high-frequency interferences and gross errors due to sampling element is obvious. Based on the characteristic of two algorithm programs, the paper discusses their applicable condition.

Keywords: power system on-line insulation monitoring, the sampled data preprocess, statistical theory, wavelet transformation.

1 Introduction

Monitoring of the insulation condition of power system apparatus is necessary in order to estimate its deterioration and maintain the reliability of electric power supply. With the transition from time-based maintenance to condition-based maintenance, the on-line insulation monitoring techniques of power apparatus is playing an important role to guarantee the safety of power delivery and find insulation faults in time. Based on the on-line insulation monitoring techniques, the condition of electrical insulation can be effectively find out, so the insidious defects will be checked out earlier, and the occurring of gusty accident may be avoided and the reliability of power system can be increased. The study on on-line insulation monitoring techniques is of theoretical and practical values.

Although considerable advantages such as laborsaving, high sensitivity, no service interruption and providing continuously trend monitoring and so on, the precision and stability of the on-line insulation monitoring result isn't satisfying. The on-line insulation monitoring system generally is located in substations or power plants,

D. Zeng (Ed.): Advances in Control and Communication, LNEE 137, pp. 751–758.
springerlink.com

where electromagnetic environment is bad. There are many factors that may influence the tested data. However, electromagnetic interferences are one of the main factors.

Along with the rapid development of computer technique, the precision of on-line insulation monitoring system is requested to be higher. In order to diminish the impulse interferences, high-frequency interferences and gross error due to sampling element, the sampled data of on-line insulation monitoring system should be preprocessed before harmonic analysis is applied to these sampled data. Based on statistical theory, an algorithm of preprocessing the sampling data in on-line insulation monitoring system is presented in this paper. In this paper, the wavelet transformation process used to preprocess the sampled data in insulation on-line monitoring system is analyzed, too. The results of two methods used to preprocess the test data are given. Their effect to decrease the impulse interferences, high-frequency interferences and gross error due to sampling element is obvious. Based on the characteristic of two algorithm programs, the paper discusses their applicable condition.

2 Description of the Algorithm of Preprocessing the Sampling Data in On-Line Insulation Monitoring System Based on Statistical Theory

The algorithm of preprocessing the sampling data in on-line insulation monitoring system based on statistical theory contains two steps: the process of picking out gross error and impulse interference points and the process of smoothing.

2.1 The Process of Picking Out Gross Error and Impulse Interference Points

The data sampling process of on-line insulation monitoring system is a sort of dynamic parametric test.

Even though the insulation has aged, the leakage current must continuously change according to the rule of insulation ageing and can't jump. So we can use the method of linear smooth move to deduce the rationality of the continuous sampling data value. The theoretical base of algorithm is that the errors of random sample data obey Gaussian distribution and Small Probability theory. For the sampling data of on-line insulation monitoring system, Small Probability theory consider that impulse interferences and gross errors due to sampling element belong in small probability events and the occurring of impulse interferences hasn't correlation. In statistical theory, to measurement of fixed value, the measured values whose sample deviation greater than triple sample standard deviation are called high abnormal values. The probability of high abnormal values occurring is 0.27%. Райта offers that high abnormal values should be picked out, which is called Райта rule or 3δ rule. In this algorithm, we try using the Райта rule that is used to preprocess fixed-value metrical data to preprocess dynamic metrical data. In order to meet Райта rule's condition, this algorithm replaces sample standard deviation with sample standard deviation's unbiased estimation. According to statistical theory this replacement is reasonable. The main aim of the process of picking out gross error points is diminishing the influence of the impulse interferences and gross error due to sampling element to precision of on-line insulation monitoring system metrical data. According to the result of preprocessing the practical sampling data, the effect is obvious.

The waveforms of on-line insulation monitoring system sampling data are often distorted and interfered. The data sampling process is equal-timed. The sampling frequency is higher. So we can consider the data sequence in short time is linear.

Given five points: P_1 (x_1, y_1), P_2 (x_2, y_2), P_3 (x_3, y_3), P_4 (x_4, y_4), P_5 (x_5, y_5) at straight-line $y = a + bx$, according to least squares theory and extreme value principle of differential and integral calculus, fit the five points with the straight-line $y = a + bx$. So we can deduce the sixth point as follow:

$$\widehat{y}_6 = \frac{1}{10}(-4y_1 - y_2 + 2y_3 + 5y_4 + 8y_5) \tag{1}$$

Equation (1) can be promoted as follow:

$$\widehat{y}_i = \frac{1}{10}(-4y_{i-5} - y_{i-4} + 2y_{i-3} + 5y_{i-2} + 8y_{i-1}) \tag{2}$$

According to the character of simple linear regression analysis, the ample standard deviation's unbiased estimation is

$$\widehat{\delta}_y = \sqrt{\frac{S_e}{n-2}} = \sqrt{\frac{\sum_{i=1}^{n}(y_i - \widehat{y}_i)^2}{n-2}} \tag{3}$$

Replaces sample standard deviation with sample standard deviation's unbiased estimation, according to Райта rule, and we can get: when

$$\left|y_i - \widehat{y}_i\right| > 3\widehat{\delta}_y \tag{4}$$

The y_i value is judged as high abnormal value and then picked out. In the sampling data sequence, the $\widehat{y}_i$ value will replace y_i value.

2.2 The Process of Smoothing

The aim of sampling data smoothing process is decreasing the high-frequency noise interferences with digital filter technique. The essence of smoothing process is taking curve fitting to the data that have been picked out gross error points. To the sampling data smoothing process of on-line insulation monitoring system, the precision of curve fitting only depend the number of fitted points. Here, we take the method of seven-pointed quadratic polynomial smoothing.

Given point i and it fore-and back points total seven points: P_{i-3} (x_{i-3}, y_{i-3}), P_{i-2} (x_{i-2}, y_{i-2}), P_{i-1} (x_{i-1}, y_{i-1}), P_i (x_i, y_i), P_{i+3} (x_{i+3}, y_{i+3}), P_{i-2} (x_{i+2}, y_{i+2}), P_{i-1} (x_{i+1}, y_{i+1}), according to least squares theory and extreme value principle of differential and integral calculus, we take quadratic polynomial $\widehat{y}_i = a_0 + a_1x + a_2x^2$ to fit the seven points. We can obtain:

$$\hat{y}_i = \frac{1}{42}(-4y_{i-3} + 6y_{i-2} + 12y_{i-1} + 14y_i + 12y_{i+1} + 6y_{i+2} - 4y_{i+3}) \quad (5)$$

When i=1, 2, 3, n-2, n-1, n, we can adjust method of giving the seven points and obtain their corresponding smoothing equation.

3 The Wavelet Transformation Algorithm of Preprocessing the Sampling Data in Power System On-Line Insulation Monitoring

Wavelet transformation is a sort of signal analysis method of time-frequency domain. It has the feature of multi-resolution representation. Wavelet transformation can decompose the signal into components of different frequency bands. The frequencies of noise interference signal are often higher, and so the components of noise interference signal will be differentiated at the Wavelet transformation of smaller scale. The impulse interference signal or gross error signal contains components of abundant frequencies. Their components will occur at some scales. When wavelet transformation is used to preprocess the sampling data of on-line insulation monitoring system, the scales must be suitable. If the scales are too small, the interference signals will still have rather large influence. If the scales are too large, the some components of real sampled data will be erased, and which will lead the distortion of initial sampling signal.

The sampled data preprocessing with wavelet transformation algorithm contains the process of eliminating the interference signals and the process of reconstructing the real signal.

At first, choose proper prototype mother wavelet function and decompose the sampling data signal with wavelet analysis. According to the theory for multi-resolution signal decomposition [4], The specific algorithm is described as follow: There is a signal $f(t) \in L^2(R)$. Here, $f(t)$ is a function of time-domain and $L^2(R)$ is the measurable Hilbert vector space, square-integral one-dimensional function. The classical norm of $f(t) \in L^2(R)$ is given by

$$\|f\|^2 = \int_{-\infty}^{+\infty} |f(x)|^2 dx \quad (6)$$

The Hilbert vector space can be decomposed as a direct sum of the orthogonal vector sub-spaces:

$$L^2(R) = \sum_{j=-\infty}^{J} W_j \oplus V_j \quad (7)$$

Here, W_j and V_j are the orthogonal vector sub-spaces, and they constitute the whole Hilbert vector space. The J is the set decomposition scales. The meaning of (7) is that the Hilbert vector space is decomposed as J of wavelet spaces W_j and scale spaces V_j. It can be shown as Fig. 1.

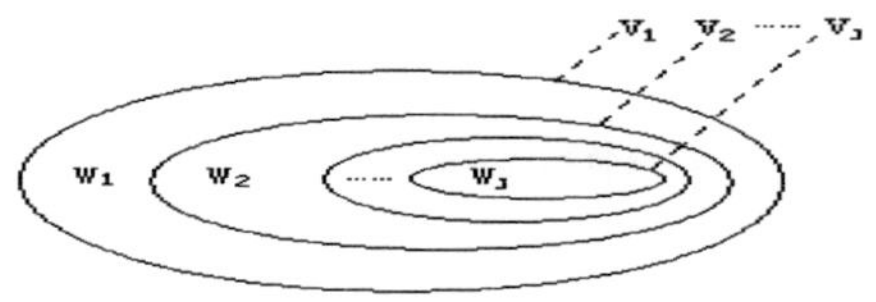

Fig. 1. Wavelet spaces Wj and scale spaces Vj schematic diagram

By applying the projection theorem [5], we can easily show that the detail signal which is given by the orthogonal projection of the original signal on the wavelet spaces W_j and scale spaces V_j at scale j. On the scale spaces V_j, the detail signal can be written:

$$f_s^j(t) = \sum_k c_{j,k}\phi_{j,k}(t)\ , \ k \in Z. \tag{8}$$

Here, $\phi(t)$ is the one-dimensional scaling function of the multiresolution and $c_{j,k}$ is the value of scaling coefficient. On the wavelet spaces W_j, the detail signal can be written:

$$f_d^j(t) = \sum_k d_{j,k}\varphi_{j,k}(t)\ , \ k \in Z. \tag{9}$$

Here, $\varphi(t)$ is the one-dimensional wavelet function of the multi-resolution and $d_{j,k}$ is the value of wavelet coefficient. The signal $f(t)$ on the whole Hilbert vector space $L^2(R)$ can be expressed as:

$$f(t) = \sum_{j=-\infty}^{J}\left(\sum_k d_{j,k}\varphi_{j,k}(t) + \sum_k c_{j,k}\phi_{j,k}(t)\right) \tag{10}$$

The decomposition process can be shown as Fig. 2.

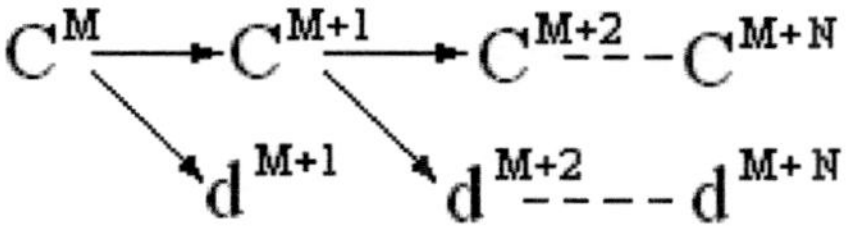

Fig. 2. Decomposition process of wavelet schematic diagram

Because the frequencies of interference signal are often higher, we can select d^{M+N}, … d^{M+1} and make them zero value or proper value [5]. Thus the influence of interference signal is eliminated. Then we need reconstruct the sampling data signal [5]. The reconstruction process is the inverse transformation of the decomposition process. The reconstruction process can be shown as Fig. 3.

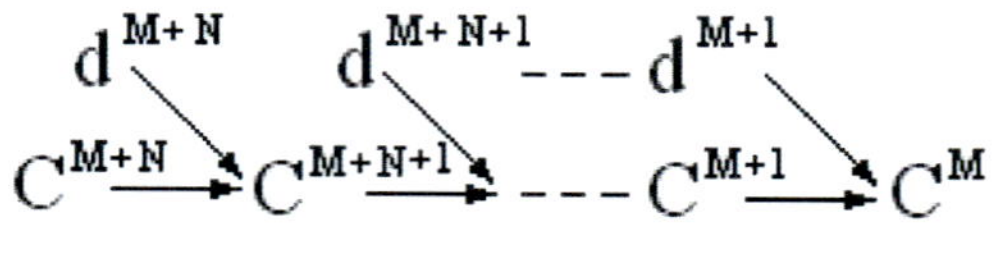

Fig. 3. Reconstruction process of wavelet schematic diagram

4 Verification of Algorithm with Test Data

Fig. 4 is the actual test waveform of MOA leakage current. The sampling rate is 25kS/s (k Sample points/sec). There are obvious impulse interferences and white noise interferences.

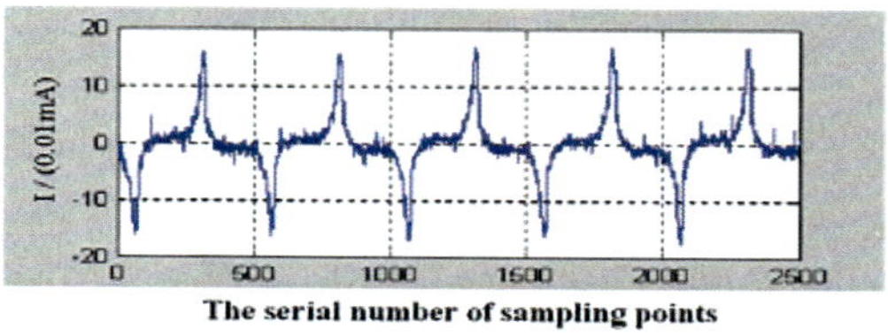

Fig. 4. Actual test waveform of MOA leakage current

Fig. 5 is the waveform, which has been processed by the process of picking out gross error points with Райта rule. The result reveals that most of gross error points and impulse interference points have been picked out and erased.

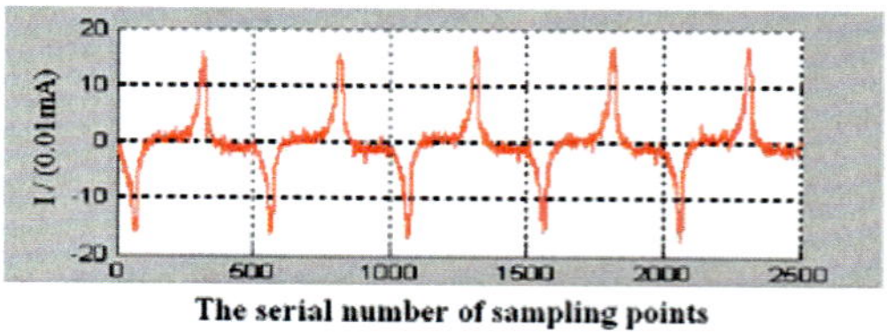

Fig. 5. Waveform processed by the picking out point algorithm based on Райта rule

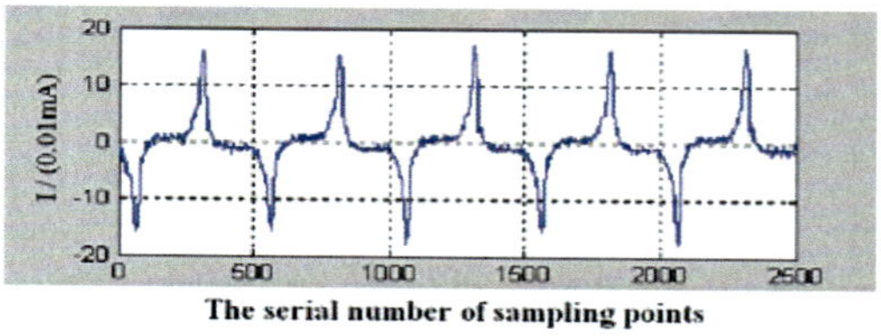

Fig. 6. Waveform processed by seven-pointed quadratic polynomial smoothing

Fig. 6 is the waveform, which has been processed by the seven-pointed quadratic polynomial smoothing. There is definite effect to eliminate the white noise interferences.

Fig. 7 is the waveform, which has been processed by the process of Daubechies4 wavelet 3 scales analysis. Using wavelet analysis, the effect of eliminating interference signals is obvious.

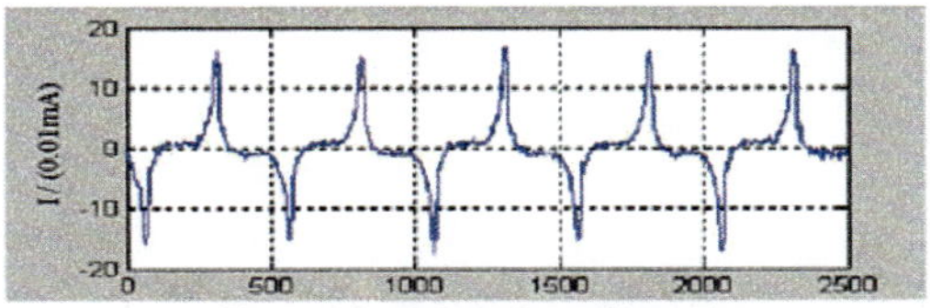

Fig. 7. Waveform processed by Daubechies4 wavelet 3 scales analysis

According to the fact of programming to realize the two algorithms, with high-level language, the operational amount of wavelet transformation algorithm is rather larger than the algorithm based on statistical theory. So the wavelet transformation algorithm offer higher requires to hardware of on-line insulation monitoring system. To the on-line insulation monitoring system based on field-bus, whose data signal processing depends on embedded CPU, the hardware and software resources are limited and the algorithm based on statistical theory is the proper choice. For conventional on-line insulation monitoring system, whose sampled analog signal is transmitted from sensor to system and then processed, the wavelet transformation algorithm can take higher precision.

5 Conclusion

An algorithm of statistical theory to preprocess the sampled data in on-line insulation monitoring system is presented in this paper. The result of preprocessing test data proves that it has better effect.

The effect of wavelet transformation algorithm is obvious. But the scales of wavelet transformation must be suitable. The wavelet transformation algorithm offer higher requires to hardware of on-line insulation monitoring system.

For the on-line insulation monitoring system based on field-bus, the algorithm based on statistical theory is the proper choice. For conventional on-line insulation monitoring system, the wavelet transformation algorithm could be taken.

References

1. Fei, Y.: Error Theory and Data Process. China Machine Press, Beijing (2000)
2. Brandt, S.: Statistical and Computational Methods in Data Analysis, 2nd edn. North-Holland Press, Amsterdam (1976)

3. Schaeffer, R.L., McClure, J.T.: Probability and Statistics for Engineers. Duxbury Press, California (1995)
4. Mallet, S.: A theory for multi-resolution signal decomposition: the wavelet representation. IEEE Transaction on Pattern Anal. Machine Intell. 11(7), 674–693 (1989)
5. Mallet, S., Hwang, W.L.: Singularity detection and processing with wavelets. IEEE Transaction on Information Theory 38(2), 617–643 (1992)

Author Index